FUNDAMENTALS OF FUZZY SETS

THE HANDBOOKS OF FUZZY SETS SERIES

Series Editors
Didier Dubois and Henri Prade
IRIT, Université Paul Sabatier, Toulouse, France

FUNDAMENTALS OF FUZZY SETS, edited by Didier Dubois and Henri Prade

MATHEMATICS OF FUZZY SETS: ***Logic, Topology, and Measure Theory***, edited by Ulrich Höhle and Stephen Ernest Rodabaugh

FUZZY SETS IN APPROXIMATE REASONING AND INFORMATION SYSTEMS, edited by James C. Bezdek, Didier Dubois and Henri Prade

FUZZY MODELS AND ALGORITHMS FOR PATTERN RECOGNITION AND IMAGE PROCESSING, by James C. Bezdek, James Keller, Raghu Krisnapuram and Nikhil R. Pal

FUZZY SETS IN DECISION ANALYSIS, OPERATIONS RESEARCH AND STATISTICS, edited by Roman Slowinski

FUZZY SYSTEMS: ***Modeling and Control,*** edited by Hung T. Nguyen and Michio Sugeno

PRACTICAL APPLICATIONS OF FUZZY TECHNOLOGIES, edited by Hans-Jürgen Zimmermann

FUNDAMENTALS OF FUZZY SETS

edited by

Didier Dubois

and

Henri Prade

IRIT, CNRS & University of Toulouse III

Foreword by Lotfi A. Zadeh

Kluwer Academic Publishers
Boston//London/Dordrecht

Distributors for North, Central and South America:
Kluwer Academic Publishers
101 Philip Drive
Assinippi Park
Norwell, Massachusetts 02061 USA
Tel: 781-871-6600
Fax: 781-871-6528
E-mail: kluwer@wkap.com

Distributors for all other countries:
Kluwer Academic Publishers Group
Distribution Centre
Post Office Box 322
3300 AH Dordrecht, THE NETHERLANDS
Tel: 31 78 6392 392
Fax: 31 78 6546 474
E-mail: orderdept@wkap.nl

Electronic Services: http://www.wkap.nl

Library of Congress Cataloging-in-Publication Data

Fundamentals of fuzzy sets / edited by Didier Dubois and Henri Prade ; foreword by Lotfi A. Zadeh.
p. cm. -- (The handbooks of fuzzy sets series ; FSHS 7)
Includes bibliographical references and index.
ISBN 0-7923-7732-X
1. Fuzzy sets. I. Dubois, Didier. II. Prade, Henri M. III. Series.

QA248.5 .F86 2000
511.3'22--dc21

99-049471

Printed on acid-free paper.

Printed in the United States of America

Contents

PART III UNCERTAINTY

FOREWORD

The seven volume Handbook of Fuzzy Sets Series is a monumental achievement. Its publication caps years of sustained effort by the principal editors Didier Dubois and Henri Prade —the walking encyclopaedias of fuzzy set theory— working in close collaboration with the editors of individual volumes, the contributors and the publisher, Kluwer. The result is a compendium which is much more than the sum of its parts; it is an integrated, authoritative and up-to-date exposition of the entire body of knowledge centered on fuzzy set theory and its wide-ranging applications. The editors and the contributors come from the ranks of leaders in their fields; the quality of exposition is uniformly high; and the organization of the Handbook reflects the exceptional mastery and deep insight of the principal editors, Didier Dubois and Henri Prade.

The vast panorama of methods, techniques and applications which is presented in the Handbook mirrors —as it should— the current perception of the topography of fuzzy set theory, its boundaries and its applications. In this connection, there is a point of semantics which is in need of clarification, namely: What is the difference, if any, between fuzzy set theory and fuzzy logic?

In some of my recent comments on this issue, I pointed out that the label "fuzzy logic" may be interpreted in two different ways: first, in a wide sense, in which case fuzzy logic (FL) is essentially coextensive with fuzzy set theory; and second, in a narrow sense in which fuzzy logic is a logical system which, in contrast to conventional logical systems, is aimed at a formalization of modes of reasoning which are approximate rather than exact.

More recently, I elaborated on my earlier suggestion by proposing a view of fuzzy logic which may be summarized as follows.

Fuzzy logic (FL), in its wide sense, has four principal facets: the logical facet, FL/L, which is basically fuzzy logic in its narrow sense; the set-theoretic facet, FL/S; the relational facet, FL/R; and the epistemic facet, FL/E. These and other facets of fuzzy logic have unsharp, overlapping boundaries.

Briefly, the logical facet, FL/L, is a logical system which underlies approximate reasoning and inference from imprecisely defined premises. The set-theoretic facet is focused on the theory of sets which have unsharp boundaries, rather than on issues which relate to logical inference. My 1965 paper on fuzzy sets is in this spirit, as are most of the papers in the realm of what is frequently referred to as fuzzy mathematics.

The relational facet, FL/R, is concerned in the main with representation and analysis of imprecise dependencies. Of central importance in FL/R are the concepts of a linguistic variable and the calculus of fuzzy if-then rules. Most of the applications of fuzzy logic in control and systems analysis relate to this facet of fuzzy logic.

The epistemic facet of fuzzy logic is focused on knowledge, meaning and imprecise information. Possibility theory is a part of this facet, while possibilistic logic is shared by the logical and epistemic facets of FL.

The core of FL, which is shared by all of its facets, is centered on two basic concepts: fuzziness and granularity. The conjunction of these concepts is fuzzy granularity or f-granularity, for short. F-granularity is a concomitant of a fundamental limitation on the cognitive ability of humans to resolve detail, process information and store data. In this perspective, FL may be viewed basically as a methodology for dealing with f-granularity in all of its manifestations, encompassing natural sciences, physical sciences, cognitive sciences, mathematics and engineering.

What I have described is a perception of the conceptual structure of fuzzy logic which is consistent in the main, but perhaps not in detail, with the organization of the Handbook. Explicitly or implicitly, f-granularity is addressed in all of its incarnations. A perception that emerges from reading the Handbook is that much of fuzzy logic may be viewed as the result of f.g-generalization (fuzzification and fuzzy granulation) of classical concepts, techniques and methodologies. The important point is that fuzzy logic complements rather than competes with existing theories. This is a point that is frequently a source of misunderstanding and misdirected criticism.

There is what I believe to be a new and important direction in FL which is in its initial stages of development and thus is not as yet reflected in the contents of the Handbook. What I have in mind is the methodology of computing with words (CW) and the computing-with-words-based methodology of the computational theory of perceptions (CTP). Basically, CW & CTP lay the groundwork for a major enlargement of the role of natural languages in FL and, eventually, in all branches of science.

Linguistic concepts, and especially the concepts of a linguistic variable and linguistic if-then rules have long played a major role in fuzzy logic and its applications. But CW moves much further in this direction by the development of a fuzzy-logic-based machinery for computing and reasoning with a much wider class of propositions drawn from a natural language. Through the use of constraint-centered semantics, such propositions are translated into what is called the generalized constraint language (GCL). In this way, reasoning and computation with propositions drawn from a natural language is reduced to constraint propagation from generalized antecedent constraints to generalized consequent constraints. The rules governing constraint propagation in CW coincide with the rules of inference in FL/L.

In the computational theory of perceptions, the objects of computation are perceptions expressed as propositions in a natural language. The importance of CTP derives from the fact that much of human reasoning is perception rather than measurement based.

An important concept which is suggested by CW is that of a precisiated natural language, PNL. Basically, PNL is a subset of a natural language which is equipped with constraint-oriented semantics and admits of translation into the generalized constraint language GCL.

The concept of a precisiated natural language lays the groundwork for a far more extensive use of natural languages in concept definition, reasoning and computation in all realms of science. Such use would represent a major paradigm shift by adding to existing theories the capability to define and compute with perceptions rather than measurements.

The development of CW and CTP, both of which are fuzzy-logic-based, has the effect of greatly amplifying the importance of fuzzy logic in the basic sciences and, especially, in applications involving the conception, design and utilization of information/intelligent systems. In this perspective, the Handbook makes a major contribution to the advancement of fuzzy logic, thereby enlarging its impact on the basic and applied sciences and providing new and powerful tools for the conception and design of systems with a much higher MIQ (Machine IQ) than those we have today.

Publication of the Handbook of Fuzzy Sets is an achievement whose importance cannot be exaggerated. The principal editors, the volume editors, the contributors and the publisher deserve our applause and appreciation.

Lotfi A. Zadeh

PREFACE

Fuzzy sets are almost thirty five years old. The novelty and the richness of the ideas carried out by the development of fuzzy sets, which have been often challenging previous views and old prevailing dogmas, have raised many doubts, criticisms, controversies cast by opponents or skeptics, especially during the first twenty years when fuzzy sets were still in their infancy or in their adolescence. Nevertheless, they have in the same time been encountering an increasing interest, and gaining more and more recognition. Each new decade has seen the blossoming of a larger set of methods and theoretical results as well as applications. This has led to a truly enormous literature, since there are presently approximately thirty thousands published papers dealing with fuzzy logic and several hundreds of books have appeared on the various facets of the theory and the methodology. There are currently several thousands of papers written every year.

Inevitably, such a gigantic amount of publications, by reporting so many achievements and applications (but sometimes also introducing more confusion or repetition than genuine progress), may leave newcomers, or even more experienced scholars, partially lost in front of this forest of results, in spite of the existence of many good monographs or edited volumes providing introductory or advanced materials to large segments of the field. Thus it was becoming more and more necessary, but also more and more challenging, to realize a general treatise covering, as much as possible, all the aspects of the development of the fuzzy set literature, and providing an organized and rather detailed view of the field and its key notions, with reference to the relevant literature.

For this reason, when Kluwer Academic Publishers came to us in the early nineties for proposing us to prepare a general Handbook of Fuzzy Sets, we enthusiastically considered the project as worth doing, although we already anticipated some of the difficulties of the enterprize. Indeed the extent and the variety of the fuzzy sets developments was forbiding us to repeat what we did once at the end of the seventies, namely writing by ourselves a monograph covering the whole field. The only way of approaching the problem was for us to organize a collective handbook, where each chapter would be prepared by one, or preferably several, recognized researchers, specialists of the covered topic. Quite rapidly we came up with a proposal structured in five volumes with a list of chapters and potential contributors, which was further expanded into six and then seven volumes, as more contributions had been accumulating. The first proposal was then refined, completed, and sometimes substantially revised, under the scientific authority and guidance of the volume editors and the feedbacks we got from the advisory board. But it is only in the mid-nineties that Kluwer gave its final green light to the project. For being honest, it is only about that time that we have started to measure the amount of difficulties for gathering about eighty tutorial introductions and surveys, in a recommended format, for each of the various areas where fuzzy set have brought contributions, in spite of the considerable dedication of the volume editors who had kindly agreed to take part to the project.

The efforts of the contributors, of the volume editors and of the series editors have finally culminated in a 7 volume handbook of about 4000 pages, covering most of the fields from the Fundamentals of Fuzzy Sets to the Practical Applications of Fuzzy Technologies, including the contributions of fuzzy sets to Mathematics, Approximate Reasoning and Information Systems, Pattern Recognition and Image Processing, Decision Analysis, Operations Research and Statistics, and Modelling and Control of Systems. Inevitably the coverage and the depth in treatment are not perfectly homogeneous through the whole set of volumes, resulting in chapters of variable lengths, somewhat mirroring the present unequal development of the different areas related to fuzzy sets. Depending on the cases, an area may have been covered by a unique, often long, chapter, or by several shorter chapters written by different authors, thus leading to differences in the granulation of the topics in some places. It is also clear that in spite of the long and sometimes huge lists of references provided with each chapter, it is now becoming impossible to cite all the existing literature, and that noticeable and interesting contributions might have been unfortunately missed. We hope that these defects, which we expect, remain limited, will not prevent the reader from finding substantial material in these Handbook volumes. We also strongly hope that the Handbooks of Fuzzy Sets Series will contribute to a better understanding, to a more informed and structured view of what has been achieved, what is going on, and what is still missing in the fuzzy sets and systems research.

Didier Dubois and Henri Prade
Toulouse

SERIES FOREWORD

Fuzzy sets were introduced in 1965 by Lotfi Zadeh with a view to reconcile mathematical modeling and human knowledge in the engineering sciences. Since then, a considerable body of literature has blossomed around the concept of fuzzy sets in an incredibly wide range of areas, from mathematics and logics to traditional and advanced engineering methodologies (from civil engineering to computational intelligence). Applications are found in many contexts, from medicine to finance, from human factors to consumer products, from vehicle control to computational linguistics, and so on.... Fuzzy logic is now currently used in the industrial practice of advanced information technology.

As a consequence of this trend, the number of conferences and publications on fuzzy logic has grown exponentially, and it becomes very difficult for students, newcomers, and even scientists already familiar with some aspects of fuzzy sets, to find their way in the maze of fuzzy papers. Notwithstanding circumstancial edited volumes, numerous fuzzy books have appeared, but, if we except very few comprehensive balanced textbooks, they are either very specialized monographs, or remain at a rather superficial level. Some are even misleading, conveying more ideology and unsustained claims than actual scientific contents.

What is missing is an organized set of detailed guidebooks to the relevant literature, that help the students and the newcoming scientist, having some preliminary knowledge of fuzzy sets, get deeper in the field without wasting time, by being guided right away in the heart of the literature relevant for her or his purpose. The ambition of the HANDBOOKS OF FUZZY SETS is to address this need. It will offer, in the compass of several volumes, a full picture of the current state of the art, in terms of the basic concepts, the mathematical developments, and the engineering methodologies that exploit the concept of fuzzy sets.

This collection will propose a series of volumes that aim at becoming a useful source of reference for all those, from graduate students to senior researchers, from pure mathematicians to industrial information engineers as well as life, human and social sciences scholars, interested in or working with fuzzy sets. The original feature of these volumes is that each chapter is written by one or several experts in the concerned topic. It provides introduction to the topic, outlines its development, presents the major results, and supplies an extensive bibliography for further reading.

The core set of volumes are respectively devoted to fundamentals of fuzzy set, mathematics of fuzzy sets, approximate reasoning and information systems, fuzzy models for pattern recognition and image processing, fuzzy sets in decision research and statistics, fuzzy systems modeling and control, and a guide to practical applications of fuzzy technologies.

D. Dubois H. Prade
Toulouse

Contributing Authors

James C. Bezdek
Computer Science Department
University of West Florida
Pensacola, FL 32514, USA

Taner Bilgic
Department of Industrial Engineering
University of Toronto
Toronto, Ontario M5S 1A4, Canada

Dionis Boixader
Sec. Matematiques i Inf., ETSAV
Pere Serra 1-15, 08190 Sant Cugat, Spain

Bernard De Baets
Department of Applied Mathematics and Computer Science
University of Gent
Krijgslaan 281 - S9, 9000 Gent, Belgium

Phil Diamond
Department of Mathematics
University of Queensland
St. Lucia, QLD 4067, Australia

Didier Dubois
Institut de Recherche en Informatique de Toulouse
Université Paul Sabatier
118 route de Narbonne, 31062 Toulouse Cedex 4, France

Janos Fodor
Department of Mathematics
Institute of Mathematics and Computer Science
University of Agricultural Sciences (GATE)
Pater Karoly u.1., 2103 Gödöllö, Hungary

Joan Jacas
Sec. Matematiques i Inf., ETSAB
Avda. Diagonal 649, 08028 Barcelona, Spain

Etienne Kerre
Fuzziness and Uncertainty Modeling
Applied Mathematics and Computer Science
University of Gent
Krijgslaan 281 - S9, 9000 Gent, Belgium

George J. Klir
Center for Intelligent Systems and Dept. of Systems Science & Industrial Eng.
Thomas J. Watson School of Engineering and Applied Science
Binghamton University - SUNY
Binghamton, NY 13902-6000, USA

Peter Kloeden
School of Computing and Mathematics
Deakin University
Australia

Radko Mesiar
Slovak Technical University Bratislava
Radlinského 11, 81368 Bratislava, Slovakia
and
UTIA AV CR , P.O. Box 18, 18204 Prague,
Czech Republic.

Hung T. Nguyen
Department of Mathematical Sciences
New Mexico State University
Las Cruces, NM 88003, USA

Walenty Ostasiewicz
Katedra Statystiki i Cybernetyki Ekonomicznej
Akademia Ekonomicsna im. O. Langegn
ul. Komandorska 118/120, 53-345 Wroclaw, Poland

Sergei Ovchinnikov
Mathematics Department
San Francisco State University
1600 Holloway Avenue, San Francisco, CA 94709, USA

Nikhil R. Pal
Electronics & Communication Science Unit
Indian Statistical Institute
Calcutta, India

Henri Prade
Institut de Recherche en Informatique de Toulouse
Université Paul Sabatier
118 route de Narbonne, 31062 Toulouse Cedex 4, France

Jordi Recasens
Sec. Matematiques i Inf., ETSAV
Pere Serra 1-15, 08190 Sant Cugat, Spain

I. Burhan Türksen
Department of Industrial Engineering
University of Toronto
Toronto, Ontario M5S 1A4, Canada

Ronald R. Yager
Machine Intelligence Institute
Iona College
New Rochelle, NY 10801, USA

ACKNOWLEDGEMENTS

As editors of the Handbooks of Fuzzy Sets Series, our deepest thanks are due to J.C. Bezdek, U. Höhle, N. Pal, S. Rodabaugh, R. Slowinski, M. Sugeno, H.J. Zimmermann who have accepted the heavy responsibility of preparing volumes of the Series. We also gladly thank the other members of the advisory board for their valuable suggestions at a very early step of the preparation of this huge projet: J. Baldwin, P. Bonissone, B. Bouchon-Meunier, A. Di Nola, D. Gabbay, J. Goguen, K. Hirota, J. Kacprzyk, E.P. Klement, G.J. Klir, R. López de Mántaras, R. Lowen, E.H. Mamdani, H.T. Nguyen, S. Orlowski, M. Roubens, E. Ruspini, E. Sanchez, P. Smets, L. Stout, M. Sugeno, H. Tanaka, E. Trillas, I.B. Türksen, P.Z. Wang, R.R. Yager, L.A. Zadeh and M. Zemankova.

The volume editors also wish to express their sincere gratitude to Alex Greene for his kind, patient and smooth cooperation in the realization of the Handbooks of Fuzzy Sets Series.

Special thanks are due to Agathe Baritaud who kindly and professionally prepared the final manuscript.

GENERAL INTRODUCTION

Didier Dubois
Henri Prade

"There is a fairly wide gap between what might be regarded as 'animate' system theorists and 'inanimate' system theorists at the present time, and it is not at all certain that this gap will be narrowed, much less closed, in the near future. There are some who feel that this gap reflects the fundamental inadequacy of conventional mathematics -the mathematics of precisely- defined points, functions, sets, probability measures, etc.- for coping with the analysis of biological systems, and that to deal effectively with such systems, which are generally orders of magnitude more complex than man-made systems, we need a radically different kind of mathematics, the mathematics of fuzzy or cloudy quantities which are not describable in terms of probability distributions. Indeed, the need for such mathematics is becoming increasingly apparent even in the realm of inanimate systems, for in most practical cases the a priori data as well as the criteria by which the performance of a man-made system are judged are far from being precisely specified or having accurately-known probability distributions".

L.A. Zadeh (1962a)
cited in (Zadeh, 1999).

When he introduced fuzzy sets thirty five years ago, L.A. Zadeh (1964), was motivated by the lack of any existing mathematical framework that could cope with the complexity of 'animate', either biological or humanistic, systems, a concern which he had started to realize a few years before (Zadeh, 1962a,b), as shown by the above citation or by his claim for a need of non-scalar-valued performance criteria, which would be less arbitrary or illusory than precisely valued ones (Zadeh, 1963). He has continued to strongly express such a view at various steps of the development of fuzzy sets with a special emphasis on man-machine systems (Zadeh, 1972, 1973, 1977, 1981, 1984, 1994, 1996, 1997, 1999), if we except the paper (Zadeh, 1969) on biological systems.

Most building blocks of the theory of fuzzy sets were proposed by Zadeh, especially fuzzy extensions of classical basic mathematical notions such as logical connectives, quantifiers, inference rules, relations, especially equivalence and ordering, arithmetic operations, etc. However the originality of the approach is to be found elsewhere, namely in the deep motivations of its author. For the first time, engineering was envisaged from the point of view of human cognition, a trend that could also be observed in other disciplines such as probability theory (with De Finetti and Ramsey's stress on subjective probability) and Artificial Intelligence, whose development was carried out in parallel with fuzzy set theory. The fuzzy systems approach shares with Artificial Intelligence the view that information possessed by humans is useful, and should be modelled and exploited so as to improve our capability of solving real problems, through a better understanding of how humans proceed. The notion of a fuzzy algorithm for parking a car (Zadeh, 1968) was a convincing illustration of this view. However the specificity of the fuzzy system approach is that it insists on a faithful modeling of human knowledge, respectful of its characteristic features: flexibility, summarization ability, granulation, gradualness. Like Artificial Intelligence, the fuzzy set methodology insisted that the quest for ever more numerical precision is partly hopeless, useless, and misleading, when it comes to understanding how complex systems work. However, contrary to the main trends in Artificial Intelligence which promoted a purely symbolic approach, the fuzzy set methodology maintained close links with numerical modeling, acknowledging that cognitive categories man uses to describe the world are not binary notions. The extent to which a statement uttered in natural language applies to a given situation is often a matter of degree. It is only if this gradualness is taken for granted that the bridge between human cognition and engineering practice based on numerical representations can be maintained. This view was expressed perhaps for the first time with full strength in Zadeh (1973).

More recently, the necessity of combining granularity and flexibility in the representation of information is the core of Zadeh (1996, 1997, 1999)'s presentation of fuzzy sets as a tool for "computing with words". This point highlights the importance of the interface between the data emanating from the physical world and the categories human beings are most comfortable with in expressing their perceptions and their knowledge. The idea that a class is not an all-or-nothing matter sometimes results from the conflict between symbolic and continuous representations of information, and sometimes because the apparent incompleteness as shown by Zadeh, each time in a specific paper. The concept of membership function is central in any fuzzy set-based method, of course, but more particularly in the field of fuzzy clustering, where membership functions are the output.

Aggregation Operators. A large body of fuzzy set-theoretic aggregation operations has been laid bare in the last twenty five years. They are presented in details in the Chapter 2 of this volume by Fodor and Yager. Fuzzy set-theoretic operations unify connectives coming from logic and trade-off operations coming from decision theory and statistics. Generalizations of set theoretic unions and intersections, variants of mean aggregations have been studied in detail at the mathematical level on an axiomatic basis. This is the theory of triangular norms, and uninorms. The diversity of connectives concerns not only the type of aggregation (conjunctive, disjunctive, trade-off). It also pertains to the possibility of weighting arguments in the aggregation, and introducing quantifiers. More recently a unified view of quantified and weighted aggregations has been proposed whereby *groups* of arguments can be weighted. Aggregation operations are widely applied in fuzzy multicriteria decision-making, fuzzy querying of information systems, and data fusion.

Measurement of Membership. Several techniques for the operational derivation of membership functions have been proposed. They rely on various interpretations of membership functions pointed out in Chapter 1. Membership grades can be related to distances to prototypes, frequencies, or utilities and costs. The elicitation of membership functions and fuzzy connectives is also a measurement problem that has not received the attention it deserves. The existing works are surveyed in Chapter 3 of this volume by Bilgic and Türksen.

Similarity and Fuzzy Orderings. Equivalence and ordering relations are very basic tools. Fuzzy relations have been used to generalize equivalence and give formal foundations to a theory of similarity. This theory is closely related to the theory of distances. Similarly, the concepts of partial ordering, preordering, linear ordering have been generalized to valued relations. These concepts are introduced in Chapter 4 of this volume by Ovchinnikov. The study of similarity relations enables a generalization of partitions to be developed, in connection with linear algebra notions such as eigenvectors, put in a non classical algebraic framework. Advanced results on fuzzy equivalence relations are surveyed in the Chapter 5 of this volume by Boixader, Jacas and Recasens. Similarity relations are important in pattern classification, and fuzzy orderings are particularly useful in decision analysis.

Fuzzy Relations Equations. Fuzzy set theory provides the opportunity of generalizing the theory of relations to many-valued relations in a general algebraic setting. Especially the max-min calculus of fuzzy relations offers a computationally nice and expressive setting for constraint propagation. Tools, both mathematical and algorithmic, for solving fuzzy relational equations have been developed. Methods for obtaining exact or approximate solutions of different types of fuzzy relational equations are reviewed in the Chapter 6 of this volume by De Baets. Fuzzy relations offer a sound theoretical basis for the study of fuzzy systems.

Possibility Theory. A new theory of uncertainty has been developed on the basis of fuzzy sets: possibility theory (Zadeh, 1978). The main purpose of possibility theory is to model incomplete information in a flexible way. Possibility theory is very similar to probability theory, but possibility theory is maxitive, not additive: the possibility of a disjunction of events is the maximum of the possibilities of each event. Moreover the uncertainty of an event is captured by two degrees: possibility and necessity, instead of one. Possibility distributions are fuzzy sets of possible mutually exclusive outcomes. The possibility of an event evaluates its degree of unsurprizingness while its necessity evaluates its degree of acceptance. Possibility theory can be a purely ordinal construct and it is then the simplest ordinal uncertainty theory, or possibility degrees can also be numerical. In this latter case they can be viewed as upper probability bounds. Numerical possibility theory is then the simplest theory of imprecise probabilities. All notions existing in probability theory (conditioning, independence, mean values, etc.) have counterparts in possibility theory. Integration in possibility theory is achieved by means of the Choquet integral on numerical universes and Sugeno integral in ordinal settings. Chapter 7 of this volume, by Dubois, Nguyen and Prade proposes an extensive introduction to the various aspects of possibility theory, trying especially to clarify the pervasive misunderstandings with respect to probability theory.

New Measures of Information. Fuzzy set theory has given the opportunity to study evaluations of the quantity of information that differ from standard probabilistic notions based on entropy. Two points of view can be envisaged: To what extent is a fuzzy set fuzzy? To what extent is a fuzzy set specific (= focuses on a small set of values)? The second point is closely connected to possibility theory. Fuzzy set inclusion also offers a tool for comparing fuzzy sets in terms of their relative specificity. Indices of specificity are obtained by fuzzy extensions of set cardinality. Chapter 8 of this volume by Klir put nonspecificity measures and other entropy-like measures for possibility theory and Shafer evidence theory in the general perspective of information theory. The first question has given rise to a literature on indices of fuzziness, which is reviewed in Chapter 9 of this volume by Pal and Bezdek. Measures of fuzziness have found natural applications in the analysis of grey images.

Fuzzy Intervals. Fuzzy intervals and numbers are fuzzy sets of the real line that generalize intervals. A calculus of fuzzy intervals that parallels the calculus of random variables has been developed under various assumptions according to whether the underlying imprecise variables are interactive or not. It actually generalizes interval computations to fuzzy sets. Fuzzy counterparts of probabilistic limit theorems exist. Fuzzy interval methods are thoroughly surveyed in the Chapter 10 by Dubois, Kerre, Mesiar and Prade, which also provides a synthesis of the abundant literature on the comparison of fuzzy numbers. Fuzzy intervals are very useful in many applications, such as decision and risk analysis, engineering design, and scheduling.

Fuzzy Interval-Valued Analysis. Metric spaces of fuzzy quantities have been defined, and serve as a basis for fuzzy analysis. This the topic of Chapter 11 of this volume by Diamond and Kloeden. Many results exist on the integration and differentiation of fuzzy-valued functions. This topic is related to the available literature on fuzzy random variables which mixes statistics and fuzzy interval computations. This latter aspect is more particularly addressed in the companion volume of the Handbooks of Fuzzy Sets Series edited by Slowinski (1998).

Approximate Reasoning and Fuzzy Rules. This is another important component of the fuzzy set approach, which explains why it is often spoken about "fuzzy logic". Fuzzy logic refers to a methodology for reasoning with pieces of knowledge expressed by means of linguistic variables, translated as fuzzy restrictions, and is not only a many-valued logic. Fundamental principles of approximate reasoning have been laid bare in terms of combination and projections of fuzzy relations. The main pattern of fuzzy reasoning is the generalized modus ponens. The many-valued extensions of implication provide a basis for modelling fuzzy rules which are basic building blocks of the fuzzy logic inferential machineries. Fuzzy rules are also instrumental in fuzzy control systems. This topic is not dealt with here, but thoroughly surveyed in the volume of the Handbooks of Fuzzy Sets Series devoted to information systems, edited by Bezdek, Dubois and Prade (1999).

The other volumes in the Handbook of Fuzzy Sets Series deal with topics ranging from pure mathematics to applications in science and engineering. Specific volumes are devoted to areas where fuzzy set-methods have been particularly successful or look very promising, such as information systems, systems engineering, pattern classification and image processing, and decision sciences.

The volume entitled *Fuzzy Sets in Approximate Reasoning and Information Systems* (Bezdek, Dubois and Prade,1999) is organized in three parts: inference, learning and data fusion, and information technology proper. In the first part, the long chapter on approximate reasoning also provides a survey of specific inference techniques such as fuzzy constraint propagation, possibilistic networks, case-based reasoning temporal reasoning, causal reasoning, etc. It is completed by two chapters on the genuinely logical aspects of fuzzy logics, namely weighted formal inference systems and the notion of fuzzy consequence relation. The second part surveys methods for extracting fuzzy rules from data, using decision trees induction or neuro-fuzzy techniques. It also shows that possibility theory and fuzzy set connectives are relevant for data fusion, where they provide very flexible fusion modes. The rest of the volume is a state of the art presentation of fuzzy approaches in database and information retrieval technologies.

The volume entitled *Fuzzy Systems: Modeling and Control*, (Nguyen and Sugeno, 1999) reflects the success encountered by fuzzy rule-based systems in systems engineering. Basic techniques for fuzzy modeling are presented, with an extensive chapter on universal approximation properties of fuzzy rule-based systems, and another chapter bridging the gap with approximate reasoning notions. The very popular topic of fuzzy control, on which so much has been written, also receives an extensive treatment, with chapters devoted to the design, the tuning and the stability of fuzzy controllers. The similarity and differences between fuzzy controllers and classical PID controllers are also explained. More chapters are devoted to data-driven learning techniques that can enhance fuzzy models or fuzzy controllers: reinforcement learning, neuro-fuzzy techniques and genetic algorithms. As opposed to rule-extraction methods, these techniques try to account for a priori fuzzy knowledge in the construction of black-box models. Lastly, some more prospective aspects of fuzzy systems analysis are covered, such as fuzzy differential inclusions and chaos theory. It is useful to point out the edited volume by Verbruggen et. al (1999), as good ancillary reading on fuzzy algorithms in systems engineering, in connection with the Fuzzy Handbook.

The volume entitled *Fuzzy Models and Algorithms for Pattern Recognition and Image Processing*, (Bezdek, Keller, Krishnapuram, and Pal, 1999) is the only monograph in the series. This huge book offers a full treatment of fuzzy information processing from the standpoint of automated classification and computer vision. Two chapters are respectively devoted to fuzzy clustering methods with objective function models, and to fuzzy relational clustering methods. The specificity of fuzzy clustering is to postpone classification decisions so as to better reflect the soft structure of the data set. An important part of the volume is devoted to the problem of classifier design, and fuzzy extensions of the kth neighbor rule. It also describes fuzzy models and algorithms inspired by neural networks, as well as fuzzy extensions of syntactic pattern recognition, and data fusion. Lastly the volume proposes an extensive survey of fuzzy sets in the field of image processing and computer vision.

The volume entitled *Fuzzy Sets in Decision Analysis, Operations Research and Statistics*, (Slowinski, 1998) gather material that pertains to the application of fuzzy sets in optimization and data analysis. The first part is devoted to preference modelling and decision analysis, and usefully complements the chapters on fuzzy relations in the present volume, especially on the topic of fuzzy orderings and their aggregations. It also explains how fuzzy set theoretic operations can be applied to multicriteria decision-making. Moreover chapters survey the application of fuzzy relations in group decision-making, and in game theory. Next, a bunch of chapters is devoted to fuzzy optimization methods: fuzzy linear and non-linear programming, fuzzy graphs and discrete problems, and fuzzy dynamic programming. The third part is devoted to data analysis methods, namely fuzzy sets in statistics, and fuzzy regression analysis. The volume also covers topics like reliability, and maintenance, where possibility theory and fuzzy intervals have been used.

The two other volumes of the series display an interesting contrast that witnesses the incredibly wide span of the area which is covered by fuzzy sets, from pure mathematics to practical applications.

The volume entitled *Mathematics of Fuzzy sets: Logic, Topology and Measure Theory*, (Höhle and Rodabaugh, 1999) should convince the skeptical readers that there is a serious mathematical side to fuzzy sets. First, a extensive chapter shows that on the side of logic proper, fuzzy sets have provided an opportunity to unify existing many-valued logic systems that have been devised well before the emergence of fuzzy sets by scholars like Lukasiewicz, and even Gödel. Ancillary readings to this chapter are the book by Hajek (1998) and the collection of papers edited by Dubois, Prade and Klement (1999). An important part of the Mathematics of Fuzzy sets volume is dedicated to a state-of-the-art and organized presentation of fuzzy topology, a field where a huge number of papers has been published in a rather disorderly fashion. These chapters try to adopt a common terminology and to clarify the basic issues, on a rather closed topic that has not yet found practical applications. This volume is completed by chapters devoted to measure-theoretic topics such as fuzzy events and conditional events, pseudo-additive measures and their applications to differential equations, and fuzzy random variables. Note that this volume covers only part of the fuzzy mathematical landscape. It does not deal with the "fuzzification" of algebraic structures. However the books by Mordeson and Malik (1998), and Mordeson and Nair (1998) fill this gap.

The volume entitled *Practical Applications of Fuzzy Technologies* (Zimmermann, 1999) is an overview of more specific applications where a proper representation of information requires fuzzy set techniques. Its aim is to illustrate the fact that a fuzzy set is a generic notion that can be used in quite unexpected areas. This volume includes a discussion of the famous fuzzy controllers from a practical perspective in the area of process control. It tells the myth from the reality. Other chapters are in fields such as engineering design (where fuzzy interval propagation plays a key-role), fault detection, quality control, robotics (especially, the famous SRI Flakey robot), production scheduling, civil engineering, ecological systems, chemistry. The field of medicine has given birth to a lot of works and two chapters survey medical expert systems and medical image processing. The book also covers other domains in human sciences such as ergonomics, psychology, and social sciences, domains related to economics such as project planning, management, finance, accounting, and actuarial sciences.

However, there remain a few specific areas where fuzzy set-based ideas have been developed that are not covered by the present volumes of the Handbooks of Fuzzy Sets Series. This is either because the area did not fit in any of the volumes due to its specificity, or because the area has not been very active for a long time. Let us mention in particular, fuzzy formal languages and automata, fuzzy switching logics (Kandel, and Lee, 1979), fuzzy algebra (pointed out above), fuzzy set-based models in quantum mechanics and mathematical physics, fuzzy hardware (Mlynek and Patyra, 1996; Patyra, 1996; Kandel and Langholz, 1997). These topics might be covered in future volumes of this Handbooks Series.

II THE ROLE OF FUZZY SETS IN INFORMATION ENGINEERING

Beside its potentially enormous economical, industrial and social impact, information has become a matter of scientific investigation of its own, standing apart from other older established fields such as physics, mathematics, biological sciences, human and social sciences. This interest is in part strongly motivated by the confluence of computer and communication technologies which allows for the collection and storage of large amounts of information and its subsequent dissemination in a very efficient way. In a recent publication (Dubois, Prade and Yager, 1997) a unified view of information sciences and engineering has been promoted that transcends boundaries existing between subfields that sometimes ignore one another, while having in common the purpose of processing some kind of information in some way, either emphasizing the computational or the representational points of view. One may define the term "information" as follows : any collection of symbols, signs or signals, produced either by the observation of natural or artificial phenomena, or by the cognitive activity of human beings or artificial machines, which can be used to understand the surrounding world, help in decision making or to communicate with one another. The concept of information processing science and engineering is an attempt to bring together under one umbrella many scientific investigations which have developed along with the emergence of the computer technology and have to do with information: automatic control, signal processing, operations research, data analysis, artificial intelligence, data bases, information retrieval, man-machine communication. Information science and engineering seem to be concerned with three main tasks (Dubois *et al.* 1997): clarifying, storing/retrieving, and exploiting information.

Clarifying information refers the process of elaborating raw data into more useful and comprehensible forms. This task ranges from enhancing information via smoothing or eliminating redundant or erroneous data, completing information via interpolation, to higher cognitive activities such as abstracting and modeling.

Retrieving information means to extract, based on certain requirements and specifications, already existing, possibly hidden, pieces of information stored in a database. The way information is retrieved is strongly dependent on the storage technology and the knowledge representation framework used.

Exploiting information means using retrieved information for further action. It encompasses a large range of activities including decision-making, optimization, design, and control. Generally in such tasks, information plays a supportive role to some sort of paradigm or model.

Let us now look at the three components of information engineering and summarize the role fuzzy sets can play in each of them.

Clarification. In clarification techniques such as clustering, recognition and classification, fuzzy set theory offers a tool for representing classes with soft boundaries as first proposed by Bellman, Kalaba and Zadeh (1966), taking advantage of the similarity notion naturally embedded in the membership functions, and the possible overlap between classes.

Fuzzy models offer not only a tool for numerical modelling with interpolation purposes, but also enable rules to be extracted from data in a flexible way, with a concern of linguistic interpretability.

In data analysis, it has become possible to exploit fuzzy/linguistic data in regression methods and other techniques such as least squares. Extensive use of possibility theory and fuzzy interval analysis is made. Statistical methods have also been extended to linguistic data.

In data fusion, be it sensor fusion, expert opinion pooling, or fuzzy database fusion, fuzzy set theoretic aggregations could play a major role due to their flexibility.

The main asset of fuzzy set theory in this side of information engineering is the smooth interface between linguistic and numerical representations of information. It may sound provocative to claim that fuzzy sets can clarify information. Yet, a linguistic summary, even if containing fuzzy predicates, is often easier to interpret than a huge data set.

Retrieval. It is often desirable to allow for flexible requests in order to more faithfully express the desires of the user of an information system. Complex, flexible queries can be modelled by fuzzy sets and their connectives. Fuzzy sets also allow for the representation of stored information, and especially the representation of the uncertainty pervading the data or their descriptions.

Information systems whose purpose is to restore information via inference techniques can take advantage of approximate reasoning tools such as possibilistic logic, fuzzy rule-based inference. Qualitative possibility theory is a natural framework for exception-tolerant reasoning. Other more specific fuzzy reasoning frameworks have been devised for temporal reasoning, order-of-magnitude reasoning, etc. Fuzzy arithmetics play a fundamental role in these more specialized tasks. Lastly, fuzzy relational methods for causal diagnosis exist.

Exploitation. When using information for action, fuzzy sets may appear in the representation of uncertainty (when making decision in uncertain environments), and in the expression of preferences in formulating objectives in optimization problems, as first stressed by Bellman and Zadeh (1970), or yet in the description of similarities (as in case-based or interpolative reasoning). Various types of problems can be concerned:

-) problems where the number of potential solutions is very small but the choice is made difficult due to many conflicting goals and/or the presence several of decision-makers. This is the realm of decision support systems for strategic decision-making. An important asset of fuzzy sets lies in the very general framework it provides that allows us to cope with the presence of multiple criteria, including the representation of complex relationships between criteria (using techniques such as OWA operators and fuzzy integrals for instance). Fuzzy relations may bridge the gap between value functions that insist on numerical, totally ordered preference models, and relational approaches that handle non transitive notions of preference and indifference and capture notions of incomparability. Lastly the framework of possibility theory offers new criteria for decision-making under uncertainty (see Chapter 7, in this volume, on this point).

-) problems where the computation of a solution is a highly combinatorial task but the local preference modeling is made very clear and simple. This is typically the case of optimization problems that occur in engineering design or large-scale systems where the difficulty lies in the intricacy of a large number of local constraints and decision variables. Here, the notion of flexible constraints in design problems is intrumental, and allows for the presence of uncertainty in the data (using possibility theory).

-) problems where the solution is a complex entity that is built in several steps because time is involved in the picture. This is the case for planification problems in robotics or manufacturing; flexible scheduling methods have been proposed, which unify the optimisation methods and the constraint-directed techniques.

-) lastly, problems where information must be exploited in a dynamic environment and reactivity is the key issue. Decision is then made in real time by exploiting some feedback. This is the realm of control problems where fuzzy set methods have been so widely applied. Here the main contribution of fuzzy sets was to provide a methodology for deriving control laws from human knowledge using fuzzy rules and interpolative reasoning. Applications of fuzzy dynamic programming are also naturally in this field.

In each of the above three components of information engineering that we have touched upon (clarifying, storing/retrieving and exploiting information), two aspects seem to be important: the representation of information and the exchange of information between man and machine. Knowledge representation has become a central topic for basic research in artificial intelligence due to the necessity for a communication language between man and computers. This communication goes in both directions: information generated by the machine should be in a form easily comprehensible by human beings while information provided by human beings should be in a format exploitable by the computer.

Logic has become the natural framework for the formalization of reasoning techniques. However classical logic alone is unable to face the major challenge of accounting for the various forms of human reasoning, necessary to clarify, retrieve and exploit information in an automated, yet human-friendly way. Moreover a great amount of information comes in the form of numerical data, as opposed to human knowledge. Numerical data processing methods, and symbolic knowledge representation and reasoning tools must thus be used conjointly, on the basis of their complementarity, and not be opposed as exclusive approaches to information engineering. This presupposes that the dogma of the monopoly of symbolic representations in Artificial Intelligence be given up, and that data-driven schools of information processing open to logical approaches, in the scope of developing software systems that not only efficiently solve problems but are capable of explaining the obtained solutions to users. Clearly, fuzzy set theory offers a significant contribution to this debate, because the reconciliation between symbolic and numerical information processing techniques is on its agenda.

III CONCLUSION: THE LEGITIMACY OF FUZZY SETS

In summary, fuzzy sets contributions to information modelling and processing can be particularly advocated in relation with the following issues:

- interfacing numerical data with linguistic labels, by making continuous quantitative scales compatible with finite vocabularies, and offering a more faithful representation of gradual properties or of classes without precisely located boundaries;

- postponing, as much as possible, arbitrary reductions of imprecision and uncertainty leading to unsupported classification decisions, thus preserving flexibilities in future choices;

- summarizing numerical data in a soft, but linguistic interpretable way;

- providing tools for modelling complex subjective evaluations by restoring commensurability between non-commensurate criteria;

- representing incomplete or imprecise information with implicit, or qualitative, levels of preference or plausibility. It paves the way to building artificial systems aware of their limited knowledge.

- approximating imprecise statistical information.

- providing a formal framework for capturing several forms of commonsense reasoning

It is interesting, after several decades of fuzzy set research, to re-examine critical statements uttered against the fuzzy set methodology, taking them seriously and checking their relevance to the current state of the art.

Kalman's fears seem to have been unwarranted. Indeed, his words express some confusion between works lacking scientific rigour and attempts at modelling fuzzy information. Fuzzy set theory does belong to the scientific tradition of rigour and hard work, as witnessed by the achievements described in the Handbooks of Fuzzy Sets Series, as well as many other books on the topic. Some have complained about the insularity of the field, and the lack of bridges with other disciplines. This situation has changed in many respects, to-date.

Some critical statements are also due to the effect produced by the word "fuzzy" which, some people, looks self-depreciative. However, the essence of fuzzy set theory is not an apology of sloppiness in science nor of a presupposed imprecision of natural language (also criticized by Arbib, 1977). It is an attempt to model the *gradualness* of linguistic categories that appear in human-originated information.

There is nothing sloppy or permissive in the notion of a gradual category, since a many-valued entity may be more informative than a binary one.

On the other hand, the idea, promoted by Bunge, that we should get rid of inexacness rather than trying to model it in a faithful way and that propositions should be ultimately true or false is still a matter of debate. Fuzzy sets take the stand that modelling lexical flexibility and accounting for incompleteness of information is a worthwhile endeavour. Rejecting it is forgetting the existence of truly gradual concepts in human cognition, and amounts to believing that information can always be made precise and certain at a low cost in practice (or to refusing to process useful information when poor). The dispute between precision searchers and imprecision modellers may be due to the existing gap between science and engineering whose purposes differ although they have to share their tools.

The remarks of S. Haack can be diversely commented upon. The paradox of precise modeling of membership to imprecise categories was also pointed out by Arbib (1977) and Parikh (1991). Although the wide use of [0, 1] as a membership scale may suggest that membership grades can be accurately measured (which they cannot, of course), numerical grades are often used as a convenient encoding of a linearly ordered scale. Continuous membership functions reflects the continuity of the space of real numbers which is often used in engineering applications. Subjective probability suffers from a similar paradox (despite well-established operational semantics), and can be efficiently applied anyway. The precision of membership grades is more a philosophical problem than a practical one. A fuzzy membership function is still an idealization, but in some cases a better one than a crisp set. The reason is that going from discontinuous representations (using crisp sets) to continuous ones (membership functions), we gain in robustness: moving the thresholds defining the support or the core of fuzzy sets only modifies membership grades slightly. Hence the precision of membership function is less important than the precision of the thresholds defining an interval representation of a linguistic category. In finite or discrete settings, no decent scholar in fuzzy set theory seriously believes in precise numerical membership grades, say, that a penguin could be a bird at level .4. Membership grades can be elements of a finite chain (a view which is compatible with basic fuzzy set connectives such as max, min operations, medians, or order-reversing maps of the scale). Moreover the context-dependent nature of fuzzy sets acknowledged by Zadeh, is often forgotten by discussants.

The claim made by Haack on the enormous complexities involved by fuzzy logic only applies to the generic reasoning machinery it would need to implement Zadeh's approximate reasoning framework using brute force methods. Only useful, computationally tractable fragments of fuzzy reasoning methods have been implemented and put at work in practical applications. One may also regret that the proportion of fuzzy set theory actually used in engineering tasks is still very little. Fuzzy controllers are based only on a very simple calculus of fuzzy rules, and, in some recent works supposedly on fuzzy control just use the name "fuzzy" because it is fashionable in some circles. Nevertheless, some not so elementary, yet tractable notions like weighted aggregations, fuzzy intervals, fuzzy random variables, have a promising future in statistics, engineering design, decision-support systems, multimedia database technology, etc. However it is much more expensive to develop, say, a fuzzy set-based software for database management, than to put a few fuzzy rules in a consumer device.

Haack is anyway right in pointing out that what is often called fuzzy logic is not really a logic in the narrow sense. Again, some fragments of approximate reasoning have received a decent logical treatment (such as possibilistic logic). But much work remains to be done to confer a genuine logical legitimacy to Zadeh's intuitions, although some logicians have started considering the matter seriously (see contributions in Dubois, Prade, and Klement, 1999).

As for the confict between probability theory and fuzzy sets, it involves two issues. One is the question of the existence of links between probabilities and fuzzy sets, for instance, the possibility of probabilistic interpretations of membership functions. The other is the legitimacy of any uncertainty management method that differs from Bayesian probability. On the first issue, the reader is referred to the Chapter 7 by Dubois, Nguyen and Prade (1999) in this volume. Many misunderstandings were due to a lack of knowledge of one field by the other. If an attitude of mutual consideration is adopted, then a lot of things can be done for the reconcilation of fuzzy sets and probability, expecially via possibility theory. However the probabilistic point of view must be the one of imprecise probability systems, and confidence intervals, hence is non-Bayesian. The relative situation between fuzzy sets and imprecise probability, random sets, likelihood functions, etc. is now clear although many details still have to be worked out.

About the objection raised by Bayesians such as Laviolette and Seaman (also before, by a famous scientist such as Lindley, 1987), it is based on the prejudiced assumption that Bayesian probability can capture all facets of uncertainty. Fuzzy sets, via possibility theory, is a non-additive uncertainty theory that accounts for incomplete information. It shares this feature with other theories of uncertainty like belief functions, imprecise probability systems, and the like. So the Bayesian objection applies to all these theories as well and brings us back to the question of the validity of modelling incompleteness, raised by Bunge. However the statement that there is more to uncertainty than Bayesians might have people believe (Klir, 1989) is more and more widely acknowledged, especially in the framework of new problems of uncertainty modeling arising in information technology

Concludingly, we hope that the Handbooks of Fuzzy Sets Series will contribute to a better understanding and to a larger recognition of 35 years of works related to fuzzy sets, offer guidance for digging in the huge specialized literature, encourage readers to apply the available results in their problems, help them further develop the field and improve the methodologies by bridging the gaps with other established fields, thus further anchoring fuzzy sets in the regular scientific tradition.

REFERENCES

Arbib M. (1977). Review article on fuzzy set theory, *Bulletin of the Amer. Mathematical Society*, **83**, 946-951.

Bellman R.E. and Zadeh L.A. (1970). Decision-making in a fuzzy environment, *Management Sciences*, **17**, 141-164.

Bellman R.E., Kalaba R. and Zadeh L.A. (1966). Abstraction and pattern classification, *J. of Mathematical Analysis and Applications*, **13**, 1-7.

Bezdek J., Dubois D. and Prade H. (Eds.) (1999). *Fuzzy Sets in Approximate Reasoning and Information Systems*, The Handbooks of Fuzzy Sets Series, Kluwer Academic Publishers, Dordrecht, The Netherlands.

Bezdek J., Keller J., Krishnapuram R. and Pal N. R. (1999). *Fuzzy Models and Algorithms for Pattern Recognition and Image Processing*, The Handbooks of Fuzzy Sets Series, Kluwer Academic Publishers, Dordrecht, The Netherlands.

Bunge M. (1983). *Epistemology and Methodology — I: Exploring the World*, Vol. 5 of Treatise on Basic Philosophy, Reidel, Dordrecht/Boston.

Dubois D., Prade H. and Klement E.P. (Eds.) (1999). *Fuzzy Sets, Logics, and Reasoning About Knowledge*, Kluwer Academic Publishers, Dordrecht, The Netherlands, to appear.

Dubois D., Prade H. and Yager R.R. (Eds.) (1997). *Fuzzy Information Engineering: A Guided Tour of Applications*, Wiley, New York.

Haack S. (1979). Do we need "fuzzy logic"?, *Int. J. of Man-Machine Studies*, **11**, 437-445.

Haack S. (1996). *Deviant Logic, Fuzzy Logic — Beyond the Formalism*, The University of Chicago Press.

Hajek P. (1998) *The Metamathematics of Fuzzy Logic*. Kluwer Academic Publishers, Dordrecht, The Netherlands.

Höhle U. and Rodabaugh S. (Eds.) (1998). *Mathematics of Fuzzy sets: Logic, Topology and Measure Theory*, The Handbooks of Fuzzy Sets Series, Kluwer Academic Publishers, Dordrecht, The Netherlands.

Kalman R.E. and Zadeh L.A. (1974). Discussion, *Man and Computer* (Marois M., ed.), North-Holland, 93-94.

Kandel A. and Langholz G. (Eds.) (1997). *Fuzzy Hardware. Architectures and Applications*, Kluwer Academic Publishers, Dordrecht, The Netherlands.

Kandel A. and Lee C.S. (1979). *Fuzzy Switching and Automata — Theory and Applications*, Crane & Russak, New York and Edward Arnold, London.

Klir G. (1989) Is there more to uncertainty than some probabilist might have us believe? *Int. J. General Systems*, 15, 347-378.

Laviolette M. and Seaman J.W., Jr (1994) The efficacy of fuzzy representations of uncertainty (with discussions). *IEEE trans. on Fuzzy Systems*, **2**, 4-42.

Lindley D.V (1987) The probability treatment of uncertainty in artificial intelligence and expert systems. Statistical Science, 2, 17-24.

Mordeson J.N. and Malik D.S (1998). *Fuzzy Commutative Algebra*, World Publ., Singapore.

Mordeson J.N. and Nair P.S. (1998). *Fuzzy Mathematics*, Physica-Verlag.

Nguyen H.T. and Sugeno M. (Eds.) (1998). *Fuzzy Systems: Modeling and Control*, The Handbooks of Fuzzy Sets Series, Kluwer Academic Publishers, Dordrecht, The Netherlands.

Parikh R. (1991). A test for fuzzy logic, *SIGACT News*, **2**, 49-50.

Patyra M.J. (Ed.) (1996). *Fuzzy Logic Hardware Implementations*, Special issue, IEEE Trans. on Fuzzy Systems, 4(4), 401-505.

Patyra M.J. and Mlynek D.M., Eds. (1996). *Fuzzy Logic — Implementation and Applications*, John Wiley & Sons, Chichester and B.G. Teubner, Stuttgart.

Slowinski R. (Ed.) (1998). *Fuzzy Sets in Decision Analysis, Operations Research and Statistics*, The Handbooks of Fuzzy Sets Series, Kluwer Academic Publishers, Dordrecht, The Netherlands.

Verbruggen H.B., Zimmermann H.-J. and Babushka R. (Eds.) (1999). *Fuzzy Algorithms for Control*, Kluwer Academic, Dordrecht, The Netherlands.

Zadeh L.A. (1962a). From circuit theory to system theory, *Proc. of Institution of Radio Engineers*, **50**, 856-865.

Zadeh L.A. (1962b). A critical view of our research in automatic control, *IRE Trans. on Automatic Control*, **AC-7**, 74.

Zadeh L.A. (1963). Optimality and non-scalar-valued performance criteria, *IEEE trans. on Automatic Control*, **8**(1), 59-60.

Zadeh L.A. (1964). Fuzzy Sets. Memo. ERL, N°64-44, University of California, Berkeley. *Information and Control*, **8**, 338-353, 1965.

Zadeh L.A. (1968) Fuzzy Algorithms, *Information and Control*, **12**, 94-102.

Zadeh L.A. (1969). Biological application of the theory of fuzzy sets and systems, *Proc. of the Symp. on the Biocybernetics of the Higher Nervous System* (Proctor L.D., ed.) Little, Brown and Co., Boston, 199-206.

Zadeh L.A. (1972). A rationale for fuzzy control, *J. Dyn. Syst. Meas. Control*, **34**, 3-4.

Zadeh L.A. (1973). Outline of a new approach to the analysis of complex system and decision processes, *IEEE Trans. on Systems, Man and Cybernetics*, **3**, 28-44.

Zadeh L.A. (1977). Linguistic characterization of preference relations as a basis for choice in social systems, *Erkenntnis*, **11**(3) 383-401.

Zadeh L.A. (1978). Fuzzy sets as a basis for a theory of possibility, *Fuzzy Sets and Systems*, **1**, 3-28.

Zadeh L.A. (1981). Possibility theory and soft data analysis, *Mathematical Frontiers of Social and Policy Sciences* (Cobb L. and Thrall R.M., eds.), Westview Press, Boulder, Colo. 69-129.

Zadeh L.A. (1984). Making computers think like people, *IEEE Spectrum*, **21**, 26-32.

Zadeh L.A. (1990). The birth and evolution of fuzzy logic, *Int. J. of General Systems*, **17**, 95-105.

Zadeh L.A. (1994). Soft computing and fuzzy logic, *IEEE Software*, **11**(6), 48-56.

Zadeh L.A. (1996). Fuzzy logic = Computing with words, *IEEE Trans. on Fuzzy Systems*, **4**, 103-111.

Zadeh L.A. (1997). Toward a theory of fuzzy information granulation and its centrality in human reasoning and fuzzy logic, *Fuzzy Sets and Systems*, **90**, 111-127.

Zadeh L.A. (1999). From computing with numbers to computing with words — From manipulation of measurements to manipulation of perceptions, *IEEE Trans. on Circuits and Systems - I: Fundamental Theory and Applications*, **45**(1), 105- 119.

Zimmermann H.-J. (Ed.) (1999). *Practical Applications of Fuzzy Technologies*, The Handbooks of Fuzzy Sets Series, Kluwer Academic Publishers, Dordrecht, The Netherlands.

I Fuzzy Sets

1 FUZZY SETS: HISTORY AND BASIC NOTIONS

Didier Dubois
Walenty Ostasiewicz
Henri Prade

Abstract. This paper is an introduction to fuzzy set theory. It has several purposes. First, it tries to explain the emergence of fuzzy sets from an historical perspective. Looking back to the history of sciences, it seems that fuzzy sets were bound to appear at some point in the 20th century. Indeed, Zadeh's works have cristalized and popularized a concern that has appeared in the first half of the century, mainly in philosophical circles. Another purpose of the paper is to scan the basic definitions in the field, that are required for a proper reading of the rest of the volume, as well as the other volumes of the Handbooks of Fuzzy Sets Series. This Chapter also contains a discussion on operational semantics of the generally too abstract notion of membership function. Lastly, a survey of variants of fuzzy sets and related matters is provided.

1.1 INTRODUCTION

The notion of a fuzzy set stems from the observation made by Zadeh (1965a) that "more often than not, the classes of ojects encountered in the real physical world do not have precisely defined criteria of membership". This observation emphasizes the gap existing between mental representations of reality and usual mathematical representations thereof, which are based on binary logic, precise numbers, differential equations and the like. Classes of objects referred to in Zadeh's citation exist only through such mental representations through natural language terms such as high temperature, young man, big size, etc., and also with nouns such as bird, chair, etc. Classical logic is too rigid to account for such categories where it appears that membership is a gradual notion rather than an all-or-nothing matter. The power of expressivity of real numbers is far beyond the limited level of precision found in mental representations. The latter are meaningful summaries of perceptive phenomenas that account for the complexity of the world. Analytical representations of physical phenomena can be faithful as models of reality, but are sometimes difficult to understand because they do not explain much by themselves, and may remain opaque to the non-specialist. Mental representations make more sense but are pervaded with vagueness, which encompasses at the same time the lack of specificity of linguistic terms, and the lack of well-defined boundaries of the class of objects they refer to. We shall then speak of fuzzy predicates, or gradual properties.

The ambition of representing human knowledge in a human-friendly, yet rigorous way might have appeared like a futile exercice not worth spending time on, and even ridiculous from a scientific standpoint, one hundred years ago. However in the meantime the emergence of computers has significantly affected the landscape of science, and we have now entered the era of information management. The development of sound theories and efficient technology for knowledge representation and automated reasoning has become a major challenge, now that many people possess computers and communicate with them in order to find information that helps them when making decisions. An important issue is to store and exploit human knowledge in various domains where little objective and precise data are available. Fuzzy set theory participates to this trend (Dubois, Prade and Yager, 1997), and, as such, has close connection with Artificial Intelligence.

Many attempts have been made, especially in this century, for augmenting the representational capabilities of logic, or for proposing non-additive models of uncertainty. One of the most radical and fruitful of these attempts was initiated by Lotfi Zadeh in 1965 with the publication of his paper "Fuzzy Sets." Starting from the idea of gradual membership, it has been the basis for both a logic of gradualness in properties and a new, particularly simple and effective, uncertainty calculus, called "Possibility Theory" by Zadeh (1978a), for handling the notions of possibility and certainty (or necessity) as gradual modalities.

When proposing fuzzy sets, Zadeh's concerns were explicitly centered on their potential contribution in the domains of pattern classification (Bellman *et al.*, 1966), processing and communication of information, abstraction and summarization (Zadeh, 1973). Although the claims that fuzzy sets were relevant in these areas appeared unsustained at the time when they were first uttered, namely in the early sixties, the future development of information sciences and engineering proved that these intuitions were right, beyond all expectations.

In the literature of fuzzy sets, the word *fuzzy* often stands for the word *vague*. Some comment on the links between vagueness and fuzziness is useful. In common use, there is a property of objects called "fuzziness"; see also Rolf (1980). From the Oxford English Dictionary we read that "fuzzy" means either not firm or sound in substance, or fringed into loose fibers. Fuzzy means also covered by fuzz, i.e., with loose volatile matter. Alike any other characteristic, "fuzzy" can be used to form a predicate of the form: "*something* is fuzzy". For example "a bear is fuzzy". It may sound strange to say that "*bald* is fuzzy", or that "*young* is fuzzy". Words (adjectives in this case) *bald* and *young* are vague (but not fuzzy in the material sense) because their meanings are not fixed by sharp boundaries. Similarly, objects are not vague.

Here however, the word "fuzzy" is applied to words, especially predicates, and is supposed to refer to the gradual nature of some of these words, which causes them to appear as vague. However, the term "vagueness" designates a much larger kind of ill-definition for words (including ambiguity), generally.

The specificity of fuzzy sets is to capture the idea of partial membership. The characteristic function of a fuzzy set, often called membership function, is a function whose range is an ordered membership set containing more that two (often a continuum of) values (typically, the unit interval). Therefore, a fuzzy set is often understood as a function. This has been a source of criticism from mathematicians (Arbib, 1977) as functions are already well-known, and a theory of functions already exists. However, the novelty of fuzzy set theory, as first proposed by Zadeh, is to treat functions as if they were subsets of their domains, since such functions are used to represent gradual categories. It means that classical set-theoretic notions like intersection, union, complement, inclusion, etc. are extended so as to combine functions ranging on an ordered membership set. In elementary fuzzy set theory, the set-union of functions is performed by taking their pointwise maximum, their intersection by their pointwise minimum, their complementation by means of an order-reversing automorphism of the membership scale, and set-inclusion by the pointwise inequality between functions. This point of view had not been envisaged earlier by mathematicians, if we except some pioneers, mainly logicians. Fuzzy set theory is indeed closely connected to many-valued logics that appeared in the thirties, if degrees of membership are understood as degrees of truth, intersection as conjunction, union as disjunction, complementation as negation and set-inclusion as implication.

This Chapter is meant to account for the history of how the notion of fuzzy sets could come to light, and it also presents a catalogue of basic notions which are presented in greater details in the other Chapters of this volume, as well as in the other volumes of the Handbooks of Fuzzy Sets Series. Section 1.2 shows that the problem of representing vagueness in logic, in physics, in linguistics, as well as the questioning of the notion of set in the twentieth century led to preliminary proposals that came close to fuzzy set theory. They make its emergence retrospectively less surprizing, if not expected. It is a way to show that fuzzy set theory is not a strange, gratuitous object that suddenly appeared out of nothing, but that it cristalized the intuitions of some leading scientists in the century. Section 1.3 presents various ways of representing a fuzzy set and provides the basic set-theoretic connectives as well as counterparts of various set-theoretic notions such as cardinality, inclusion and the like. It describes canonical tools for extending many mathematical notions to fuzzy sets. It also introduces special types of fuzzy sets useful in applications, like convex fuzzy sets, a noticeable example being fuzzy intervals. Section 1.4 overviews offsprings of the notion of a fuzzy set, that have been developed further on for their own sake, such as fuzzy relations, and set-functions for which fuzzy sets play the role of density. Section 1.5 is a repertory of variants of fuzzy sets found in the literature, where the membership function is changed into more elementary entities (a mere ordering relation, for instance), or more complex entities (for instance, when membership grades become functions themselves). It also describes some types of non-classical sets that have common features with, although different from, fuzzy sets. Finally, Section 1.6 discusses interpretive issues. Indeed a set is a very abstract notion, and it is very difficult to use pure set-theoretic intuitions in order to build membership functions in practice. The most popular interpretations of fuzzy sets in terms of similarity, uncertainty and preference profiles are reviewed. These concrete views of membership functions are those found in the current practice of fuzzy set-based methods in applications.

Two anthologies of Lotfi Zadeh's papers exist (Yager *et al.*, 1987, Klir and Bo Yuan, 1996). The first published book ever, specifically devoted to fuzzy sets is the one (in French) by Kaufmann (1973, translated into English in 1975), closely followed by a mathematical treatise by Negoita and Ralescu (1975a), based on a 1974 monograph in Romanian. Basic generic books on fuzzy sets are the ones by by Dubois and Prade (1980), Kandel (1986), Novak (1986), Klir and Bo Yuan (1995), among others. The book by Kruse *et al.* (1994) is more focused on foundations, in connection with probabilistic underpinnings. Introductory, mathematically oriented, monographs are the ones of Gottwald (1993), Lowen (1996), Nguyen and Walker (1996), while introductions which are more concerned with methodological issues and applications are those of Zimmerman (1985), Terano, Asai and Sugeno (1987), Klir and Folger (1988), Klir *et al.* (1997), Pedrycz and Gomide (1998). For some basic fuzzy papers of the first twenty five years, and references of other books, see also the anthology edited by Dubois, Prade and Yager (1993).

1.2 THE HISTORICAL EMERGENCE OF FUZZY SETS

About a hundred years ago, the American philosopher Charles Peirce was one of the first scholars in the modern age to point out, and to regret, that "Logicians have too much neglected the study of *vagueness*, not suspecting the important part it plays in mathematical thought." (Peirce, 1931). This point of view was also expressed some time later by Bertrand Russell (1923). Discussions on the links between logic and vagueness are not unusual in the philosophical literature in the first half of the century (Copilowish, 1939; Hempel, 1939). Even Wittgenstein (1953) pointed out that concepts in natural language do not possess a clear collection of properties defining them, but have extendable boundaries, and that there are central and less central members in a category. In spite of the considerable interest for multiple-valued logics raised in the 1930s by Jan Łukasiewicz (1910a, b; 1920, 1930) and his school who developed logics with intermediary truth value(s), it was the American philosopher Max Black (1937) who first proposed so-called "consistency profiles" (the ancestors of fuzzy membership functions) in order to "characterize vague symbols." The generalization of the traditional characteristic function has been first considered by H. Weyl (1940), who explicitly replaces it by a continuous characteristic function. The same kind of generalization was further proposed in 1951 by Kaplan and Schott (1951). They suggested calculi for generalized characteristic functions of vague predicates, and the basic fuzzy set connectives already appeared in these works. Strangely enough it is the mathematician of probabilistic metric spaces, Karl Menger (1951a), who, in 1951, was the first to use the term "ensemble flou" (the French counterpart of "fuzzy set") in the title of a paper of his.

This section gives some details on works published by what can be considered as forerunners of fuzzy sets. Some aspects of the early developments are described in more details by Gottwald (1984) and Ostasiewicz (1991, 1992b). The basic philosophical view presented in this section is the one of Ostasiewicz (1992a). Other points of view on the epistemology of fuzzy sets can be found in Toth (1987, 1992, 1997), von Furstenberg (1990).

1.2.1 Fuzzy-ism

The term "fuzzy set" has become a fashionable (and is still often ill-regarded) slogans used in scientific circles and sometimes even in everyday life. It is very much advertised, often ill defined, sometimes completely misused and misunderstood. The doctrine of fuzzy-ism has been properly defined by M. M. Gupta (1977) as "a body of concepts and techniques aimed at providing a systematic framework for dealing with the *vagueness* and *imprecision* inherent in human *thought* processes" (italics added by the authors of this Chapter). The three basic keywords, the three pillards of this doctrine which form its philosophical basis, are therefore following: thinking, vagueness, and imprecision.

i) The process of thinking i.e., mental creation, is apparently not intrinsically connected with linguistic expressions; only for the exchange of thought (the communication of ideas) do we need spoken language or its written equivalent (see Fraenkel *et al.*, 1973). Thinking hence refers to such processes or constructions in the mind, which, only after being cast in a *linguistic* form, become liable of analysis and logical tests.

ii) The two other terms: vagueness and imprecision, on which the whole doctrine of fuzzy-ism is founded, are considered here as empirical phenomena, and should be treated therefore as characteristics of language, and more generally of information.

Since the earliest antiquity, different attempts to cope with these phenomena have been undertaken. Various logical calculi were developed to reason with loose concepts, and different formal tools were proposed to deal with imprecise orders of magnitude. Fuzzy set theory seems to be one of the many links in a long chain of various tools invented with the aim of dealing with different aspects of language and knowledge.

There are two different kinds of theories of perception and thought (see Rolf (1981)): Some assume that the content of perception is never identical with the perceived entity. Others however assert that this content is identical with the object perceived, provided that perception is veridical. Identity theories were strongly criticized by B. Russell (1923), who clearly distiguished the properties of words from the properties of things. In this now famous paper, he writes that vagueness and precision are characteristics which can only belong to a language and that "apart from representation there can be *no such thing as vagueness or precision"* (italics are added). This means that vagueness is a characteristic of words, not of things. The argument that "things might actually be vague, as well as being vaguely described, is not properly intelligible" (see Garret, 1991; Zemach, 1991).

1.2.2 Philosophical Background

The ambition of fuzzy set theory is to provide a formal setting for incomplete and gradual information, as expressed by people in natural language. There is a very long tradition of philosophical interest in ambiguity and imprecision of knowledge. These notions, as some others like inexacteness, vagueness, uncertainty, etc., are apprehended very differently by scholars. There is however a quite general agreement in considering all such notions as relevant for representational systems, of which a language is a typical example. From this point of view, an information system can be considered as just another example thereof.

World and Words. The distinction between what pertains to language from what pertains to the rest of the world is very old. Among the classical ontological points of view on the external world, namely, holistic, individualistic and systemic, this section adopts the third one. This means that the world (external, or real world) is interpreted as a *system.* Under the notion of system one usually means a more or less autonomous aggregate of interconnected items (for more details see Bunge, 1967). This point of view stems from the so-called discrete ontology, according to which a world is considered as a collection of *objects* (also called components, elements, things, entities, items, etc.), which have some *properties* and some *relations* that hold between them (see Hasenjaeger, 1972; Narens, 1981).

Let us divide all properties into two groups:

- classificatory properties,also called sortal concepts, and
- measurable properties.

Measurable properties are in turn further divided into two groups:

- comparative or intensional properties, and
- metric or extensional properties, also called magnitudes.

In natural language, objects are represented by proper names, properties are expressed by adjectives and nouns, whereas prepositions and verbs tend to express relations between two or more things (see Russell, 1934). In artificial languages, for example in a language baed on first order logic, objects are represented by singular names, whereas properties and relations are represented by predicates. Problems of the connections between the real world, reasoning, and representation in semiotic systems are not discussed here. For some details and references see

Andreoli (1956), Birkhoff (1969), Kemeny (1963) and particularly Quine (1960, 1981).

Vagueness. An essential feature of vagueness is that it concerns the *meaning* of signs of a language. Vagueness is treated as particular case of, or as one of the types of, ambiguity. Any notion is said to be vague when its *meaning* is not fixed by sharp boundaries. There is a graceful degradation between situations to which a vague notion fully applies and situations to which it does not apply at all. There is thus a whole range of situations to which a vague notion partially apply. This is the *gradualness* phenomenon, also called "membership gradience" by Lakoff (1987). It is very important to emphasize that such gradualness, also called *fuzziness* here, is an essential characteristic of the concept of vagueness, which differs from uncertainty. Many authors have misinterpreted gradualness, considering that full applicability of a vague notion to a particular situation may be *dubious* (see Rescher, 1964). However such doubt may arise only in the case one is forced to decide if a vague notion applies or not to a particular situation, that is, if we are obliged to act as if the notion under concern were all-nothing. Hence we consider that gradualness is a primitive aspect of vagueness and that classification doubt is only a result of gradualness.

Vagueness, restricted to its fuzziness side, can be accounted for by attaching to any situation a *grade of applicability* of a given concept to it. A proposition is said to be vague if it contains gradual predicates. A characterictic feature of such propositions is that they may be neither true nor false when applied to depict a given situation, but in-between (Copilowish, 1939). Namely they can be true *to a degree.* This notion of partial truth is not only a convenient fiction. Symbolic forms of such degrees of truth can actually be found in natural languages, namely linguistic hedges like "very", "rather", "almost not", etc., which are very common. If somebody qualifies as "very young", it means that the truth level of the statement " this person is young" is "very", which means that this person has a high degree of membership to the class of "young people". This type of linguistic hedges apply only to gradual predicates. It provides a methodology for spotting such predicates. For instance, "single" is not a gradual predicate because "very single" is not an expression of common usage. See Lakoff (1972) for a study of linguistic hedges, putting together linguistic and fuzzy sets.

Modeling the gradual features of predicates enable some paradoxes of classical logic to be solved, such as the Sorites paradox:

> If a heap of sand is *small*, adding a single sand particle to it leaves it *small.*
> A heap of sand with a single particle is *small*
>
> ---
>
> Hence all heaps of sand are *small.*

The paradox stems from the all-or-nothing treatment of the gradual predicate *small.* The paradox disappears once it is acknowledged that, as particles of sand are added to the small heap, the degree of truth of the proposition "the heap of sand is small" decreases little by little (Goguen, 1969).

The question is however: why is there fuzziness and ambiguity in all languages? It seems that there are at least two reasons for this state of facts.

A first clue is that any language, or more generally any representational system is discrete, and the external world seems to be continuous (*natura non facit saltis*). This gap between discrete representations and continuous perceptions explains, according to Aristotle, the prevalence of ambiguity in languages.

"Since we cannot introduce the realities themselves into our discussion, but have to use words as symbols for them, we suppose that what follows in the words will follow in the realities too, like people reckoning with counters. But it is not the same. For whereas words and the quantity of sentences are limited, realities are unlimited in number. It is therefore inevitable for the same sentence and the one word to mean more than one thing" (translated by Robinson, 1941).

In particular, a certain number of words refer to supposedly continuous numerical scales. Consider the word *young* for instance, applied to humans. It is natural to think that the more fine-grained the scale of age (going from years to months, weeks, days etc.), the more difficult is it for a person who uses the term *young* in a given context, to fix an age threshold below which *young* fully applies, and above which it does not at all. There is a conflict between the linguistic representation of the age scale (it is a finite term set, for instance {*young, mature, old*}) and its numerical representation (say, the real interval [0, 140] years for humans). Il seems attractive to try and cope with the rigid nature of the term set by attaching continuous consistency profiles (Black, 1937) to each term. Each profile represents the range of ages covered by a term, along with the degree of applicability of this term to each age. The range of functions representing consistency profiles, which Zadeh (1965a) calls membership functions, may contain as many elements as in the age scale, hence be continuous itself. Terms like *young* will be called gradual (or fuzzy) predicates.

A second reason for the pervasive vagueness of languages and the gradualness of symbols denoting classes of objects may also be due to the fact that natural language tolerates the presence of exceptions. For instance, among objects people call *birds,* some fly, other don't, because some even have no wings. In other words, among birds, some are more typical than others. Lakoff (1987) speaks of "centrality", « the idea that some members in a category can be better members of that category than others». So, the class of birds is not just a set, it is a set partially ordered by a relation of compared typicality. Grades of applicability, or degrees of membership, if any, evaluate how normal, or central, individuals are in the class. But then it is not always so natural to represent degrees of membership numerically. The consistency profile is then only a partially arbitrary encoding of the partial ordering relation *more typically bird than.*

There is a dispute among scholars of vagueness as to whether a membership threshold exists for vaguely defined classes. Let us consider one of the oldest broadly discussed examples of vagueness, namely the notion *bald*. A man without hair on the head is bald, while a very hirsute man is not. In many cases however, applicability of the word *bald* is partial, since a person may have more or less hair on the head. However, if we are forced to decide whether to say such a person to whom some hair is missing is *bald* or not, we shall be hesitant about where to set a *baldness/non baldness* threshold. The key issue when debating about vagueness is whether such threshold, even if ill-located, exists or not. In the fuzzy set tradition, it is assumed that it does not exist. Some threshold appears only when a fuzzy category is made crisp by forcing individuals into a black-or-white membership decision. However, if this threshold separating membership from non-membership exists, then the excluded middle and contradiction laws will hold for vague concepts, and membership grades only stem from a noisy threshold (Fine, 1975). If this theshold does not exist, then these basic laws of logic will not necessarily hold.

Imprecision. Imprecision is also characteristic of language, and more particularly pertains to so-called measurable concepts, and particularly metric properties. Because such concepts are represented by numbers, one often uses the term of *numerical imprecision*. This imprecision stems from the fact that any measurement yields results of limited accuracy. It is possible to speak only of a one-many correspondence between the results of measurement and real numbers.

The idea of using intervals instead of single numbers (Chwistek, 1948) has been developed and extended in various direction by Mellor (1965, 1966), who has among other things defined the *amount of imprecision* in the concept to be equal to the length of an assigned interval. In logic, imprecision appears under the form of disjunctions: the statement that a proposition p OR q is true contains imprecision because it is not known whether p is true, q is true, or both. This situation is often called *incompleteness*. So in the general case, imprecision is represented by sets, viewed as a *disjunction* of elements (not a conjunction, as usual with sets). For instance, an interval representing an ill-known value contains elements which are mutually exclusive: only one such element is the value of interest. We shall also speak of *disjunctive sets*.

An attempt to develop a formal theory enabling manipulations with imprecise or indeterminate quantities has been undertaken very early by R. Young (1931). This pioneering work is continued in present days in the form of the so-called interval mathematics. The idea of a variable which assumes individually considered values, has been replaced by R. Young by a concept of many-valuedness. This concept was conceived as a set of values but considered collectively. The theory of such sets has been developed under the name *mereology* by S. Lesniewski (1916, 1992)

A related issue is the one of indistinguishability. Poincaré (1905) stressed that even if it is impossible to distinguish a physical quantity A from another quantity B, and quantity B from quantity C, we may sometimes distinguish A from C. He had espressed this raw result of experience by the often quoted relation:

$$A = B, B = C \text{ and } A < C$$

which has been regarded as formalizing the idea of the physical continuum. This result contradicts classical logic, and it is called therefore the Poincaré's paradox. However, it make sense if A = B is interpreted as d(A, B) < ε for some distance function d and some threshold ε. Then any observed value A = *a* is imprecise and should be interpreted as an interval I(A) of center *a* and width ε, and A = B just means that I(A) and I(B) intersect. In 1908, E. Borel has critically discussed this problem relating it to the afore-mentioned sand heap paradox (see Borel, 1956).

K. Menger (1951b) suggested a different solution to this paradox. He proposed to describe the equality to the two raw results by " associating with A and B a number, namely, the probability of finding A and B indistinguishable...". Nowadays, imprecise quantities are also modeled by fuzzy subsets of the real numbers giving rise to the development of the so-called fuzzy arithmetic (See Chapter 10 by Dubois, Kerre, Mesiar and Prade in this volume).

Uncertainty. While imprecision and vagueness refer to the contents of a piece of information expressed in some language, uncertainty refers to the ability of an agent to claim whether a proposition holds or not. Modeling uncertainty refers to the quantification of partial belief regarding propositions. Uncertainty about Boolean propositions can be gradual as is the case with probability theory where partial belief ranges on the unit interval. However, in propositional logic, uncertainty is three-valued: one can be sure about a proposition, sure about its negation, or unsure about both. Explicitly expressing and reasoning about these three situations requires the setting of modal logic (Hughes and Cresswell, 1968), where certainty is captured by the modal necessity and reflects provability, while the lack of certainty is captured by the modal possibility, and reflects logical consistency.

However the type of uncertainty modelled by probability is different from the type of uncertainty modelled by propositional logic. Probability theory often models uncertainty resulting from conflicting, precisely observed, pieces of information. This is usually the case in statistics where a random experiment, run several times, does not produce the same outcomes. Propositional logic only accounts for uncertainty stemming from incomplete or imprecise information. When such information is fuzzy, a model of gradual uncertainty is derived, which extends the one of propositional logic, in that such uncertainty is also due to a lack of information (and not too large a quantity of conflicting precise information as with frequentist probability). So fuzzy sets leads to a gradual theory of uncertainty which differs in its purpose from the one of frequentist probability. This is possibility theory, where gradual notions of possibility and necessity are present (see Dubois and Prade, 1988, and Chapter 7 by Dubois, Nguyen and Prade in this volume).

This discussion points out that uncertainty differs from imprecision and vagueness and only result from them. In order to see better the differences between the three notions discussed here, let us consider the following assertions about some car:

- This car is between 10 and 15 years old (pure imprecision).
- This car is very big (imprecision and vagueness).
- This car was probably made in Germany (uncertainty).

In the first case there is a lack of knowledge, due to a lack of ability to measure or to evaluate numerical features. In the second case there is a lack of precise definition of the notion *big* and the modifier *very* indicates a rough degree of "bigness", and the third case expresses uncertainty about a well-defined proposition, perhaps based on statistics.

1.2.3 From Logic to Fuzzy Logics

There is a bewildering Babel of tongues as to what logic is about (see Cohen and Nagel, 1962). Different schools, the traditional, the linguistic, the psychological, the epistemological, and the mathematical, speak different languages, and each regards the other as not really dealing with logic at all. The founder of logic as a distinct discipline, is Aristotle (384-322 B.C.), and the view created and developed by him is called today the traditional or Aristotlelian logic. However, this view, whereby logic is a formal approach to human reasoning, is the one underlying recent works in Artificial Intelligence, especially people interested in knowledge representation and reasoning about knowledge. This is also the point of view of fuzzy logic.

Aristotle and his disciples have developed two parts of logic: the logic of terms, and the syllogisms. The logic of propositions is ascribed to Chrisippus and his school, i.e., Stoics. After the Great Greeks, the most significant contribution to logic has been done by Gottlob Frege (1848-1925). His works deal with the question of the soundness of the deductive method. Another important contribution has been made by Jan Lukasiewicz (1878-1956) who is considered as the main founder of multi-valued logic. And multi-valued logic is to fuzzy set theory what classical logic is to set theory.

The earliest reference to the three-valued logic is "the farewell lecture" delivered by Lukasiewicz (1920) on March 7, 1918 in the Warsaw University on the occasion of taking duties in Polish government. In this lecture, he announced the following: "that system, which is worked out in detail last summer, is as coherent and self-consistent as Aristotle logic, and is much richer in laws and formulae. That new logic, by introducing the concept of objective possibilities, destroys the former concept of science, based on necessity". This new system has been published for the first time in Polish in 1920b (see the English version in Lukasiewicz (1970). This system has been axiomatized by Wajsberg (1931). See Gottwald (1999) in the Handbooks of Fuzzy Sets Series, for more details).

Independently from Lukasiewicz and starting from different premises another system of multi-valued logic was discovered by Emil Post (1921). In his work, Post has proved, among other things, that every sentence of m-valued logic can be interpreted as an ordered set of m-1 sentences of usual logic. In the theory of fuzzy sets, the analog of this theorem exists under the form of a representation of fuzzy sets by a family of usual sets.

Since these works by Lukasiewicz and Post, many kinds of multivalued logics were discovered and studied (see for example, Rescher (1969), Hajek (1998), Gottwald (1999).

Graded Truth. Classically, the concept of truth is defined as adequation to fact. The basic principle of classical logic known as the Principle of Bivalence asserts the following:

Every proposition is either true or false.

Introducing this principle, which asserts that all propositions must be either true or false, Aristotle wanted to characterise propositions as opposed to other kinds of sentences like questions, orders, etc. This concept of truth has been formally developed in detail for the first time in the late twenties by Alfred Tarski (1940).

Lukasiewicz (1910a) called this Principle of Bivalence the *Aristotlelian prejudice*, and decided to put the end to it once and for all. This means that it should be replaced by the more general Principle of Valence, which says the following:

Every proposition has a truth value.

Lukasiewicz had characterized propositions as those sentences which predicate something about something and hence assert something. Such sentences are not only either true or false but they can have an intermediary truth value (expressed by a number from the unit interval). The idea that truth is a matter of degree is also advocated in the philosophical works of Bradley (1914) ("All truths and all errors in my view may be called relative, and the difference in the end between them is a matter of degree"); see (Haack, 1996) for a critical discussion.

Trying to develop a quantitative concept of truth, H. Reichenbach has proposed his "probability logic" in which the alternative true-false is replaced by a continuous scale of truth values. H. Reichenbach (1949) begins with a simple illustrative example of the statement "I shall hit the center". As a measure of the degree of truth of this statement, H. Reichenbach proposes to measure the distance of the hit to the center and to take the truth-value as equal to 1/(1+r). This can be done of course *after* the shot. Unfortunately we cannot find advice in his work as how to evaluate the truth of the above sentence *before* the shot. However, it is easy to figure out retrospectively that this method is actually evaluating the fuzzy proposition "I shall hit *close to* the center". It can only be done after the hit. Quantifing the proposition before the hit is a matter of belief assessment, not of measuring graded truth.

Herman Weyl (1940) also devised a logic where any proposition is assigned a number from a fixed subset L of the unit interval. The only condition which this subset L must fulfill is that it must be closed under the operation 1-x. The operations on the set L interpreted as logical connectives *and*, *or*, *not*, and *implies*, are defined as follows:

$$\text{a } \textit{and} \text{ b} = \min(a,b),$$
$$\text{a } \textit{or} \text{ b} = \max(a,b),$$
$$\textit{not} \text{ a} = 1 - a,$$
$$\text{a } \textit{implies} \text{ b} = 1 - a + \min(a, b) = \min(1, 1 - a + b).$$

H. Weyl (1946) has also introduced a calculus of vague predicates treated as functions defined on fixed universe of discourse U, with values in the unit interval. Operations on such predicates $f : U \rightarrow [0,1]$ have been defined as follows:

$$f \cap g = \min(f, g) \text{ (conjunction);}$$
$$f \cup g = \min(f, g) \text{ (disjunction);}$$
$$f^c = 1 - f \text{ (negation).}$$

Clearly, this is one ancestor of the fuzzy set calculus. However, one of the approaches discussed by him for interpreting these connectives again considers truth values as probabilities. As shown below, this interpretation is dubious, first because probabilities are not compositional, but also because probability and truth address different issues.

Truth Versus Modalities. The logical origins of fuzzy set theory are pervaded with such a confusion between the truth values of fuzzy sentences and modalities, especially probabilities. This is due to a misunderstanding regarding the law of excluded middle and the law of non-contradiction, and the connections between many-valued logics and modal logics. The principle of bivalence, formulated and strongly defended by Chrisippus and his school in antique Greece, was for instance questioned by Epicureans, and even rejected by them in the case of propositions referring to future contingencies.

Let us take an example considered already by Aristotle, namely the proposition:

"There will be a sea battle to-morrow (p) and
there will not be a sea battle to-morrow (¬p)"

This proposition "p and ¬p" is ever false, because of the non-contradiction law and the proposition "p or ¬p" is ever true, because *tertium non datur*. But we may *fail to know* if the proposition "there will be a sea battle to-morrow" is true and if the proposition "there will not be a sea battle to-morrow" is true. In this case, at least intuitively, it seems reasonable to say that it is *possible* that there will be a sea battle to-morrow but at the same time, it is *possible* that there will not be a sea battle to-morrow. There has been a recurrent tendency, until the twentieth century many-valued logic tradition, to claim the failure of the bivalence principle on such grounds, and to consider the *possible* as a third truth value. However the proposition *possible* p is not the same as p, and *possible* ¬p is not the negation of *possible* p. Hence the fact that the proposition

possible p and *possible* ¬p

may be true does not question the law of non-contradiction. On the contrary, vague or fuzzy propositions are ones such that, due to the gradual boundary of their sets of models, proposition "p and ¬p" is not completely false in some interpretations. This is why Moisil (1972) speaks of fuzzy logics as Non-Chrisippean logics.

In fact, besides the modality *possible*, Aristotle already considered other so-called modal expressions like *impossible*, and *necessary*, giving the base for the so-called modal logic, for which the negation of *possible* p is *impossible* p, which is considered as equivalent to *necessary* ¬p.

Modal logic and multivalued logic are closely related, both have at least the same philosophical roots (Katsoff, 1937). First of all, the development of both of them has been influenced by the principle of bivalence. Second, the probabilistic tradition of considering equipossibility as identical to equiprobability, has misled researchers to view many-valued logics as a setting for probabilistic reasoning.

In 1786 J. J. Ebert has defined the modality as an explicate determination of the *grade* of intensity to which some property is affirmed or negated. The first significant work on modal logic has been published in 1897 by H. MacColl (1897, 1900), who introduced different modal sentences like *known to be true*, *known to be false*, *doubtful* and *neither known to be true nor known to be false*. In his work, MacColl discussed the following modalities: *true*, *false*, *certain*, *imposible* and *variable*, i.e., *neither certain nor impossible* (see Becker, 1930). This idea had been rudimentary realized for the first time probably by B. Bolzano (1837) (see also Lukasiewicz, 1930).

B. Bolzano had introduced a new logical concept, which he called "the validity of proposition". Let us consider propositions of the type "x is A", where one variable is declared to be changeable and the other (i.e., A) not. For example in the proposition "the entity Caius is mortal", the name "Caius" is treated as one that can be changed at will and replaced by new names. By changing this name we obtain propositions some of which are true, and some false. The quantity of the true and the false propositions in certain circumstances can be computed. The ratio of the number of the true propositions obtainable from a given proposition by changing the values of certain variables, to the number of all propositions obtainable in this way is the "validity" of a proposition.

Suppose for example that the whole universe of discourse is the set of six numbers {1,2, ...,6}. Furthermore assume that in proposition "5 is divisible by 3" the value 5 can be changed, then the validity of this proposition is equal to 2/6. However, again here, the main idea is to assert the belief in a certain particular proposition as induced from a proportion of true similar propositions in a population of examples. It is not connected to the notion of graded truth (which remains binary for propositions of the form "m is divisible by n)".

Many attempts were further on undertaken to import probability theory into realm of logic or to reconstruct logic in accordance with the probabilistic laws by means of many-valued logics. In 1932 Z. Zawirski argued that "la logique nouvelle n'est qu'une nouvelle manière d'éclairer les formules du calcul des probabilités". He distinguished two kinds of multi-valued logics: qualitative and metrical. For more details, see Zawirski (1932, 1934).

In 1934 H. Reichenbach introduced the notion of "probability proposition" to which he assigns the probability interpreted as a grade of truth. The probability proposition has the form $A \rightarrow_p B$ of an implication and is treated as a relation between two classes: the class A which is called the reference class, and class B, the attribute class. The expression" $A \rightarrow_p B$" means "the probability of the attribute B that is concerned with reference to A is equal to p". It is important to emphasize the conditional form of probability expression, since "no one would say that the probability of finding a die on the table with face 1 up has the value 1/6, if the die had not be thrown" (Reichenbach, 1949).

Graded truth and Probability: Clarifying the Case. Very early, when many-vaued logics came to light, some scholars in the foundations of probability became aware that probabilities differ from what logicians call truth-values. De Finetti (1936), witnessing the emergence of many-valued logics (especially the works of Lukasiewicz), pointed out that uncertainty, or partial belief, as captured by probability, is a meta-concept with respect to truth degrees, and goes along with the idea that a proposition, in its usual acceptance, is a binary notion. On the contrary, the notion of partial truth, as put forward by Lukasiewicz (1930), leads to changing the notion of proposition proper. The definition of a proposition is a matter of convention. To quote De Finetti (our translation):

> «Propositions are assigned two values, true or false, and no other, not because there "exists" an a priori truth called "excluded middle law", but because we call "propositions" logical entities built in such a way that only a yes/no answer is possible... A logic, similar to the usual one, but leaving room for three or more [truth] values, cannot aim but at compressing several ordinary propositions into an single many-valued logical entity, which may well turn out to be very useful...».

In this statement, one can perceive the idea of a fuzzy set as a collection of non-fuzzy level-cuts. It also puts it very clearly that many-valued logics, hence fuzzy sets, deal with many-valuedness in a logical format, not with uncertainty or probability. On the contrary, uncertainty pertains to the beliefs held by an agent, who is not totally sure whether a proposition of interest is true or false, without questioning the fact that ultimately this proposition cannot be but true or false. To quote De Finetti (1936) again:

> «Even if, in itself, a proposition cannot be but true or false, it may occur that a given person does not know the answer, at least at a given moment. Hence for this person, there is a third attitude in front of a proposition. This third attitude does not correspond to a third truth-value distinct from yes or no, but to the doubt between the yes and the no (as people, who, due to incomplete or undeciferable information, appear as of "unknown sex" in a given statistics, do not constitute a third sex. They only form the group of people whose sex is unknown».

Probabilistic logic, contrary to multiple-valued logic is not substituted to binary logic. It is only superposed to it. However this point is not always clearly made by the forefunners of many-valued logics. Carnap (1945) also points out the difference in nature between truth-values and probability values (hence degrees thereof), precisely because "true" (resp: false) is not synonymous to "known to be true" (resp: known to be false), that is to say, verified (resp: falsified). He criticizes Reichenbach on his claim that probability values should supersede the two usual truth-values. Quoting Carnap:

> «A given sentence is often neither verified nor falsified; nevertheless it is either true or false, whether anybody knows it or not. In this way an inadvertent confusion of "true" and "verified" may lead to doubts about the validity of the principle of excluded middle... I agree with Reichenbach that a concept referring to an absolute and unobtainable maximum should be replaced by a concept referring to a high degree in a continuous scale. However what is superseded by "highly probable"... is the concept of "confirmed to the highest degree or verified", and not the concept "true"».

The above discussion should convince the reader that, despite some early claims, many-valued logics, that form the logical underpinnings of fuzzy set theory, have strictly nothing to say about uncertainty and partial belief, contrary to probability. In their essence, fuzzy sets grasp the idea of partial truth, gradedness in the compatibility of a statement and a state of facts. The use of fuzzy sets as a tool for describing ill-known information (as done in possibility theory, see Chapter 7 by Dubois *et al.* in this volume) only comes as a by-product of the use of sets and intervals for the representation of incompleteness and inaccurateness.

1.2.4 From Sets to Fuzzy Sets

The notion of set is one of the most important one, used frequently in every day life as well as mathematics. In mathematics this term is so important, that it is argued that every mathematician must know something about set theory. Already in 1666, Leibniz has defined a notion of a set as "any number of things whatever may be taken simultaneously and yet treated as a whole". (See Kruze-Blachowicz, 1982). Georg Cantor (1845-1918), Kronecker's disciple, made an attempt to base the whole mathematics on the theory of sets. He gave the following definition: "by a set we understand every collection made into a whole of definite, distinct objects of our intuition or of our thought. (See Murawski, 1994). Cantor's theory proved to be inconsistent. The notion of "set" was however intuitively appealing and the theory of sets quite nice and powerful. Many attempts to improve this theory instead of rejecting it were undertaken. For those who criticized and wanted to reject this theory, D. Hilbert (1862-1943) gave his now famous sentence: "nobody could expel us from this paradise".

The most important proposals aiming to overcome the inconsistency of Cantor's theory are due to E. Zermelo (1871-1953), B. Russell (1872-1970) and S. Lesniewski (1916, 1992). Russell and Zermelo have introduced some restrictions of different characters to the unlimited freedom in Cantor's Paradise. Zermelo's method has been improved by A. Fraenkel, and now this system of set theory is known as Zermelo-Fraenkel theory of sets, or shortly as ZF system. Within this system, but not generalizing it, fuzzy sets theory is developed, as well as as the theory of rough sets (Pawlak, 1982, 1991). Lesniewski has proposed a completely different system, for the foundations of mathematics, but unrelated to fuzzy sets.

Cantor vs. Lesniewski. Sets in Cantor's sense differ essentially from sets in Lesniewski's sense. Cantorian sets as *abstract* entities are conceived as an aggregation of things treated as *one* abstract object. Therefore one speaks about the sets-as-one view. Philosophically, these sets are conceived as Platonic ideas, they exist independently of our thought, "whiteness" exists independently of white objects. A collection of seven given pairs of shoes is not to be identified with the collection of the actual fourteen shoes, they are differents sets. This of course may be seen as intuition-defying.

Sets in Lesniewski's theory are treated collectively as *material* (concrete, physical) objects. In this case one speaks about the set-as-many. Philosophically, Lesniewski's theory could be considered as a nominalistic theory of sets, according to the basic doctrine of this theory, "whiteness" does not exist, only white objects do.

In natural language, a distinction is sometimes made between these two approaches, by using different nouns. For example the term "soldiers" is conceived as a set-as-one, but "army" is a set-as-many.

In the case of a "set-as-thing" (set-as-one) the most important relation is the relation of elementhood or membership, in the case of a "set of" (set-as-many) the fundamental relation in the relation of being part. One soldier is an element of a set "soldiers", any part of a set "army" is an army too.

For a set-as-one the following properties hold:

i) $x \neq \{x\}$;
ii) if $x \in X$ and $X \in Y$ then $x \notin Y$; ($\in$, denoting "is member of" is not transitive);
iii) $\varnothing$ is a very important set;

on the contrary, for sets-as-many, it holds:

i) x={x};
ii) if $x \angle X$ and $X \angle Y$ then $x \angle Y$; ($\angle$, denoting "is part of" is transitive);
iii) ∅ does not exist, it has no concrete meaning.

Boole. One often assumes that the empirical world consists of a finite collection of objects, with visible empirical properties and with visible empirical relations holding among them (see Narens, 1981). As early as in 1854, G. Boole assumed the existence of the above mentioned two classes: the class of all properties and the class of all objects. He established two laws connecting properties with objects (see Boole, 1854).

First of all he has defined an algebra on the class of all properties with respect to the words: *and, or, not, implies*. This algebra is called to-day a Boolean algebra.

The second law asserts the existence of an isomorphism between the algebra of properties and the algebra defined on the family of all sets corresponding to properties. Hence G. Boole assumed that to any property A corresponds the set Â of all objects possessing this property.

Suppose that the expression "x is A" means that an object x has a property A; then the set Â of all objects having the property A is:

$$\hat{A}=\{x|\ x \text{ is } A\}.$$

If X denotes the universe of all objects, then, subset Â can be defined by means of the so-called characteristic function, which has been introduced in 1936 by Ch. de la Vallée Poussin.

$$f_{\hat{A}}(x) = \begin{cases} 1, \text{ if } x \in \hat{A} \\ 0, \text{ if } x \notin \hat{A} \end{cases}$$

Defining proper algebraic operations on the set of such functions, one arrives at the following diagram of isomorphisms:

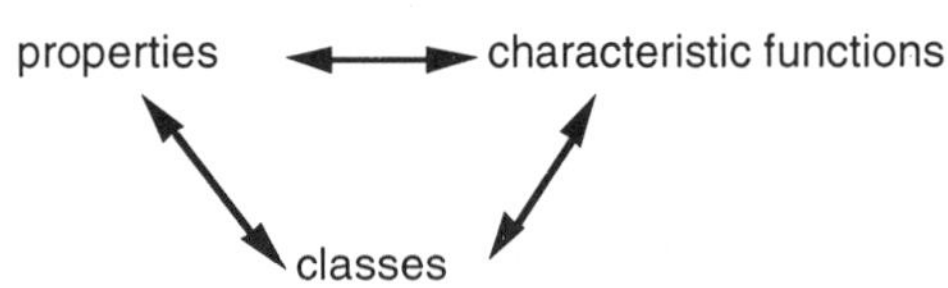

It is interesting to note, that the calculus of classes which forms a Boolean algebra is a part of general theory of sets, and that this part is made of axioms of which do not contain the primitive term "∈". On the other hand, a Boolean algebra defined on a class of properties is the strongest structure as far as connectives *and* and *implies* are concerned.

Kaplan and Schott. As early as in 1946, the philosopher Abraham Kaplan argued in favor of the usefulness of the classical calculus of sets for practical applications. The essential novelty he introduces with respect to the Boolean calculus consists in «entities which have a degree of vagueness characteristic of actual (empirical classes)» (see Kaplan, 1946).

Such calculus has been presented by A. Kaplan and H. Schott (1951), and has been called the calculus of empirical classes (CEC). Instead of notion of "property", Kaplan and Schott prefer to use the term *profile* defined as a type of quality (such as red, green, etc.). This means that a profile could be a simple property like *red*, *green*, etc. or it could be a complex property like *red and 20 cm long*, *green and 2 years old*, etc.

They have replaced the classical characteristic function by an *indicator* which takes on values in the unit interval. These values are called the weight from a given profile to a specified class. In the work of Kaplan and Schott, the notion of "empirical class" corresponds to the notion of "fuzzy set", and a value in the range of the generalized characteristic function (indicator, in their terminology) is called by Kaplan and Schott a "degree of membership" (Zadehian grade of membership). The notation $w_\lambda(x, A) = \alpha$ states that the weight from the profile λ, characterizing object x to the class A is α. Subsequently, a simplified notation w(x, A) will be used, and according to modern terminology these weights will be also called grades of membership. Indicators of profiles are now called membership functions of fuzzy sets.

The above diagram for the classical calculus of classes, can take the new following form:

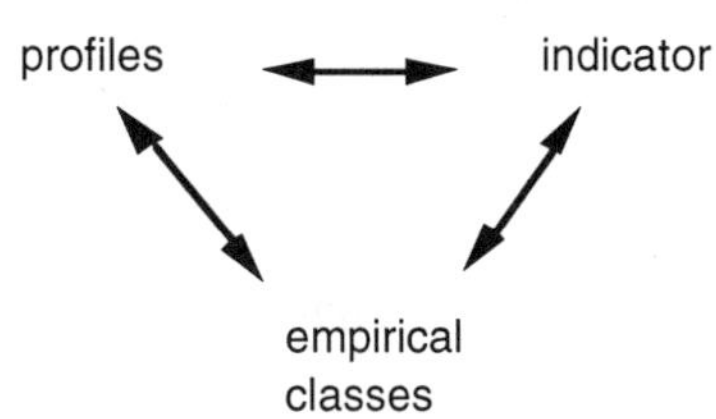

Kaplan and Schott have defined various operations for the empirical classes. Some of them are briefly reviewed in this paragraph. The basic notions are defined as follows:

1. Class sum : $w(x, A \cup B) = \max(w(x,A), w(x, B))$.
2. Class product: $w(x, A \cap B) = \min(w(x,A), w(x, B))$.
3. Class contrary: $w(x, A^c) = 1 - w(x,A)$.

It is stated that these operations fulfill the following conditions: commutativity, associativity, De Morgan laws, the laws of tautology, the two distributions laws, the laws of double negation and of transposition. This means that on the set of all empirical classes form a De Morgan algebra, a structure, introduced for the first time by the Romanian logician G. Moisil (1935).

In order to have the counterpart to some classical definitions, particularly concerning the null (N) and universal classes (U), they introduce quasi-null (ν) and quasi-universal (υ) classes, defined as follows:

1. Null class	$w(x,N) = 0$, for all x.
2. Universal class	$w(x, U) = 1$, for all x.
3. Quasi-null classes	$w(x, \nu) \leq 1/2$, for all x.
4. Quasi-universal classes	$w(x, \upsilon) \geq 1/2$, for all x.

The inclusion for empirical classes is defined as follows:

$$A \subseteq B \Leftrightarrow A = A \cap B.$$

This is equivalent to $A \subseteq B \Leftrightarrow w(x, A) \leq w(x, B)$ for all x. Using these definitions the following properties were established:

1. $A \cup N = A$.
2. $A \cup A^c$ is quasi-universal.
3. $A \cap A^c$ is quasi-null.
4. $A \subseteq B \Rightarrow A^c \cup B$ is quasi-universal.
5. $A \subseteq B \Rightarrow A \cap B^c$ is quasi-null.

In addition to the inclusion ordering given by relation $\subseteq$, a different ordering, called w-sharpening, is defined as follows:

$$A \geq B \Leftrightarrow \begin{cases} w(x, A) \geq w(x, B) \text{ for } w(x, B) \geq 1/2 \\ w(x, B) \geq w(x, A) \text{ for } w(x, B) < 1/2. \end{cases}$$

Then A is closer to a standard class than B, if $A \neq B$. An idealization A of a class A is defined as follows:

$$A^* = \begin{cases} 1 \text{ if } w(x, A) \geq 1/2 \\ 0 \text{ if } w(x, A) < 1/2. \end{cases}$$

Using this notion, the meaning of a term is understood as follows. Two terms are said to have the same meaning, if the corresponding classes (i.e., their extensions) are equal, and these terms have *substantially* the same meaning if the idealizations of corresponding classes are equal. The classical calculus of classes is properly considered by Kaplan and Schott as an *idealization* which disregards the vagueness characterizing symbols in their actual use.

In order to establish connections with the general theory of sets, Kaplan and Schott have modified the principle of extensionality. For this reason they have defined membership at degree α, symbolized by $\in^{\alpha}$, as the relation of an object to a class when the weight from the profile characterizing this object to the class is α. The principle of extensionality is defined as follows:

$$A = B \text{ if and only if } x \in^{\alpha} A \Leftrightarrow x \in^{\alpha} B, \forall x.$$

This calculus is very close to the theory of fuzzy sets of Zadeh. The work by Kaplan and Schott which contains many results, independently rediscovered by L. Zadeh (1965a) in his first paper (some of them being even more deeply developed), nevertheless remained confidential. For the applications of this calculus, see Schutz (1959).

Menger. On the contrary, the paper by K. Menger (1951) is well known in the fuzzy set community; the recognition it received is perhaps higher than it deserves, within the fuzzy set history. K. Menger only suggested the necessity of developing a theory in which the element-set-relation is replaced by the probability of an element belonging to a set. For that reason he has defined a notion «ensemble flou» (the paper is originally in French) which he has translated by himself into English as *hazy set* (Menger, 1966).

K. Menger (1951b, 1966), like H. Weyl, suggested to replace *qualities*, treated as monadic relations in classical sense, by *quantities*, treated as functions of one variable. A hazy set A is defined therefore as a function p_A: X→ [0, 1] and a value $p_A(x)$ is interpreted now as the probability that element x belongs to a set A. Unfortunately, the intuition motivating this probability has not been explained, nor the modes of determining it. It is not clear either what does "element belongs to a set" means as an event. Menger (1966) acknowledged some relationship between his idea of "cloud-like" points, or "hazy set", and the idea, recently expressed at that time, of a fuzzy set, anyway emphasizing the differences between the two concepts.

Menger (1951a, 1951c) considered what is now known in the fuzzy set literature as max-product transitive fuzzy relations, but with a probabilistic interpretation. His work in probabilistic geometry deals with triangular inequality when the location of points are associated with probability distribution functions. It was motivated by the non-transitivity (in a strict sense) of equality in a physical continuum, recalled above in Section 1.2.2, as discussed by Poincaré (1905). Moreover Menger's work on probabilistic metric spaces also led to the introduction of so-called triangular norms and co-norms, which have been extensively studied by B. Schweizer and A. Sklar (1983), and which later have turned out to be also basic operators for fuzzy sets (see Chapter 2 by Fodor and Yager in this volume).

Klaua. In fact the idea of graded membership was suggested by Menger himself to the (East) German mathematician D. Klaua who, between 1965 and 1972, developed an extensive theory of many-valued sets that has much in common with fuzzy set theory, albeit with a more mathematically-oriented treatment (e.g., Klaua, 1966). See Gottwald (1984) for an extensive account of the often forgotten works by Klaua (and for their genesis), along with a full bibliography.

NB. The French linguist Yves Gentilhomme (1968) coined the term "ensemble flou" (the French counterpart for "fuzzy set"), without being aware of Zadeh's (1965a) fuzzy sets. Gentilhomme used this term for naming a pair of nested (classical) subsets, (E_1,E_2) with $E_1 \subseteq E_2$, in order to distinguish between "central interpretations" (in E_1) of the concept thus described, and "peripheral" ones (in E_2 - E_1).

1.3 BASIC NOTIONS OF FUZZY SET THEORY

This section provides basic definitions of fuzzy set theory and its main connectives. More details on fuzzy set connectives are given in Chapter 2 of this volume. The emphasis here is put on the various representations of a fuzzy set, that are instrumental when extending formal notions from sets to fuzzy sets. The problem of comparing fuzzy sets is discussed and methods for building comparison indices are surveyed. This section finally addresses the properties of fuzzy sets induced by the structural properties of the referential, such as convexity, and distances between fuzzy sets.

1.3.1 Representations of a Fuzzy Set

Membership Functions. L. A. Zadeh (1965a) has given in his now famous paper the following definition: A fuzzy set is a class with a continuum of membership grades. So, a fuzzy set (class) F in a referential U is characterized by a membership function which associates with each element u in U a real number in the interval [0, 1]. The value of the membership function at element u represents the "grade of membership"of u in F. A fuzzy set F is thus defined as a mapping

$$F: U \rightarrow [0, 1],$$

and it is a kind of generalization of the traditional characteristic function of a subset A

$$A: U \rightarrow \{0, 1\}.$$

Note that fuzzy sets are actually fuzzy subsets of U, as emphasized by Kaufmann (1975). There is a tendency now to identify the theory of fuzzy sets with a theory of generalized characteristic functions. This is why in the following we shall denote the membership grade of u to a fuzzy set F as F(u) instead of $\mu_F(u)$. In particular, F(u) = 1 reflects full membership of u in F, while F(u) = 0 expresses absolute non-membership in F. Usual sets can be viewed as special cases of fuzzy sets where only full membership and absolute non-membership are allowed. They are called crisp sets, or Boolean sets. When $0 < F(u) < 1$, one speaks of partial membership. For instance, the term *young* (for ages of humans) pictured on Figure 1.1 applies to a 30-year old individual only at degree 0.5. A fuzzy set can be also denoted as a set of pairs made of an element of u and its membership grade when positive: $\{(u, F(u)), u \in (0, 1]\}$. The set of fuzzy subsets of U is denoted $\mathcal{F}(U)$.

The membership function attached to a given word (such as *young*) depends on the contextual intended use of the word; a young retired person is certainly older than a young student, and the idea of what is a young student also depends on the user. However, in the different contexts, *young* will be understood as a gradual property generally.

Membership degrees are fixed only *by convention*, and the unit interval as a range of membership grades, is arbitrary. The unit interval is natural for modelling membership grades of fuzzy sets of real numbers. The continuity of the membership scale reflects the continuity of the referential. Then a membership degree can be viewed as a degree of proximity between u and the prototypes of F, that is, the elements v such that F(v) = 1. The membership grade decreases as elements are located farther from such prototypes. For instance, Figure 1.1 presents a plausible representation of the predicate young (for human beings in a given context). This representation points out that there is no precise threshold between ages that qualify as *young* and ages that qualify as not young. More precisely there is a gap between protopypes of *young* and proptypes of *not young*. It is clear that fuzzy sets can offer a natural interface between linguistic representations and numerical representations. Of course, membership grades never appear as such in natural languages. It has been pointed out in Section 1.2.2 that, in natural language, gradual predicates are those to which linguistic hedges such as "very" can be applied. Such linguistic hedges are the trace of gradual membership in natural language. Clearly the numerical membership grade corresponding to "very" is itself ill-defined. It is a fuzzy set of membership degrees as suggested by Zadeh (1972). He suggested to build the membership function of "very young" from the one of "young and the one of very, by letting *very-young* = *very(young(·))*. So, fuzzy membership grades model linguistic hedges that modify the membership function (See also Bouchon-Meunier *et al.*, 1999).

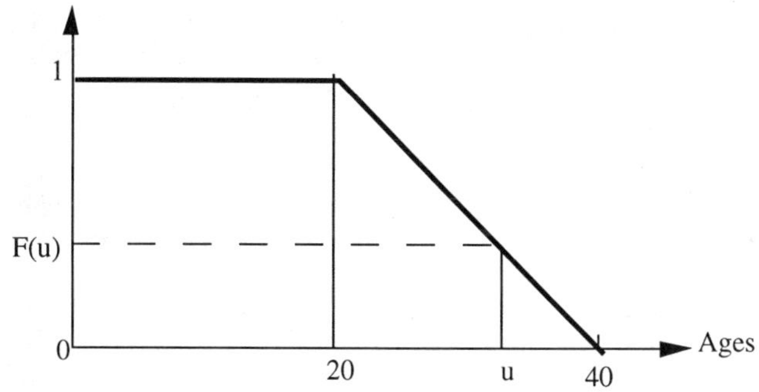

Figure 1.1. Representing *young* in : "a young person"

However if the referential set U is a finite set of objects then the use of the unit interval as a set of membership grades is more difficult to justify. A finite totally ordered set L will then do. It results from a partitioning of elements of U with respect to a fuzzy set F, each class in the partition gathering elements with equal membership, and the set of classes being ordered from full membership to non-membership.

It should be emphasized that the fuzziness of a property is not viewed as a defect in the linguistic expression of knowledge (e.g., lack of precision, sloppiness, limitation of the natural languages), but rather as a way of expressing gradualness. In that sense, fuzzy sets do not have exactly the same concern as other approaches to vagueness. For instance, K. Fine (1975) proposes that statements about a vague predicate be taken to be true if and only if they hold for all possible ways of making the predicate precise. It enables the classical logical relationships between a vague predicate A and its negation not-A to be preserved. In contrast, fuzzy set applications take advantage of the limited overlapping between A and not-A, and more generally, of this kind of overlapping in fuzzy partitions, in order to have a smooth transition from one class to another.

Parikh (1983) questions the possibility of precisely assessing degrees of truth for a vague predicate. In practice, however membership degrees have mainly an ordinal meaning. In other words it is the ordering induced by the membership degrees between the elements that is meaningful, rather than the exact value of the degrees. This is in agreement with the qualitative nature of the most usual operations that are used on these degrees (min, max and the complementation to 1 as an order-reversing operation in [0,1], as recalled in Sections 1.2.4, and 1.3.4, below). Obviously a fuzzy membership function will depend on the context in various ways. First, the universe of discourse (i.e., the domain of the membership function) has to be defined (e.g., *tallness* is not the same thing for a man or for a tree). Second, it may depend on the other classes which are used to cover the domain. For instance, with respect to a given domain, *young* does not mean exactly the same thing if the remaining vocabulary includes only *old*, or is richer and contains both *mature* and *old.* Lastly, a fuzzy membership function may vary from one person to another. However, what is really important in practice is to correctly represent the pieces of knowledge provided by an expert and capture the meaning he intends to give to his own words. Wether there is a universal consensus on the meaning of a linguistic expression like *young man* is another matter.

Level Cuts. The view of a fuzzy set as a membership function can be dubbed the *vertical view* of fuzzy sets, in reference to Figure 1.1 Another possible and very convenient view is to consider a fuzzy set as a nested family of classical subsets, via the notion of level-cut. The α-level cut F_α of a fuzzy set F is the set $\{u \in U: F(u) \geq \alpha\}$, for $1 \geq \alpha > 0$. The idea is to fix a positive threshold α and to consider as members of the set the elements with membership grades above the threshold. Moving the threshold in the unit interval, the family of crisp sets $\{F_\alpha: 1 \geq \alpha > 0\}$ is generated. This is the *horizontal view* of a fuzzy set.

For $\alpha = 1$, the *core* $\overset{\circ}{F}$ of F is obtained. It gathers the prototypes of F. Letting α vanish, the *support* s(F) of F is obtained. It contains elements with positive membership grades, those which belong to some extent to F. Note that the support is different from $F_0 = U$. Gentilhomme (1968)'s "ensembles flous" were fuzzy sets with only a core and a support.

The set of level-cuts of F is nested in the sense that (Figure 1.2)

$$\alpha > \beta \text{ implies } F_\alpha \subseteq F_\beta. \tag{1.1}$$

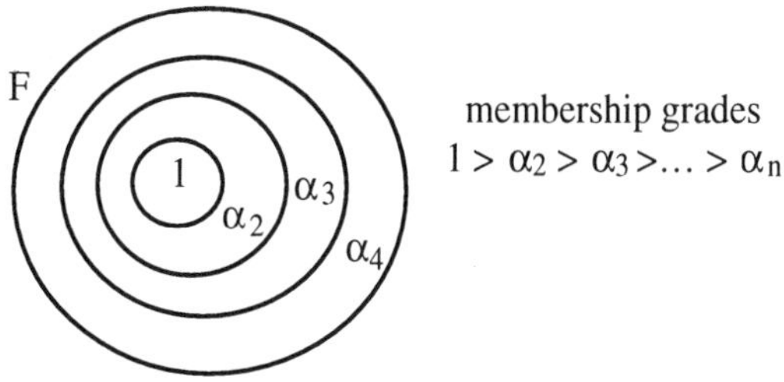

Figure 1.2. The horizontal view of fuzzy sets: level-cuts

One more property, namely *upper continuity*, is verified by level-cuts, namely, for any non-decreasing sequence sequence $\alpha_1 \leq \alpha_2 \leq \ldots$ whose limit is α,

$$F_\alpha = \lim_{n \to \infty} \cap_{i=1,n} F_{\alpha_i}. \tag{1.2}$$

In particular, $F_\alpha = \cap_{\beta < \alpha} F_\beta$. Going from the level-cut representation to the membership function and back is easy. The membership function can be recovered from the level-cut as follows:

$$\begin{aligned} F(u) &= \sup\{\alpha : u \in F_\alpha\} \\ &= \sup_{\alpha \in [0,1]} \min(\alpha, F_\alpha(u)) \end{aligned} \tag{1.3}$$

with $F_\alpha(u) = 1$ if $u \in F_\alpha$ and 0 otherwise.

Conversely, given an indexed nested family $\{A_\alpha : 1 \geq \alpha > 0\}$ such that $A_0 = U$ and (1.1), (1.2) hold, then there is a unique fuzzy set F whose level-cuts are precisely $F_\alpha = A_\alpha$ for each $\alpha \in [0, 1]$. It is given by (1.3) with F_α changed into A_α. This representation theorem was obtained by Negoita and Ralescu (1975a, b); see also Negoita and Ralescu (1987). In the case of a finite family of nested sets indexed by elements of a finite chain, the representation theorem holds without using property (1.2). (1.3) can be generalized by means of some distortion function $\eta: [0, 1] \to [0, 1]$, and F^η can be defined as $F^\eta(u) = \sup\{\eta(\alpha) : u \in F_\alpha\}$. Ralescu (1992) shows that a representation theorem still makes sense.

There are other kinds of representations of fuzzy sets, for instance using strict level-cuts $\underline{F}_\alpha = \{u \in U: F(u) > \alpha\}$, $1 > \alpha \geq 0$. Note that the support of F is $\underline{F}_0$.

Then $\underline{F}_\alpha = \bigcup_{\beta > \alpha} \underline{F}_\beta$ and

$$\begin{aligned} F(u) &= \inf\{\alpha : u \notin \underline{F}_\alpha\} \\ &= \inf_{\alpha \in [0, 1]} \max(\alpha, \underline{F}_\alpha(u)) \end{aligned} \tag{1.4}$$

Note that level cuts are far from containing all sets nested between the core and the support of F. Yet any such set may be considered as a possible Boolean representative of F. This requirement is fulfilled by assigning to each set A the degree to which it contains all representative elements of F and only them, say

$$\begin{aligned} r(A) &= \inf_{u \in A} F(u) \text{ if } \mathring{F} \subseteq A \\ &= 0 \text{ otherwise.} \end{aligned}$$

Clearly the following properties hold:

$r(F_\alpha) = \alpha$;

$r(A) > 0$ if and only if $\mathring{F} \subseteq A \subseteq s(F)$;

$A \subseteq B$ implies $r(A) \geq r(B)$.

Convex Combinations of Sets. In the following, let us assume the set of membership grades is the unit interval. First assume that the referential set U is finite. Then the set of distinct level-cuts of F is finite too. Let $F_1 \subseteq \ldots \subseteq F_n$ be these level-cuts, where F_1 is the core of F and F_n is supposed to be the support of F. Assign to F_i the weight $\alpha_i = \sup\{\alpha : F_\alpha = F_i\}$. Clearly $1 = \alpha_1 > \alpha_2 > \ldots > \alpha_n > 0 = \alpha_{n+1}$ (see Figure 1.2).

Define the set of positive weights $\{p_i\}_{i = 1,\ldots,n}$ as $p_i = \alpha_i - \alpha_{i+1}$. Then the sum $\sum_{i = 1,\ldots,n} p_i = 1$. What is obtained is a convex combination of sets in the sense that

$$F(u) = \sum_{i: u \in F_i} p_i \tag{1.5}$$

Note the similarity of (1.5) with (1.3). So, any fuzzy set on a finite set can be modelled by a weighted family of nested sets of the form $\{(F_i, p_i)\}_{i = 1,\ldots, n}$ with $\sum_{i = 1,\ldots,n} p_i = 1$, and conversely, if the family of sets is nested: $F_1 \subseteq \ldots \subseteq F_n$. This type of weighted family of sets is called a random set (Robbins, 1944; Bronowski and Neyman, 1944; Kendall, 1974; Matheron, 1975), because each weight can be interpreted as the probability that $F = F_i$. More precisely, it is a nested random set. Nevertheless this probabilistic interpretation of the convex combination must be taken with caution. It may mean that an unknown Boolean set is underlying F, whose boundaries are unknown but do exist (Fine, 1975;

Williamson, 1994). On the contrary, the Zadehian view is that precise boundaries of F do *not* exist because this boundary is gradual. Choosing one interpretation may have consequence on the proper use of fuzzy sets.

More generally in the infinite case the set of weights can be replaced by the Lebesgue measure λ on the unit interval. and (1.5) becomes

$$F(u) = \lambda\{\alpha: u \in F_\alpha\} = \int_0^1 F_\alpha(u)\, d\alpha.$$

The view of fuzzy sets as random sets has been considered independently by several authors like Fortet and Kambouzia (1976), Goodman (1982) who points out that (1.5) defines the membership function as a one-point coverage function of a random set, Wang (1983) and Wang and Sanchez (1982) (eq. (1.5) is said to compute the falling shadow of the random set), and Orlov (1975, 1978, 1980).

More recently, Orlovski (1990, 1995) has generalized this view in the following way: Let U be a set of objects and $\mathcal{P}$ be a set of Boolean properties. Let Γ be a set-valued mapping from $\mathcal{P}$ to U, such that $\Gamma(p)$ is the set of objects which verify property p. Assume that there is a monotonic set function g on $\mathcal{P}$ that quantifies the importance of sets of properties. Namely g(A) is the importance of the set A of properties. Then the membership grade F(u) is defined as $F(u) = g(\{p : u \in \Gamma(p)\})$.

Points in the Hypercube. In the finite case, namely, if $U = \{u_1, \ldots, u_n\}$, a fuzzy set can be encoded as a point in the unit hypercube $[0, 1]^n$ (Kosko, 1990, 1991). It enables geometric metaphors of fuzzy set notions which may be of tutorial value.

1.3.2 Scalar Characteristics of a Fuzzy Set

A fuzzy set can be evaluated from several points of view, namely

-) to what extent a fuzzy set is not empty? and the dual question: and to what extent it covers the universe?
-) how large is a fuzzy set? (= does it contain many elements?)
-) how fuzzy is a fuzzy set?

Height. The answer to the first question can be answered by computing the *height* of a fuzzy set, namely:

$$ht(F) = \sup_u F(u).$$

A fuzzy set is said to be normal if and only if $\exists u, F(u) = 1$. It implies that $ht(F) = 1$ (not conversely, in general). Otherwise the fuzzy set can be considered as empty to some extent, as evaluated by $1 - ht(F)$. In particular $ht(\emptyset) = 0$: the empty set has a constant membership function equal to 0. Conversely the extent to which F covers the referential is measured by the *plinth* of F, namely:

$$\mathrm{pl}(F) = \inf_u F(u).$$

Cardinality. The size (or scalar cardinality) of a fuzzy set is evaluated via a real-valued extension of cardinality :

$$|F| = \sum_{u \in U} F(u).$$

De Luca and Termini (1972) introduced this definition and called it the *power* of a fuzzy set. It evaluates "how many" elements F contains. This definition presupposes i) that membership grades are numerical; ii) that the support of F is finite. In the case of infinite support one may use the integral of the membership function on its support, when it exists:

$$|F| = \int_U F(u)\, du.$$

This definition may look paradoxical since the result is generally not an integer. However if the fuzzy set is used to describe an uncertainty distribution, then this evaluation does reflect how little specific this distribution is. See Klir's Chapter 8 in this volume for this view of cardinality -like indices for fuzzy sets. Moreover the scalar cardinality of F can be understood as an average cardinality as explain below. Note that the relation "larger than" is well expressed by the fuzzy set inclusion, and the fuzzy cardinality is coherent with fuzzy inclusion. Scalar cardinality can evaluate the surface of a fuzzy set of a plane for instance. There is a full book by Wygralak(1996) devoted to fuzzy cardinality

Indices of Fuzziness. Typical of fuzzy sets is that for some elements, the membership decision, insofar as this decision must be made, is difficult or uncertain. Mesuring this difficulty can be addressed by another ordering on fuzzy sets (Mukaidono, 1975). F is said to be at least as sharp as G if and only if

$$F \geq G \Leftrightarrow \forall u, \begin{cases} \text{either } F(u) \geq G(u) \geq 1/2 \\ \text{or } 1/2 \geq G(u) \geq F(u). \end{cases}$$

F is said to be sharper than G if and only if $F \geq G$ and $F \neq G$. A slight variant of this ordering has already been suggested by Kaplan and Schott (see above) before the term fuzzy set was coined. This ordering reflects the idea that the closer to 0.5 is F(u), the more uncertain is any decision regarding actual membership versus non-membership. The sharpest fuzzy sets are the usual crisp sets. The least sharp one is unique and is such that F(u) = .5, ∀u. An idealization of F is any crisp set $A \geq F$. It is not unique because if F(u) = 0.5 for some u, then an idealization of F indifferently obeys $F^*(u) = 1$ or $F^*(u) = 0$. However if $F(u) \neq 0.5$ for all u, then the idealization is unique, and is the 0.5-cut of F.

Some remarks seem to be useful:

-) This definition presupposes that there is a middle point in the membership scale (0.5, for the unit interval). For finite membership scales this is not obvious.

-) The least sharp fuzzy set is rather empty. There are many minimally sharp normal fuzzy sets.

-) Clearly if G is less sharp than F then deciding whether u belongs to G or not is more difficult than to F.

The sharpness relation underlies a requirement of coherence between membership grades to F and G: If F is sharper that G then for any u membership to F cannot be but confirming the tendency given by G. One may extend this ordering by getting rid of this constraint. Namely, F is said to be at least as clear as (= not more fuzzy than) G if and only if $\forall u$, $|F(u) - 0.5| \geq |G(u) - 0.5|$, which is denoted $F \succcurlyeq G$. Clearly if $F \underline{\succ} G$, then $F \succcurlyeq G$, but not conversely.

Indices of fuzziness (and there are many of them) evaluate to which extent a set is fuzzy, in accordance with the above relations of comparative fuzziness. There are three ideas leading to indices of fuzziness:

-) The idea of uncertainty of membership decision leads to the entropic view of De Luca and Termini (1972):

$$e(F) = -\sum_{u \in U} (F(u) \operatorname{Ln} F(u) + (1 - F(u))\operatorname{Ln}(1 - F(u)))$$

Let $U = \{u_1, \ldots, u_n\}$. This expression is nothing but the Shannon entropy of the joint probability on $\{0, 1\}^n$, whose marginals are $P_i(1) = F(u_i)$, $P_i(0) = 1 - F(u_i)$.

-) The idea of distance to any idealization of F as proposed by Kaufmann (1975)

$$\text{fuz}(F) = 1 - \frac{1}{|U|} \sum_{u \in U} |F(u) - F^*(u)|.$$

-) The idea of similarity between a fuzzy set and its complement (see Section 1.4.1 below), as proposed by Yager (1979), namely the more a set is fuzzy the less it can be told apart from its complement.

Chapter 9 by Pal and Bezdek in this volume gives a more detailed survey of indices of fuzziness.

1.3.3 Extension Principles

The main interest of the level-cut representation is to be very handy when extending set-theoretic notions to fuzzy sets. First any set-to-set function can be in principle extended level cut per level cut, provided that what is obtained is a family of level cuts. Then any usual point-to -point function can be lifted to a fuzzy-set-to-fuzzy-set function on this basis. Lastly, set-to-point functions (set functions for short) can also use this method, either yielding a fuzzy-valued function, or a scalar-valued function. It is much more difficult to extend set-functions to fuzzy set-valued arguments under the vertical view. When a definition making sense for fuzzy sets can be expressed in terms of level cuts, this definition is said to be cut-worthy (Bandler and Kohout, 1993). See De Baets and Kerre (1994) for a survey of of fuzzy concepts defined via cuts.

Set-To-Set Functions The basic extension scheme is as follows. Suppose some function f from the family of subsets of U to the family of subsets of V is given, and F is a fuzzy subset of U. How to define $f(\mathrm{F})$ for the fuzzy subset F? The answer is: compute $f(\mathrm{F}_\alpha)$ for all level-cuts, and build $f(\mathrm{F})$ via (1.3) as

$$f(\mathrm{F})(\mathrm{v}) = \sup\{\alpha : \mathrm{v} \in f(\mathrm{F}_\alpha)\}. \tag{1.6}$$

However, this method will be fully satisfactory only if the family $\{f(\mathrm{F}_\alpha): 1 \geq \alpha > 0\}$ verifies (1.1) and (1.2), i.e., is properly nested and upper continuous. Then we can be sure that $f(\mathrm{F}_\alpha) = [f(\mathrm{F})]_\alpha$, that is, the mapping f applied to level-cuts produces level cuts of a fuzzy set. Often, the only way of getting such a canonical extension is by restricting f to special functions and F to special fuzzy sets. An alternative view is to give up vertical view of fuzzy sets and the use of membership functions, and exploit a family of indexed sets obeying the nestedness condition only. This is the path followed by Herencia (1998) who calls such families graded sets.

Extending Usual Functions. This technique can be applied to extend usual functions to fuzzy set-valued arguments. Namely, let f be a function from U to V. Given a fuzzy set F on U, the fuzzy set $f(\mathrm{F})$ on V can be obtained in two steps. First extend f to set-valued arguments. This is standard in mathematics. Namely, for any subset A of U, $f(\mathrm{A}) = \{f(\mathrm{u}) : \mathrm{u} \in \mathrm{A}\}$. The second step is to apply (1.6) to the extended function. Note that in that case it is obvious that $\mathrm{A} \subseteq \mathrm{B}$ implies $f(\mathrm{A}) \subseteq f(\mathrm{B})$. Hence the set $\{f(\mathrm{F}_\alpha): 1 \geq \alpha > 0\}$ verifies (1.1). In the finite case, it ensures that $f(\mathrm{F}_\alpha) = [f(\mathrm{F})]_\alpha$, and (1.6) reduces to the usual form of the extension principle (Zadeh, 1965a):

$$\begin{aligned} f(\mathrm{F})(\mathrm{v}) &= \sup\{\mathrm{F}(\mathrm{u}) : \mathrm{v} = f(\mathrm{u})\} \\ &= 0 \text{ if } \{\mathrm{v} = f(\mathrm{u})\} = \emptyset. \end{aligned} \tag{1.7}$$

On continuous domains, the property $f(F_\alpha) = [f(F)]_\alpha$ may fail to hold and some conditions must be added. Nguyen (1978) has shown that a sufficient condition is that the supremum be attained in (1.7). This is the case for fuzzy intervals, where f is supposed to be continuous and the membership function $F(\cdot)$ upper semi-continuous (see Chapter 10 of this volume)). However, for strict level cuts $f(\underline{F}_\alpha) = [\underline{f(F)}]_\alpha$ is valid for set-valued extensions of usual functions.

From Set Functions to Fuzzy-Valued Functions. If the above conditions are not met, applying (1.6) may still be interesting to compute. For instance if f is now an injective set-function, that is, a function from 2^U to V, the image of set F_α is a point $f(F_\alpha) \in V$, and $f(F)$ is a fuzzy subset of V such that $f(F)(v) = \alpha$ for $v = f(F_\alpha)$. Note that (1.4) does not lend itself to this kind of extension.

If the function is not injective, then (1.6) selects the highest membership grade for any value v. This approach has been adopted by Zadeh (1983) for computing the *fuzzy cardinality* of a fuzzy set, where $f(A) = |A|$ for any subset of a finite set U. What is obtained for $|F|$ is a fuzzy subset of integers :

$$|F|(n) = \max\{\alpha : n = |F_\alpha|\}.$$

This definition can also be applied to relative cardinality. It also leads to the notion of fuzzy quantifiers (see Chapter 2 by Fodor and Yager in this volume).

However in (Dubois and Prade, 1985b), it is pointed our that this extension of scalar evaluations of sets to fuzzy sets has drawbacks. If U is finite then the support of $|F|$, namely $\{|F_\alpha|, 1 > \alpha \geq 0\}$ may contain holes, i.e., integers n, $|F_1| < n < |s(F)|$ such that $|F|(n) = 0$ where one whould expect a non-increasing membership function, i.e., if $m > n$ then $|F|(m) \leq |F|(m)$. This regularity is recovered if the level cut representation of fuzzy sets is extended to all subsets of U nested between the support and the core of F as suggested in Section 1.3.1. Then $|F|(n)$ given by the above formula is improved into $|F|(n) = \max\{r(A) : n = |A|\}$ which has the expected non-increasing behavior. This approach may be applied to compute fuzzy -valued extensions of any scalar evaluation $f(A)$ of sets to fuzzy sets.

Scalar Fuzzy Set-Functions. The representation in terms of convex combination of sets provides a systematic way of extending set-functions into scalar evaluations of fuzzy sets. Especially, for a function f from from 2^U to $\mathbb{R}$, the image of the level-cut F_α is a real number $f(F_\alpha)$ and $f(F)$ is a real number of the form

$$f(F) = \int_0^1 f(F_\alpha)\, d\alpha.$$

It is the average value of f on the random set underlying F. In the finite case, and in accordance with (1.5), what is found is

$$f(F) = \sum_{i = 1,\ldots, n} f(F_i)p_i. \tag{1.8}$$

In some cases the vertical and the horizontal view yields the same results. For instance, this approach recovers the height of a fuzzy set (then $f(A) = 0$ if A is empty and 1, otherwise), its plinth ($f(A) = 1$ if $A = U$ and 0 otherwise), but not the indices of fuzziness since their value is 0 for crisp sets. If f is the cardinality, then, (1.8) is also in full agreement with the notion of scalar cardinality of a fuzzy set in Section 1.3.2, and we do have that $\sum_{u\in U} F(u) = \sum_{i = 1,\ldots, n} |F_i|p_i$. More generally, the probability of a fuzzy event has been defined by Zadeh (1968) as follows, using the integral of the membership function over a measurable set U:

$$P(F) = \int_U F(u)\, dP(u)$$

It turns out that $P(F)= \int_0^1 P(F_\alpha)d\alpha$ as first pointed out by Höhle (1976). Similarly, Smets (1981) has introduced the belief degree in a fuzzy event in a similar way, on a finite set U. A belief function can be defined by means of a family of weighted sets $\{(E_i,m_i)\}_{i = 1,\ldots, n}$ with $\sum_{i = 1,\ldots,n} m_i = 1$. The weight m_i is the probability that the set E_i represents an as accurate as possible description of the state of affairs. The degree of belief in an event A is the amount of support given to A by the body of information$\{(E_i,m_i)\}_i$, namely, $Bel(A) = \sum_{E_i\subseteq A} m_i$. The degree of plausibility of an event A is the amount of support not given to A^c namely, $Pl(A) = 1- Bel(A^c) = \sum_{E_i\cap A\neq\emptyset} m_i$. The degree of belief in a fuzzy event F can be equivalently defined in terms of the membership function or in terms of level cuts:

$$Bel(F) = \int_0^1 Bel(F_\alpha)d\alpha = \sum_{i = 1,\ldots, n} m_i \min_{u\in E_i} F(u)$$

$$Pl(F) = \int_0^1 Pl(F_\alpha)d\alpha = \sum_{i = 1,\ldots, n} m_i \max_{u\in E_i} F(u)$$

These are known as Choquet integrals (Grabisch *et al.*, 1995). However the range of applicability of (1.8) is very large. For instance, if a set represents an area for which the notion of perimeter makes sense, say $per(F) \in \mathbb{R}$, then (1.8) provides a scalar evaluation of the perimeter of a fuzzy set F, while in the vertical view it is much more difficult to define it using the membership function $F(\cdot)$.

A natural qualitative substitute to (1.8) is Sugeno integral (Sugeno, 1974, 1977):

$$\oint_F \circ f = \sup_{\alpha\geq 0} \min(\alpha, f(F_\alpha)). \tag{1.9}$$

where f is an inclusion- monotonic set-function, often called fuzzy measure, such that if T is a subset of T' then $g(T) \leq g(T')$, ranging on the same scale as the degrees of membership. In the finite setting, on $\{u_1,\ldots, u_n\}$:

$$\int F \circ f = \max_{j=1,n} \min(F(u_j), f(\{u_1,\ldots, u_j\})$$
$$= \min_{j=1,n} \max(F(u_j), f(\{u_1,\ldots, u_{j-1}\}),$$

where the elements of U are named such that $F(u_1) \geq F(u_2) \geq \ldots \geq F(u_n)$. $\int F \circ f$ is actually the median of $2n - 1$ terms $\{F(u_1), F(u_2) \ldots F(u_n), g(\{u_1\}),\ldots, g(\{u_1,\ldots, u_{n-1}\})\}$(see for instance Kandel and Byatt, 1978; Dubois and Prade, 1980a), as opposed to the convex combination of sets view, which yields average values. Sugeno integrals are the basis for defining possibility and necessity of fuzzy events.

1.3.4 Basic Connectives

Having at our disposal a representation of fuzzy classes in the form of membership functions (supposed to take value in the unit interval unless otherwise specified), and their representations in terms of classical sets, it is natural to try and do with fuzzy sets what is done with sets. Zadeh, like Kaplan and Schott, has proposed extensions of inclusion, equality, complementation, union and intersection to fuzzy sets. A point worth mentioning about fuzzy set combination, especially in the numerical setting, is the existence of a panoply of different families of operators, including various extensions of logical ones like conjunctions and disjunctions, together with averaging and compensatory operations.

Inclusions. The inclusion of fuzzy sets F and G can be defined in two basic ways:

i) The most usual way, proposed by Zadeh (1965a), consists in considering that F is included in G as soon as each element belongs to G at least as much as to F, in other words:

$$F \subseteq G \Leftrightarrow \forall u \in U, F(u) \leq G(u) \quad (1.10)$$

ii) A stronger view of inclusion is to consider that all elements belonging whatsoever to F must be prototypes of G. Strong inclusion of F into G is then defined as the inclusion of the support of F in the core of G (Dubois and Prade, 1980a):

$$F \subseteq_s G \Leftrightarrow s(F) \subseteq \mathring{G} \quad (1.11)$$

The first definition of inclusion leads to define equality between fuzzy sets by the equality of their membership functions (via double inclusion). This definition, although mathematically sound may look very demanding in practice since membership functions may not be precisely known, or may be partially arbitrary. The strong inclusion leads to consider that genuinely fuzzy sets can never be considered as perfectly equal: only usual sets can be equal, since $F \subseteq_S G$ and $G \subseteq_S F \Leftrightarrow \forall u \in U, F(u) = G(u) \in \{0, 1\}$.

Intersection and Union. Other set-theoretic notions have been extended to fuzzy sets. Intersection and union of fuzzy sets can be introduced via level-cuts using the canonical extension (1.3) as follows:

$$(F \blacklozenge G)_\alpha = F_\alpha \blacklozenge G_\alpha \text{ for } \blacklozenge = \cap \text{ and } \cup.$$

This approach yields a genuine family of level-cuts, and the following well-known fuzzy connectives

$$F \cap G(u) = \min(F(u), G(u)),$$
$$F \cup G(u) = \max(F(u), G(u)),$$

proposed by Zadeh (1965a). In the same paper, another definition of intersection and union was proposed, namely using a product and the probabilistic sum respectively:

$$F \cap_p G(u) = F(u) \cdot G(u),$$
$$F \cup_p G(u) = F(u) + G(u) - F(u) \cdot G(u).$$

These operations can be obtained under the random set view of fuzzy sets, considering the one-point coverage function of the intersection of independent nested random sets representing F and G. For instance, if $F = \{(F_i, p_i)\}_{i = 1,\ldots, n}$ and $G = \{(G_j, q_j)\}_{j = 1,\ldots, m}$ then,

$$F \cap_p G(u) = \textstyle\sum_{u \in F_i \cap G_j} p_i \cdot q_j;$$
$$F \cup_p G(u) = \textstyle\sum_{u \in F_i \cup G_j} p_i \cdot q_j.$$

Note that in the latter random set-like extensions, level-cuts of various levels F_α and G_β are combined, while the max-min connectives are obtained by considering the same thresholdings for F and G.

The third pair of fuzzy intersection and union was first suggested by Giles (1976). They are linear connectives up to a truncation, in accordance with Lukasiewicz implication:

$$F \cap_L G(u) = \max(0, F(u) + G(u) - 1);$$
$$F \cup_L G(u) = \min(1, F(u) + G(u)).$$

These connectives can again be recovered in a random set view of fuzzy sets. However since independence of random sets is no longer assumed, one must put conditions on the weights of the joint convex combination of sets $\{(F_i \times G_j, r_{ij})\}_{i,j=1,\ldots,n}$ with marginals defined by $\{(F_i, p_i)\}_{i=1,\ldots,n}$ and G =

$\{(G_j, q_j)\}_{j=1,\ldots,m}$. For instance, the linear connectives are recovered as soon as $r_{ij} = 0$, $\forall$ i, j such that $F_i \cup G_j \neq U$, which corresponds to maximal exclusiveness between F and G (see Goodman and Nguyen,1985; Dubois and Prade, 1989).

All the above connectives coincide with classical ones when applied to usual Boolean sets. One may define extreme (often called drastic) forms of fuzzy set-theoretic connectives by restricting to these limit conditions:

$$
\begin{aligned}
F \cap_d G(u) &= \min(F(u), G(u)) \text{ if } \max(F(u), G(u)) = 1 \\
&= 0 \text{ otherwise;} \\
F \cup_d G(u) &= \max(F(u), G(u)) \text{ if } \min(F(u), G(u)) = 0 \\
&= 1 \text{ otherwise.}
\end{aligned}
$$

More generally, fuzzy set-theoretic intersections can be modelled by so-called triangular norms and fuzzy set-theoretic unions by co-norms, as first proposed by Alsina *et al.* (1983), Dubois and Prade (1980a, b). There is a consensus among researchers on the t-norm/conorm based setting for fuzzy sets. Triangular norms and conorms are semigroup operations on the unit interval, respecting the natural ordering structure (so that if $F \subseteq G$ then $F \cap H \subseteq G \cap H$ may hold); (see Chapter 2, in this volume, for details). This requirement implies that the fuzzy unions and intersections should be monotonic (non-decreasing). Triangular norms come from the literature in stochastic geometry, rooted in the works of Menger. The basic triangular norms are precisely the four above fuzzy intersections introduced above. In particular the minimum is the greatest triangular norm (hence the most liberal fuzzy intersection), and the drastic intersection is the smallest triangular norm. Similarly the maximum operation yields the smallest conorm acting as fuzzy union operation, and the drastic union is the most encompassing one.

Complementation. Fuzzy set complementation is often defined by $F^c(u) = 1 - F(u)$ $\forall u \in U$. It is the most straightforward way of expressing the obvious requirement that the more u belongs to F the less it belongs to its complement F^c, and the other requivement that negation cancels itself by repeating. Under these requirements, and the limit condition that fuzzy complementation should coincide with its crisp counterpart on crisp sets, continuous complementation operations are all of the form 1 –a, up to an isomorphism (Trillas, 1979). However it is not directly justified by the level-cut approach since Under this definition $(F^c)_\alpha = (\underline{F}_{1-\alpha})^c$. Indeed the family of the complements of level-cuts of a fuzzy set $(F_\alpha)^c$ does not form a family of level-cuts with respect to this indexing. However this complementation is a by-product of the convex combination view, since, in the discrete case, for instance, F^c corresponds to the convex combination $(F_i{}^c, p_i)\}_{i=1,\ldots,n}$. Variants of fuzzy complementations are surveyed in Chapter 2 in this volume.

Structural Properties. Naive set theory is based on a specific algebraic structure of subsets equipped with intersection, union and complementation:

Boolean algebras. It would be good if the family of fuzzy subsets of a set U could be a Boolean algebra under a suitable definitions of fuzzy connectives. However this is impossible, mathematically: gradual membership and compositionality of membership grades are incompatible with the Boolean structure. As a consequence, some properties of Boolean algebras must be deleted for fuzzy sets.

The first debatable properties of sets, when considering fuzzy sets are the laws of contradiction (a proposition cannot be both true and false) and excluded middle (a proposition is either true or false): $F \cap F^c = \emptyset$ and $F \cup F^c = U$. Indeed since a fuzzy set has a gradual boundary it seems plausible that there should be some overlap between F and its complement and that together they do not fill the whole universe. Choosing the min-max system with the complementation $1 - (\cdot)$ (up to an isomorphism) is the only way of preserving all properties of the Boolean structure, but these two laws (Bellman and Giertz, 1973; Fung and Fu, 1975) :

Associativity:	$(F \cap G) \cap H = F \cap (G \cap H)$;
	$(F \cup G) \cup H = F \cup (G \cup H)$;
Commutativity:	$F \cap G = G \cap F$; $F \cup G = G \cup F$;
Identity:	$F \cap U = F$; $F \cup \emptyset = F$;
Absorption:	$F \cap \emptyset = \emptyset$; $F \cup U = U$;
Idempotence:	$F \cap F = F$; $F \cup F = F$;
De Morgan Laws:	$(F \cap G)^c = F^c \cup G^c$; $(F \cup G)^c = F^c \cap G^c$;
Mutual distributivity:	$F \cup (G \cap H) = (F \cup G) \cap (F \cup H)$;
	$F \cap (G \cup H) = (F \cap G) \cup (F \cap H)$;
Involution:	$(F^c)^c = F$.

This structure is that of a distributive pseudo-complemented lattice. Although the laws of contradiction and excluded middle fail they hold to a limited extent since $(F \cap F^c)_\alpha = \emptyset$ for $\alpha > 0.5$ and $(F \cup F^c)_\alpha = U$ for $\alpha < 0.5$. Note that the fuzzy set inclusions are recovered from the connectives as follows:

$$F \subseteq G \Leftrightarrow F \cup G = G \Leftrightarrow F \cap G = F;$$
$$F \subseteq_S G \Leftrightarrow F^c \cup G = U \Leftrightarrow F \cap G^c = \emptyset.$$

Keeping the excluded middle and contradiction laws implies the deletion of idempotence and mutual distributivity (Dubois and Prade, 1980b). The most natural fuzzy set algebra is then based on the linear connectives: there does hold $F \cap_L F^c = \emptyset$ and $F \cup_L F^c = U$. These connectives indeed enforce maximal exclusiveness between conjunctively combined sets. Under such connectives, the family of fuzzy sets form a so-called MV-algebra (Chang, 1954) which is very popular in current logico-algebraic developments of fuzzy set theory (Belluce *et al.*, 1991; Gottwald, 1999). The characteristic properties of this structure can be described only using the union $\cup_L$ and the fuzzy complementation, the intersection being recovered using De Morgan laws (Mundici, 1986): associativity and commutativity, identity and absorption laws of $\cup_L$, involution of the complementation, and

$$\emptyset^c = U;$$
$$(F^c \cup_L G)^c \cup_L G = (G^c \cup_L F)^c \cup_L F.$$

Moreover the max-min fuzzy set structure (hence the two fuzzy set inclusions) is recovered as:

$$F \cup G = (F \cap_L G^c) \cup_L G \text{ and } F \cap G = (F \cup_L G^c) \cap_L G.$$

The fuzzy set structure based on the product and the probabilistic sum is weaker than the two other ones since not only do the excluded middle and contradiction laws fail, but the idempotence and mutual distributivity fail too. This structure has been studied only recently (Hajek *et al.*, 1997). Again the max-min structure can be recovered as a by-product.

Other Aggregation Operations. Fuzzy set theory goes beyond classical set theory not only by generalizing basic set-theoretic connectives like unions and intersections, but also by proposing new kinds of set-theoretic connectives which have no Boolean counterparts. For instance, it makes sense to compute a new fuzzy set from F and G by taking the arithmetic mean of the membership grades F(u) and G(u), $\forall\, u \in U$, (on the unit interval), or the median of F(u), G(u) and a prescribed membership level α. These operations are well-known in statistics and decision theory but are not usually interpreted as set-theoretic operations or logical connectives expressing, say, a compensatory "AND". For instance, someone looking for *cheap AND luxury cars* does not really try to perform a logical intersection between the set of *cheap cars* and the set of *luxury* ones (most probably this intersection is empty). Such a linguistic AND most plausibly accepts compensation, trade-off between the two criteria and more naturally translates into an arithmetic mean. Such a connective has no room in classical logic and only fuzzy set theory can put it in such a perspective. It is possible to combine usual sets using arithmetic mean but the result is not a set any more. The mean operations for combining fuzzy sets are all operations lying between minimum and maximum. Note that some of them coincide with Boolean intersection (geometric means) or Boolean unions (De Morgan duals of geometric means). Families of means were first introduced in fuzzy set theory by Dujmovic (1974), and systematically studied by Dyckoff and Pedrycz (1983). See Chapter 2 in this volume for more details. Trying to preserve associativity for families of operations lying between min and max is almost impossible. The only such continuous associative operation was found by Fung and Fu (1975). It is of the form of a qualitative order-statistics connective

$$M_\alpha(F(u), G(u)) = \text{median}\{F(u), G(u), \alpha\}.$$

There are other classes of symmetric fuzzy set-theoretic operations that have absolutely no classical counterparts nor exist in other disciplines, like symmetric sums, introduced by Silvert (1979). They are monotonic, commutative and invariant under De Morgan laws : $(F \blacklozenge G)^c = F^c \blacklozenge G^c$. The arithmetic mean and $M_{1/2}(F(u), G(u))$ are examples of such symmetric sums that are idempotent. Other less common symmetric sums are associative on (0, 1), like $\frac{a \cdot b}{a \cdot b + (1-a) \cdot (1-b)}$ and even less usual non-associative ones such as $\frac{\min(a, b)}{1 - |a - b|}$ and $\frac{\max(a, b)}{1 + |a - b|}$. See Chapter 2 of this volume, for more details on this family. Symmetric sums look tailored to design meaningful fuzzy voting procedures (where the meaning of the end-points of the rating scale should have no influence on the way preference profiles, modelled by fuzzy sets of preferred candidates, are combined); see (Dubois and Koning, 1991).

This variety of combination operations may help solve difficulties one may encounter using only maximum and minimum, like the following: Smith and Osherson (1984) give the example that "a perfectly brown apple" seems to be more typical of the conjunction *brown apple* than of either *brown* or *apple*. However, this example also points out that when we are interested in a class C, which is made of borderline elements of two other classes, say A and B, we may have to renormalize the result of the combination of the membership functions of A and B in order to obtain the one of C.

Commensurability. It is important to notice that the membership scale is supposed to be common to the various fuzzy sets that are involved in a given aggregation. This is the *commensurability* assumption, which underlies all the fuzzy set literature and has been found debatable by some opponents to the theory (French, 1984). Indeed, in applications, the fuzzy sets may refer to non-commensurate entities (the height and the weight). The role of membership grades is precisely to restore some form of commensurability between entities that are usually unrelated, but are related, in a given context, at the syntactic level (e.g., when *fast* cars of *moderate* price are selected in a database). It is to be considered as a convenient and reasonable assumption in many situations. Deleting this assumption would lead to difficulties when combining membership functions, namely it is usually impossible to merge complete preorderings, as estabished by Arrow (1963) in social choice theory. The commensurability hypothesis on the degrees of satisfaction with respect to various fuzzy sets makes it possible to combine fuzzy sets in a meaningful way. General principles for fuzzy set membership function elicitation resort to existing methods in measurement theory (Krantz *et al.*, 1971). See also Section 1.6.2, and Chapter 3 of this volume.

1.3.5 Set-Theoretic Comparisons Between Fuzzy Sets

Fuzzy sets can be compared as to their relative positions in the referential. However if U is not equipped with a topological structure of any kind, only set-theoretic comparisons can be envisaged. Namely given two sets A and B, one may address the following issues:

i) Is the intersection between A and B empty or not?

ii) Is A a subset of B?

iii) Are A and B equal?

Comparing Crisp Sets. Questions (i), (ii) and (iii) can be answered in a yes- or no manner for crisp sets. However, using cardinality, one may also quantify to what extent A and B come close to having an empty intersection, to satisfying an inclusion relationship, or yet to being equal. In the following we shall denote O(A,B), I(A, B) and E(A, B) degrees of intersection (overlap), inclusion and equality between A and B respectively.

Boolean-valued indices are as follows:

$$O(A,B) = 1 \text{ if and only if } A \cap B \neq 0,$$
$$I(A, B) = 1 \text{ if and only if } A \subseteq B,$$
$$E(A, B) = 1 \text{ if and only if } A = B.$$

Consistency. The natural extension of the Boolean overlap index is the consistency (Zadeh, 1978a)

$$O(F, G) = ht(F \cap G) = \sup_{u \in U} \min(F(u), G(u)) \tag{1.12}$$

Then O(F, G) = 1 if there exists a prototype element of both F and G, and 0 if F and G are completely disjoint. This view of overlap is clearly not based on counting. The minimum can be changed into another fuzzy conjunction.

Inclusion Indices and Implications. For the Boolean inclusion index, fuzzy extensions have been proposed by Bandler and Kohout (1980a):

$$I(F, G) = \inf_{u \in U} F(u) \rightarrow G(u), \tag{1.13}$$

where → is a fuzzy implication. Fuzzy implications play an important role in fuzzy set theory, for representing fuzzy rules. Let us briefly summarizes the main families (see Bouchon-Meunier *et al.*, 1999, for more details). There are basically two families, the "strong" and residuated implication functions (S-implications and R-implications for short) respectively defined as follows (Trillas and Valverde, 1985).

1) S-implications are of the form $a \rightarrow b = S(1 - a, b)$, where S is a fuzzy disjunction operation. They are a copy of material implication "not A or B". The name of *strong* implication, when S = max, or probabilistic sum, refers to the strong fuzzy set inclusion (1.11). The two basic strong implication operations are

Kleene-Dienes implication	$\max(1 - a, b)$ (for S = max);
Reichenbach implication	$1 - a + ab$ (for S = probabilistic sum).

Using these implications in (1.13), and denoting I_S the strong inclusion indices, $I_S(F, G) = 1$ if and only if $F \subseteq_s G$ and in particular, for genuine fuzzy sets

$I_S(F, F) \neq 1$. What is really evaluated is the plinth of $F^c \cup G$. In particular $I_S(F, G) = 0$ only if a prototype of F is totally rejected by G. Note that, as expected, $I_S(F, G) = 1 - O(F, G^c)$.

2) *R-implications* are a generalization of the residuation operations in lattices (Birkhoff, 1967). Implication is then envisaged in relation to a deduction process, where A implies B means that B is at least as true as A. It also amounts to preserving the deduction theorem in many-valued logic (from A and B, conclude C if and only if from A, it follows that B implies C), which writes, with the minimum as conjunction operation, for $\alpha, \beta, \gamma \in [0, 1]$:

$$\min(\alpha, \beta) \leq \gamma \text{ if and only if } \alpha \leq \beta \to \gamma$$

Hence $\beta \to \gamma = \sup\{\alpha \in [0, 1]: \min(\alpha, \beta) \leq \gamma\}$, that is:

$$\beta \to \gamma \begin{array}{l} = 1 \text{ if } \beta \leq \gamma \\ = \gamma \text{ otherwise} \end{array} \qquad \text{(Gödel implication)}$$

Using product instead of min yields:

$$\beta \to \gamma \begin{array}{l} = \min(1, \gamma/\beta) \\ = 1 \text{ if } \beta = 0 \end{array} \qquad \text{(Goguen implication)}$$

Using residuated implications in the inclusion index (then denoted I_R) is in agreement with Zadeh's inclusion of fuzzy sets since $I_R(F, G) = 1$ if and only if $F \subseteq G$. Moreover, $I_R(F, G) = 0$ as soon as there is an element in the support of F which is not in the support of G. Hence $I_R(F,G)$ is more liberal with residuated implications than with strong implications when close to 1, and less liberal when close to 0. The related index $O(F, G) = 1 - I_R(F, G^c)$ can still be viewed as a degree of overlap but it is not symmetric.

The other commonly used implication is Lukasiewicz implication $\min(1, 1 - a + b)$ which is both strong (for S = bounded sum) and residuated (with respect to the linear conjunction). Then $I_L(F, G) = 1$ if and only if $F \subseteq G$, and 0 only if a prototype of F is totally rejected by G.

Willmott (1986) has pointed out that inclusion indices stated above satisfy the following transitivity property, for three fuzzy sets F, G, H: $I(F, H) \geq I(F, G) + I(G, H) - 1$. However, known results on fuzzy relations after Valverde (1985) indicate that for $a \to b$ = Gödel implication $I_R(F, H) \geq \min(I_R(F, G), I_R(G, H))$, and for $a \to b$ = Goguen implication $I_R(F, H) \geq I_R(F, G) \cdot I_R(G, H)$ holds.

Fuzzy Set Differences. The fuzzy set difference can be defined by means of the complement of implication operations $(F \setminus G = (F \to G)^c$. The following fuzzy set-theoretic differences can then be derived, among others:

From Kleene-Dienes implication : $F \setminus G(u) = \min(F(u), 1 - G(u))$;

From Lukasiewicz implication : $F \setminus G(u) = \max(0, F(u) - G(u))$ known as the bounded difference;

From Gödel implication: $F \setminus G(u) = 0$ if $F(u) \leq G(u)$ and $1 - G(u)$ otherwise.

Fuzzy symmetric difference operations can be derived by letting $F \Delta G = (F \to G)^c \cup (G \to F)^c$. Noticeably the following operations are obtained

With Kleene-Dienes implication :

$$F\Delta G(u) = \max(\min(F(u), 1 - G(u), \min(G(u), 1 - F(u)))$$

With Lukasiewicz implication : $F\Delta G(u) = |F(u) - G(u)|$

With Gödel implication:

$$F\Delta G(u) = 0 \text{ if } F(u) = G(u) \text{ and } 1 - \min(F(u), G(u)) \text{ otherwise.}$$

Equality Indices. Evaluations of fuzzy set equality can be achieved as the degree to which both F is included in G and G is included in F. So a natural definition is based on inclusion indices:

$$E(F, G) = \min(I(F, G), I(G, F)).$$

Due to the use of minimum, these indices can be written as

$$E(F, G) = \inf_{u \in U} F(u) \leftrightarrow G(u)$$

where $a \leftrightarrow b = \min(a \to b, b \to a)$ is a degree of equality between membership values. Namely, with

Kleene-Dienes implication: $a \leftrightarrow b = \min(\max(1-a, b), \max(1 -b, a)))$;

Godel implication: $a \leftrightarrow b = 1$ if $a = b$ and $\min(a, b)$ otherwise;

Lukasiewicz implication: $a \leftrightarrow b = 1 - |a - b|$.

Clearly $1 - a \leftrightarrow b$ is a symmetric difference operation.

NB. Indices of fuzziness can be defined as $O(F, F^c)$ or $E(F, F^c)$ since $O(F, F^c) = E(F, F^c) = 0$ when F is not fuzzy, and F is the least sharp fuzzy set if and only if

$F = F^c$ (Yager, 1979). Kosko (1986) has also suggested the form $I(F \cup F^c, F \cap F^c)$ (see Chapter 9 by Pal and Bezdek, in this volume).

Counting-Based Comparisons. Another family of comparison indices is based on counting. These indices can be absolute or relative. In the non-fuzzy case, absolute indices can be

$$OC(A, B) = |A \cap B|;\ IC(A, B) = |A^c \cup B|;$$

$EC(A, B) = |(A^c \cup B) \cap (B^c \cup A)|$ or $\min(|A^c \cup B|, B^c \cup A|)$ (these options are not equivalent since the former expression is smaller that the latter).

Absolute comparison indices for fuzzy sets based on counting can be obtained by changing sup in the definition of O(F, G), and inf in the definition of I(F, G) into the sum. Using

$$EC(F, G) = \Sigma_{u \in U} F(u) \leftrightarrow G(u),$$

with $a \leftrightarrow b = 1 - |a - b|$, it is obvious that $1 - EC(F, G)$ is the Hamming distance between fuzzy sets $\Sigma_{u \in U} |F(u) - G(u)|$ (Kaufmann, 1975). Moreover $EC(F, F^c)$ is exactly the index of fuzziness fuz(F) of Kaufmann (1975). Obvious extensions of the Hamming distance are the Euclidean distance between fuzzy sets, as well as all indices of the form $(\Sigma_{u \in U} |F(u) - G(u)|^p)^{1/p}$ for $p \geq 1$. Willmott (1986) has studied transitivity of such indices of inclusion and equality (normalised versions thereof). For many of them the property $I(F, H) \geq I(F, G) + I(G, H) - 1$ holds, depending on the normalisation factor chosen.

Relative counting-based indices of overlap, inclusion and equality of fuzzy sets can be obtained by using scalar cardinality of fuzzy sets in the following expressions

$$ROC(F, G) = \frac{|F \cap G|}{|F| \cdot |G|} \text{ (overlap index)}$$

$$RIC(F, G) = \frac{|F \cap G|}{|F|} \text{ (inclusion index)}$$

$$REC(F, G) = \frac{|F \cap G|}{|F \cup G|} \text{ (equality index)}.$$

The above definitions presuppose that F and G be different from Ø. Note that if product is used instead of min, ROC(A, B) is similar to the Bhattacharyya index for comparing two probability distributions, easily adapted to compare fuzzy sets, namely (Sahnoun *et al.*, 1991)

$$OC(A, B) = \sum_{u \in U} \left(\frac{F(u) \cdot G(u)}{|F| \cdot |G|}\right)^{\frac{1}{2}}.$$

Murthy et al. (1985) have also adapted the probabilistic notions to the comparison of fuzzy sets, using a standardized squared Euclidean distance, that is the counterpart of a correlation coefficient.

Index RI(F, G), originally proposed by Sanchez (1979), also cited in Dubois and Prade (1980, 1982), has been popularized by Kosko (1991). The index RE(F, G) is an extension of the Jaccard index, proposed in Dubois and Prade (1982), which differs from min(RIC(F, G), RIC(G, F)) = $\frac{|F \cap G|}{\max(|F|, |G|)} \geq \frac{|F \cap G|}{|F \cup G|}$, generally. Note that RIC(F, G) also acts as an overlap measure since it is 0 only if F and G are disjoint, and 1 only if F = G. RIC(F, G) = 0 also characterizes disjoint sets, but is not symmetric.

For surveys, on similarity indices, see Zwick *et al.* (1987) who report on a comparative psychometric experiment indicating that the counting-based indices seem to be less faithful to human perception of similatity that those focusing on some level-cut. Especially, 1 - O(F, G) performs well.

Building Comparison Indices. A general framework for building fuzzy set comparison indices has been proposed by Dubois and Prade (1982a). The above three kinds of indices have been considered: overlap indices (called partial matching), inclusion indices and similarity indices (evaluating equality between fuzzy sets). All indices rely on a scalar evaluator of a fuzzy set, which assigns to each fuzzy set F in $\mathcal{F}$(U) a number e(F) in the unit interval. Basically, e(F) is the height, the plinth or the relative cardinality of a fuzzy set. However more general evaluators could be envisaged especially, using a weight distribution over elements of U so as to give more importance to some elements than to others, using a Sugeno integral (1.9) or a Choquet integral (see Grabisch *et al.,* 1995). One only requests that an evaluator obey the following properties:

$$e(\emptyset) = 0; \ e(U) = 1;$$
$$\text{if } F \subseteq G \text{ then } e(F) \leq e(G).$$

An existential evaluator is one such that e(F) = 0 implies F = Ø (like the height). A universal evaluator is one such that e(F) = 1 implies F = U (like the plinth). The relative cardinality is both existential and universal. Axioms for comparison indices have been proposed by Dubois and Prade (1982a), and Sinha and Dougherty (1993) specifically for inclusion. The following freely borrows from both.

First, any set-theoretic comparison index should be invariant up to a permutation of elements of U:

Invariance: If σ denotes a bijection of U to itself then $C(F, G) = C(F \circ \sigma, G \circ \sigma)$ for C = I, O, E.

Axioms more specific to each type of comparison indices are as follows:

An *overlap index* is a mapping O(F, G) from $\mathcal{F}(U)^2$ to the unit interval such that

O1) O(F, G) = 0 if and only if F and G have disjoint support.
O2) O(F, G) = 1, whenever $F \subseteq_S G$ or $G \subseteq_S F$.
O3) O(F, G) = O(G, F) (symmetry).
O4) If $F \subseteq G$ then $O(H, F) \leq O(H, G)$.

A typical overlap index for comparing normal fuzzy sets is $O(F, G) = e(F \cap G)$ for an existential evaluator (like the height) and the fuzzy intersection being min or product. It satisfies a stronger property than O2: O(F, G) = 1, if and only if $\exists$ u, F(u) = 1 and G(u) = 1. For subnormal fuzzy sets, O2 can be violated, because in the conditions specified by this axiom, $e(F \cap G) \leq \min(e(F), e(G))$ which may be less than 1. The counting-based index ROC(F, G) does not verify property O2.

A *inclusion index* is a mapping I(F, G) from $\mathcal{F}(U)^2$ to the unit interval, such that

I1) I(F, G) = 1 if and only if $F \subseteq G$.

I2) I(F, G) = 0 if $\exists$ u, F(u) = 1 and G(u) = 0.

I3) If $F \subseteq G$ then $I(H, F) \leq I(H, G)$.

I4) If $F \subseteq G$ then $I(F, H) \geq I(G, H)$.

A consequence of the above axioms is the following bounding properties

$$
\begin{aligned}
I(F \cup G, H) &\leq \min(I(F, H), I(G, H)) \\
I(H, F \cap G) &\leq \min(I(H, F), I(H, G)) \\
I(H, F \cup G) &\geq \max(I(H, F), I(H, G)) \\
I(F \cap G, H) &\geq \max(I(F, H), I(G, H))
\end{aligned}
$$

Such axioms are freely inspired from (Sinha and Dougherty, 1993) where equality instead of some of the above inequalities is requested. A typical inclusion index for comparing normal fuzzy sets is $I(F, G) = e(F \rightarrow G)$ for a universal evaluator (like the plinth) and a residuated implication (since $a \rightarrow b = 1$ for $a \leq b$, and $1 \rightarrow 0 = 0$). It takes the form (1.13) for a residuated implication.

The counting-based index RIC(F, G) violates I2 and is 0 only in case of disjoint supports. It is as much a non-symmetric overlap index as an inclusion index. Sinha and Dougherty also find it natural to request

I5) $I(F, G) = I(G^c, F^c)$.

However it presupposes that the implication is reciprocal, which, leaves room only for Lukasiewicz implication in $e(F \to G)$.

For strong inclusion indices, I1 should be replaced by

SI1) $I_S(F, G) = 1$ if and only if $F \subseteq_S G$.

A typical strong inclusion index or comparing normal fuzzy sets is $O(F, G) = e(F^c \cup G)$ for a universal evaluator (like the plinth) and the fuzzy disjunction being max or the probabilistic sum. It takes the form of (1.13) for a strong implication. Note that I5 will be satisfied by strong implication indices, generally.

An equality index is a mapping $E(F, G)$ from $\mathcal{F}(U)^2$ to the unit interval, such that

E1) $E(F, G) = 1$ if and only if $F = G$.
E2) $E(F, G) = 0$ if F and G have disjoint support.
E3) $E(F, G) = E(G, F)$ (symmetry).

A typical equality index family for comparing normal fuzzy sets is of the form $E(F, G) = \min(I(F, G), I(G, F))$ as studied above. Another family can be built from a symmetric difference as $E(F, G) = n(e(F \Delta G))$, where n is a fuzzy complementation operation (decreasing function) and an existential operation. The counting-based index REC(F, G) also verifies these properties.

One might as well request properties like:

$E(F, G) = 1$ implies $E(F \cup H, G \cup H) = 1$ and $E(F \cap H, G \cap H) = 1$.

This is satisfied by $E(F, G) = 1 - \inf_{u \in U} |F(u) - G(u)|$ and by REC(F, G), with a proper handling of the empty set case. See (Pappis and Karacapilidis, 1993), (Chen *et al.*, 1995) (Wang *et al.*, 1995) for a studies of properties of some of these comparisons indices.

In his well-known paper on similarity, Tversky (1977) comes up via an axiomatic approach to the following general forms of similarity index between crisp sets (A\B denotes set-difference $A \cap B^c$):

$$C(A, B) = q \cdot e(A \cap B) + a \cdot e(A \setminus B) + b \cdot e(B \setminus A) \text{ (absolute);}$$

$$RC(A, B) = \frac{e(A \cap B)}{q \cdot e(A \cap B) + a \cdot e(A \setminus B) + b \cdot e(B \setminus A)} \text{ (relative).}$$

Bouchon-Meunier *et al.* (1996) have tried to extend this approach to the comparison of fuzzy sets, using evaluations of intersection and differences of fuzzy sets. They propose comparison indices of the form $f(e(F \cap G), e(F \backslash G), e(G \backslash F))$. According the the properties of function f, various comparison indices, especially equality and inclusion, can be retrieved. Fonck *et al.* (1998) build measures of similarity between fuzzy sets from the characterization of aggregation operations applied to fuzzy relations. The search for a unified, axiomatically founded, framework for the definition of comparison indices for fuzzy sets is still going on.

Compatibility. The comparison between fuzzy sets may be more fully described by means of a fuzzy-valued compatibility index COMP(F ; G) introduced by Zadeh (1978b):

$$\mathrm{COMP}(F\,;\,G)(t) = \begin{cases} \sup_{u:F(u)=t} G(u) \\ 0 \text{ if } F^{-1}(t) = \emptyset \end{cases} \quad \forall\, t \in [0,1]. \tag{1.14}$$

COMP(F ; G) is the fuzzy set of possible values of the degree of membership in F of an ill-located element of U whose possible values are restricted by the fuzzy set G. It is an imprecise membership grade, computing by applying the extension principle to the membership function $F(\cdot)$ itself. Namely COMP(F ; G) = F(G). For F = G, and with continuous membership functions that are onto, COMP(F ; G)= τ, a fuzzy set of the unit interval such that $\tau(u) = u$ for $u \in [0, 1]$. For $F = G^c$, $\mathrm{COMP}(F\,;\,G^c) = \phi$, a fuzzy set of the unit interval such that $\phi(u) = 1 - u$ for $u \in [0, 1]$. The fuzzy sets τ and ϕ are usually interpreted as fuzzy truth-values "true" and "false". Moreover

$G \subseteq_s F$ if and only if $\mathrm{COMP}(F\,;\,G) = \{1\}$,
$G \subseteq_s F^c$ if and only if $\mathrm{COMP}(F\,;\,G) = \{0\}$.
$G \subseteq F$ if and only if $\mathrm{COMP}(F\,;\,G)(u) \le \tau(u)$ for $u \in [0, 1]$

Hence COMP(F ; G) is a kind of general inclusion index.

1.3.6 Fuzzy Sets on Structured Referentials

So far, nothing has been assumed about the referential set U that carries the fuzzy sets. However in practice this set will have a structure of its own, and there will be some interplay between this structure and the fuzzy sets on U. New properties of fuzzy sets will stem from the structure that equips U.

Convex Fuzzy Sets. A very common case is when U is a separable complete metric space, like the set $\mathbb{R}^n$ of real vectors. Using cut-worthy definitions, a lot of properties making sense for subsets of $\mathbb{R}^n$ will also make sense for fuzzy sets thereof: convexity, star-shapedness, connectedness, etc. Namely a fuzzy set will have property P if and only if its level-cuts have this property in the classical sense (see Brown, 1971). One of the early definitions is the one of a convex fuzzy set, one whose level-cuts are convex. This definition can be equivalently spelled out in terms of membership functions (Zadeh, 1965a):

F is convex if and only if $\forall$u, v, w, where w lies in the segment bounded by u and v, $F(w) \geq \min(F(u), F(v))$.

It is also possible to write w as $\alpha \cdot u + (1 - \alpha) \cdot v$, where $\alpha \in [0, 1]$. Intersection of fuzzy sets (by the minimum) preserves convexity. A stronger form of convexity is obtained by requesting a strict inequality in the definition. As for boundedness, the level-cut definition (level-cuts of F are bounded sets) is weaker than the requirement that the support of F be bounded. Convex fuzzy sets and their topological implications have been further studied by Lowen (1980) and Drewniak (1987), among others.

The most popular and useful example of convex fuzzy set is the *fuzzy interval*, extensively studied in Chapter 10 of this volume, where a full bibliography is provided. A fuzzy interval is a fuzzy set of the real line whose level-cuts are intervals. Very often, it is required that the level-cuts be closed intervals, which means that the membership function is upper semi-continuous.

Fuzzy Algebraic Structures. More generally, if the set U is equipped with some operation * then a fuzzy set F will be closed under * if its level-cuts are closed. In terms of the membership function, it comes down again to define the closure of F with respect to * as:

$$\forall u, v,\ F(u * v) \geq \min(F(u), F(v)).$$

The first examples of such a fuzzy extension of algebraic concepts are the fuzzy groups and fuzzy ideals of Rosenfeld (1971). Since then, a considerable literature on fuzzy algebraic structures has been published (see (Mordeson and Malik, 1998), for a thorough organized survey). This kind of definition differs from the idea of extending the operation * to fuzzy set arguments F and G, defining F*G as

$$F * G\,(w) = \sup\{\min(F(u), G(v)) : u * v = w\},$$

which comes down to lifting the operation * to the sets of fuzzy sets of U. This is what is done in fuzzy interval analysis, with addition, multiplication etc. of fuzzy intervals. Operations on fuzzy intervals lead to consider max-based "convolutions" of pairs of functions, which have been previously considered by Bellman and Karush (1961), in the framework of dynamic programming. Generally, not all properties of * remain on $\mathcal{F}(U)$ although some (like associativity of *) go through.

Some fuzzy extensions of classical notions of algebraic structures have proved to be practically useful and mathematically fruitful, contrary to the statement of MacLane (1983) talking about "extravagant claims for applications". For instance fuzzy intervals, have proved to be helpful in the formulation and the solving of optimization and uncertainty propagation problems. The field of image processing has also considerably benefited from fuzzy structured sets, for extending tools for the analysis of regions, shapes, to grey-level areas (see Rosenfeld, 1979, 1984; Bezdek *et al.*, 1999); this is noticeably the case with fuzzy mathematical morphology, for instance, Bloch and Maitre (1994, 1995), De Baets *et al.* (1994), Sinha ad Dougherty (1995).

However, it seems that many papers in fuzzy algebra remain straightforward and lack both motivations and applications, because the extension of many notions is so simple to carry out using level-cuts that it can be developed for its own sake without a significant effort. Head (1995) has emphasized that many results in fuzzy algebra can be subsumed by simple, generic meta-theorems.

Distances Between Fuzzy Sets. When the referential U is equipped with a metric structure of its own, comparing the relative position of fuzzy sets can be achieved by defining some distance between fuzzy sets. This is quite different from comparing fuzzy sets from a set-theoretical point of view, as done in the previous section. Namely, the above indices would rate equally for disjoint sets of a metric space no matter how far they are located from each other, contrary to distance indices. While comparison indices take a vertical view (comparing membership grades for each element in the referential), distance indices take an horizontal view, integrating distances between elements in each level cut. Let d be the distance function on U, such that

$d(u, v) = 0$ if and only if $u = v$ (separability);
$d(u, v) = d(v, u)$;
$d(u, v) \leq d(u, w) + d(w, v)$.

In order to extend this distance to fuzzy sets, two options are possible: one may look for a scalar or a fuzzy evaluation. One of the extension principles mentioned in Section 1.3.3 must be used.

Consider two subsets A and B of U. The set-valued extension of d is defined by

$$D(A, B) = \{d(u, v) : u \in A, v \in B\}.$$

It only depicts the set of actual distances between an element of A and an element of B. The value $\inf D(A, B)$ is the minimal distance separating A and B, while $\sup D(A, B)$ is the diameter of the union $A \cup B$. Let us call it the *descriptive distance*.

The fuzzy descriptive distance between fuzzy sets F and G has been defined by Dubois and Prade (1980a): it is the fuzzy set D(F, G) such that

$$\begin{aligned} D(F, G)(a) &= \sup\{\min(F(u), G(v)) : d(u, v) = a\} \\ &= \sup\{\alpha, a \in D(F_\alpha, G_\alpha)\}. \end{aligned}$$

Alternatively, the minimal distance can be extended separately as $\inf D(F, G)(a) = \sup\{\alpha, a = \inf D(F_\alpha, G_\alpha)\}$. What is obtained is an decreasing membership function. This notion is useful when studying the geometry of images (Rosenfeld, 1984; Dubois and Jaulent, 1987).

The most standard scalar extension of distance between A and B is the Hausdorff distance

$$d_H(A, B) = \max(\inf_{u \in A} \sup_{v \in B} d(u, v), \inf_{v \in B} \sup_{u \in A} d(u, v)).$$

This distance is based on the following idea: for each element in A look for the most remote element in B, then check for the element in A for which the distance to the most remote element in B is minimal. The same is done exchanging B and A and the longest distance of the two components is kept. Note that d_H is still a metric, while D is not. Namely $D(A, A) \neq \{0\}$ while $d_H(A, B) = 0$ if and only if A = B. Moreover triangular inequality is basically lost for D, although some properties can be derived from the ones of d (see Dubois and Prade, 1980a).

The Hausdorff distance between fuzzy sets can be either fuzzy or scalar. Scalar distances that retain good properties can be defined by merging the values $\{d_H(F_\alpha, G_\alpha), \alpha \in (0, 1]\}$. For instance

$$d_H^{\infty}(F, G) = \sup\{d_H(F_\alpha, G_\alpha), \alpha \in (0, 1]\};$$

$$d_H^{1}(F, G) = \int_0^1 d_H(F_\alpha, G_\alpha)d\alpha\,.$$

These notions have been introduced by Puri and Ralescu (1983), and Goetschel and Voxman (1983). The major reference on this topic is the book by Diamond and Kloeden (1994) (See also Chapter 11 by Diamond and Kloeden in this volume). Zwick *et al.* (1987) reviews many early proposals of such distances, which they test along with set-theoretic equality indices, for their psychological plausibility. Lowen and Peeters (1998) provide an extensive study of a class of distances between fuzzy sets for which the triangle inequality is not fulfilled.

Variants of such distances could be envisaged by merging the values $\{d_H(F_\alpha, G_\beta), \alpha, \beta \in (0, 1]\}$, in the scope of the random set approach. For instance

$$d_H^{1}(F, G) = \int_0^1\int_0^1 d_H(F_\alpha, G_\beta)d\alpha d\beta,$$

as proposed (in the discrete case) by Dubois and Jaulent (1987). See also Bloch and Maitre (1994). Besides fuzzy Hausdorff distances are defined in Dubois and Prade (1983). The idea is basically to draw the diagram of $\{d_H(F_\alpha, G_\alpha), \alpha \in (0, 1]\}$, plotting $\{d_H(F_\alpha, G_\alpha)$ against α, as in the case of the fuzzy valued cardinality. Unfortunately, there is no monotony of $d_H(F_\alpha, G_\alpha)$ with α. One may also use a Sugeno-like integral so as to catch the most representative value of $\{d_H(F_\alpha, G_\alpha), \alpha \in (0, 1]\}$ provided that the distance function does range in the unit interval.

1.4 NOTIONS DERIVED FROM FUZZY SETS

The notion of fuzzy sets has given birth to various developments that are closely related to fuzzy set theory, but go beyond the simple idea of a membership function, the set-theoeretic operations, and fuzzy set comparison indices. Here, we review two essential such derived notions: moving from one referential to several ones, sets become relations. And fuzzy sets give birth to a formal framework for valued relations where the intensity of the link between elements is modelled. Next, fuzzy sets can be lifted to set-functions that differ from probability measures. The idea is that given a membership function F(·) and a crisp set A, to attach to set A a degree that is the aggregatiom of membership grades in A. Following this path enables fuzzy sets to become the basis of alternative, non-probabilistic uncertainty theories.

1.4.1 Fuzzy Relations

A fuzzy relation is a fuzzy subset R of a Cartesian product of sets. For instance, consider a fuzzy set of 3-tuples (u, v, w) ∈ U×V×W. Membership degrees R(u,v,w) estimate the strength of the link between elements u, v and w. This notion was anticipated by Menger (1951a, c) although he tried to interpret it in the framework of probability theory.

Fuzzy relations can model gradual n-ary predicates, possibly by suitable combinations, using fuzzy set-theoretic operations, of gradual unary predicates. For instance, a *stout* man means somebody who is *short and broad-shouldered.* If simple membership functions of *short* and *broad-shouldered* are known, it makes sense to model *stout* as a fuzzy relation on the referential U×V, where U is the human height scale, and V is a scale for shoulder width. This relation is obtained by aggregating the two fuzzy sets *short* and *broad-shouldered* by means of some fuzzy conjunction. More complex concepts like *comfort*, *"good customer,* etc. may, if they are hierarchically decomposed into simple fuzzy concepts (Zimmermann, 1985), be modelled by a fuzzy relation on a multidimensional scale. Fuzzy relations, obtained by combination of fuzzy sets, thus offer a general setting for multifactorial evaluation, out of a logical expression describing the complex concept of interest.

Composition of Fuzzy Relations. A very important special case of fuzzy relations is the fuzzy binary relation, which links two elements together. A fuzzy binary relation R between U and V can also be viewed as a fuzzy set valued function which associates to each element u ∈ U a fuzzy set R(u) on V, with membership function R(u, ·), called the fuzzy image of u in V. Conversely, the fuzzy set $R^{-1}(v)$ with membership function R(·, v), is called the fuzzy inverse image of v in U.

Given a relation R_1 relating two sets U and V, and another relation R_2 relating U and V, it is possible to define the composition of these relations, denoted $R_1 \circ R_2$. It is a relation linking U and W directly. This composition is defined as follows: in order to relate $u \in U$ and $w \in W$, select an intermediary element $v \in V$ and define the link strength between u and w through v as a function of the link strengths between u and v, and between v and w, at most the weakest of both. Then vary the intermediary element v in V, and consider v such that the link strength between u and w through v is maximal. This is the direct link strength between u and w. Mathematically, the most currently found fuzzy relational composition is the max-min composition which reads $R = R_1 \circ R_2$ where (Zadeh, 1975a):

$$R(u,w) = \sup_v \min(R_1(u,v), R_2(v,w)).$$

This composition reduces to the classical composition of relations (u is related to w if and only if there exists v such that u is linked to v, and v is linked to w). In the above equation, the minimum operation can be changed into any fuzzy intersection operation without altering these semantics. Yet other types of variants of fuzzy relational products exist (Bandler and Kohout, 1980b, Dubois and Prade, 1992a; De Baets and Kerre 1993). They can be obtained by exploiting comparison indices between the fuzzy set of images of u in V, namely, $R_1(u)$, and the fuzzy inverse image of w in V. Namely $R(u,w) = O(R(u), R^{-1}(w))$ (index of overlap), $I(R(u), R^{-1}(w))$ (index of inclusion of R(u) in $R^{-1}(w)$, or $E(R(u), R^{-1}(w))$ (index of equality).

The image of a fuzzy set F of U in V through R is $R(F) = F \circ R$, a particular case of the above composition where

$$R(F)(v) = \sup_u \min(F(u), R(u,v)).$$

This definition of R(F) can be called the upper image of F through R since it excludes only those elements in V which have no link with any element in U. The lower image of a non empty fuzzy set F through R is the fuzzy set of elements in V which are related to all elements in F:

$$R_*(F)(v) = \inf_u \max(1 - F(u), R(u,v)) = (F \circ R^c)^c(v)$$

Note that $R(F \cup G) = R(F) \cup R(G)$, and $R_*(F \cap G) = R_*(F) \cap R_*(G)$.

Fuzzy Cartesian Products. A particular case of fuzzy relation is the fuzzy Cartesian product, namely $R = F \times G$, where F is a fuzzy subset of U, G is a fuzzy subset of V, and $R(\cdot, \cdot) = \min(F(\cdot), G(\cdot))$. There is a precise reason why the minimum operation is used here. The relation R can be projected on U or V, thus yielding the projections $R^{\downarrow U}$ and $R^{\downarrow V}$ defined as:

$$R^{\downarrow U}(u) = \sup_v R(u,v); R^{\downarrow V}(v) = \sup_u R(u,v).$$

In terms of composition, $R^{\downarrow U} = R \circ V$ and $R^{\downarrow V} = U \circ R$. Generally the inclusion $R \subseteq R^{\downarrow U} \times R^{\downarrow V}$ holds, for any fuzzy relation. A fuzzy relation is separable if and only if it coincides with the Cartesian product of its projections, in other words iff $R = R^{\downarrow U} \times R^{\downarrow V}$. Note that if $R = F \blacklozenge G$ for another fuzzy intersection operator, where F is a fuzzy subset of U, G is a fuzzy subset of V, then this is a weak form of separability. However the property $R = R^{\downarrow U} \blacklozenge R^{\downarrow V}$ will hold only if F and G are normal fuzzy sets. Separable fuzzy relations are the key-notion of the fuzzy relational calculus, that underlies the theory of approximate reasoning (Zadeh, 1979; Bouchon-Meunier *et al.*, 1999) and fuzzy constraint propagation (Dubois et al, 1996), as well as fuzzy arithmetics (see Chapter 10 by Dubois *et al.* in this volume).

Extension Principles for Functions of Several Variables. The extension principles described in Section 1.3.3 still make sense when changing the fuzzy set into a fuzzy relation. Especially it enables functions of several variables to be lifted to function with fuzzy set arguments. In particular, when the fuzzy relation is separable, the variables are said to be noninteractive (Zadeh, 1975b). Namely, let f be a function from U×V to W. Let F and G be fuzzy sets on U and V respectively. Define $R = F \blacklozenge G$. then the fuzzy image or (F, G) through f is defined by

$$f(F, G)(w) = \sup\{F(u) \blacklozenge G(v): u * v = w\}$$

This is the basic building block of fuzzy interval analysis (see Chapter 10 in this volume).

Other types of extension principles can be considered. For instance if F and G are represented as a convex combination of sets of the form $\{(F_i \times G_j, r_{ij})\}_{i,\, j = 1,\dots, n}$ with marginals $\{(F_i, p_i)\}_{i = 1,\dots, n}$ and $G = \{(G_j, q_j)\}_{j = 1,\dots, m}$, then it makes sense of define the image of (F, G) via f as the convex combination $\{(f(F_i, G_j), r_{ij})\}$ (Dubois and Prade, 1991). The case of independent fuzzy sets is when $r_{ij} = p_i \cdot q_j$. However the obtained family of sets may no longer be nested. One can still get a normalized fuzzy set using the one-point coverage function (1.5). These types of operations have been considered by, e.g., Bloch and Maitre (1994) for the purpose of fuzzy mathematical morphology. However these kinds of fuzzy set propagation have received little attention to-date.

Fuzzy Relational Equations. A research topic has focused the interest of many a researcher, starting with Sanchez (1976), namely how to find all solutions R_2 to fuzzy relational equations of the form $R_1 \circ R_2 = R_3$. A major finding has been the existence of a unique largest solution, which contains all other ones in the sense of fuzzy inclusion, when such solutions exist. This solution is precisely of the form

$$R_{2*}(v, w) = \inf_u R_1(u, v) \rightarrow R_3(u,w)$$

for a residuated implication. It is closely related to lower images via a fuzzy relation. A related problem is the study of the set of eigenvectors and eigenvalues of a fuzzy relation on $U \times U$, that is the solutions F of $R \circ F = F$, where F is a fuzzy set on U (Sanchez, 1978). A basic book on this topic is the one of Di Nola *et al.* (1989). See Chapter 6 by De Baets in this volume.

These considerations come close to similar ones in linear algebra (a binary fuzzy relation on finite sets can be encoded as a matrix). However here, the underlying algebra is exotic, and non-linear in the traditional sense, generally. This kind of structure where maximum plays the role of addition and some fuzzy conjunction plays the role of product has been studied in other fields under various names: minimax algebras in operations research (Cuninghame-Green, 1991), incline algebras in social sciences (Cao *et al.*, 1984). It is also a special form of a ring-like structure (without invertibility) called dioïd. See Gondran (1976) for the max-min fuzzy relational algebra, and Gondran (1979) for a more general setting. Such structures popped up in graph theory and are exploited in the theory of discrete-event systems (Bacelli et al, 1992).

In fuzzy set theory, fuzzy relational equations are exploited for deriving mathematical models of fuzzy rules and posing problems in abductive reasoning for the purpose of automated diagnosis (see, e.g., Bouchon-Meunier *et al.*, 1999). Fuzzy relations are also the basic tool for modelling what Zadeh (1965b) has called fuzzy systems (i.e., systems that are too complicated or ill-known, so that only vague descriptions thereof make sense or are accessible). A fuzzy system is a model where the input-output relation is one-to-many, and where some outputs are more plausible than others, for a given input (and/or state). As such, and similarly to linear systems, fuzzy systems take advantage of the algebraic non-classical framework of fuzzy relations, that plays a role similar to linear algebra for linear systems (Pedrycz, 1989).

Similarity Relations. Two kinds of fuzzy binary relations linking elements of the same set U have been particularly studied: similarity relations (also called indistinguishability relations, or yet fuzzy equality relations), and fuzzy ordering (or preference) relations. A similarity relation is an extension of the equivalence relation, enjoying gradual versions of reflexivity, symmetry and transitivity. They can be defined via level-cuts. See Bandler and Kohout (1988, 1993), and De Baets and Kerre (1994) for studies of the properties of binary fuzzy relations that can be defined via level-cutting. The most common definitions are as follows:

reflexivity : $S(u, u) = 1, \forall u$;
symmetry : $S(u, u') = S(u', u), \forall u, \forall u'$;
min-transitivity : $S(u, u'') \geq \min(S(u, u'), S(u', u'')), \forall u, u', u''$.

The latter property reads $S \circ S \subseteq S$, using fuzzy relational composition. The degree $S(u, u')$ evaluates the proximity, or similarity between elements u and u'.

Actually the degree $1 - S(u, u')$ is a kind of normalized distance. More precisely, when S is min-transitive, $1 - S$ is an ultrametric, well-known in taxonomy. Then, the level cuts of S form a nested family of equivalence relations (Zadeh, 1971a). The fuzzy equivalence class of u is of the form $[u]_S = \{(u, S(u, v)), v \in U\}$.

Using the linear intersection operation instead of min in the transitivity, $1 - S$ obeys the triangular inequality (up to truncation effects), and comes very close to a usual notion of distance. A simple way to get a fuzzy relation satisfying transitivity in the sense of the linear intersection is to start with n classical equivalence relations $S_1, \ldots, S_n$ on X and to define (Bezdek and Harris, 1978)

$$S(u, v) = \Sigma_{i=1,n}\, \alpha_i \cdot S_i(u, v) \tag{1.15}$$

where $\Sigma\alpha_i = 1$ and $\alpha_i > 0, \forall i$. When U has cardinality more than 3, this form of transitivity is more general than convex decomposability into equivalence classes. However the problem of finding a criterion to determine if a linearly transitive similarity relation decomposes into (1.15) is solved by Kasumov (1996).

Using a more general fuzzy conjunction to define the transitivity of a fuzzy relation, the notion of distance can thus be generalized.

The converse problem has been solved by Valverde (1985), i.e., given a family of fuzzy sets on X that represents clusters of similar elements, find the underlying similarity relation. Let $F_1, \ldots, F_n$ be fuzzy sets on U. Then the fuzzy relation S defined by

$$S(u, v) = \min_{i=1,n} (\max(F_i(u), F_i(v)) \rightarrow \min(F_i(u), F_i(v))), \tag{1.16}$$

is a max-* transitive similarity relation for a suitable continuous intersection operation *; namely, $\alpha \rightarrow \lambda = \sup\{\lambda \in [0, 1] : \alpha * \lambda \leq \beta\}$. Moreover if the family of fuzzy sets is the fuzzy quotent set $U/S = \{[u]_S \mid u \in U\}$ for a given max-* transitive similarity relation S, then (1.16) applied to this family produces S again.

The application of (1.16) to the 3 basic kinds of transitive similarity relation gives fuzzy relations computed via fuzzy equivalence connectives: :

– max-min transitivity

$$\begin{aligned} S(u,v) &= \min_{i: F_i(u) \neq F_i(v)} \min(F_i(u), F_i(v)). \\ &= 1 \text{ otherwise} \end{aligned}$$

– max-product transitivity (Ovchinnikov (1982))

$$S(u,v) = \min_{i=1,n} \min\left(\frac{Fi(u)}{Fi(v)}, \frac{Fi(v)}{Fi(u)}\right).$$

– max-linear transitivity

$$S(x,v) = \min_{i=1,n} 1 - |F_i(x) - F_i(v)|.$$

For more details on fuzzy equivalence relations, see Chapters 4 and 5 by Ovchinnikov, and by Boixader *et al.*, respectively, in this volume. Similarity relations are useful in automated classification (Bezdek *et al.*, 1999), and some forms of approximate reasoning such as interpolative and case-based reasoning (see, e.g., Bouchon-Meunier *et al.*, 1999).

Fuzzy Partitions. The notion of a fuzzy partition was first proposed by Ruspini (1969). A fuzzy partition is a family of fuzzy sets $\{F_1,\ldots, F_n\}$ such that $\forall i = 1, \ldots, n$, $F_i \neq \emptyset$, $F_i \neq U$, and

$$\forall u \in U, \Sigma_{i = 1,\ldots,n} F_i(u) = 1. \tag{1.17}$$

An obvious example of fuzzy partition is the pair (F, F^c) made of a non-empty fuzzy set and its complement. For crisp sets, the definition is equivalent to the usual one for non-degenerate partitions. This definition has been widely used in pattern classification. It is closely related to the relation linking the likelihood functions of a family of exhaustive and mutually exclusive classes C_i, namely $\Sigma_{i=1,\ldots,n} P(C_i | u) = 1$, interpreting the degree of membership of an object to a fuzzy set as the probability of classifying this object using the name of this fuzzy set (Hisdal, 1989; Mabuchi, 1992). In particular this definition makes sense only for numerical membership functions. Kasumov (1996) has proved that a fuzzy partition $\{F_1,\ldots, F_n\}$ à la Ruspini is decomposable as a convex sum of Boolean partitions if and only if the fuzzy cardinalities $|F_i| \geq 1$, for all i.

A fuzzy partition satisfies the usual properties of a partition with respect to the linear fuzzy connectives, namely, $F_i \cap_L F_j = \emptyset$ if $i \neq j$ and $F_1 \cup_L \ldots \cup_L F_n = U$. However the above definition is more strict. To recover (1.17), one must change the disjointness condition into $F_i \cap_L (\cup_{L\, j \neq i} F_j) = \emptyset$ (Butnariu, 1983)) Note that we cannot change $\cup_L$ and $\cap_L$ into other pairs of fuzzy connectives given in Section 1.3.4 because the definition degenerates and yields Boolean partitions.

However this definition is not *fully* in accordance with the usual notion of partition, nor does it yield the fuzzy equivalence classes of a similarity relation.

First, nothing prevents from having $F_i = F_j$ for $i \neq j$ in the partition. For instance if $F = F^c$, all elements in F having membership .5. This is strange. More generally it may happen that F_i is at least partially contained in the union of other fuzzy classes. This phenomenon is studied by Klement and Moser (1997). It has led these authors to use geometric methods so as to minimize redundancy in fuzzy partitions. Other try to add conditions of linear independence between the membership functions $F_i(\cdot)$, see for instance Mundici (1998).

In the case of the minimum operation, the fuzzy equivalence classes of a similarity relations do not form a fuzzy partition in the sense of Ruspini. Namely there exists k fuzzy sets $\{G_1,..., G_k\}$ such that each G_l is of the form $[u]_S$ for some u, and $G_1 \cup G_2 \cup ... G_k = U$ (the usual fuzzy union) and $G_i \cap G_j$ is subnormalized for $i \neq j$. In other words the cores of the G_i's form a standard partition. A less committing definition of a fuzzy partition, not contradicting the one coming from similarity relations is thus the following (Dubois and Prade, 1992b): A weak fuzzy partition is a family of distinct fuzzy sets $\{F_1,..., F_n\}$ such that their supports cover U and their cores do not overlap.

Some authors have tried to study fuzzy partitions from the standpoint of fuzzy relations, following the work of Valverde (1985). Höhle(1988) has proposed axioms than define fuzzy equivalence classes of a max-* transitive similarity relation S. In Dubois and Prade (1992), based on these works, the following definition was proposed: F_i is a fuzzy equivalence class of S if and only if the three properties are valid

-) F_i is a normal fuzzy set
-) $F_i o S \subseteq F_i$ (extensionality)
-) $F_i \times F_i \subseteq S$ (singleton condition)

The first condition ensures a non-degenerated notion. The second is a closure condition: elements belonging to F_i are similar to each other and dissimilar to other elements in U. The third condition views S as formed of its equivalent classes (here the Cartesian product is in the sense of *). More works along these lines are those of De Baets and Mesiar (1997), and Klawonn and Castro (1995). See also Chapter 5 by Boixader *et al.*, in this volume.

Fuzzy Orderings. A fuzzy partial preordering relation R is one that is reflexive and transitive (Zadeh, 1971). The semantics or the relation is then in terms of preference and R(u, u') evaluates to what extent element u is preferred to, or dominates, element u', in the wide sense. Specific properties can be added namely:

Antisymmetry : $\min(R(u, u'), R(u', u)) = 0, \forall u, \forall u'$;
Completeness : $\max(R(u, u'), R(u', u)) = 1, \forall u, \forall u'$.

Adding antisymmetry makes it a fuzzy partial ordering expressing strict dominance. Adding completeness makes it a fuzzy preordering relation excluding incomparability, and generalizing usual preordering relations. Adding both makes it a linear ordering relation that is no longer fuzzy (Chakraborty and Das, 1985).

Among problems that have focused the attention of researchers, one is the one of decomposing a fuzzy preordering relation R into a similarity relation S, and a fuzzy ordering relation P. In the classical case it can be done in a unique way and $R = S \cup P \cup P^{-1}$. This question has been examined first by Orlovski (1978) and more recently by Fodor and Roubens (1994), and many others since then. See Perny and Roubens (1998) for a survey on this topic.

Another issue is the condition of antisymmetry which has been criticized as being too strong. Especially, residuated fuzzy implications should define fuzzy orderings on the unit interval. However it is not true that $\min(a \to b, b \to a) = 0$, generally. Bodenhofer (1998) has proposed to introduce a notion of antisymmetry with respect to a similarity relation S, assumed to be min-transitive here for simplicity. A fuzzy ordering P is then defined as a fuzzy preordering that is S-reflexive ($S \subseteq R$) and S-antisymmetric ($\min(R(u, u'), R(u', u)) \leq S(u, u')$). This relaxed framework enables complete fuzzy orderings (and residuated fuzzy implications) to be genuine fuzzifications of standard linear orderings.

The question of transitivity-preserving aggregation of fuzzy preorderings has been also considered, in the scope of social choice. Generalized version of Arrow's impossibility theorem on the aggregation of usual preordering relations have been obtained. This question is surveyed by Fodor *et al.* (1998) in the setting of multicriteria decision-making, and by Kacprzyk and Nurmi (1998) in the scope of group decision theory and voting. These works are particularly relevant for decision-making problems and are thus dealt with in the volume of the Handbooks of Fuzzy Sets Series devoted to this topic; see Slowinski (1998).

1.4.2 Possibility Measures and Other Fuzzy Set-Based Functions

An expression such as "X is F" where X is a variable and F a fuzzy set (e.g., "age is young") may be used in two different types of situations, which both take advantage of the fuzziness of F. On the one hand the expression "X is F" can take place in a situation where the *value of X is precisely known* and we estimate the extent to which this value is compatible with the label F (whose meaning obviously depends on the context). In this case we are interested in the *gradual*, *soft* nature of the qualification stated by "X is F". For instance, one looks for somebody who should be young and we estimate to what extent a person satisfies this requirement or can be qualified as *young* (e.g., John, whose age is known to be 32, can be considered as *young* to the degree 0.8 in a fixed context).

However, the expression "X is F" may in other situations mean "*all that is known* about the value of X, is that X is F" (without knowing the value of X precisely in this case). This corresponds to a situation of *incomplete* information (pervaded with imprecision and uncertainty), a situation in which *we can only order the possible values* of X according to their level of plausibility or possibility. When a fuzzy set is used to represent what is known about the value of a single-valued variable, the degree attached to a value expresses the level of possibility that this value is indeed *the value* of the variable. The fuzzy set F is then interpreted as a *possibility distribution* (Zadeh, 1978a), which expresses various shades of plausibility on the possible values of the ill-known variable X. Note that distinct values may simultaneously have a degree of possibility equal to 1.

Possibility and Necessity. In the case of incomplete information, we can compute to what extent a statement like "the value of X is in subset A", is consistent with the available information "X is F". This is estimated by means of a possibility measure Π(A), and a necessity measure N(A) (Zadeh, 1978; Dubois and Prade, 1980, 1988).

When A is an ordinary subset, Π(A) is defined as the maximum of F(·) over A; namely Π(A) = O(A, F). Clearly, $\Pi(A \cup B) = \max(\Pi(A), \Pi(B))$. Besides, we shall be all the more certain that X takes its value in some subset A as all the values outside A have a smaller degree of possibility; this can be estimated by means of a dual measure of necessity $N(A) = 1 - \Pi(A^c)$ (A^c is the complement of A). This duality between possibility and necessity measures (Dubois and Prade, 1980) is a graded version of the one existing in modal logic (Hughes and Cresswell, 1968), which expresses a relation between the necessary and the possible, already advocated by Aristotle and his school; indeed some linkage between possibility theory and some modal logic systems have been pointed out (e.g., Fariñas del Cerro and Herzig, 1991). Necessity degrees N(A) express how certain the proposition "X is A" is, in the face of the possibility distribution induced by F. More precisely it expresses to what extent "X is A" is implied by "X is F", namely $N(A) = I_S(F, A)$, using (1.13) with a Kleene-Dienes implication. Clearly, $N(A \cap B) = \min(N(A), N(B))$.

Going further, given a fuzzy set G playing the role of a fuzzy event, and a fuzzy set F playing the role of a possibility distribution, possibility and necessity measures can be defined for fuzzy events as follows:

$$\Pi_F(G) = O(G, F); \; N_F(G) = I_S(F,G).$$

These evaluations, respectively the consistency of F and G (1.12) and the degree of strong inclusion of F in G still verify the duality condition $N_F(G) = 1 - \Pi_F(G^c)$. They have been exploited as measures of fuzzy pattern matching in computer science applications (Cayrol *et al.*, 1982; Dubois and Prade, 1988). They are also special kinds of Sugeno integrals, namely

$$\Pi_F(G) = \oint G \circ \Pi \text{ (Dubois and Prade, 1980);}$$

$$N_F(G) = \oint G \circ N \text{ (Grabisch } et\ al.\text{, 1992).}$$

Moreover $\Pi_F(G)$ and $N_F(G)$ are also the upper and lower expectations with respect to the fuzzy set $\tau(u) = u$ for $u \in [0, 1]$ in the sense of Sugeno integral of the fuzzy-valued compatibility index COMP(F ; G). Namely, $\Pi_F(G)$ and $N_B(A)$ can be recovered from COMP(F ; G) as follows (Baldwin and Pilsworth, 1979; Yager, 1983; Dubois and Prade, 1985a)

$$\Pi_F(G) = \sup_{t \in [0, 1]} \min(t, \mathrm{COMP}(F ; G)(t))$$

$$N_B(A) = \inf_{t \in [0, 1]} \max(t, 1 - \text{COMP}(F ; G)(t))$$

The English economist G. L. S. Shackle is a particularly important precursor of possibility theory. In a series of works starting in the late 1940s, Shackle (1949, 1955, 1961, 1979) has proposed to replace probability with a new concept in order to have a more satisfactory model of uncertainty in economics. He introduced "degrees of potential surprise," which estimate to what extent events are impossible, and used the maximum and minimum operations for combining the degrees. He also exploits a potential surprise curve which up to a complementation to 1 is nothing but a possibility distribution in Zadeh's sense. Moreover, he discussed at length "uncertainty as probability" vs. "uncertainty as possibility." In spite of the interest of some logicians in his work (e.g., Hamblin, 1959), the mathematics of Shackle's theory were never completely worked out, but find a faithful echo within possibility theory.

Independently, the philosopher L.J. Cohen (1973) advocated the minimum operation for estimating the so-called grade of inductive support attached to the conjunction of two propositions from the grades associated with each of them. These degrees are accounted for by necessity measures. Some years later the philosopher N. Rescher (1976) proposed a theory of plausible reasoning where propositions are organized in layers of various "plausibility" degrees. These degrees, which in our terminology are rather "certainty degrees," are propagated by means of inference rules which are similar to the ones appearing in the syntactic part of possibilistic logic, a logic whose semantics completely agree with possibility theory (see Dubois *et al.*, 1994). The consequence principle that is used goes back to Theophrastus, a disciple of Aristotle, and expresses that the certainty attached to a conclusion cannot be weaker than that of the weakest premisse essential to drawing the conclusion. Formally, this inference rule can be expressed in terms of a necessity measure N in the form of an inequality $N(B) \geq \min(N(A), N(A^c \cup B))$, where $A^c \cup B$ is the material implication (i.e., $A \rightarrow B$). This inequality can be obtained as a consequence of the min-decomposability of necessity measures for the conjunction.

Shafer (1976)'s belief functions exploits random sets for representing imprecise and uncertain information. The special case of a Shafer's (1976) consonant plausibility function (when the focal elements which give birth to this set function are nested), is mathematically identical to Zadeh's (1978a) possibility measure. Shafer's contour functions (corresponding to the plausibility of singletons) can be viewed as a fuzzy set membership function, a point first made by Kampé de Fériet (1980). Shilkret's (1971) maxitive measures are another example of possibility-like measures, introduced in this case from a purely mathematical point of view. See Chapter 7 in this volume, for a detailed presentation of possibility theory.

Pseudo-Additive Set-Functions. More generally, a fuzzy set can represent an uncertainty distribution which gives birth to more general set-functions than both possibility and probability measures. Namely given a triangular co-norm $\perp$, that is, a monotonically non-decreasing semi-group of the unit interval with identity 0 and absorbing element 1, a pseudo-additive measure (Dubois and Prade, 1982b; Pap, 1994; Weber, 1984) is a numerical set-function g on a *finite* set U such that

$$g(U) = 1 \; ; \; g(\emptyset) = 0;$$
$$\text{if } A \cap B = \emptyset, \; g(A \cup B) = g(A) \perp g(B)$$

A pseudo-additive measure is totally characterized by a fuzzy set F on U such that $F(u) = g(\{u\})$, since

$$g(A) = \perp_{u \in A} F(u).$$

The fuzzy set must satisfy the condition $\perp_{u \in U} F(u) = 1$, that results from $g(U) = 1$.

When the co-norm is the maximum, possibility measures are recovered and F must be a normal fuzzy set. Then the dual set function $g^c(A) = 1 - g(A^c)$ is a necessity measure. When the co-norm is the bounded sum, g is a truncated additive set-function since $g(A) = \min(1, \Sigma_{u \in A} F(u))$. The condition $g(U) = 1$ implies $\Sigma_{u \in U} F(u) \geq 1$. The set function is a probability measure only if F coincides with a probability density, namely, $\Sigma_{u \in U} F(u) = 1$. This condition is however never required in the framework of fuzzy sets, since this condition corresponds, generally, to a partially empty set ($ht(F) < 1$). When the co-norm is the probabilistic sum, g is isomorphic to a non-finite additive measure and F is again a normal fuzzy set.

Such generalized set-function are the basis of pseudo-analysis (Pap, 1997).

Also very related to this notion of pseudo-additive set-function is the introduction, by Kampé de Fériet, Forte and Benvenuti (1969), of generalized information measures which are no longer necessarily additive, but only decomposable with respect to a fuzzy disjunction (a triangular co-norm, including the case of max operation).

1.5 GENERALISATIONS AND VARIANTS OF FUZZY SETS

The notion of a fuzzy set, as proposed by Zadeh (1965a), namely a function from U to the unit interval that generalizes a characteristic function of a subset, is appealing enough and calls for further generalizations. As already pointed out, the use of the unit interval as a set of membership grades is a pure matter of convention. It looks natural especially when the referential U is the set of real numbers. However, if U is a finite set this convention sounds much more arbitrary.

In the literature, there has been a lot of attempts to go beyond simple membership functions. One can classify these works as follows. A first idea is to change the set of membership values. Some proposals along this line build an abstraction of the membership function by turning the unit interval into a suitable algebraic structure, or even preserving only the ordinal nature of the membership function. Other works use another set of numbers different from the unit interval. A second idea is to account for the ill-defined precision of membership grades. Then membership grades become intervals, or even probability distributions. Iterating the construct leads to view a membership grade as a fuzzy set of the membership scale. Conversely, a third idea is to consider that elements of the referential are fuzzy sets, and especially form a fuzzy partition of another referential. It leads to an joint use of fuzzy sets and rough sets that acccounts for fuzziness and granularity.

1.5.1 L-Fuzzy Sets

For mathematicians, it is important to build a model of a fuzzy set where only the useful properties of the membership scale are kept. This concern was first laid bare by Goguen (1967). Different ways of generalization of the notion of characteristic function had already been proposed by H. Rasiowa and by E. Szpilrajn. Szpilrajn (1936) considered functions from the referential to $\{0, 1\}^\infty$. H. Rasiowa (1964) used functions $f:X \to L$, where L is any logical algebra.

Let L be a membership scale. The expected properties of such a set are as follows:

-) L is equipped with a partial ordering structure >, so as to make sense of gradual intensity of membership and fuzzy set inclusion.

-) L must possess a unique maximal element 1 and a unique minimal element 0, so as to represent the referential U and the empty set Ø.

-) L must be a lattice (Birkhoff, 1967) so that the intersection and union of fuzzy sets may be defined using the joint $\wedge$ and the meet operation $\vee$ of the lattice.

So the weakest structure to be thought of for L is that of a complete lattice $(L, \wedge, \vee, 0, 1)$, and a L-fuzzy set is a membership function from U to L, such that

$$F \cap G(u) = F(u) \wedge G(u);$$
$$F \cup G(u) = F(u) \vee G(u).$$

It is well known that the structure of the sets of fuzzy subsets of U is precisely the same as the structure of L.

A useful additional property is to let $\{\lambda \in L : \alpha \wedge \lambda \leq \beta\}$ have a greatest lower bound $\alpha \to \beta$ for all $\alpha, \beta \in L$. Then L is a Brouwerian lattice, also called Heyting algebra (the latter only requires a lattice with a bottom element). One may also dually require that $\{\lambda \in L : \alpha \vee \lambda \geq \beta\}$ have a least upper bound $\beta \,(-)\, \alpha$ for all α, $\beta \in L$ (dual Brouwerian lattice). The operation $\alpha \to \beta$ models implication and is very useful from a logical point of view. In particular a form of fuzzy complementation can then be defined as $F^c(u) = F(u) \to 0$. However it is not involutive and F^c is a crisp set, even if F is fuzzy. The operation $\beta \,(-)\, \alpha$ can model set-difference and yet another fuzzy complementation is $F^c(u) = 1 (-) F(u)$, again yielding crisp sets and being not involutive. A complete lattice that is Brouwerian and dually Brouwerian is a distributive lattice where $\wedge$ and $\wedge$ are mutually distributive. Adding the requirement of a totally ordered membership set leads to the abstraction of the unit interval equipped with max and min in the infinite case, and to a finite chain of membership grades otherwise.

The usual way to get an involutive complementation is to equip L with a decreasing (order-reversing) involutive bijection n, such that $\beta > \alpha$ implies $n(\alpha) > n(\beta)$ and $n(0) = 1$. This is called peudo-complementation in the lattice. One may require that L be a general Boolean lattice, to get all usual properties while retaining the idea of partial membership (Brown, 1971). However this construct is not so easily found in practice. Adding the requirement of a totally ordered membership set leads indeed to a collapse on usual sets.

Another idea due to Goguen (1969) is to equip $(L, \wedge, \vee, 0, 1)$ with an extra semi-group conjunction-like operation $*$ monotonic with respect to the ordering, and get a complete lattice ordered semigroup (closg). Examples of such structures are obtained when L is the unit interval and $*$ is the product or Lukasiewicz conjunction. With the latter operation, L is an MV-algebra (Chang, 1954, Mundici, 1986). In each case, the lattice structure is recovered from the properties of this new operation. More recently, all such structures for building fuzzy sets unified by Höhle (1995) under the name GL-monoid, subsuming complete Heyting algebras and MV-algebras. Much of the mathematics of fuzzy sets are developed in such abtstract algebraic setting (See Höhle and Rodabaugh, 1999).

Measures of fuzziness of L-fuzzy sets are studied by De Luca and Termini (1974) and Yager(1980).

1.5.2 Fuzzy Sets as Ordering Relations

The theory of fuzzy sets often rely on two rather strong assumptions: membership grades in a fuzzy set are numbers (in the unit interval); membership grades to different fuzzy sets are commensurate. However the basic idea of grade of membership just suggests that referring to some concept F, an object u can be more compatible to F than another element u'. *Bird* is a fuzzy concept insofar as swallows are better examples of birds than penguins. So the most primitive object one can think of form representing the membership in F is a partial ordering relation $\geq_F$ on a subset of U that represents the support of F.

On such a basis let us examine the two assumptions: The first assumption, cardinality, is not natural when the referential is finite, because from a measurement theoretic point of view it is difficult, if not impossible to represent an ordering on a finite set in a unique way, where numbers really make sense (e.g., they can be added, multiplied...). For instance what is the real-valued membership grade of a penguin to the class of birds? This difficulty is alleviated by the use of more abstract L-fuzzy sets. For instance, birds species can be classified into categories corresponding to ranked levels of birdiness and L may thus be a finite chain.

However, the second assumption, commensurability, remains with L-fuzzy sets, since the membership scale remain common for all fuzzy sets. However useful and natural it may look in many application contexts, it may also be challenged in other contexts. One may not find a sentence like "this object is as round as it is tall" very appealing. Hence the only way out is to adopt the view of a membership function as a partial ordering. However the price of this shift is our unability to define meaningful set-theoretic connectives due to the lack of common membership scale for the fuzzy sets.

An example given by French (1986) illustrates the point: Let U = {u, v} and F, G, ordinal fuzzy sets such that $u >_F v$ and $v \geq_G u$. Suppose we represent $\geq_F$ by a regular membership funcfion F, such that F(u) = 0.9 > F(v) = 0.1, and $\geq_G$ by another membership function G such that G(u) = 0.2 < G(v) = 0.8. Then $F \cap G(u) = 0.2 > F \cap G(v) = 0.1$. Hence using this representation the intersection relation $\geq_{F \cap G}$ should satisfy $u >_{F \cap G} v$. Now suppose we had represented the membership relations by means of other membership functions F' and G' defined by: F'(u) = 0.8 > F'(v) = 0.7, and G'(u) = 0.1 < G'(v) = 0.6. then we obtain $F' \cap G'(u) = 0.1 < F' \cap G'(v) = 0.6$., that contradicts the ordering $\geq_{F \cap G}$.

Some authors have adopted an ordinal view of graded membership and tried to cope with this difficulty. Finch (1981) represents membership functions by complete preorderings. He tries to define membership grades in terms of proportion of dominated elements, and discussses the compatibility of this approach with Zadeh connectives. Basu *et al.* (1992) introduce, in the scope of choice theory, a more complex ordinal setting called soft set-theoretic structure, where it is possible to control the commensurability assumption. The idea is to define a set of soft sets Ψ on a finite set U as a primitive objet. It is a set with cardinality $|\Psi| \geq 2^{|U|}$. A one-to-one function $C: 2^U \to \Psi$ identifies the crisp sets in Ψ. Relative membership is captured by a partial preordering relation $\geq$ on $U \times \Psi$, such that $(u, F) \geq (v, G)$ means that u belongs to F at least as much as v belongs to G. Additional conditions are given so that membership to a crisp set is always at least as high as membership in any soft set, the latter being at least as high as the non-membership to a crisp set. Moreover the restriction of the membership relation obtained by fixing a soft set and moving the elements must be an ordering. In such framework, the membership relation can be configurated so that it forbids commensurability between membership levels to two different soft sets, or on the contrary, to allow it fully (if the relation $\geq$ is complete). Basu *et al.* (1992) study the correspondence between soft sets and L-fuzzy sets.

Bollman-Sdorra *et al.* (1993) propose a measurement-theoretic setting whose properties enable the basic fuzzy set system, that is, the membership function, the minimum and maximum operations for intersection and union connectives, to be meaningful. More on measurement issues appear in Chapter 3 of this volume, by Bilgic and Türksen.

When fuzzy sets are modelled by complete preordering relations, they can be used as ordinal possibility distributions which are the basis of qualitative possibility theory (e.g.Dubois and Prade, 1991; see also Chapter 7 in this volume).

1.5.3 Toll Sets

The introduction of the notion of fuzzy set by Zadeh (1965a) was mainly motivated by the problem of modelling linguistically expressible categories with unsharp boundaries. Then starting with the usual convention for set characteristic functions: zero (0) for non-membership, one (1) for membership, it leads to take the open interval (0,1) as a possible set of intermediary degrees of membership. Until now, in spite of the theoretical interest of L-fuzzy sets, [0,1]-valued fuzzy sets are often considered in practice. In (Dubois and Prade, 1993), the following questions are raised: Is there any other numerical scale of interest when generalizing classical subsets by extending the possible range of their characteristic functions? Has this change of scale an effect on the practical interpretation of the corresponding generalized subsets, as wall as on the mathematical structure with which we can equip them ? It seems that indeed the semantics of membership grades has influence on the type of structure that looks natural for fuzzy sets.

More particularly, the use of the real scale $\mathbb{R}^{+*} = [0,+\infty) \cup \{+\infty\}$ can be used, underlying a cost interpretation. Namely a zero cost will coincide with the idea of an undebatable, unquestionable membership, while an infinite cost will represent an impossible membership, an absolute non-membership. A membership grade can then be easily interpreted in terms of the "toll" to pay in order to let element u become a member of the considered subset. This is why we propose to call "toll set" this kind of generalized subset, rather than "fuzzy set" (which does not suggest a cost interpretation). Let us denote T a toll set and its ψ_T tool function.

In this framework, the grade of membership of John to the "fuzzy set" of *young* people, ψ_{young}(John) would be understood as the price that John must pay, or would have to pay, to be called a young man (like a club-membership fee). When $\psi_T(u) = 0$, it means that there is no obstacle for element u to belong to T (free membership) ; when $\psi_T(u) = +\infty$, it is forbidden for u to belong to T, since only finite costs may be paid. This interpretation suggests that

$$\forall u, \min(\psi_T(u), \psi_{T^c}(u)) = 0 \qquad (1.18)$$

since any u should belong for free to a class or to its complement, and if you have to pay for being T, this is because you are rather not T, i.e., T^c ; moreover both T and T^c cannot be toll-free, i.e.

$$\psi_T(u) = 0 \Rightarrow \psi_{T^c}(u) > 0 \qquad (1.19)$$

i.e., $\max(\psi_T(u),\psi_{T^c}(u)) > 0$. Particularly, $\psi_U(u) = 0$, $\forall$ u $\in$ U, and $\psi_\emptyset(u) = +\infty$, $\forall$ u $\in$ U (the referential U is free, and it is impossible to belong to the empty set).

The union of toll sets is defined by the min operation, i.e.

$$\psi_{S\cup T}(u) = \min(\psi_S(u),\psi_T(u))$$

since to be a member of S or T, u has just to "pay" the cheapest fare. Thus (1.18) expresses the law of excluded middle $T \cup T^c = U$. Several choices are possible for the operation $*$ defining the intersection. It should obey the following inequalities:

$$\max(\psi_S(u),\psi_T(u)) \le \psi_{S\cap T}(u) = \psi_S(u) * \psi_T(u) \le \psi_S(u) + \psi_T(u). \qquad (1.20)$$

The lower bound in (1.20) corresponds to the maximal possible "discount" for joint membership (the agent only pays the higher toll for u) while the upper bound corresponds to no discount at all : the agent has to pay both tolls for u to belong to S and T. Note that (1.19) is thus some kind of law of non-contradiction, but not completely fulfilled since $T \cap T^c \neq \emptyset$. Inclusion is defined, like for ordinary fuzzy sets (except that the scale works in the opposite way), by

$$S \subseteq T \Leftrightarrow \psi_S \ge \psi_T.$$

Clearly using the one-to-one mapping $\alpha = e^{-\beta}$ from $\mathbb{R}^{+*}$ to [0,1], we can compare toll sets with usual fuzzy sets : ($\mathbb{R}^{+*}$, min, $*$) is mapped on ([0,1], max, $\perp$) with $\min(a,b) \ge a \perp b \ge a \cdot b$. Note that due to (1.18)-(1.19) the complementation of toll sets is not compositional but should obey the constraint

$$\psi_{\bar{T}}(x) \begin{cases} > 0 \text{ if } \psi_T(x) = 0 \\ = 0 \text{ if } \psi_T(x) > 0. \end{cases}$$

The following complementation obeys this requirement :

$$\psi_{\bar{T}}(x) = \begin{cases} +\infty & \text{if } \psi_T(x) = 0 \\ 0 & \text{if } \psi_T(x) > 0 \end{cases}$$

which has a "para-consistent" flavor (indeed $T \cap T^c \neq \emptyset$ for some T, i.e., it is allowed for an element u to belong to T and to somewhat belong to T^c, provided that the corresponding cost is finite). As we can see, changing the membership scale modifies the interpretation and leads to privilege operations on $\mathbb{R}^{+*}$ which are not necessarily the counterparts (in the sense of a correspondence such as $\alpha = e^{-\beta}$) of the most standard fuzzy set-theoretic operations used on the scale [0,1]. In particular the above toll set complementation is not mapped to the involution $1 - \alpha$, and rather corresponds to the dual-Brouwerian lattice negation $1(-)\alpha$, mentioned in Section 1.5.1. The mathematically natural involution on $\mathbb{R}^{+*}$ namely $1/\psi_S(u)$ makes no sense in terms of cost.

Note that so-called indicator functions used in convex analysis (Rockafellar, 1970) are not based on the usual binary membership scale {0, 1}. The indicator function ψ_T of a classical subset $T \subseteq U$ is defined by

$$\psi_T(x) = \begin{cases} 0 & \text{if } x \in T \\ +\infty & \text{if } x \notin T \end{cases}$$

When U is $\mathbb{R}^n$, T is closed (respectively : convex) if its indicator is lower semicontinuous (respectively : convex). Aubin (1990a, b) has introduced extended indicator functions ψ_T taking their values in $\mathbb{R}^{+*}$. They are precisely toll sets since ψ_T is regarded as a "cost" or "penalty function". Based on this notion, Aubin (1990b) has also defined toll set-valued maps in order to deal with toll differential inclusions (differential inclusions generalize differential equations to set-valued maps).

1.5.4 Interval--Valued Fuzzy Sets

An interval-valued fuzzy set is a fuzzy set whose membership function is many-valued and forms an interval in the membership scale. This idea is the simplest one that captures the imprecision of membership grades. It was first proposed by Sambuc (1975), under the name Φ-fuzzy set. Assume F(u) is an interval $[F_*(u), F^*(u)]$. Then obvious extensions of the basic fuzzy connectives are as follows:

$$F \cap G(u) = [\min(F_*(u), G_*(u)), \min(F^*(u), G^*(u))]; \tag{1.21}$$

$$F \cup G(u) = [\max(F_*(u), G_*(u)), \max(F^*(u), G^*(u))]; \tag{1.22}$$

$$F^c(u) = [1 - F^*(u), 1 - F_*(u)]. \tag{1.23}$$

The "ensembles flous" of Gentillomme (1968) can as well be viewed as crisp interval-valued sets wher membership of peripheral elements of F is [0, 1]. Interval-valued fuzzy sets appear in the literature under various guises.

A first example is the idea of an ill-defined set due to the lack of knowledge of its elements. Given a collection of objects Ω, for which we have some (possibly incomplete) knowledge about the value of an attribute X, represented by means of a possibility distribution F^ω for each object ω, the set of objects such that "X is A" (A may be itself fuzzy in the general case) is an ill-known set described by a pair of fuzzy sets, namely the fuzzy set F^* of objects which more or less possibly satisfy the requirement "X is A" and the fuzzy set F_* of objects which more or less necessarily (i.e., certainly) satisfy it. Namely, $F^*(\omega) = \Pi(A)$ and $F_*(\omega) = N(A)$. Since $N(A) > 0$ implies $\Pi(A) = 1$, the latter fuzzy set F_* is strongly included in the former F^*. Interval-valued fuzzy sets generated in this way and satisfying this constraint are called twofold fuzzy sets (Dubois and Prade, 1987).

Another natural way of generating interval-valued fuzzy sets was noticed by Turksen and Yao (1984). In classical logic, connectives can be equivalently expressed in normal forms: conjunctions of disjunctions (conjunctive normal form = CNF) and disjunctions of conjunctions (disjunctive normal form = DNF). Moving to fuzzy logic, these forms are no longer equivalent, and, generally, the conjunctive normal form gives membership grades greater than the disjunctive form. For instance, with crisp sets $A \cup B = (A \cap B^c) \cup (A \cap B) \cup (A^c \cap B)$. Using the idempotent family of fuzzy connectives, note that, generally:

$$\max(a, b) > \max(\min(a, 1 - b), \min(a, b), \min(1 - a, b),$$

This property carries over to more general families of connectives (Türksen, 1986, 1995). Hence when performing some fuzzy set-theoretic aggregation of regular membership functions, the simultaneous use of CNF and DNF forms of the fuzzy connectives yields an interval-valued fuzzy set.

The idea of sub-definite sets (suggested by Narin'yani, 1980) presupposes that membership of an element to a set does not determine its non-membership. This author developed a calculus of such non-classical sets for knowledge representation purposes. A subdefinite set S is a pair (A^+, A^-) of disjoint subsets, a subset A^+ of elements which definitely belong to S and a subset A^- of elements which definitely do not belong to it, togehter with some constraints on the cardinality of these subsets.

This view has been taken over in the fuzzy set field by Atanassov (1986) who proposes the notion of intuitionistic fuzzy sets. An intuitionistic fuzzy set IF is defined by a pair of membership functions (F^+, F^-) where $F^+(u)$ is the degree of membership of u in IF and $F^-(u)$ is its degree of non-membership. The two membership functions are supposed to verify the constraint $F^+(u) + F^-(u) \leq 1$. The name "intuitionistic" stems from this inequality which is supposed to express a rejection of the excluded middle law, like in intuitionistic logic. Intuitionistic fuzzy set-theoretic operations are proposed as follows

$$IF \cap IG(u) = (\min(F^+(u), G^+(u)), \max(F^-(u), G^-(u)));$$
$$IF \cup IG(u) = (\max(F^+(u), G^+(u)), \min(F^-(u), G^-(u)));$$
$$IF^c(u) = (1 - F^-(u), 1 - F^+(u)).$$

On this basis many subsequent papers have been written, developing set-theoretic as well as logical aspects of the theory. Atanassov also sugests links to modal logic, defining "necessary"IF as F^+ and "possible"IF as $1 - F^-$. See Atanassov (1999) for a monograph on this approach. Some remarks are easily made on such an attempt. First, while the theory is supposed to be intuitionistic, the complementation of intuitionistic fuzzy sets is involutive, which is unexpected. Moreover denoting $F_* = F^+$ and $F^* = 1 - F^-$, it becomes clear that an intuitionistic fuzzy set is just another encoding of an interval-valued fuzzy set. Yet other concepts equivalent to interval-valued fuzzy sets are grey sets of (Deng, 1989) and vague sets of Gau and Buehrer (1993). See Bustince and Burillo (1996) for a full discussion of these notions.

1.5.5 Type 2 Fuzzy Sets

Type 2 fuzzy sets are fuzzy sets whose membership functions are fuzzy-valued, that is for any u in U, F(u) is a fuzzy set of membership grades, a fuzzy subset of [0, 1]. This notion was suggested by Zadeh (1971b). Mizumoto and Tanaka (1976) were the first to consider them in detail. The appeal of this notion is due to the modelling of imprecision of membership grade. Indeed one of the first critical view-points on fuzzy set was to lay bare the paradox of the impossibility of modeling a vague concept via a precise membership function (Arbib, 1977). Of course a type 1 (= usual) fuzzy set is a particular case of type two fuzzy set.

Despite its natural appeal, the notion of type 2 fuzzy sets is tricky. First, one must be careful with the definition of type 2 fuzzy connectives. Suppose the usual max and min-based union and intersection on type 1 fuzzy sets is used. Noticing that the sets of fuzzy sets of [0, 1] form a distributive lattice, one may be tempted to use the meet and the joint of this lattice, namely

$$F \cap G(u) = F(u) \cap G(u);$$
$$F \cup G(u) = F(u) \cup G(u)$$
$$G^c(u) = G(u)^c.$$

However these definitions do not collapse with the usual fuzzy set connectives, since for instance, applying the above defined union connective to regular fuzzy sets, one obtains $F \cup G(u) = \{F(u), G(u)\}$. What must be done is to extend the minimum and the maximum (and more generally, all fuzzy set connectives) to fuzzy set arguments using the extension principle. For instance (Mizumoto and Tanaka, 1976; Dubois and Prade, 1979):

$$F \cap G(u)\ (\alpha) = \sup\{\min(F(u)\ (\beta), G(u)\ (\gamma)), \min(\beta, \gamma) = \alpha\};$$
$$F \cup G(u)\ (\alpha) = \sup\{\min(F(u)\ (\beta), G(u)\ (\gamma)), \max(\beta, \gamma) = \alpha\}.$$
$$G^c(u) = 1 - G(u).$$

In the particular case when F(u) is an interval $[F_*(u), F^*(u)]$, then we retrieve the above operations of interval-valued fuzzy sets (1.21, 1.22, 1.23). Another difficulty with type 2 fuzzy sets is that they are subject to a regression to infinity. One may recursively apply the rationale of type 2 fuzzy sets, and assume that fuzzy-valued membership grades are themselves type 2 fuzzy sets. The fuzzy subsets of U then become type 3 fuzzy sets. So type M fuzzy sets for any $m > 1$ can be defined, and it is not clear when to stop. A way out of this problem is to give up the idea of membership functions and to use for instance semi-sets of Vopenka (1979), as advised by Novak (1992). However the notion of membership function remains a more easy object to handle, despite its possible philosophical paradoxes, which do not prevent their use in practice.

1.5.6 Probabilistic Extensions of Fuzzy Sets

There are two main probabilistic extensions of fuzzy sets: the probabilistic sets of Hirota (1981), and the random fuzzy sets first introduced by Féron (1976, 1981).

A probabilistic set is obtained by randomizing membership grades of each element of U separately. More specifically a probabilistic set is defined by a function F from $U \times \Omega$ to [0, 1], where $F(u, \cdot)$ is a random variable on the probability space $(\Omega, \mathcal{B}, P)$. The probability distribution P is supposed not to depend on F, and models the imprecision of the membership function. The idea is that when asking several people for membership grades F(u), they may come up with different answers. The probabilistic component takes this variability into account. The average value of the random variable $F(u, \cdot)$ can be taken as the actual membership grade. However standard deviation and other moments can be computed. A full account of probabilistic sets can be found in the monograph by Czogala and Hirota (1986). A similar concept is proposed by Suzuki (1993), also in the scope of providing consensus between expert opinions on totally ordered referentials. The theory of expertons outlined by Kaufmann (1988) also pertains to the same issue.

A random fuzzy set is the obvious fuzzy extension of a random set, namely a weighted family of fuzzy sets $\{(F_i, p_i)\}_{i = 1,\ldots, n}$ with weights such that $\Sigma_{i = 1,\ldots,n}\, p_i = 1$. This type of construct was proposed in the setting of probability spaces under the name fuzzy random variables (Kwakernaak (1978, 1979; Puri and Ralescu, 1986). See Gebhardt *et al.* (1998)'s and Ralescu (1999)'s Chapters in the Handbook of Fuzzy Sets for extensive surveys on this notion, respectively from a statistical and mathematical point of view. The book by Kruse and Meyer (1987) also constitutes a basic reference. Compared with a probabilistic set, a random fuzzy set considers the membership fuction as a lumped random object, while probabilistic sets view membership grades as separate random variables (especially if the probability measure P depends on the element u).

Random fuzzy sets can be the basis for a generalized theory of belief functions. Namely, $(F_i, p_i)\}_{i = 1,\ldots, n}$ is viewed as defining a set of fuzzy focal elements. The degree of belief in a fuzzy event, based on fuzzy focal elements can be defined from two points of view: the first one is to exploit degrees of comparison between fuzzy sets introduced in Section 1.3.5: degree of inclusion $I(F_i, F)$ for the belief function, and degree of overlap $O(F_i, F)$ for the plausibility function (Yager, 1982; Dubois and Prade, 1985b; Ishizuka *et al.*, 1982):

$$\mathrm{Bel}(F) = \Sigma_{i = 1,\ldots, n}\, m_i\, I(F_i, F);$$
$$\mathrm{Pl}(F) = \Sigma_{i = 1,\ldots, n}\, m_i\, O(F_i, F).$$

The second approach (Yen, 1990, 1992) extends Smets (1981) 's definition of the degree of belief in a fuzzy event (Section 1.3.3), now considering fuzzy focal elements as convex combinations of sets, i.e., $F_i = (F_{ij}, q_{ij})\}_{j = 1,\ldots, m}$. It gives

$$\mathrm{Bel}(F) = \Sigma_{i = 1,\ldots, n}\, m_i\, \Sigma_{j = 1,\ldots, m}\, q_{ij}\, \min_{u \in F_{ij}} F(u).$$
$$\mathrm{Pl}(F) = \Sigma_{i = 1,\ldots, n}\, mi\, \Sigma_{j = 1,\ldots, m}\, qij\, \max_{u \in F_{ij}} F(u).$$

More recent works along this line appear in Denoeux (1997, 1998, 1999). They introduce, respectively, interval-valued and fuzzy-valued belief structures.

1.5.7 Level 2 Fuzzy Sets

The dual construct to a type 2 fuzzy set is that of a level 2 fuzzy set. A level 2 fuzzy set is a type 1 fuzzy subset of a family of fuzzy subsets Zadeh (1972). This time, the membership grade F(u) remains point-valued, but element u is considered as a fuzzy set.

This notion can be regarded as yet another baroque construct. However, it turns out that it is much more used than type 2 fuzzy sets. First, Goguen (1974) has pointed out that concepts form fuzzy hierarchies. For instance, "light color" is a fuzzy subset of colours, each of which is a fuzzy subset of wavelengths. Also the result of a fuzzy inference process, via a fuzzy input and a system of fuzzy rules yields a level two fuzzy set attaching a weight to the fuzzy conclusion of each rule (Bouchon-Meunier *et al.*, 1999).

Moreover the notions of overlap and inclusion indices lead to view each fuzzy set F as a fuzzy set of the fuzzy power set. Namely $\forall G \in \mathcal{F}(U)$, define the two level 2 fuzzy sets $\tilde{F}^*$ and $\tilde{F}_*$ as follows:

$$\tilde{F}^*(G) = O(F, G); \tag{1.24}$$
$$\tilde{F}_*(G) = I(G, F). \tag{1.25}$$

This definition may be useful when restricting it to special families of fuzzy subsets of U, for instance a fuzzy partition. Then one may build approximate representations of fuzzy sets that may be useful because more economic from a computer storage point of view (Willaeys and Malvache, 1979).

An interesting issue is how to reconstruct a type 1 fuzzy set on U from a level 2 fuzzy set on a family of subsets of U. If $\mathcal{F} = \{G_1,\ldots, G_k\}$ is such a family, and the level two fuzzy set is F on $\{G_1,\ldots, G_k\}$, then Zadeh (1972) has proposed to define the fuzzy set H on U by

$$H(u) = \max_{i = 1,\ldots, k} F(G_i)\cdot G_i(u).$$

The problem of handling level 2 fuzzy sets and moving from level 1 to level 2 representations and back is closely related to the problem of fuzzy granulation and fuzzy granular computing, pointed out by Zadeh (1997) as one of the basic purposes of fuzzy set theory. It also comes close to the theory of rough sets. Many problems in this area remain open.

1.5.8 Fuzzy Rough Sets and Rough Fuzzy Sets

Although fuzzy similarity relations capture the idea of indiscernibility, indiscernibility is not the fundamental idea underlying the concept of a fuzzy set. This notion is captured by rough set theory (Pawlak, 1982, 1991) where equivalence relations describe partitions made of classes of indiscernible objects. The problem of rough set theory is to describe subsets of a set by means of equivalence classes under the form of upper and lower approximations.

Examples.

1. *Measurement scale* : U = [0, 2.5] is a human size scale between 0 and 2.5 meters, that allows for infinite precision. In practice, only millimeters can be measured, i.e.a set of adjacent intervals, whose representatives are of the form n/1000 with $0 \leq n \leq 2500$, n integer. In usual communication between individuals on this matter, the implicit representation of this scale is even coarser: only centimeters (or inches) make sense.

2. *Information system* : U is a set of item identifiers (objects), $\mathcal{A}$ is a set of attributes a, which map elements of U to suitable scales. An equivalence relation R is defined by x R y if and only if $\forall a \in \mathcal{A}$, a(u) = a(v). An equivalence class $[u]_R$ denotes the class of objects which have the same description as u in terms of attributes in $\mathcal{A}$. This example is given by Pawlak (1982).

3. *Image processing* U is a rectangle screen, i.e., a Cartesian product $[0,a] \times [0,b]$, X/R is a discretization grid into pixels, $[(x,y)]_R$ being the pixel that contains a point (x,y) in X. This is the 2-dimensional version of example 1.

Let A be a subset of U and R be an equivalence relation on U. The main question addressed by rough sets (Pawlak (1982)) is : how to represent A by means of the set of equivalence classes U/R ? Denote $[u]_R$ the equivalence class of $u \in U$. A *rough set* is a pair of subsets $R^*(A)$ and $R_*(A)$ of U/R that approach as close as possible A from outside and inside respectively :

$$R^*(A) = \{[u]_R \mid [u]_R \cap A \neq \emptyset, u \in U\}; \tag{1.26}$$
$$R_*(A) = \{[u]_R \mid [u]_R \subseteq A, u \in U\}. \tag{1.27}$$

$R^*(A)$ (resp.: $R_*(A)$) is called the upper (resp. : lower) approximation of A by R. $R^*(A)$ contains $R_*(A)$. When $R^*(A) \neq R_*(A)$, it means that due to the indiscernibility of elements in X, A cannot be perfectly described. More precisely, the set difference $R^*(A) - R_*(A)$ is a rough (imprecise) description of the boundary of A by means of "granules" of X/R.

Rough Sets Versus Fuzzy Sets. One might be tempted to view a rough set as a special kind of fuzzy set with three levels of membership, that is, with a core and a support. However as pointed out by Pawlak (1985), the fuzzy intersection and union operations cannot be applied to combine rough sets since, although $R^*(A \cup B) = R^*(A) \cup R^*(B)$ and $R_*(A \cap B) = R_*(A) \cap R_*(B)$ hold, inclusions $R^*(A \cap B) \subseteq R^*(A) \cap R^*(B)$ and $R_*(A) \cup R_*(B) \subseteq R_*(A \cup B)$ only are valid. This is because a rough set is not just a pair of nested subsets made of unions of equivalence classes. It is really a 3-uple (A, $R^*(A)$, $R_*(A)$), and some constraints relate the two sets in the interval-valued representation. More discussions on the difference between fuzzy sets and rough sets are in Wygralak (1989), Chanas and Kuchta (1992), Lin (1998a).

In (Dubois and Prade, 1990; 1992b) we pointed out that indiscernibility and fuzziness are distinct facets of imperfect knowledge. Indiscernibility refers to the granularity of knowledge, which affects the definition of universes of discourse. Fuzzy set theory rather relies on ordering relations that express intensity of membership. Fuzziness is present because categories of natural language are often gradual notions and refer to sets with smooth boundaries. In the above example of measurement, fuzziness stems from the fact that the continuous nature of the measurement scale suggest gradual boundaries for subsets. Indiscernibility in human perception lead people to granulate information (Zadeh, 1997), that is, they use a discrete set of chunks. In the example of image processing, rough set theory is about the size of the pixels, fuzzy set theory is about the existence of more than two levels of gray.

Although the starting points of fuzzy set and rough set theories are clearly distinct, they display a natural complementarity, and rough sets can be extended by either by coarsening a fuzzy set via an equivalence relation, or by turning equivalence relations into fuzzy similarity relations. People do granulate information, but they do it in a soft way, because they cannot fix precise threshold. Hence the marriage between rough sets and fuzzy sets is naturally expected.

Approximation of Fuzzy Sets by Equivalence Relations. The idea of upper and lower approximations of sets (in terms of approximate covering by equivalence classes), underlying rough sets, can be extended to fuzzy sets. Let U be a set, R be an equivalence relation on U and F be a fuzzy set in U. Let $\{X_1,\ldots, X_k\}$be the set of equivalence classes of R. The upper and lower approximations $R^*(F)$ and $R_*(F)$ of a fuzzy set F by R are fuzzy sets of U/R with membership functions defined by

$$R^*(F)(X_i) = \sup\{F(u) \mid u \in X_i\}; \tag{1.28}$$
$$R_*(F)(X_i) = \inf\{F(u) \mid u \in X_i\}. \tag{1.29}$$

where $R^*(F)(X_i)$ (resp. : $R^*(F)(X_i)$) is the degree of membership of X_i in $R^*(F)$ (resp. : $R_*(F)$). $(R^*(F), R_*(F))$ is called a rough fuzzy set. Moreover these two last equations are basic in C-calculus (Caianiello 1973; 1987), whose main applications are in image processing, and which can be retrospectively viewed in this manner.

Blurring Sets by Similarity Relations. Similarity relations can account for the lack of strict transitivity of perceived equality that bothered Poincaré (1905) and motivated Menger (1951c) (see Section 1.2). Given a fuzzy similarity relation R on U, counterparts of (1.26) and (1.27) in this setting allow for a description of any set A by means of of an upper and a lower approximation $R^*(A)$ and $R_*(A)$, which are fuzzy sets defined as follows:

$$R^*(A)(u) = \sup_{v \in A} R(u, v);$$
$$R_*(A)(u) = \inf_{v \notin A} 1 - R(u, v).$$

However in the spirit of information granulation (Zadeh, 1997), these approximations should be viewed as level 2 fuzzy sets of a fuzzy partition $\Phi = \{F_1, \ldots, F_k\}$ of U. Then

$$\Phi^*(F)(F_i) = \sup_{v \in A} F_i(v); \tag{1.30}$$
$$\Phi_*(F)(F_i) = \sup_{v \notin A} 1 - F_i(v). \tag{1.31}$$

What is obtained is a fuzzy rough set. Both (1.26) and (1.27), and (1.30) and (1.31) are subsumed by (1.24) and (1.25), for the purpose of approximate description of a fuzzy set by means of fuzzy granules that are fuzzy clusters of similar elements of U. More works on fuzzy rough sets are in Nakamura and Gao (1991), Lin (1998b).

1.6 SEMANTICS AND MEASUREMENT OF FUZZY SETS

A set is a rather abstract notion. Sets may be useful for many purposes and set membership does not mean the same thing at the operational level in each and every context. The same thing occurs with membership grades and fuzzy sets whose presentation is often mathematically oriented (like in Section 1.3), hence remote from applied settings. What fuzzy sets mean is left to the sagacity of users. As a consequence, when we scan the fuzzy set literature, including Zadeh's own papers, there is no uniformity in the interpretation of what a membership grade means. This situation has caused many a critique by fuzzy set opponents, and also many a misunderstanding within the field itself. Most negative statements expressed in the literature turn around the question of interpreting and eliciting membership grades.

Our claim is that, far from being a weakness, the existence of several understandings of what a membership grade may mean proves the potential richness of the concept of fuzzy set, and explains its pervasiveness in many unrelated fields of investigation. However, beyond the success of fuzzy logic in engineering problems, it seems that the condition for an improved recognition of fuzzy sets by the scientific community is that the various semantics of fuzzy sets be articulated in a clear way. It seems to be a crucial step in order for instance to start considering the basic question of a measurement theory for membership functions, a topic which only very few fuzzy research scholars have considered (with some noticeable exceptions such as Türksen (1991), see Chapter 3 by Bilgic and Türksen in this volume).

From that point of view the situation of fuzzy sets should be compared with the situation of probability theory. The reasons why probability theory is widely recognized as a major scientific achievement are threefold: i) probability theory is based on an impressive mathematical construct; ii) probability theory has proved useful in many applications; iii) probability theory has clear, well understood (although not unique) semantics. After 35 years of fuzzy set research, much has been done towards fulfilling the first and the second condition. The mathematics of fuzzy sets have drastically progressed, in connection with logic, algebra, analysis, and measure theory etc. Figure 1.3 describes how fuzzy set theory is embedded in the landscape of classical mathematics. Moreover there has been application of fuzzy sets in pure mathematics: category theory, topology, non-additive measure theory, and analysis (see Höhle and Rodabaugh, 1999). No serious scholar can nowadays claim that fuzzy sets are mathematically unsound (although, like in all fields, weak papers and wrong papers have been published).

Similarly, with the success of fuzzy systems analysis and control, fuzzy classification methods, fuzzy image analysis, fuzzy data bases and the like (covered by the Handbooks of Fuzzy Sets Series), the practical usefulness of fuzzy sets in the engineering field becomes a matter of fact. However the questions of the semantics and the empirical foundations and the measurement of fuzzy sets remain partially unresolved. Note that even in probability theory, debates on the semantics are not settled, since there exists frequentist ("objectivist") views of probability that are still in conflict with subjectivist views (based on cost and betting behavior), as well as some other views (see T. Fine, 1973; Hacking, 1975). However, for each semantic approach to probability, there exists a coherent and extensive explanation that justifies why the laws of probability theory should be adhered to, and thought experiments that explain how degrees of probability can be obtained. This is perhaps still the weak point of fuzzy set theory at the present time despite the existing literature on the elicitation of membership grades. This section, based on a previous position paper (Dubois and Prade, 1997) suggests that the situation in that respect is likely to change in the future and it gives several directions of future research on the foundations of fuzzy sets, further elaborated in Chapter 3 of this volume.

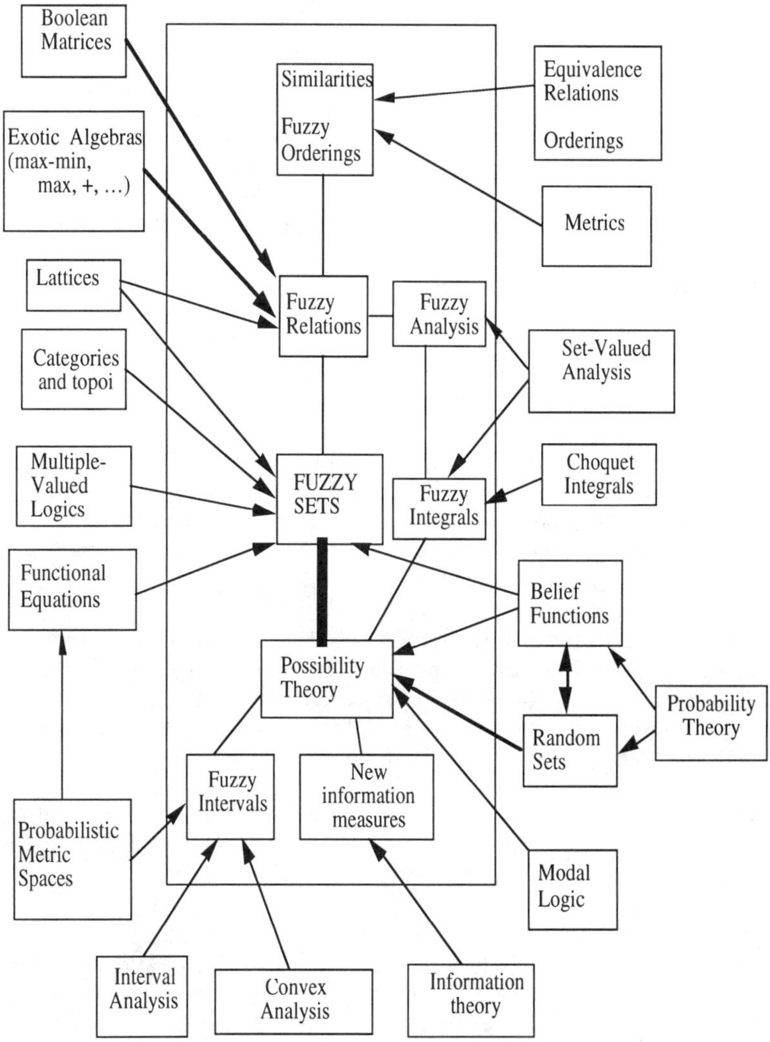

Figure 1.3. Mathematical environment of Fuzzy set Theory

1.6.1 What Membership Grades May Mean

Fuzzy sets seem to be relevant in three types of information-driven tasks where graded membership plays a role: classification and data analysis (Bezdek Keller *et al.*, 1999), decision-making problems (Slowinski, 1998) and approximate reasoning (Bouchon-Meunier *et al.*, 1999). Of course this typology of problems is not exhaustive, since for instance it does not explicitly mention fuzzy modeling and control (Nguyen and Sugeno, 1998; Verbruggen *et al.*, 1999). However the latter involves classification and data analysis and a form of approximate reasoning. More generally most other tasks addressed by means of fuzzy sets (for instance database querying) involve one or several of the above ones and participate to a general field of investigation one may called *fuzzy information engineering* (Dubois, Prade and Yager, 1987). The three basic tasks, that have been investigated by many researchers, actually correspond and/or exploit three semantics of the membership grades, respectively in terms of similarity, preference and uncertainty. Indeed, considering the degree of membership F(u) of an element u in a fuzzy set F, defined on a referential U, one can find in the literature, three interpretations of this degree.

Degree of Similarity. Historically, the oldest interpretation of membership grades views F(u) as the degree of proximity of u from prototype elements of F. It goes back to Bellman, Kalaba and Zadeh (1966) which advocated the interest of the fuzzy set concept in pattern classification from the start. This view is particularly at work in clustering analysis (Bezdek, *et al.*, 1999), regression analysis and the like (Diamond and Tanaka, 1998), where the problem is that of abstracting a representation from a set of data by exploiting proximity between pieces of information. It is also at work in fuzzy control techniques, where the similarity degrees between the current situation and the prototypical ones described in the condition parts of the rules, are the basis for the interpolation mechanism between the conclusions (Nguyen and Sugeno, 1998; Verbruggen *et al.,* 1999).

Degree of Preference. F represents a set of more or less preferred objects (or values of a decision variable x) and F(u) represents an intensity of preference in favor of object u, or the feasibility of selecting u as a value of x. Fuzzy sets then represent criteria or flexible constraints. This view is the one later put forward by Bellman and Zadeh (1970); it has given birth to an abundant literature on fuzzy optimisation, especially fuzzy linear programming, and decision analysis (Slowinski, 1998). Approximate reasoning is then concerned with the propagation of preferences when several constraints (which may be fuzzy) relate the variables. Applications pertain to engineering design and scheduling problems.

Degree of Uncertainty. This interpretation was implicitly proposed by Zadeh (1978a, 1979) when he introduced possibility theory and developed his theory of approximate reasoning. The degree of membership F(u) can be viewed the degree of plausibility that a parameter x has value u, given that all that is known about it is that "x is F". Then the values encompassed by the support of the membership function are mutually exclusive, and the membership degrees rank these values in terms of their respective plausibility. Set functions called possibility and necessity measures can be derived so as to rank-order events in terms of unsurprizingness and acceptance, respectively. They are non-additive, as opposed to probabilities. This view has been used in expert systems, and artificial intelligence (Bouchon-Meunier *et al.*, 1999)

Note that the word possibility, now tightly linked with fuzzy sets, can convey an epistemic meaning (F then describes the more or less plausible values of x) or a physical meaning (F(u) being the degree of ease of having x = u). Viewing F(u) as a degree of uncertainty only refers to the epistemic interpretation. The physical meaning of possibility actually has more to do with feasibility. In his seminal paper, Zadeh (1978a) does not fully stick to the epistemic view and the proposed example of "Hans eating more or less eggs for breakfast" rather refers to ease of attainment. The epistemic view of possibility theory in terms of potential surprize is found in the pioneering works of Shackle (1963).

Example. It is noticeable that the three semantics of fuzzy sets appear in the works of Zadeh and that he was the first to propose each of them. In order to highlight the case for these three semantics, let us take a simple example.

For instance, one might be interested in *classifying* cars of known dimensions in the categories of "big cars" "regular cars" and "small cars". It is clear that this is a matter of degree. To compute the degree of membership of each car to the category "big", one may choose a prototype of a big car like a Mercedes, and construct a measure of distance between our car under concern and this Mercedes. The greater the distance between our car and the Mercedes, the less similar they are. One may consider that the membership grade of our car to the fuzzy set of "big cars" is a decreasing function of this distance.

In contrast, one may be interested in *buying* a big car. Then the membership grade of a given tentative car to the class of big cars now reflects our degree of satisfaction with this particular car, according to the criterion "size", so membership now reflects preference. Note that here the choice of the car is ours. In other words the variable whose value is the name of the chosen car is controllable.

A very different situation is when somebody says (s)he just saw a big car and reports it. In this situation, we would like to figure out *what is known* about the car. the membership grade of a given tentative car to the class of big cars now reflects our degree of plausibility that this kind of car is the same as the one seen by the person. When this membership degree is high, our confidence that we know which car it is may still be low, especially if there are several alternatives. However if this membership degree is low then the car can be rejected as a very implausible candidate. Now the degree of membership pertains to uncertainty, not preference. In this case, the choice of the car is not ours: a big car passed by and it is what it is. In other words, the variable whose value is the name of the big car is now uncontrollable.

The existence of more than one semantics can also be encountered with fuzzy relations. For instance if R is a preference relation on U, R(u,v) may either reflect the intensity of preference (to what extent u is preferred to v) or the uncertainty about preference (how sure it is that u is preferred to v). However in the following we restrict the discussion to fuzzy sets.

1.6.2 Measuring Membership Grades

In order to operationally evaluate degrees of membership there are two basic attitudes.

The Ordinal Approach. A first attitude is not to assume that a membership grade can be represented by a number. Although fuzzy sets have been introduced with a numerical flavor, a membership function is indeed not necessarily mapped on a set of numbers, but an ordered set such as a complete lattice is enough (Goguen, 1967). The only problem is then to prescribe which membership set makes sense in the given context. In some cases a linearly ordered set of qualitative levels will do, in other cases only a partially ordered set makes sense. A membership function can even be more abstractly construed as an ordering relation $\geq_F$, attached to a predicate F, where u $\geq_F$ u' means that u is more F than u'. In that case a fuzzy set F is the pair (S_F, $\geq_F$) where S_F is the support of the fuzzy set and $\geq_F$ orders the elements of the support in terms of intensity of membership (Section 1.5.2). A fuzzy relation R on U × U can also be viewed as a ternary relation (on U^3), i.e., a collection {$\geq_u$, u ∈ U} of binary relations that are complete preorderings. Then R(u, u') ≥R(u, u") can represent a situation where u' $\geq_u$ u", which reads u' is closer to u than u". These structures are common in conditional logics of counterfactuals (Lewis, 1973).

The Cardinal Approach: Distance. If membership grades are supposed to be numerical then there should be some operational definition of these numbers. There are basically three kinds of quantities that can help measuring fuzzy set membership: distance, frequency and cost. Distance is obvious when a membership function is interpreted in terms of similarity. Namely if the referential set U is equipped with a distance d then F(u) can be computed by means of a decreasing function of the distance d(u, u*) of u to the prototype element u* of F (if unique). This has been very often done in clustering and more generally in many applications of fuzzy sets for the definition of fuzzy numbers. Ruspini (1989) has strongly emphasized the metric semantics of fuzzy sets. However the measurement of similarity is not a very frequent issue in the past literature, if one excepts works by Tversky (1977). Similarity apparently has a bad reputation in the scientific world and has as such received less attention than uncertainty and preference, although the popularity of case-based reasoning in Artificial Intelligence (Kolodner, 1993) has put forward the key role of similarity in cognitive tasks, and the lack of operational foundations for this concept. The advent of case-based reasoning seems to offer more opportunity of application of fuzzy logics (see Bouchon-Meunier *et al.* (1999), for a survey of fuzzy case-based reasoning).

The Cardinal Approach: Frequency. Uncertainty is often measured in terms of frequency of observed situations in a random experiment. But this approach leads to probability theory when the repeated observations are precise. In that restricted sense, probability assignments to the elements of U might be viewed as special membership functions such that the sum of membership grades is 1 (fuzzy singletons; see Section 1.4.2, and Kosko, 1990). However, when the situations observed in a repeated way are imprecise, more general kinds of membership functions are obtained. This approach casts fuzzy sets in the context of ramdom sets. See Section 1.3.1, eq. (1.5). The degree of membership F(u) can then be computed as the proportion of observations that do not rule out the situation u. Then the membership function can be interpreted as a "plausibility" function, not a probability function since F(u) = 1 just means that u is ruled out by no observation, and that other situations u' can get membership one (even all of them, if the observations were so imprecise as to be non-informative). In such a context, the membership function shares with the probability assignment the fact that the values u such that F(u) > 0 represent the mutually exclusive value of some unknown parameter. However the membership function represents only a part of the information contained in the imprecise observations: the whole information can be retrieved only if the observations were precise (this is standard probability theory) or nested (this is possibility theory, developed in Chapter 7 of this volume).

Another probabilistic approach to the measurement of membership grades is to use likelihood functions, namely F(u) can be measured by P('F' | u), i.e., the proportion of times the situation u has been tagged F in a random experiment where individuals presented to situation u are asked to put the tag 'F' on u or not. This type of experiments is very common in the early fuzzy set literature (Hersch *et al.*, 1979) and has been systematically envisaged by Hisdal (1989) and Mabuchi (1992). The recent proposal by Paris (1997) also tries to justify fuzzy connectives along the same lines. The membership degree is based on counting arguments in favor of considering element u as an F, as opposed to the arguments that go against it. The likelihood view can be related to the random set view in a simple way. See Chapter 7 by Dubois *et al.*, in this volume for an extensive comparison between fuzzy sets and probability. That chapter suggests that disputes between the two fields are partially unfounded.

The Cardinal Approach: Cost. The relationship between membership degrees and cost has been seldom considered. F(u) could reflect the price somebody would agree to pay for making u a prototype of F. Then, the set of membership grades is no longer the unit interval but the non-negative reals, were 0 corresponds to full membership and non-membership is described by $+\infty$. Such fuzzy sets are toll sets of Section 1.5.3. Clearly, relating membership functions to cost sounds natural when they encode preference profiles. They become very similar to utility functions, for which there exist extensive measurement theories. This direction has been rather seldom envisaged in the fuzzy community, if we except scholars like Giles (1988).

Metric Views of Preference and Plausibility. Note that the similarity semantics of fuzzy sets can serve as a basis for the estimation of the two others. Preference and plausibility are derived notions, mainly because of their epistemic nature. Preference is often a function of cost; but it may also reflect similarity with respect to an ideal object and this view is often found in the literature on optimization (especially goal programming). Besides, a new approach to decision theory has been proposed by Gilboa and Schmeidler (1995) and called case-based decision theory. Preference beween acts are expressed by a counterpart of expected utility where the probability function is changed into a similarity relation that evaluates the resemblance between the current situation and stored past situations where decisions were made.

Plausibility may also be viewed as similarity with respect to a maximally plausible situation. This idea is expressed in the logics of counterfactuals by Lewis(1973), where the plausibility of an imaginary situation is evaluated by the degree of similarity to the current situation. The idea of relating similarity and uncertainty also pervades the probabilistic literature since the likelihood functions attached to classes are directly computed by means of distances to cluster centers when Gaussian functions are used. Least square methods also have probabilistic semantics. So there exists some links between similarity and uncertainty although they have been little explored in a systematic way.

1.6.3 The Semantic Meaningfulness of Fuzzy Logic

To summarize the discussion, membership functions in application-oriented tasks are no longer abstract set-theoretic notions, but can and should be naturally related to basic measurement of quantities such as distance, frequency and cost. However the calculus of fuzzy sets has been developed in the abstract framework of algebra, without paying attention to the semantics of membership functions. As a consequence, there exists a very nice and mathematically sound framework for combining fuzzy sets, but it is often presented without casting it into an interpretive framework. The risk is to leave the user with no guidelines about how to apply fuzzy set theory, and to use semantically meaningless operations. Of course the calculus of fuzzy sets can also be applied in a purely data-driven way, the relevant fuzzy set operation being selected via a learning method, for instance by putting fuzzy set-theoretic operations inside a neural net-like structure (Pedrycz, 1995, for instance). In general the identification of membership functions and fuzzy connectives could be both data- and knowledge-driven. However the question of the semantic meaningfulness of fuzzy set theory is an interesting issue as such.

Let us insist again on the two basic assumptions of fuzzy set theory: commensurability and compositionality.

Commensurability, Again. Degrees of membership to fuzzy sets pertaining to unrelated concepts are supposed to be*commensurate*. This is done by resorting to a common membership scale (that need not be numerical). This commensurability assumption is often taken for granted and never emphasized in the fuzzy literature (See Section 1.5.2). Yet, it is better to state it clearly in the face of newcomers to the theory, as it sheds light on potential limitations of the fuzzy approach and helps locating it within alternative settings such as measurement theory and multiple-criteria decision analysis. The commensurability assumption is really the one that allows fuzzy set theory to work.

Compositionality. The compositionality assumtion presupposes that the degree of membership to a compound fuzzy set is a function of the membership degrees to each of the components. This assumption is not always fully in accordance with the meaning of the membership functions. For instance, assume now the membership function is described in terms of similarity. Namely let A be a subset of prototypes of F and define the membership function of the fuzzy set F in terms of similarity to a prototype as follows:

$$F(u) = \max_{u' \in A} S(u, u').$$

F is the image of A through the similarity relation S on U and is denoted $A \circ S$. It is the upper part of a fuzzy rough set (Section 1.5.8). One might be tempted to use fuzzy set-theoretic operations to combine such fuzzy sets, requiring the usual compositionality assumptions. Unfortunately such a calculus cannot be compositional. Namely given B the set of prototypes of another fuzzy set G, it is easy to see that we cannot in general accept that $F \cap G = \min(F(\cdot), G(\cdot))$ nor any connective different from min. The reason is that generally $(A \cap B) \circ S \subset (A \circ S) \cap (B \circ S)$. For instance the former can be empty while the latter is not. However the equality holds with union instead of intersection, and the equality for intersection may hold if F and G are defined on different sets, S is defined on their Cartesian product and is itself a Cartesian product of similarity relations. There is also a problem with fuzzy complementation (see Dubois *et al.,* 1995). So the max-min theory of fuzzy sets cannot precisely account for this particular similarity-based semantics of F. The max-min fuzzy set theory appears as an approximation (since under this scheme $F \cap G \leq \min(F(\cdot), G(\cdot))$ holds).

The advantage of casting fuzzy set-theoretic operations in a semantic framework is to be able to lay bare operational assumptions behind a given aggregation operation. For instance fuzzy set intersection using product can be explained by intersecting independent underlying random sets, and again this is an approximation of the full random set intersection. Kruse, Gebhardt and Klawonn (1994) try to advocate probabilistic underpinnings of possibility theory, including Zadeh's extension principle in a systematic way. Fuzzy set-theoretic operations can thus be justified in terms of random sets, or likelihood functions, as already advocated (see also Chapter 7 by Dubois *et al.*). In this interpretive setting, compositionality is not universal, but underlies precise prerequisites. Fuzzy set-theoretic operations in the context of uncertainty are useful in data fusion (See Dubois, Prade and Yager, 1999).

It seems that it is when membership functions account for preference that the compositionality assumption of fuzzy set theory makes no problem so far. There is a huge body of fuzzy set-theoretic aggregations that can be used when the membership functions represent utility-like functions. Even multiattribute utility theory leads to compositional methods when criteria to be merged are assumed to be independent. Cost-based explanations of some aggregation operations are also outlined in Section 1.5.3. However, with an unbounded scale such as $[0, +\infty)$, it is hard to intuitively justify an involutive negation. The framework of fuzzy sets for multiple attribute evaluation has the potential to account for dependency among attributes (Grabish **et al.**,1995). However there is still a large amount of work to be done in order to justify in an axiomatic format the fuzzy set aggregation operators in terms of the decision-maker's attitudes about trading-off between criteria. See Slowinski (1998) on the fuzzy set-theoretic approach to criteria aggregation.

Uncertainty and Preference: The Analogy. Interestingly, possibility theory itself lends itself to both uncertainty-based and preference-based semantics (see Chapter 7 of this volume, by Dubois *et al.*). Namely a possibility distribution $\pi = F(\cdot)$ on a set U can represent a preference distribution that models a fuzzy constraint or a plausibility profile that describes more or less normal states of the world. The necessity function induced by $F(\cdot)$, i.e., $N(A) = \inf_{u \notin A} 1 - F(u)$ for any subset A, can be viewed either as the degree of acceptance of A (uncertainty semantics), or the degree of priority of a constraint of the form $x \in A$ (since stating that $N(A) = \alpha$ means that the degree of satisfaction of any solution $x = u \notin A$ is at most $1 - \alpha$). The preference-based semantics of possibility theory is systematically exploited in fuzzy constraint satisfaction (Dubois, Fargier and Prade, 1996). In that scope, fuzzy arithmetics can be viewed as a generalization of constraint propagation. One may call it preference propagation, as used in engineering design (See Antonsson and Sebastian, 1999). It has nothing to do with risk analysis based on fuzzy ill-known quantities even if they are formally expressed by the same equations. Maxmin fuzzy arithmetics for preference propagation can be justified on the basis of an assumption of lack of compensation between preference levels, while the possibilistic calculus of fuzzy quantities is justified quite differently either in terms of random set computations (Dubois and Prade, 1991), or, in an ordinal setting, using a minimal commitment assumption that leads to the combination of possibility distributions using the minimum operation. Putting together uncertainty and preference profiles in the setting of possibility theory leads to a qualitative decision theory that parallels expected utility theory (Dubois and Prade, 1995), and gives axiomatic foundations to possibilistic counterparts of expected utility functions (see Chapter 7 in this volume for details).

1.6.4 Membership Grades: Truth Values or Uncertainty Degrees

There is some confusion in the fuzzy literature on the potential of fuzzy sets for handling uncertainty. As seen in Section 2, this confusion has been around for a long time and has hampered the sound development of many-valued logics and their use in knowledge representation. In the more recent history of fuzzy sets, this state of confusion can be exemplified in some texts by fuzzy set proponents claiming that probability theory models randomness while fuzzy set models subjective uncertainty (hence ignoring subjective probability). It is also present in the expert systems literature where certainty factors have been confused with membership grades, and are assumed to be compositional. It also pervades the literature antagonistic to fuzzy sets where the compositinality of conjunction, disjunction and negation in fuzzy logic is considered as mathematically inconsistent (see Elkan (1994) for the latest restatement of this fallacy). And the debate is still going on. See (Hajek and Paris, 1997), for instance.

In fact, insofar as fuzzy sets are used only to model gradualness in properties, membership grades may model degrees of truth of fuzzy propositions, not degrees of uncertainty of propositions. More precisely membership grades reflect the compatibility between the precise description of a state of affairs and a fuzzy proposition. In the scope of knowledge representation, choosing a truth set is a matter of convention like choosing the range of a variable, while uncertainty is a meta-notion which reflects incomplete or contradictory knowledge. In classical logic the convention is that truth is binary. Fuzzy set theory (and before, multiple-valued logics) has modified this convention. This shift in convention does not entitle degrees of truth to be systematically interpreted as degrees of uncertainty. Moreover a degree of truth is an abstract notion (like a membership grade) that cannot be directly measured as such. Many-valued truth simply corresponds to the idea of a non-Boolean variable on an ordered range, which can represent any gradual quantity.

In contrast, uncertainty, viewed as partial belief, is a more particular notion and degrees of belief or plausibility can be attached to a *non-fuzzy* proposition in order to model the fact that it is not known whether this proposition is true or false (See Section 1.2.3). Uncertainty is at the meta-level with respect to truth and is related to practical situations where information is missing or is contradictory. Degrees of belief are not compositional. In classical logic truth is binary (true or false) while uncertainty is ternary (surely true, surely false, or unknown). In uncertainty theories (probability, belief functions etc.), truth is usually binary (crisp propositions) while uncertainty takes on all values in the unit interval. In fuzzy set theory, truth is many-valued but there is no uncertainty insofar as the element, the membership grade of which is computed, is precisely located. Knowing that a bottle is precisely half full, we can say that it is half true that the bottle is full, which does not mean at all that the probability that the bottle is full is one half (in this latter case, it is possible that the bottle is in fact empty for instance). This metaphor has been used by several authors. Fuzzy-truth values (which Zadeh (1979b) has claimed to be typical of fuzzy logic) put the two views together: they are possibility distributions that describe partially unknown non-binary truth-values (see Bouchon-Meunier *et al.*, 1999).

As pointed out above, a degree of membership F(u) can sometimes been used as a degree of uncertainty, namely a degree of plausibility, depending on the available information. When it is attached to the fuzzy proposition 'X is F' and it is known that X = u for a non-Boolean variable u, F(u) is a "degree of truth" of the fuzzy proposition, and is just a way of semantically encoding the assignment X = u in the language containing predicate F. As a degree of uncertainty, F(u) is attached to the non-fuzzy proposition p = 'X = u', which is now a Boolean proposition, when all that is known is that the value of X is somewhere in the support of F; then F(u) is interpreted as the degree of possibility that p is true (Zadeh, 1978a).

This discussion (see also Bouchon-Meunier *et al.*, 1999) leads to two entirely different extensions of classical logic that exploits the notion of a fuzzy set:

• *many-valued logics* are exclusively devoted to the handling of vague propositions $\tilde{p}$. The underlying algebraic structure is weaker than a Boolean algebra, and can be consistent with truth-values $t(\tilde{p})$ that lie in the unit interval and remain truth-functional. However since, as seen above, it sounds difficult to come up with a full fledged semantics for fuzzy set-theoretic operations in terms of similarity or uncertainty, the question of the operational meaning of many-valued logic systems of Lukasiewicz and Gödel, as revived by Pavelka (1979), Novák (1990), and Hájek (1998) remains open. Maybe truth-values are best interpreted in terms of preference or cost (unless the logic manipulates predicates like *probable*, interpreted as a gradual property).

• *possibilistic logic* that is built on top of classical logic, and where each crisp proposition is attached a degree of certainty N(p) such that N(p) = 1 if and only if p is surely true and N(p) = 0 expresses the complete lack of certainty that p is true (either p is false when $N(\neg p) = 1$, or it is unknown if p is true or false then $N(\neg p) = 0$). The degree N(p) is compositional *for conjunction only* ($N(p \wedge q) = \min(N(p),N(q))$, and $N(p \vee q) \geq \max(N(p), N(q))$ generally). For instance, if $q = \neg p$, $p \vee q$ is tautological, hence surely true ($N(p \vee q) = 1$), but p may be unknown ($N(p) = N(\neg p) = 0$). Moreover $N(\neg p) = 1 - \Pi(p)$ where $\Pi(p)$ is the degree of possibility of proposition p. Functions N and Π stem from the existence of a fuzzy set of more or less possible worlds, one of which is the actual world. It is described by means of a possibility distribution π on the interpretations of the language, and N(p) is computed as the degree of impossibility of the proposition $\neg p$. Possibilistic logic is either a logic of uncertainty (Dubois *et al.*, 1994) or a problem solving tool for prioritized constraints (Lang, 1991).

In summary, possibilistic logic describes incomplete, more or less entrenched, pieces of knowledge about where the actual world is, or more or less prioritary constraints expressing where the best solution to a problem is to be found; so a possibilistic knowledge base corresponds to a single fuzzy subset of possible worlds. In many-valued logics fuzzy sets describe the flexible extensions of vague predicates, but it is not always clear how to understand the operational meaning of truth values appearing in the language. As a consequence, many-valued logics have seldom be applied to knowledge-representation problems in a convincing way, while possibilistic logic turns out to be very close to exception-tolerant reasoning and partial constraint satisfaction problems in Artificial Intelligence (See Bouchon-Meunier *et al.*, 1999)).

1.6.5 Towards Membership Function Measurement

This section has tried to advocate the importance of future research in the operational semantics of fuzzy sets because it may greatly help getting a deeper understanding of the now impressive mathematical apparatus of fuzzy set theory, while making the field better established as a sound methodology for solving information engineering problems. It is also important so as to keep fuzzy set theory in the tradition of cognitive modelling, which was its original purpose. Fuzzy controllers, and fuzzy rule-based modelling which have become the most popular and visible side of applied fuzzy set theory, are only the emerged part of the fuzzy iceberg, and as time passes this technology seem to borrow less and less to fuzzy set theory itself, and mainly becomes a tool for approximating functions.

Besides, due to a significant progress in the mathematics of fuzzy sets and the successes in applications, the fuzzy field is now mature enough to start in-depth investigations on the links with other established fields, such as probability theory, decision theory, measurement theory, and the like. It seems that there might be at least four quite different directions to be followed. The current state of the art is covered in more details in Chapter 3 of this volume by Bilgic and Türksen.

i) *Probabilistic semantics of fuzzy sets based on random sets and likelihood functions*. The potential of possibility theory for the approximate representation of uncertainty sounds promising. This direction has already been significantly investigated, but much is left to be done before Bayesians and other kinds of statisticians start considering fuzzy sets as a natural and fruitful concept;

ii) *The formalization of similarity*. This is a direction where much practical work has been done, viz. the field of pattern classification and case-based reasoning, but where the foundations are little known. Distances have been studied from a purely mathematical point of view. No strong scientific tradition on similarity measurement exists like in probability theory for instance. But there are some formal philosophic studies studying the links between truth and similarity, that are not so well-known by fuzzy set people, e.g., Niiniluoto (1987);

iii) *Numerical preference modeling*. There is both a huge literature outside fuzzy sets in multiple criteria decision-making and a very big amount of tools for criteria aggregation in fuzzy set theory. However there is little work bridging the gap between fuzzy sets and conjoint measurement, while this has been for a long time a natural open question (partially addressed by researchers like Türksen (1991), and Bollman-Sdorra *et al.* (1993)). The situation seems to be better with relational methods in preference modeling where there is a very active field on fuzzy relations clearly positioned with respect to the non-fuzzy tradition (Fodor and Roubens, 1994; Slowinski, 1998);

iv) *Qualitative approaches to reasoning and decision*. Since membership grades may be elements of a lattice, and be nonnumerical, there is a whole avenue of research dealing with qualitative handling of similarity, preference and uncertainty which closely relates fuzzy set and possibility theory to symbolic artificial intelligence. This link between fuzzy sets and symbolic AI is as strong as the one between neural nets and fuzzy rule-based systems, using numerical membership functions. For instance, possibility theory emerges from the study of revision of logical theories, since the epistemic entrenchment relations of Gärdenfors (1988) are representable by necessity measures (Dubois and Prade, 1991b) and possibilistic inference is nonmonotonic; see Benferhat *et al.* (1997). See Bouchon-Meunier *et al.*, (1999) for a survey on possibilistic reasoning. The main reason for this link to current trends in Artificial Intelligence is that possibility theory is the simplest comparative uncertainty theory, namely the only one where an ordering on a set of realizations induces an ordering on events.

1.7 CONCLUSION

In this introductory chapter, some emphasis has been put on the historical and interpretive sides of fuzzy set theory, beyond the organized review of technical basic notions. Hopefully, a brief, but encompassing overview of fuzzy set research in the twentieth century has been provided. Looking back in time, what is really amazing is the diversity of fields, where intuitions about fuzziness were expressed and more or less formalized, and the number of scientists who participated to the emergence of the fuzzy set concept. Also it is surprizing to see how long it took before such a simple, although powerful, idea of graded membership, could be cast into a proper, widely accepted mathematical model, due to the far-ranged vision, the tenacity, and the numerous seminal papers of Lotfi Zadeh. As already pointed out, this is no surprize if the emergence and the blossoming of fuzzy set theory takes place in the era of information technology, since this approach proposes some of the new formal tools it needs to be properly developed, in particular for interfacing numerical data and symbolic knowledge, and for coping with uncertainty partial belief and incomplete knowledge in a flexible way.

Many issues, regarding the handling of gradual categories in automated reasoning, the treatment of uncertainty in possibilistic reasoning, and the measurement of gradual membership from the three points of view mentioned in Section 6, that is similarity, uncertainty and preference, remain open. More time is needed before fuzzy set theory is finalized its its formal and empirical developments, and fully recognized in science, although much has been achieved in 35 years. However the idea that information management is a full-fledged domain of scientific investigation, distinct from mathematics, physics, etc. is itself not yet widely accepted. Fuzzy set theory is part of this debate.

Now that many basic formal notions are by and large explored, the next step is really to anchor fuzzy set theory in the scientific tradition, by actively bridging the remaining gaps to more established fields that have, in the recent past, made some relevant contributions to the area of human information processing.

REFERENCES

Alsina C., Trillas E. and Valverde L. (1983). On some logical connectives for fuzzy sets theory, *J. of Math. Anal. & Appl.*, **93**, 15-26.

Andreoli G. (1956). La logica del finito e del concreto, *Giornale di matematiche*, **LXXXIV**, 19-48.

Antonsson E.K. and Sebastian H.J. (1999). Fuzzy sets in engineering design, *Practical Applications of Fuzzy Technologies* (Zimmermann H.J., ed.), The Handbooks of Fuzzy Sets Series, Kluwer Academic Publ.

Arbib (1977). Review article on fuzzy set theory, *Bull. Amer. Math. Soc.*, **83**, 946-951.

Arrow K.J. (1963). *Social Choice and Individual Values*, Wiley, New York.

Atanassov K. (1986). Intuitionistic fuzzy sets, *Fuzzy sets and Systems*, **20**, 87-96.

Atanassov K. (1999). *Intuitionistic Fuzzy Sets*, Physica-Verlag, Heidelberg.

Aubin J.P. (1990a). A survey of viability theory, *SIAM J. Control and Optimization*, **28**, 749-788.

Aubin J.P. (1990b). Fuzzy differential inclusions, *Problems of Control and Information Theory*, **19**(1), 55-67.

Baccelli F., Cohen G. Olsder G.J. and Quadrat J.P. (1992). *Synchronization and Linearity: An Algebra for Discrete-Event Systems*, J. Wiley, New-York.

Baldwin J.F. and Pilsworth B.W. (1979). Fuzzy truth definition of possibility measure for decision classification, *Int. J. of Man-Machine Studies*, **11**, 447-463.

Bandler W. and Kohout L.J. (1980a). Fuzzy power sets and fuzzy implication operators, *Fuzzy Sets and Systems*, **4**, 13-30

Bandler W. and Kohout L.J. (1980b). Fuzzy relational products as a tool for analysis and synthesis of the behaviour of complex natural and artificial systems, *Fuzzy Sets: Theory and Applications to Policy Analysis and Information Systems* (Wang P.P. and Chang S.K., eds.), Plenum Publ., 341-367.

Bandler W. and Kohout L.J. (1988). Special properties, closures and interiors of crisp and fuzzy relations, *Fuzzy sets and Systems*, **26**, 317-332.

Bandler W. and Kohout L.J. (1993). Cuts commute with closures, *Fuzzy Logic: State of the Art* (Lowen R. and Roubens M., eds., Kluwer Academic Publ., Dordrecht, 161-168.

Basu, K., Deb R. and Pattanaik P.K. (1992). Soft sets: An ordinal formulation of vagueness with some applications to the theory of choice, *Fuzzy Sets and Systems*,**45**, 45-58.

Becker O. (1930). Zur Logik der Modalitaeten, *Jahrbuch f. Phil. und phaeenomenologische Forschung*, **11**, 497-548.

Bellman R.E. and Giertz M. (1973). On the analytic formalism of theory of fuzzy set, *Information Science*, **5**, 149-157.

Bellman R.E. and Karush W. (1961). On a new functional transform in analysis: The maximum transform, *Bull. Amer. Math. Soc.*, **67**, 501-503.

Bellman R.E., Kalaba R. and Zadeh L.A. (1966). Abstraction and pattern classification, *J. Math. Anal. Appl.*, **13**, 1-7.

Bellman R. and Zadeh L.A. (1970) Decision making in a fuzzy environment, *Management Science*, **17**, B141-B164.

Belluce L.P., Di Nola A. and Sessa S. (1991). Triangular norms MV — Algebras and fuzzy set theory, *Math. Japonica*, **36**, 481-487.

Bezdek J., Keller J., Krishnapuram R. and Pal N.R. (1999). *Fuzzy Models and Algorithms for Pattern Recognition and Image Processing*, The Handbooks of Fuzzy Sets Series, Kluwer Academic Publishers, Dordrecht, The Netherlands.

Bezdek J.C. and Harris J.D. (1978). Fuzzy partitions and relations: An axiomatic basis for clustering, *Fuzzy Sets and Systems*, **1**, 111-127.

Bezdek J.C. and Harris J.D. (1979) Convex decompositions of fuzzy partitions, *J. Math. Anal. & Appl.*, **67**, 490-512.

Bilgic T. and Türksen I.B. (1999). Measurement of membership functions: Theoretical and empirical work, *Fundamentals of Fuzzy Sets* (Dubois D. and Prade H. eds.), Kluwer Acad. Publ., 1999. This volume.

Birkhoff G. (1967). *Lattice theory*, Providence, Rhode Island.

Birkhoff G. (1969). Mathematics and psychology, *SIAM Review*, **11**, 429-469.

Black M. (1937). Vagueness, *Phil. of Science*, **4**, 427-455.Reprinted in *Language and Philosophy: Studies in Method*, Cornell University Press, Ithaca and London, 1949, 23-58. Also in *Int. J. of General Systems*, **17**, 1990, 107-128.

Bloch I. and Maitre H. (1994) Fuzzy mathematical morphology, *Annals of Mathematics and Artificial Intelligence*, **10**, 55-84.

Bloch I. and Maitre H. (1995) Fuzzy mathematical morphologies: A comparative study, *Pattern Recognition*, **28**, 1341-1387.

Bodenhofer U. (1998) *A similarity-based generalizatton of fuzzy orderings*, Ph.D Thesis, J. Kepler Universität, Linz, Austria.

Boixader D., Jacas J.and Recasens J. (1999) Fuzzy Equivalence Relations: Advanced Material. In:*Fundamentals of Fuzzy Sets* (Dubois D. and Prade H., eds.), Kluwer Acad. Publ., 1999. This volume.

Bollman-Sdorra P., Wong S.K.M. and Yao Y.Y. (1993). A measure-theoretic axiomatization of fuzzy sets, *Fuzzy Sets and Systems*, **60**, 295-307.

Bolzano B. (1837). *Wissennschaftslehre vol. II.*, Sulzbach.

Boole G. (1854). *An Investigation of the Laws of Thought*, MacMillan. Reprinted by Dover,New York.

Borel E. (1956). *Probabilité et certitude*, Presses Universitaires de France, Paris.

Bouchon-Meunier B., Dubois D, Godo L. and Prade H. (1999). Fuzzy sets and possibility theory in approximate and plausible reasoning, *Fuzzy Sets in Approximate Reasoning and Information Systems*, The Handbooks of Fuzzy Sets Series, Kluwer Academic Publishers, Dordrecht, The Netherlands.

Bouchon-Meunier B., Rifqi M. and Bothorel S. (1996). Towards general measures of comparison of objects, *Fuzzy Sets and Systems*, **84**, 143-153.

Bradley F.H. (1914) *Essays on Truth and Reality*. Oxford University Press, Oxford.

Bronowski J. and Neyman J. (1944). The variance of the measure of a two-dimensional random set, *Ann. Math. Stat.*, **16**, 330-341.

Brown J.G. (1971). A note on fuzzy sets, *Information and Control*, **18**, 32-39.

Bunge M. (1967). *Scientific research I. The search for systems; II. The search for truth*, Springer Verlag, Berlin.

Bustince H. and Burillo P. (1996). Vague sets are intuitionistic fuzzy sets, *Fuzzy Sets and Systems*, **79**, 403-405.

Butnariu D. (1983). Additive fuzzy measures and integrals, *J. Math Anal. & Appl.*, **93**, 436-452.

Caianiello E.R. (1973). A calculus for hierarchical systems, *Proc. of the 1st Inter. Conf. on Pattern Recognition*, Washington, DC, 1-5.

Caianiello E.R. (1987). C-calculus: an overview, *Topics in the General Theory of Structures* (Caianiello E.R. and Aizerman M.A., eds.), D. Reidel, Dordrecht, 163-173.

Cao Z.Q., Kim K.H. and Roush F.W. (1984). *Incline Algebra and Applications*, Ellis Horwood, Chichester, UK.

Carnap R. (1945). Two concepts of probability, *Philosophy and Phenomenological Research*, **5**, 513-532.

Cayrol M., Farreny H. and Prade H. (1982). Fuzzy pattern matching, *Kybernetes*, **11**, 103-116.

Chakraborty M.K. and Das M. (1985). Reduction of fuzzy strict order relations, *Fuzzy Sets and Systems*, **15**, 33-44.

Chanas S. and Kuchta D. (1992). Further remarks on the relation between rough and fuzzy sets, *Fuzzy Sets and Systems*, **47**, 391-394.

Chang C.C. (1958). Algebraic analysis of many valued logics, *Trans. on Amer. Society of Mathematics*, **87**, 467-490.

Chen S.M., Yeh M.S. and Hsiao P.Y. (1995). A comparison of similarity measures of fuzzy values, *Fuzzy Sets and Systems*, **72**, 79-89.

Chwistek L. (1948). *The limits of science*, Routledge and Kegan Paul, London.

Cohen L.J. (1973). A note on inductive logic, *The J. of Philosophy*, **LXX**, 27-40.

Cohen M.R. and Nagel E. (1962). *An introduction to logic*, Harcourt Brace and World.

Copilowish I.M. (1939). Border-line cases, vagueness, and ambiguity, *Phil. of Science*, **6**, 181-195.

Cuninghame-Green R.A. (1991). Minimax algebra and applications, *Fuzzy Sets and Systems*, **41**, 251-267.

Czogala E. and Hirota K. (1986). *Probabilistic Sets: Fuzzy and Stochastic Approach to Decision, Control and Recognition Processes*, ISR 91, Verlag TÜV Rheinland, Köln, Germany.

De Baets B. and Kerre E (1993). Fuzzy relational compositions, *Fuzzy Sets and Systems*, **60**, 109-120.

De Baets B. and Kerre E (1994). The cutting of compositions, *Fuzzy Sets and Systems*, 62, 295-309.

De Baets B., Kerre E. and Gupta M.M. (1994). The fundamentals of fuzzy mathematical morphology — Part 1: Basic concepts; Part 2: Idempotence, convexity and decomposition, *Int. J. General Systems*, **23**, Part 1: 155-171; Part 2: 307-322.

De Baets B. and Mesiar R. (1998). $\mathcal{T}$-partitions, *Fuzzy Sets and Systems*, **97**, 211-223.

De Finetti B. (1936). La logique de la probabilité, *Actes Congrès Int. de Philos. Scient.*, Paris 1935, Hermann et Cie Editions, Paris, IV1- IV9.

De Luca A. and Termini S. (1972). A definition of non-probailistic entropy in the setting of fuzzy sets theory, *Information & Cont.*, **24**, 301-312.

De Luca A. and Termini S. (1974). Entropy of L-fuzzy sets, *Information and Control*, **24**, 55-73.

Deng J.L. (1989). Introduction to grey system theory, *J. Grey Systems*, **1**, 1-24.

Denoeux T. (1997). Reasoning with imprecise belief structures, Technical report Heudiasyc 97/45, Université de Compiegne, France. To appear in *Int. Journal of Approximate Reasoning*.

Denoeux T. (1998). Allowing imprecision in belief representation using fuzzy-valued belief structures, *Int. Conf. on Information Processing and Management of Uncertainty in Knowledge-based Systems (IPMU'98)*, Vol. 1, Paris, 48-55.

Denoeux T. (1999). Modeling vague beliefs using fuzzy-valued belief structures, *Fuzzy Sets and Systems*, to appear.

Di Nola A., Sessa S., Pedrycz W. and Sanchez E. (1989). *Fuzzy Relation Equations and their Applications to Knowledge Engineering*, Kluwer Academic Publ., Dordrecht.

Diamond P. and Kloeden P. (1994). *Metric spaces of fuzzy sets*, World Scientific, Singapore.

Diamond P. and Tanaka H. (1998). Fuzzy regression analysis, *Fuzzy Sets in Decision Analysis, Operations Research and Statistics* (Slowinski R., ed.), The Handbooks of Fuzzy Sets Series, Kluwer Academic Publ., Boston, 349-387.

Drewniak J. (1987). Convex and strongly convex fuzzy sets, *J. of Math. Anal. & Appl.*, **126**, 292-300.

Dubois D., Esteva F., Garcia P., Godo L. and Prade H. (1995). Similarity-based consequence relations, *Symbolic and Quantitative Approaches to Reasoning and Uncertainty (Proc. of the Europ. Conf. on ECSQARU'95, Fribourg, Switzerland, July 1995)* (Froidevaux C. and Kohlas J., eds.), Lecture Notes in Artificial Intelligence, Vol. 946, Springer Verlag, Berlin, 171-179.

Dubois D., Fargier H. and Prade H. (1996). Possibility theory in constraint satisfaction problems: Handling priority, preference and uncertainty, *Applied Intelligence*, **6**, 287-309.

Dubois D. and Jaulent M.C. (1987). A general approach to parameter evaluation in fuzzy digital pictures, *Pattern Recognition Letters*, **6**, 251-259.

Dubois D., Kerre E., Mesiar R. and Prade H. (1999). Fuzzy interval analysis, *Fundamentals of Fuzzy Sets* (Dubois D. and Prade H., eds.), Kluwer Acad. Publ., 1999. This volume.

Dubois D. and Koning J.L. (1991). Social choice axioms for fuzzy set aggregation, *Fuzzy Sets and Systems*, **43**, 257-274.

Dubois D., Lang J. and Prade H. (1994). Possibilistic logic, *Handbook of Logic in Artificial Intelligence and Logic Programming, Vol. 3* (Gabbay D.M., Hogger C.J., Robinson J.A. and Nute D., eds.), Oxford University Press, 439-513.

Dubois D., Nguyen H.T. and Prade H. (1999). Possibility theory, probability and fuzzy sets: Misunderstandings, bridges and gaps, *Fundamentals of Fuzzy Sets* (Dubois D. and Prade H., eds.), Kluwer Acad. Publ., 1999. This volume.

Dubois D. and Prade H. (1979). Operations in a fuzzy-valued logic, *Information and Control*, **43**(2), 224–240.

Dubois D. and Prade H. (1980a). *Fuzzy Sets and Systems — Theory and Applications*, Academic Press, New York.

Dubois D. and Prade H. (1980b). New results about properties and semantics of fuzzy-set-theoretic operators, *Fuzzy Sets: Theory and Applications to Policy Analysis and Information Systems* (Wang P.P. and Chang S.K., eds.), Plenum Publ., 59-75.

Dubois D. and Prade H. (1982a). A unifying view of comparison indices in a fuzzy set-theoretic framework, *Recent Development in Fuzzy Set and Possibility Theory* (Yager R.R., ed.), Pergamon Press, 3-13.

Dubois D. and Prade H. (1982b). A class of fuzzy measures based on triangular norms, *Int. J. General Systems*, **8**, 43-61.

Dubois D. and Prade H. (1983). On distances between fuzzy points and their use for plausible reasoning, *IEEE Int. Conf. on Systems Man and Cybernetics*, Bombay and New Delhi, IEEE, Pistacaway, NJ, 300-303.

Dubois, D. and H. Prade. (1985a). Evidence measures based on fuzzy information, *Automatica*, **21**(5), 547-562.

Dubois, D. and H. Prade. (1985b). Fuzzy cardinality and the modeling of imprecise quantification, *Fuzzy Sets and Systems*, **16**, 199-230.

Dubois D. and Prade H. (1987). Twofold fuzzy sets and rough sets: Some issues in knowledge representation, *Fuzzy Sets and Systems*, **23**, 3-18.

Dubois D. and Prade H. (1988). possibility theory

Dubois D. and Prade H. (1989). Fuzzy sets, probability and measurement, *Europ. J. of Operational Research*, **40**, 135-154.

Dubois D. and Prade H. (1990). Rough fuzzy sets and fuzzy rough sets, *Int. J. of General Systems*, **17**, 191-209.

Dubois D. and Prade H. (1991a). Random sets and fuzzy interval analysis, *Fuzzy Sets and Systems*, **42**, 87-101.

Dubois D. and Prade H. (1991b). Epistemic entrenchment and possibilistic logic, *Artificial Intelligence*, **50**, 223-239.

Dubois D. and Prade H. (1992a). Upper and lower images of a fuzzy set induced by a fuzzy relation: Applications to fuzzy inference and diagnosis, *Information Sciences*, **64**:203-232.

Dubois D. and Prade H. (1992b). Putting rough sets and fuzzy sets together, *Intelligent Decision Support — Handbook of Applications and Advances of the Rough Sets Theory* (Slowinski R., ed.), Kluwer Academic Publ., Dordrecht, 203-250.

Dubois D. and Prade H. (1993). Toll sets and toll logic, *Fuzzy Logic: State of the Art* (Lowen R. and Roubens M., eds.), Kluwer Academic Publ., Dordrecht, 169-177.

Dubois D. and Prade H. (1995). Possibility theory as a basis for qualitative decision theory, *Proc. of the International Joint Conference on Artificial Intelligence(IJCAI 95)*, Montreal, 1924-1930.

Dubois D. and Prade H. (1997). The three semantics of fuzzy sets, *Fuzzy Sets and Systems*, **90**, 141-150.

Dubois D., Prade H. and Yager R.R. (Eds.) (1993). *Readings in Fuzzy Sets and Intelligent Systems*, Morgan Kaufmann, San Mateo, CA.

Dubois D., Prade H. and Yager R.R. (Eds.) (1997). *Fuzzy Information Engineering: A Guided Tour of Applications*, John Wiley & Sons, New York.

Dubois D., Yager R.R. and Prade H. (1999). Merging fuzzy information, *Fuzzy Sets in Approximate Reasoning and Information Systems*, The Handbooks of Fuzzy Sets Series, Kluwer Academic Publishers, Dordrecht, The Netherlands.

Dujmovic J.J. (1974). Weighted conjunctive and disjunctive means and their application in system evaluation, *Publ. de la Fac. d'Electrotechnique de l'Univ. à Belgrade*, **461**, 147-158.

Dyckhoff H. and Pedrycz W. (1984). Generalized means as model of compensatory connectives, *Fuzzy Sets and Systems*, **14**:143-154.

Eisler R. (1899). Wörterbuch der Philosophischen Begriffe und Ausdrücke, Berlin.

Elkan E. (1994). The paradoxical success of fuzzy logic, (with replies) *IEEE Expert*, **August**, 3-49.

Fariñas del Cerro L. and Herzig A. (1991). A modal analysis of possibility theory, *Proc. of the Inter. Workshop on Fundamentals of Artificial Intelligence Research (FAIR'91)* (Jorrand Ph. and Kelemen J., eds.), Lecture Notes in Computer Sciences, Vol. 535, Springer Verlag, Berlin, 11-18.

Féron R. (1976). Ensembles aléatoires flous, *C.R. Acad. Sci.* (Paris), **Série A**, 303-306.

Féron R. (1981). Probabilistic and statistical study of random fuzzy sets whose referential is $\mathbb{R}^N$, *Applied Systems and Cybernetics–Vol. VI: Fuzzy Sets and Fuzzy Systems, Possibility Theory and Special Topics in Systems Research*, (Lasker G.E., ed.), Pergamon Press, New York, 2831-2836.

Finch P.D. (1981). Characteristics of interest and fuzzy sets, *Information Sciences*, **24**, 121-134.

Fine K. (1975). Vagueness, truth and logic, *Synthese*, **30**:265-300.

Fine T. (1973). *Theories of probability*, Academic Press, New York.

Fodor J., Orlowski S., Perny P. and Roubens M. (1998) The use of fuzzy preference models in multiple criteria choice, ranking and sorting. In: *Fuzzy Sets in Decision Analysis, Operations Research and Statistics* (Slowinski R., ed.), The Handbooks of Fuzzy Sets Series, Kluwer Academic Publishers, Dordrecht, 69-101.

Fodor J. and Roubens M. (1994). *Fuzzy Preference Modelling and Multicriteria Decision Support*, Kluwer Academic Publ., Dordrecht.

Fodor J. and Yager R.R. (1999). Fuzzy set-theoretic operators and quantifiers. In:*Fundamentals of Fuzzy Sets* (Dubois D. and Prade H., eds.), Kluwer Acad. Publ., 1999. This volume.

Fonck P., Fodor J. and Roubens M. (1988). An application of aggregation procedures to the definition of measures of similarity between fuzzy sets, *Fuzzy Sets and Systems*, **97**, 67-74.

Fortet R. and Kambouzia M. (1976). Ensembles aléatoires et ensembles flous, *Publications Econométriques*, **IX**(1), 1-23.

Fraenkel A.A., Bar-Hillel Y.A. and Levy A. (1973). *Foundations of Set Theory*, North-Holland.

French S. (1984). Fuzzy decision analysis: Some criticisms, *TIMS/Studies in the Management Sciences*, **20**, 29-44.

French S. (1986). *Decision Theory: An Introduction to the Mathematics of Rationality*, John Wiley & Sons, New York.

Fung L.W. and Fu K.S. (1975). An axiomatic approach to rational decision-making in a fuzzy environment, *Fuzzy Sets and Their Application to Cognitive and Decision Processes* (Zadeh L.A., Fu K.S., Tanaka K. and Shimura M., eds.), Academic Press, New York, 227-256.

von Furstenberg G.M. (Ed.) (1990). *Acting under uncertainty*, Kluwer, Dordrecht.

Gärdenfors P. (1988). *Knowledge in Flux*, MIT Press, New York.

Garret B. (1991). *Vague identity and vague objects*, *Nous*, **25**,341-351.

Gau W.L. and Buehrer D.J. (1993). Vague sets, *IEEE Trans. on Systems, Man and Cybernetics*, **23**, 610-614.

Gentilhomme Y. (1968). Les ensembles flous en linguistique, *Cahiers de Linguistique Théorique et Appliquée* (Bucarest), **5**, 47-63.

Gebhardt J., Gil M.A. and Kruse R. (1998). Fuzzy set-theoretic methods in statistics, *Fuzzy Sets in Decision Analysis, Operations Research and Statistics* (Slowinski R., ed.), The Handbooks of Fuzzy Sets Series, Kluwer Academic Publ., Boston, 311-347.

Gilboa and Schmeidler (1995). Case-based decision. *The Quarterly J. of Economics*, **Aug.**, 607-639.

Giles R. (1976). Lukasiewicz logic and fuzzy theory, *Int. J. of Man-Machine Studies*, **8**, 313-327.

Giles R. (1988). The concept of grade of membership, *Fuzzy Sets and Systems*, **25**, 297-323.

Goetschel R. Jr. and Voxman W. (1983). Topological properties of fuzzy numbers, *Fuzzy Sets and Systems*, **10**, 87-99.

Goguen J.A. (1967). L-fuzzy sets, *J. Math. Anal. Appl.*, , 8:145-174.

Goguen J.A. (1969). The logic of inexact concepts, *Synthese*, **19**, 325-373.

Goguen, J.A. (1974). Concept representation in natural and artificial languages: Axioms, extensions and applications for fuzzy sets, *Int. J. of Man-Machine Studies*, **6**, 513-561.

Gondran M. (1976). Valeurs propres et vecteurs propres en classification hiérarchique, *RAIRO Informatique Théorique*, **10**(3), 39-46.

Gondran M. (1979). Les éléments p-réguliers dans les dioïdes, *Discrete Mathematics*, **25**:33-39.

Goodman I.R. (1982). Fuzzy sets as equivalence classes of random sets, *Fuzzy Sets and Possibility Theory* (Yager R.R., Ed.), Pergamon Press, Oxford, 327-342.

Goodman I.R. and Nguyen H.T. (1985). *Uncertainty Models for Knowledge-Based Systems*, North-Holland, Amsterdam.

Gottwald S. (1984). Fuzzy sets theory: Some aspects of the early development, *Aspects of vagueness* (Skala H.J., Termini S. and Trillas E., eds.), D. Reidel, 13-29.

Gottwald S. (1993). *Fuzzy Sets and Fuzzy Logic — Foundations of Applications from a Mathematical Point of View*, Vieweg, Wiesbaden, Germany.

Gottwald S. (1999) Many-valued logic and fuzzy set theory, *Mathematics of Fuzzy sets: Logic, Topology and Measure Theory* (Höhle U. and Rodabaugh S., eds.), The Handbooks of Fuzzy Sets Series, Kluwer Academic Publishers, Dordrecht, The Netherlands, 5-90.

Grabisch M., Murofushi T. and Sugeno M. (1992). Fuzzy measure of fuzzy events defined by fuzzy integrals, *Japanese J. of Fuzzy Theory and Systems*, **3**:251-264. Also in *Fuzzy Sets and Systems,* **50**, 1992, 293-313.

Grabisch M., Nguyen H.T. and Walker E.A. (1995). *Fundamentals of Uncertainty Calculi with Applications to Fuzzy Inference*, Kluwer Academic Publ., Dordrecht.

Gupta M.M. (1977). Fuzzy-ism, the first decade, *Fuzzy Automata and Decision Processes* (Gupta M.M., Saridis G.N. and Gaines B.R., eds.), North-Holland, Amsterdam, 5-10.

Haack S. (1979). Do we need "fuzzy logic"?, *Int. J. of Man-Machine Studies*, **11**, 437-445.

Haack S. (1996). *Deviant Logic, Fuzzy Logic — Beyond the Formalism*, The University of Chicago Press.

Hacking I. (1975). *The Emergence of Probability*, Cambridge University Press.

Hajek P. (1998) *The Metamathematics of Fuzzy Logics*, Kluwer Academic, Dordrecht.

Hajek P. Godo L and Esteva F. (1996) A complete many-valued logic with product conjunction, *Archive for Mathematical Logic*, **35**, 191-208.

Hajek P. and Paris J.(1997) A dialogue on fuzzy logic, *Soft Computing*, **1**, 3-5.

Hamblin C.L. (1959). The modal "probably", *Mind*, **LXVIII**(270), 234-240.

Hasenjaeger C. (1972). *Introduction to the basic concepts and problems of modern logic*, D. Reidel.

Head T. (1995). A metatheorem for deriving fuzzy theorems from crisp versions, *Fuzzy Sets and Systems*, **73**, 349-358.

Hempel C.G. (1939). Vagueness and logic, *Phil. of Science*, **6**, 163-180.

Herencia J.A. (1998). Graded sets, points and numbers, *Mathware & Soft Computing*, **5**, 189-199.

Hersh H.M., Caramazza A. and Brownell H.H. (1979). Effects of context on fuzzy membership functions, *Advances in Fuzzy Set Theory and Applications* (Gupta M.M., Ragade R.R. and Yager R.R., eds.), North-Holland, Amsterdam, 389-408.

Hirota K. (1981). Concepts of probabilistic sets, *Fuzzy Sets and Systems*, **5**, 31-46.

Hisdal E. (1988a). Are grades of membership probabilities?, *Fuzzy Sets and Systems*, **25**, 325-348.

Hisdal E. (1988b). The philosophical issues raised by fuzzy set theory, *Fuzzy Sets and Systems*, **25**, 349-356.

Höhle U. (1976). Maβe auf unscharfen Mengen, *Zeitschrift für Wahrscheinlichkeits-theorie und verwandte Gebiete*, **36**, 179-188.

Höhle, U. (1988) Quotients with respect to similarity relations, *Fuzzy Sets and Systems*, **27**, 31-44.

Höhle U. (1995). Commutative, residuated L-monoids, *Non-Classical Logics and Their Applications to Fuzzy Subsets* (Höhle U. and Klement P., eds.), Kluwer Academic Publ., 53-196.

Höhle U. and Rodabaugh S. (Eds.) (1999). *Mathematics of Fuzzy sets: Logic, Topology and Measure Theory*, The Handbooks of Fuzzy Sets Series, Kluwer Academic Publishers, Dordrecht, The Netherlands.

Höhle U. (1998). Many-valued equalities, singletons and fuzzy partitions, *Soft Computing*, **2**, 134-140.

Höhle U. and Rodabaugh S. (Eds.) (1999). *Mathematics of Fuzzy sets: Logic, Topology and Measure Theory*, The Handbooks of Fuzzy Sets Series, Kluwer Academic Publishers, Dordrecht, The Netherlands.

Hughes G.E and Cresswell M.J. (1968). *An Introduction to Modal Logic*, Methuen, London.

Ishizuka M., Fu K.S. and Yao J.T.P. (1982). Inferences procedures and uncertainty for the problem-reduction method, *Information Science*, **28**, 179-206.

Kacprzyk J. and Nurmi H. (1998). Group decision making under fuzziness, In: *Fuzzy Sets in Decision Analysis, Operations Research and Statistics* (Slowinski

R., ed.), The Handbooks of Fuzzy Sets Series, Kluwer Academic Publishers, Dordrecht, 103-136.

Kampé de Fériet J. (1980). Une interprétation des mesures de plausibilité et de crédibilité au sens de G. Shafer et de la fonction d'appartenance définissant un ensemble flou de L. Zadeh, *Publ. I.R.M.A. de Lille I*, **2**(6), II.01-II.20. Translated into "Interpretation of membership functions of fuzzy sets in terms of plausibility and belief," *Fuzzy Information and Decision Processes* (Gupta M.M. and Sanchez E., eds.), North-Holland, Amsterdam, 1982, 93-98.

Kampé de Fériet J., Forte B. and Benvenuti P. (1969). Forme générale de l'opération de composition continue d'une information, *C.R. Académie des Sciences de Paris*, **t. 269**, Ser. A, 529-534.

Kandel A. (1986) *Fuzzy Mathematical Techniques with Applications*, Addison Wesley Publ. Corp., Reading.

Kandel A. and Byatt W.J. (1978). Fuzzy sets, fuzzy algebra, and fuzzy statistics, *Proc. of the IEEE*, **66**(12), 1619-1639.

Kaplan A. (1946). Definition and specification of meanings, *J. Phil.*, **43**, 281-288.

Kaplan A. and Schott H.F. (1951). A calculus for empirical classes, *Methods*, **III**, 165-188.

Kasumov N. (1996). Metric properties of fuzzy partitions, *Fuzzy Sets and Systems*, **81**, 365-378.

Katsoff H. (1937). Modality and probability, *Phil. Rev.*, **46**, 78-85.

Kaufmann A. (1975). *Introduction to the Theory of Fuzzy Subsets–Vol. 1: Fundamental Theoretical Elements*, Academic Press, New York.

Kaufmann A. (1988). Theory of expertons and fuzzy logic, *Fuzzy Sets and Systems*, **28**, 295-304.

Kemeny J.G. (1963). Analyticity versus fuzziness, *Synthese*, **15**, 57-80.

Kendall D.G. (1974). Foundations of a theory of random sets, *Stochastic Geometry* (Harding E.F. and Kendall D.G., eds.), Wiley, New York, 322-376.

Klaua D. (1966). Grundbegriffe einer mehrwertigen Mengenlehre, *Monatsber. Deut. Acad. Wiss. Berlin,* **8**, 782-802.

Klawonn F. and Castro J.L. (1995). Similarity in fuzzy reasoning, *Mathware and Soft Computing,* **2**, 336-350.

Klement E.P. and Moser B. (1997). On the redundancy of fuzzy partitions, *Fuzzy Sets and Systems*, **85**, 195-201.

Klir G.J. and Folger T.A. (1988). *Fuzzy Sets, Uncertainty, and Information*, Prentice Hall, Englewook Cliffs, NJ.

Klir G.J., St Clair U.H. and Yuan B. (1997). *Fuzzy Set Theory — Foundations and Applications*, Prentice Hall, Englewook Cliffs, NJ.

Klir G.J. and Yuan B. (1995). *Fuzzy Sets and Fuzzy Logic — Theory and Applications*, Prentice Hall, Upper Saddle River, NJ.

Klir G.J. and Yuan B. (1996). *Fuzzy Sets, Fuzzy Logic, and Fuzzy Systems — Selected Papers by Lotfi A. Zadeh*, World Scientific, Singapore.

Kolodner J. (1993). *Case-Based Reasoning*, Morgan Kaufmann, San Mateo, CA.

Kosko B. (1986). Fuzzy entropy and conditioning, *Information Sciences*, **40**:165-174.

Kosko B. (1990). Fuzziness vs. probability, *Int. J. of General Systems*, **17**, 211-240.

Kosko B. (1991). *Neural Networks and Fuzzy Systems*, Prentice-Hall, Englewood Cliffs, NJ.

Krantz D.H., Luce D.D., Suppes P. and Tversky A. (1971). *Foundations of Measurement–Vol. I: Additive and Polynomial Representations*, Academic Press, New York.

Kruse R., Gebhardt J. and Klawonn F. (1994). *Foundations of Fuzzy Systems*, Wiley, New York.

Kruse R. and Meyer K. (1987). *Statistics with Vague Data,* D. Reidel, Dordrecht.

Kruze-Blachowicz K. (1982). Leibniz's Demonstratio Existentiae Dei ad mathematicam certitudinem exacta: A logical analysis, *Studies in Logic, Grammar and Rethoric*, Bialystok, 43-55.

Kwakernaak H. (1978). Fuzzy random variables — PartI: Definitions and theorems; Part II: Algorithms and examples for the discrete case, *Information Sciences*, Part I: **15**, 1-29; Part II: **17**, 253-278.

Lakoff G. (1972). A study in meaning criteria and the logic of fuzzy concepts, *Papers from the 8th Regional Meeting Chicago Linguistic Society*, Chicago Linguistic Soc., Chicago. Reprinted in *J. of Philosophical Logic*, 2, 458-508.

Lakoff G. (1987). *Women, Fire, and Dangerous Things — What Categories Reveal about the Mind*, University of Chicago Press, Chicago.

Lang J. (1991). Possibilistic logic as a logical framework for min-max discrete optimization problems and prioritized constraints, *Proc. of the Inter Workshop on Fundamentals of Artificial Intelligence Research (FAIR'91)*, Lecture Notes in Computer Science, Vol. 535, Springer Verlag, Berlin, 112-126.

Lesniewski S. (1916). Podstawy ogolnej teorii mnogosci, Prace Polskiego Kola Naukowego w Moskwie, Nr. 2, Moscow.

Lesniewski S. (1992). *Collected works*, Polish Scientific Publishers, Warszawa.

Lewis D. (1973). *Counterfactuals*, Basil Blackwell.

Lin T.Y. (1998a). Sets with partial membership: A rough set view of fuzzy sets, *Proc.IEEE Int. Conf. on Fuzzy Systems*, Anchorage, Al., 785-790.

Lin T.Y. (1998b). Fuzzy partitions — Part I: Rough sets view of fuzzy sets, *Proc 7th Int. Conf. on Information Processing and Management of Uncertainty in Knowledge-based Systems (IPMU'98)*, Paris, Editions Medicales et Scientifiques, 1167-1174.

Lowen R. (1980). Convex fuzzy sets, *Fuzzy Sets and Systems*, **3**, 291-310.

Lowen R. (1996). *Fuzzy Set Theory — Basic Concepts, Techniques and Bibliography*, Kluwer Academic Publ., Dordrecht.

Lowen R. and Peeters W. (1998). Distances between fuzzy sets representing grey level images, *Fuzzy Sets and Systems*, **99**, 135-149.

Lukasiewicz J. (1910a). O zasadzie sprzecznosci u Arystotelesa. Studium Krytyczne (On the principle of contradiction in Aristotle. A critical study) (new edition by PWN 1987).

Lukasiewicz J. (1910b). O zasadzie wylaczonego srodka, *Przeglad Filozoficzny*, **13**, 372-373.

Lukasiewicz J. (1920). O logice trojwartosciowej, *Ruch Filozoficzny*, **5**, 170-171.

Lukasiewicz J. (1930). Philosophical remarks on many-valued systems of propositional logic, Reprinted in *Selected Works* (Borkowski, ed.), Studies in

Logic and the Foundations of Mathematics, North-Holland, Amsterdam, 1970, 153-179.
Lukasiewicz J. (1970). *Selected Works*, Borkowski, ed., Studies in Logic and the Foundations of Mathematics, North-Holland, Amsterdam.
Mabuchi S. (1992). An interpretation of membership functions and the properties of general probabilistic operators as fuzzy set operators — Part I: Case of type 1 fuzzy sets, *Fuzzy Sets and Systems,* **49**:271-283.
MacColl H. (1897). Symbolic reasonning II, *Mind*, **6**, 494-510.
MacColl H. (1900). Symbolic reasoning III, *Mind*, **9**, 190-203.
MacLane S. (1983). The health of mathematics, *The Mathematical Intelligencer*, **5**, 53-55.
Matheron G. (1975). *Random sets and integral geometry*, J. Wiley.
Mellor D.H. (1965). Experimental error and deducibility, *Phil. of Science*, **32**, 102-122.
Mellor D.H. (1966). Inexactness and explanation, *Phil. of Science*, **33**, 345-359.
Mellor D.H. (1967). Imprecision and explanation, *Phil. of Science*, **34**, 1-10.
Menger K. (1951a). Ensembles flous et fonctions aleatoires, *Comptes Rendus de l'Académie des Sciences de Paris,* **232**, 2001-2003.
Menger K. (1951b). Probabilistic geometry, *Proc. N.A.S.*, **37**, 226-229.
Menger K. (1951c). Probabilistic theories of relations, *Proc. Natl. Acad. Sci.*, **37**, 178-180.
Menger K. (1966). Positivistic geometry, *Proc. Mach Symposium*. Reprinted as "Geometry and positivism: A probabilistic microgeometry," *Selected Papers in Logic and Foundations, Didactics, Economics*, D. Reidel, Dordrecht, 1979, Chapter 20, 225-234.
Mizumoto M. and Tanaka K. (1976). Some properties of fuzzy sets of type 2, *Information and Control,* **31**, 312-340.
Moisil G. (1935). Recherches sur l'algébre de la logique, *Annales Scientifiques de l'Université de Jassy*, **XXII**, 1-118.
Moisil G. (1972). La logique des concepts nuancés, *Essais sur les Logiques Non Chrysippiennes*, Editions Acad. Repub. Soc. Roum, Bucharest, 157-163.
Mordeson J.N. and Malik D S (1998). *Fuzzy Commutative Algebra.* World Publ.
Mordeson J.N. and Nair P.S. (1998). *Fuzzy Mathematics*, Physica-Verlag, Singapore.
Mukaidono M. (1975). On some preperties of fuzzy logic, *Systems, Computers, Control*, **6**(2), 36-43.
Mundici D. (1986). Interpretation of AF C*-algebras in Lukaziaewicz sentential calculus, *J. Functional Analysis*, **65**, 15-63.
Mundici D. (1998). Nonboolean partitions and their logic, *Soft Computing*, **2**, 18-22.
Murawski R. (1994). *Filozofia matematyki, antologia tekstow klasycznych*, Wyd. naukowe UAM, Poznan.
Murthy C.A., Pal S.K. and Dutta Majumder D. (1985). Correlation between two fuzzy membership functions, *Fuzzy Sets and Systems*, **17**, 23-38.
Nakamura, A. and Gao, J.M. (1991) 'A logic for fuzzy data analysis', *Fuzzy Sets and Systems* **39,** 127-132.

Narens L. (1981). A general theory of ratio scalability with remarks about the measurement-theoretic concept of meaningfulness, *Theory and Decision*, **13**, 1-70.

Narin'yani A.S. (1980). Sub-definite sets: New data-type for knowledge representation, (In Russian) Memo n° 4-232, Computing Center, Novosibirsk (26 pages).

Negoita C.V. and Ralescu D.A. (1975a). *Applications of Fuzzy Sets to Systems Analysis*, Interdisciplinary Systems Research Series, Vol. 11, New York: Birkhaeuser, Basel & Stuttgart and Halsted Press.

Negoita C.V. and Ralescu D.A. (1975b). Representation theorems for fuzzy concepts, *Kybernetes*, **4**, 169-174.

Negoita, C.V. and Ralescu D.A. (1987). *Simulation, Knowledge-Based Computing, and Fuzzy Statistics*, Van Nostrand Reinhold, New York.

Nguyen H.T. (1978). A note on the extension principle for fuzzy sets, *J. Math. Anal. Appl.*, **64**, 369-380.

Nguyen H.T. and Walker E.A. (1996). *A First Course in Fuzzy Logic*, CRC Press, New York.

Niiniluoto I. (1987). *Truthlikeness,* D. Reidel, Dordrecht.

Novak V. (1986). *Fuzzy Sets and their Applications*, Adam Hilger, Bristol, UK.

Novak V. (1990). On the syntactico semantical completeness of first order fuzzy logic, *Kybernetika,* Part I: **25**, 47-66; Part II: **25**, 134-154.

Novak V. (1992). *The Alternative Mathematical Model of Linguistic Semantics and Pragmatics*, Plenum Press, New. York.

Orlov A.I. (1975). Foundations of fuzzy sets (generalization of Zadehian concept), *Algorithms of multidimensional statistical analysis and their applications*, CEMI AN SSSR, Moscow 1980, 287-308 (in Russian).

Orlov A.I. (1978). Fuzzy and random sets, *Prikladnoï Mnogomiernii Statisticheskii Analyz* (Nauka, Moscow), 262-280 (in Russian).

Orlov A.I. (1980). Random sets with independent elements (Lucyeans) and their applications, *Algorithms and Programms for Applied Statistical Analysis* (Ayvasian S.A. and Enyyukov I.C., eds.), Nauka, Moscow, 287-308 (in Russian).

Orlovski S. (1978). Decision-making with a fuzzy preference relation, *Fuzzy Sets and Systems*, **1**, 155-168.

Orlovski S. (1990). Decomposition of properties and definition of fuzzy set, *Fuzzy Sets and Systems*, **37**, 53-64.

Orlovski S. (1995). *Calculus of Decomposable Properties, Fuzzy Sets, and Decisions*, Allerton Press.

Ostasiewicz W. (1991). Pioneers of fuzziness, *Busefal*, **46**, 4-15.

Ostasiewicz W. (1992a). Uncertainty and vagueness, *Advances in fuzzy sets and applications* (Tofan I., Gil Aluja J., Costinescu O. and Teodorescu H.N., Eds.), Iasi.

Ostasiewicz W. (1992b). Half a century of fuzzy sets. *Supplement to Kybernetika* 28:17-20 (Proc. of the Inter. Symp. on Fuzzy Approach to Reasoning and Decision Making, Bechyne, Czechoslovakia, June 25-29, 1990).

Ovchinnikov, S.V. (1982) 'On fuzzy relational systems', in Proc. 2d World Conf. on Mathematics at the Service of Man, Las Palmas, Spain, pp. 566-569.

Ovchinnikov, S.V. (1999) An Introduction to Fuzzy Relations. In:*Fundamentals of Fuzzy Sets* (Dubois D. and Prade H., eds.), Kluwer Acad. Publ., 1999. This volume.

Pal N. and Bezdek J.C. (1999). Quantifying different facets of fuzzy uncertainty, *Fundamentals of Fuzzy Sets* (Dubois D. and Prade H., eds.), Kluwer Acad. Publ., 1999. This volume.

Pap E. (1994). *Null-additive Set Functions*, Kluwer Academic Publ., Dordrecht.

Pap E. (1997). Pseudo-analysis as a mathematical base for soft computing, *Soft Computing*, **1**, 61-68.

Pappis C.P. and Karacapilidis N.I. (1993). A comparative assessment of measures of similarity of fuzzy values, *Fuzzy Sets and Systems*, **56**, 171-174.

Parikh R. (1983). The problem of vague predicates, *Language, Logic and Method*, (Cohen R.S. and Wartopsky M.W., eds.), D. Reidel, Dordrecht, 241-261.

Paris J.B. (1997). The semantics of fuzzy logics, *Soft Computing* , **1**, 143-147.

Pavelka J. (1979). On fuzzy logic, *Zeitschr. f. Math. Logik und Grundlagen d. Math.,* **25**, Part I: 45-52; Part II: 119-134; Part III: 447-464.

Pawlak Z. (1982) Rough sets, *Int. J. of Computer & Information Sciences*, **11**(5), 341-356.

Pawlak Z. (1985). Rough sets and fuzzy sets, *Fuzzy Sets and Systems*, **17**, 99-102.

Pawlak Z. (1991). *Rough Sets — Theoretical Aspects of Reasning about Data*, Kluwer Academic Publ., Dordrecht.

Pedrycz W. (1989). *Fuzzy Control and Fuzzy Systems*, Research Studies Press, Taunton, UK.

Pedrycz W. (1995). *Fuzzy Sets Engineering,* CRC Press, Boca Raton, FL.

Pedrycz W. and Gomide F. (1998). *An Introduction to Fuzzy Sets — Analysis and Design*, MIT Press, Cambridge, Ma.

Peirce C.S. (1931). *Collected Papers of Charles Sanders Peirce*, C. Hartshorne and P. Weiss, eds., Harvard University Press, Cambridge, MA.

Perny P. and Roubens M. (1998). Fuzzy preference modelling. In: *Fuzzy Sets in Decision Analysis, Operations Research and Statistics* (Slowinski R., ed.), The Handbooks of Fuzzy Sets Series, Kluwer Academic Publishers, Dordrecht, 3-30.

Poincaré H. (1905). *La Valeur de la Science*, Flammarion, Paris.

Post E.L. (1921). Introduction to a general theory of elementary propositions, *American J. of Mathematics,* **43**, 165-185.

Puri M. and Ralescu D. (1983). Differentials of fuzzy functions, *J. Math. Anal. Appl.*, **91**, 552-558.

Puri M.L. and Ralescu D. (1986). Fuzzy random variables, *J. of Math. Anal. & Appl.*, **114**, 409-422.

Quine W.V. (1960). *Word and object*, The MIT Press, Cambridge.

Quine W.V. (1981). *Theories and things*, The Belknap Press of Harvard University Press, Cambridge.

Ralescu D.A. (1992). A generalization of the representation theorem, *Fuzzy Sets and Systems*, **51**, 309-311.

Ralescu D.A. (1999). Fuzzy random variables revisited, *Mathematics of Fuzzy sets: Logic, Topology and Measure Theory* (Höhle U. and Rodabaugh S., eds.), The Handbooks of Fuzzy Sets Series, Kluwer Academic Publishers, Dordrecht, 700-710.

Rasiowa H. (1964). A generalization of a formalized theory of fields of sets on non-classical logics, *Rozprawy Matematyczne*, **42**, 3-29.
Rasiowa H. and Nguyen C.H. (1992). LT-fuzzy sets, *Fuzzy Sets and Systems*, **47**, 323-339.
Reichenbach H. (1949). *The theory of probability*, University of California Press.
Rescher N. (1964). *Introduction to logic*, St. Martins Press, New York.
Rescher N. (1969). *Many-Valued Logics*, McGraw-Hill, New-York.
Rescher N. (1976). *Plausible Reasoning: An Introduction to the Theory and Practice of Plausibilistic Inference*, Van Gorcum, Assen/Amsterdam.
Robbins H.E. (1944). On the measure of random set, *Ann. of Math. Stat.,* **15**, 70-74.
Robinson R. (1941). III.- Ambiguity, *Mind*, **50**, 140-155.
Rockafellar R.T. (1970) *Convex Analysis*. Princeton University Press, Princeton, NJ.
Rolf B. (1980). A theory of vagueness, *Jour. of philosophical logic*, **9**, 315-325.
Rolf B. (1981). *Topics on vagueness*, Ph. D. Thesis, University of Lund, Sweden.
Rosenfeld A. (1971). Fuzzy groups, *J. Math. Anal. Appl.*, **35**, 512-517.
Rosenfeld A. (1979). Fuzzy digital topology, *Information and Control*, **40**, 76-87.
Rosenfeld A. (1984). The fuzzy geometry of image subsets, *Pattern Recognition Letters*, **12**, 311-317.
Ruspini E.H. (1969). A new approach to clustering, *Information Control,* **15**, 22-32.
Ruspini E.H. (1989). On the semantics of vague knowledge, *Revue Internationale de Systemique,* **3**, 387-420.
Russell B. (1923). Vagueness, *Austr. J. of Philosophy*, **1**, 84-92.
Russell B. (1934). *The problem of philosophy*, Williams & Norgate, London.
Sahnoun Z., DiCesare F. and Bonissone P.P. (1991). Efficient methods for computing linguistic consistency, *Fuzzy Sets and Systems*, **39**, 15-26.
Sambuc R. (1975). *Fonctions Φ-floues. Application à l'aide au diagnostic en pathologie thyroidienne*, Ph. D. Thesis Univ. Marseille, France.
Sanchez E. (1976). Resolution of composite fuzzy relation equations, *Information and Control,* **30**, 38-48.
Sanchez E. (1977). Inverses of fuzzy relations — Application to possibility distributionsand medical diagnosis, *Fuzzy Sets and Systems*, **2**, 75-86.
Sanchez E. (1978). Resolution of eigen fuzzy set equations, *Fuzzy Sets and Systems,* **1**(1), 69-74.
Schutz W.C. (1959). Some applications of the logical calculus for empirical classes for social methodology, *Psychometrica*, **24**, 69-87.
Schweizer B. and Sklar A. (1983). *Probabilistic Metric Spaces*, North-Holland, Amsterdam.
Shackle G.L.S. (1949). *Expectation in Economics*, Cambridge University Press, Cambridge, UK.
Shackle G.L.S. (1955). *Uncertainty in Economics*, Cambridge University Press, Cambridge, UK.
Shackle G.L.S. (1961). *Decision Order & Time: In Human Affairs*, Cambridge University Press, Cambridge, UK (2nd edition, 1969).

Shackle G.L.S. (1979). *Imagination and the Nature of Choice*, Edinburgh University Press, Edinburgh, UK.

Shafer G. (1976). *A Mathematical Theory of Evidence*, Princeton University Press, Princeton, NJ.

Shilkret N. (1971). Maxitive measure and integration, *Indag. Math.*, **33**, 109-116.

Silvert W. (1979). Symmetric summation: A class of operations on fuzzy sets, *IEEE Trans. on Systems, Man and Cybernetics,* **9**(10), 657-659.

Sinha D. and Dougherty E.R. (1993). Fuzzification of set inclusion: Theory and applications, *Fuzzy Sets and Systems*, **55**, 15-42. Erratum, **55**, 15-42.

Sinha D. and Dougherty E.R. (1995). A general axiomatic theory of intrinsically fuzzy mathematical morphologies, *IEEE Trans. on Fuzzy Systems*, **3**,389-403.

Slowinski R. (Ed.) (1998). *Fuzzy Sets in Decision Analysis, Operations Research and Statistics*, The Handbooks of Fuzzy Sets Series, Kluwer Academic Publishers, Dordrecht, The Netherlands.

Smets P. (1981). The degree of belief in a fuzzy event, *Information Sciences,* **25**, 1-19.

Smith E.E. and Osherson D.N. (1984). Conceptual combination with prototype concepts, *Cognitive Science,* **8**, 337--361.

Szpîlrajn E. (1936). Sur l'équivalence des suites d'ensembles of l'équivalence des fonctions, *Fundamenta Mathematicae*, **26**, 302-326.

Sugeno M. (1974). *Theory of fuzzy integrals and its application*, Doctoral Thesis, Tokyo Institute of Technology.

Sugeno M. (1977). Fuzzy measures and fuzzy integrals — A survey, *Fuzzy Automata and Decision Processes* (Gupta M.M., Saridis G.N. and Gaines B.R., eds.), North-Holland, Amsterdam, 89-102.

Suzuki H. (1993). Fuzzy sets and membership functions, *Fuzzy Sets and Systems*, **58**, 123-132.

Tarski A. (1940). *Introduction to logic*, Oxford University Press.

Terano T., Asai K. and Sugeno M. (1987). *Fuzzy Systems theory and its Applications*, Academic Press.

Toth H. (1987). From fuzzy-set theory to fuzzy set-theory: some critical remarks on existing concepts, *Fuzzy Sets and Systems*, **23**, 219-237.

Toth H. (1992). Reconstruction possibilities for fuzzy sets: Towards a new level of understanding?, *Fuzzy Sets and Systems*, **52**, 283-303.

Toth H. (1997). Fuzziness: From epistemic considerations to terminological clarification, *Int. J. of Uncertainty, Fuzziness and Knowledge-Based Systems*, **5**, 481-503.

Trillas E. (1979). Sobre funciones de negación en la teoría de conjuntos difusos (in spanish), *Stochastica*, **III**, 1, 47-60. English version: On negation functions in fuzzy set theory, *Advances of Fuzzy Logic* (Barro S., Bugarín A. and Sobrino A., eds.), Publicación de la Universidade de Santiago de Compostela, 1998, 31-40.

Trillas E. and Valverde L. (1985). On implication and indistinguishability in the setting of fuzzy logic, *Management Decision Support Systems using Fuzzy Sets and Possibility Theory* (Kacprzyk J. and Yager R. R., eds.), Verlag TÜV Rheinland, Köln, 198-212.

Turksen I.B. (1991). Measurement of membership functions and their acquisition, *Fuzzy Sets and Systems,* **40**, 5-38.

Turksen I.B. and Yao D.D.W. (1984). Representation of connectives in fuzzy reasoning: The view through normal forms, *IEEE Trans. on Systems, Man and Cybernetics,* **14**:146-151.

Türksen I. B. (1986). Interval valued fuzzy sets based on normal forms, *Fuzzy Sets and Systems*, **20**, 191-210.

Türksen I.B. (1995). Fuzzy normal forms, *Fuzzy Sets and Systems*, **69**, 319-346.

Tverski A. (1977). Features of similarity, *Psychol. Review,* **84**, 327-352.

Valverde, L. (1985) 'On the structure of F-indistinguishability operators', *Fuzzy Sets and Systems*, **17**, 313-328.

Verbruggen H.B., Zimmermann H.J. and Babuska R. (1999). *Fuzzy Algorithms for Control*, Kluwer Academic Publ., Boston.

Vopenka P. and Hajek P. (1972). *The theory of semisets*, Academia, Prague.

Wajsberg M (1931). Aksjomatyzacja trojwartosciowego rachunku zdan, *Comp. Rend. Scanc. Soc. Csi. Lett. Varsovic, Classe III*, **24**, 126-148.

Wang P.Z. (1983). From the Fuzzy Statistics to the Falling Random Subsets, *Advances in Fuzzy Sets, Possibility Theory and Applications* (Wang P.P., eds.), Plenum Press, New York, 81-96.

Wang P.Z. and Sanchez E. (1982). Treating a fuzzy subset as a projectable random set, *Fuzzy Information and Decision Processes* (Gupta M.M. and Sanchez E., eds.), North-Holland, Amsterdam, 213-220.

Wang X., Baets B. D. and Kerre E. (1995). A comparative study of similarity measures, *Fuzzy Sets and Systems*, **73**, 259-268.

Weber S. (1984). $\perp$-decomposable measures and integrals for Archimedean t-conorms $\perp$, *J. Math. Anal. Appl.*, **101**, 114-138.

Weyl H. (1940). The ghost of modality, *Philosophical essays in memory of Edmund Husserl*, Cambridge, MA, 278-303.

Weyl H. (1946). Mathematic and logic, *Amer. Math. Month.,* **53**, 2-13.

Wierman M.J. (1997). Extending set functions and relations, *Int. J. of Genral Systems*, **26**, 91-96.

Willaeys D. and Malvache N. (1981). The use of fuzzy sets for the treatment of fuzzy information by computer, *Fuzzy Sets and Systems*, **5**, 323-328.

Williamson T. (1994). *Vagueness.* Routledge, London.

Willmott R. (1986). On the transitivity of containment and equivalence in fuzzy power set theory, *J. of Mathematical Analysis and Applications,* **120**, 384-396.

Wittgenstein L. (1953). *Philosophical Investigations*, Macmillan, New York.

Wygralak M. (1989). Rough sets and fuzzy sets — Some remarks on interrelations, *Fuzzy Sets and Systems*, **29**, 241-243.

Wygralak M. (1996). *Vaguely Defined Objects — Representations, Fuzzy Sets and Nonclassical Cardinality Theory*, Kluwer Academic Publ., Dordrecht.

Yager R.R. (19). measyrement of fuzzy

Yager R.R. (1979). On the measure of fuzziness and negation–Part I: Membership in the unit interval, *Int. J. of General Systems,* **5**, 221-229.

Yager R.R. (1980). On the measure of fuzziness and negation–Part II: Lattices, *Information and Control,* **44**, 236-260.

Yager R.R. (1982). Generalized probabilities of fuzzy events from fuzzy belief structures, *Information Sciences*, **28**, 45-62.

Yager R.R. (1983). Some relationships between possibility, truth and certainty, *Fuzzy Sets and Systems*, **11**, 151-156.

Yager R.R., Ovchinnikov S., Tong R.M.. and Nguyen H.T. (Eds.) (1987). *Fuzzy Sets and Applications: Selected Papers by L.A. Zadeh*, Wiley, New York.

Yen J. (1990). Generalizing the Dempster-Shafer theory to fuzzy sets, *IEEE Trans. on Systems, Man and Cybernetics*, **20**, 559-570.

Yen J. (1992). Computing generalized belief functions for continuous fuzzy sets, *Int. J. of Approximate Reasoning*, **6**, 1-31.

Young C. (1931). The algebra of many-valued quantities, *Mathematische Annalen*, **104**, 260-290.

Zadeh L.A. (1965a). Fuzzy sets, *Information Control*, **8**, 338-353.

Zadeh L.A. (1965b). Fuzzy sets and systems, *System Theory* (Fox J., ed.), Microwave Research Institute Symposia Series XV, Polytechnic Press, Brooklyn, NY, 29-37. Reprinted in *Int. J. of General Systems,* **17**, 1990, 129-138.

Zadeh L.A. (1968). Probability measures of fuzzy events, *J. Math. Anal. & Appl.,* **23**, 421-427.

Zadeh L.A. (1971a). Similarity relations and fuzzy orderings, *Information Sciences,* 177-200.

Zadeh L.A. (1971b). Quantitative fuzzy semantics, *Information Sciences,* **3**, 159-176.

Zadeh L.A. (1972). A fuzzy-set-theoretic interpretation of linguistic hedges, *J. of Cybernetics,* **2**, 4-34.

Zadeh L.A. (1973). Outline of a new approach to the analysis of complex systems and decision processes, *IEEE Trans. on Systems, Man and Cybernetics,* **3**, 28-44.

Zadeh L.A. (1975a). Calculus of fuzzy restrictions, *Fuzzy Sets and their Applications to Cognitive and Decision Processes* (Zadeh L.A., Fu K.S., Tanaka K. and Shimura M., eds., Academic Press, New York, 1-39.

Zadeh L.A. (1975b). The concept of a linguistic variable and its application to approximate reasoning, *Information Sciences,* Part 1: **8**, 199-249, Part 2: **8**, 301-357, Part 3: **9**, 43-80.

Zadeh L.A. (1978a). Fuzzy sets as a basis for a theory of possibility, *Fuzzy Sets and Systems*, **1**, 3-28.

Zadeh L.A. (1978b). PRUF: A meaning representation language for natural languages, *Int. J. of Man-Machine Studies,* **10**, 395-460.

Zadeh L.A. (1979a). Fuzzy sets and information granularity, *Advances in Fuzzy Set Theory and Applications* (Gupta M.M., Ragade R.K. and Yager R.R., eds.), North-Holland, Amsterdam, 3-18.

Zadeh L.A. (1979b). A theory of approximate reasoning, *Machine Intelligence, Vol. 9* (JHayes J.E., Michie D. and Mikulich L.I., eds.), Elsevier, New York, 149-194.

Zadeh L.A. (1983). A computational approach to fuzzy quantifiers in natural languages, *Computer & Mathematics with Applications,* **9**(1), 149-184.

Zadeh L.A. (1997). Toward a theory of fuzzy information granulation and its centrality in human reasoning and fuzzy logic, *Fuzzy Sets and Systems*, **90**, 111-127.

Zawirski Z. (1932). Les logiques nouvelles et le champ de leur application, *Revue de métaphysique et de morale.*

Zawirski Z. (1934). *Stosunek logiki wielowartosciowej do rachunku prawdopodobienstwa*, Polskic Towarzystwo Przyjaciol Nauk, Poznan.

Zemach E.M. (1991). Vague objects, *Nous*, **3**, 323-340.

Zimmermann H.J. (1985). *Fuzzy Sets Theory — and its Applications*, Kluwer/Nijhoff Publ., Boston. Second revised edition, Kluwer Academic Publ., Boston, 1991. Third revised edition, 1996.

Zwick R., Carlstein E. and Budescu D.V. (1987). Measures of similarity among fuzzy concepts: A comparative analysis, *Int. J. of Approximate Reasoning*, **1**, 221-242.

2 FUZZY SET-THEORETIC OPERATORS AND QUANTIFIERS

János Fodor

Ronald R. Yager

Abstract: This chapter summarizes main ways to extend classical set-theoretic operations (complementation, intersection, union, set-difference) and related concepts (inclusion, quantifiers) for fuzzy sets. Since these extensions are mainly pointwisely defined, we review basic results on the underlying unary or binary operations on the unit interval such as negations, t-norms, t-conorms, implications, coimplications and equivalences. Some strongly related connectives (means, OWA, weighted, and prioritized operations) are also considered, emphasizing the essential differences between these and the formerly investigated operator classes. We also show other operations which have no counterpart in the classical theory but play some important role in fuzzy sets (like symmetric sums, weak t-norms and conorms, compensatory AND).

2.1 INTRODUCTION

It is needless to emphasize the importance of the seminal paper by Lotfi Zadeh (Zadeh, 1965) which gave us an extremely powerful tool for representing different kinds of human knowledge. At the same time, it started the investigation of an interesting problem class on the proper extension of classic set-theoretic operators to the field of fuzzy subsets. The aim of the present chapter is to review the main results related to such extensions. In addition, an overview of similar problems concerning extensions of quantifiers is also given.

Let X be a given set. In classical set theory intersection, union and complement of subsets of X are defined in a unique way. If $A, B \subseteq X$ are crisp sets then for any $a \in X$ we have $a \in A \cap B$ if and only if $a \in A$ **AND** $a \in B$,

$a \in A \cup B$ if and only if $a \in A$ **OR** $a \in B$, $a \in A^c$ if and only if **NOT** $a \in A$. The uniqueness is due to the fact that **AND, OR** and **NOT** are two-valued logic operations. For example, if P_1 and P_2 are propositions being either **TRUE** or **FALSE** then P_1 **AND** P_2 is **TRUE** if and only if both P_1 and P_2 are **TRUE**. We can use a *valuation* of any proposition P in the following natural way:

$$v(P) := 1 \quad \text{if and only if} \quad P \quad \text{is } \mathbf{TRUE},$$
$$v(P) := 0 \quad \text{if and only if} \quad P \quad \text{is } \mathbf{FALSE}.$$

In fact, this valuation corresponds to the *characteristic function* $\chi_A : X \to \{0,1\}$ of a subset $A \subseteq X$ when the proposition '$a \in A$' is considered for any element $a \in X$. That is, we have

$$\chi_A(a) = \begin{cases} 1 & \text{if } a \in A \\ 0 & \text{if } a \notin A \end{cases} \qquad (a \in X).$$

By using characteristic functions and the usual connectives used in Boolean logic ($\wedge$, $\vee$ and $\neg$), set-theoretic operations can be represented for all $a \in X$ as follows:

$$\begin{aligned} \chi_{A\cap B}(a) &:= \chi_A(a) \wedge \chi_B(a), \\ \chi_{A\cup B}(a) &:= \chi_A(a) \vee \chi_B(a), \\ \chi_{A^c}(a) &:= \neg\chi_A(a). \end{aligned} \tag{2.1}$$

Expressions in (2.1) imply that classical set-theoretic operations are defined pointwise. Note that we use the symbol $:=$ indicating that the item on its left-hand side is equal, by definition, to the already known item on its right-hand side. Also, $\overline{m,n} := \{m, m+1, \ldots, n-1, n\}$ is used for any $m, n \in \mathbb{N}$, $m < n$.

As it is well-known, classical subsets of a given non-empty set X endowed with set-theoretic operations form a *Boolean algebra* structure (Sikorski, 1964). That is, conditions listed in Table 2.1 are satisfied for all $A, B, C \subseteq X$. Note that the list could be shortened because some properties are consequences of the others.

Properties B1 – B9 are respectively called *idempotency*, *commutativity*, *associativity*, *absorption*, *distributivity*, *identity*, *complementarity*, *involution*, and *De Morgan's laws*. Moreover, the complementarity condition B7 is usually called the *law of the excluded middle* when union is used (i.e., $A \cup A^c = X$), while it is called the *law of contradiction* in case of intersection (i.e., $A \cap A^c = \emptyset$).

If we consider fuzzy sets on X with membership values in the closed unit interval $[0,1]$, the extension of classical set-theoretic operations is not so obvious. Good overview papers have been published about this problem (Czogała and Drewniak, 1984; Dubois and Prade, 1984b; Dubois and Prade, 1985; Klement, 1981b; Yager, 1991a). Denote the set of fuzzy subsets of X by $\mathcal{F}(X)$.

In any case, we suppose (except if it is otherwise stated) that these extended operations can be defined pointwise. That is, there exist three functions $n : [0,1] \to [0,1]$, $T : [0,1]^2 \to [0,1]$ and $S : [0,1]^2 \to [0,1]$ such that

$$A^c(a) \quad := \quad n\left(A(a)\right),$$

Table 2.1 Properties of Boolean algebras

B1.	$A \cup A = A,$	$A \cap A = A$
B2.	$A \cup B = B \cup A,$	$A \cap B = B \cap A$
B3.	$A \cup (B \cup C) = (A \cup B) \cup C,$	$A \cap (B \cap C) = (A \cap B) \cap C$
B4.	$A \cup (A \cap B) = A,$	$A \cap (A \cup B) = A$
B5.	$A \cup (B \cap C) = (A \cup B) \cap (A \cup C),$	$A \cap (B \cup C) = (A \cap B) \cup (A \cap C)$
B6.	$A \cup \emptyset = A,$	$A \cap X = A$
B7.	$A \cup A^c = X,$	$A \cap A^c = \emptyset$
B8.	$(A^c)^c = A$	
B9.	$(A \cup B)^c = A^c \cap B^c,$	$(A \cap B)^c = A^c \cup B^c$

$$\begin{aligned} (A \cap_T B)(a) &:= T\,(A(a), B(a))\,, \\ (A \cup_S B)(a) &:= S\,(A(a), B(a))\,, \end{aligned} \tag{2.2}$$

where $A, B \in \mathcal{F}(X)$ and we denote a fuzzy subset and its *membership function* with the same symbol. Therefore, desired properties of fuzzy set-theoretic operations can be obtained via the corresponding properties of the above functions n, T and S.

To determine appropriate classes of functions n, T and S, we try to preserve as many of the above properties B1–B9 as possible in the fuzzy case.

2.2 COMPLEMENTATION

Let X be a given set and $A \in \mathcal{F}(X)$. The *complement* of A, denoted by A^c, is pointwisely defined by a function $n : [0,1] \to [0,1]$ as in (2.1). As a minimal requirement, complementation of fuzzy subsets should coincide with the usual complementation when n is applied to characteristic functions of crisp subsets of X. In addition, when the grade of membership of an element $a \in X$ in A is greater than that of another element $b \neq a$ in A, then the grade of membership of b in A^c cannot be higher than that of a in A^c. More formally, we propose the following definition.

Definition 1 A function $n : [0,1] \to [0,1]$ satisfying conditions

$$n(0) = 1, \quad n(1) = 0 \tag{2.3}$$

and

$$x, y \in [0,1],\ x \leq y \quad \Longrightarrow \quad n(x) \geq n(y) \qquad (n \text{ is nonincreasing}) \tag{2.4}$$

is called a *negation*. ■

Obviously, thus defined function class of negations is too broad both for theoretical and practical purposes. Further properties can be proposed, as it was done in (Bellman and Giertz, 1973):

$$x, y \in [0,1],\ x < y \quad \Longrightarrow \quad n(x) > n(y) \qquad (n \text{ is strictly decreasing}), \tag{2.5}$$

$$n \quad \text{is continuous}, \tag{2.6}$$

$$n(n(x)) = x \quad \text{for all} \quad x \in [0,1] \qquad (n \text{ is involutive}). \tag{2.7}$$

According to these conditions, two subclasses of negations are defined now.

Definition 2 A negation is called *strict* if it is also strictly decreasing and continuous (that is, n satisfies (2.5) and (2.6)). A strict negation is said to be *strong* if it is involutive too (i.e., (2.7) also holds.) ■

Let us consider some examples to illustrate the different classes.

The function $n_i : [0,1] \to [0,1]$ defined by

$$n_i(x) := \begin{cases} 1 & \text{if } x = 0 \\ 0 & \text{if } x > 0 \end{cases}$$

is said to be the *intuitionistic* negation, see (Yager, 1979; Yager, 1980b). This is not strictly increasing, not continuous and not involutive. Hence, it is neither a strict nor a strong negation.

Exactly the same is true for the *dual intuitionistic* negation defined by

$$n_{di}(x) := \begin{cases} 1 & \text{if } x < 1 \\ 0 & \text{if } x = 1 \end{cases},$$

see (Ovchinnikov, 1983).

Despite the lack of nice properties, these two negations are nevertheless important from at least one theoretical point of view. Namely, these are (together with involutions) "optimal" in the sense that they are notions nearest to the crisp one (more explanation and results can be found in (Ovchinnikov, 1983)). In addition, the intuitionistic and the dual intuitionistic negation can be regarded as two extreme cases of negations. Indeed, it is easy to see that for any negation n we have

$$n_i \leq n \leq n_{di}.$$

An example of a strict negation which is not strong can be given by $n(x) := 1 - x^2$ ($x \in [0,1]$). Obviously, this function is not involutive.

The *standard negation* is simply

$$N(x) := 1 - x, \quad x \in [0,1].$$

This was proposed in the first paper on fuzzy sets by Zadeh (Zadeh, 1965) to define the complement of a fuzzy subset of X. Clearly, this negation is strong.

A family of strong negations including the standard one is defined by Sugeno (Sugeno, 1977) (under the name λ-complement) as follows:

$$N_\lambda(x) := \frac{1-x}{1+\lambda x}, \quad \lambda > -1, \quad x \in [0,1].$$

Because any strict negation n is a strictly decreasing and continuous function, one can define its inverse n^{-1}, which is also a strict negation and in general differs from n. Clearly, we have $n^{-1} = n$ if and only if n is involutive. This means that the graph of the function n is symmetric with respect to the line $\{(x,y) \in [0,1]^2 \mid x = y\}$.

Another important property of a strict negation n is that there exists a unique value $0 < s_n < 1$ such that $n(s_n) = s_n$. The value s_n can be interpreted as a *membership crossover point* (see (Dubois and Prade, 1985)). That is, $A(a) > s_n$ if and only if $A^c(a) < s_n$ (A^c is defined via n). Notice that we also have $n^{-1}(s_n) = s_n$.

In general, n denotes a strict negation and N means a strong negation in the sequel.

2.2.1 *Representation of negations*

Trillas published a representation theorem for strong negations (Trillas, 1979). According to his result, any strong negation can be obtained from the standard negation by a suitable transformation, by the help of an automorphism of the unit interval. Let $a, b \in \mathbb{R}$, $a < b$. A continuous, strictly increasing function $\varphi : [a,b] \to [a,b]$ with boundary conditions $\varphi(a) = a, \varphi(b) = b$ is called an *automorphism* of the interval $[a,b] \subset \mathbb{R}$.

Now we are able to cite the result of (Trillas, 1979) in a slightly modified form as follows.

Theorem 1 *A function $N : [0,1] \to [0,1]$ is a strong negation if and only if there exists an automorphism φ of the unit interval such that for all $x \in [0,1]$ we have*

$$N(x) = \varphi^{-1}(1 - \varphi(x)). \blacksquare \qquad (2.8)$$

In such a case $N_\varphi(x)$ denotes $\varphi^{-1}(1 - \varphi(x))$ ($x \in [0,1]$) and is called a *φ-transform* of the standard negation.

2.2.2 *Other important results*

1. Clearly, the representation of strong negations is not unique. The following statement can be proved (see (Fodor and Roubens, 1994)).

Proposition 1 *Suppose that N is a strong negation. Then we have*

$$N(x) = \varphi^{-1}(1 - \varphi(x)) = \psi^{-1}(1 - \psi(x))$$

with automorphisms φ, ψ *of the unit interval if and only if there exists an automorphism* η *of the interval* $[-\frac{1}{2}, \frac{1}{2}]$ *such that* $\eta(-x) = -\eta(x)$ *and*

$$\psi(x) = \frac{1}{2} + \eta\left(\varphi(x) - \frac{1}{2}\right). \blacksquare$$

2. The result of Trillas (Trillas, 1979) (see Theorem 1 above) is generalized to $\mathcal{F}_J(X)$ (the set of fuzzy subsets of X with membership values in a subset J of $[0,1]$) by Esteva and Domingo (Esteva and Domingo, 1980).

3. For sake of theoretical completeness, it is worth mentioning that there is a more general approach to complementations of fuzzy sets. Lowen (Lowen, 1978) defined a complement of a fuzzy subset as follows. Let $\{N_a\}_{a \in X}$ be a family of strong negations and A be a fuzzy subset of X. Then define A^c by

$$A^c(a) := N_a(A(a)), \quad a \in X.$$

That is, in this case it is assumed that the negation depends also on the point $a \in X$. Fuzzy complements in Lowen's sense are called *variable pointwise involutions* by Ovchinnikov (Ovchinnikov, 1983).

As we have mentioned, Ovchinnikov (Ovchinnikov, 1983) studied negations which are "optimal" in the sense that they are notions nearest to the crisp one. He proved that there are only three types of optimal negations: (i) involutions, (ii) the intuitionistic negation, (iii) the dual intuitionistic negation. In addition, the most general form of an involution in $\mathcal{F}(X)$ is given by a variable pointwise involution of Lowen (Lowen, 1978).

4. An operator closely related to the complement of a fuzzy subset is that of the *antonym*. Antonyms relate concepts such as tall–short, big–small, old–young, while complement relates tall–not tall, short–not short, etc. For more details we refer to the papers (Ovchinnikov, 1981; Trillas and Riera, 1981).

2.3 INTERSECTION AND UNION

Let A, B fuzzy subsets of X. Their *intersection* and *union* can be defined pointwise, by using functions $T : [0,1]^2 \to [0,1]$ and $S : [0,1]^2 \to [0,1]$ as follows:

$$\begin{aligned} (A \cap_T B)(a) &:= T(A(a), B(a)) \quad \text{for all } a \in X \\ (A \cup_S B)(a) &:= S(A(a), B(a)) \quad \text{for all } a \in X. \end{aligned}$$

The following result shows that it is impossible to keep the Boolean algebra structure of classical subsets of X when fuzzy subsets are considered. For more details and proofs see (Dubois and Prade, 1980b).

Theorem 2 *Suppose* A *is a proper fuzzy subset of* X *and its complement* A^c *is defined via a strong negation* N.

(a) If the law of the excluded middle $A \cup_S A^c = X$ holds then the union is no longer idempotent (that is, $A \cup_S A \neq A$).

(b) If the law of contradiction $A \cap_T A^c = \emptyset$ is valid then the intersection cannot be idempotent (i.e., $A \cap_T A \neq A$). ∎

Moreover, if $\cap_T$ and $\cup_S$ are mutually distributive then neither the law of excluded middle nor the law of contradiction can be valid. Hence, one is obliged to choose between these two laws on one hand, and the mutual distributivity and idempotency on the other hand (see also Theorem 3 below).

2.3.1 Triangular norms and conorms

The properties concerning operations $\cap_T$ and $\cup_S$ separately are expressed as follows, in terms of functions T and S. In each property the quantification 'for all $x, y, z \in [0,1]$' is included but not written.

Identity law:

$$\textbf{T1.}\ T(x,1) = x \quad (A \cap X = A), \tag{2.9}$$

$$\textbf{S1.}\ S(x,0) = x \quad (A \cup \emptyset = A). \tag{2.10}$$

Commutativity:

$$\textbf{T2.}\ T(x,y) = T(y,x) \quad (A \cap B = B \cap A), \tag{2.11}$$

$$\textbf{S2.}\ S(x,y) = S(y,x) \quad (A \cup B = B \cup A). \tag{2.12}$$

Associativity:

$$\textbf{T3.}\ T(x,T(y,z)) = T(T(x,y),z) \quad (A \cap (B \cap C) = (A \cap B) \cap C), \tag{2.13}$$

$$\textbf{S3.}\ S(x,S(y,z)) = S(S(x,y),z) \quad (A \cup (B \cup C) = (A \cup B) \cup C). \tag{2.14}$$

The following condition is natural, although it has nothing to do with algebraic structures, especially with Boolean algebras.

Monotonicity: for all $x, y, u, v \in [0,1]$, $x \leq u, y \leq v$,

$$\textbf{T4.}\ T(x,y) \leq T(u,v), \tag{2.15}$$

$$\textbf{S4.}\ S(x,y) \leq S(u,v). \tag{2.16}$$

Notice that the only difference between properties of T and S is given by the identity law.

Definition 3 A function $T : [0,1]^2 \to [0,1]$ is called a *triangular norm*(*t-norm* for short) if and only if it satisfies conditions T1–T4.

A function $S : [0,1]^2 \to [0,1]$ is a *triangular conorm* (*t-conorm* for short) if and only if it satisfies conditions S1–S4. ■

Notice that continuity of a t-norm and a t-conorm is not taken for granted. In addition, these conditions do not even imply that all t-norms, as two-place functions, are measurable (see (Klement, 1982) for constructing a counter-example).

Conditions (2.9), (2.15) and (2.10), (2.16) respectively imply

$$T(0,x) = 0, \quad S(1,x) = 1 \quad \text{for all} \quad x \in [0,1]. \tag{2.17}$$

Also, by (2.9), (2.15) (respectively, by (2.10), (2.16)) we have that

$$T(x,y) \leq \min(x,y), \quad S(x,y) \geq \max(x,y) \tag{2.18}$$

for any t-norm T and t-conorm S ($x, y \in [0,1]$). Moreover, the extension of a t-norm (t-conorm) for more than two arguments is unique since associativity is satisfied. It can also be proved that conditions T1–T4 for defining t-norms (resp. S1–S4 for t-conorms) are independent.

It is worth noting that the class of t-norms was introduced in the theory of statistical (probabilistic) metric spaces as a tool for generalizing the classical triangular inequality by Menger (Menger, 1942) (see also (Schweizer and Sklar, 1983)).

From an algebraic point of view, $([0,1],T)$ and $([0,1],S)$ are commutative ordered semigroups for any t-norm T and t-conorm S. The order can either be the usual ordering or the reverse ordering of real numbers. The main difference between $([0,1],T)$ and $([0,1],S)$ is that 1 is the unit element of the first and a zero element of the second one, while 0 is the unit element of $([0,1],S)$ and is the zero element of $([0,1],T)$.

Clearly, for every t-norm T and strong negation N, the operation S defined by

$$S(x,y) := N(T(N(x),N(y))), \quad x,y \in [0,1] \tag{2.19}$$

is a t-conorm. In addition, in this case $T(x,y) = N(S(N(x),N(y)))$ ($x, y \in [0,1]$). Then S is called the *N-dual t-conorm* to T, and T is called the *N-dua* *t-norm* to S. In case of the standard negation (i.e., when $N(x) := 1 - x$ fo $x \in [0,1]$) we simply speak about dual t-conorm (resp. t-norm). Obviously equality (2.19) expresses the De Morgan's law in the fuzzy case.

Definition 4 A triplet (T,S,N) is called a *De Morgan triplet* if and only if T is a t-norm, S is a t-conorm, N is a strong negation and they satisfy (2.19). ■

There are several examples of De Morgan triplets. We list here the mos frequently used and (either from a theoretical or from a practical point c view) important ones. In all cases we consider the standard negation.

• ***The minimum and the maximum.*** In his classical paper Zadeh (Zadeh, 1965) proposed to use

$$\min(x,y), \quad \max(x,y)$$

as models for defining intersection and union of fuzzy subsets ($x, y \in [0,1]$). As it was already mentioned, 'min' is the greatest t-norm and 'max' is the least t-conorm (see (2.18)). The special role of 'min' and 'max' will be studied in the next section.

• ***The product and the probabilistic sum.*** Another example is given by the product t-norm Π and its dual t-conorm Π' called probabilistic sum

$$\Pi(x,y) := xy, \quad \Pi'(x,y) := x + y - xy, \qquad (x, y \in [0,1]),$$

see e.g. (Goguen, 1969). Note that the use of the product and its dual has nothing to do with probability theory.

• ***The Łukasiewicz t-norm and t-conorm.*** The t-norm W introduced by Łukasiewicz in 1931 (see (Łukasiewicz, 1970)) and its dual t-conorm W' is defined by

$$W(x,y) := \max\{x+y-1, 0\}, \quad W'(x,y) := \min\{x+y, 1\} \qquad (x, y \in [0,1]).$$

These are very often called *bold intersection* and *bounded sum*, respectively (see e.g. Giles (1976)).

• ***The nilpotent minimum and maximum.*** A new t-norm was found by Perny (Perny, 1992), which was independently rediscovered by Fodor (Fodor, 1995), and called *nilpotent minimum* and *nilpotent maximum*, respectively. These differ essentially from the others, and are defined by the following formulas:

$$\min_0(x,y) := \begin{cases} \min(x,y) & \text{if } x+y > 1 \\ 0 & \text{otherwise} \end{cases},$$

and

$$\max_1(x,y) := \begin{cases} \max(x,y) & \text{if } x+y < 1 \\ 1 & \text{otherwise} \end{cases}.$$

• ***The drastic product and sum.*** The weakest t-norm Z and strongest t-conorm Z' can be defined by using conditions (2.9) and (2.10), respectively, as follows:

$$Z(x,y) := \begin{cases} \min(x,y) & \text{if } \max(x,y) = 1 \\ 0 & \text{otherwise} \end{cases},$$

and

$$Z'(x,y) := \begin{cases} \max(x,y) & \text{if } \min(x,y) = 0 \\ 1 & \text{otherwise} \end{cases}.$$

Thus, as one can easily check, for any t-norm T and t-conorm S we have

$$Z \le T \le \min, \quad \max \le S \le Z'.$$

Moreover, $W \le \Pi$ ($W' \ge \Pi'$) and $W \le \min_0$ ($W' \ge \max_1$), but Π and $\min_0$ (resp. Π' and $\max_1$) are not comparable in this sense.

2.3.2 The special role of minimum and maximum

In the literature of fuzzy sets, a key role is played by operations minimum and maximum as models of intersection and union, respectively. One reason behind this role in practical problems is the simplicity of 'min' and 'max'. In addition, we would like to enlighten this role by a number of characterizations of these two operators. Apart from properties T1–T4 for 'min' and S1–S4 for 'max', they also satisfy the following conditions.

Distributivity: For all $x, y, z \in [0, 1]$,

$$S(x, T(y, z)) = T(S(x, y), S(x, z)), \tag{2.20}$$

$$T(x, S(y, z)) = S(T(x, y), T(x, z)). \tag{2.21}$$

Continuity:

$$T \text{ and } S \text{ are continuous.} \tag{2.22}$$

Strict monotonicity on the diagonal: for $x, y \in [0, 1]$,

$$x < y \quad \text{implies} \quad T(x, x) < T(y, y) \text{ and } S(x, x) < S(y, y). \tag{2.23}$$

Idempotency: for all $x \in [0, 1]$,

$$T(x, x) = x, \quad S(x, x) = x. \tag{2.24}$$

Absorption: for all $x, y \in [0, 1]$,

$$T(x, S(x, y)) = x, \quad S(x, T(x, y)) = x. \tag{2.25}$$

Non-compensation: for all $x, y, z \in [0, 1]$,

$$x < y < z \quad \text{imply} \quad T(x, z) \neq T(y, y), \ S(x, z) \neq S(y, y). \tag{2.26}$$

It is an important and interesting problem to find out which of these properties are necessary and sufficient to conclude that $T = \min$ and $S = \max$. Answers are given in the following statement. We emphasize again that the strongest conditions implying $T = \min$ and $S = \max$ are either distributivity (2.20), (2.21) or idempotency (2.24).

Theorem 3 *Let T and S be any binary operations on $[0,1]$. Then the following assertions are equivalent:*

(i) $T = \min$ *and* $S = \max$.

(ii) *T and S are commutative, associative, nondecreasing, continuous binary operations satisfying also conditions (2.18), (2.23), and $T(1,1) = 1$, $S(0,0) = 0$.*

(iii) *T and S are commutative, associative, continuous, nondecreasing, idempotent binary operations satisfying (2.17).*

(iv) *T and S are associative, nondecreasing binary operations satisfying also (2.18) and (2.25).*

(v) *T and S are nondecreasing, idempotent operations satisfying (2.18).*

(vi) *T and S are associative, nondecreasing binary operations satisfying also (2.18), (2.20), (2.21), and (2.23).*

(vii) *T and S are continuous, nondecreasing, idempotent, non-compensative operations with $T(x,1) = T(1,x) = x$, $S(x,0) = S(0,x) = x$.*

(viii) *T and S are nondecreasing operations satisfying (2.20), (2.21) and $T(x,1) = T(1,x) = x$, $S(x,0) = S(0,x) = x$.*

(ix) $(A \cap_T B)_\alpha = A_\alpha \cap B_\alpha$, $(A \cup_S B)_\alpha = A_\alpha \cup B_\alpha$ *for all fuzzy sets $A, B \in \mathcal{F}(X)$.(Here $\alpha \in [0,1]$ and the crisp set A_α is the α-cut of A defined as usual, by $A_\alpha := \{a \in X \mid \mu_A(a) \geq \alpha\}$).* ■

One can find the above equivalent statements and their proofs in the literature as follows. Equivalence of (i) and (ii) was proved by Bellman and Giertz (Bellman and Giertz, 1973). Fung and Fu (Fung and Fu, 1975) showed that (i) is equivalent to (iii). The proof of equivalence of (i), (iv), (v) and (vi) can be found in (Hamacher, 1978). Bellman and Zadeh (Bellman and Zadeh, 1977) proved that (i) $\iff$ (vii), and Alsina *et al.* (Alsina et al., 1983) showed the equivalence between (i) and (viii). Finally, the last characterization, which is really in the spirit of fuzzy sets, was given by Klement (Klement, 1981a).

Further characterizations of 'min' ('max') will be given in terms of fuzzy inclusions later in this chapter.

2.3.3 Continuous Archimedean t-norms and t-conorms

A broad class of problems consists of the representation of multi-place functions in general by composition of "simpler" functions and functions of fewer variables (see (Ling, 1965) for a brief survey), such as

$$K(x,y) = g(f(x) + f(y)), \tag{2.27}$$

where K is a two-place function and f, g are real functions. In that general framework, the representation of (two-place) associative functions by appropriate one-place functions is a particular problem. It was Abel who first obtained such a representation i (Abel, 1826), assuming also commutativity, strict monotonicity and differentiability.

Since Abel's result, a lot of contributions have been made to representations of associative functions (and generally speaking, of abstract semigroups). It was proved by Ling (Ling, 1965) that, for example, the minimum cannot be represented in the form (2.27) assuming either f, g to be continuous or strictly decreasing functions. This negative result indicates that we have to consider a suitable subclass of all the t-norms. The properties of these t-norms are given in the following definition. For any $x \in [0, 1]$ and any associative binary operation K on [0,1], denote $x_K^{(n)}$ the nth power of x defined by $x_K^{(n)} := K(\underbrace{x, \ldots, x}_{n-\text{times}})$.

Definition 5 A t-norm T (resp. a t-conorm S) is said to be

(a) *continuous* if T (resp. S) as a function is continuous on the unit interval;

(b) *Archimedean* if $\lim_{n\to\infty} x_T^{(n)} = 0$ (resp. $\lim_{n\to\infty} x_S^{(n)} = 1$) for all $x \in]0, 1[$. ■

Note that the definition of the Archimedean property is borrowed from the theory of semigroups. If a t-norm T is Archimedean then it satisfies the condition

$$T(x, x) < x \quad \text{for all } x \in]0, 1[. \tag{2.28}$$

If a triangular norm T is continuous then it is Archimedean if and only if the condiiton (2.28) is fulfilled. Similarly, if a t-conorm S is Archimedean then it satisfies

$$S(x, x) > x \quad \text{for all } x \in]0, 1[. \tag{2.29}$$

If a t-conorm S is continuous then it is Archimedean if and only if (2.29) holds. Note that inequalities (2.28) and (2.29) are usually used to define Archimedean property for t-norms and t-conorms, respectively. Generally speaking, $T(x, x) < x$ does not imply the Archimedean property for non-continuous t-norms (see (Schweizer and Sklar, 1983) for a counterexample).

Considering the previous examples, the t-norm 'min' is continuous but not Archimedean. Z is not continuous, but it is Archimedean since $Z(x, x) = 0$ for all $x \in]0, 1[$. Π and W are continuous and Archimedean. $\min_0$ is neither Archimedean nor continuous. However, it is left-continuous.

We state here the representation theorem of continuous Archimedean t-norms and t-conorms attributed very often to Ling (Ling, 1965). In fact, her main theorem can be deduced from previously known results on topological semigroups, see (Faucett, 1955; Mostert and Shields, 1957; Paalman – De Miranda, 1964). Nevertheless, the advantage of Ling's approach is twofold: treating two different cases in a unified manner and establishing elementary proofs.

Theorem 4 *A t-norm T is continuous and Archimedean if and only if there exists a strictly decreasing and continuous function $f : [0,1] \to [0,\infty]$ with $f(1) = 0$ such that*

$$T(x,y) = f^{(-1)}(f(x) + f(y)), \tag{2.30}$$

where $f^{(-1)}$ is the pseudoinverse of f defined by

$$f^{(-1)}(x) := \begin{cases} f^{-1}(x) & \text{if } x \leq f(0) \\ 0 & \text{otherwise} \end{cases}$$

Moreover, representation (2.30) is unique up to a positive multiplicative constant. ■

We say that T *is generated by* f if T has representation (2.30). In this case f is said to be an *additive generator* of T.

In our examples, an additive generator of W is given by $f_W(x) := 1 - x$. Then $f_W^{(-1)}(x) = \max\{1 - x, 0\}$. Π is generated by $f_\Pi(x) := -\log(x)$ with $f_\Pi^{(-1)}(x) = f_\Pi^{-1}(x) = \exp(-x)$.

Theorem 5 *A t-conorm S is continuous and Archimedean if and only if there exists a strictly increasing and continuous function $g : [0,1] \to [0,\infty]$ with $g(0) = 0$ such that*

$$S(x,y) = g^{(-1)}(g(x) + g(y)), \tag{2.31}$$

where $g^{(-1)}$ is the pseudoinverse of g defined by

$$g^{(-1)}(x) := \begin{cases} g^{-1}(x) & \text{if } x \leq g(1) \\ 1 & \text{otherwise} \end{cases}$$

Moreover, representation (2.31) is unique up to a positive multiplicative constant. ■

Let S be a continuous Archimedean t-norm. We say that S *is generated by* g if S has representation (2.31). In this case g is said to be an *additive generator* S.

An additive generator of W' is given by $g_{W'}(x) := x$. Then $g_{W'}^{(-1)}(x) = \min\{x, 1\}$. Π' is generated by $g_{\Pi'}(x) := -\log(1 - x)$ with $g_{\Pi'}^{(-1)}(x) = g_{\Pi'}^{-1}(x) = 1 - \exp(-x)$.

Remark that Aczél published the representation of strictly increasing, continuous and associative two-place functions on open or half-open real intervals (Aczél, 1948; Aczél, 1949; Aczél, 1966). This was the starting point to be generalized by Ling (Ling, 1965).

One can see by the previous examples that there is an essential difference between the Łukasiewicz t-norm W and the product Π. Indeed, for W there exist $x, y > 0$ such that $W(x,y) = 0$, while this is impossible for Π. We formulate this property of W in the following definition.

Definition 6 We say that a t-norm T has *zero divisors* if and only if there exist $x, y \in]0,1[$ such that $T(x,y) = 0$. T is said to be *positive* if $x, y > 0$ imply $T(x,y) > 0$. T is called *strict* if it is a continuous and strictly increasing function in each place on $]0,1[^2$. Finally, T is called *nilpotent* if and only if T is continuous, Archimedean with zero divisors. ■

It is obvious that min and Π are positive t-norms, while W, $\min_0$ and Z have zero divisors. The only strict t-norm (resp. t-conorm) among the previous examples is Π (resp. Π'), while the only nilpotent t-norm (resp. t-conorm) is W (eresp. W'). It is easy to see that a continuous Archimedean t-norm T is positive if and only if it is strict.

The representation theorem of t-norms (resp. t-conorms) does not indicate any condition on the value of a generator function at 0 (resp. at 1). On the basis of this value, one can classify continuous Archimedean t-norms (resp. t-conorms) as it is stated in the following proposition.

Theorem 6 *Let T be a continuous Archimedean t-norm with additive generator f, and S be a continuous Archimedean t-conorm with additive generator g. Then*

(a) T is nilpotent if and only if $f(0) < +\infty$;

(b) T is strict if and only if $f(0) = \lim_{x \to 0} f(x) = +\infty$;

(c) S is nilpotent if and only if $g(1) < +\infty$;

(d) S is strict if and only if $g(1) = \lim_{x \to 1} g(x) = +\infty$. ■

Ling (Ling, 1965) proved also that every continuous Archimedean t-norm (resp. t-conorm) is the limit of a pointwise convergent sequence of strict t-norms (resp. t-conorms).

2.3.3.1 Representation of nilpotent t-norms and t-conorms.

Using the general representation theorem of continuous Archimedean t-norms, we can give another form of representation for a class of continuous t-norms with zero divisors. More exactly, for continuous t-norms T such that $T(x, n(x)) = 0$ holds with a strict negation n for all $x \in [0,1]$. Such t-norms are Archimedean, as it was proved in (Ovchinnikov and Roubens, 1992). The following theorem is established after (Ovchinnikov and Roubens, 1991).

Theorem 7 *A continuous t-norm T is such that $T(x, n(x)) = 0$ holds for all $x \in [0,1]$ with a strict negation n if and only if there exists an automorphism φ of the unit interval such that*

$$T(x,y) = \varphi^{-1}(\max\{\varphi(x) + \varphi(y) - 1, 0\}) \qquad (2.32)$$

and

$$n(x) \le \varphi^{-1}(1 - \varphi(x)). \ ■$$

In other words, this result says that any nilpotent t-norm is isomorphic to (i.e., is a φ-transform of) the Łukasiewicz t-norm. In addition, if $T = W_\varphi$, then the greatest strict negation n satisfying $T(x, n(x)) = 0$ for all $x \in [0,1]$ is strong, and is equal to N_φ.

For t-conorms, one can state the following form of the general representation theorem.

Theorem 8 *A continuous t-conorm S satisfies condition $S(x, n(x)) = 1$ for all $x \in [0,1]$ with a strict negation n if and only if there exists an automorphism φ of the unit interval such that*

$$S(x,y) = \varphi^{-1}(\min\{\varphi(x) + \varphi(y), 1\}) \tag{2.33}$$

and

$$n(x) \geq \varphi^{-1}(1 - \varphi(x)). \blacksquare$$

In the light of the representation theorem, there does not exist additive generator (i.e., a continuous, strictly decreasing function) either of $\min_0$ or Z. However, the weakest t-norm Z has a representation (see again (Ling, 1965)) in the form of

$$Z(x,y) = g(f(x) + f(y)),$$

for example, with the following functions:

$$f(x) := \begin{cases} 2-x & \text{if } x < 1 \\ 0 & \text{if } x = 1 \end{cases},$$

$$g(x) := \begin{cases} 1 & \text{if } 0 \leq x \leq 1 \\ 2-x & \text{if } 1 \leq x \leq 2 \\ 0 & \text{otherwise} \end{cases}.$$

Notice that f is decreasing but discontinuous while g is continuous and noninreasing.

2.3.3.2 Representation of strict t-norms and t-conorms.

Turning to strict t-norms and t-conorms, it is obvious that any strict t-norm and strict t-conorm is Archimedean. The following form of representation can be found in (Schweizer and Sklar, 1983).

Theorem 9 *A continuous t-norm T is strict if and only if there exists an automorphism φ of the unit interval such that*

$$T(x,y) = \varphi^{-1}(\varphi(x)\varphi(y)) \quad (x, y \in [0,1]). \blacksquare \tag{2.34}$$

An automorphism φ such that representation (2.34) holds is called a *multiplicative generator* of T.

Concerning t-conorms, we have the following result.

Theorem 10 *A continuous t-conorm S is strict if and only if there exists an automorphism φ of the unit interval such that*

$$S(x,y) = \varphi^{-1}(\varphi(x) + \varphi(y) - \varphi(x)\varphi(y)) \quad (x, y \in [0,1]). \blacksquare \tag{2.35}$$

Therefore, any continuous strict t-conorm is isomorphic to the probabilistic sum.

2.3.3.3 General form of continuous t-norms and t-conorms. Let us consider continuous t-norms and t-conorms without any additional property. Suppose that $\{[a_m, b_m]\}$ is a countable family of non-overlapping, closed, proper subintervals of [0,1], denoted by $\mathcal{I}$. With each $[a_m, b_m] \in \mathcal{I}$ associate a continuous Archimedean t-norm T_m. Let T be a function defined on $[0,1]^2$ by

$$T(x,y) := \begin{cases} a_m + (b_m - a_m)T_m\left(\frac{x-a_m}{b_m-a_m}, \frac{y-a_m}{b_m-a_m}\right) & \text{if } (x,y) \in [a_m, b_m]^2 \\ \min(x,y) & \text{otherwise} \end{cases} \tag{2.36}$$

In this case T is called the *ordinal sum* of $\{([a_m, b_m], T_m)\}$ and each T_m is called a *summand.*

Similarly, with each $[a_m, b_m] \in \mathcal{I}$ associate a continuous Archimedean t-conorm S_m. Let S be a function defined on $[0,1]^2$ by

$$S(x,y) = \begin{cases} a_m + (b_m - a_m)S_m\left(\frac{x-a_m}{b_m-a_m}, \frac{y-a_m}{b_m-a_m}\right) & \text{if } (x,y) \in [a_m, b_m]^2 \\ \max(x,y) & \text{otherwise} \end{cases} \tag{2.37}$$

In this case S is called the *ordinal sum* of $\{([a_m, b_m], S_m)\}$ and each S_m is called a *summand.*

Assume that T is a continuous t-norm. Then, one of the following three cases is valid for T:

1. $T = \min$;

2. T is Archimedean;

3. there exist a family $\{([a_m, b_m], T_m)\}$ such that T is the ordinal sum of this family,

see (Mostert and Shields, 1957; Ling, 1965). It has also been proved that a continuous t-conorm S is

1. either the maximum;

2. or Archimedean;

3. or there exist a family $\{([a_m, b_m], S_m)\}$ such that S is the ordinal sum of this family.

The following interesting result of (Jenei and Fodor, 1998) shows an important relationship between continuous, and continuous Archimedean t-norms.

Theorem 11 *Every continuous t-norm is a uniform limit of a sequence of continuous Archimedean t-norms.*

2.3.4 Parametered families of t-norms and t-conorms

In this section we list a number of parametered families of t-norms and t-conorms (with generators, if they exist). Very often, as the parameter tends to the limits of its range, we obtain t-norms (resp. t-conorms) 'min' and Z (resp. 'max' and Z') as extreme members of these families. This is due to a result proved by Dombi (Dombi, 1982b). We recall only the case of t-norms here.

Theorem 12 *Suppose T is a continuous Archimedean t-norm with additive genarator f. For any $\lambda \in]0,\infty[$ the function f^λ is also an additive generator of a t-norm. If T_λ is the t-norm generated by f^λ then for all $x, y \in [0,1]$ we have*

$$\lim_{\lambda\to\infty} T_\lambda(x,y) = \min(x,y),$$

and

$$\lim_{\lambda\to 0} T_\lambda(x,y) = Z(x,y). \blacksquare$$

Now we can turn to particular families of operations.

2.3.4.1 Frank family. Let $s > 0, s \neq 1$ be a real number. Define a parametric family of continuous Archimedean t-norms in the following way:

$$T^s(x,y) := \log_s \left(1 + \frac{(s^x - 1)(s^y - 1)}{s - 1}\right).$$

We can extend this definition for $s = 0$, $s = 1$ and $s = \infty$ by taking limits. Thus

$$\begin{aligned} T^0(x,y) &:= \lim_{s\to 0} T^s(x,y) = \min\{x,y\}, \\ T^1(x,y) &:= \lim_{s\to 1} T^s(x,y) = xy, \\ T^\infty(x,y) &:= \lim_{s\to\infty} T^s(x,y) = \max\{x+y-1,0\}. \end{aligned}$$

Each T^s is a strict t-norm for $s \in]0,\infty[$. The corresponding additive generators f_s ($s \in]0,\infty[$) are given by

$$f_s(x) = \begin{cases} -\log x & \text{if } s = 1 \\ -\log \frac{s^x - 1}{s-1} & \text{if } s \in]0,\infty[,\ s \neq 1 \end{cases}.$$

The family $\{T^s\}_{s\in[0,\infty]}$ is called the *Frank family* of t-norms, see (Frank, 1979). Very often, it is also called the family of *fundamental t-norms* because of the role it plays in several applications. Note that members of this family are decreasing functions of the parameter s (see e.g. (Klement et al., 1997)).

The De Morgan law enables us to define the Frank family of t-conorms $\{S^s\}_{s\in[0,\infty]}$ by

$$S^s(x,y) := 1 - T^s(1-x, 1-y)$$

for any $s \in [0,\infty]$.

In (Frank, 1979) one can find the following interesting characterization of these parametrized families.

Theorem 13 *A continuous t-norm T and a continuous t-conorm S satisfy the functional equation*

$$T(x,y) + S(x,y) = x + y \tag{2.38}$$

if and only if

(a) there is a number $s \in [0,\infty]$ such that $T = T^s$ and $S = S^s$,

or

(b) T is representable as an ordinal sum of t-norms, each of which is a member of the family $\{T_s\}$, $0 < s \leq \infty$, and S is obtained from T via equation (2.38). ■

We remark that these families will play a central role in the investigation of the contraposition law for QL-implications later in this chapter.

2.3.4.2 Hamacher family. Let us define three parametrized families of t-norms, t-conorms and strong negations, respectively, as follows.

$$\begin{aligned} T_\alpha(x,y) &:= \frac{xy}{\alpha + (1-\alpha)(x+y-xy)}, \quad \alpha \geq 0, \\ S_\beta(x,y) &:= \frac{x+y+(\beta-1)xy}{1+\beta xy}, \quad \beta \geq -1, \\ N_\gamma(x) &:= \frac{1-x}{1+\gamma x}, \quad \gamma > -1. \end{aligned}$$

Hamacher (Hamacher, 1978) proved the following characterization theorem.

Theorem 14 *(T, S, N) is a De Morgan triplet such that*

$$\begin{aligned} T(x,y) = T(x,z) \quad &\Longrightarrow \quad y = z, \\ S(x,y) = S(x,z) \quad &\Longrightarrow \quad y = z, \\ \forall z \leq x \quad \exists y, y' \quad &\text{such that} \quad T(x,y) = z, S(z,y') = x \end{aligned}$$

and T and S are rational functions if and only if there are numbers $\alpha \geq 0, \beta \geq -1$ and $\gamma > -1$ such that $\alpha = \frac{1+\beta}{1+\gamma}$ and $T = T_\alpha, S = S_\beta$ and $N = N_\gamma$. ■

The above family of t-norms and t-conorms is called the *Hamacher family* of t-norms and t-conorms, respectively. Note that each member of these families is a strict t-norm resp. a strict t-conorm. Additive generators f_α of T_α are given as follows:

$$f_\alpha(x) = \begin{cases} \frac{1-x}{x} & \text{if } \alpha = 0 \\ \log\left(\frac{\alpha+(1-\alpha)x}{x}\right) & \text{if } \alpha > 0 \end{cases}.$$

Members of the Hamacher family of t-norms are decreasing functions of the parameter α.

Another characterization of the Hamacher family of t-norms with positive parameter has been obtained by Fodor and Keresztfalvi (Fodor and Keresztfalvi, 1994) as the only solutions of a functional equation

$$T\left(x, 1 - \frac{T(x, 1-y)}{x}\right) = xy, \quad x, y \in [0,1],$$

where T is a t-norm such that the function φ defined by $\varphi(x) := \left[\frac{\partial T(x,y)}{\partial y}\right]_{y=0}$, $x \in [0,1]$, is a multiplicative generator of T (that is, $T(x,y) = \varphi^{-1}(\varphi(x)\varphi(y))$, $x, y \in [0,1]$).

2.3.4.3 Sugeno–Weber family. The following t-norms and t-conorms were suggested by Weber (Weber, 1983) for modelling intersection and union of fuzzy set. For $\lambda > -1$ and $x, y \in [0,1]$, define

$$\begin{aligned} T_\lambda(x,y) &:= \max\left\{\frac{x+y-1+\lambda xy}{1+\lambda}, 0\right\}, \\ S_\lambda(x,y) &:= \min\{x+y+\lambda xy, 1\}. \end{aligned}$$

Additive generators f_λ of T_λ can be defined as follows:

$$f_\lambda(x) := \begin{cases} 1-x & \text{if } \lambda = 0 \\ 1 - \frac{\log(1+\lambda x)}{\log(1+\lambda)} & \text{otherwise} \end{cases}.$$

Note that $\{T_\lambda\}_{\lambda>-1}$ are increasing functions of the parameter λ. Additive generators of S_λ are given by $g_\lambda(x) = 1 - f_\lambda(x)$, $x \in [0,1]$. The t-conorms S_λ were suggested as addition rules by Sugeno (Sugeno, 1974) for so-called λ-fuzzy measures.

2.3.4.4 Schweizer–Sklar family. There are several particular families of parametric t-norms and t-conorms introduced by Schweizer and Sklar, see (Schweizer and Sklar, 1961; Schweizer and Sklar, 1963). Define T_p for $p \in]-\infty, \infty[$, $x, y \in [0,1]$ as follows:

$$T_p(x,y) := (\max\{x^{-p} + y^{-p} - 1, 0\})^{-1/p}.$$

Note that T_p is strict for $p \in [0, \infty[$ and is nilpotent for $p \in]-\infty, 0[$. This family is increasing in the parameter p. Additive generators of T_p can be defined by

$$f_p(x) := \begin{cases} -\log x & \text{if } p = 0 \\ \frac{x^{-p}-1}{p} & \text{otherwise} \end{cases}.$$

2.3.4.5 Yager family. The following family of nilpotent t-norms and t-conorms introduced by Yager (Yager, 1980a) is one of the most frequently applied one. For $p \in]0, \infty[$ and $x, y \in [0, 1]$, define

$$T_p(x,y) := \max\left\{1 - ((1-x)^p + (1-y)^p)^{1/p}, 0\right\},$$

$$S_p(x,y) := \min\left\{(x^p + y^p)^{1/p}, 1\right\}.$$

The corresponding additive normed generators are given by $f_p(x) := (1-x)^p$ and $g_p(x) := x^p$ for $p \in]0, \infty[$, $x \in [0, 1]$. It is interesting to note that $\lim_{p \to +0} T_p = W$. The family $\{T_p\}$ is strictly increasing in p.

2.3.4.6 Dombi family. The following family of strict t-norms and t-conorms was introduced by Dombi (Dombi, 1982b). Define for $\lambda \in]0, \infty[$ and $x, y \in]0, 1[$

$$T_\lambda(x,y) := \frac{1}{1 + \left(\left(\frac{1-x}{x}\right)^\lambda + \left(\frac{1-y}{y}\right)^\lambda\right)^{1/\lambda}}$$

$$S_\lambda(x,y) := 1 - T_\lambda(1-x, 1-y).$$

The corresponding additive generators f_λ and g_λ are given by

$$f_\lambda(x) := \left(\frac{1-x}{x}\right)^\lambda, \quad g_\lambda(x) := \left(\frac{x}{1-x}\right)^\lambda.$$

The family $\{T_\lambda\}$ is increasing in λ. Notice also that $g_\lambda(x) f_\lambda(x) = 1$ for $x \in]0, 1[$, $\lambda \in]0, \infty[$.

2.3.4.7 Dubois–Prade family. All the above families were continuous and Archimedean (with possible exceptions for some limit values of the parameters). The following family of t-norms introduced by Dubois and Prade (Dubois and Prade, 1980b) is not Archimedean. Define T_γ for $\gamma \in [0, 1]$ and $x, y \in [0, 1]$ as follows:

$$T_\gamma(x,y) := \frac{xy}{\max(x, y, \gamma)}.$$

Note that $T_0 = \min$, $T_1 = \Pi$. For $\gamma \in]0, 1[$, T_γ is the ordinal sum of $\{([0, \gamma], \Pi)\}$; that is, the only summand is the product t-norm.

2.3.4.8 Perny-Fodor family. A new family of t-norms and t-conorms was found by Perny (Perny, 1992), and rediscovered by Fodor (Fodor, 1995) during the study of contraposition law of fuzzy implications. Let N be any strong negation and define T_N and S_N for $x, y \in [0,1]$ by

$$\begin{aligned} \min_N(x,y) &:= \begin{cases} \min(x,y) & \text{if } x > N(y) \\ 0 & \text{otherwise,} \end{cases}, \\ \max_N(x,y) &:= \begin{cases} 1 & \text{if } x \geq N(y) \\ \max(x,y) & \text{otherwise,} \end{cases}. \end{aligned}$$

Each T_N is left-continuous and each S_N is right-continuous. Note that the extension of T_N for more than two arguments is given by

$$T_N(x_1,\ldots,x_n) = \begin{cases} \min(x_1,\ldots,x_n) & \text{if } x_i > N(x_j) \text{ when } i \neq j \\ 0 & \text{otherwise,} \end{cases}$$

Particular parametric families of T_N and S_N can be obtained via parametrization of the strong negation N.

Finally, the interested reader can find much more details about t-norms and related concepts in the forthcoming book (Klement et al., 1998).

2.3.5 Complementation defined from intersection and union

In classical set theory, the complement A^c of a crisp subset A of X can be interpreted as follows:

(i) A^c is the largest among crisp subsets B of X such that $A \cap B = \emptyset$;

(ii) A^c is the smallest among crisp subsets B of X such that $A \cup B = X$.

If $A \in \mathcal{F}(X)$ then we can try to use the above way of thinking to define the complement of A. That is, let A^c be defined as follows, according to either (i) or (ii):

$$A^c(a) := \sup\{ z \in [0,1] \mid T(A(a),z) = 0 \}, \tag{2.39}$$

$$A^c(a) := \inf\{ z \in [0,1] \mid S(A(a),z) = 1 \} \tag{2.40}$$

where T is a t-norm and S is a t-conorm. Notice that it suffices to deal only with the first case (2.39), because any t-conorm S can be written as $S(x,y) = 1 - T(1-x, 1-y)$ with a suitable t-norm T.

Basically, two main cases should be considered.

1. T is a positive t-norm. Then $T(x,z) = 0$ if and only if $x = 0$ or $z = 0$. Thus, in this case (2.39) implies that

$$A^c(a) = \begin{cases} 1 & \text{if } A(a) = 0 \\ 0 & \text{if } A(a) > 0 \end{cases} \quad \text{for all } a \in X.$$

Using the intuitionistic negation n_i introduced in Section 2.2, one can recognize that

$$A^c(a) = n_i(A(a)) \quad \text{for all } a \in X.$$

2. T is a t-norm with zero divisors. In this case we can distinguish two basic classes.

(a) $T = Z$ (the drastic t-norm)

Then (2.39) gives the following complement:

$$A^c(a) = \begin{cases} 1 & \text{if } A(a) < 1 \\ 0 & \text{if } A(a) = 1 \end{cases} \qquad \text{for all } a \in X.$$

That is, the complement is generated by the dual intuitionistic negation n_{di}:

$$A^c(a) = n_{di}\,(A(a)) \quad \text{for all } a \in X.$$

(b) $T = W_\varphi$ (a φ-transform of the Łukasiewicz t-norm)

In this case $T(x, y) = \varphi^{-1}\,(\max\{\varphi(x) = \varphi(y) - 1, 0\})$, whence (2.39) implies

$$A^c(a) = \varphi^{-1}(1 - \varphi(A(a))) = N_\varphi(A(a)) \text{ for all } a \in X.$$

That is, the complement is generated by the strong negation N_φ (the φ-transform of the standard negation). Note that $T = \min_{0,\varphi}$ leads to the same complement.

As a conclusion, these forms of complementation obtained from intersection and union also support the statement of Ovchinnikov (Ovchinnikov, 1983), cited in Section 2.2, on "optimal" negations.

2.4 INCLUSION AND DIFFERENCE

Suppose X is a crisp set and consider two fuzzy sets A, B on X. In his classical paper, Zadeh (Zadeh, 1965) proposed the following definition of fuzzy sets:

$$A \subseteq B \quad :\Longleftrightarrow \quad A(a) \leq B(a) \text{ for all } a \in X. \tag{2.41}$$

This is a straigthforward extension of ordinary inclusion of crisp sets. Although it is used extensively both in theoretical and practical problems, it is pointed out that this inclusion is a crisp binary relation between fuzzy sets and there is a need for a comprehensive study for soft models of inclusion. Clearly, any crisp binary relation on $\mathcal{F}(X)$ is a discontinuous mapping from $\mathcal{F}(X) \times \mathcal{F}(X)$ to [0,1].

In classical set theory there is a close link between set inclusion and set difference expressed as follows:

$$A \subseteq B \quad \Longleftrightarrow \quad \text{there does not exist } a \in A \setminus B$$

for crisp subsets A, B of X. However, at this point we would like to emphasize that set inclusion is a binary *relation* defined on subsets of a given set while set difference is a binary *operation* on the same universe. Nevertheless, it suffices to study only one of the two notions (we choose fuzzy inclusion), and then to apply the above connection to the other (i.e., to difference of fuzzy sets).

Because of the correspondence between operations in fuzzy logic and fuzzy set theory, it is clear that, semantically, fuzzy inclusions are close to fuzzy implications. For this reason, first we study fuzzy implications. The precise connection between fuzzy inclusions and fuzzy implications will be shown later.

2.4.1 *Fuzzy implications*

There are several approaches to implications. We define a fuzzy implication as a connective. In a very broad sense, any function $I : [0,1]^2 \to [0,1]$ such that it is antitone/monotone and preserves the values of the crisp implication on $\{0,1\}$ can be considered as a fuzzy implication (see (Kitainik, 1993)).

Definition 7 A function $I : [0,1]^2 \to [0,1]$ is called a *fuzzy implication* if and only if it satisfies the following properties:

I1. $I(0,0) = I(0,1) = I(1,1) = 1$; $I(1,0) = 0$.

I2. If $x \leq z$ then $I(x,y) \geq I(z,y)$ for all $y \in [0,1]$.

I3. If $y \leq t$ then $I(x,y) \leq I(x,t)$ for all $x \in [0,1]$. ■

The reason behind I1 is obvious, while a fuzzy implication is required to be antitone/monotone (i.e., I2 and I3 should be satisfied) because it measures that the consequent is more true than the antecedent (see (Smets and Magrez, 1987)).

Clearly, a fuzzy implication I has the following properties (as a consequence of the definition):

I4. $I(0,x) = 1$ for all $x \in [0,1]$.

I5. $I(x,1) = 1$ for all $x \in [0,1]$.

One can easily construct the smallest and the greatest fuzzy implication, denoted by I_0 and I_1, respectively. In fact, we have for all $x, y \in [0,1]$ that

$$I_0(x,y) = \begin{cases} 1 & \text{if } x = 0 \text{ or } y = 1 \\ 0 & \text{otherwise} \end{cases},$$

$$I_1(x,y) = \begin{cases} 0 & \text{if } x = 1 \text{ and } y = 0 \\ 1 & \text{otherwise} \end{cases}.$$

From an axiomatic point of view, further properties may be required for a fuzzy implication that can be important also in some applications:

I6. $I(1,x) = x$. (Trillas and Valverde, 1981)

I7. $I(x, I(y,z)) = I(y, I(x,z))$. (Trillas and Valverde, 1981)

I8. $x \leq y$ if and only if $I(x,y) = 1$. (Gaines, 1976)

I9. $N(x) := I(x, 0)$ is a strong negation.

I10. $I(x, y) \geq y$. (Yager, 1980a)

I11. $I(x, x) = 1$. (Bandler and Kohout, 1980)

I12. $I(x, y) = I(N(y), N(x))$ with a strong negation N.

I13. I is a continuous function.

Property I6 yields that tautology cannot justify anything. Condition I7 is called the exchange principle, and is based on the following equivalence:

$$\text{"if } P_1 \text{ then (if } P_2 \text{ then } P_3\text{)"} \iff \text{"if } (P_1 \textbf{ AND } P_2) \text{ then } P_3\text{"}.$$

I8 expresses that implication defines an ordering, I9 reflects that $P \to Q = \neg P$ if $v(Q) = \text{FALSE}$. I10 is the numerical counterpart of $P \to (Q \to P)$. I11 is called identity principle and it yields that $P \to P$ is always true. I12, the contraposition law (or in other words, the contrapositive symmetry), expresses a relationship between modus ponens and modus tollens, see (Smets and Magrez, 1987). It turns out to be a key property with respect to fuzzy inclusions as we will show later in this section. In general, this is a strong condition, see (Fodor, 1995). I13 prevents implication from reacting in a chaotic way to a small change of the truth value of either the antecedent or the consequent. This is also a fairly restrictive condition.

We would like to illustrate some interdependencies among these axioms now. For more details see (Fodor and Roubens, 1994).

Theorem 15 *Let $I : [0,1]^2 \to [0,1]$ be any function which fulfils conditions I3, I7, I8. Then I satisfies also I1, I2, I4, I5, I6, I10 and I11.*

At this point we would like to emphasize that the contraposition law is not implied by conditions I3, I7 and I8.

2.4.2 Fuzzy implications defined by t-norms, t-conorms and negations

To be consistent, implications and conjunctions (or implications and disjunctions) cannot be studied independently. Thus, we introduce two particular classes of fuzzy implications based on t-norms, t-conorms and negations. These were identified in (Trillas and Valverde, 1981; Trillas and Valverde, 1985), (Prade, 1982), (Weber, 1983), (Dubois and Prade, 1984a; Dubois and Prade, 1984b). Suppose that T is a t-norm, S is a t-conorm and N is a strong negation.

Definition 8 An *S-implication* associated with a t-conorm S and a strong negation N is defined by

$$I_{S,N}(x, y) := S(N(x), y) \quad (x, y \in [0,1]). \tag{2.42}$$

An *R-implication* associated with a t-norm T is defined by

$$I_T(x,y) := \sup\{z|T(x,z) \leq y\} \quad (x,y \in [0,1]). \blacksquare \tag{2.43}$$

The idea behind the definition of $I_{S,N}$ is obvious from the crisp case: $P \to Q = \neg P \vee Q$. One can justify expression for I_T by the following classical set-theoretic identity :

$$A \cup B = \overline{A \setminus B} = \cup\{Z \mid A \cap Z \subseteq B\},$$

where $\setminus$ denotes set-difference operator. This identity has something to do with the residuation concept in algebra (see references cited above). Note that residuals of uninorms have been studied in (De Baets and Fodor, 1997a; De Baets and Fodor, 1997b).

It is easy to see that both $I_{S,N}$ and I_T satisfy properties I1–I3 for any t-norm T, t-conorm S and strong negation N, thus they are fuzzy implications. Note that if T is an Archimedean t-norm with additive generator f then

$$I_T(x,y) = f^{-1}(\max\{f(y) - f(x), 0\}) \qquad (x,y \in [0,1]).$$

For the sake of completeness we mention a third type of implications used in quantum logic and called *QL-implication* defined as follows:

$$I_{T,S,N}(x,y) := S(N(x), T(x,y)) \qquad (x,y \in [0,1]).$$

For the idea behind QL-implications, see (Dubois and Prade, 1991). In general, $I_{T,S,N}$ violates property I2, so it is not a fuzzy implication in the sense of Definition 7. Conditions under that I2 is satisfied can be found in (Fodor, 1995).

Now we characterize S-implications (see also (Trillas and Valverde, 1981; Trillas and Valverde, 1985; Fodor and Roubens, 1994)).

Theorem 16 *A fuzzy implication is an S-implication with an appropriate t-conorm S and a strong negation N if and only if I satisfies I6,I7 and I12.* ■

Now we characterize implications which can be defined as R-implications based on left-continuous t-norms (see also (Fodor, 1991a; Fodor, 1993; Fodor and Roubens, 1994; Miyakoshi and Shimbo, 1985).

Theorem 17 *A function $I : [0,1]^2 \to [0,1]$ is an R-implication based on a left-continuous t-norm if and only if I satisfies conditions I2, I7, I8 and the function $x \mapsto I(a,x)$ is right-continuous for any fixed $a \in [0,1]$.* ■

We would like to emphasize that left-continuous t-norms play a key role from several points of view. It is worth noting that a t-norm T is left-continuous if and only if the following condition is satisfied :

$$T(x,z) \leq y \Longleftrightarrow I_T(x,y) \geq z. \tag{2.44}$$

As it was pointed out in (Höhle, 1988), a rather general acceptable algebraic structure of the valuation set (instead of the unit interval) consists of an *integral cl-monoid* (see also (Birkhoff, 1967) for more details). Integral cl-monoids present common framework for complete Heyting algebras (see (Birkhoff, 1967)) and for left-continuous t-norms on the unit interval. In addition, they form a basic structure in which abstract ideal theory takes place.

For a long time, the nilpotent minimum seemed to be the only t-norm (up to an automorphism) which is left-continuous and not continuous on the unit interval. A rather general picture of such t-norms can be found in a recent PhD thesis (Jenei, 1997a), together with some other very important results on t-norms and related algebraic structures (Jenei, 1996; Jenei, 1997b; Jenei, 1998b; Jenei, 1998a; Jenei, 1998c).

In (Dubois and Prade, 1984c) it was shown that S- and R-implications can be merged into a single family if the class of triangular norms is enlarged to non-commutative and non-associative conjunctions (for results concerning the general case, see (Fodor, 1991a; Fodor, 1993)). In fact, let T be a t-norm, and let $\mathcal{S}_n$ and $\mathcal{R}$ be two transformations of T into an S-implication and an R-implication, respectively. That is, we define

$$x\mathcal{S}_n(T)y \;:=\; N(T(x,N(y))), \tag{2.45}$$

$$x\mathcal{R}(T)y \;:=\; \sup\{z\in[0,1] \mid T(x,z)\le y\}, \tag{2.46}$$

where N is a strong negation, $x, y \in [0,1]$.

Obviously, $\mathcal{S}_n \circ \mathcal{S}_n(T) = T$ for any t-norm T (here $\circ$ denotes composition) and $\mathcal{S}_n \circ \mathcal{R}(T)$ is a (non-commutative) conjunction T^*, which is not a t-norm in general. It can be proved (see (Dubois and Prade, 1984c; Fodor, 1991a)) that we have

$$\mathcal{S}_n \circ \mathcal{R} \circ \mathcal{S}_n \circ \mathcal{R}(T) = T \tag{2.47}$$

for a t-norm T if and only if T is left-continuous. In other words, the following diagram (Figure 2.4.2) commutes for a t-norm T if and only if T is left-continuous.

It is possible to extend classical inference properties to the fuzzy case. For example, consider I12, the contraposition law. Clearly, it has a one-to-one analog in ordinary set theory: $A \subseteq B$ if and only if $B^c \subseteq A^c$. However, the Modus Ponens expressed as

$$T(x, I(x,y)) \le y \quad \text{for all } x, y \in [0,1],$$

where T is a t-norm and I is a fuzzy implication, does not seem to have any set-theoretical interpretation. Instead, the connection between inclusion and intersection in ordinary set theory is given by

$$A \subseteq B \cap C \quad \Longleftrightarrow \quad A \subseteq B,\ A \subseteq C,$$

see (Giles, 1976) for motivations.

Figure 2.1 S- and R-implications in a common framework

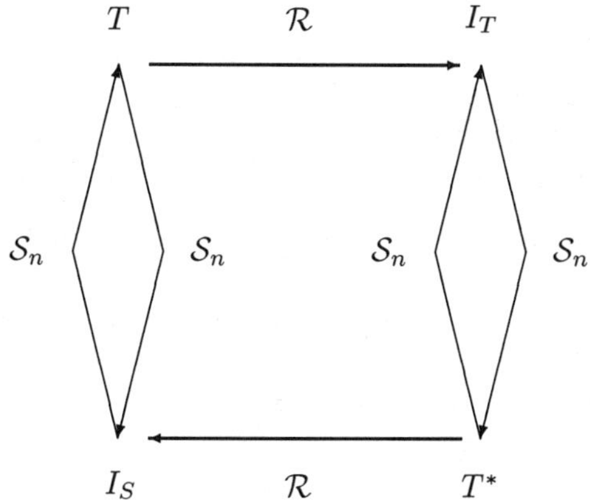

In the fuzzy case, this formula can be transformed into

$$I(x, T(y, z)) = T(I(x, y), I(x, z)), \quad (x, y, z \in [0, 1]), \tag{2.48}$$

which is the distributivity property of I. Triangular norms with distributive residuated implications are characterized in the following theorem, due to (Kitainik, 1993).

Theorem 18 *(a) The minimum is the only left-continuous t-norm possessing distributive R-implication.*

(b) The minimum is the only t-norm (irrespective of continuity) for which R-implication satisfies both distributivity and the Modus Ponens.

(c) For Archimedean t-norms, if a fuzzy implication satisfies distributivity and the contraposition law with respect to a strong negation, the implication cannot be continuous. ∎

Note that statements (a) and (b) give new characterizations of the minimum. Concerning the contraposition law, the following result was proved in (Fodor, 1995).

Theorem 19 *Let T be a left-continuous t-norm and N be a strong negation. Then the R-implication I_T satisfies the contraposition law with respect to N if and only if*

$$I_T(x, y) = N(T(x, N(y))) \quad \textit{for all } x, y \in [0, 1].$$

If T satisfies I12 with N, then

$$N(x) = I_T(x, 0) \quad \textit{for all } x \in [0, 1]$$

and

$$T(x, y) = 0 \quad \textit{if and only if} \quad x \leq N(y). \blacksquare$$

Therefore, the most general class of fuzzy implications (defined by t-norms, t-conorms and strong negations) satisfying the contraposition law consists of S-implications. Indeed, an R-implication can possess contraposition law with respect to a strong negation N if and only if it is at the same time an S-implication, where S is the N-dual t-conorm of T.

For continuous t-norms, we have a unicity result (up to an automorphism of the unit interval) as follows (see (Fodor and Roubens, 1994)).

Theorem 20 *Let T be a continuous t-norm and N be a strong negation. The R-implication I_T satisfies the contraposition law with respect to N if and only if there is an automorphism φ of the unit interval such that*

$$\begin{aligned} T(x,y) &= \varphi^{-1}\left(\max\{\varphi(x)+\varphi(y)-1,0\}\right), \\ N(x) &= \varphi^{-1}\left(1-\varphi(x)\right). \end{aligned}$$

In this case I_T is given by

$$I_T(x,y) = \varphi^{-1}\left(\min\{1-\varphi(x)+\varphi(y),1\}\right). \blacksquare \tag{2.49}$$

When I_T is supposed to be continuous, one can represent I_T in the form of (2.49), as it was proved in (Smets and Magrez, 1987). With a reduced number of conditions, the following result is true ((Fodor and Roubens, 1994)).

Theorem 21 *A function $I : [0,1]^2 \to [0,1]$ is such that I3, I7, I8 and I13 are satisfied if and only if there exists an automorphism φ of the unit interval such that*

$$I(x,y) = \varphi^{-1}(\min\{1-\varphi(x)+\varphi(y),1\}). \blacksquare \tag{2.50}$$

For sake of completeness, we recall a result on QL-implications that satisfy contraposition law with respect to a certain strong negation (for proof and more details see (Fodor, 1995)). Note first that if $I_{T,S,N}$ satisfies contraposition law with respect to a strong negation N_1 then we have $S(x, N_1(x)) = 1$ for all $x \in [0,1]$. If, in addition, S is continuous then

$$S(x,y) = \varphi^{-1}(\min\{\varphi(x)+\varphi(y),1\}) \tag{2.51}$$

and $N_1(x) \geq N(x) = \varphi^{-1}(1-\varphi(x))$. Supposing equality here (that is, $N_1 = N$), the following result is true.

Theorem 22 *Suppose S is a continuous t-conorm with representation (2.51), $N(x) = \varphi^{-1}(1-\varphi(x))$ and T is a t-norm which is either Archimedean or idempotent. Then the QL-implication $I_{T,S,N}$ satisfies the contraposition law with respect to N if and only if there exists an $s \in [0,+\infty]$ such that*

$$T(x,y) = \varphi^{-1}(T^s(\varphi(x),\varphi(y)),$$

where T^s is a member of the Frank family with parameter $s \in [0, +\infty]$.

In this case $I_{T,S,N}$ is an S-implication based on the t-conorm $\varphi^{-1}(S^{1/s}(\varphi(x), \varphi(y)))$, with $S^{1/s}(x, y) := 1 - T^{1/s}(1 - x, 1 - y)$, where we employ $1/0 := +\infty$ and $1/(+\infty) := 0$. ■

Therefore, under the above-mentioned conditions, a QL-implication can satisfy the contraposition law if and only if it is at the same time an S-implication with an appropriate t-conorm S. In case of $\varphi(x) = x$ for all $x \in [0, 1]$ in the above theorem, QL-implications having the contraposition law with respect to the standard negation are given as follows:

$$I^s(x, y) := \min\{1 - x + T^s(x, y), 1\}, \quad s \in [0, +\infty].$$

Hence, for example, we obtain

$$I^0(x, y) = \max\{1-x, y\}, \quad I^1(x, y) = 1-x+xy, \quad I^{+\infty}(x, y) = \min\{1-x+y, 1\}.$$

2.4.3 *Negations defined by implications*

As we have seen, several types of negations can be introduced in fuzzy set theory. The link between fuzzy implications and negations can be expressed by requiring that

$$n(x) := I(x, 0) \quad \text{for all } x \in [0, 1],$$

be a negation, where I is a fuzzy implication. This is motivated by the corresponding classical rule. Then one can prove the following results (see (Fodor and Roubens, 1994; Wangming, 1992)).

Suppose that $I : [0, 1]^2 \to [0, 1]$ is a function satisfying I3, I7, I8 and define $n(x) := I(x, 0)$ ($x \in [0, 1]$). Then (a) n is a negation; (b) $x \leq n(n(x))$ for all $x \in [0, 1]$; (c) $n(n(n(x))) = n(x)$ for all $x \in [0, 1]$. If, in addition, n is continuous then it is involutive. Thus, under the above conditions, n cannot be a noninvolutive strict negation: it is either discontinuous or a strong negation. If n is continuous then I fulfils I12 with the previous n.

For a positive t-norm T (like min or Π), the negation obtained via R-implication is not continuous at all. In fact, we have that $I(x, 0) = n_i(x)$.

Continuity of the implication is sufficient but not necessary to obtain strong negation via residuation. As an example, consider the nilpotent minimum, and

$$I_{\min_N}(x, y) := \begin{cases} 1 & \text{if } x \leq y \\ \max(N(x), y) & \text{otherwise} \end{cases},$$

where N is a strong negation. Then $I_{\min_N}$ is not continuous but $I_{\min_N}(x, 0) = N(x)$ is a strong negation.

Closing this subsection, in Table 2.4.3 we list the most common implications which are based on t-norms, t-conorms and the standard strong negation (for more details see e.g. (Dubois and Prade, 1991; Fodor, 1995; Goguen, 1969)). These implications are usually called after the person who discovered and published them (i.e., Kleene–Dienes, Rechenbach, Łukasiewicz, Gödel, Goguen, Zadeh, and Fodor, respectively).

Table 2.2 Fuzzy implications

Form	*Type*	*Properties*
$\max(1-x, y)$	S with $S = \max$ QL with $S = W'$, $T = W$	1–7, 9, 10, 12, 13
$1 - x + xy$	S with $S = \Pi'$	1–7, 9, 10, 12, 13
$\min\{1 - x + y, 1\}$	S with $S = W'$ R with $T = W$ QL with $S = W'$, $T = \min$	1-13
$\begin{cases} 1 & \text{if } x \le y \\ y & \text{otherwise} \end{cases}$	R with $T = \min$	1–8, 10, 11
$\min\{y/x, 1\}$	R with $T = \Pi$	1–8, 10, 11
$\max(1-x, \min(x, y))$	QL with $S = \max$, $T = \min$	2, 3, 5, 6, 9, 13
$\begin{cases} 1 & \text{if } x \le y \\ \max(1-x, y) & \text{if } x > y \end{cases}$	R with $T = \min_0$ S with $T = \max_1$ QL with $T = \min$, $S = \max_1$	1–12

2.4.4 *Axioms for fuzzy inclusions*

There is a nice axiomatic approach to fuzzy inclusions proposed by Kitainik (Kitainik, 1986; Kitainik, 1987b; Kitainik, 1987a; Kitainik, 1993) (see also (Sinha and Dougherty, 1993)). Generally speaking, a fuzzy inclusion is considered as a fuzzy binary relation on the set of all fuzzy subsets $\mathcal{F}(X)$ of X, satisfying four axioms. Three of them were studied in the previous parts of this section, while the fourth one reflects the fact that the degree of inclusion of fuzzy sets A and B does not depend on the permutation of points of X. In this subsection we denote the inclusion of Zadeh by $\subseteq$.

Definition 9 A *fuzzy inclusion* on a set X is defined as a fuzzy binary relation *inc* on $\mathcal{F}(X)$ satisfying the following axioms:

INC 1. (Contraposition) $inc(A, B) = inc(B^c, A^c)$ for all $A, B \in \mathcal{F}(X)$.

INC 2. (Distributivity) $inc(A, B \cap C) = inc(A, B) \wedge inc(A, C)$ for all $A, B, C \in \mathcal{F}(X)$.

INC 3. (Symmetry) $inc(A,B) = inc(\sigma A, \sigma B)$ for all $A, B \in \mathcal{F}(X)$, where the fuzzy set σA is defined by $\sigma A(x) := A(s(x))$ with a one-to-one mapping $s : X \to X$.

INC 4. (Heritage) Applying *inc* to crisp sets, it coincides with the usual set inclusion.

■

Clearly, the following crisp interrelations for $A, B, C \in X$ motivate the above definition:

$$\begin{aligned} A \subseteq B \quad &\Longleftrightarrow \quad B^c \subseteq A^c, \\ A \subseteq (B \cap C) \quad &\Longleftrightarrow \quad (A \subseteq B) \wedge (A \subseteq C), \\ A \subseteq B \quad &\Longleftrightarrow \quad \sigma A \subseteq \sigma B. \end{aligned}$$

The binary fuzzy relation *inc* can also be interpreted as an inclusion grade (see (Dubois and Prade, 1980a)). First we list some elementary properties of fuzzy inclusions as follows.

Theorem 23 *For any fuzzy inclusion inc on $\mathcal{F}(X)$ the following conditions are also satisfied:*

(i) $inc(A \cup B, C) = inc(A, C) \wedge inc(B, C)$,

(ii) inc is $\subseteq$-antitone with respect to its first argument and $\subseteq$-monotone with respect to its second argument.

(iii) $x \neq y$ *implies* $inc(\alpha\chi_{\{x\}}, (\beta\chi_{\{y\}})^c) = 1$ *for all* $\alpha, \beta \in [0,1]$, $x, y \in X$.

(iv) $inc(A,B) = \inf_{x \in X} inc\left(A(x)\chi_{\{x\}}, (B^c(x)\chi_{\{x\}})^c\right)$.

(v) The set of fuzzy inclusions inc is closed with respect to α-cuts.

■

Now we show the basic relationship between fuzzy implications and inclusions. The following theorem can easily be derived from Theorem 6.3 in (Kitainik, 1993) and the above Theorem 23.

Theorem 24 *A binary fuzzy relation R defined on $\mathcal{F}(X)$ is a fuzzy inclusion on $\mathcal{F}(X)$ if and only if there is a contrapositive fuzzy implication I on X such that*

$$R(A,B) = \inf_{x \in X} I(A(x), B(x)) \qquad (2.52)$$

for all fuzzy sets A, B on X. ■

On the basis of results on the contraposition law (see Theorems 19 and 22 and remarks there), the natural class of fuzzy implications underlying fuzzy set inclusion is the class of S-implications. That is, a typical fuzzy inclusion is defined as follows:

$$inc(A, B) := \inf_{x \in X} S\left(N(A(x)), B(x)\right),$$

where S is a t-conorm and N is a strong negation.

There are some natural properties of fuzzy inclusions that can be required on the basis of the corresponding properties of crisp inclusion, as we define them now.

Definition 10 A fuzzy inclusion *inc* is called

- *reflexive* if and only if $inc(A, A) = 1$ for all $A \in \mathcal{F}(X)$.
- *antisymmetric* if and only if $A, B \in \mathcal{F}(X)$, $A \neq B$ imply

$$\min\{inc(A, B), inc(B, A)\} = 0.$$

- *transitive* if and only if $\min\{inc(A, B), inc(B, C)\} \leq inc(A, C)$ for all $A, B, C \in \mathcal{F}(X)$. ■

In the following theorem we state some characterizations of the inclusion proposed by Zadeh ((Zadeh, 1965)), proved in (Kitainik, 1993).

Theorem 25 *The fuzzy inclusion $\subseteq$ of Zadeh is*

(a) the smallest reflexive fuzzy inclusion;

(b) the greatest antisymmetric fuzzy inclusion;

(c) the only reflexive, antisymmetric and transitive fuzzy inclusion.

■

In particular, in (Kitainik, 1993) it is proved that there does not exist a fuzzy inclusion possessing reflexivity, transitivity and continuity. That is, well-defined algebraic and topological properties of fuzzy inclusions are against each other. Therefore, there is no reason to criticize the lack of softness of Zadeh's inclusion: any transitive and reflexive fuzzy inclusion, satisfying the above axiomatics, is necessarily discontinuous.

2.4.5 Difference of fuzzy sets

As we have pointed out, there is a basic connection between inclusion and difference of crisp sets $A, B \in X$ expressed as follows:

$$A \subseteq B \quad \Longleftrightarrow \quad \text{there is no } x \in A \setminus B. \tag{2.53}$$

This link enables us to define the corresponding connection between inclusion and difference of fuzzy subets of X. If $A, B \in \mathcal{F}(X)$ and N is a strong negation then

$$inc(A, B) = N\left(\sup_{x \in X}(A \setminus B)(x)\right) \tag{2.54}$$

can be considered as an extension of (2.53) for the fuzzy case. Starting from this equality, we immediately obtain that

$$inc(A, B) = \inf_{x \in X} N((A \setminus B)(x)).$$

Comparing this equation with (2.52) and applying it for special fuzzy sets $\alpha(x) := \alpha\chi_{\{x\}}$, $\beta(x) = \beta\chi_{\{x\}}$, we easily get for all $x \in X$ that

$$(A \setminus B)(x) = N(I(A(x), B(x))) = N(I(B^c(x), A^c(x))). \tag{2.55}$$

Here we used that I in (2.52) should satisfy the contraposition law.

By equation (2.55), we can introduce a family of binary operations that can be considered as underlying operations for defining fuzzy set difference. Since these new operations are linked to implications like t-conorms to t-norms, we will call them *coimplications*. This term was introduced in (Fodor and Roubens, 1994) in a slightly different form.

Definition 11 A function $I_c : [0,1]^2 \to [0,1]$ is called *coimplication* if it satisfies the following conditions:

COIMP 1. $I_c(0,0) = I_c(1,0) = I_c(1,1) = 0$; $I_c(0,1) = 1$.

COIMP 2. If $x \leq z$ then $I_c(x,y) \geq I_c(z,y)$ for all $y \in [0,1]$.

COIMP 3. If $y \leq t$ then $I_c(x,y) \leq I_c(x,t)$ for all $x \in [0,1]$. ■

The simple connection between implications and coimplications is given by the following proposition.

Theorem 26 *A function $I_c : [0,1]^2 \to [0,1]$ is a coimplication if and only if the function*

$$I(x,y) = N(I_c(N(x), N(y))) \tag{2.56}$$

is an implication for any strong negation N. ■

Therefore, there is no need to deal with coimplications in details since results on implications can be applied by the duality expressed by (2.56).

For example, coimplications corresponding to R- and S-implications are defined respectively as

$$I_c(x,y) = \inf\{z \mid S(x,z) \geq y\} \tag{2.57}$$

and

$$I_c(x,y) = T(N(x), y). \tag{2.58}$$

Notice that these two classes have been investigated in (Dubois and Prade, 1984b). To see that both are connectives which support set-difference, consider the following classical identity (X is a given set, $A, B \subseteq X$):

$$B \setminus A = B \cap A^c = \cap\{Z \mid A \cup Z \supseteq B\}.$$

Particular coimplications can easily be obtained by implications listed in Table 2.4.3. Further properties of coimplications can be required on the basis of conditions I6–I13 for implications.

2.5 EQUIVALENCE

In the spirit of the previous sections, we give the definition of an *equivalence* as a binary operation on the unit interval (see (Fodor and Roubens, 1994)). We refer also to (Höhle and Stout, 1991; Höhle, 1992; Klawonn and Kruse, 1993).

Definition 12 A function $E : [0, 1]^2 \to [0, 1]$ is called *equivalence* if it satisfies the following conditions:

$$E(x, y) = E(y, x) \quad \forall x, y \in [0, 1], \tag{2.59}$$
$$E(0, 1) = E(1, 0) = 0 \tag{2.60}$$
$$E(x, x) = 1 \quad \forall x \in [0, 1] \tag{2.61}$$
$$x \le x' \le y' \le y \Rightarrow E(x, y) \le E(x', y'). \blacksquare \tag{2.62}$$

The following proposition shows a close link between equivalences and implications, as it is in the classical case.

Theorem 27 *The following statements are equivalent:*
(a) $E : [0, 1]^2 \to [0, 1]$ *is an equivalence*
(b) there exists a fuzzy implication I *having also property* I11 *such that*

$$E(x, y) = \min\{I(x, y), I(y, x)\} \quad (x, y \in [0, 1]). \tag{2.63}$$

(c) there exists a fuzzy implication I *having also property* I11 *such that*

$$E(x, y) = I(\max(x, y), \min(x, y)). \blacksquare \tag{2.64}$$

Notice the natural appearance of R-implications in equivalences. In fact, if one would like to use S-implications then I11 implies

$$S(N(x), x) = 1 \quad \text{for all } x \in [0, 1],$$

that is, the law of excluded middle must be satisfied by S and N. For a continuous t-conorm S this condition implies the existence of an automorphism

φ of the unit interval such that S is the φ-transform of the Lukasiewicz t-conorm. In this case, using $N(x) = \varphi^{-1}(1 - \varphi(x))$, both R- and S-implications coincide. That is, there is no chance to use an S-implication different from the corresponding R-implication in formula (2.63) or (2.64).

In further investigations of equivalences, there may be considered several other conditions to be satisfied (we refer to the listed properties of fuzzy implications).

2.6 UNINORMS

In (Yager and Rybalov, 1996a; Fodor et al., 1997; Yager, 1994; Yager and Rybalov, 1996b) Yager, Rybalov and Fodor introduced a unifying generalization of t-norms and t-corms called *uninorms.* A uninorm allows the neutral (identity) element to be any number in the unit interval. Triangular norms and conorms are special cases of this with neutral element 1 and 0, respectively. More formally, the following definition can be proposed.

Definition 13 A function $U : [0,1]^2 \to [0,1]$ is called a *uninorm* if it is increasing, associative and commutative and there exists $e \in [0,1]$ such that

$$U(x,e) = x \qquad (x \in [0,1]).\blacksquare$$

The element e corresponding to a uninorm U is clearly unique and is called the *neutral element* (or identity element) of U. In what follows we clarify the picture about the structure of uninorms (Fodor et al., 1997).

Proposition 2 *Consider a uninorm U with neutral element e.*

(i) If $e \in]0,1]$, then the binary operator T_U defined by

$$T_U(x,y) := \frac{U(ex,ey)}{e}$$

is a t-norm.

(ii) If $e \in [0,1[$, then the binary operator S_U defined by

$$S_U(x,y) := \frac{U(e+(1-e)x, e+(1-e)y) - e}{1-e}$$

is a t-conorm.

The structure of a uninorm with neutral element $e \in]0,1[$ on the squares $[0,e]^2$ and $[e,1]^2$ is therefore closely related to t-norms and t-conorms. For $e \in]0,1[$, we denote by ϕ_e and ψ_e the linear transformations defined by $\phi_e(x) = \frac{x}{e}$ and $\psi_e(x) = \frac{x-e}{1-e}$. According to Proposition 2, to any uninorm U with neutral element $e \in]0,1[$, there corresponds a t-norm T and a t-conorm S such that:

$$U(x,y) = \begin{cases} \phi_e^{-1}(T(\phi_e(x), \phi_e(y))) & \text{for } (x,y) \in [0,e]^2, \\ \psi_e^{-1}(S(\psi_e(x), \psi_e(y))) & \text{for } (x,y) \in [e,1]^2. \end{cases}$$

Concerning the other parts of the unit square, we have the following proposition.

Proposition 3 *Consider a uninorm U with neutral element e, then we have*

$$\min(x,y) \leq U(x,y) \leq \max(x,y) \qquad ((x,y) \in [0,e] \times [e,1] \cup [e,1] \times [0,e]).$$

The values of any uninorm at the points $(0,1)$ and $(1,0)$ of the unit square must be equal either to 0, or to 1. Thus, the following statement holds.

Proposition 4 *For any uninorm U, one of the following two cases always hold:*

(i) U is a conjunctive (or andlike) uninorm: $U(0,1) = U(1,0) = 0$;

(ii) U is a disjunctive (or orlike) uninorm: $U(0,1) = U(1,0) = 1$.

2.6.1 Important classes of uninorms

2.6.1.1 The classes $U_{\min}$ and $U_{\max}$. According to Proposition 3, a uninorm U with neutral element e acts on $[0,e] \times [e,1]$ and $[e,1] \times [0,e]$ as a *mean operator*. The following two propositions investigate the extreme cases where it acts as the minimum operator or as the maximum operator on $[0,e[\times]e,1] \cup]e,1] \times [0,e[$.

Theorem 28 *A binary operator U is a conjunctive uninorm with neutral element $e \in]0,1]$ such that $U(\cdot,1)$ is continuous on $[0,e[$ if and only if there exists a t-norm T and a t-conorm S such that*

$$U(x,y) = \begin{cases} \phi_e^{-1}(T(\phi_e(x), \phi_e(y))) & \textit{if } (x,y) \in [0,e]^2 \\ \psi_e^{-1}(S(\psi_e(x), \psi_e(y))) & \textit{if } (x,y) \in [e,1]^2 \\ \min(x,y) & \textit{otherwise} \end{cases} .$$

For obvious reasons, the class of uninorms characterized by Theorem 28 is denoted by $U_{\min}$. A similar class of disjunctive uninorms, denoted $U_{\max}$, can also be characterized similarly. The only difference is that we obtain max instead of min in the above expression.

For a given t-norm T and t-conorm S, Theorem 28 shows how to construct a conjunctive uninorm that has T and S as underlying t-norm and t-conorm. In case of a strict t-norm and strict t-conorm, yet another construction exists, as will be shown in the next subsection.

2.6.1.2 Representable uninorms. In analogy to the representation of continuous Archimedean t-norms and t-conorms in terms of additive generators, Fodor, Yager and Rybalov have investigated the existence of uninorms with a similar representation in terms of a single-variable function (Fodor et al., 1997). This search turns out to lead back to the class of *aggregative operators* introduced by Dombi in (Dombi, 1982a). This work is also closely related to that of Klement, Mesiar and Pap on associative compensatory operators (Klement et al., 1996).

Proposition 5 *Consider* $e \in]0,1[$ *and a strictly increasing continuous* $[0,1] \to \overline{\mathbb{R}}$ *mapping* h *with* $h(0) = -\infty$, $h(e) = 0$ *and* $h(1) = +\infty$. *The binary operator* U *defined by*

$$U(x,y) := h^{-1}(h(x) + h(y)) \qquad ((x,y) \in [0,1]^2 \setminus \{(0,1),(1,0)\}),$$

and either $U(0,1) = U(1,0) = 0$ *or* $U(0,1) = U(1,0) = 1$, *is a uninorm with neutral element* e.

The mapping h in the foregoing proposition is called an *additive generator* of the uninorm U. The following result characterizes those uninorms for which such an additive generator exists.

Theorem 29 *Consider a uninorm* U *with neutral element* $e \in]0,1[$. *There exists a strictly increasing continuous* $[0,1] \to \overline{\mathbb{R}}$ *mapping* h *with* $h(0) = -\infty$, $h(e) = 0$ *and* $h(1) = +\infty$ *such that*

$$U(x,y) = h^{-1}(h(x) + h(y)) \qquad ((x,y) \in [0,1]^2 \setminus \{(0,1),(1,0)\})$$

if and only if

(i) U *is strictly increasing and continuous on* $]0,1[^2$;

(ii) there exists a strong negation N *with fixpoint* e, *i.e.* $N(e) = e$, *such that*

$$U(x,y) = N(U(N(x),N(y))) \qquad ((x,y) \in [0,1]^2 \setminus \{(0,1),(1,0)\}).$$

Uninorms characterized by Theorem 29 are called *representable*. The strong negation corresponding to a representable uninorm U with additive generator h, as mentioned in condition (ii) of Theorem 29, is denoted by N_U, and is given by $N_U(x) = h^{-1}(-h(x))$. This condition indicates the self-duality of U w.r.t. N_U (excluding the two problematic corner points). Note that for any representable uninorm U it holds that $U(x,0) = 0$ for all $x \in [0,1[$, and $U(x,1) = 1$ for all $x \in]0,1]$.

From Proposition 5 it follows that any representable uninorm comes in a conjunctive and a disjunctive version, i.e. there always exist two representable uninorms that only differ at the points $(0,1)$ and $(1,0)$. The disjunctive version of a conjunctive representable uninorm U will be denoted U^*. Note that the partial mapping $U(\cdot,1)$ is not continuous at 0 and that the partial mapping $U^*(\cdot,0)$ is not continuous at 1. It is immediately clear that the classes $\mathcal{U}_{\min}$ and $\mathcal{U}_{\max}$ do not contain representable uninorms. Note that the additive generator of a representable uninorm is unique up to a positive multiplicative constant.

Representable uninorms are almost continuous, i.e. continuous except the points $(0,1)$ and $(1,0)$. Moreover, the strict increasingness implies that any representable uninorm U with neutral element e satisfies

$$\begin{aligned} U(x,x) &< x \quad \text{for all } x \in]0,e[, \\ U(x,x) &> x \quad \text{for all } x \in]e,1[, \end{aligned}$$

an obvious generalization of the Archimedean property of continuous t-norms and t-conorms. Moreover, the strict increasingness and continuity imply that the t-norm and t-conorm underlying a representable uninorm are both *strict*. Moreover, given a strict t-norm T with an additive generator f and a strict t-conorm S with an additive generator g, we can always construct a representable uninorm U with desired neutral element $e \in]0,1[$ that has T and S as underlying t-norm and t-conorm. It suffices to consider as additive generator the mapping h defined by

$$h(x) := \begin{cases} -f\left(\dfrac{x}{e}\right) & \text{if } x \leq e, \\ g\left(\dfrac{x-e}{1-e}\right) & \text{if } x \geq e. \end{cases}$$

On the other hand, the following proposition indicates that representable uninorms are in some sense also generalizations of nilpotent t-norms and nilpotent t-conorms.

Proposition 6 *Consider a representable uninorm U with neutral element e, then we have*

$$U(x, N_U(x)) = N_U(e) = e \qquad (x \in]0,1[).$$

As an example, consider the $[0,1] \to \overline{\mathbb{R}}$ mapping h defined by

$$h(x) = \log \frac{x}{1-x},$$

then the corresponding conjunctive representable uninorm U with neutral element $\frac{1}{2}$ is given by

$$U(x,y) = \begin{cases} \dfrac{xy}{(1-x)(1-y)+xy} & \text{if } (x,y) \in [0,1]^2 \setminus \{(0,1),(1,0)\}, \\ 0 & \text{otherwise} \end{cases}.$$

The interested reader can find some results on *idempotent uninorms* in (De Baets, 1998). Residual operators of uninorms (these are very close to fuzzy implications and coimplications) are also studied in (De Baets and Fodor, 1997a; De Baets and Fodor, 1998).

2.7 MEAN AGGREGATION OPERATORS

A close look at t-norms indicates two very important properties associated with these operators. The first is *anti-monotonicity in cardinality*:

$$T(x_1, x_2, \ldots, x_n) \geq T(x_1, x_2, \ldots, x_n, x_{n+1}) \qquad (x_1, \ldots, x_{n+1} \in [0,1]).$$

As can be observed, this property implies as we add arguments to the aggregation the overall score tends to decrease. In particular, we can't ever increase the aggregated value, we only can decrease it. The second property, closely related to the first, is *zero fixation*,

$$T(x_1, x_2, \ldots, x_n) = 0 \quad \text{if there is } x_i = 0 \ (i \in \overline{1,n}).$$

This property is closely related to the fact that $T \leq \min$. We can also see that the t-conorm exhibits analogous properties. A t-conorm operator is *monotonic in cardinality*: $S(x_1, x_2, \ldots, x_n) \leq S(x_1, x_2, \ldots, x_n, x_{n+1})$ $(x_1, \ldots x_{n+1} \in [0,1])$. Thus adding arguments generally results in an increase in the aggregated value. In particular, we can't ever decrease the aggregated value, we only can increase it. The t-conorm has a *one fixation* property, $S(x_1, x_2, \ldots, x_n) = 1$ if any of the $x_i = 1$ $(i \in \overline{1,n})$. For many applications, especially in information fusion, these characteristics are too restrictive. In these applications involving aggregation of fuzzy sets the type of aggregation required is neither the pure *anding* of the t-norm nor the pure *oring* of the t-conorm, with respective fixations and monotonicities under the addition of arguments, but the type of aggregation operator desired lies somewhere between these two extremes. In some sense, this is true for uninorms. Uninorms exhibit both anti-monotonicity and monotonicity in cardinality depending upon the value of the identity e:

$$\begin{aligned} U(x_1, x_2, \ldots, x_n) &\leq U(x_1, x_2, \ldots, x_n, x_{n+1}) && \text{if } x_{n+1} > e \\ U(x_1, x_2, \ldots, x_n) &\geq U(x_1, x_2, \ldots, x_n, x_{n+1}) && \text{if } x_{n+1} < e \\ U(x_1, x_2, \ldots, x_n) &= U(x_1, x_2, \ldots, x_n, x_{n+1}) && \text{if } x_{n+1} = e. \end{aligned}$$

Thus we see that we can both increase and decrease the aggregated value by the addition of arguments.

The associativity property in addition to allowing us to define these operators just in terms of the binary case allows for the simple calculation of multiple arguments.

In this section, we shall introduce another type of aggregation operators which are called the *mean aggregation operators.*

Definition 14 A mapping $G : [0,1]^n \to [0,1]$ is called a *mean aggregation operator* if it satisfies the following axiomatic conditions:

M1. Commutativity: The aggregation is indifferent to the ordering of the arguments.

M2. Monotonicity: $G(x_1, x_2, \ldots, x_n) \geq G(y_1, y_2, \ldots, y_n)$ if $x_i \geq y_i$ for all $i \in \overline{1,n}$.

M3. Idempotency: If $x_i := c$ for $i \in \overline{1,n}$, $c \in \mathbb{R}$, then $G(x_1, x_2, \ldots, x_n) = c$.

One implication of these conditions is that the mean aggregation always lies between the min and max of the arguments.

As can easily be established the arithemtic mean $\widehat{x} := \dfrac{1}{n}\sum_{j=1}^{n} x_j$ is a mean operator $(x_i \in \mathbb{R},\ i \in \overline{1,n})$. The max, min and median are also averaging operators of the above type.

The property of associativity is generally not available for mean operators. As discussed in (Dubois and Prade, 1985), the idempotency requirement generally conflicts with associativity (see (Fung and Fu, 1975; Fodor, 1996) for further results). The lack of associativity has some important implications. One is that the definition for the aggregation with argument of size $n \in \mathbb{N}$ is not implicit in the definition of the binary case. With t-norms and t-conorms we of course have the associativity property and use it to extend the domain of these operators from their original definition in terms of two arguments to the case of n arguments. In the case of mean operators the definition of G must be such that it covers all cases of $n \in \mathbb{N}$. With associativity not available for relating the definition of G at different dimensions some imperative must be applied to hold our definition together as we add elements to an aggregation. In (Yager, 1993) Yager suggested one criterion which he called the consistency condition for attaining this goal. An averaging operator is said to be *consistent* if the following condition is satisfied by G for all $n \in \mathbb{N}$ and all arguments:

$$\text{if } x_{n+1} \geq G(x_1, x_2, \ldots, x_n) \text{ then } G(x_1, x_2, \ldots, x_n, x_{n+1}) \geq G(x_1, x_2, \ldots, x_n);$$

$$\text{if } x_{n+1} \leq G(x_1, x_2, \ldots, x_n) \text{ then } G(x_1, x_2, \ldots, x_n, x_{n+1}) \leq G(x_1, x_2, \ldots, x_n).$$

We see that this condition effectively says that *if we add some element to our aggregation our new aggregated value should move in the direction of this element.*

A closely related property to these is the property which is called *self identity* in (Yager, 1993; Yager, 1995). An operator G is said to posses the self identity property if for all arguments we have:

$$\text{If } e := G(x_1, x_2, \ldots, x_n) \text{ then } G(x_1, x_2, \ldots, x_n, e) = G(x_1, x_2, \ldots, x_n).$$

The self identity condition says that adding an element to an aggregation whose value is the same as the aggregation leads to no change in the aggregate value. It can be shown that if we impose the self identity condition on an aggregation that is monotonic we attain the consistency condition described above.

A further property that one can associate with a mapping G is *naturalness.* By this we mean the condition that for all $x \in [0, 1]$

$$G(x) = x.$$

Thus, naturalness implies that one argument aggregations evaluate to the one argument.

We see that self identity and naturalness implies idempotency. From naturalness $G(x) = x$ in addition self identity implies $G(x, x) = x$. Continuing to apply self identity we get idempotency.

We now can define a special class of mean operators which we call *self consistent mean operators.* Before introducing these operators we discuss the concept of a *bag.* A bag associated with the set $[0, 1]$ is any collection of elements drawn from $[0, 1]$. A bag is different from a set in that it allows multiple copies of the same element. We shall use $\mathcal{B}^{[0,1]}$ to indicate the set of all bags associated with the interval $[0,1]$. Essentially we can view $\mathcal{B}^{[0,1]}$ to consist of all the possible arguments that can be applied to our aggregation. For a given bag $B \in \mathcal{B}^{[0,1]}$ we indicate the dimension of B, $\mathrm{Dim}(B)$, as the number of elements in B. We emphasize that $\mathcal{B}^{[0,1]}$ contains bags of all different dimensions, and clearly we have

$$\mathcal{B}^{[0,1]} = \bigcup_{n \geq 1} [0,1]^n.$$

Definition 15 A mapping $G : \mathcal{B}^{[0,1]} \rightarrow [0,1]$ is called a *self consistent mean* operator if G satisfies the following conditions:

1. **Naturalness:** $G(x) = x$.

2. **Commutativity:** The ordering of the arguments don't matter.

3. **Monotonicity:** For bags of the same dimension condition M2 applies.

4. **Self Identity.**

It can be easily shown that for any fixed dimension $n \in \mathbb{N}$, a self consistent mean operator satisfies M1–M3 and is thus a mean operator.

In (Blom, 1979; Lemke et al., 1983; Dyckhoff and Pedrycz, 1984; Kaymak and Van Nauta Lemke, 1993) a parameterized family of mean operators, called the *root-power mean operator*, is described (see also (Aczél, 1966)). The root-power mean operator is defined as

$$G_\alpha(x_1, \ldots, x_n) := \left(\frac{x_1^\alpha + x_2^\alpha + \ldots + x_n^\alpha}{n} \right)^{\frac{1}{\alpha}}, \tag{2.65}$$

where the parameter $\alpha \in]-\infty, \infty[$. It can be shown that for any α, G_α is a self consistent mean.

This family reduces to a number of interesting special cases of mean operators for particular values of the parameter α. For $\alpha \rightarrow \infty$ this operator approaches the max operator, $G_\infty(x_1, \ldots, x_n) = \max\{x_1, \ldots, x_n\}$. For $\alpha \rightarrow -\infty$ it tends to the min operator, $G_{-\infty}(x_1, \ldots, x_n) = \min\{x_1, \ldots, x_n\}$. For $\alpha = 1$ we obtain the *arithmetic mean*, $G_1(x_1, \ldots, x_n) = \dfrac{x_1 + x_2 + \ldots + x_n}{n}$. For $\alpha = -1$ we obtain the *harmonic mean*, $G_{-1}(x_1, \ldots, x_n) = \dfrac{n}{\frac{1}{x_1} + \frac{1}{x_2} + \ldots + \frac{1}{x_n}}$. For $\alpha \rightarrow 0$ the family G_α tends to the *geometric mean*, $G_0(x_1, \ldots, x_n) = (x_1 x_2 \ldots x_n)^{\frac{1}{n}}$.

For further results on general means we refer to (Fodor and Roubens, 1994; Fodor and Roubens, 1995).

2.8 ORDERED WEIGHTED AVERAGING OPERATORS

In (Yager, 1988) Yager introduced a family of mean-like operators called the *ordered weighted averaging (OWA) operators.* We shall see that the OWA aggregation operators allow us to easily adjust the degree of *anding* and *oring* implicit in an aggregation. In this section we introduce the basic concepts associated with the OWA operators (Yager, 1988).

Definition 16 An OWA operator of dimension n is a mapping

$$f : \mathbb{R}^n \to \mathbb{R},$$

that has an associated n-vector W

$$W := \begin{bmatrix} w_1 \\ w_2 \\ \vdots \\ w_n \end{bmatrix}$$

such that

1. $w_i \in [0, 1]$,
2. $\sum_{i=1}^{n} w_i = 1$,

where $f(x_1, \ldots, x_n) := \sum_{j=1}^{n} w_j y_j$ with y_j the jth largest of the x_i's. ■

Since we shall view the OWA weights as a column vector we shall find it convenient to refer to the weights with the low indices as weights at the top and those with the higher indices with weights at the bottom.

A fundamental aspect of this operation is the re-ordering step, in particular an aggregate x_i is *not* associated with a particular weight w_i but rather a weight is associated with a particular *ordered position* of aggregate. This ordering step introduces a nonlinearity into the aggregation process.

Example 1 Assume

$$W := \begin{bmatrix} .4 \\ .3 \\ .2 \\ .1 \end{bmatrix}.$$

Consider the aggregation $f(.7, 1, .3, .6)$. In this case carrying out the reordering process we get that $y_1 = 1$, $y_2 = .7$, $y_3 = .6$ and $y_4 = .3$. Then performing the aggregation $\sum_{j=1}^{4} w_j y_j$ we get

$$f(.7, 1, .3, .6) = (.4)(1) + (.3)(.7) + (.2)(.6) + (.1)(.3) = .76. \quad ■$$

It is noted that different OWA operators are distinguished by their weighting vector W. Three important special cases of OWA aggregations can be pointed out.

1. F^*: In this case the weighting is denoted as W^* and is defined such that $w_1 := 1$ and $w_j := 0$ for $j \in \overline{2, n}$.

2. F_*: In this case the weighting is denoted as W_* and is defined such that $w_n := 1$ and $w_j := 0$ for $j \in \overline{1, n-1}$.

3. F_A: In this case the weighting is denoted as W_A and is defined such that $w_j := \frac{1}{n}$ for all $j \in \overline{1, n}$.

It can easily be seen that $F^*(x_1, \ldots, x_n) = \max\{x_1, \ldots, x_n\}$, $F_*(x_1, \ldots, x_n) = \min\{x_1, \ldots, x_n\}$ and $F_A(x_1, \ldots, x_n) = \frac{1}{n}\sum_{i=1}^{n} x_i$.

Another important example of OWA operators is the median aggregator. We define W_{med} as follows:

$$\text{for } n \text{ odd} \left\{ \begin{array}{c} w_{\frac{n+1}{2}} := 1 \\ w_i := 0 \text{ for all other } i \end{array} \right\},$$

$$\text{for } n \text{ even} \left\{ \begin{array}{c} w_{\frac{n}{2}} := w_{\frac{n}{2}+1} := .5 \\ w_i := 0 \text{ for all other } i \end{array} \right\}.$$

The resulting F is called the median aggregator.

Another interesting class of OWA operators is defined by the following weighting vector

$$w_k := 1,$$
$$w_j := 0 \quad \text{for } j \neq k.$$

This form of aggregation selects the kth largest element in the argument as its aggregated value.

A number of important properties can be associated with the OWA operators. For any OWA operator F we obviously have

$$F_*(x_1, \ldots, x_n) \leq F(x_1, \ldots, x_n) \leq F^*(x_1, \ldots, x_n).$$

Thus the upper and lower star OWA operators are its boundaries. From the above it becomes clear that for any F

$$\min\{x_1, \ldots, x_n\} \leq F(x_1, \ldots, x_n) \leq \max\{x_1, \ldots, x_n\}.$$

The OWA operator is *commutative*. Let $< x_1, \ldots, x_n >$ be a bag of aggregates and let $< t_1, \ldots, t_n >$ be any permutation of the x_i's, then for any OWA operator F

$$F(x_1, \ldots, x_n) = F(t_1, \ldots, t_n).$$

A third characteristic associated with these operators is *monotonicity*. Assume x_i and u_i are a collection of aggregates, $i \in \overline{1, n}$ such that for each i, $x_i \geq u_i$ then for any OWA aggregation F

$$F(x_1, \ldots, x_n) \geq F(u_1, \ldots, u_n).$$

Another characteristic associated with these operators is *idempotency*. If $x_i := x$ for all $i \in \overline{1, n}$ then for any OWA operator

$$F(x_1, \ldots, x_n) = x.$$

From the above we can see the OWA operators have the basic properties associated with mean aggregation operators. In addition, the following property, called **stability for positive linear transformations with the same unit and independent zeroes (SPLU)**, was observed to hold for any OWA operator in (Fodor et al., 1995):

$$F(rx_1 + t_1, \ldots, rx_n + t_n) = rF(x_1, \ldots, x_n) + T(t_1, \ldots, t_n)$$

for all $(x_1, \ldots, x_n) \in \mathbb{R}^n$, all $r > 0$, all $(t_1, \ldots, t_n) \in \mathbb{R}^n$. Moreover, this condition is satisfied also for the reordered values $y_1 \geq y_2 \geq \ldots \geq y_n$ of $(x_1, \ldots, x_n)$. The following result, among others, was proved in (Fodor et al., 1995).

Theorem 30 *The class of ordered weighted averaging operators corresponds to the aggregators which satisfy the properties of commutativity, monotonicity, idempotency, and SPLU for ordered values.*

As we see from the above the OWA operator allows us, by appropriate choice of the weighting vector, to move continuously from the *and* (min) to *or* (max) type aggregation. In order to classify these OWA operators in regards to their location on this continuum, a measure of **orness**, associated with any vector W, can be introduced. The **orness** measure is defined as

$$\mathbf{orness}(W) := \frac{1}{n-1} \sum_{i=1}^{n} (n-i) \cdot w_i.$$

We note that for any W the $\mathbf{orness}(W)$ is always in the unit interval. Furthermore it is noted that the nearer W is to an *or* the closer its measure is to one while the nearer it is to an *and* the closer its measure is to zero. Supporting this it can be easily shown that

1. $\mathbf{orness}(W^*) = 1$.

2. $\mathbf{orness}(W_*) = 0$.

3. $\mathbf{orness}(W_A) = .5$.

Generally it should be appreciated that an OWA operator with much of the non-zero weights at near the top will be an *orlike* operator, $\mathbf{orness}(W) \geq .5$. At the other extreme when the weights are non-zero near the bottom the OWA operator will be *andlike*, $\mathbf{orness}(W) \leq .5$.

The following theorem formally expresses this characteristic.

Theorem 31 *Assume W and W' are two n-dimensional OWA vectors such that*

1. $w_i = w'_i$ *for* $i \neq j$ *or* $i \neq k$,

2. $w_j = w'_j + \Delta$,

3. $w_k = w'_k - \triangle$,

where $\triangle > 0$, $j > k$. *Then* $\mathbf{orness}(W) \geq \mathbf{orness}(W')$.

This theorem shows that as we move weight up the vector we increase the $\mathbf{orness}(W)$, while moving weight down causes us to decrease $\mathbf{orness}(W)$.

Except for the extreme cases of $\mathbf{orness}(W) = 1$ and 0, which are obtained by W^* and W_* respectively, numerous different vectors W can attain a given measure of **orness**. For example both

$$\begin{bmatrix} 0 \\ 0 \\ 1 \\ 0 \\ 0 \end{bmatrix}, \quad \text{and} \quad \begin{bmatrix} .2 \\ .2 \\ .2 \\ .2 \\ .2 \end{bmatrix}$$

attain a measure of **orness** of 0.5.

We should note that in the case of the root-power mean, G_α, introduced in the preceding section the value α plays a role similar to the orness measure just introduced. However this observation allows us to make an important distinction between the generalized means and the OWA operators. Given a dimension $n \in \mathbb{N}$, for each value of α there is only one available generalized mean aggregator while for the OWA operators for a fixed dimension $n \in \mathbb{N}$ and a given measure of **orness** there is a multiple number of weighting vectors that can be used.

To help further distinguish amongst the different OWA aggregators, a measure called the dispersion (or entropy) of an OWA vector W is defined as

$$\mathrm{Disp}(W) := -\sum_{i=1}^{n} w_i \ln w_i.$$

The measure of dispersion can be used to indicate the degree to which we use all the information contained in the argument, $(x_1, \ldots, x_n)$ when calculating the aggregated value $F(x_1, \ldots, x_n)$. In particular, the larger $\mathrm{Disp}(W)$, the more of the information provided by the argument we use. To illustrate this consider the case of

$$W_1 := \begin{bmatrix} 0 \\ 0 \\ 1 \\ 0 \\ 0 \end{bmatrix}, \quad \text{and} \quad W_2 := \begin{bmatrix} .2 \\ .2 \\ .2 \\ .2 \\ .2 \end{bmatrix}.$$

Both of these calculate the same degree of **orness**, .5. However we readily see that W_1 just uses the information contained in one argument value, the third largest, whereas W_2 uses the information in all the arguments equally. This observation is supported by our measure of dispersion where $\mathrm{Disp}(W_1) = 0$ and $\mathrm{Disp}(W_2) = \ln 5$.

If F is an OWA aggregation with weights w_i ($i \in \overline{1,n}$), the dual of F, denoted $\widehat{F}$, is an OWA aggregation of the same dimension with weights $\widehat{w_i} := w_{n-i+1}$. We can easily see that if F and $\widehat{F}$ are duals then

1. $\text{Disp}(\widehat{F}) = \text{Disp}(F)$,

2. $\textbf{orness}(\widehat{F}) = 1 - \textbf{orness}(F)$.

Thus if F is orlike its dual is andlike.

The great generality that the OWA operators offer us bring with them the issue of selecting the weights that are to be used in a particular application. A number of approaches are available to accomplish this task (Yager and Filev, 1994a). A very interesting approach to the generation of the OWA weights was suggested by O'Hagan ((O'Hagan, 1990)). This approach makes use of the measures of *orness* and *entropy* introduced earlier. O'Hagan suggested using the following procedure for calculating the weights associated with an OWA aggregation. The first step is to choose a desired level of *orness*, α, to be associated with the aggregation. The second step is then to select from among the weighting vectors that attain this level of *orness* the one that has the maximum dispersion (entropy). As suggested by O'Hagan we can formulate this procedure in terms of the following mathematical programming problem:

$$\textbf{Max:} \; -\sum_{i=1}^{n} w_i \ln w_i$$

Subject to:

$$\begin{aligned}\frac{1}{n-1}\sum_{i=1}^{n}(n-i)\cdot w_i &= \alpha,\\ \sum_{i=1}^{n} w_i &= 1,\\ w_i &\in [0,1].\end{aligned}$$

This family of OWA operators are called the **ME-OWA** operators where the **ME** stands for Maximum Entropy.

Another family of OWA weights were introduced in (Yager and Filev, 1994b). These OWA operators are called S-OWA operators. These operators are notable for their ease of generation from the degree of *orness* as well as the simplicity of their resulting aggregation process. More specifically, this class consists of two families of S-OWA operators, the first being *orlike* and the second being *andlike.*

The *orlike* S-OWA operators, denoted F_{SO}, are defined by a family of OWA weights such that with $\alpha \in [0,1]$ we have

$$\begin{aligned}w_1 &:= \frac{1}{n}(1-\alpha)+\alpha,\\ w_i &:= \frac{1}{n}(1-\alpha), \quad i=2,\ldots,n.\end{aligned}$$

Using these weights result in an aggregation of the form

$$F_{SO}(x_1,\ldots,x_n) = \alpha \max\{x_1,\ldots,x_n\} + (1-\alpha)\frac{1}{n}\sum_{i=1}^{n} x_i, \tag{2.66}$$

which is a weighted average of the max and the arithmetic mean of the aggregates. We note that when $\alpha := 0$ we get $\frac{1}{n}\sum_{i=1}^{n} a_i$ and when $\alpha := 1$ we get $\max(a_i)$.

The *orness* measure of this aggregation can be shown to be

$$\mathbf{orness}(F_{SO}) = \frac{\alpha+1}{2}.$$

We see that, since $\alpha \in [0,1]$, the orness value lies in $[.5, 1]$, thus F_{SO} is indeed an *orlike* measure. Furthermore, the "orness" increases as α increases, when $\alpha := 1$, we get $\mathbf{orness}(F_{SO}) = 1$ and when $\alpha := 0$, we get $\mathbf{orness}(F_{SO}) = .5$.

The *andlike* S-OWA operators, denoted F_{SA}, are defined by the weights

$$\begin{aligned} w_i &:= \frac{1}{n}(1-\beta) \quad i = 1,\ldots,n-1, \\ w_n &:= \frac{1}{n}(1-\beta) + \beta, \end{aligned}$$

where $\beta \in [0,1]$. Using these weights we get

$$F_{SA}(x_1,\ldots,x_n) = \beta \min\{x_1,\ldots,x_n\} + \frac{1}{n}(1-\beta)\sum_{i=1}^{n} x_i.$$

Using F_{SA} we are getting a weighted average of the min and the arithmetic mean. The associated measure of *orness* is

$$\mathbf{orness}(F_{SA}) = \frac{1-\beta}{2},$$

which lies between 0 and .5.

Given a value of *orness*, λ, it is easy to generate the weights associated with the appropriate aggregate:

1) if $\lambda \geq .5$ use an *orlike* S-OWA operator with $\alpha := 2\lambda - 1$,

2) if $\lambda < .5$ use an *andlike* S-OWA with $\beta := 1 - 2\lambda$.

Once having obtained the weights the calculation for a given aggregate triple is very simple, just requiring the summation of the aggregates plus the finding of either the max or min of the aggregate.

Another approach to obtaining the OWA weights is to learn the weights from observations on the argument and aggregated value. This approach is discussed in detail in (Filev and Yager, 1994; Filev and Yager, 1997).

2.9 QUANTIFIERS

In classical logic, *propositions* are considered and are required to be either *true* or *false*. Such propositions are expressed in some language, and are of the form

$$x \text{ is } P,$$

where x (the subject of the sentence) is an element of a given set X and P is a predicate characterizing a property. For instance, "Budapest is the capital of Hungary" is a proposition in which "Budapest" is the subject (a particular city), and "the capital of Hungary" is a predicate characterizing the property of being a city where the government of Hungary is carried on. Obviously, this proposition is true.

We use the general form "x is P", where $x \in X$ for a given nonempty set X. Then P can be considered as a function defined on X, being a proposition for each $x \in X$. Such a function is called a *predicate* and the particular proposition assigned to a specified $x \in X$ is denoted by $P(x)$.

We can quantify the applicability of a predicate (with respect to the domain X). In two-valued logic, two types of quantification are mainly used: existential and universal quantification.

Existential quantification of a predicate P on the domain X means "There exists an $x \in X$ such that x is P", or equivalently, "Some $x \in X$ are P". Such an *existential quantifier* is denoted by $\exists$.

Universal quantification of a predicate P on the domain X is given by the statement "For every $x \in X$, x is P", or equivalently, "All $x \in X$ are P". The symbol $\forall$ is used for a *universal quantifier.* Notice that a quantified predicate is a proposition (does not depend on a particular element $x \in X$).

Possible generalizations of existential and universal quantifiers are summarized by Ostasiewicz (Ostasiewicz, 1988). The key issue in the selection of those elements in X which satisfy property represented by P. More formally, define the subset X_P of X by

$$X_P := \{x \in X \mid P(x)\}.$$

Clearly, the universally quantified predicate "All $x \in X$ are P" is equivalent to

$$X_P \in K_\forall, \quad \text{where } K_\forall := \{X\},$$

and the existentially quantified predicate "Some $x \in X$ are P" can be expressed as

$$X_P \in K_\exists, \quad \text{where } K_\exists := 2^X \setminus \emptyset.$$

It can also be seen that other quantifiers such as "No $x \in X$ is P" or "Some $x \in X$ are not P" have equivalent forms as

$$X_P \in K_{\text{No}}, \quad \text{with } K_{\text{No}} := \{\emptyset\},$$

and

$$X_P \in K_{\text{Sn}}, \quad \text{where } K_{\text{Sn}} := 2^X \setminus X,$$

respectively.

Therefore, it is natural to define any quantifier Q by selecting some subsets of X to form the set K_Q, and using

$$\text{“}Q\ x \in X \text{ are } P\text{”} \quad :\Longleftrightarrow \quad X_P \in K_Q.$$

For more details on this approach and its extensions for the fuzzy case, see (Ostasiewicz, 1988).

As an example for the use of K_Q, consider collective decision-making, where K_Q (a set of some subsets of X) is called a majority space, and elements of K_Q are called coalitions (see e.g. (Vilenkin and Shreider, 1977)). Suppose $d_1, d_2, \ldots d_m$ are individual decisions, for example $d_i \in \{0,1\}^n$ for some $n \in \mathbb{N}$. Having a fixed majority space K_Q, the social decision d can be obtained by the majority rule as follows:

$$d = \bigvee_{C \in K_Q} \left(\bigwedge_{i \in C} d_i \right).$$

For instance, if $n = 1$ then $d = 1$ if and only if there exists some coalition C in K_Q such that all members of C agree with a decision (i.e., $d_i = 1$ for all $i \in C$).

Turning back to quantification, anyone can observe that there is a gap between classical (existential and universal) quantifiers and quantification used in natural discourse, such as “most”, “few”, “almost all”. As Hintikka (Hintikka, 1978) noted, “there are all sorts of subtle problems about natural–language quantifiers which have no counter part in logicians' so called quantification theory”. A solution to this problem was suggested by Zadeh (Zadeh, 1983), who introduced *linguistic quantifiers*, providing a more flexible and natural representation of human knowledge. The link between linguistic quantifiers and OWA operators is summarized next.

2.10 LINGUISTIC QUANTIFIERS AND OWA OPERATORS

As we have indicated the OWA operators provide a machinery for implementing aggregations which lie between the *and* and the *or*. Thus if $A_1, \ldots, A_n$ are a collection of fuzzy subsets of X and F_W is an OWA operator based on the weighting vector W then we can construct a fuzzy subset D of X such that

$$D(x) := F_W(A_1(x), A_2(x), \ldots, A_n(x)).$$

We shall denote this as $D = F_W(A_1, \ldots, A_n)$. For the special case of *and* we use W_* (min) and membership in D effectively requires membership in *all* of the A_i. For the special case of *or* we use W^* (max) and membership in D effectively requires membership in *at least one* of the A_i. Other choices of W essentially model membership requirements lying between these two extremes.

There exists in natural language a class of words (concepts) which convey aggregation policies that include these two extreme cases as well as those that lie between these them. Zadeh ((Zadeh, 1983)) called these concepts *linguistic quantifiers*. Examples of these words are *almost all, few, many, most, nearly half, at least twenty percent* as well as the the extreme cases of *all* and *their exists* (at least one). We note that these words which are conveying information about the proportions. Very ingeniously Zadeh suggested in (Zadeh, 1983) that linguistic quantifiers which say something about the proportion can be represented as fuzzy subsets of the unit interval. Thus if Q is a linguistic quantifier, such as *most*, then Q can be represented as a fuzzy subset Q of $[0, 1]$, where for each $r \in [0, 1]$, $Q(r)$ indicates the degree to which the proportion r satisfies the concept denoted by Q. Thus if Q is *most* then if $Q(.8) = 1$ we would be saying that 80% is completely compatible with the idea conveyed by the linguistic quantifier *most* while if $Q(.5) = .7$ we are indicating that the proportion 50% is only .7 compatible with the concept of *most.*

Having provided for a representation of linguistic quantifiers in terms of fuzzy sets let us characterize some special classes of quantifiers.

I. We shall say a linguistic quantifier Q is *regular non-decreasing* if

1. $Q(0) = 0$;

2. $Q(1) = 1$;

3. if $r_1 > r_2$ then $Q(r_1) \geq Q(r_2)$.

We note that these linguistic quantifiers are characterized by a situation in which the satisfaction to the concept of the quantifier increases as the proportion increases. These quantifiers exhibit a kind of monotonicity. Among this class of linguistic quantifiers are concepts like *at least* α, *most* and *all.*

II. We shall call a quantifier Q *regular non-increasing* if

1'. $Q(0) = 1$;

2'. $Q(1) = 0$;

3'. if $r_1 < r_2$ then $Q(r_1) \geq Q(r_2)$.

These linguistic quantifiers are characterized by a situation in which the satisfaction to the concept of the quantifier decreases as the proportion increases. These quantifiers exhibit a kind of anti-monotonicity. Among this class of linguistic quantifiers are concepts like *at most* α and *few.*

III. We shall call a quantifier Q *regular unimodal* if

1". $Q(0) = Q(1) = 0$;

2". $Q(r) = 1$ for $r \in [a, b]$;

3". if $r_2 \leq r_1 \leq a$ then $Q(r_1) \geq Q(r_2)$;

4". if $r_2 \geq r_1 \geq b$ then $Q(r_2) \leq Q(r_1)$.

Among this class of quantifiers are concepts such as *about* α.

Figure 2.10 shows these different types of quantifier.

Figure 2.2 Types of linguistic quantifiers

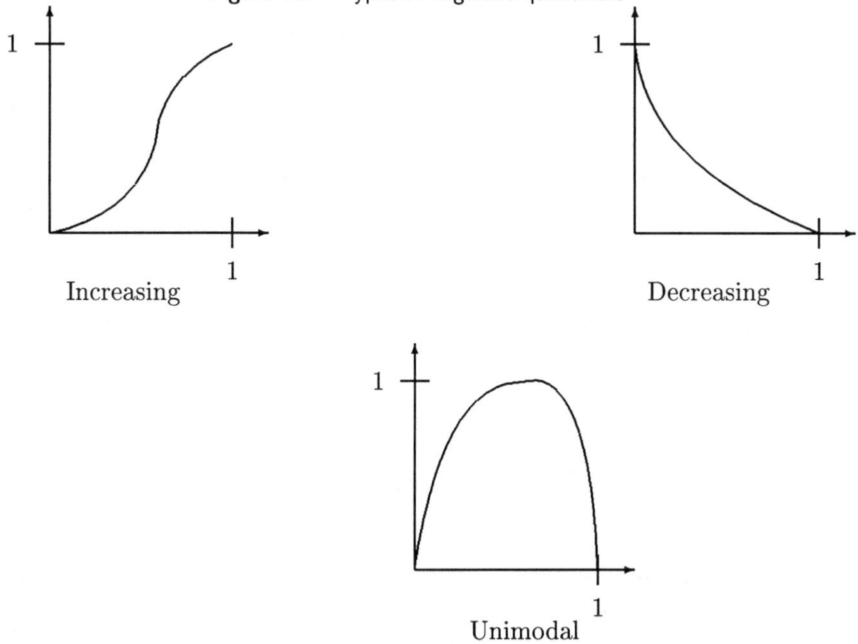

Linguistic quantifiers can be used in the aggregation of fuzzy subsets, this process is called *quantifier guided aggregation.* Assume $A_1, A_2, \ldots, A_n$ are a collection of fuzzy subsets over the set X. Let Q be a proportion quantifier of the type we just described. We shall use the notation

$$D = Q(A_1, A_2, \ldots, A_n)$$

to indicate the fuzzy subset D of X is formed in a manner that the membership grade $D(x)$ is obtained by requiring that x is a member of Q of the A_i. This type of aggregation is particularly useful in applications of fuzzy set theory to multi-criteria decision making, information retrieval and pattern recognition. In these applications we use the fuzzy subset A_i to represent a condition associated with an object being a solution to the problem. Rather then requiring *all* the conditions to be satisfied we only need Q of the conditions satisfied. As we shall see the OWA aggregation operators play a central role in implementing this type of aggregation.

First we shall investigate the two classic quantifiers *all* and *their exists.* It is well know that these quantifiers correspond to the connectives *and* (min) and *or* (max). The quantifier, *for all,* Q_* is defined such that

$$Q_*(r) := Q_\forall := \begin{cases} 0 & \text{for } r < 1, \\ 1 & \text{for } r = 1. \end{cases}$$

For the quantifier *there exists* we have

$$Q^*(r) := Q_\exists := \begin{cases} 0 & \text{for } r = 0, \\ 1 & \text{for } r > 0. \end{cases}$$

We note that these quantifiers are examples of what we called regular nondecreasing quantifiers. Furthermore we see that for any regular nondecreasing quantifier Q it is the case that

$$Q^*(r) \geq Q(r) \geq Q_*(r),$$

thus all regular nondecreasing quantifiers lie between the *and* and *or*. Since the OWA operators provide for aggregations between these two extremes it appears natural that these operators should be useful for implementing quantifier guided aggregations in the case of regular nondecreasing quantifiers. In the following we describe the procedure for accomplishing task which was introduced in (Yager, 1988).

To calculate $D = Q(A_1, A_2, \ldots, A_n)$ we proceed as follows.

1. Obtain the OWA weighting vector W associated with the fuzzy subset Q representing the quantifier Q,

$$w_i := Q\left(\frac{i}{n}\right) - Q\left(\frac{i-1}{n}\right), \quad i = 1, \ldots, n.$$

2. For each $x \in X$ calculate

$$D(x) := F_W(A_1(x), A_2(x), \ldots, A_n(x)),$$

that is $D(x)$ is the OWA aggregation of the memberships of the $A_i(x)$ using the weights found in step 1.

The following Figure 2.2 illustrates the process of obtaining the weights.

The regularity and the increasing nature of Q guarantees the satisfaction of the two conditions:

1. $\sum_{i=1}^{n} w_i = 1$;

2. $w_i \in [0, 1]$.

Let us look at the weights generated from some basic types of quantifiers. Consider first the quantifier *for all*, Q*. Using our method for generating weights

$$w_i^* := Q_*\left(\frac{i}{n}\right) - Q_*\left(\frac{i-1}{n}\right),$$

Figure 2.3 Obtaining weights from a quantifier

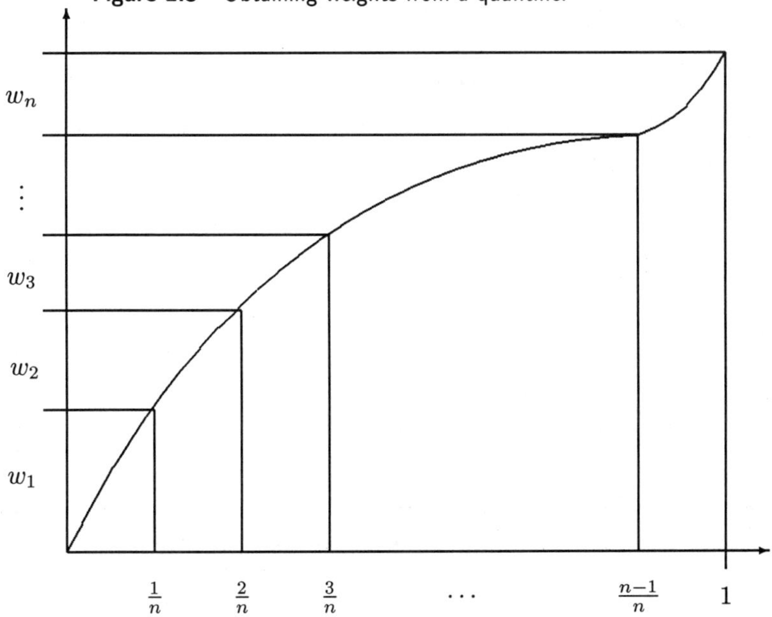

we get

$$w_i = 0 \quad i < n,$$

$$w_i = 1 \quad i = n.$$

This is exactly what we previously denoted as W_* and is the min operator. Consider next the quantifier *their exists*, Q^*. Using our weight generating method we get

$$w_1 = 1,$$

$$w_i = 0 \quad i \neq 1.$$

This is exactly what we denoted as W^* and is the max operator. Consider next the quantifier defined by $Q(r) := r$ $(r \in [0, 1])$, this is an identity or linear type quantifier.

In this case we get

$$w_i = Q\left(\frac{i}{n}\right) - Q\left(\frac{i-1}{n}\right) = \frac{i}{n} - \frac{i-1}{n} = \frac{1}{n}.$$

This gives us the pure averaging OWA aggregation operator (the arithmetic mean).

We now turn to the implementation of quantifier guided aggregation when the quantifier is regular nonincreasing. We recall that a regular *non-increasing* quantifier such as *few* or *not more than* α, is representable by a fuzzy subset of the unit interval having three distinguishing properties:

1. $Q(0) = 1$;
2. $Q(1) = 0$;
3. if $r_1 < r_2$ then $Q(r_1) \geq Q(r_2)$.

Formally we define a fuzzy subset $\widehat{Q}$ called the *antonym* of Q (see also section 2.2 in this chapter) via the following relationship between their membership grades: $\widehat{Q}(r) := Q(1-r)$. We first note that

$$\widehat{Q}(0) = Q(1-0) = Q(1) = 0.$$

In addition

$$\widehat{Q}(1) = Q(1-1) = Q(0) = 1.$$

Finally if $r_1 < r_2$ then

$$\widehat{Q}(r_1) = Q(1-r_1) \leq Q(1-r_2) \leq \widehat{Q}(r_2).$$

Thus when Q is a regular non-increasing quantifier its antonym $\widehat{Q}$ is a regular non-decreasing quantifier. Having introduced this antonym of Q we can now provide a procedure for evaluating

$$D := Q(A_1, A_2, \dots, A_n)$$

when Q is represented by a regular non-increasing quantifier Q.

1. Obtain the OWA weighting vector $\widehat{W}$ associated with the fuzzy subset $\widehat{Q}$, the antonym of Q

$$\widehat{w_i} := \widehat{Q}\left(\frac{i}{n}\right) - \widehat{Q}\left(\frac{i-1}{n}\right), \quad i \in \overline{1,n}.$$

2. For each $x \in X$ calculate

$$D(x) := F_{\widehat{W}}(1 - A_1(x), 1 - A_2(x), \dots, 1 - A_n(x)),$$

that is $D(x)$ is the OWA aggregation of the memberships of the negation of the $A_i(x)$.

We now turn to the case where the guiding quantifier is a unimodal one. We first note that any unimodal quantifier Q can be expressed in terms of a regular nondecreasing quantifier Q_1, and a regular nonincreasing quantifier Q_2. Assume $a, b \in [0,1]$ are the beginning and end of the interval where Q is equal to one, i.e.,

$$\begin{aligned} Q(x) &< 1 \quad (0 \leq x < a), \\ Q(x) &= 1 \quad (a \leq x \leq b), \\ Q(x) &< 1 \quad (b < x \leq 1). \end{aligned}$$

We now define Q_1 as $Q_1(x) := Q(x)$ for $0 \leq x < a$ and $Q_1(x) := 1$ for $a \leq x \leq 1$. It is easy to see that Q_1 is regular non-decreasing. We define Q_2 as $Q_2(x) := 1$ for $0 \leq x < b$ and $Q_2(x) := Q(x)$ for $b \leq x \leq 1$. We can easily show that Q_2 is regular non-increasing. We can now express

$$Q(x) = Q_1(x) \wedge Q_2(x) = \min[Q_1(x), Q_2(x)].$$

We now can use these two quantifiers to evaluate

$$D = Q(A_1, A_2, \ldots, A_n)$$

when Q is represented by a regular unimodal quantifier Q.

1. Calculate $D_1 := Q_1(A_1, A_2, \ldots, A_n)$;

2. Calculate $D_2 := Q_2(A_1, A_2, \ldots, A_n)$;

3. $D := D_1 \cap D_2$.

In the preceding we have shown how OWA operators can be used to implement *quantifier guided aggregations*. Implicit in this discussion is the development of a procedure of obtaining the OWA weighting vectors from linguistic quantifiers.

2.11 WEIGHTED UNIONS AND INTERSECTIONS

We now turn the issue of implementing the union and intersection operation when the associated fuzzy subsets have weights associated with them. In the following we shall assume we have a collection of fuzzy subsets $A_1, A_2, \ldots, A_n$ of X. Associated with each fuzzy subset A_i is a weight $\alpha_i \in [0,1]$ $(i \in \overline{1,n})$ corresponding to the importance of the fuzzy subset. The larger the α_i the more important the associated fuzzy subset in the aggregation. It is assumed that zero weight means no importance. Let $R : [0,1]^n \to [0,1]$ indicate an aggregation operation, either intersection or union, our problem is to find the fuzzy subset D, the weighted aggregation, denoted

$$D := R(f(A_1, \alpha_1), f(A_2, \alpha_2), \ldots, f(A_n, \alpha_n)).$$

The general procedure for accomplishing the task is to first transform the original fuzzy subsets, based upon their associated weights, into modified fuzzy subsets, which we shall denote as $B_i := f(A_i, \alpha_i)$, and then perform the appropriate aggregation on these modified fuzzy subsets. Letting $f : [0,1]^2 \to [0,1]$ be the *importance transformation function* we obtain new fuzzy subsets B_i, where

$$B_i(x) := f((A_i(x), a_i)) \quad (i \in \overline{1,n}),$$

and then calculate $D := R(B_1, B_2, \ldots, B_n)$. If R is the intersection operation and T is the *and* operation being used we get

$$D(x) := T(B_1(x), B_2(x), \ldots, B_n(x)).$$

If R is the union operation and S is the *or* operation being used we get

$$D(x) := S(B_1(x), B_2(x), \dots, B_n(x)).$$

Our focus in this section will be on the construction of the importance transformation function f. While the form of f will depend upon whether the operation R is a union or intersection the required characteristics of f is the same for both cases. In the following we describe these basic characteristics.

1. The relationship between membership grades should be maintained:
if $A_i(x) \geq A_i(y)$ then $B_i(x) \geq B_i(y)$.

2. Importance of one is the nominal case:
$A_i(x) = f(A_i(x), 1)$.

3. Zero importance should result a modified fuzzy subset that does not effect the aggregation.

4. $B_i(x)$ should change monotonically as α_i moves from zero to one.

It is in the process of satisfying the third requirement, and implicitly the fourth, that mandates different forms of f for the union and intersection. We recall that for the intersection (t-norm aggregations) the value one is the identity ($T(x, 1) = x$), thus a membership grade of one doesn't effect the intersection operation. Because of this we require that when R is an intersection we should have zero importance transforming any membership grade to one, thus in the case of intersection

$$f(A_i(x), 0) := 1.$$

For the case of union (t-conorm aggregations) the value zero is the identity ($S(x, 0) = x$), thus a membership grade of zero doesn't effect the union operation. Because of this we require that when R is a union we should have zero importance transforming any membership grade to zero, thus in the case of union

$$f(A_i(x), 0) := 0.$$

This distinction causes us to have different importance transformation operators for union and intersection. In the following we shall denote the intersection transformation as h and the union transformation as g. We now shall introduce some importance transformation operators which can be used to transform A_i into B_i.

In (Yager, 1987a) Yager suggested a general class of importance transformation operators that can be used for the intersection operation:

$$B_i(x) := S(A_i(x), \bar{\alpha}_i),$$

in the above S is any t-conorm and $\bar{\alpha}_i := 1 - \alpha_i$. Among this class of transformation operators are the following

1. $B_i(x) := \bar{\alpha}_i \vee A_i(x)$ (here $S := \max$);
2. $B_i(x) := \bar{\alpha}_i + \alpha_i A_i(x)$ (here $S(a,b) := a + b - ab$);
3. $B_i(x) := (1 + A_i(x) - \alpha_i) \wedge 1$ (here $S(a,b) := \min(a+b,1)$).

Observe that in each case $B_i(x) := I(\alpha_i, A_i(x))$ with a fuzzy implication I (for more details and general investigations see (Fodor and Roubens, 1994)). In (Yager, 1977) Yager introduced a importance transformation which is not from this class:

4. $B_i(x) := (Ai(x))^{\alpha_i}$.

For the union operation a general class of importance transformation operators is given by

$$B_i(x) := T(A_i(x), \alpha_i),$$

where T is any t-norm. Among this class of operators are the following

1': $B_i(x) := \alpha_i \wedge A_i(x)$ (here $T := \min$);

2': $B_i(x) := \alpha_i \cdot A_i(x)$ (here $T(a,b) := ab$);

3': $B_i(x) := (A_i(x) + \alpha_i - 1) \vee 0$ (here $T(a,b) := \max(a+b-1,0)$).

It is easy to see that in each case we have $B_i(x) := I_c(1 - \alpha_i, A_i(x))$ with an appropriate coimplication I_c (see again (Fodor and Roubens, 1994)). An additional operator not from this class is

4': $B_i(x) := 1 - (\bar{A}_i(x))^{\alpha_i}$.

2.12 PRIORITIZED FUZZY OPERATIONS

In defining the basic set operations, union and intersection, we required that the operations were monotonic in the sense that if $A_1 \subset A_2$ and $B_1 \subset B_2$ then

$$A_1 \cap B_1 \subset A_2 \cap B_2 \quad \text{and} \quad A_1 \cup B_1 \subset A_2 \cup B_2.$$

In trying to develop aggregation operations that enable us to do default reasoning or commonsense reasoning, we are forced to give up this condition. A whole group of logics called non-monotonic logics have been suggested for capturing commonsense reasoning. In (Yager, 1987b; Yager, 1991b), Yager has introduced such a nonmonotonic logic for the theory of approximate reasoning. As we shall see this will lead us to consider a special class of prioritized aggregation operators.

The essential issue which one must deal with in this endeavor can be expressed in the following. Assume we have two fuzzy subsets A and B. The conjunction of these propositions induces a fuzzy subset $D := A \cap B$. Of particular significance is the observation that if A and B are conflicting then $D = \emptyset$. In trying to capture commonsense knowledge, a different result is required. In the commonsense reasoning environment, we desire that the conjunction of these two propositions be such that if their intersection is empty, they are conflicting, we revert to the prioritized fuzzy set, which we shall assume is A, otherwise we take their intersection. In order to emulate this type of reasoning, Yager (Yager, 1991b) introduced a non monotonic intersection.

Definition 17 Assume A and B are two fuzzy subsets of X. We define the basic *prioritized intersection* operator η where

$$D := \eta(A, B),$$

and where D is also a fuzzy subset of X such that

$$A(x) \wedge \{B(x) \vee (1 - \text{Poss}[B|A])\} := D,$$

where $\text{Poss}[B|A] = \max_x[A(x) \wedge B(x)]$.

We note that the max and min used in the definition can be replaced by any t-conorm and t-norm.

A first observation we make is that this operator is *not* pointwise. We see that $D(x)$ is not just a function of $A(x)$ and $B(x)$ but, through the $\text{Poss}(B|A)$ term, depends upon A and B for all values in X.

The second observation we make is that this operation is **em** commutative, the ordering of the arguments effect the value it assumes. To see this easily consider the case where $A \cap B = \emptyset$. In this case we get different answers if we consider A or B first. In particular,

$$\eta(A, B) = A, \quad \eta(B, A) = B.$$

Thirdly we note that this operator is *nonmonotonic.* Consider the following three sets

$$\begin{aligned} A &:= \{x_1, x_2, x_3, x_4\}, \\ B_1 &:= \{x_2, x_5\}, \\ B_2 &:= \{x_5\}. \end{aligned}$$

In this case

$$D_1 := \eta(A, B_1) = \{x_2\}, \quad D_2 := \eta(A, B_2) = \{x_1, x_2, x_3, x_4\},$$

thus even though $B_2 \subset B_1$, it is the case that $D_1 \subset D_2$.

It is important to emphasize that in the special case when $\text{Poss}[B|A] = 1$ this operation reduces to the ordinary intersection, $\eta(A, B) = A \cap B$. Thus this basic nonmonotonic operation can be used in situations in we want the ordinary intersection when the arguments are not conflicting.

The kind of noncommutativity displayed here introduces the concept of priority amongst the arguments in the aggregation.

Since this operator is not associative we must describe the process for combining n fuzzy subsets under this operator ($n \in \mathbb{N}$). Assume $\{A_i\}_{i\in\overline{1,n}}$ is a collection of fuzzy subsets where it is assumed that A_i has priority over A_j if $i < j$ ($i, j \in \overline{1,n}$). Our process for aggregating these n fuzzy sets will consist of aggregating them in order of their priorities. We shall let D_i indicate the aggregated up to and including A_i. The value of D_i is calculated using our nonmonotonic intersection situation as follows

$$D_i := \eta(D_{i-1}, A_i)$$

with the initializing value $D_0 := X$.

A more general version of this prioritized operator can be obtained if we use t-norm and t-conorm operators instead of min and max.

Definition 18 Assume A and B are two fuzzy subsets of X. The *generalized prioritized intersection operation* η_{TS} is denoted as $\eta_{TS}(A, B) := E$ where E is also a fuzzy subset of X with membership grades

$$E(x) := T(A(x), S((1 - \text{Poss}[B|A], B(x))),$$

where T is a t-norm and S is a t-conorm. ■

By selecting different manifestations of T and S we can obtain different formulations for the prioritized intersection. Let us look at some manifestations of this operator. We shall consider three t-norms and three t-conorms:

$$\begin{aligned} T_1(a,b) &:= 0 \vee (a+b-1), \\ T_2(a,b) &:= ab, \\ T_3(a,b) &:= \min(a,b), \end{aligned}$$

and

$$\begin{aligned} S_1(a,b) &:= \max(a,b), \\ S_2(a,b) &:= a+b-ab, \\ S_3(a,b) &:= 1 \wedge (a+b). \end{aligned}$$

Using these operators we can form nine different intersection operators. Let us denote these as η_{ij} when $T := T_i$ and $S := S_j$ (in the following we use P to denote $\text{Poss}[B|A]$ and $\bar{P}$ to denote $1 - \text{Poss}[B/A]$).

$$\begin{aligned} E_{11}(x) &= 0 \vee (A(x) + (\bar{P} \vee B(x)) - 1), \\ E_{12}(x) &= 0 \vee (A(x) + B(x) \cdot P - P) = 0 \vee (A(x) - P\bar{B}(x)), \\ E_{13}(x) &= 0 \vee (A(x) + (1 \wedge (B(x) + \bar{P})) - 1), \\ E_{21}(x) &= A(x)(\bar{P} \vee B(x)), \\ E_{22}(x) &= A(x)(\bar{P} + B(x)P), \\ E_{23}(x) &= A(x)(1 \wedge (\bar{P} + B(x))), \\ E_{31}(x) &= A(x) \wedge (B(x) \vee \bar{P}), \\ E_{32}(x) &= A(x) \wedge (\bar{P} + B(x)P), \\ E_{33}(x) &= A(x) \wedge (\bar{P} + B(x)). \end{aligned}$$

In (Yager, 1997b) Yager used fuzzy modeling techniques (Yager and Filev, 1994a) to develop a class of prioritized intersections of the form

$$E(x) := \text{Poss}[B|A]T(A(x), B(x)) + (1 - \text{Poss}[B|A])A(x),$$

where T is any t-norm operator.

In (Yager, 1991b), Yager also introduced a nonmonotonic union.

Definition 19 Assume A and B are two fuzzy subsets of X. We define the basic *prioritized union* Ψ such that $\Psi(A, B) := E$, where E is also a fuzzy subset of X such that

$$E(x) := A(x) \vee (B(x) \wedge \text{Poss}[\bar{A}|\bar{B}]).$$

We can easily see that this operator is non-commutative, non-pointwise and non-monotonic. It reduces to the ordinary union if $\text{Poss}[\bar{A}|\bar{B}] = 1$. Furthermore we can show that Ψ and η are related by De Morgan's law.

2.13 OTHER AGGREGATION OPERATORS ON FUZZY SETS

In this section we shall briefly describe some other aggregation operators used on fuzzy sets.

2.13.1 Symmetric sums

In (Silvert, 1979) Silvert introduced a class of operators called *symmetric sums.* Assume A and B are fuzzy subsets over the space X. The fuzzy subset F on X, denoted $F := A * B$, is called the *symmetric sum of A and B* if $F(a) := A(a) * B(a)$, where $a \in X$, and $*$ is a binary operator having the following properties ($x, y \in [0, 1]$):

1. $0 * 0 = 0$ and $1 * 1 = 1$;
2. commutativity: $x * y = y * x$;
3. $*$ is continuous and monotonic;
4. $1 - (x * y) = (1 - x) * (1 - y)$.

Silvert has shown that any symmetric sum can be expressed in the form

$$x * y = \left(1 + \frac{g(1 - x, 1 - y)}{g(x, y)}\right)^{-1}, \quad x, y \in [0, 1],$$

where g is any increasing real mapping which is positive, continuous and has $g(0, 0) = 0$.

It can easily be shown that if $g(x, y) := x + y$ then $x * y = \dfrac{x + y}{2}$. More generally g can be any t-norm or t-conorm. In the case where g is the min we get $x * y := \dfrac{\min(x, y)}{1 - |x - y|}$, and when g is the max we get $x * y := \dfrac{\max(x, y)}{1 + |x - y|}$.

The following properties can be shown to hold for this operator

1. $x * (1 - x) = 0.5$ for $0 < x < 1$;
2. $0 * 1 = 0.5$ except if $g(0, x) = 0$ in which case $0 * 1$ is undefined;
3. The only associative symmetric sum having idempotency is $\text{med}_{1/2}$, where med_α is defined by

$$\text{med}_\alpha(x, y) := \begin{cases} y & \text{if } x \leq y \leq \alpha, \\ \alpha & \text{if } x \leq \alpha \leq y, \text{ or } y \leq \alpha \leq x, \\ x & \text{if } \alpha \leq x \leq y. \end{cases}$$

For further results related to the median see (Fodor, 1996).

Dombi (Dombi, 1982a) provided a generalization of these operators in which he replaced the standard negation with any strong negation. He required $N(x * y) = N(x) * N(y)$ where N is any strong negation ($x, y \in [0, 1]$). Further details about symmetric sums can be found in (Silvert, 1979; Dubois and Prade, 1985).

2.13.2 *Weak t-norms*

In (Yager, 1997a) Yager introduced class of operators which provide aggregations lying between the classical t-norms and t-conorms. He called these operators *weak t-norms*, which might cause some misunderstanding since under the same name (weak t-norms) Fodor defined a different class (Fodor, 1991b).

Definition 20 (Fodor) A function $w : [0, 1]^2 \to [0, 1]$ is called a *weak t-norm* if it is monotone and satisfies the following conditions: $w(x, 1) \leq x$, $w(1, y) = y$ for all $x, y \in [0, 1]$. ■

Fodor applied this notion to the definition of fuzzy strict preferences. In the rest of this chapter we will exclusively use the following notion of a weak t-norm.

Definition 21 (Yager) An aggregation operator $f : [0, 1]^n \to [0, 1]$ is called a *weak t-norm* operator if it is commutative, monotone, auasi-associative, satisfies zero fixation, and $f(1, 1, \ldots, 1) = 1$. ■

Definition 22 (Yager) An aggregation operator $g : [0, 1]^n \to [0, 1]$ is called a *weak t-conorm* operator if it is commutative, monotone, quasi-associative and satisfies one fixation, and $g(0, 0, 0, ..., 0) = 0$. ■

In the above, quasi-associativity is understood as follows. Let $\perp_1, \ldots, \perp_q$ be a collection of associative operators. An aggregation operator $\text{Agg}(x_1, \ldots, x_n)$ is called *quasi-associative* if $\text{Agg}(x_1, \ldots, x_n) := H(A_1, \ldots, A_p)$ where $A_i = \perp_i(x_1, \ldots, x_n)$ and H is a mapping from $[0, 1]^p$ into the unit interval.

We can easily show that duality exists between weak t-norms and weak t-conorms. In (Yager, 1997a) it is shown that the weak t-norm operators are characterized by the follow aggregation principle:

R1: If *any* of the arguments have a *low* value use a *t-norm aggregation.*

R2: If *all* the arguments have *high* values use a *t-conorm aggregation.*

The weak t-conorms are characterized by the following principle:

R3: If *all* of the arguments have a *low* value then use a *t-norm aggregation.*

R4: If *any* of the arguments have a *high* value then use a *t-conorm aggregation.*

2.13.3 Compensatory operators

Closing this section, we recall operations called 'compensatory AND', proposed by Zimmermann and Zysno (Zimmermann and Zysno, 1980) on the basis on some experimental results. For a theoretical study of these operators we refer to (Dombi and Zysno, 1982). Consider fuzzy sets A, B on X and let $\gamma \in [0,1]$. Two particular forms (exponential and convex linear) of compensatory AND are defined as follows (see (Turksen, 1992)):

$$\begin{aligned} \mu_{A\theta B}(x) &:= [T(\mu_A(x), \mu_B(x))]^{1-\gamma}[S(\mu_A(x), \mu_B(x))]^{\gamma}, \\ \mu_{A\vartheta B}(x) &:= (1-\gamma)T(\mu_A(x), \mu_B(x)) + \gamma S(\mu_A(x), \mu_B(x)), \end{aligned}$$

where T is a t-norm and S is a t-conorm. We can show that θ is a weak t-norm in the sense of Yager.

On the other hand, Turksen (Turksen, 1986) introduced so-called interval valued fuzzy sets. He observed that conjunctive and disjunctive normal forms (CNF(.) and DNF(.), respectively) are different in fuzzy logic: for a certain class of t-norms and t-conorms we have DNF(.) $\subseteq$ CNF(.). For the above forms of compensatory AND, Turksen (Turksen, 1992) investigated whether it is true that

$$A\theta B \in [\mathrm{DNF}(A \text{ AND } B), \mathrm{CNF}(A \text{ AND } B)],$$

and

$$A\theta B \in [\mathrm{DNF}(A \text{ OR } B), \mathrm{CNF}(A \text{ OR } B)]$$

for certain values of the compensation parameter $\gamma \in [0,1]$, and the same problem for $A\vartheta B$ $(A, B \in \mathcal{F}(X))$.

In (Yager, 1997a) Yager introduced a general class of weak t-norm operators as

$$F(x_1, x_2, \ldots, x_n) := (1-\alpha)T(x_1, x_2, \ldots, x_n) + \alpha S(x_1, x_2, \ldots, x_n),$$

where $\alpha := \min\{x_1, \ldots x_n\}$, $x_i \in [0,1]$ for $i \in \overline{1,n}$. We see that F is a weighted average of a t-norm and t-conorm. However, it should be pointed out that this is a non-linear weighted average since α depends upon the arguments to be aggregated. Thus we see these weak t-norms provided for aggregations between t-norms and t-conorms. We can express F as $F = T + \alpha(S - T)$. We see that for α small the t-norm component predominates while as α gets bigger the t-conorm component predominates. So that this operator moves from t-norm to t-conorm depending upon the value of the minimal arguments. We also note that if T is the associated t-norm then

$$F(x_1, x_2, \dots, x_n) \geq T(x_1, x_2, \dots, x_n),$$

thus the aggregation is always at least as large as the value obtained by the associated t-norm.

In (Yager, 1997a) a general class of weak t-conorm operators was introduced as $G(x_1, x_2, \dots, x_n) := (1 - \beta)T(x_1, \dots x_n) + \beta S(x_1, \dots, x_n)$ where $\beta := \max\{x_1, \dots x_n\}$.

Acknowledgments

The work of J. Fodor has been partially supported by the National Scientific Research Fund (OTKA) under grant T025163.

References

Abel, N. H. (1826). Untersuchung der fuctionen zweier unabhängig veränderlichen grössen x und y wie $f(x, y)$, welche die eigenschaft haben, dass $f(z, f(x, y))$ eine symmetrische function von x, y und z ist. *J. Reine Angew. Math.*, 1:11–15. also as Oeuvres Complètes de N.H. Abel, Vol. 1, Christiana, 1881, 61–65.

Aczél, J. (1948). Über eine klasse von funktionalgleichungen. *Comment. Math. Helv.*, 54:247–256.

Aczél, J. (1949). Sur les opérations définies pour des nombres réels. *Bull. Soc. Math. de France*, 76:59–64.

Aczél, J. (1966). *Lectures on Functional Equations and Applications.* Academic Press, New York.

Alsina, C., Trillas, E., and Valverde, L. (1983). On some logical connectives for fuzzy sets theory. *J. Math. Anal. Appl.*, 93:15–26.

Bandler, W. and Kohout, L. (1980). Fuzzy power sets and fuzzy implication operators. *Fuzzy Sets and Systems*, 4:13–30.

Bellman, R. and Giertz, M. (1973). On the analytic formalism of the theory of fuzzy sets. *Inform. Sci.*, 5:149–156.

Bellman, R. and Zadeh, L. (1977). Local and fuzzy logic. In Dunn, J. and Epstein, G., editors, *Modern Uses of Multiple-valued Logic*, pages 103–156. D. Reidel, Dordrecht.

Birkhoff, G. (1967). *Lattice Theory.* American Mathematical Society, Providence, Rhode Island, third edition.

Blom, W. B. A. (1979). *Goals and Decisions in a Fuzzy Environment (in Dutch).* Department of Electrical Engineering, Delft University of Technology.

Czogała, E. and Drewniak, J. (1984). Associative monotonic operations in fuzzy set theory. *Fuzzy Sets and Systems*, 12:249–269.

De Baets, B. (1998). Idempotent uninorms. *European Journal of Operational Research.* (to appear).

De Baets, B. and Fodor, J. (1997a). On the structure of uninorms and their residual implicators. In Gottwald, S. and Klement, E., editors, *18th Linz Seminar on Fuzzy Set Theory: Enriched Lattice Structures for Many-Valued and Fuzzy Logics*, pages 81–87. (February 25–March 1, 1997), Johannes Kepler Universität, Linz, Austria.

De Baets, B. and Fodor, J. (1997b). Residual operators of representable uninorms. In Zimmermann, H.-J., editor, *Proceedings of the 5th European Congress on Intelligent Techniques and Soft Computing*, pages 52–56. (September 8-12, 1997, Aachen, Germany), Verlag Mainz, Wissenschaftsverlag, Aachen.

De Baets, B. and Fodor, J. (1998). Residual operators of uninorms. *Soft Computing.* (to appear).

Dombi, J. (1982a). Basic concepts for a theory of evaluation: the aggregative operator. *European Journal of Operations Research*, 10:282–293.

Dombi, J. (1982b). A general class of fuzzy operators, the De Morgan class of fuzzy operators and fuzziness measures induced by fuzzy operators. *Fuzzy Sets and Systems*, 8:149–163.

Dombi, J. and Zysno, P. (1982). Comments on the γ-model. In Trappl, R., editor, *Cybernetics and Systems Research*, pages 711–714. North-Holland, Amsterdam.

Dubois, D. and Prade, H. (1980a). *Fuzzy Sets and Systems: Theory and Applications.* Academic Press, New York.

Dubois, D. and Prade, H. (1980b). New results about properties and semantics of fuzzy set-theoretic operators. In Wang, P. and Chang, S., editors, *Fuzzy Sets: Theory and Applications to Policy Analysis and Information Systems*, pages 59–75. Plenum, New York.

Dubois, D. and Prade, H. (1984a). Fuzzy logics and the generalized modus ponens revisited. *Int. J. Cybernetics and Systems*, 15:293–331.

Dubois, D. and Prade, H. (1984b). Fuzzy set-theoretic differences and inclusions and their use in the analysis of fuzzy equations. *Control and Cybernetics*, 13:129–145.

Dubois, D. and Prade, H. (1984c). A theorem on implication functions defined from triangular norms. *Stochastica*, VIII:267–279.

Dubois, D. and Prade, H. (1985). A review of fuzzy set aggregation connectives. *Inform. Sci.*, 36:85–121.

Dubois, D. and Prade, H. (1991). Fuzzy sets in approximate reasoning, part 1: Inference with possibility distributions. *Fuzzy Sets and Systems*, 40:143–202.

Dyckhoff, H. and Pedrycz, W. (1984). Generalized means as model of compensative connectives. *Fuzzy Sets and Systems*, 14:143–154.

Esteva, F. and Domingo, X. (1980). Sobre funciones de negacion en $[0, 1]$. *Stochastica*, IV:141–166.

Faucett, W. M. (1955). Compact semigroups irreducibly connected between two idempotents. *Proc. Amer. Math. Soc.*, 6:741–747.

Filev, D. P. and Yager, R. R. (1994). Learning OWA operator weights from data. In *Proceedings of the Third IEEE International Conference on Fuzzy Systems*, pages 468–473, Orlando.

Filev, D. P. and Yager, R. R. (1997). On the issue of obtaining OWA operator weights. *Fuzzy Sets and Systems.* (to appear).

Fodor, J. (1991a). On fuzzy implication operators. *Fuzzy Sets and Systems*, 42:293–300.

Fodor, J. (1991b). Strict preference relations based on weak t-norms. *Fuzzy Sets and Systems*, 43:327–336.

Fodor, J. (1993). A new look at fuzzy connectives. *Fuzzy Sets and Systems*, 57:141–148.

Fodor, J. (1995). Contrapositive symmetry of fuzzy implications. *Fuzzy Sets and Systems*, 69:141–156.

Fodor, J. (1996). An extension of Fung–Fu's theorem. *Int. J. of Uncertainty, Fuzziness and Knowledge-Based Systems*, 4:235–243.

Fodor, J. and Keresztfalvi, T. (1994). Characterization of the Hamacher family of t-norms. *Fuzzy Sets and Systems*, 65:51–58.

Fodor, J., Marichal, J.-L., and Roubens, M. (1995). Characterization of the ordered weighted averaging operators. *IEEE Transactions on Fuzzy Systems*, 3:236–240.

Fodor, J. and Roubens, M. (1994). *Fuzzy Preference Modelling and Multicriteria Decision Support.* Theory and Decision Library, Series D, Volume 14. Kluwer Academic Publishers, Dordrecht.

Fodor, J. and Roubens, M. (1995). On meaningfulness of means. *J. Comput. Appl. Math.*, 64:103–115.

Fodor, J., Yager, R., and Rybalov, A. (1997). Structure of uninorms. *Int. J. of Uncertainty, Fuzziness and Knowledge-Based Systems*, 5(4):411–427.

Frank, M. J. (1979). On the simultaneous associativity of $f(x,y)$ and $x+y-f(x,y)$. *Aeq. Math.*, 19:194–226.

Fung, L. W. and Fu, K. S. (1975). An axiomatic approach to rational decision-making in a fuzzy environment. In L.A. Zadeh, K.S. Fu, K. T. and Shimura, M., editors, *Fuzzy Sets and their Applications to Cognitive and Decision Processes*, pages 227–256. Academic Press, New York, San Francisco, London.

Gaines, B. (1976). Foundations of fuzzy reasoning. *Int. J. Man-Machine Stud.*, 8:623–668.

Giles, R. (1976). Łukasiewicz's logic and fuzzy set theory. *Int. J. Man-Machine Stud.*, 8:313–327.

Goguen, J. A. (1969). The logic of inexact concepts. *Synthese*, 19:325–373.

Hamacher, H. (1978). *Über logische Aggrationen nichtbinär explizierter Entscheidungskriterien; Ein axiomatischer Beitrag zur normativen Entscheidungstheorie.* Rita G. Fischer Verlag, Frankfurt.

Hintikka, R. (1978). Quantifiers in logic and quantifiers in natural languages. In Korner, S., editor, *Philosophy of Logic.* Basil Blackwell, Oxford.

Höhle, U. (1988). Quotients with respect to similarity relations. *Fuzzy Sets and Systems*, 27:31–44.

Höhle, U. (1992). M-valued sets and sheaves over integral commutative CL-monoids. In Rodabaugh, S., Klement, E., and Höhle, U., editors, *Applications of Category Theory to Fuzzy Subsets*, pages 33–72. Kluwer Academic Publishers, Dordrecht.

Höhle, U. and Stout, L. (1991). Foundations of fuzzy sets. *Fuzzy Sets and Systems*, 40:257–296.

Jenei, S. (1996). On discontinuous triangular norms. *BUSEFAL*, 66:38–42.

Jenei, S. (1997a). *Characterization and limit theorems on triangular norms.* PhD thesis, Eötvös Loránd University, Budapest.

Jenei, S. (1997b). New family of girard monoids on intervals via annihilation. *BUSEFAL*, 69:94–101.

Jenei, S. (1998a). Fibred triangular norms. *Fuzzy Sets and Systems.* (to appear).

Jenei, S. (1998b). New family of triangular norms via contrapositive symmetrization of residuated implications. *Fuzzy Sets and Systems.* (to appear).

Jenei, S. (1998c). On archimedean triangular norms. *Fuzzy Sets and Systems.* (to appear).

Jenei, S. and Fodor, J. (1998). On continuous triangular norms. *Fuzzy Sets and Systems.* (to appear).

Kaymak, U. and Van Nauta Lemke, H. R. (1993). A parametric generalized goal function for fuzzy decision making with unequally weighted objectives. In *Proceedings of the Second IEEE International Conference on Fuzzy Systems*, pages 1156–1160, San Francisco.

Kitainik, L. (1986). Axiomatics and properties of fuzzy inclusions. *Scientific Works of the Institute for System Studies*, 10:97–107. (in Russian).

Kitainik, L. (1987a). Comparative study of fuzzy inclusions. *Scientific Works of the Institute for System Studies*, 14:83–92. (in Russian).

Kitainik, L. (1987b). Fuzzy inclusions and fuzzy dichotomous decision procedures. In Kacprzyk, J. and Orlovski, S., editors, *Optimization Models using Fuzzy Sets and Possibility Theory*, pages 154–170. D. Reidel,, Dordrecht.

Kitainik, L. (1993). *Fuzzy Decision Procedures with Binary Relations.* Kluwer Academic Publishers, Dordrecht.

Klawonn, F. and Kruse, R. (1993). Equality relations as a basis for fuzzy control. *Fuzzy Sets and Systems*, 54:147–156.

Klement, E., Mesiar, R., and Pap, E. (1996). On the relationship of associative compensatory operators to triangular norms and conorms. *International Journal of Uncertainty, Fuzziness and Knowledge-Based Systems*, 4:129–144.

Klement, E., Mesiar, R., and Pap, E. (1997). A characterization of the ordering of continuous t-norms. *Fuzzy Sets and Systems*, 86:189–195.

Klement, E., Mesiar, R., and Pap, E. (1998). Triangular norms. (to appear).

Klement, E. P. (1981a). An axiomatic theory of operations on fuzzy sets. Technical Report Institutsbericht 18, Institut für Mathematik, Johannes Kepler Universität Linz.

Klement, E. P. (1981b). Operations on fuzzy sets and fuzzy numbers related to triangular norms. In *Proceedings of the Eleventh International Symposium on Multiple-valued Logic*, pages 218–225.

Klement, E. P. (1982). Operations on fuzzy sets – an axiomatix approach. *Inf. Sci.*, 27:221–232.

Lemke, H. R. V. N., Dijkman, T. G., Haeringen, H. V., and Pleeging, M. (1983). A characteristic optimism factor in fuzzy decision making. In *Proc. IFAC Symposium on Fuzzy Information, Knowledge Representation and Decision Analysis*, pages 283–288, Marseille, France.

Ling, C. H. (1965). Representation of associative functions. *Publ. Math. Debrecen*, 12:189–212.

Lowen, R. (1978). On fuzzy complements. *Inform. Sci.*, 14:107–113.

Łukasiewicz, J. (1970). *Selected Works.* Reidel, Amsterdam.

Menger, K. (1942). Statistical metric spaces. *Proc. Nat. Acad. Sci. USA*, 28:535–537.

Miyakoshi, M. and Shimbo, M. (1985). Solutions of composite fuzzy relational operations with triangular norms. *Fuzzy Sets and Systems*, 16:53–63.

Mostert, P. S. and Shields, A. L. (1957). On the structure of semigroups on a compact manifold with boundary. *Ann. of Math.*, 65:117–143.

O'Hagan, M. (1990). Using maximum entropy-ordered weighted averaging to construct a fuzzy neuron. In *Proceedings 24th Annual IEEE Asilomar Conf. on Signals, Systems and Computers*, pages 618–623, Pacific Grove, Ca.

Ostasiewicz, W. (1988). Formalization of quantifier phrases. In Gupta, M. and Yamakawa, T., editors, *Fuzzy Logic in Knowledge-Based Systems, Decision and Control*, pages 271–286. North-Holland, Amsterdam.

Ovchinnikov, S. (1981). Representations of synonymy and antonymy by automorphisms in fuzzy set theory. *Stochastica*, V:95–107.

Ovchinnikov, S. (1983). General negations in fuzzy set theory. *J. Math. Anal. Appl.*, 92:234–239.

Ovchinnikov, S. and Roubens, M. (1991). On strict preference relations. *Fuzzy Sets and Systems*, 43:319–326.

Ovchinnikov, S. and Roubens, M. (1992). On fuzzy strict preference, indifference and incomparability relations. *Fuzzy Sets and Systems*, 47:313–318.

Paalman – De Miranda, A. B. (1964). Topological semigroups. Technical report, Mathematisch Centrum, Amsterdam.

Perny, P. (1992). *Modélisation, agrégation et exploitation des préférences floues dans une problématique de rangement.* PhD thesis, Université Paris-Dauphine.

Prade, H. (1982). *Modèles mathématiques de l'imprécis et de l'incertain en vue d'applications au raisonnement naturel.* PhD thesis, Université P. Sabatier, Toulouse.

Schweizer, B. and Sklar, A. (1961). Associative functions and statistical triangle inequalities. *Publ. Math. Debrecen*, 8:169–186.

Schweizer, B. and Sklar, A. (1963). Associative functions and abstract semigroups. *Publ. Math. Debrecen*, 10:69–81.

Schweizer, B. and Sklar, A. (1983). *Probabilistic Metric Spaces,*. North-Holland, Amsterdam.

Sikorski, R. (1964). *Boolean Algebras.* Springer, Berlin.

Silvert, W. (1979). Symmetric summation: a class of operations on fuzzy sets. *IEEE Transactions on Systems, Man and Cybernetics*, 9:659–667.

Sinha, D. and Dougherty, E. (1993). Fuzzification of set inclusion: theory and applications. *Fuzzy Sets and Systems*, 55:15–42.

Smets, P. and Magrez, P. (1987). Implication in fuzzy logic. *Int. J. of Approximate Reasoning*, 1:327–347.

Sugeno, M. (1974). *Theory of fuzzy integrals and its applications.* PhD thesis, Tokyo Institute of Technology, Tokyo.

Sugeno, M. (1977). Fuzzy measures and fuzzy initegrals: a survey. In M.M. Gupta, G. S. and Gaines, B., editors, *Fuzzy Automata and Decision Processes*, pages 89–102. North-Holland, Amsterdam.

Trillas, E. (1979). Sobre funciones de negación en la teoría de conjuntos difusos. *Stochastica*, III:47–60.

Trillas, E. and Riera, T. (1981). Towards a representation of "synonyms" and "antonyms" by fuzzy sets. *BUSEFAL*, 5:42–68.

Trillas, E. and Valverde, L. (1981). On some functionally expressable implications for fuzzy set theory. In *Proc. of the 3rd Intern. Seminar on Fuzzy Set Theory*, pages 173–190. Johannes Kepler Universität.

Trillas, E. and Valverde, L. (1985). On implication and indistinguishability in the setting of fuzzy logic. In Kacprzyk, J. and Yager, R., editors, *Management Decision Support Systems using Fuzzy Sets and Possibility Theory*, pages 198–212. Verlag TÜV Rheinland, Köln.

Turksen, I. (1986). Interval valued fuzzy sets based on normal forms. *Fuzzy Sets and Systems*, 20:191–210.

Turksen, I. (1992). Interval-valued fuzzy sets and 'compensatory AND'. *Fuzzy Sets and Systems*, 51:295–307.

Vilenkin, N. and Shreider, Y. (1977). Majority spaces and quantifier "majority". *Semiotika i Informatika*, 8:45–82. (in Russian).

Wangming, W. (1992). Representation of Archimedean implications. (unpublished).

Weber, S. (1983). A general concept of fuzzy connectives, negations and implications based on t-norms and t-conorms. *Fuzzy Sets and Systems*, 11:115–134.

Yager, R. (1979). On the measure of fuzziness and negation, part I: membership in the unit interval. *Internat. J. General Systems*, 5:221–229.

Yager, R. (1980a). An approach to inference in approximate reasoning. *Int. J. Man-Machine Stud.*, 13:323–338.

Yager, R. (1980b). On the measure of fuzziness and negation II. lattices. *Information and Control*, 44:236–260.

Yager, R. (1988). On ordered weighted averaging aggregation operators in multi-criteria decision making. *IEEE Transactions on Systems, Man and Cybernetics*, 18:183–190.

Yager, R. (1991a). Connectives and quantifiers in fuzzy sets. *Fuzzy Sets and Systems*, 40:39–75.

Yager, R. (1993). MAM and MOM operators for aggregation. *Information Sciences*, 69:259–273.

Yager, R. (1995). A unified approach to aggregation based upon MOM and MAM operators. *International Journal of Intelligent Systems*, 10:809–855.

Yager, R. and Rybalov, A. (1996a). Uninorm aggregation operators. *Fuzzy Sets and Systems*, 80:111–120.

Yager, R. R. (1977). Multiple objective decision making using fuzzy sets. *International J. of Man-Machine Studies*, 9:375–382.

Yager, R. R. (1987a). A note on weighted queries in information retrieval systems. *J. of the American Society of Information Sciences*, 38:23–24.

Yager, R. R. (1987b). Using approximate reasoning to represent default knowledge. *Artificial Intelligence*, 31:99–112.

Yager, R. R. (1991b). Nonmonotonic set theoretic operations. *Fuzzy Sets and Systems*, 42:173–190.

Yager, R. R. (1994). Misrepresentations and challenges: A response to Elkan. *IEEE Expert*, August:41–42.

Yager, R. R. (1997a). On a class of weak triangular norm operators. *Information Sciences*, 96:47–78.

Yager, R. R. (1997b). Structures for prioritized fusion of fuzzy information. Technical Report MII-1719, Machine Intelligence Institute, Iona College, New Rochelle, NY.

Yager, R. R. and Filev, D. P. (1994a). *Essentials of Fuzzy Modeling and Control.* John Wiley, New York.

Yager, R. R. and Filev, D. P. (1994b). Parameterized "andlike" and "orlike" OWA operators. *International Journal of General Systems*, 22:297–316.

Yager, R. R. and Rybalov, A. (1996b). Uni-norms: a unification of t-norms and t-conorms. In *Proceedings of Nafips Conference*, pages 50–54, Berkeley.

Zadeh, L. (1965). Fuzzy sets. *Information and Control*, 8:338–353.

Zadeh, L. (1983). A computational approach to fuzzy quantifiers in natural languages. *Computing and Mathematics with Applications*, 9:149–184.

Zimmermann, H.-J. and Zysno, P. (1980). Latent connectives in human decision making. *Fuzzy Sets and Systems*, 4:37–51.

3 MEASUREMENT OF MEMBERSHIP FUNCTIONS: THEORETICAL AND EMPIRICAL WORK

Taner Bilgiç
I. Burhan Türkşen

Abstract: This chapter presents a review of various interpretations of the fuzzy membership function together with ways of obtaining a membership function. We emphasize that different interpretations of the membership function call for different elicitation methods. We try to make this distinction clear using techniques from measurement theory.

3.1 INTRODUCTION AND PREVIEW

Since Zadeh (1965) introduced the notion of fuzzy sets one of the main difficulties has been with the *meaning and measurement* of membership functions. Particularly, lack of a consensus on the meaning of membership functions has created some confusion. This confusion is neither bizarre nor unsound. After all fuzzy sets are totally characterized by their membership functions and in order to diffuse this cloud of confusion and for a sound theory of fuzzy sets a rigorous semantics together with practical elicitation methods for membership functions are necessary.

There are various interpretations as to where fuzziness might stem from. Depending on the interpretation of fuzziness one subscribes to, the meaning attached to the membership function changes. It is the objective of this chapter to review the various interpretations, theoretical models for various interpretations, and elicitation methods of membership functions.

We first start with the formal (i.e., mathematical) definition of a membership function[1]. A fuzzy (sub)set, say F, has a membership function μ_F[2] defined as a function from a well defined universe (the referential set), X, into the unit interval as: $\mu_F : X \to [0, 1]$ (see Figure 3.1).

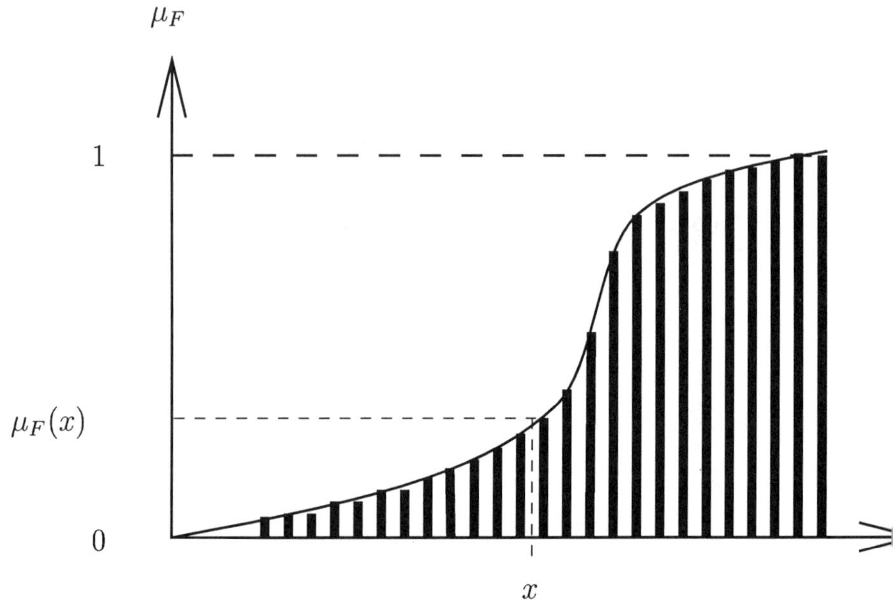

Figure 3.1 A membership function

Hence, the vague predicate "John (x) is tall (T)" is represented by a number in the unit interval, $\mu_T(x)$. There are several possible answers to the question "What does it mean to say $\mu_T(x) = 0.7$?"

likelihood view 70% of a given population declared that John is tall (see Section 3.2.1).

random set view 70% of a given population described "tall" as an interval containing John's height (see Section 3.2.2).

similarity view John's height is away from the prototypical object which is truly "tall" to the degree 0.3 (a normalized distance, see Section 3.2.3).

utility view 0.7 is the *utility* of *asserting* that John is tall (see Section 3.2.4).

measurement view When *compared* to others, John is taller than some and this fact can be encoded as 0.7 on some scale (see Section 3.2.5).

In Section 3.2, each of the interpretations mentioned above for the concept of grade of membership are taken up one-by-one, emphasizing their differences as well as similarities. Consequences of subscribing to a certain interpretation are also highlighted. It is important to realize that these interpretations are hypotheses about where *fuzziness* stems from and each interpretation suggests a calculus for manipulating membership functions. The calculus of fuzzy sets as described by Zadeh (1965) and his followers are sometimes appropriate as the calculus of fuzziness but sometimes inappropriate depending on the interpretation. After presenting a critical discussion of each interpretation, we give a measurement-theoretic interpretation, on the basis of which lies the comparative view. Measurement-theoretic view is covered in some depth in Section 3.2.5 as it sheds light to the foundations of every other interpretation.

In Section 3.3, we summarize experimental research conducted with the aim of constructing the membership functions and/or verifying the axioms of the fuzzy set theory. This section mentions various methods of obtaining membership functions as they are used in experimental research of fuzzy set theory. The interplay of the elicitation method and the assumed interpretation of fuzzy sets is emphasized.

We close the chapter with a quick summary of the review in Section 3.4.

In general, the aim of this chapter is to make the reader familiar with different (sometimes competing) interpretations of fuzzy sets and methods of elicitation for practical applications. The controversies and arguments, as well as further references are usually given as end notes for the interested reader to follow up.

3.2 INTERPRETATIONS OF GRADE OF MEMBERSHIP

When someone is introduced to the fuzzy set theory, the concept of graded membership sounds fairly intuitive since this is just an extension of a well known concept: membership in a set to "graded membership" in a set. However, the second step is quite mind boggling:

"How can graded membership be measured?"

This question has been considered, albeit in the context of many-valued logics, by many people from different disciplines[3]. Although more than two thousand years ago, Aristotle commented on an "indeterminate truth value", the interest in formal aspects of many-valued logics has started in early 1900's (McCall & Ajdukiewicz 1967, Rosser & Turquette 1977). But the *meaning* of multiple truth values has not been explained to satisfaction. For some, this is sufficient to discard many-valued logics all together (Kneale 1962, French 1984). On the other hand, the intellectual curiosity never let go of the subject (Scott 1976).

Anyone who is to use fuzzy sets must answer the following questions:

1. What does graded membership mean?

2. How is it measured?

3. What operations are meaningful to perform on it?

To answer the first question one has to subscribe to a certain view of fuzziness. Mainly there has been two trends in the interpretations of fuzziness: those who think that fuzziness is subjective as opposed to objective and those who think that fuzziness stems from the individual, as opposed to from a group of people (or sensors, etc.) as depicted in Figure 3.2.

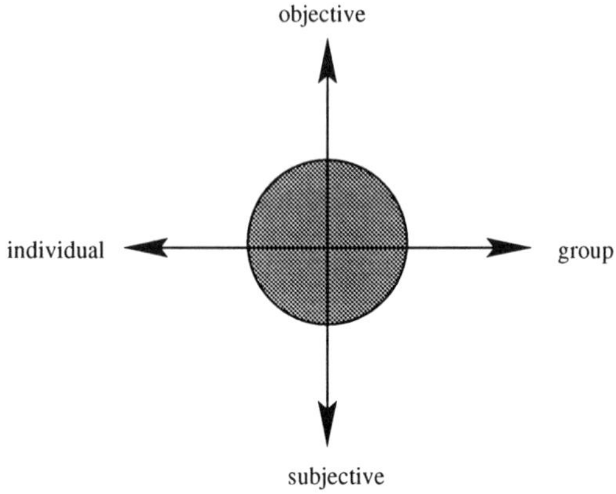

Figure 3.2 Main interpretations of fuzziness

We first summarize some of these views and then present a measurement-theoretic development which sheds light into all three questions.

Both the likelihood and the random set views of the membership function implicitly assume that there are *more than one* evaluator or experiments are *repeated*. Therefore, if one thinks of membership functions as "meaning representation", they come close to the claim that "meaning is essentially objective" and fuzziness arises from inconsistency or error. On the other hand, during the initial phases of the development of fuzzy sets, it has been widely accepted that membership functions are *subjective* and context dependent (Zadeh 1965, 1975). The similarity and utility views of the membership function differ from the others in their espousing a subjective interpretation. The measurement view is in the connection of subjective and objective views in the sense that the problem can be defined in both ways depending on the observer(s) who is (are) making the comparison. The comparisons can be results of subjective evaluations or results of "precise" (or idealized) measurements.

3.2.1 The likelihood view

Particularly Hisdal (1985, 1988) advocates a *likelihood* view of the concept of a grade of membership[4]. Hisdal's work is a typical example of creating a *descriptive* (interpretive is the word she uses) theory of fuzzy sets.

Hisdal's TEE (Threshold, Error, assumption of Equivalence) model for membership functions considers several *sources* of fuzziness:

1. Errors in measurement (which are essentially statistical in nature).

2. Incomplete information.

3. Interpersonal contradictions.

Furthermore, Hisdal argues that, *for a singe subject* equipped with perfect information (about, say, John's height) there cannot be any fuzziness! In this case John would be called 'tall' if his height is greater than a well-defined threshold or not.

Hisdal considers three ways of eliciting the membership function:

1. Label experiment: John is tall where tall $\in$ {very tall, tall, short}.

2. Yes-No experiment: Is John tall?

3. Grade of membership experiment: What is the degree of belonging of John to the set of tall people?

Essentially, Hisdal's argument goes as follows: when asked to what degree a person is tall, the subject who has an imprecise measurement of the person's height (say she knows that it is within 180 ± 5 cm.) constructs an *error curve* to quantify the possible error around 180 cm. Then, she would argue as: "in 75% of the cases the subject's height lie in my quantization interval for 'tall' ". Therefore

$$\mu_{\text{tall}}(180cm) = P(\text{tall}|x = 180cm) = 0.75$$

where $P(\text{tall}|x = 180cm)$ denotes the likelihood that the label "tall" is assigned to the subject in question given that the subject's height is 180 cm.

Hisdal postulates that a subject's grade of membership is the modification of her (crisp) answers to the Label or Yes-No experiments *modified* by her *estimate* of the error curve.

These arguments essentially apply to the first type of fuzziness: errors in measurement[5]. Hisdal also goes on to explore the connectives which are valid under this interpretation. She argues that min and max are not always appropriate as conjunction and disjunction, respectively. This should come as no surprise since in this probabilistic setting, we know that joint probabilities are not always simply recovered from the constituent probabilities.

Like Hisdal, Mabuchi (1992) also defines membership function as likelihood but assumes that the likelihood is determined by the number of *evidence grounds* of all necessary grounds which support an attribute's membership. The construction of membership functions follows closely that of a checklist paradigm (Bandler & Kohout 1986, Kohout & Bandler 1993) and max-min calculus is not supported as disjunction and conjunction.

Thomas (1979, 1995) elaborates on a likelihood semantics for fuzzy sets. Thomas subscribes to the philosophical point of view that "meaning is essentially objective and is a convention among the users of a language." Furthermore, Thomas also opposes to the determinism and idealism that measurement theory imposes. He claims that measurement is essentially a vague process. These views lead Thomas to equate the likelihood function to the membership function and hence provide a likelihood semantics for fuzzy set theory. Like Hisdal, Thomas also finds out that the min-max calculus as proposed by Zadeh is not sufficient in this framework.

3.2.2 Random set view

The definition given for a membership function in Section 3.1 is called the *vertical* representation (Dubois & Prade 1989). One can also view the membership function *horizontally* where a fuzzy set F on a referential set X is represented in terms of its "level-cuts" (see Figure 3.3):

$$\{F_\alpha : \alpha \in (0, 1]\}$$

where $F_\alpha = \{x : \mu_F(x) \geq \alpha\}$. Zadeh (1971) defines the membership function as:

$$\mu_F(x) = \sup\{\alpha \in (0, 1] : x \in F_\alpha\}.$$

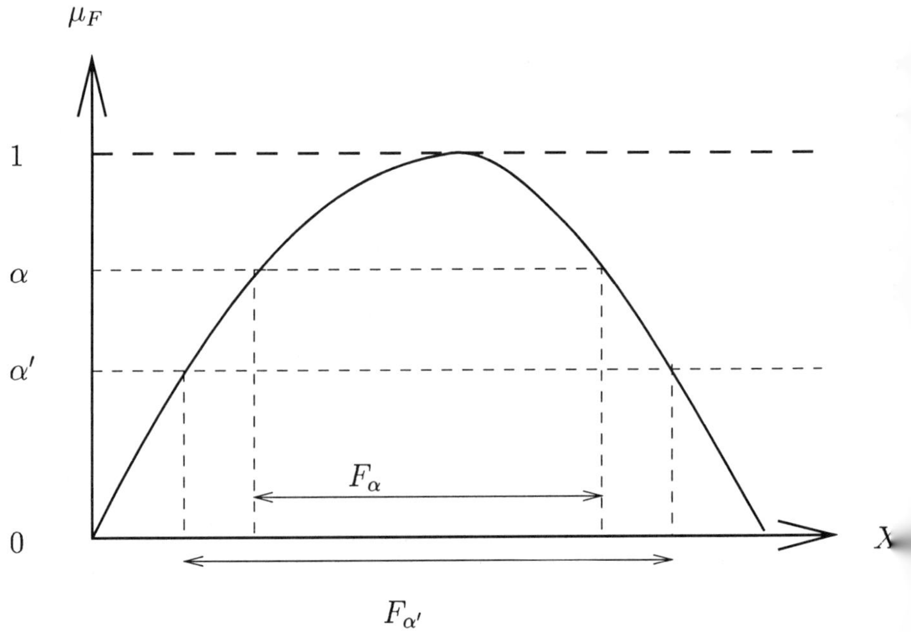

Figure 3.3 The horizontal representation of a membership function

In this representation we view the membership function as a nested family of level-cuts[6]. Note that each level-set, F_α, is a set in the classical sense.

The membership function can also be represented as an integral:

$$\mu_F(x) = \int_0^1 \mu_{F_\alpha}(x) d\alpha$$

where $F_\alpha(x) = 1$ if $x \in F_\alpha$ and 0 otherwise (Dubois & Prade 1989). Of course, one must make continuity and measurability assumptions so that the integral is well-defined. In that sense, it is a specialized view of the membership function (see

Section 3.2.5 for more details on continuity and Archimedean assumptions). However, in this view, the membership function can be viewed as a uniformly distributed random set consisting of the Lebesgue measure on $[0, 1]$ and the set-valued mapping $F_\alpha : (0, 1] \to X$.

Therefore, for $\mu_F(x) = 0.7$, the random set view is: 70% of the population defined an *interval* on the referential set, X, as an interval containing x (e.g. $x =$ John's height) on the basis of an evaluation F (e.g. tallness). The remaining 30% defined intervals which *excluded* x to be F.

3.2.3 Similarity View

The similarity view of the membership functions naturally arise in prototype theory where membership is a notion of being *similar* to a *representative* of the category (Rosch & Mervis 1975, Lakoff 1987). In that sense, membership function measures the *degree of similarity* of an element to the set in question. It is assumed that there exists a *perfect* example of the set (or the category) that belongs to the set to the full degree. Others belong to the set to a degree measured by their relative *distance* to the perfect example.

Osherson & Smith (1981, 1982) claim that in rating the prototypicality of a stimulus in some category, subjective probability is sufficient. They critically discuss the (in)adequacy of fuzzy operators like conjunction and implication. There is agreement on the need to account gradients in category membership, however, a major task is to validate to what extent do people use gradient category membership in day-to-day cognitive tasks. There are two theories on the subject:

- People assign borderline stimuli to the category in which it has the greatest membership. Thereby economizing on cognitive effort (Lakoff 1973).
- People are mainly "thresholders" assigning an item to a category if the stimuli are greater than a certain threshold (Kochen 1975).

Pipino, van Gich & Tom (1981) report evidence on both types of behaviour for the same cognitive tasks.

Kempton (1981) reviews and defends the similarity view by extending traditional cognitive anthropological folk classification methods using methods based on fuzzy set theory.

Zysno (1981) discusses scaling the membership function. His approach stems from a similarity view of the membership function. He takes a single human being as the *measurement device* and assumes that fuzziness arises from the insufficient cognitive abilities of this person who is faced with the task of "comparing the object with a certain prototype or imaginable ideal." This approach naturally leads to a notion of *distance* which Zysno relates to the membership function as:

$$\mu(x) = \frac{1}{1 + f(d_x)}$$

where d_x is the distance of object x from the ideal and $f(d_x)$ is a certain function of the distance. Then he proposes alternatives for the function $f(d_x)$ and empirically evaluates them.

A theoretical discussion of this method is given by Zimmermann & Zysno (1985) in which a mathematical model of a membership function is defined on the real line where membership is a function of the distance between an object and an ideal object. The proposed model is also extensively tested in an experimental setting.

However, when one talks about distance, there is a more substantial assumption that comes with it: a metric space. Many empirical researchers take the notion of a metric space on which a distance measure can be defined for granted. What needs to be discussed is the properties of this metric space and its correspondence to psychophysical laws.

Zwick, Carlstein & Budescu (1987) study different measures of similarity among fuzzy concepts. They consider similarity measures defined both on metric spaces and in a set theoretic manner. They consider 19 such measures and try to experimentally verify the performance of these measures. They report that all measures successfully yield similar and dissimilar concepts in a crude manner. However, when the aim is to distinguish between degrees of similarities metric measures perform better. From a measurement-theoretic point of view this is expected. In a crude categorization, one is interested only about similar objects not with *how similar* they are. This can be done on an ordinal scale. However, when the aim is to find out how similar the concepts are then one needs to move to interval and ratio scales on which metric measures are meaningful and carry more information.

Ruspini (1991) also considers a similarity semantics for fuzzy set theory. He builds the notion of similarity on a certain modal logic in which the accessibility relation of the modal logic is a similarity relation in a metric space. This method provides a formalism of fuzzy set theory based on the possible world semantics of modal logics (Chellas 1980)[7].

The notion of similarity seems to require some sort of topology (or a metric space) on which it can be defined (Tversky 1977). Most often than not, someone who subscribes to the similarity view implicitly assumes a continuous, metric space. However this assumption needs a critical examination before committing to the similarity view. Possibly the best way to put similarity view under a critical framework is via Measurement Theory (c.f. Section 3.2.5). We will have more to say on this point in that section.

3.2.4 View from utility theory

Giles (1988) offer a decision-theoretic interpretation for the membership function. He argues that a set is equivalent to a *property* and claims that a sound meaning to the membership function can be given by considering it *together* with the problem of *fuzzy reasoning*. Hence, he is following the path of logic to come up with a semantic theory for the concept of a graded truth value.

He defines a fuzzy sentence as "a sentence to which we attribute a degree of belief" which is a function of *possible worlds* (or possible states of mind, nature etc.). Giles considers sentences that are *asserted* rather than merely uttered. For example, when one asserts that "John is tall" Giles assumes that there exists a pay-off function related to this assertion. This pay-off function offers more if the statement is closer to truth (e.g., you are more credible among your peers if you speak the truth!). This assumption drives the *utility theory* approach to the semantics of the membership function. (Giles 1988, p. 304) claims that "Utilities of assertions carry the meanings of assertions". Of course, when defined this way, the utility approach yields a membership function on an interval scale (unique up to positive linear transformations) (Fishburn 1970).

He dismisses the effects of context on the meaning of membership function (as in a child who is 160 cm in height is a tall child but not a tall person) by claiming that one should try to represent the *average* meaning of the assertion in a *normal* society.

When viewed this way, the connectives for the theory are no more truth functional! Therefore, none of the triangular norms and conorms (cf. Chapter 2 of this handbook) are candidates for disjunction and conjunction. A semantic approach to fuzzy set theory, as Giles attempted, defies the *algebraic* view in which classical connectives are simply extended to many-valued settings in a truth functional way.

3.2.5 View from measurement theory

The three problems posed at the beginning of Section 3.2 can be taken up by using the techniques of measurement theory (Krantz, Luce, Suppes & Tversky 1971, Roberts 1979, Narens 1985, Suppes, Krantz, Luce & Tversky 1989, Luce, Krantz, Suppes & Tversky 1990). Measurement theory is concerned with *representation* and *meaningfulness* of the particular representation.

Most often than not, mathematical structures are investigated in their own right without any application in mind. This is essential to the growth of human mathematical enterprise. On the other hand, fuzzy set theory is an *empirical science*; it claims that it represents a natural phenomena that we observe in our everyday lives: fuzziness. Therefore a theory of fuzzy sets must have clear correspondences in our daily lives as well. Measurement theory bridges that gap by taking a phenomenon which can be modelled as an algebraic structure (e.g. fuzziness) and by mapping it into a *numerical* structure (hence capturing the essential core of the *measurement* process). In such a theory, one can discuss the representation of a qualitative structure and the meaningfulness of such a representation[8].

We utilize measurement theory to *formally*[9] answer the general questions that we posed at the beginning of this chapter. The aim is to *bring to the surface* the axioms that are implicit in subscribing to a particular interpretation of membership functions and to discuss them as to their testability, validity and/or intuitiveness.

We start with Saaty's (1986) formulation who espouses a ratio scale. After discussing the assumptions of this formulation, we argue that ratio scales are neither *natural* nor acceptable for membership functions as Saaty claims them to be.

We then differentiate between two related but *different* measurement problems: *(i)* membership measurement (Section 3.2.5) and *(ii)* property ranking (Section 3.2.5). We argue that, although the first problem received much attention in the literature the second one is closely related to the question of "which connectives to use?". Then, we combine the two problems to arrive at an axiomatization for fuzzy set theory (Section 3.2.5)[10]. In the Appendix to this chapter, we give definitions, and representation and uniqueness results for ordered algebraic structures which can be taken to model fuzzy set theory. These results provide a technical background to facilitate the discussion of the measurement-theoretic frameworks for fuzzy set theory in Sections 3.2.5–3.2.5. The reader who is interested in technical details can use the Appendix as a reference.

Saaty's measurement on a ratio scale. Saaty (1986) thinks that it is hopeless to seek to determine the membership function through direct scaling (fundamental measurement) and espouses a derived scale. He proceeds with assuming a finite set $\mathcal{A}$ with n elements called the alternatives and a set of criteria, $\mathcal{C}$. He defines a binary relation on $\mathcal{A}$ denoted by $\succ_c$ and read as "more preferred than with respect to criteria c". Then he *assumes* that $\succ_c$ is complete and there *exists* a representation, say P_c for $\succ_c$ and continues to *add* axioms on this *fundamental* scale! The first axiom he adds is what he calls the *reciprocal axiom* which states:

$$P_c(A_i, A_j) = 1/P_c(A_j, A_i)$$

for all A_i and A_j in $\mathcal{A}$. This implies that the *initial fundamental scale*, P_c, is measured at least on a *ratio scale*. Saaty does not show how this can be done and, therefore, implicitly takes all the axioms that lead to a ratio scale for granted.

However, the basic tenet of measurement theory (as in any axiomatization of a theory) is to lay out assumptions as clear as possible so that their validity can be tested and/or they can be accepted on normative grounds. By assuming that such preference comparisons can be measured on a ratio scale without justifying this position does not provide a sound formalism for fuzzy set theory[11].

Membership measurement. There are two important (but different) measurement problems in fuzzy set theory. The first kind deals with measuring the degree of membership of *several subjects or objects* in a single fuzzy set. This problem has been studied in (Yager 1979, Norwich & Türkşen 1982b, Norwich & Türkşen 1984, Türkşen 1991, Bollmann-Sdorra, Wong & Yao 1993, Bilgiç 1995) among others.

In this problem, there is a single fuzzy set (or a fuzzy term), F, and a finite number of agents, objects, elements, etc. in A. The question is: "to what degree an agent from A belongs to fuzzy set F?". The first measurement problem takes the relation "an agent is more F than another agent", where F is a fuzzy term (typically an adjective). The important thing to notice is the fact that this representation compares several agents over a *single* fuzzy term, F. The resulting representation measures the degree to which each agent belongs to the fuzzy set F.

To capture this graded membership concept, consider a binary relation, $\succsim_F$ on A with the following interpretation:

$$a \succsim_F b \Longleftrightarrow a \text{ belongs to } F \text{ at least as much as } b \text{ belongs to } F$$

or equivalently,

$$a \succsim_F b \Longleftrightarrow a \text{ is at least as } F \text{ as } b \text{ is } F.$$

Examples of such sentences are:

- Mary is at least as intelligent as John
- The new generation is less political than the old one

- This task is more important than the other

Norwich & Türkşen(1982b, 1984) consider the relation, $\succsim_F$, and discuss its possible representations.

The structure is assumed to be bounded (i.e., there exists elements a^- and a^+ in A such that for all other a in A, $a^+ \succsim_F a \succsim_F a^-$). This amounts to assuming that there are agents for which the fuzzy term F is completely true or completely false. If F is "tall", for example, there may be some elements of A for which "tall" and "not tall" are completely unambiguous. However, if F is "cute" or "bright" etc. it is very hard to find agents who are absolutely "cute" or "not cute".

If the boundary condition is accepted, then it is clear that an ordinal scale representation exists for $\langle A, \succsim_F \rangle$ once the transitivity and connectedness of $\succsim_F$ are accepted. The ordinal scale for measuring fuzziness allows only the comparison of membership values and no other numerical manipulation.

Norwich & Türkşen (1982b) also consider the difference measurement problem. Discussing the axioms of an algebraic difference structure $\langle A \times A, \succsim_F \rangle$ for membership in a fuzzy set, they argue that membership can be measured on an interval scale. The extra information required in this case is to be able to compare *pairs* of agents (e.g., John is taller than James than Jason is taller than Jack.).

The interval scale uniqueness for this representation guarantees that averaging is a meaningful operation as well as comparison but the notion of an *origin* is still arbitrary (i.e., the bounds do *not* invariably map to the same numbers in the numerical domain.

On the basis of the above arguments it can be shown that a function, $\mu_F : A \to [0, 1]$ exists such that:

$$a \succsim_F b \Longleftrightarrow \mu_F(a) \geq \mu_F(b).$$

The uniqueness is either ordinal or interval depending on the axioms that are accepted and $\mu_F(a)$ measures the degree with which a belongs to F. In the case of an interval scale, the following representation also holds:

$$(a, b) \succsim_F (c, d) \Longleftrightarrow \mu_F(a) - \mu_F(b) \geq \mu_F(c) - \mu_F(d).$$

In (Norwich & Türkşen 1984, Section 2.2) it is argued that ratio scales are not likely to arise in fuzzy set theory. They suggest that there is a difficulty with the *concatenation* of psychological attributes and hence no operational meaning can be assigned to the concatenation operator (addition) of *extensive* measurement which results in a ratio scale[12].

Property ranking. A complementary problem is to consider *many* fuzzy terms for a *single* agent. This way one can introduce the connectives into the measurement-theoretic setting.

The justification of using different operators as connectives for fuzzy sets are considered by many authors. Bellman & Giertz (1973) give the first axiomatic justification for using min and max operators followed by Fung & Fu (1975). Yager (1979) also considers the same problem and proposes axioms from which min and max follow as the only solutions. The introduction of triangular norms, conorms and parametrized families of functions as possible candidates of connectives triggered more research

(Yager (1980), Dubois & Prade (1980, 1982), Weber (1983), Alsina (1985) among others).

Bollmann-Sdorra *et al.* (1993) take up the same problem and clearly distinguish the membership measurement and property ranking problems. Bollmann-Sdorra *et al.* justify the use of min and max operators as intersection and union, respectively.

Bilgiç (1995) and Bilgiç & Türkşen (1995a) consider the same problem. They start with a very weak algebraic structure for which only an ordinal scale representation exists. Then, they continue to add axioms on this structure on an "as-needed" basis and show how ratio and absolute scale representations are obtained[13]. At each addition of a new axiom they discuss the testability and validity of the new axiom for fuzzy set theory.

This approach is in accord with some linguistic theories where it is claimed that although adjectives (like tall, clever etc.) linguistically precede their comparative forms (taller, cleverer etc.), the comparative forms precede the simple forms *logically* (Sapir 1944, Palmer 1981). In particular, Kamp (1975, p. 127) argues that:

> ... when we learn a language like English we learn the meanings of individual adjectives and, moreover, the semantic function which this comparative-forming operation performs, *in general*, so that we have no difficulty in understanding, on first hearing, the meaning of the comparative of an adjective which we had thus far only encountered in the positive. If this is so then the meaning of an adjective must be such that the comparative can be understood as a semantic transformation of that meaning into the right binary relation.

This amounts to saying that the meaning of adjectives (fuzzy terms) requires its comparative form for formal analysis; an idea perfectly reflected in measurement theory.

It should be stressed that the problem as defined in this manner *does not assume* that the subject is evaluated on a scale for each individual fuzzy term. Hence, the disjunction in fuzzy set theory is *not* the combination of two fuzzy scales from the membership measurement problem of Section 3.2.5. Such a formulation is briefly reviewed in Section 3.4 which uses conjoint measurement techniques.

The qualitative relation of concern is[14]:

$$F_i \succsim_a F_j \iff \text{an agent } a \text{ is more } F_i \text{ than s/he is } F_j.$$

Examples of such expressions are:

- John is taller than he is clever.
- Inventory is higher than it is low.
- Coffee is at least as unhealthy as it is tasty.
- Her last novel is more political than it is confessional.

It should be noticed that there is a *single subject* in such sentences. Therefore, the treatment of this section excludes those sentences like "I am taller than you are strong". General two-subject sentences are discussed in Section 3.2.5.

The problem is formalized by imposing an algebraic, qualitative structure $\langle \mathcal{F}_a, \succsim_a, \oplus_a \rangle$ where $\mathcal{F}_a$ is a (countable) set of fuzzy terms related to a *single subject* a:

$$\mathcal{F}_a = \{F_1, F_2, F_3, \ldots\},$$

$\succsim_a$ is an ordering of those terms and $\oplus_a$ is intended to model the disjunction of two fuzzy terms[15].

While membership measurement can at best be performed on an interval scale, *formally* as strong as absolute scale representations exist for the property ranking problem (cf. Theorem 1 in the Appendix).

The formal existence of scales that are stronger than interval scales poses the problem of coming up with a *natural origin*. However, since the ranking of properties of an individual is highly a subjective act of the observer, there cannot be universally accepted bounds on the measurement scale (Norwich & Türkşen 1984). This makes all the scales resulting from the measurement, *relative* to the observer and amounts to concluding that the formally attainable absolute or ratio scales are not likely to arise in the framework of fuzzy set theory (Bilgiç & Türkşen 1995a).

Of all the axioms imposed on the fuzzy set structure, the boundaries and the Archimedean axiom are the most problematic (cf. Appendix to this chapter for formal definitions of these axioms).

The boundary axiom asserts the existence of (unique) *maximal and minimal* fuzzy terms and requires that for a given subject, there exists a fuzzy term which is more suitable to the subject than any other fuzzy term. Although this sounds reasonable for some cases, it is unlikely that it will hold for *all* fuzzy terms.

The Archimedean axiom (cf, Definition 3 in the Appendix) means that if Mary is cleverer than she is angry, there should be a *finite* amount of anger which, when attributed to Mary, will make her at least as angry as she is clever. Admittedly this might hold for *some* cases. However consider the statement "alcohol intake is more hazardous to liver than it is enjoyable." Archimedean axiom states that a finite amount of joy would make alcohol intake more enjoyable than it is hazardous. Is this quantity really finite?

It is important to notice that all the representations that are discussed in this section are "two-way" representations in the sense that the conditions to be imposed on the qualitative structure should be both *necessary and sufficient* for a representation on the numerical domain. If "one-way" representations are considered, then many of the problematic structural axioms like Archimedean and continuity that are necessary for the representation in the real line can be omitted. However, this approach does not give rise to the fruitful concept of *meaningfulness*.

If one only considers one-way representations then one can use t-conorms for all representations since they satisfy all the conditions of the theorems and *more*. However, such usage must be made with care. The numbers resulting from using a t-conorm as the disjunction can only be compared to each other in the ordinal sense and no other arithmetic operation (however tempting) can be performed on them unless one is ready to accept (or empirically verify) the Archimedean and continuity axioms.

If everyone cannot agree on the bounds of the system, the ratio and absolute scale representations that arise may be considered to be *relative* to the observer.

On the other hand, without the Archimedean axiom, there does not seem to be a representation that has a uniqueness stronger than ordinal scale.

All the representation theorems yield a numerical function, say ν_a, that takes on values in the unit interval:

$$F_i \succsim_a F_j \Longleftrightarrow \nu_a(F_i) \geq \nu_a(F_j).$$

This function represents the ordering of attributes (ranking of properties) that can be associated to a single subject a and *does not necessarily* represent the *belonging* of the subject to the fuzzy set F.

Simultaneous measurement of membership and property ranking. It is almost impossible to treat separately the measurement of membership functions and the selection of connectives to be used in the combination of membership functions.

The scales resulting from membership measurement and property ranking problems do not necessarily measure the same entity. The degree with which a subject belongs to a certain fuzzy set may not be equal to the degree that fuzzy set is associated with him/her[16].

One way to pursue is to *combine* the membership measurement and property ranking problems by introducing a new structure where the resulting measurement scale necessarily measures the membership degree in a fuzzy set. This path is followed by (Bollmann-Sdorra *et al.* 1993, Bilgiç 1995, Bilgiç & Türkşen 1995b)[17].

Bollmann-Sdorra *et al.* (1993) consider an ordered algebraic structure which justify the use of *min* as conjunction and *max* as disjunction. Their representation is ordinal.

Bilgiç & Türkşen (1995b) introduce two ways this combination can be made. In the first one, the two different problems are simply cast into a bounded semigroup structure (cf. Appendix). The consequences of this model are analyzed. It is argued that since accepting the Archimedean axiom (cf. Definition 3 in the Appendix) can be very hard for some fuzzy terms, the ratio scale representations are not likely to arise.

Apparently, in the literature of the ordered algebraic structures, there are very strong (absolute scale) representations for similar structures that one can use to represent a fuzzy set structure (cf. Appendix). Figure 3.4 summarizes the representation theorem presented in the Appendix (Theorem 1)[18].

However, representations stronger than ordinal scales require unacceptable structural axioms like the strong monotonicity (SM) and the Archimedean (Ar) conditions (cf. Definition 3 in the Appendix).

If one gives up the Archimedean axiom and endows the structure with other axioms (particularly the Idempotency axiom, cf. Figure 3.4), some weak representations can be obtained. These are ordinal scale representations and particularly the function *max* as originally suggested by Zadeh can be recovered as the unique disjunction satisfying some reasonable axioms.

However, ordinal scale representations exist at the cost of accepting that the equivalence of two fuzzy propositions is *transitive*. This fact is known to yield a class of paradoxes usually called the heaps paradox or sorites. The equivalence of fuzzy terms is not necessarily transitive. In order to introduce the *intransitivity* of equivalence in fuzzy set theory Bilgiç & Türkşen (1995b) propose another model in which the weak ordering axiom is replaced by the "interval ordering" axiom which results in a threshold representation. Unfortunately, this representation has some peculiar uniqueness characteristics[19].

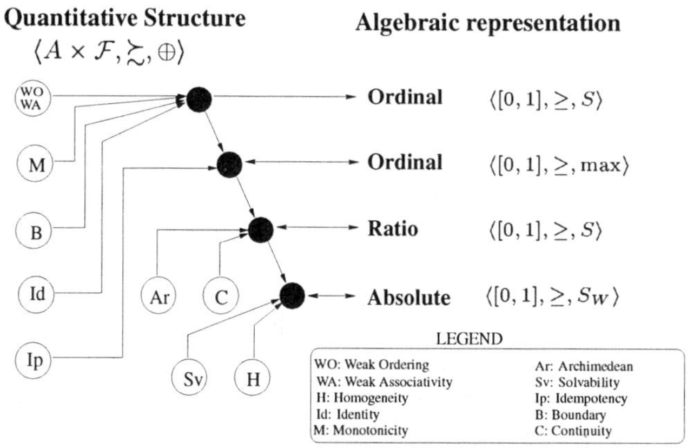

Figure 3.4 Summary of representations

The measurement models show that measurement of membership functions in fuzzy set theory is *formally* possible. However, the acceptability of each formal model needs to be critically analyzed.

On the other hand, once the Archimedean axiom is omitted it is not possible to come up with representations stronger than ordinal scale. This suggests that if one is not ready to accept the Archimedean axiom the only meaningful operation that can be performed on the measurement scale is the comparison (i.e., the results are on an ordinal scale). Any other arithmetic operation is simply meaningless. This suggests that triangular norms and conorms can be used to model connectives in fuzzy set theory but their results should not be attached any cardinal significance.

Archimedean axiom is necessary when a representation into real numbers is sought. If a representation into field extensions of the real number system is considered, then Archimedean condition is not necessary anymore (Narens 1985). Such a path has been followed by Nola & Gerla (1986). However, this approach has consequences of a philosophical nature as to why degrees of truth are not real numbers and why do they require non-standard analysis (Robinson 1966). This is still an open question.

If one accepts the Archimedean axiom and continuity, the results about the additive generators of Archimedean triangular norms and conorms simply state that using these in the unit interval and using addition on the extended real numbers amount to the same thing. Therefore, using an Archimedean triangular norm or a conorm as conjunction or disjunction of fuzzy sets is simply a matter of preference or convenience. For disjunction, one can equivalently use addition on extended reals. Therefore, the justification of continuous, Archimedean triangular norms and conorms require some unnatural structural axioms and in the end they amount to ordinary, additive extensive measurement. In order to move away from additivity one either has to give up the Archimedean axiom which as we have seen does not lead to strong representations (but nevertheless recovers max) or give up some other structural axiom. Luce *et al.* (1990) show that if one accepts the Archimedean axiom then associativity (in presence of positivity) amounts to additivity. Hence, the first candidate to give up seems to be the associativity (Bilgiç 1996). The attempts of Fodor (1993) and Fodor & Kereszfalvi (1995) to generalize triangular norms by dropping associativity require a measurement-theoretic discussion.

As for giving up the continuity, the main thrust behind the concept of a fuzzy set is continuous gradation. Hence moving away from continuity would be against the very nature of fuzziness.

Conjoint Measurement. One particular difficulty with the notion of fuzzy terms is that they can be *composed of* two or more simple terms (e.g. the term 'comfort' may constitute two parts 'humidity' and 'temperature'). Türkşen (1991) applies *conjoint measurement* techniques to this problem with the aim of showing that the scale for the higher order term (comfort) is constructed from the individual scales (humidity and temperature) using triangular norms and conorms. Under the assumption of *decomposability* Türkşen shows that all the triangular norms and conorms preserve the monotonicity of the building scales.

Türkşen (1991) considered the problem of *justifying* triangular norms, conorms, and parametrized families of them as valid under conjoint measurement. In so doing, he considers only "one-way" representations. To find both necessary and sufficient conditions that the main operations such as triangular norms, conorms and the parametrized families must follow and to discuss their acceptability in fuzzy set theory requires further investigation.

A critique of measurement theory. The measurement-theoretic discussion presented in this section relies on the *representational* view of measurement. There are three main objections to the representational theory of measurement (RTM) from psychologists and philosophers alike (Michell 1990, Henry E. Kyburg 1992):

1. RTM cannot be applied
   - because it uses an axiomatic approach, and
   - because it cannot give a satisfactory account of actual scientific measurement practice.
2. Errors of measurement cannot be incorporated into the RTM framework
3. Metaphysical objections:
   - RTM is wrongly *liberal* in what it accepts as measurement.
   - Whether numbers are primitive or obtained as results of a measurement process (realism on numbers or not).

RTM is motivated by the belief that numbers are not *real* and tries to show how numbers find their way into measurement (Michell 1990). The role of numbers in measurement is to *represent*. However, Michell (1990) espouses another view. He agrees with the methods of RTM in mapping an empirical system to a numerical system but, in his view, numbers are physical quantities rather than a mere abstraction. He considers that a numerical system which satisfies the properties of real numbers as the *only* possible candidate for measurement. And hence other numerical systems (ordinal scales for example) cannot be considered to yield a measurement scale.

Luce (1996) counters these criticisms and defends the RTM from several vantage points. He points out the main objective of the RTM as:

> "Contemporary" measurement theory builds on top of the classical view and introduces the, now well established, notion of *scale types*. Its purpose is primarily to inform the sciences—especially the behavioral and social ones—of the full range of measurement possibilities and to provide concrete contributions to those sciences (Luce 1996, p. 79).

Although there are valid points in the critique of the RTM, particularly about representing the errors of measurement, RTM is, by far, the most widely accepted account of measurement (see also Bilgiç & Türkşen (1997), in this connection).

3.3 ELICITATION METHODS

In the experimental research concerning membership functions, we detect two main trends: one branch tries to experimentally *validate* the postulates of fuzzy set theory. The second branch adopts a certain interpretation of the concept of the 'grade of membership' and then seeks ways of its elicitation. Since in both cases one needs ways of eliciting membership functions, we will summarize elicitation methods together with experimental validation results.

Mainly, there appears to be six methods used in experiments with the aim of constructing membership functions (Norwich & Türkşen 1982a, Chameau & Santamarina 1987a, Türkşen 1991):

- polling: do you agree that John is tall? (Yes/No).
- direct rating (point estimation): classify color A according to its darkness, classify John according to his tallness. In general, the question is: "How F is a?".
- reverse rating: identify the person who is tall to the degree 0.6? In general, identify a who is F to the degree $\mu_F(a)$.
- interval estimation (set valued statistics): give an interval in which you think color A lies, give an interval in which you think the height of John lies.
- membership function exemplification: what is the degree of belonging of color A to the (fuzzy) set of dark colors? What is the degree of belonging of John to the set of tall people? In general, "To what degree a is F?".
- pairwise comparison: which color, A or B, is darker (and by how much?)

3.3.1 Polling

In polling one subscribes to the point of view that fuzziness arise from *interpersonal* disagreements. The question "Do you agree that a is F?" is asked to *different* individuals. The answers are polled and an average is taken to construct the membership function.

Hersh & Carmazza (1976) use this approach[20] in their experiment 1. They presented their subjects a phrase (like small, very small, large, very large etc.) and then showed 12 squares in random order. The subjects responded by 'yes' or 'no' depending on whether they think that the phrase applies to the shown square or not. The subjects were also shown the 12 squares at the beginning of the experiment in ascending order so that every subject was operating within approximately *the same context*. Although the experimental results justify $(1 - \mu(x))$ for the connective NOT, the 'squaring the membership function' is not justified for VERY. Instead of squaring, VERY seems to *shift* the membership function (a result also reported in (Zysno 1981, Norwich & Türkşen 1984, Chameau & Santamarina 1987a)). They also report a reasonably good match between the membership function of "either small or large" and the *maximum* of the membership functions of small and large.

Polling is also one of the natural ways of eliciting membership functions for the likelihood interpretation.

3.3.2 Direct Rating

Direct rating seems to be the most straightforward way to come up with a membership function. This approach subscribes to the point of view that fuzziness arise from *individual subjective* vagueness[21]. The subject is required to classify a with respect to F *over and over again* in time. The experiment has to be carefully designed so that it will be harder for the subject to remember past answers. Hersh & Carmazza (1976) use this approach in their experiment 2. Essentially, this is a repetition of their experiment 1 for a single subject repeated over days. They report similar results for negation, disjunction and emphasizing (VERY) for each individual as in the group membership experiment. Stemming from responses of one particular subject they differentiate between *linguistic* and *logical* interpretations of membership functions.

One can also use direct rating to *compare* the answers of a subject against a pre-defined membership function Türkşen (1988, 1991). In that case the subject is asked "How F is a?" and the experimenter has a perfect knowledge of the evaluation for F (e.g. subject's height is known to the experimenter but not to the subject). The same question is asked to the same subject over and over again, and the membership is constructed using the assumption of probabilistic errors and by estimating a few key parameters as is usual for this type of construction.

Chameau & Santamarina (1987a) also discuss this method (which they call membership exemplification, however we reserve that term for another method). They use *several* subjects and aggregate their answers as opposed to asking a single subject same questions over and over again as is done in Norwich & Türkşen (1984) experiments. Norwich & Türkşen (1982c) and Chameau & Santamarina (1987a) report that this method results in membership functions with a wider spread (more fuzzy) when compared to polling and pairwise comparison.

Thole, Zimmermann & Zysno (1979) consider the measurement of membership functions and the justification of connectives within an empirical setting based on measurement theory. They argue that since the numerical membership scale is *bounded* the scale *has to be* an absolute scale. They mention the biases that one can have in direct elicitation methods, particularly the "end effect" which is a common problem in all bounded scales (including probability). They opt for an indirect elicitation method in the spirit of measurement theory and admit that only interval scales can be constructed with indirect methods. Then they suggest a combination of a direct rating technique and a Thurstonian scaling method. This way they create two scales and then try to identify the relationship of the two scales. They also report that neither the minimum nor the product operators are adequate representations of conjunction[22].

Nowakowska (1977) considers scaling of membership functions. He uses direct rating and poses the problem in terms of psychophysical scaling. He clarifies the assumptions to be made in order to perform direct rating.

Direct rating methods require evaluations to be measured on at least interval scales (Türkşen 1988, 1991).

3.3.3 Reverse Rating

In this method, the subject is given a membership degree and then asked to identify the object for which that degree corresponds to the fuzzy term in question (Türkşen 1988, 1991). This method can be used for individuals by repeating the same question for the same membership function as well as for a group of individuals.

Once the subject's (or subjects') responses are recorded the conditional distributions can be taken to be normally distributed and the unknown parameters (mean and variance) can be estimated as usual. This method also requires evaluations to be made on at least interval scales.

Chameau & Santamarina (1987a) consider reverse rating as a valuable tool to *verify* the membership function obtained by using another approach rather than an acquisition method.

3.3.4 Interval Estimation

Interval estimation subscribes to the *random set*-view of the membership function (cf. Section 3.2.2). The subject is asked to give an *interval* that describes the Fness of a. Let I_i be the set-valued observation (the interval) and m_i the frequency with which I_i is observed. Then $R = (I_i, m_i)$ defines a random set in the sense of Section 3.2.2 (Dubois & Prade 1989, 1991). Notice that this method is more appropriate to situations where there is a clear linear ordering in the measurement of the fuzzy concept like in tallness, heat, time, etc.

Chameau & Santamarina (1987a) find this approach of elicitation particularly advantageous over polling and direct rating in which the answer mode is necessarily crisp (Yes/No). Interval estimation is a relatively simple way of acquiring the membership function and it results in membership functions that are "less fuzzy" (the spread is narrower) when compared to direct rating and polling.

Interval estimation subscribes to the *uncertainty* view of membership functions as opposed to the *vagueness* view and in that sense it brings the issues of uncertainty modelling using fuzzy set theory, random sets, possibility measures and their relations to probability theory (Dubois & Prade 1993). Dubois & Prade (1986) and Civanlar &

Trussel (1986) describe statistical methods to come up with the membership function while subscribing to the random-set interpretation.

Zwick (1987) also consider the random set interpretation and uses the law of comparative judgments to assess the membership function.

Recently, set-valued statistics has been proposed as a more "natural" way of handling data than point-valued statistics. The set-valued statistics methods also assume a random-set or likelihood interpretation of the membership function. Such methods are analyzed in Kruse & Meyer (1987) in some detail.

Li & Yen (1995, Chapter 4) discuss various methods of elicitation methods. Their main assumption states fuzziness as a form of uncertainty. As a result of this view the first three methods they propose are suitable for the random set view of the fuzzy sets. They ask their subjects for intervals which include the fuzzy concept at hand. They do not differentiate between single subject versus multiple subject fuzzy concepts. And they do not discuss the issue of commensurability between two measurements they obtain for different agents and different fuzzy concepts.

3.3.5 Membership Exemplification

In terms of membership function exemplification, Hersh & Carmazza (1976) performed a test for the direct elicitation of the membership function. Hersh & Carmazza ordered twelve squares in ascending order and indicated each square with an ordinal number. They asked the subjects "write the number(s) which is appropriate for "large", "very large", "small" etc. The results are at variance with direct rating and polling most likely because there is no *repetition* in this elicitation method to normalize the effects of error or "noise".

Kochen & Badre (1974) also report experimental results on exemplification (they call it *anchoring*) which makes the resulting membership functions more precise (than without exemplification).

Zysno (1981) uses exemplification in an empirical setting. He asks 64 subjects from 21 to 25 years of age to rate 52 different statements of age with respect to one of the four sets: very young man, young man, old man, and very old man. He utilizes a scale from 0 to 100 to collect the answers. Since, at the outset, he has some hypothesis on the nature of the membership function (see Section 3.2.3), he mainly tries to test those hypotheses.

Kulka & Novak (1984) test for whether people use min-max or algebraic product and sum when combining fuzzy concepts. They use the method of exemplification (using ellipses and rectangles) in their experiments. Their conclusions are weak and they call for more experiments to come up with stronger conclusions.

However, the use of computer graphics to give an example membership function to be modified by the subject greatly enhanced this procedure as is usually witnessed in commercial applications of "fuzzy expert system shells".

3.3.6 Pairwise Comparison

Kochen & Badre (1974) report experimental results for the "precision" of membership functions using pairwise comparison method. They report experimental evidence for the precision of *greater, very much greater* and *much greater* in decreasing precision, in that order. How the addition of *very* makes the adjective more precise is highlighted.

Oden (1979) discusses the use of fuzzy set theory in psycholinguistic theories. He considers *comparisons* of the form: "which is a better example of a bird: an eagle or a pelican?", and after the answer to this question (say, an eagle is chosen): "How *much more* of a bird is an eagle than a pelican?". But, by asking for the strength of the preference directly, Oden falls for a pitfall for which many researchers have been cautioning us for a long time (see e.g. (Luce & Raiffa 1957, p. 32, Fallacy 3)).

Chameau & Santamarina (1987a) also use the same pairwise comparison technique and report it to be as robust as polling and direct rating. Following Saaty (1974), they require the subjects to provide pairwise comparisons *and the strength* of preference. This yields a non-symmetric full matrix of relative weights. The membership function is found by taking the components of the eigenvector corresponding to the maximum eigenvalue. The values are also *normalized*. Chameau & Santamarina (1987a) also find the requirement that evaluations be on a ratio scale to be unnatural (cf. Section 3.2.5).

However, they espouse a "comparison-based point estimation" which determines the position of a set of stimuli on the reference axis by pairwise comparison and the membership is calculated by *aggregating* the values provided by several subjects. Although the subjects of Chameau & Santamarina (1987a) experiments ranked this method almost as good as the interval estimation method (which was ranked as the best method), this method also needs the unfortunate assumption of a ratio scale. Furthermore, pairwise comparison requires many comparison experiments in a relatively simple domain.

3.3.7 Fuzzy Clustering Methods

Most of the elicitation techniques described so far subscribe to the subjective interpretation of the membership function. However, a particularly important concern in practice is approaches to construct membership functions from a *given set of data*. When adequate amount of data is already collected and a model based on fuzzy sets is required to analyze the problem, the membership functions must be constructed from the available data.

Fuzzy clustering techniques (Ruspini 1969, Bezdek & Harris 1979, Sugeno & Yasukawa 1993) are appropriate tools for such an analysis. Usually the analysis proceeds as follows (Nakanishi, Turksen & Sugeno 1993):

- Apply clustering on the output data and then project it into the input data, generate clusters and select the variables associated with input-output relations. The clustering method determines clusters on the data space based on Euclidean norm.

- Form the membership functions for the variables selected (i.e., determine the shape of the membership functions). There is a procedure (Nakanishi *et al.* 1993, Section 2.3.1) that lets one to select four parameters which completely characterizes trapezoidal membership functions[23].

- Select the input variables by dividing the data into three subgroups. Use two groups in the model building for selection of effective (important) variables and cross validation for data set independence. The third group is used as the test data to validate the goodness of the model.

Since most fuzzy clustering techniques are based on Euclidean norm, formation of fuzzy clusters, and hence membership functions with non-Euclidean norms requires further investigation.

3.3.8 Neural-fuzzy Techniques

With the advance in neural network models and their applications to machine learning a considerable interest has been on unifying neural networks and fuzzy set theory (Rocha 1982, Takagi 1990, Kosko 1991). In this section we briefly mention several studies in which neural networks have been utilized to come up with membership functions from a given set of data.

Takagi & Hayashi (1991) discuss a Neural Network that generates nonlinear, *multi-dimensional* membership functions which is a membership function generating module of a larger system that utilized fuzzy logic. They claim that the advantage of using nonlinear multi-dimensional membership functions is in its effects in reducing the number of fuzzy rules in the rule base.

Yamakawa & Furukawa (1992) present an algorithm for learning membership functions using a model of the *fuzzy neuron*. Their method uses example-based learning and optimization of cross-detecting lines. They assign *trapezoidal* membership functions and automatically come up with its parameters. The context is handwriting recognition. They also report some computational results for their algorithm.

On the experimental side, Erickson, Lorenzo & Woodbury (1994) claim that fuzzy membership functions and fuzzy set theory *better explain* the classification of taste responses in brain stem. They analyze previously published data and allow each neuron to belong to *several* classifications *to a degree*. This degree is measured by the neuron's response to the stimuli. They show that their model based on fuzzy set theory explains the data better than other statistical models.

Furukawa & Yamakawa (1995) describe two algorithms that yield membership functions for a fuzzy neuron and their application to recognition of hand writing. The crossing points of two (trapezoidal) membership functions are *optimized* for the task at hand.

3.3.9 General Remarks

In general, Chameau & Santamarina (1987a, 1987b) report good agreement between direct rating, interval estimation and membership exemplification with the comment that, in most of the cases, fuzzy sets obtained by exemplification method are wider (fuzzier) than the ones obtained by other methods. The main difficulty with the point-estimation method is the contradiction between fuzziness of the perception and the crispness of the response mode. This difficulty is overcome by the interval estimation method which in turn needs a minimum number of assessors or assessments. However, Chameau & Santamarina (1987a) report that as low as five assessments are sufficient. Exemplification yields membership functions without further processing which is an advantage. The way they carry out the pairwise comparison method assumes a ratio scale for the measurements which is hardly justified (cf. Section 3.2.5).

The assessors that took part in the experiments of Chameau & Santamarina (subjectively) rated the interval estimation method as the best in terms of expected consistency and expected quality. The age of the assessors affected their response, particularly in the "old-not old" task.

In all of the elicitation methods, Chameau & Santamarina obtain the membership functions based on *averaging or aggregation* of the responses from several assessors. In that sense they do not subscribe to the individualistic interpretation of fuzziness. Chameau & Santamarina justify this approach by assuming that fuzziness *is a property of the phenomenon* rather than a property attributed by the observer, which may be disputed.

One important issue in constructing membership functions is the *context*. It has always been emphasized that fuzzy set theory and particularly the membership functions are context dependent. Hersh, Carmazza & Brownell (1979) discuss the effects of *changing context* in determining the membership functions, a problem which eluded linguists for a long time (Kamp 1975). They report that the frequency of occurrence of the elements does not effect the location and form of the membership function (i.e., if one asks "Is John tall?" over and over again the resulting membership is consistently of the same form). On the other hand, the number of unique elements significantly affects the form of the membership function (i.e., the answer to the question "Is John tall?" varies when there is only one more person to consider versus when there is more than one more person to consider. The context changes!). This result tends to suggest that the membership function is not only a function of the object from the universe of discourse but the discourse as well. That is, if X is the universe of discourse and $x \in X$, the membership function for the fuzzy subset F of X is actually of the form:

$$\mu_F(x) = f_F(x, X).$$

Dombi (1990) takes up the problem of determining the membership function. He focuses on providing a theoretical basis for membership construction which is relevant to the problem at hand, described with only a few parameters that are *meaningful*. He avoids discussing problems of context dependency, measurement and uncertainty.

He postulates five axioms of which the most unnatural is the fourth one: μ is a rational function of polynomials of the following form

$$\mu(x) = \frac{a_0 x^n + a_1 x^{n-1} + \cdots + a_n}{A_o x^m + A_1 x^{m-1} + \cdots + a_m} \qquad (m \neq 0).$$

Furthermore, Dombi requires that a membership function be such that $(n + m)$ is minimal.

Based on these assumptions he derives a *mathematical form* for the membership function and verifies his result using the empirical data obtained by Zimmermann & Zysno (1985).

He also discusses the connections of the proposed form to logical connectives. Particularly, he considers aggregation operators (which generalize triangular norms) and derives expressions for the conjunction and disjunction.

This work is interesting in the sense that it starts the analysis from the membership function, postulates a *form* for membership function and then *derives* conjunction and disjunction. On the other hand, the axioms (or constraints, rather) are not further discussed for their rationality.

Chen & Otto (1995) consider constructing *continuous* membership functions from a given set of discrete points (evaluations). The rationale is that: one can only answer *finite* amount of questions from which a continuous membership function has to be constructed. They propose that the membership function can be constructed on an interval scale. When one requires membership functions to be continuous and convex[24], curve fitting methods might yield membership functions that are outside the unit interval and non-convex. In order to obtain continuous membership functions that are invariably bounded, convex and continuous Chen & Otto propose a *constrained interpolation* method. This method, which minimally considers the semantic issues, is a powerful way of obtaining a continuous membership function from discrete data, a problem frequently encountered in practice.

Kruse, Gebhardt & Klawonn (1994, Section 2.8.3) briefly discuss elicitation of membership functions. They acknowledge the elicitation as a major problem in fuzzy set theory. They maintain that the fuzzy concepts can only be measured on an ordinal scale and they emphasize that "the initial choice of precise numbers for membership functions is not important". They claim that the fuzzy methods are *insensitive* to precise measurement of membership functions and what is more important is the *qualitative properties* of the phenomena modelled using fuzzy sets.

3.4 SUMMARY

Membership functions are the building blocks of fuzzy set theory. We outlined five possible interpretations of the membership function and critically discussed each interpretation. We believe that a satisfactory treatment of the meaning of membership function can be given within measurement theory. The view from measurement theory provides insights into the foundations of the fuzzy set theory. In that framework one can lay down the axioms and thoroughly discuss them as to their applicability to the theory.

The results of the measurement-theoretic investigation indicates some difficulties in the foundations related to boundaries and to some of the *cardinal scales* (ratio and absolute scales). It is unlikely that the bounds on the membership scale are *universal*. In asserting this we do not only think that no consensus on bounds is available but also there are no normative grounds on which the bounds *should* exist. This observation brings about the *subjectivity* of membership evaluations. Furthermore, the requirements of ratio and absolute scales seem to be too strict for membership functions. Hence, the only *meaningful* operation on membership functions turn out to be the comparison (ordinal scale).

On the experimental side, there has been numerous studies for eliciting membership functions. These are mainly characterized by whether they espouse an *individual* or a *group* interpretation of fuzziness and whether they espouse an objective or a subjective view of fuzziness. Experimental results for both interpretations are reported using various methods of elicitation. The experimental results are also available for various interpretations discussed in Section 3.2. The results of the measurement-theoretic investigation impose a constraint on the elicitation method: the method should not be *intrusive* in the sense that it should not require arithmetic manipulations on the raw membership evaluation.

Furthermore, *automatic* elicitation methods like fuzzy clustering methods and fuzzy-neural techniques stem from a practical need of eliciting membership functions from a set of *given data*. These are considered more *objective* ways of elicitation when compared to others.

Notes

1. This is the definition given in Zadeh (1965). Various other formalisms also appear in the literature and many of those will be reviewed in this chapter.

2. Although this is the original notation, the recent trend is to use $F(.)$ instead of μ_F to avoid heavy mathematical notation. Since we are not going to use this notation extensively in formulas in this chapter, we stick with the older notation.

3. A many-valued (actually infinite-valued) logic based on fuzzy set theory exists and is called *fuzzy logic* (Goguen 1969, Zadeh 1975, Giles 1976). This is analogous to the one between two-valued set theory and first order predicate logic. Hence, the fuzzy set intersection and union correspond to connectives AND and OR of fuzzy logic, respectively. In this study, the terms union or disjunction of fuzzy sets and intersection or conjunction of fuzzy sets are used interchangeably.

4. Gaines (1978) also espouses a similar view in the context of the discussion "Are membership functions probabilities?". This debate is still alive (Laviolette & J. W. Seaman 1994).

5. For a different perspective see Henry E. Kyburg (1992) where errors of measurement are treated probabilistically in a Bayesian framework.

6. The term "nested" is used in the sense that whenever $\alpha \geq \alpha'$ then $F_\alpha \subseteq F_{\alpha'}$.

7. See also Klir (1994) in this connection.

8. The approach to measurement theory in this section is purely *representational and formal* as is exemplified and widely accepted in the references mentioned. However, see Section 3.4 for different views on measurement theory.

9. This "formality" of the measurement-theoretic approach is somewhat disliked by experimental psychologists (Smithson 1987, pp. 78–79).

10. These measurement problems should not be confused with a totally different measurement problem: measurement of *fuzzy measures*. That problem is studied elsewhere (Suppes 1974, Dubois 1986, Dubois 1988). There the aim is to consider an *algebra* of subsets of a set and their representation. That type of a measurement problem is more akin to the measurement theoretic representations of probabilities and highlights the formal differences between fuzzy set theory, probability theory and fuzzy measure theory (Dubois & Prade 1989).

11. Saaty used the same technique in his Analytic Hierarchy Process (AHP) Saaty (1977, 1980). However, the validity of the scale for AHP remained problematic to date (Triantaphyllou & Mann 1990, Lambert 1992, Murphy 1993, Saaty 1994). Particularly, Lambert (1992) concludes that (although based on ratio scales) the methods with which AHP come up with membership functions (priorities) are only "rough cuts".

12. In the Appendix, we show another representation theorem in which ratio scale formally arises and discuss its applicability to fuzzy set theory in Sections 3.2.5 and 3.2.5.

13. This is done mainly by following each item in Theorem 1 of the Appendix.

14. One might argue that the qualitative relation to be defined here is a "crisp" one. Then the sentence "John is cleverer than he is tall" simply means "John is clever but not tall" without showing any sign of *degrees* of tallness or cleverness. We do not subscribe to this point of view and assume that when one utters single subject sentences of the sort to be described, one has a degree of belonging in mind which we attempt to measure (Bolinger 1972).

15. Actually, one might consider a richer structure in which there is room for conjunction, negation, and implication as well. But we do not do that here for the following reasons: *(i)* conjunction and implication can easily be derived from a disjunction via the negation, and *(ii)* the representation that we have for negation functions is only *ordinal* (Trillas 1979). See also MV-algebras and their representations in this connection (Chang 1958, Chan 1959, Nola 1993).

16. Consider John, who is a basketball player, and consider his tallness. Among all the other attributes of John, let his tallness be the least that one can associate with him. Therefore, $\nu_{John}(tall) = 0$. But when compared to other people $\mu_{tall}(John) > 0$.

17. See also (de Cooman 1991, de Cooman & Kerre 1994) in this connection where the aim is not guided by measurement theory but similar ideas arise within their characterization of fuzzy sets.

18. Notice that in Figure 3.4 there is no mention of an *interval scale*. An interval scale requires a different algebraic structure $\langle A_1 \times A_2, \succsim, \oplus \rangle$ (positive difference structure) with slightly different axioms (Krantz *et al.* 1971, p. 147).

19. Another representation in which the membership functions are point-valued but the connectives are vague linguistic operators also generates interval-valued representations. These representations turn out to be ordinal as developed in Bilgiç (1995) based-on the idea of an "interval-valued fuzzy set" (Türkşen 1986).

20. See also Labov (1973) for the same approach.

21. This approach is in tune with the claim that the vagueness in verbal concepts is an integral part of the concept rather than a result of summing of variable responses over individuals (McCloskey & Glucksberg 1978).

22. This finding enables Zimmermann & Zysno (1980) to come up with a "compensatory AND" operator which is a parametric operator that covers minimum and product. Relationship of this operator to interval-valued fuzzy sets is shown in (Türkşen 1992).

23. There has been a recent interest in using trapezoidal membership functions in fuzzy set theory mainly because their special structure yields more efficient computations. By definition, triangular membership functions are special cases of trapezoidal membership functions. However, trapezoidal functions are not as general as continuous spline functions. See e.g. (Pedrycz 1994) for a motivation to use triangular membership functions in fuzzy control.

24. A membership function, μ, (over the real numbers) is called *convex* if and only if for all $x, y \in \mathbb{R}$ and for all $\lambda \in [0,1]$, $\mu(\lambda x + (1-\lambda)y) \geq \min\{\mu(x), \mu(y)\}$.

References

Aczél, J. (1966). *Lectures on Functional Equations and Applications*, Academic Press, New York.

Alsina, C. (1985). On a family of connectives for fuzzy sets, *Fuzzy Sets and Systems* **16**: 231–235.

Bandler, W. & Kohout, L. J. (1986). The use of checklist paradigm in inference systems, *in* C. V. Negoita & H. Prade (eds), *Fuzzy Logic in Knowledge Engineering*, Verlag TÜV, Rheinland, Köln, chapter 7, pp. 95–111.

Bellman, R. & Giertz, M. (1973). On analytic formalism of the theory of fuzzy sets, *Information Sciences* **5**: 149–156.

Bezdek, J. C. & Harris, J. D. (1979). Convex decompostions of fuzzy partitions, *Journal of Mathematical Analysis and Applications* **6**: 490–512.

Bilgiç, T. (1995). *Measurement-Theoretic Frameworks for Fuzzy Set Theory with Applications to Preference Modelling*, PhD thesis, University of Toronto, Dept. of Industrial Engineering Toronto Ontario M5S 1A4 Canada.

Bilgiç, T. (1996). The archimedean assumption in fuzzy set theory, *in* M. H. Smith, M. A. Lee, J. Keller & J. Yen (eds), *Proceedings of NAFIPS '96*, IEEE Press, pp. 299–303.

Bilgiç, T. & Türkşen, I. (1995a). Measurement-theoretic justification of fuzzy set connectives, *Fuzzy Sets and Systems* **76**(3): 289–308.

Bilgiç, T. & Türkşen, I. B. (1995b). Measurement-theoretic frameworks for fuzzy set theory, *Working notes of the IJCAI-95 workshop on "Fuzzy Logic in Artificial Intelligence"*, pp. 55–65. 14th International Joint Conference on Artificial Intelligence Montréal, Canada.

Bilgiç, T. & Türkşen, I. B. (1997). Elicitation of membership functions: How far can theory take us?, *Proceedings of FUZZIEEE'97*, IEEE Press, Barcelona.

Bolinger, D. L. M. (1972). *Degree words*, Vol. 53 of *Janua linguarum. Series maior*, The Hague Mouton.

Bollmann-Sdorra, P., Wong, S. K. M. & Yao, Y. Y. (1993). A measurement-theoretic axiomatization of fuzzy sets, *Fuzzy Sets and Systems* **60**(3): 295–307.

Chameau, J. L. & Santamarina, J. C. (1987a). Membership functions part I: Comparing method of measurement, *International Journal of Approximate Reasoning* **1**: 287–301.

Chameau, J. L. & Santamarina, J. C. (1987b). Membership functions part II: Trends in fuzziness and implications, *International Journal of Approximate Reasoning* **1**: 303–317.

Chan, C. C. (1959). A new proof of the completeness of the lukesiewicz axioms, *Transactions of Maerican Mathematical Society* **93**: 74–80.

Chang, C. C. (1958). Algebraic analysis of many-valued logics, *Transactions of American Mathematical Society* **88**: 467–490.

Chellas, B. F. (1980). *Modal logic: an introduction*, Cambridge University Press, Cambridge, UK.

Chen, J. E. & Otto, K. N. (1995). Constructing membership functions using interpolation and measurement theory, *Fuzzy Sets and Systems* **73**: 313–327.

Civanlar, M. R. & Trussel, H. J. (1986). Constructing membership functions using statistical data, *Fuzzy Sets and Systems* **18**: 1–14.

de Cooman, G. (1991). Evaluation sets and mappings-the order-theoretic aspect of the meaning of properties, *in* E. E. Kerre (ed.), *Introduction to the Basic Principles of Fuzzy Set Theory and Some of its Applications*, Universteit Gent, Communication and Cognition, Gent, Belgium, pp. 159–213.

de Cooman, G. & Kerre, E. E. (1994). Order norms on bounded partially ordered sets, *The Journal of Fuzzy Mathematics* **2**: 281–310.

Dombi, J. (1990). Membership function as an evaluation, *Fuzzy Sets and Systems* **35**: 1–22.

Dubois, D. (1986). Belief structures, possibility theory and decomposable confidence measures on finite sets, *Computers and Artificial Intelligence* **5**(5): 403–416.

Dubois, D. (1988). Possibility theory: Searching for normative foundations, *in* B. R. Munier (ed.), *Risk, Decision and Rationality*, D. Reidel Publishing Company, pp. 601–614.

Dubois, D. & Prade, H. (1980). *Fuzzy sets and systems : theory and applications*, Academic Press, New York.

Dubois, D. & Prade, H. (1982). A class of fuzzy measures based on triangular norms: a general framework for the combination of information, *International Journal of General Systems* **8**: 43–61.

Dubois, D. & Prade, H. (1986). Fuzzy sets and statistical data, *European Journal of Operational Research* **25**: 345–356.

Dubois, D. & Prade, H. (1989). Fuzzy sets, probability and measurement, *European Journal of Operational Research* **40**: 135–154.

Dubois, D. & Prade, H. (1991). Random sets and fuzzy interval analysis, *Fuzzy Sets and Systems* **42**: 87–101.

Dubois, D. & Prade, H. (1993). Fuzzy sets and probability: Misunderstandings, bridges and gaps, *Second IEEE International Conference on Fuzzy Systems*, IEEE, pp. 1059–1068. San Fransisco, California March 28–April 1, 1993.

Erickson, R. P., Lorenzo, P. M. D. & Woodbury, M. A. (1994). Classification of taste responses in brain stem: Membership in fuzzy sets, *Journal of Neurophysiology* **71**(6): 2139–2150.

Fishburn, P. C. (1970). *Utility theory for decision making*, Operations Research Society of America. Publications in operations research, no. 18, Wiley, New York.

Fodor, J. C. (1993). A new look at fuzzy connectives, *Fuzzy Sets and Systems* **57**: 141–148.

Fodor, J. C. & Keresztfalvi, T. (1995). Nonstandard conjunctions and implications in fuzzy logic, *International Journal of Approximate Reasoning* **12**: 69–84.

French, S. (1984). Fuzzy decision analysis: Some criticisms, *in* H. Zimmermann, L. Zadeh & B. Gaines (eds), *Fuzzy Sets and Decision Analysis*, North Holland, pp. 29–44.

Fuchs, L. (1963). *Partially Ordered Algebraic Systems*, Pergamon Press, London.

Fung, L. W. & Fu, S. K. (1975). An axiomatic approach to rational decision making in a fuzzy environment, *in* H. J. Zimmermann, L. A. Zadeh & B. R. Gaines (eds), *Fuzzy Sets and Their Application to Cognitive and Decision Processes*, Academic Press, New York, pp. 227–256.

Furukawa, M. & Yamakawa, T. (1995). The design algorithms of membership functions for a fuzzy neuron, *Fuzzy Sets and Systems* **71**(3): 329–343.

Gaines, B. R. (1978). Fuzzy and probabilistic uncertainty logics, *Information and Control* **38**(2): 154–169.

Giles, R. (1976). Łukasiewicz logic and fuzzy set theory, *International Journal of Man-Machine Studies* **8**(3): 312–327.

Giles, R. (1988). The concept of grade of membership, *Fuzzy Sets and Systems* **25**: 297–323.

Goguen, J. (1969). The logic of inexact concepts, *Synthese* **19**: 325–373.

Henry E. Kyburg, J. (1992). Measuring errors of measurement, *in* C. W. Savage & P. Ehrlich (eds), *Philosophical and Foundational Issues in Measurement Theory*, Lawrence Erlbaum Associates Inc. Publishers, New Jersey, USA, chapter 5, pp. 75–92.

Hersh, H. & Carmazza, A. (1976). A fuzzy set approach to modifiers and vagueness in natural language, *Journal of Experimental Psychology: General* **105**(3): 254–276.

Hersh, H., Carmazza, A. & Brownell, H. H. (1979). Effects of context on fuzzy membership functions, *in* M. M. Gupta, R. M. Ragade & R. R. Yager (eds), *Advances in Fuzzy Set Theory*, North-Holland, Amsterdam, pp. 389–408.

Hisdal, E. (1985). Reconciliation of the Yes-No versus grade of membership dualism in human thinking, *in* M. M. Gupta, A. Kandel, W. Bandler & J. B. Kiszka (eds), *Approximate Reasoning in Expert Systems*, North-Holland, Amsterdam, pp. 33–46. Also in Dubois, D., H. Prade, and R. Yager eds., Readings in Fuzzy Sets for Intelligent Systems, Morgan Kaufmann, pp. 854-860, 1993.

Hisdal, E. (1988). Are grades of membership probabilities?, *Fuzzy Sets and Systems* **25**: 325–348.

Kamp, J. A. W. (1975). Two theories about adjectives, *in* E. L. Keenan (ed.), *Formal Semantics of Natural Language*, Cambridge University Press, London, pp. 123–155.

Kempton, W. (1981). *The folk classification of ceramics: A study of cognitive prototypes*, Academic Press, New York.

Klir, G. J. (1994). Multivalued logics versus modal logics: alternative frameworks for uncertainty modeling, *in* P. P. Wang (ed.), *Advances in Fuzzy Theory and Technology*, Duke University, Durham, North Carolina U.S.A., pp. 3–48. Vol. 2.

Kneale, W. C. (1962). *The development of logic*, Oxford , Clarendon Press, England.

Kochen, M. (1975). Applications of fuzzy sets in psychology, *in* L. A. Zadeh, K. S. Fu, K. Tanaka & M. Shimura (eds), *Fuzzy Sets and thei applications to cognitive and decision processes*, Academic Press, pp. 395–408.

Kochen, M. & Badre, A. N. (1974). On the precision of adjectives which denote fuzzy sets, *Journal of Cybernetics* **4**: 49–59.

Kohout, L. J. & Bandler, W. (1993). Interval-valued systems for approximate reasoning based on checklist paradigm semantics, *in* P. P. Wang (ed.), *Advances of Fuzzy Set Theory*, Bookwrigths Press, Durham, North Carolina USA, pp. 167–194. Volume 1.

Kosko, B. (1991). *Neural Networks and Fuzzy Systems*, Prentice-Hall, Englewood Cliffs, NJ.

Krantz, D. H., Luce, R. D., Suppes, P. & Tversky, A. (1971). *Foundations of Measurement*, Vol. 1, Academic Press, San Diego.

Kruse, R., Gebhardt, J. & Klawonn, F. (1994). *Foundations of Fuzzy Systems*, John Wiley and Sons, New York.

Kruse, R. & Meyer, K. D. (1987). *Statistics with Vague data*, Theory and Decision Library: Series B Mathematical and Statistical Methods, D. Reidell publishing company, Dordrech Holland.

Kulka, J. & Novak, V. (1984). Have fuzzy operators a psychologocal correspondence?, *Studia Psychologica* **26**: 131–140.

Labov, W. (1973). The boundaries of words and their meanings, *in* C. J. Bailey & R. W. Shuy (eds), *New Ways of Analyzing Variation in English*, Georgetown University Press, Washington.

Lakoff, G. (1973). Hedges: A study in meaning criteria and the logic of fuzzy concepts, *Journal of Philosophical Logic* **2**: 458–508.

Lakoff, G. (1987). *Women, Fire, and Dangerous Things: What categories reveal about the mind*, The University of Chicago Press, Chicago.

Lambert, J. (1992). The fuzzy set membership problem using the hierarchy decision method, *Fuzzy Sets and Systems* **48**(3): 323–330.

Laviolette, M. & J. W. Seaman, J. (1994). The efficacy of fuzzy representations of uncertainty, *IEEE Transactions on Fuzzy Systems* **2**(1): 4–15. With comments from D. Dubois and H. Prade, E. Hisdal, G. J. Klir, B. Kosko, N. Wilson, D. V. Lindley.

Li, H. X. & Yen, V. C. (1995). *Fuzzy Sets and Fuzzy Decision-Making*, CRC Press, Boca Raton.

Ling, C. H. (1965). Representation of associative functions, *Publicationes Mathematicae Debrecen* **12**: 189–212.

Luce, R. D. (1996). The ongoing dialog between empirical science and measurement theory, *Journal of Mathematical Psychology* **40**: 78–98.

Luce, R. D. & Raiffa, H. (1957). *Games and Decisions*, Wiley, New York.

Luce, R., Krantz, D., Suppes, P. & Tversky, A. (1990). *Foundations of Measurement*, Vol. 3, Academic Press, San Diego, USA.

Mabuchi, S. (1992). An interpretation of membership functions and the properties of general probabilistic operators as fuzzy set operators .1. case of type-1 fuzzy-sets, *Fuzzy Sets and Systems* **49**(3): 271–283.

McCall, S. & Ajdukiewicz, K. (eds) (1967). *Polish logic, 1920-1939*, Oxford, Clarendon P. papers by Ajdukiewicz [and others]; with, an introduction by Tadeusz Kotarbinski, edited by Storrs McCall, translated by B. Gruchman [and others].

McCloskey, M. E. & Glucksberg, S. (1978). Natural categories: Well defined or fuzzy sets?, *Memory and cognition* **6**: 462–472.

Michell, J. (1990). *An introduction to the logic of psychological measurement*, L. Erlbaum Associates, Hillsdale, N.J.

Murphy, C. K. (1993). Limits on the analytic hierarchy process from its consistency index, *European Journal of Operational Research* **65**: 138–139.

Nakanishi, H., Turksen, I. B. & Sugeno, M. (1993). A review and comparison of six reasoning methods, *Fuzzy Sets and Systems* **57**: 257–294.

Narens, L. (1985). *Abstract Measurement Theory*, MIT Press, Cambridge, Mass.

Nola, A. D. (1993). MV algebras in the treatment of uncertainty, *in* R. Lowen & M. Roubens (eds), *Fuzzy logic : state of the art*, Kluwer Academic Publishers, Dordrecht ; Boston, pp. 123–131.

Nola, A. D. & Gerla, G. (1986). Nonstandard fuzzy sets, *Fuzzy Sets and Systems* **18**: 173–181.

Norwich, A. M. & Türkşen, I. B. (1982a). The construction of membership functions, *in* R. R. Yager (ed.), *Fuzzy Sets and Possibility Theory: Recent Developments*, Pergamon Press, New York, pp. 61–67.

Norwich, A. M. & Türkşen, I. B. (1982b). The fundamental measurement of fuzziness, *in* R. R. Yager (ed.), *Fuzzy Sets and Possibility Theory: Recent Developments*, Pergamon Press, New York, pp. 49–60.

Norwich, A. M. & Türkşen, I. B. (1982c). Stochastic fuzziness, *in* M. M. Gupta & E. Sanchez (eds), *Approximate Reasoning in Decision Analysis*, North-Holland, pp. 13–22.

Norwich, A. M. & Türkşen, I. B. (1984). A model for the measurement of membership and the consequences of its empirical implementation, *Fuzzy Sets and Systems* **12**: 1–25.

Nowakowska, M. (1977). Methodological problems of measurement of fuzzy concepts in the social sciences, *Behavioral Science* **22**: 107–115.

Oden, G. C. (1979). Fuzzy propositional approach to psyholinguistic problems: an application of fuzzy set theory in cognitive science, *in* M. M. Gupta, R. K. Ragade & R. R. Yager (eds), *Advances in Fuzzy Set Theory and Applications*, North-Holland, Amsterdam, pp. 409–420.

Osherson, D. N. & Smith, E. E. (1981). On the adequacy of prototype theory as a theory of concepts, *Cognition* **9**: 35–58.

Osherson, D. N. & Smith, E. E. (1982). Gradedness and conceptual combination, *Cognition* **12**: 299–318.

Palmer, F. (1981). *Semantics*, 2 edn, Cambridge University Press, New York.

Pedrycz, W. (1994). Why triangular membership functions?, *Fuzzy Sets and Systems* **64**(1): 21–30.

Pipino, L. L., van Gich, J. P. & Tom, G. (1981). Experiments in the representation and manipulation of labels of fuzzy sets, *Behavioral Science* **26**: 216–228.

Roberts, F. (1979). *Measurement Theory*, Addison Wesley Pub. Co.

Robinson, A. (1966). *Non-standard Analysis*, North-Holland Pub. Co., Amsterdam.

Rocha, A. F. (1982). Basic properties of neural circuits, *Fuzzy Sets and Systems* **7**: 109–121.

Rosch, E. & Mervis, C. B. (1975). Family resemblances: Studies in the internal structure of categories, *Cognitive Pschology* **7**: 573–605.

Rosser, J. B. & Turquette, A. R. (1977). *Many-Valued Logics*, Greenwood Press, Westport Connecticut.

Ruspini, E. H. (1969). A new approach to clustering, *Information Control* **15**(1): 22–32.

Ruspini, E. H. (1991). On the semantics of fuzzy logic, *International Journal of Approximate Reasoning* **5**(1): 45–88.

Saaty, T. L. (1974). Measuring the fuzziness of sets, *Journal of Cybernetics* **4**: 53–61.

Saaty, T. L. (1977). A scaling method for priorities in hierarchical structures, *Journal of Mathematical Psychology* **15**: 234–281.

Saaty, T. L. (1980). *The Analytic Hierarcy Process*, McGraw-Hill, New York.

Saaty, T. L. (1986). Scaling the membership function, *European Journal of Operational Research* **25**: 320–329.

Saaty, T. L. (1994). Homogeneity and clustering in AHP ensures the validity of the scale, *European Journal of Operational Research* **72**: 598–601.

Sapir, E. (1944). Grading: a study in semantics, *Philosophy of Science* **11**: 93–116.

Schweizer, B. & Sklar, A. (1983). *Probabilistic Metric Spaces*, North-Holland, Amsterdam.

Scott, D. (1976). Does many-valued logic have any use?, *in* S. Körner (ed.), *Philosophy of logic*, Camelot Press, Southampton, Great Britain, chapter 2, pp. 64–95. with comments by T.J. Smiley, J.P. Cleave and R. Giles. Bristol Conference on Critical Philosophy, 3d, 1974.

Smithson, M. (1987). *Fuzzy Set Analysis for Behavioral and Social Sciences*, Recent Research in Psychology, Springer-Verlag, Berlin, Germany.

Sugeno, M. & Yasukawa, T. (1993). A fuzzy logic based approach to qualitative modelling, *IEEE Transactions on Fuzzy Systems* **1**(1): 1–24.

Suppes, P. (1974). The measurement of belief, *Journal of Royal Statistical Society Series B* **36**(2): 160–175.

Suppes, P., Krantz, D., Luce, R. & Tversky, A. (1989). *Foundations of Measurement*, Vol. 2, Academic Press, San Diego.

Takagi, H. (1990). Fusion technology of fuzzy theory and neural network: Survey and future directions, *Proc. of International Conference on Fuzzy Logic and Neural Networks*, Iizuka, Japan, pp. 13–26.

Takagi, H. & Hayashi, I. (1991). Nn-driven fuzzy reasoning, *International Journal of Approximate Reasoning* **5**(3): 191–213. (Special Issue of IIZUKA'88).

Thole, U., Zimmermann, H. J. & Zysno, P. (1979). On the suitability of minimum and product operators for the interpretation of fuzzy sets, *Fuzzy Sets and Systems* **2**: 167–180.

Thomas, S. F. (1979). *A theory of semantics and possible inference, with application to decision analysis*, PhD thesis, University of Toronto.

Thomas, S. F. (1995). *Fuzziness and Probability*, ACG Press, Wichita Kansas USA.

Triantaphyllou, E. & Mann, S. H. (1990). An evaluation of the eigenvalue approach for determining the membership values in fuzzy sets, *Fuzzy Sets and Systems* **35**: 295–301.

Trillas, E. (1979). Sobre funciones de negación en la teoría de conjuntos difusos, *Stochastica* **3**: 47–59.

Türkşen, I. B. (1986). Interval valued fuzzy sets based on normal forms, *Fuzzy Sets and Systems* **20**: 191–210.

Türkşen, I. B. (1988). Stochasic fuzzy sets: a survey, *Combining Fuzzy Imprecision with Probabilistic Uncertainty in Decision Making*, Springer-Verlag, New York, pp. 168–183.

Türkşen, I. B. (1991). Measurement of membership functions and their assessment, *Fuzzy Sets and Systems* **40**: 5–38.

Türkşen, I. B. (1992). Interval valued fuzzy sets and 'Compensatory AND', *Fuzzy Sets and Systems* **51**: 295–307.

Tversky, A. (1977). Features of similarity, *Psychological Review* **84**(4): 327–352.

Weber, S. (1983). A general concept of fuzzy connectives, negations and implications based on t-norms and t-conorms, *Fuzzy Sets and Systems* **11**: 115–134.

Yager, R. R. (1979). A measurement–informational discussion of fuzzy union and intersection, *International Journal of Man–Machine Studies* **11**: 189–200.

Yager, R. R. (1980). On a general class of fuzzy connectives, *Fuzzy Sets and Systems* **4**: 235–242.

Yamakawa, T. & Furukawa, M. (1992). A design algorithm of membership functions for a fuzzy neuron using example-based learning, *Proceedings of the First IEEE Conference on Fuzzy Systems*, San Diego, pp. 75–82.

Zadeh, L. A. (1965). Fuzzy sets, *Information Control* **8**: 338–353.

Zadeh, L. A. (1971). Similarity relations and fuzzy orderings, *Information Sciences* **3**: 177–200.

Zadeh, L. A. (1975). Fuzzy logic and approximate reasoning, *Synthese* **30**: 407–428.

Zimmermann, H.-J. & Zysno, P. (1980). Latent connectives in human decision making, *Fuzzy Sets and Systems* **4**: 37–51.

Zimmermann, H. J. & Zysno, P. (1985). Quantifying vagueness in decision models, *European Journal of Operational Research* **22**: 148–154.

Zwick, R. (1987). A note on random sets and the Thurstonian scaling methods, *Fuzzy Sets and Systems* **21**: 351–356.

Zwick, R., Carlstein, E. & Budescu, D. V. (1987). Measures of similarity among fuzzy concepts: A comparative analysis, *International Journal of Approximate Reasoning* **1**(2): 221–242.

Zysno, P. (1981). Modeling membership functions, *in* B. B. Rieger (ed.), *Empirical Semantics I*, Vol. 1 of *Quantitative Semantics, Vol. 12*, Studienverlag Brockmeyer, Bochum, pp. 350–375.

Appendix: Ordered algebraic structures and their representations

In this section basic definitions, and representation and uniqueness results for algebraic structures, called *ordered semigroups*, are given. These results form the basis of discussions on measurement-theoretic semantics that are discussed in Sections 3.2.5 and 3.2.5.

Mainly, the results of Fuchs (1963) and Schweizer & Sklar (1983) are translated in terms of ordered algebraic structures as is customary in the measurement theory literature.

Definition 1 *The algebraic structure $\langle A, \oplus \rangle$ where A is a nonempty set and $\oplus$ is a binary operation on A is called a* semigroup *if and only if $\oplus$ is associative (i.e., for all $a, b, c \in A$, $(a \oplus b) \oplus c = a \oplus (b \oplus c)$). If there exists $e \in A$ such that for all $a \in A$, $e \oplus a = a \oplus e = a$ the structure $\langle A, \oplus, e \rangle$, is called a* semigroup with identity e *or a* monoid. *Finally, $\langle A, \oplus, e \rangle$ is a* group *if and only if it is a semigroup with identity e and any element of A has an inverse:for all $a \in A$, there exists $b \in A$ such that $a \oplus b = b \oplus a = e$.*

When the algebraic structure is also endowed with an ordering, $\succsim$, we obtain *ordered algebraic structures*.

Definition 2 *Let A be a non-empty set, $\succsim$ a binary relation on A and $\oplus$ a binary operation on A. $\langle A, \succsim, \oplus \rangle$ is an* ordered structure *if and only if the following axioms are satisfied:*

> *(weak ordering) $\succsim$ is connected and transitive,*
>
> *(monotonicity) for all $a, b, c, d \in A$, $a \succsim c$ and $b \succsim d$ imply $a \oplus b \succsim c \oplus d$.*

The asymmetric part ($\succ$) and the symmetric complement ($\sim$) of any relation $\succsim$ are defined as usual: $a \succ b$ if and only if $a \succsim b$ and not $b \succsim a$ and $a \sim b$ if and only if $a \succsim b$ and $b \succsim a$.

Adding more properties to an ordered algebraic structure results in specializations of the concept. In this paper, we only consider ordered *semigroups* (where the concatenation is associative). These are summarized in the following definition:

Definition 3 *Let $\mathcal{A} = \langle A, \succsim, \oplus \rangle$ be an ordered algebraic structure such that $\langle A, \oplus \rangle$ is a semigroup. Then $\mathcal{A}$ is called an* ordered semigroup. *Furthermore, it is said to be:*

Weakly Associative *(WA) $a \oplus (b \oplus c) \sim (a \oplus b) \oplus c$.*

Solvable *(Sv) iff whenever $a \succ b$ then there exists $c \in A$ such that $a \succsim b \oplus c$.*

Strongly Monotonic *(SM) iff whenever $a \succsim b$ then $a \oplus c \succsim b \oplus c$ then $c \oplus a \succsim c \oplus b$.*

Homogeneous *(H) iff whenever $a \succsim b$ if and only if $a \oplus c \succsim b \oplus c$ if and only if $c \oplus a \succsim c \oplus b$.*

Idempotent *(Ip) iff for all $a \in A$, $a \oplus a \sim a$.*

Bounded *(B) iff there exist* u *and* e *in* A *such that: for all* $a \in A$, $u \succ e$, $u \succsim a$ *and* $a \succsim e$.

Archimedean *(Ar) iff for any* $a, b \in A$ *there exists a positive integer* m *such that* $a^{(m)} \succ b$ *where* $a^{(m)}$ *is recursively defined as* $a^1 = a$, $a^{(m)} = a \oplus a^{(m-1)}$.

Continuous *iff* $\oplus$ *is continuous as a function of two variables, using the order topology on its range and the relative product topology on its domain.*

By a representation of an ordered algebraic structure, we mean a real valued function that maps the ordered algebraic structure, $\langle A, \succsim, \oplus \rangle$ to a numerical structure, $\langle X, \geq, S \rangle$, where X is a subset of $\mathbb{R}$, $\geq$ is the natural ordering of real numbers and $S : X \times X \to X$ is a function. Since we focus on ordered semigroups, in the resulting representation, S is necessarily associative.

The boundary condition, asserts the existence of a minimal and a maximal element in set A. Hence, given the weak ordering and the boundaries, one can replace the set A by the familiar interval notation $[e, u]$

The following lemma demonstrates some of the consequences of axioms imposed on a bounded ordered semigroup (Schweizer & Sklar 1983).

Lemma 1 *Let* $\mathcal{A} = \langle A, \succsim, \oplus \rangle$ *be a bounded ordered semigroup with bounds* e *and* u*.Then* $\mathcal{A}$ *also satisfies the following conditions for all* $a, b \in A$:

(i) $a \oplus b \succsim sup(a, b)$,

(ii) $u \oplus a \sim a \oplus u \sim u$,

(iii) $a \oplus a \succsim a$.

Schweizer & Sklar (1983, Section 5.3) define a *function* on a closed real interval endowed with the natural ordering, $\geq$. Here, a more abstract structure is considered but their results carry over to our setting without modification since our relation, $\succsim$, is transitive and connected and hence $\sim$ is an equivalence.

Representation theorems with varying uniqueness characteristics can be given for ordered semigroups. These are summarized in the following:

Theorem 1 *The algebraic structure* $\langle A, \succsim, \oplus \rangle$ *is:*

(i) a bounded ordered semigroup *if and only if there exists* $\gamma : [e, u] \to X \triangleq [\underline{x}, \overline{x}]$ *such that,* $a \succsim b \iff \gamma(a) \geq \gamma(b)$, $\gamma(e) = \underline{x}$, $\gamma(u) = \overline{x}$, *and* $\gamma(a \oplus b) = S(\gamma(a), \gamma(b))$ *where* $X \triangleq [\underline{x}, \overline{x}]$ *is a closed subset of* $\mathbb{R}$ *and* S *is an associative, monotonic function such that* $S : [\underline{x}, \overline{x}] \times [\underline{x}, \overline{x}] \to [\underline{x}, \overline{x}]$ *which has* $\underline{x}$ *as its identity. Furthermore,* γ' *is another representation if and only if there exists a strictly increasing function* $\phi : [\underline{x}, \overline{x}] \to [\underline{x}', \overline{x}']$ *with* $\phi(\underline{x}) = \underline{x}'$ *and* $\phi(\overline{x}) = \overline{x}'$ *and such that for all* $x \in [\underline{x}, \overline{x}]$, $\gamma'(x) = \phi(\gamma(x))$ *(ordinal scale).*

(ii) a bounded idempotent semigroup *if and only if all the conditions in (i) are satisfied and* $S = \max$.

(iii) a continuous Archimedean bounded ordered semigroup *if and only if the conditions of (i) are satisfied with* $X = [0,1]$, *and there exists a strictly increasing continuous function,* $g : [0,1] \to \bar{\mathbb{R}}^+ = [0,\infty]$ *with* $g(0) = 0$, *such that for all* $x, y \in [0,1]$, $S(x,y) = g^{[-1]}(g(x) + g(y))$, *where* $g^{[-1]}$ *is the pseudo-inverse of g given by:* $g^{[-1]}(\alpha) = g^{-1}(\min\{\alpha, g(1)\})$. *Furthermore, g is unique up to a positive constant (ratio scale).*

(iv) a solvable homogeneous Archimedean strongly monotonic ordered semigroup *if and only if it is isomorphic to a sub semigroup of* $\langle \mathbb{R}^+, \geq, + \rangle$. *Moreover, two such isomorphisms are unique up to a positive constant (ratio scale).*

(v) a solvable Archimedean strongly monotonic ordered semigroup *if and only if it is isomorphic to a sub semigroup,* $\langle [0,1], \geq, S_W \rangle$ *where* $S_W(x,y) = \min\{x + y, 1\}$ *for all* $x, y \in [0,1]$ *and two such isomorphisms are necessarily equivalent (absolute scale).*

First two parts of Theorem 1 can easily be proven (see Bilgiç & Türkşen (1995a) and Bilgiç (1995) for details), part three is Ling's (1965) representation theorem for a continuous Archimedean triangular norm (see Schweizer & Sklar (1983) for historical comments on this representation and Aczél (1966) for a proof) and parts four and five are from Fuchs (1963).

II Fuzzy Relations

4 AN INTRODUCTION TO FUZZY RELATIONS

Sergei Ovchinnikov

Abstract. This chapter presents an introduction to the theory of fuzzy relations. First, proximity and similarity relations, their classes and fuzzy partitions are introduced and their main properties are investigated. Then, main classes of fuzzy orderings are defined and classified with respect to the duality relations. Transitivity properties of fuzzy orderings and the Ferrers property of induced fuzzy orderings are established. Finally, we present elementary versions of two representation theorems.

4.1 INTRODUCTION

Crisp relations play an important role in mathematics and its applications to other sciences. In the area of pure mathematics, most mathematical structures can be, and very often are, described in terms of relations, see Bourbaki (1974), Summary of Results. The notions of equivalence and ordering relations are used practically in all fundamental mathematical constructions; the theory of ordered sets itself is a big and rapidly developing branch of pure mathematics. In general system theory, an abstract system is defined as a relation among possible inputs and outputs, see Mesarovic and Takahara (1989). Binary and general multidimensional crisp relations have also found natural applications in modeling various concepts inherent to 'soft' sciences like psychology, sociology, linguistic, etc. Many models in measurement and decision-making theories are based on the notion of a relation. Relations are also widely used in different branches of computer science.

The idea to use numbers to describe relations is not a new one. In response to Poincaré's critical thoughts regarding the "physical continuum" (Poincaré, 1912),

Karl Menger suggested to use probabilistic indistinguishability relations as "a more realistic theoretical description of the equality of two elements", see Menger (1951). General probabilistic relations are important tools in measurement, decision-making, and choice theories, see Suppes et al. (1989), section 16.4.2.

Fuzzy set theory suggests a different approach to modeling relations. While a crisp relation represents just the presence or absence of interaction or association among elements of given sets, the values of the membership function of a *fuzzy relation* represent *degrees* of association. Thus crisp relations are simply a special case of fuzzy relations, like crisp sets are a restricted case of fuzzy sets.

Lotfi Zadeh introduced the notion of a fuzzy binary relation in his first paper on fuzzy sets in Zadeh (1965), and defined such important concepts as similarity relations and fuzzy orderings in Zadeh (1971). The concept of a fuzzy relation is one of the most fundamental notions in fuzzy set theory and is a major mathematical tool in numerous application areas.

In this chapter we introduce basic definitions and properties of fuzzy relations and describe main classes of fuzzy binary relations. The reader should be aware of the fact that even in the classical theory of binary relations there is no generally accepted terminology and definitions of major classes of binary relations vary from one source to another. The situation is more complicated in a fuzzy environment: two properties could define quite different classes of fuzzy binary relations but just the same class when restricted on crisp relations. In addition, there is a legion of models for basic logical connectives in fuzzy set theory making it extremely flexible when modeling relations. With one exception (Section 6), we adopt in this chapter the classical models $\mathrm{MIN}(x, y)$, $\mathrm{MAX}(x, y)$, and $x \mapsto 1 - x$, $x, y \in [0, 1]$ for logical connectives AND, OR, and negation, respectively. The symbols $\wedge$ and $\vee$ will be used as synonyms for MIN (infimum) and MAX (supremum).

The chapter is organized as follows.

Basic definitions, general properties, and operations on fuzzy relations are introduced in Section 4.2. We use conventions introduced in Zadeh (1971).

In sections 4.3 and 4.4 we present basic facts about proximity and similarity relations together with related concepts of fuzzy coverings and fuzzy partitions. These relations have numerous applications in such areas as pattern recognition, classifications, numerical taxonomy, cluster analysis, etc. Proximity and similarity relations are also excellent models for (usually nontransitive) indifference and incomparability relations in the area of preference modeling.

Major classes of fuzzy orderings are introduced in Section 4.5. These relations play an important role in the areas of decision-making and optimization in a fuzzy environment.

Transitive fuzzy binary can be expressed in terms of families of positive real functions on the universe. These results are commonly known as representation theorems. In Section 4.6 we formulate two representation theorems in the particular case when the product of two numbers is employed as a model for the logical connective AND. One of these theorems can be regarded as a fuzzy analog of Szpilrajn's theorem in the theory of ordered sets.

We prove of all theorems and give references to the original publications. For further readings we suggest books by Klir and Foger (1988) and Zimmerman (1988) where the reader will find good elementary introduction to the theory of fuzzy relations together with examples of applications. On a more advanced level, the theory of fuzzy binary relations is presented in Fodor and Roubens (1994). The first four chapters of this book cover in a very comprehensive way the contemporary theory of fuzzy binary relations. The rest of this book is concerned with applications to the aggregation problem, ranking procedures, and multiple criteria decision-making. These three books also have extensive bibliographies. For comprehensive references to the studies in the area of fuzzy sets prior to 1980, the reader is referred to Dubois and Prade (1980).

4.2 BASIC CONCEPTS

Let $\mathbf{X}_1 \times \mathbf{X}_2 \times \ldots \times \mathbf{X}_n$ be the Cartesian product of sets $\mathbf{X}_1, \mathbf{X}_2, \ldots, \mathbf{X}_n$. A *fuzzy relation* R on $\mathbf{X}_1 \times \mathbf{X}_2 \times \ldots \times \mathbf{X}_n$ is a fuzzy set defined on $\mathbf{X}_1 \times \mathbf{X}_2 \times \ldots \times \mathbf{X}_n$. Elements of this Cartesian product are n-tuples $(x_1, x_2, \ldots, x_n)$ where $x_i \in \mathbf{X}_i$ for all i, $1 \leq i \leq n$. For any given n-tuple $(x_1, x_2, \ldots, x_n)$, $R(x_1, x_2, \ldots, x_n)$ is the value of the membership function of R representing the degree to which elements of this n-tuple are connected.

In the particular case when $n = 2$, a *fuzzy binary relation* R *on* $\mathbf{X} \times \mathbf{Y}$ is a fuzzy set on the Cartesian product $\mathbf{X} \times \mathbf{Y}$. If $\mathbf{Y} = \mathbf{X}$, then R is a fuzzy set defined on $\mathbf{X} \times \mathbf{X}$. In this case we say that R is a *fuzzy binary relation on* $\mathbf{X}$.

Suppose $\mathbf{X} = \{x_1, x_2, \ldots, x_n\}$ and $\mathbf{Y} = \{y_1, y_2, \ldots y_m\}$ are finite sets and R is a fuzzy binary relation defined on $\mathbf{X} \times \mathbf{Y}$. A convenient way of representing R involves a *relation matrix* (r_{ij}) defined by

$$r_{ij} = R(x_i, y_j)$$

for all i and $j, 1 \leq i \leq n, 1 \leq j \leq m$.

Example 4.1. Suppose $\mathbf{X} = \{x_1, x_2, x_3, x_4, x_5\}$. We define a fuzzy binary relation on $\mathbf{X}$ by

$$R = \begin{bmatrix} 1 & 0.8 & 0.2 & 0.6 & 0.6 \\ 0 & 1 & 0 & 0 & 0.6 \\ 0 & 0 & 1 & 0 & 0.5 \\ 0 & 0 & 0 & 1 & 0.6 \\ 0 & 0 & 0 & 0 & 1 \end{bmatrix}.$$

Then, for instance,

$$R(x_1, x_4) = 0.6,\ R(x_3, x_5) = 0.5,\ R(x_5, x_5) = 1, R(x_4, x_3) = 0.$$

The *domain* of a fuzzy binary relation R on $\mathbf{X} \times \mathbf{Y}$ is denoted by $\operatorname{dom} R$ and is defined by

$$\operatorname{dom} R(x) = \bigvee_{y \in \mathbf{Y}} R(x, y)$$

for all $x \in \mathbf{X}$. Likewise, the *range* of R is denoted by $\operatorname{ran} R$ and is defined by

$$\operatorname{ran} R(y) = \bigvee_{x \in \mathbf{X}} R(x, y)$$

for all $y \in \mathbf{Y}$.

The *height* of R is defined by

$$\operatorname{hgt}(R) = \bigvee_{y \in \mathbf{Y}} \bigvee_{x \in \mathbf{X}} R(x, y).$$

Thus defined height is exactly the height of R considered as a fuzzy set on $\mathbf{X} \times \mathbf{Y}$.

Since fuzzy relations are just fuzzy sets defined on Cartesian products, all properties and operations introduced for fuzzy sets can be used for fuzzy relations. In particular, the notions of a *union* and *intersection* are defined in the usual way, and a fuzzy relation R on $\mathbf{X}$ can be represented by its *resolution form*

$$R = \bigvee_{\alpha \in [0,1]} \alpha R_\alpha,$$

where R_α is an *α-level set* of the fuzzy binary relation R.

The *composition* operation is, probably, the most important operation on fuzzy binary relations. Suppose P and Q are fuzzy binary relations on $\mathbf{X} \times \mathbf{Y}$ and $\mathbf{Y} \times \mathbf{Z}$, respectively. The *max-min composition* $R = P \circ Q$ of P and Q is defined as a fuzzy binary relation on $\mathbf{X} \times \mathbf{Z}$ by

$$R(x, z) = \bigvee_{y \in \mathbf{Y}} P(x, y) \wedge Q(y, z).$$

In a more general setting, the *max-* $*$ *composition* is defined by substituting the $\wedge$ operation in the above formula with another binary operation $*$ on the unit interval. Very often triangular norms are used for this purpose. For example, the *max-product* composition $P \odot Q$ is defined by

$$(P \odot Q)(x, z) = \bigvee_{y \in \mathbf{Y}} P(x, y) \cdot Q(y, z),$$

when the product t-norm is used instead of $\wedge$.

Example 4.2. The following matrix equations illustrate the max-min and max-product compositions for the same pair of fuzzy binary relations:

$$\begin{bmatrix} 0.3 & 0.5 & 0.8 \\ 0 & 0.7 & 1 \\ 0.4 & 0 & 0.5 \end{bmatrix} \circ \begin{bmatrix} 0.9 & 0 & 0.7 & 0.7 \\ 0.3 & 0.2 & 0 & 0.9 \\ 1 & 0 & 0.5 & 0.5 \end{bmatrix} = \begin{bmatrix} 0.8 & 0.2 & 0.5 & 0.5 \\ 1 & 0.2 & 0.5 & 0.7 \\ 0.5 & 0 & 0.5 & 0.5 \end{bmatrix}$$

$$\begin{bmatrix} 0.3 & 0.5 & 0.8 \\ 0 & 0.7 & 1 \\ 0.4 & 0 & 0.5 \end{bmatrix} \odot \begin{bmatrix} 0.9 & 0 & 0.7 & 0.7 \\ 0.3 & 0.2 & 0 & 0.9 \\ 1 & 0 & 0.5 & 0.5 \end{bmatrix} = \begin{bmatrix} 0.8 & 0.1 & 0.4 & 0.45 \\ 1 & 0.21 & 0.5 & 0.63 \\ 0.5 & 0 & 0.28 & 0.28 \end{bmatrix}$$

Essentially, we use the classical "row by column" rule to compute the composition matrix, but, for instance, operations $\wedge$ and $\vee$ substitute for multiplication and addition, respectively, in the case of max-min composition.

The max-min composition is an associative operation, i.e., $R \circ (S \circ T) = (R \circ S) \circ T$. Thus we can define $R^n = R^{n-1} \circ R$ for $n > 1$, where, by definition $R^1 = R$. Generally speaking, $R \circ S \neq S \circ R$.

In what follows we consider only fuzzy binary relations on a single set $\mathbf{X}$. These relations play a very important role in fuzzy set theory and its applications. Various combinations of main properties of fuzzy binary relations define specific classes of these relations. Examples of these classes include similarity relations and various fuzzy orderings.

In this section we introduce three most important properties of fuzzy binary relations; more properties will be discussed later in this chapter. Let R be a fuzzy binary relation on $\mathbf{X}$. The three properties under consideration are:

Reflexivity: $R(x, x) = 1$

Symmetry: $R(x, y) = R(y, x)$

Transitivity: $R(x, y) \wedge R(y, z) \leq R(x, z)$

for all $x, y, z \in \mathbf{X}$. The transitivity property can be written in the following equivalent form using max-min composition operation:

Transitivity: $R \circ R \subseteq R$.

If R is transitive and reflexive fuzzy binary relation, then $R \circ R = R$.

Sometimes a binary operation different from $\wedge$ is used in the definition of the transitivity property. For instance, the product-transitivity property is defined by

$$R(x, y) \cdot R(y, z) \leq R(x, z)$$

for all $x, y, z \in \mathbf{X}$.

The *transitive closure* R^* of a fuzzy binary relation R is defined by

$$R^* = R \cup R^2 \cup \ldots \cup R^n \cup \ldots = \bigcup_{i=1}^{\infty} R^i,$$

where, by definition, $R^1 = R$. The transitive closure of R is the smallest transitive fuzzy binary relation containing R. R is transitive if and only if $R^* = R$.

The properties of reflexivity, symmetry, and transitivity are obvious generalizations of the classical concepts. They coincide with the classical definitions if R is a binary relation. The reader should be aware of the fact that the fuzzy set theory provides very flexible tools for generalizing classical concepts. Properties that are quite different in a fuzzy environment could be equivalent when restricted to classical sets. For instance the class of fuzzy binary relations on $\mathbf{X}$ satisfying

$$R(x, x) \geq 0.5, \quad \text{for all } x \in \mathbf{X}$$

is much wider than the class of reflexive fuzzy binary relations. On the other hand, the above condition defines just reflexive relations when restricted to binary relations.

In the rest of this chapter, we assume that $\mathbf{X}$ is a finite set.

4.3 COVERINGS AND PROXIMITY RELATIONS

The notion of a similarity relation is a special case of a more general notion of a *proximity* relation (also known as a *compatibility* or *tolerance* relation). Similarity relations and their properties are studied in Section 4.4. In this section we introduce the notions of a fuzzy covering and proximity relation and discuss their properties.

Definition 4.1. Let U be a fuzzy set on $\mathbf{X}$. A family of fuzzy sets $\Sigma = \{P_i\}_{i \in J}$ is said to be a *fuzzy covering* of U if

$$U = \bigcup_{i \in J} P_i.$$

This definition is a natural generalization of the classical one. Note, that each P_i is a fuzzy subset of U.

Definition 4.2. Let U be a fuzzy set with universe $\mathbf{X}$. R is said to be a *proximity relation on* U, if

$$R(x, y) = R(y, x), \tag{4.1}$$

$$R(x, y) \leq R(x, x) \wedge R(y, y), \tag{4.2}$$

$$R(x, x) = U(x), \tag{4.3}$$

for all x and y in $\mathbf{X}$.

Note, that in the case when $U = \mathbf{X}$, a proximity relation is just a reflexive and symmetric fuzzy binary relation on $\mathbf{X}$.

For any given fuzzy covering $\Sigma = \{P_i\}_{i \in J}$ of U we define a fuzzy binary relation R_Σ by

$$R_\Sigma(x, y) = \bigvee_{i \in J} \{P_i(x) \wedge P_i(y)\}$$

Lemma 4.1. R_Σ is a proximity relation on U.

Proof. We prove (4.2), since (4.1) and (4.3) are obviously satisfied. Since each P_i is a subset of U, we have

$$R_\Sigma(x, y) = \bigvee_{i \in J} \{P_i(x) \wedge P_i(y)\} \leq U(x) \wedge U(y) = R_\Sigma(x, x) \wedge R_\Sigma(y, y)$$

□

The notion of a *class* of a fuzzy binary relation R plays an important role in the theory of proximity and similarity relations.

Definition 4.3. Let R be a fuzzy binary relation on $\mathbf{X}$. A fuzzy set K is said to be a *pre-class* of R if $K(x) \wedge K(y) \leq R(x, y)$ for all x and y in $\mathbf{X}$. Maximal (with respect to the inclusion relation, $\subset$) pre-classes of R are called *classes* of R. We denote the set of all classes of R by Σ_R.

The notion of a class is well-defined, since the set of all pre-classes of a given fuzzy binary relation R is an inductive partially ordered set, see Birkhoff (1979), VII, 7). We denote the set of all classes of R by Σ_R.

The assignment $\Sigma \mapsto R_\Sigma$ defines a function $\mathbb{F}$ from the set of all fuzzy coverings of U to the set of all proximity relations on U. The following theorem shows that $\mathbb{F}$ is a surjective (onto) function.

Theorem 4.1. (Yeh and Bang, 1975, Ovchinnikov and Riera, 1982). Let R be a proximity relation on a fuzzy set U. There exists a covering Σ of U such that

$$R = R_\Sigma.$$

Proof. Consider the family $\{K_{\{a,b\}}\}$ of fuzzy sets defined by

$$K_{\{a,b\}} = \begin{cases} R(a, b), & \text{if } x \in \{a, b\}, \\ 0, & \text{otherwise.} \end{cases}$$

for all $a, b \in \mathbf{X}$. We have

$$K_{\{a,b\}}(a) \wedge K_{\{a,b\}}(a) = R(a,b) \leq R(a,a), \text{ by (4.3)},$$
$$K_{\{a,b\}}(a) \wedge K_{\{a,b\}}(b) = R(a,b),$$
$$K_{\{a,b\}}(b) \wedge K_{\{a,b\}}(a) = R(a,b) = R(b,a), \text{ by (4.2)},$$
$$K_{\{a,b\}}(b) \wedge K_{\{a,b\}}(b) = R(a,b) \leq R(b,b), \text{ by (4.3)},$$
$$K_{\{a,b\}}(x) \wedge K_{\{a,b\}}(y) = 0 \text{ if } x \notin \{a,b\} \text{ and } y \notin \{a,b\}.$$

Therefore, $K_{\{a,b\}}$ is a pre-class of R. Let Σ be a set of classes such that for any $\{a,b\}$ there is a class in Σ containing $K_{\{a,b\}}$. Then, for any $K \in \Sigma$, $K(a) \wedge K(b) \leq R(a,b)$ and there is K such that $K(a) \wedge K(b) = R(a,b)$. Indeed, let $K \in \Sigma$ be a class containing $K_{\{a,b\}}$. Then,

$$R(a,b) = K_{\{a,b\}}(a) \wedge K_{\{a,b\}}(b) \leq K(a) \wedge K(b) \leq R(a,b).$$

Hence,

$$\bigvee_{K \in \Sigma} \{K(a) \wedge K(b)\} = R(a,b) \text{ for all } a, b \in \mathbf{X}.$$

Since R is a proximity relation on U, we have from the previous formula

$$\bigvee_{K \in \Sigma} K(a) = U(a) \text{ for all } a \in \mathbf{X}.$$

i.e., Σ is a cover of U and $R_\Sigma = R$. □

Corollary 4.1. If $\Sigma = \Sigma_R$ is the set of all classes of R, then $R_{\Sigma_R} = R$.

The following example shows that function $\mathbb{F}$ is not a one-to-one function.

Example 4.3. Let $\mathbf{X} = \{x_1, x_2, x_3\}$ and $U = \mathbf{X}$. Consider two coverings of $\mathbf{X}$ defined by the tables

$$\Sigma_1 = \begin{array}{cccc} & P_1 & P_2 & P_3 \\ x_1 & 1 & \alpha & \alpha \\ x_2 & \alpha & 1 & \gamma \\ x_3 & \beta & \gamma & 1 \end{array} \quad \text{and} \quad \Sigma_2 = \begin{array}{cccc} & P_1 & P_2 & P_4 \\ x_1 & 1 & \alpha & \beta \\ x_2 & \alpha & 1 & \alpha \\ x_3 & \beta & \gamma & 1 \end{array}$$

where $\alpha < \beta < \gamma$ (for example, $P_2(x_3) = \gamma$ and $P_4(x_1) = \beta$). It is easy to verify that

$$R_{\Sigma_1} = R_{\Sigma_2} = \begin{bmatrix} 1 & \alpha & \beta \\ \alpha & 1 & \gamma \\ \beta & \gamma & 1 \end{bmatrix}.$$

$R = R_{\Sigma_1} = R_{\Sigma_2}$ is obviously a proximity relation with the following classes

$$\Sigma_R = \begin{array}{c|cccc} & P_1 & P_2 & P_3 & P_4 \\ x_1 & 1 & \alpha & \alpha & \beta \\ x_2 & \alpha & 1 & \gamma & \alpha \\ x_3 & \beta & \gamma & 1 & 1 \end{array}$$

Hence, we have at least three different covers Σ_1, Σ_2, and Σ_R such that

$$R = R_{\Sigma_1} = R_{\Sigma_2} = R_{\Sigma_R}.$$

Although $\mathbb{F}$ defined by $\Sigma \mapsto R_\Sigma$ is not a bijective mapping, the techniques developed above provide an important tool for studying proximity relations.

4.4 SIMILARITY RELATIONS AND FUZZY PARTITIONS

In this section we study a special class of proximity relations - similarity relations. In the rest of the chapter we assume that $U = \mathbf{X}$ where $\mathbf{X}$ is a finite set.

The notion of a similarity relation was introduced in Zadeh (1971) as a generalization of the classical notion of equivalence relations.

Definition 4.4. A fuzzy binary relation S on $\mathbf{X}$ is a *similarity relation* on $\mathbf{X}$ if and only if it satisfies the following properties:

$$S(x,x) = 1 \qquad \textit{reflexivity}$$

$$S(x,y) = S(y,x) \qquad \textit{symmetry}$$

$$S(x,y) \wedge S(y,z) \leq S(x,z) \qquad \textit{transitivity}$$

for all $x, y, z \in \mathbf{X}$.

Obviously, similarity relations are just transitive proximity relations on $\mathbf{X}$. Note, that transitivity and reflexivity properties imply

$$\bigvee_{t \in \mathbf{X}} \{S(x,t) \wedge S(t,y)\} = S(x,y) \tag{4.4}$$

for all $x, y \in \mathbf{X}$.

Example 4.4. Let $\mathbf{X} = \{x_1, x_2, x_3\}$. Suppose a proximity relation S on $\mathbf{X}$ is defined by the following matrix

$$S = \begin{bmatrix} 1 & \alpha & \gamma \\ \alpha & 1 & \beta \\ \gamma & \beta & 1 \end{bmatrix},$$

where $\alpha, \beta, \gamma \in [0,1]$. This relation is transitive if and only if

$$\beta \wedge \gamma \leq \alpha, \ \ \alpha \wedge \beta \leq \gamma, \ \ \alpha \wedge \gamma \leq \beta.$$

These conditions are satisfied if and only if two of the numbers α, β, γ are equal and not greater than the third number. For instance,

$$S = \begin{bmatrix} 1 & \beta & \gamma \\ \beta & 1 & \beta \\ \gamma & \beta & 1 \end{bmatrix},$$

where $\beta \leq \gamma$, is always a similarity relation.

Let Σ be a fuzzy covering of $\mathbf{X}$ and R_Σ a proximity relation associated with Σ. Is R_Σ a similarity relation on $\mathbf{X}$? The answer to this question is found in the following theorem.

Theorem 4.2. (Ovchinnikov and Riera, 1982). Let $\Sigma = \{P_i\}_{i \in J}$ be a fuzzy covering of $\mathbf{X}$. R_Σ is a similarity relation on $\mathbf{X}$ if and only if for any $P_i, P_j \in \Sigma$ there is $P_k \in \Sigma$ such that

$$\operatorname{hgt}(P_i \cap P_j) \wedge P_i(x) \wedge P_j(y) \leq P_k(x) \wedge P_k(y), \tag{4.5}$$

for all $x, y \in \mathbf{X}$, where $\operatorname{hgt}(P_i \cap P_j)$ is the height of the intersection $P_i \cap P_j$.

Proof. Suppose Σ satisfies (4.5). It suffices to prove transitivity of R_Σ. We have

$$\begin{aligned} R_\Sigma(x,y) \wedge R_\Sigma(y,z) &= \left\{ \bigvee_{i \in J} [P_i(x) \wedge P_i(y)] \right\} \wedge \left\{ \bigvee_{j \in J} [P_j(y) \wedge P_j(z)] \right\} \\ &= \bigvee_{i,j \in J} [P_i(x) \wedge P_i(y) \wedge P_j(y) \wedge P_j(z)] \\ &\leq \bigvee_{i,j \in J} [\operatorname{hgt}(P_i \cap P_j) \wedge P_i(x) \wedge P_j(z)] \\ &\leq \bigvee_{k \in J} [P_k(x) \wedge P_k(z)] = R_\Sigma(x,z). \end{aligned}$$

Let R_Σ be a similarity relation on X. Then, for all $x, y, t \in \mathbf{X}$,

$$R_\Sigma(x,t) \wedge R_\Sigma(t,y) \leq R_\Sigma(x,y),$$

or, equivalently,

$$\left\{ \bigvee_{i \in J} [P_i(x) \wedge P_i(t)] \right\} \wedge \left\{ \bigvee_{j \in J} [P_j(t) \wedge P_j(y)] \right\} \leq \bigvee_{k \in J} [P_k(x) \wedge P_k(y)],$$

which implies

$$\bigvee_{i,j \in J} [P_i(x) \wedge P_i(t) \wedge P_j(t) \wedge P_j(y)] \leq \bigvee_{k \in J} [P_k(x) \wedge P_k(y)],$$

for all $x, y, t \in \mathbf{X}$. Hence, for any given P_i and P_j in Σ there is P_k in Σ such that

$$P_i(x) \wedge P_i(t) \wedge P_j(t) \wedge P_j(y) \leq P_k(x) \wedge P_k(y)$$

for any $t \in \mathbf{X}$. Therefore,

$$\begin{aligned} h(P_i \cap P_j) \wedge P_i(x) \wedge P_j(y) &= \bigvee_{t \in \mathbf{X}} [P_i(x) \wedge P_i(t) \wedge P_j(t) \wedge P_j(y)] \\ &\leq P_k(x) \wedge P_k(y). \end{aligned}$$

Thus (4.5) is satisfied. □

Our next theorem shows that the classes of a similarity relation S on $\mathbf{X}$ considered as a proximity relation are exactly similarity classes of S introduced in Zadeh (1971) as follows.

Definition 4.5. Let S be a similarity relation on $\mathbf{X}$. For any given $a \in \mathbf{X}$, a *similarity class* of a is a fuzzy set $S[a]$ on $\mathbf{X}$ defined by the membership function $S[a](x) = S(a, x)$ for all $x \in \mathbf{X}$.

Theorem 4.3. (Ovchinnikov and Riera, 1982). A fuzzy set P is a similarity class of a similarity relation S if and only if it is a class of S considered as a proximity relation.

Proof. We note first that any similarity class of S is a pre-class of S. Indeed,

$$S[a](x) \wedge S[a](y) = S(a, x) \wedge S(a, y) = S(x, a) \wedge S(a, y) \leq S(x, y),$$

since S is a symmetric and transitive fuzzy binary relation.

Let P be a class of S and a be an element of $\mathbf{X}$ such that $P(x) \leq P(a)$ for all x. (Such an element a exists, since $\mathbf{X}$ is assumed to be a finite set.) We have

$$P(x) = P(a) \wedge P(x) \leq S(a, x) = S[a](x).$$

Since $S[a]$ is a pre-class of S, we conclude that $P = S[a]$.

Let $S[a]$ be a similarity class of S and P be a class of S containing $S[a]$. Then $P(x) \geq S[a](x)$ for all $x \in \mathbf{X}$. In particular, $P(a) = 1$, since S is a reflexive relation. We have

$$S[a](x) \leq P(x) = P(a) \wedge P(x) \leq S(a, x) = S[a](x)$$

for all $x \in \mathbf{X}$. Therefore $S[a] = P$. □

Generally speaking it is possible that $S[a] = S[b]$ for $a \neq b$. The following theorem establishes a necessary and sufficient conditions for two similarity classes to be identical.

Theorem 4.4. $S[a] = S[b]$ if and only if $S(a,b) = 1$.

Proof. Suppose $S[a] = S[b]$. Then

$$S(a,b) = S[a](b) = S[b](b) = S(b,b) = 1.$$

Conversely, suppose $S(a,b) = 1$. Then, by transitivity of S,

$$S[a](x) = S(a,x) \geq S(a,b) \wedge S(b,x) = S(b,x) = S[b](x).$$

Similarly, $S[b](x) \geq S[a](x)$. □

Corollary 4.2. If $S[a] \neq S[b]$, then $\operatorname{hgt}(S[a] \cap S[b]) < 1$.

Proof. Suppose $\operatorname{hgt}(S[a] \cap S[b]) = 1$. Then there is $x \in \mathbf{X}$ such that

$$S[a](x) = S[b](x) = 1.$$

Then,

$$S(a,b) \geq S(a,x) \wedge S(x,b) = S[a](x) \wedge S[b](x) = 1$$

implying $S[a] = S[b]$. □

Let I be an equivalence relation on $\mathbf{X}$. The equivalence classes of I form a partition of $\mathbf{X}$, i.e., $\mathbf{X}$ can be represented as the union of disjoint nonempty equivalence classes. In the case of similarity relations, although two distinct similarity classes could have a nonempty intersection, their α-level sets, for a given α, form a partition of $\mathbf{X}$. Indeed, let S be a similarity relation on $\mathbf{X}$. Then, for any given $\alpha \in [0,1]$, S_α is an equivalence relation on X and $aS_\alpha b$ if and only if $S(a,b) \geq \alpha$. For a given $a \in \mathbf{X}$,

$$S[\,a\,]_\alpha = \{x \,|\, S[\,a\,](x) \geq \alpha\} = \{x \,|\, a\, S_\alpha\, x\}.$$

Therefore, α-level sets of similarity classes of S are exactly equivalence classes of the α-level set of S. Thus, for a given $\alpha \in [0,1]$, the α-level sets of similarity classes of S form a partition of $\mathbf{X}$ which is called an α-partition of $\mathbf{X}$. These partitions are nested in the sense that an α-partition is a refinement of a β-partition if and only if $\alpha \geq \beta$. The following example illustrates some of the above concepts and introduces an important notion of a partition tree associated with a given similarity relation.

Example 4.5. (Zadeh, 1971). Let S be a fuzzy binary relation on

$$\mathbf{X} = \{x_1, x_2, x_3, x_4, x_5, x_6\}$$

defined by

$$S = \begin{bmatrix} 1 & 0.2 & 1 & 0.6 & 0.2 & 0.6 \\ 0.2 & 1 & 0.2 & 0.2 & 0.8 & 0.2 \\ 1 & 0.2 & 1 & 0.6 & 0.2 & 0.6 \\ 0.6 & 0.2 & 0.6 & 1 & 0.2 & 0.8 \\ 0.2 & 0.8 & 0.2 & 0.2 & 1 & 0.2 \\ 0.6 & 0.2 & 0.6 & 0.8 & 0.2 & 1 \end{bmatrix}.$$

It is readily verified that S is a similarity relation on **X**. There are five distinct similarity classes of S: $S[x_1] = S[x_3]$, $S[x_2]$, $S[x_4]$, $S[x_5]$, and $S[x_6]$. Since there are only four different values of the membership function $S(x, y)$, there are four distinct α-level sets to consider. Namely, $S_{0.2}$, $S_{0.6}$, $S_{0.8}$, and $S_{1.0}$. The nested family of partitions generated by distinct α-level sets of S can be represented in the form of a partition tree, as shown on Figure 4.1. The concept of a partition tree may be considered as a generalization of the concept of a quotient set of an equivalence relation.

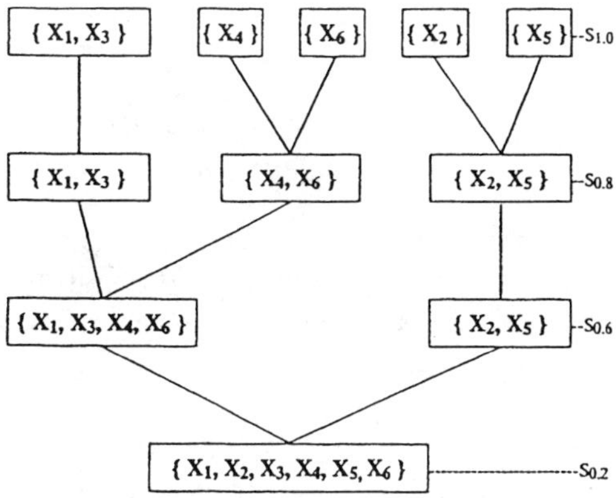

Figure 4.1

These observations lead us to the following definition.

Definition 4.6. A fuzzy cover Σ of **X** is said to be a *fuzzy partition* of **X** if there is a similarity relation S on X such that Σ is the set of all distinct similarity classes of S.

In order to describe fuzzy partitions independently, we introduce the following notations.

Let Π_Σ be the set of all different 1.0-level sets of elements of Σ. If Π_Σ is a partition of **X**, then, for any $a \in$ **X**, there is an element $[a] \in \Sigma$ such that $a \in [a]_{1.0}$, i.e., $[a](a) = 1$.

Theorem 4.5. (Ovchinnikov, 1981). Let Σ be a finite cover of X. Σ is a fuzzy partition of **X** if and only if

(i) Π_Σ is a partition of **X**, and

(ii) $\text{hgt}([\,a\,] \cap [\,b\,]) = [\,a\,](b) \wedge [\,b\,](a)$, for all a and b in X.

Proof. (a) Suppose Σ is a fuzzy partition of **X**, i.e., Σ is the set of similarity classes of some similarity relation S. Then Π_Σ is the set of all distinct 1.0-level sets of similarity classes of S. Since $S[a](a) = 1$ for any $a \in \mathbf{X}$, elements of Π_Σ are nonempty sets with union **X**. By Corollary 4.2, intersections of different sets in Π_Σ are empty. Therefore, Π_Σ is a partition of **X**.

For any element $a \in \mathbf{X}$, we have $[a] = S[a]$. Indeed, $[a]$ is a similarity class of S. Therefore, $[a] = S[x]$ for some $x \in \mathbf{X}$. We have $S(x, a) = S[x](a) = [a](a) = 1$, and, by Theorem 4.4, $S[x] = S[a]$. Further, using symmetry of S and (4.4),

$$
\begin{aligned}
h([\,a\,] \cap [\,b\,]) &= \bigvee_{x \in \mathbf{X}} \{[a](x) \wedge [b](x)\} \\
&= \bigvee_{x \in \mathbf{X}} \{S[a](x) \wedge S[b](x)\} \\
&= \bigvee_{x \in \mathbf{X}} \{S(a, x) \wedge S(b, x)\} \\
&= S(a, b) = S(a, b) \wedge S(b, a) \\
&= a[b] \wedge b[a].
\end{aligned}
$$

(b) Let Σ be a fuzzy cover satisfying conditions (i) and (ii) of the theorem. We define a fuzzy binary relation S on X by $S(a, b) = [a](b)$. We need to prove that S is a reflexive, symmetric, and transitive fuzzy binary relation.

Obviously, $S(a, a) = [a](a) = 1$ for all $a \in \mathbf{X}$. Hence, S is a reflexive fuzzy binary relation.

We have, by (ii),

$$
[a](t) \wedge [b](t) \leq a\,[b] \wedge b[a] \quad \text{for all } t \in \mathbf{X}. \tag{4.5}
$$

Substituting $t = a$ and $t = b$, we obtain $b[a] \leq a[b]$ and $a[b] \leq b[a]$, respectively. Therefore, $a[b] = b[a]$, and S is a symmetric fuzzy binary relation.

Finally, by symmetry of S and (4.5),

$$
S(x, t) \wedge S(t, y) = [x](t) \wedge [y](t) \leq [x](y) \wedge [y](x) = S(x, y),
$$

i.e., S is a transitive fuzzy binary relation. □

4.5 FUZZY ORDERINGS

In general a *fuzzy ordering* is a fuzzy binary relation satisfying the transitivity property. Since similarity relations are transitive and symmetric, we usually assume that fuzzy orderings satisfy some kind of asymmetry property.

For a given fuzzy binary relation R on **X**, we define the following fuzzy relations associated with R:

Inverse relation: $R^{-1}(x,y) = R(y,x)$,

Complement relation: $\overline{R}\,(x,y) = 1 - R(x,y)$,

Symmetric part: $R^{s}(x,y) = R(x,y) \wedge R(y,x)$,

Dual relation: $R^{d}(x,y) = 1 - R(y,x)$,

for all $x, y \in \mathbf{X}$. In the framework of preference modeling, if R is considered as a weak preference relation, then R^{s} and R^{d} are meaningful models for the indifference relation and the strict preference associated with R. It is readily verified that $R^{s} = R \cap R^{-1}$, $R^{d} = \overline{R^{-1}}$, and $(R^{d})^{d} = R$, i.e., relations R and R^{d} are duals of each other.

In order to introduce particular classes of fuzzy orderings, we define the following properties in addition to the properties of reflexivity, symmetry, and transitivity. We shall say that a fuzzy binary relation R on $\mathbf{X}$ is

Irreflexive if $R(x,x) = 0$,

Asymmetric if $R(x,y) \wedge R(y,x) = 0$,

Weakly asymmetric if $R(x,y) \wedge R(y,x) < 1$,

Negatively transitive if $R(x,y) \vee R(y,z) \geq R(x,z)$,

Complete if $R(x,y) \vee R(x,y) > 0$,

Strongly complete if $R(x,y) \vee R(y,x) = 1$,

for all $x, y, z \in \mathbf{X}$.

The correspondences between properties of mutually dual relations are demonstrated in Table 4.1.

Relation	*Dual relation*
reflexive	irreflexive
symmetric	symmetric
asymmetric	strongly complete
transitive	negatively transitive
negatively transitive	transitive
complete	weakly asymmetric

Table 4.1. Properties of dual fuzzy binary relations

matrices:

$$P = \begin{bmatrix} 0 & 1-\alpha & 1-\gamma \\ 0 & 0 & 1-\beta \\ 0 & 0 & 0 \end{bmatrix} \quad \text{and} \quad I = \begin{bmatrix} 1 & \alpha & \gamma \\ \alpha & 1 & \beta \\ \gamma & \beta & 1 \end{bmatrix}.$$

In addition to the defining properties, fuzzy orderings introduced above have some other important properties. For instance, all relations in the left column in Figure 4.2 are irreflexive and all relations in the right column are reflexive. It is less obvious that weak and linear orderings are transitive relations as the following theorem asserts.

Theorem 4.6. Let R be a strongly complete and transitive fuzzy binary relation. The dual relation R^{d} associated with R is a transitive fuzzy binary relations.

Proof. We need to prove that

$$R^{\mathrm{d}}(x, y) \wedge R^{\mathrm{d}}(y, z) \leq R^{\mathrm{d}}(x, y)$$

for all $x, y, z \in \mathbf{X}$. R^{d} is a negatively transitive and asymmetric fuzzy binary relation, since R^{d} is a dual of R and R is transitive and strongly complete. Therefore,

$$\begin{aligned} R^{\mathrm{d}}(x, y) \wedge R^{\mathrm{d}}(y, z) &\leq R^{\mathrm{d}}(x, y) \wedge \left[R^{\mathrm{d}}(y, x) \vee R^{\mathrm{d}}(x, z)\right] \\ &= \left[R^{\mathrm{d}}(x, y) \wedge R^{\mathrm{d}}(y, x)\right] \vee \left[R^{\mathrm{d}}(x, y) \wedge R^{\mathrm{d}}(x, z)\right] \\ &= R^{\mathrm{d}}(x, y) \wedge R^{\mathrm{d}}(x, z) \leq R^{\mathrm{d}}(x, z). \end{aligned}$$

□

The same assertion is true for the symmetric part of a given transitive fuzzy binary relation.

Theorem 4.7. Let R be a transitive fuzzy binary relation. The symmetric part R^{s} of R is a transitive fuzzy binary relation.

Proof. We have

$$\begin{aligned} R^{\mathrm{s}}(x, y) \wedge R^{\mathrm{s}}(x, z) &= R(x, y) \wedge R(y, x) \wedge R(y, z) \wedge R(z, y) \\ &= R(x, y) \wedge R(y, z) \wedge R(z, y) \wedge R(y, x) \\ &\leq R(x, z) \wedge R(z, x) = R^{\mathrm{s}}(x, z) \end{aligned}$$

for all $x, y, z \in \mathbf{X}$.

□

The rows of this table can be read backwards, since the dual relation of a dual relation is the relation itself. For instance, if a given fuzzy binary relation is negatively transitive, then the dual relation is transitive.

Various combinations of the above properties define major classes of fuzzy orderings. We say that a fuzzy binary relation R is a

partial ordering	if R is asymmetric and transitive,
weak ordering	if R is asymmetric and negatively transitive,
linear ordering	if R is a complete weak ordering,
quasi-transitive relation	if R is dual to a partial ordering,
complete quasi-ordering	if R is dual to a weak ordering,
reflexive linear ordering	if R is dual to a linear ordering.

According to the correspondences in Table 4.1, a fuzzy binary relation R is a *quasi-transitive relation* if R is strongly complete and negatively transitive, a *complete quasi-ordering* if R is strongly complete and transitive, and a *reflexive linear ordering* if R is strongly complete, weakly asymmetric, and transitive.

Classes of fuzzy orderings are presented in the form of a diagram in Figure 4.2.

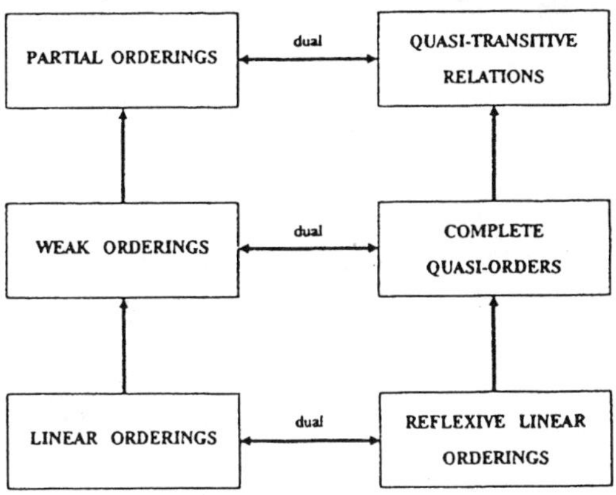

Figure 4.2

Example 4.6. Let $\mathbf{X} = \{x_1, x_2, x_3\}$. We define a fuzzy binary relation R on $\mathbf{X}$ by

$$R = \begin{bmatrix} 1 & 1 & 1 \\ \alpha & 1 & 1 \\ \gamma & \beta & 1 \end{bmatrix},$$

where α, β, γ are numbers from the open unit interval $(0, 1)$. Thus defined R is a strongly complete fuzzy binary relation on $\mathbf{X}$. The dual relation $P = R^{\mathrm{d}}$ and the symmetric part $I = R \cap R^{-1}$ associated with R are represented by the following

Let us consider some examples of fuzzy orderings in the right column of Figure 4.2.

Example 4.7. Let R be a fuzzy binary relation on $\mathbf{X} = \{x_1, x_2, x_3\}$ defined by

$$R = \begin{bmatrix} 1 & 1 & 1 \\ \alpha & 1 & 1 \\ \gamma & \beta & 1 \end{bmatrix},$$

where $\alpha, \beta, \gamma \in [0, 1]$. We have

$$R \circ R = \begin{bmatrix} 1 & 1 & 1 \\ \alpha & 1 & 1 \\ \gamma & \beta & 1 \end{bmatrix} \circ \begin{bmatrix} 1 & 1 & 1 \\ \alpha & 1 & 1 \\ \gamma & \beta & 1 \end{bmatrix} = \begin{bmatrix} 1 & 1 & 1 \\ \alpha \vee \gamma & 1 & 1 \\ \gamma \vee (\alpha \wedge \beta) & \beta \vee \gamma & 1 \end{bmatrix}.$$

One can readily verify that R is transitive, i.e., $R \circ R \subseteq R$ if and only if

$$\gamma = \alpha \wedge \beta. \tag{4.6}$$

Obviously, R is a strongly complete fuzzy binary relation on $\mathbf{X}$. R is a reflexive linear order if and only if $\alpha, \beta, \gamma \in (0, 1)$ and (4.6) holds.

Example 4.8. Consider now a fuzzy binary relation R on $\mathbf{X} = \{x_1, x_2, x_3\}$ defined by

$$R = \begin{bmatrix} 1 & 1 & 1 \\ 1 & 1 & 1 \\ \alpha & \alpha & 1 \end{bmatrix},$$

where $\alpha \in [0, 1]$. By (4.6), R is a complete quasi-ordering which is not a reflexive linear ordering, since the weak asymmetry property is not satisfied. Thus the class of all reflexive orderings is a proper subclass of all complete quasi-orderings.

Example 4.9. Suppose R on $\mathbf{X} = \{x_1, x_2, x_3\}$ is given by

$$R = \begin{bmatrix} 1 & 1 & 1 \\ \alpha & 1 & 1 \\ \gamma & \beta & 1 \end{bmatrix},$$

where $0 \leq \alpha < \gamma < \beta < 1$. By (4.6), thus defined R is not transitive. Let us prove that it is a negatively transitive fuzzy binary relation, i.e., that

$$R(x, y) \vee R(y, z) \geq R(x, z),$$

for all $x, y, z \in \mathbf{X}$. The only nontrivial case is $x = x_3, y = x_2,$ and $z = x_1$ and the previous inequality is easily verified. Relation R provides an example of a quasi-transitive relation which is not a complete quasi-order.

Examples 4.7 - 4.9 show that the vertical arrows on Figure 4.2 indeed indicate proper inclusions.

In many applications, crisp ordering relations are often characterized by the assignment of numerals to objects, attributes, and events according to certain rules (see, for instance, Roberts (1979)). For example, to express a strict preference relation P on $\mathbf{X}$ one would like to assign a real number $u(x)$ to each $x \in \mathbf{X}$ in such a way that xPy if and only if $u(x) > u(y)$. This function u is usually called an *ordinal utility function*.

In a fuzzy environment one may wish to use fuzzy numbers, or, in a more general setting, fuzzy sets defined on the set $\mathbb{R}$ of all reals, as values of ordinal utility functions. Suppose that, for any $x \in \mathbf{X}$, a fuzzy set U is defined on $\mathbb{R}$ by its membership function $U(x,t)$, $x \in \mathbf{X}$, $t \in \mathbb{R}$. This membership function defines a fuzzy binary relation on $\mathbf{X} \times \mathbb{R}$. We shall assume that, for any $x \in \mathbf{X}$, $U(x,t)$ is a membership function of a normal fuzzy set on $\mathbb{R}$, i.e.,

$$\text{for any } x \in \mathbf{X} \text{ there is } t \in \mathbb{R} \text{ such that } U(x,t) = 1.$$

This condition is definitely satisfied in the case of fuzzy numbers. We denote by L the usual ordering $\leq$ on the set of reals $\mathbb{R}$. L is a crisp binary relation on $\mathbb{R}$ defined by the characteristic function

$$L(s,t) = \begin{cases} 1 & \text{if } s \leq t, \\ 0 & \text{otherwise}, \end{cases}$$

for all $s, t \in \mathbb{R}$. For a given U, we define the *induced fuzzy binary relation* R on $\mathbf{X}$ by

$$R = U^{-1} \circ L \circ U$$

(Ovchinnikov, 1991). The membership function of R is given by

$$R(x,y) = \bigvee_{s \leq t} \{U(x,s) \wedge U(y,t)\}$$

for all $x, y \in \mathbf{X}$. It is easy to verify that thus defined R is a strongly complete fuzzy binary relation. Actually, the induced fuzzy binary relation is the Baas-Kwakernaak index (Baas and Kwakernaak, 1977) which is regarded as "very natural in the framework of fuzzy set theory" (Dubois and Prade, 1982). Note that induced fuzzy binary relations are not necessarily transitive relations.

To characterize fuzzy induced binary relations we introduce a new property generalizing the classical concept of interval ordering (Roberts, 1979). We say that a fuzzy binary relation R on $\mathbf{X}$ is *Ferrers* if

$$R(x,y) \wedge R(u,v) \leq R(x,v) \vee R(u,y) \tag{4.7}$$

for all $x, y, u, v \in \mathbf{X}$. Like the symmetry property, the Ferrers property is self-dual, i.e., if R is Ferrers then R^{d} is Ferrers.

Definition 4.7. A fuzzy binary relation R on $\mathbf{X}$ is said to be

a) an *interval ordering* if it is asymmetric and Ferrers, and

b) a *reflexive interval ordering* if it is strongly complete and Ferrers.

Thus defined interval orderings are dual to each other. Now, we have the following theorem.

Theorem 4.8. (Ovchinnikov, 1991). The induced fuzzy binary relation is a reflexive interval ordering.

Proof. It suffices to prove that R is Ferrers, i.e., that

$$R(a,b) \wedge R(c,d) \leq R(a,d) \vee R(c,b)$$

for all a, b, c, and d in $\mathbf{X}$, or, equivalently

$$\left\{\bigvee_{x\leq y} F(a,x) \wedge F(b,y)\right\} \wedge \left\{\bigvee_{z\leq t} F(c,z) \wedge F(d,t)\right\}$$
$$\leq \left\{\bigvee_{u\leq v} F(a,u) \wedge F(d,v)\right\} \vee \left\{\bigvee_{q\leq s} F(c,q) \wedge F(b,s)\right\}.$$

It is sufficient to prove that, for any given $x \leq y$ and $z \leq t$ there exist $u \leq v$ and $q \leq s$ such that

$$F(a,x) \wedge F(b,y) \wedge F(c,x) \wedge F(d,t)$$
$$\leq [F(a,u) \wedge F(d,v)] \vee [F(c,q) \wedge F(b,s)].$$

Let us consider two cases.

(i) $x \leq t$. Then, by choosing $u = x$ and $v = t$, we have

$$F(a,x) \wedge F(b,y) \wedge F(c,x) \wedge F(d,t) \leq F(a,x) \wedge F(d,t)$$
$$\leq [F(a,x) \wedge F(d,t)] \vee [F(c,q) \wedge F(b,s)].$$

(ii) $x > t$. Now we choose $q = z$ and $s = y$. Then,

$$F(a,x) \wedge F(b,y) \wedge F(c,x) \wedge F(d,t) \leq F(b,y) \wedge F(c,z)$$
$$\leq [F(a,u) \wedge F(d,v)] \vee [F(c,z) \wedge F(b,y)].$$

□

Although the induced fuzzy orderings are not necessarily transitive, their duals are transitive as the following theorem asserts.

Theorem 4.9. (Ovchinnikov, 1991). Let R be an irreflexive Ferrers fuzzy binary relation. Then R is transitive.

Proof. Let us substitute $c = b$ and $d = c$ in (4.7). We have

$$R(a,b) \wedge R(b,c) \leq R(a,c) \vee R(b,b) = R(a,c),$$

for all a, b, and c in $\mathbf{X}$. □

Suppose R is irreflexive and transitive. Then, for any $x, y \in \mathbf{X}$,

$$R(x,y) \wedge R(y,x) \leq R(x,x) = 0.$$

Thus R is asymmetric and therefore is a partial ordering. Since the class of crisp interval orderings is a proper subclass of partial orderings, the same fact holds in the fuzzy case. On the other hand, the following theorem asserts that weak orderings are special case of interval orderings.

Theorem 4.10. Let R be a weak ordering on $\mathbf{X}$. Then R is an interval ordering.

Proof. We need to prove that

$$R(a,b) \wedge R(c,d) \leq R(a,d) \vee R(c,b)$$

for all a, b, c, and d in $\mathbf{X}$. It suffices to prove that one of the numbers on the left side of the previous inequality is less than one of the numbers on the right side. We have, by negative transitivity of R,

$$R(a,b) \leq R(a,c) \vee R(c,b) \quad \text{and} \quad R(c,d) \leq R(c,a) \vee R(a,d).$$

By asymmetry of R, either $R(a,c) = 0$, or $R(c,a) = 0$. Hence, either $R(a,b) \leq R(c,b)$ or $R(c,d) \leq R(a,d)$.

□

Theorem 4.10 implies that interval orderings and reflexive interval orderings can be included in our diagram of fuzzy orderings as shown on Figure 4.3.

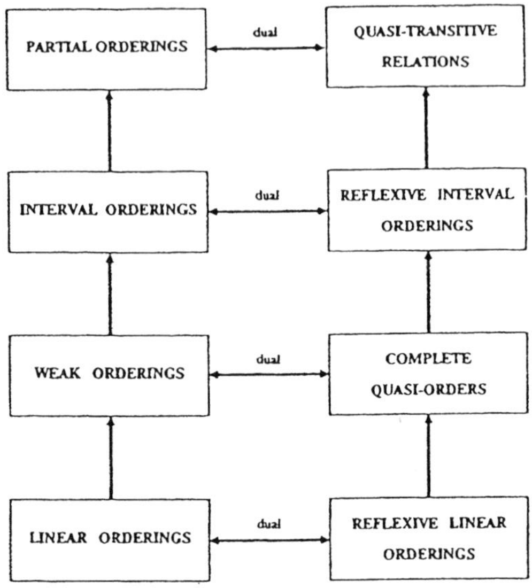

Figure 4.3

4.6 REPRESENTATION THEOREMS

In many instances membership functions of general $*$-transitive relations on a given set **X** can be expressed in terms of real valued functions on **X**. In this section we consider only product-transitive fuzzy binary relations. The reader is referred to Valverde (1985) and Fodor and Roubens (1994) where more general results are found.

Definition 4.8. A fuzzy binary relation R on **X** is $\odot$*-transitive* (*max-product transitive*) if and only if

$$R(x, y) \cdot R(y, z) \leq R(x, z)$$

for all $x, y, z \in$ **X**, or, equivalently, $R \odot R \subset R$ (see Section 4.2).

By definition, a reflexive and $\odot$-transitive fuzzy binary relation on **X** is a *fuzzy preordering* on **X**. A symmetric preordering is a $\odot$*-similarity relation* on **X**.

Consider the following examples of a fuzzy preordering and a $\odot$-similarity relation.

Example 4.10. Let $\mathbf{X} = \{x_1, x_2, x_3\}$. Define a fuzzy binary relation R on $\mathbf{X}$ by

$$R = \begin{bmatrix} 1 & 0.4 & 0.2 \\ 1 & 1 & 0.5 \\ 0.4 & 0.2 & 1 \end{bmatrix}.$$

This reflexive relation is a fuzzy preordering. Indeed,

$$\begin{bmatrix} 1 & 0.4 & 0.2 \\ 1 & 1 & 0.5 \\ 0.4 & 0.2 & 1 \end{bmatrix} \odot \begin{bmatrix} 1 & 0.4 & 0.2 \\ 1 & 1 & 0.5 \\ 0.4 & 0.2 & 1 \end{bmatrix} = \begin{bmatrix} 1 & 0.4 & 0.2 \\ 1 & 1 & 0.5 \\ 0.4 & 0.2 & 1 \end{bmatrix},$$

hence R is $\odot$-transitive. Consider now another fuzzy binary relation S on $\mathbf{X}$ defined by

$$S = \begin{bmatrix} 1 & 0.4 & 0.2 \\ 0.4 & 1 & 0.5 \\ 0.2 & 0.5 & 1 \end{bmatrix}.$$

This symmetric and reflexive relation is a $\odot$-similarity relation on $\mathbf{X}$. Indeed,

$$\begin{bmatrix} 1 & 0.4 & 0.2 \\ 0.4 & 1 & 0.5 \\ 0.2 & 0.5 & 1 \end{bmatrix} \odot \begin{bmatrix} 1 & 0.4 & 0.2 \\ 0.4 & 1 & 0.5 \\ 0.2 & 0.5 & 1 \end{bmatrix} = \begin{bmatrix} 1 & 0.4 & 0.2 \\ 0.4 & 1 & 0.5 \\ 0.2 & 0.5 & 1 \end{bmatrix},$$

thus R is $\odot$-transitive.

The following two theorems are representation theorems for fuzzy preorderings and $\odot$-similarity relations.

Theorem 4.11. (Ovchinnikov, 1984). A fuzzy binary relation R on $\mathbf{X}$ is a fuzzy preordering if and only if there is a family $\mathcal{F}$ of positive real functions on $\mathbf{X}$ such that

$$R(x, y) = \inf_{f \in \mathcal{F}} \min \left\{ \frac{f(x)}{f(y)}, 1 \right\} \tag{4.8}$$

for all $x, y \in \mathbf{X}$.

Proof. (i) Suppose R is defined by (4.8) for a given family of positive functions on $\mathbf{X}$. Then

$$\begin{aligned} R(x, y) \cdot R(y, z) &= \inf_{f \in \mathcal{F}} \min \left\{ \frac{f(x)}{f(y)}, 1 \right\} \cdot \inf_{f \in \mathcal{F}} \min \left\{ \frac{f(y)}{f(z)}, 1 \right\} \\ &\leq \inf_{f \in \mathcal{F}} \min \left\{ \frac{f(x)}{f(y)} \cdot \frac{f(y)}{f(z)}, 1 \right\} = R(x, z). \end{aligned}$$

Clearly, $R(x, x) = 1$ for all $x \in \mathbf{x}$.

(ii) Let R be a fuzzy preordering on $\mathbf{X}$. We define a family $\mathcal{F}$ of positive functions on $\mathbf{X}$ by

$$\mathcal{F} = \{f_{a,n}\}_{a\in\mathbf{X},\, n=1,2,\ldots}$$

where

$$f_{a,n}(x) = \max\left\{R(x,a), \frac{1}{n}\right\}.$$

Then, for any given $x, y \in \mathbf{X}$,

$$\begin{aligned}\inf_{a,n} \min\left\{\frac{f_{a,n}(x)}{f_{a,n}(y)}, 1\right\} &\leq \inf_{n} \min\left\{\frac{f_{y,n}(x)}{f_{y,n}(y)}, 1\right\} \\ &= \inf_{n} \min\{f_{y,n}(x), 1\} \\ &= R(x,y).\end{aligned}$$

On the other hand, using max-product transitivity, we have

$$\begin{aligned}R(x,y)\cdot f_{a,n}(y) &= R(x,y)\cdot \max\left\{R(y,a), \frac{1}{n}\right\} \\ &= \max\left\{R(x,y)\cdot R(y,a), R(x,y)\cdot\frac{1}{n}\right\} \\ &\leq \max\left\{R(x,a), \frac{1}{n}\right\} \\ &= f_{a,n}(x).\end{aligned}$$

Hence,

$$R(x,y) \leq \frac{f_{a,n}(x)}{f_{a,n}(y)},$$

which implies

$$R(x,y) \leq \inf_{a,n} \min\left\{\frac{f_{a,n}(x)}{f_{a,n}(y)}, 1\right\}.$$

Thus (4.8) is proven. □

A similar argument proves the following theorem.

Theorem 4.12. (Ovchinnikov, 1984). A fuzzy binary relation S on $\mathbf{X}$ is a $\odot$ similarity relation if and only if there is a family $\mathcal{F}$ of positive functions on $\mathbf{X}$ such that

$$S(x,y) = \inf_{f\in\mathcal{F}} \min\left\{\frac{f(x)}{f(y)}, \frac{f(y)}{f(x)}\right\} \tag{4.9}$$

for all $x, y \in \mathbf{X}$.

In the following examples we apply these theorems to the relations from Example 4.10.

Example 4.11. Consider R defined by

$$R = \begin{bmatrix} 1 & 0.4 & 0.2 \\ 1 & 1 & 0.5 \\ 0.4 & 0.2 & 1 \end{bmatrix}$$

on $\mathbf{X} = \{x_1, x_2, x_3\}$. Let $\mathcal{F} = \{f_1, f_2\}$ be a family of two functions on $\mathbf{X}$ defined by the following table

	f_1	f_2
x_1	2	5
x_2	5	10
x_3	10	2

A direct calculation shows that these two functions indeed generate R according to (4.8).

Example 4.12. Consider S defined by

$$S = \begin{bmatrix} 1 & 0.4 & 0.2 \\ 0.4 & 1 & 0.5 \\ 0.2 & 0.5 & 1 \end{bmatrix}$$

on $\mathbf{X} = \{x_1, x_2, x_3\}$. Let f be a function on $\mathbf{X}$ defined by

	f
x_1	2
x_2	5
x_3	10

Obviously, $R(x,y) = \min\left\{\frac{f(x)}{f(y)}, \frac{f(y)}{f(x)}\right\}$.

Theorems 4.12 and 4.13 do not claim the uniqueness of representations in (4.8) and (4.9). Moreover, there are infinitely many families generating the same relation. For instance, if all functions in the given family are multiplied by the same positive constant, the new family generates the same relation. On the other hand, the number of functions in $\mathcal{F}$ (or, better say, the cardinality of $\mathcal{F}$) cannot be chosen arbitrarily. For instance, in Example 4.11, R cannot be represented by a family consisting of a single function. Indeed, if a fuzzy preordering R is generated by a single function, it must be a strongly complete fuzzy binary relation according to (4.8). The minimal cardinality of all families $\mathcal{F}$ representing the same fuzzy preordering R (respectively, $\odot$-similarity relation S) is called the *dimension* of R (respectively, S). A complete characterization of one-dimensional fuzzy preorderings and $\odot$-similarity relations is found in Ovchinnikov (1984).

REFERENCES

Baas, S. M. and Kwakernaak, H. (1977). Rating and ranking of multiple aspect alternatives using fuzzy sets, *Automatica,* **13**, 47-58.

Birkhoff, G. (1979). *Lattice theory*, American Mathematical Society Colloquium Publications, Volume XXV, Providence, RI.

Bourbaki, N. (1974). *Elements of mathematics: theory of sets* , Addison-Wesley, Reading, MA.

Dubois, D. and Prade, H. (1980). *Fuzzy sets and systems: theory and applications*, Academic Press, New York.

Dubois, D. and Prade, H. (1982). The use of fuzzy numbers in decision analysis, in: *Fuzzy Information and Decision Processes*, eds. Gupta, M.M. and Sanchez, E., North-Holland, Amsterdam, 309-321.

Fodor, J. and Roubens, M. (1994). *Fuzzy preference modeling and multicriteria decision support*, Kluwer Academic Publishers, Dordrecht, Boston, London.

Klir, G.J. and Folger, T.A. (1988). *Fuzzy sets, uncertainty, and information*, Prentice Hall, Englewood Cliffs, NJ.

Menger, K. (1951). Probabilistic theories of relations, *Proceedings of the National Academy of Science*, **37**, 178-180.

Mesarovic, M. D. and Takahara, Y. (1989). *Abstract system theory,* Lecture Notes in Control and Information Sciences, v. 116.

Ovchinnikov, S. V. (1981) Structure of fuzzy binary relations, *Fuzzy Sets and Systems,* **6**, 169-195.

Ovchinnikov, S. and Riera, T. (1982). On fuzzy classifications, *Fuzzy Sets and Possibility Theory: Recent Developments*, ed.Yager, R.R., Pergamon Press, New York, 120-132.

Ovchinnikov, S.V. (1984). Representations of transitive fuzzy relations, *Aspects of Vagueness*, eds. Skala, H.J., Termini, S., and Trillas, R., D. Reidel Publishing Company, Dordrecht Boston Lancaster, 105-118.

Ovchinnikov, S. (1991). The duality principle in fuzzy set theory, *Fuzzy Sets and Systems*, **42**, 133-144.

Poincaré, H. (1912). *La science et l' hypothèse*, Paris.

Roberts, F.S. (1979). *Measurement theory*, Addison-Wesley, Reading, MA.

Suppes, P., Kranz, D.M., Luce, R.D., and Tversky, A. (1989). *Foundations of measurement*, Volume II, Academic Press, San Diego.

Valverde, L. (1985). On the structure of F-indistinguishability operators, *Fuzzy Sets and Systems*, **17**, 313-328.

Yeh, R.T. and Bang, S.Y. (1975) Fuzzy relations, fuzzy graphs and their applications in clustering analysis, *Fuzzy Sets and Their Applications to Cognitive and Decision Processes*, eds. Zadeh, L.A., Fu, K.S., Tanaka, K., and Shimura, M., Academic Press, New York, 125-149.

Zadeh, L.A. (1965). Fuzzy sets, *Information and Control*, **8**, 338-353.

Zadeh, L.A. (1971). Similarity relations and fuzzy orderings, *Information Sciences,* **3**, 177-200.

Zimmerman, H.-J. (1991). *Fuzzy set theory and its applications*, 2nd edition, Kluwer Academic Publishers, Boston, Dordrecht, London.

5 FUZZY EQUIVALENCE RELATIONS: ADVANCED MATERIAL

Dionis Boixader

Joan Jacas

Jordi Recasens

Abstract: This Chapter presents an overview of the different aspects of the concept of fuzzy equivalence relation (FER) as the extension to the fuzzy framework of the classical idea of equivalence. In this setting, new concepts like generator, dimension and base arise naturally. On the other hand, these relations can be dualy related with some kind of generalized metrics that allows a metric-like study of their properties. This chapter starts introducing some general ideas extending, for any triangular continuous norm, the concept of similarity relation already presented in Chapter 4. Then, we explain different methods for its effective construction. The relationship between fuzzy equivalence relations and generalized metrics is also studied. Next, based on the representation theorem, the concepts of generator, dimension and base are introduced. The structure of the generators set is studied and some procedures for calculating bases are presented.

.1. INTRODUCTION

ll branches of human knowledge have the need to distinguish the objects of ıeir universe of discourse, therefore, some kind of equality must be defined ı them. From the classical point of view, this equality must be an equiva- ›nce relation in the set X of the objects. Nevertheless, an increasing number f problems arising mainly from the so called Soft Sciences, are unsolvable in ıis classical context since the equalities defined in these fields, are not transitive. From this standpoint, a new approach is needed in order to cope with the idea of degree of equality or similarity that leads to a more general definition of equality, strongly related to the concept of neighbourhood and, therefore, introducing some topological background. The main difficulty in defining this generalized equality lies in the modelization of the transitive property. In this context, the transitivity must be understood as a threshold of the value of equality between two objects x and z, $E(x,z)$, given the values between the pairs x, y and y, z, $E(x,y)$ and $E(y,z)$ respectively. In any case, the transitivity can be modelized by the formula $E(x,y) * E(y,z) \leq E(x,z)$ where $*$ stands for an operation in an ordered set $(L, \leq)$ of the values of E, that is necessarily associative. Therefore, $(L,*)$ must be a semigroup. In the sequel, $*$ will represent a continuous t-norm and consequently, L the unit interval, and the associated generalized equality will be termed fuzzy equivalence relation (FER). Depending on the context and authors these relations are also termed similarity relations, fuzzy equalities, indistinguishabilities. The selection of a t-norm for modelizing the operation $*$ is due, not only to a technical reason (t-norms are a well-known set of operations) [Alsina et al. 1983; Schweizer and Sklar 1983] but also among others, to two important facts: firstly, crisp equivalence relations are transitive with respect to any t-norm $*$ (in this sense FER are a generalization of equivalence relations) and secondly, if $(*, \perp, n)$ is a De Morgan triple [García and Valverde 1989] that modelizes the three basic connectives of a multivalued logic, $\rightarrow_*$ the residuated implication associated with $*$ [Trillas and Valverde 1984; Valverde 1985; Godo et al. 1991], $p(x,y)$ is the proposition "x and y are indistinguishable" and we define $E(x,y) = v(p(x,y))$, (the truth-value of p) then the transitivity of E is equivalent to the affirmation of the proposition "if $p(x,y)$ and $p(y,z)$ then $p(x,z)$" that is:

$$v((p(x,y) \cap p(y,z)) \rightarrow p(y,z)) = 1 \text{ is equivalent to } E(x,y) * E(y,z) \leq E(x,z).$$

From this point of view, FER are the equivalences associated with this type of logic. Moreover, in [Höhle 1992] it is pointed out that this kind of generalized equality is the natural tool to introduce the M-sets as fuzzy sets with extensional properties. In this sense, fuzzy equivalences are basic elements in a sound foundation of the Fuzzy Sets Theory and they also play an important role in the construction of a clear model for some concepts applied in fuzzy control [Kruse et al. 1994]. Finally, generalized equalities are also a cornerstone in the modern theory of Fuzzy Cluster Analysis [Jacas and Valverde 1990; Ruspini 1982].

This Chapter, after some preliminaries, presents different aspects of FER using the Representation Theorem as the main tool (section 2). The relation between FER and a special kind of generalized metrics is presented in section 3. Then, we investigate the structure of the set of generators from the algebraic and topological points of view (section 4). In section 5, we deal with the problem of searching a base for a given FER.

In this section we introduce some basic concepts related to FER that will be used in the chapter. Troughout this chapter, E will represent a FER with respect to a continuous triangular norm $*$ (t-norm for short), $\to_*$ its residuation defined by

$$x \to_* y = \sup\{\alpha \in [0,1] \mid (x * \alpha) \leq y\}.$$

X will be a non empty set, $\mathcal{F}(X)$ the set of fuzzy sets on X and $h \in \mathcal{F}(X)$ one its elements. In order to simplify the notation, the symbol $\to_*$ will be substituted by $\to$ unless otherwise stated.

Remark. It is worth noting that $([0,1], \leq, *)$ is a residuated lattice, and the operation $\to$ defined above is the corresponding residuation w.r.t. the t-norm $*$. Further, in a logical context, $\to$ may be interpreted as the implication $\to_*$ based on the t-norm $*$.

Definition 5.1. *Given a t-norm $*$, a FER E on a set X is a fuzzy relation on X that satisfies*

(1) $E(x,x) = 1\ \forall x \in X$ *(Reflexivity)*

(2) $E(x,y) = E(y,x)\ \forall x,y \in X$ *(Symmetry)*

(3) $E(x,y) * E(y,z) \leq E(x,z)\ \forall x,y,z \in X$ *($*$-transitivity).*

In section 2 it is proved that any FER can be constructed starting from a family of fuzzy sets, and different ways of constructing these relations are presented.

Depending on the t-norm used to modelize the transitivity, we can find in the literature different names such as similarity relations for $*$=minimum [Zadeh 1971], probabilistic relation for $*$=product [Menger 1951], likeness relations for $*$=Lukasiewicz t-norm [Bezdek and Harris 1978]. All these relations were summarized by E. Trillas [Trillas 1982] (following Menger's ideas) in 1982 under the name of indistinguishability operators since they are the key idea in order to solve the Poincaré paradox [Poincaré 1902] in the sense that in a physical context, equal means indistinguishable and therefore equalities are not transitive in the classical sense and we can find situations where $A = B$, $B = C$ and $A \neq C$.

5.2. HOW TO BUILD FUZZY EQUIVALENCE RELATIONS

There are different ways to generate FER depending on the way the data is given and on the use you want to make of it.

One of the most common tools used in the construction of FER is the Max-$*$ product which allows us to construct the transitive closure of a given reflexive and symmetric fuzzy relation (also called tolerance relation). Another way to generate FER is by using the Representation Theorem [Ovchinnikov 1984; Valverde 1985; Jacas 1990] that allows the construction of these relations starting on from a family of fuzzy sets. For instance, if we have a set X of objects to be classified and some fuzzy subsets of X expressing the degree of fulfilment of these objects to some criteria, the theorem provides a method to generate a FER on X in a natural way.

Other methods are less useful [Fodor and Roubens 1994] and here we will only present the way to build decomposable relations for its interest in Mamdani's approach to fuzzy control.

Let us start with the transitive closure of fuzzy relations.

Definition 5.2. *Let $*$ be a t-norm and R, S two fuzzy relations in a set X. The Max-$*$ (or Sup-$*$) product $R \circ S$ of R and S is the fuzzy relation on X defined by*

$$(R \circ S)(x,y) = \sup_{z \in X} (R(x,z) * S(z,y)) \quad \forall x, y \in X$$

Due to the associativity of Max-$*$ product, we can define for each $n \in N$ the power R^n of a fuzzy relation R recursively:

$$\begin{aligned} R^1 &= R \\ R^{n+1} &= R \circ R^n \quad \forall n \in N \end{aligned}$$

Definition 5.3. *The *-transitive closure (or *-closure) $\overline{R}$ of a fuzzy relation R on a set X is defined by*

$$\overline{R} = \sup_{n \in N} R^n.$$

Example 5.1. Let R be the relation given by the matrix

$$R = \begin{pmatrix} 1 & 0.3 & 0.5 & 0.7 \\ 0.3 & 1 & 0.2 & 0.8 \\ 0.5 & 0.2 & 1 & 0.3 \\ 0.7 & 0.8 & 0.3 & 1 \end{pmatrix}$$

The transitive closures $\overline{R}_{\text{Ł}}, \overline{R}_{\Pi}, \overline{R}_{\text{Min}}$ of R with respect to the t-norms of Łukasiewicz, product and minimum respectively are

$$\overline{R}_{\text{Ł}} = \begin{pmatrix} 1 & 0.5 & 0.5 & 0.7 \\ 0.5 & 1 & 0.2 & 0.8 \\ 0.5 & 0.2 & 1 & 0.3 \\ 0.7 & 0.8 & 0.3 & 1 \end{pmatrix} \qquad \overline{R}_{\Pi} = \begin{pmatrix} 1 & 0.56 & 0.5 & 0.7 \\ 0.56 & 1 & 0.28 & 0.8 \\ 0.5 & 0.28 & 1 & 0.35 \\ 0.7 & 0.8 & 0.35 & 1 \end{pmatrix}$$

$$\overline{R}_{\text{Min}} = \begin{pmatrix} 1 & 0.7 & 0.5 & 0.7 \\ 0.7 & 1 & 0.5 & 0.8 \\ 0.5 & 0.5 & 1 & 0.5 \\ 0.7 & 0.8 & 0.5 & 1 \end{pmatrix}$$

Proposition 5.1. *If R is a reflexive and symmetric fuzzy relation on a finite set X of cardinal n, then $\overline{R} = R^{n-1}$.*

Proposition 5.2. *Let R be a reflexive and symmetric fuzzy relation on a set X. $R = \overline{R}$ if and only if $\forall x, y, z \in X$, $R(x,y) * R(y,z) \leq R(x,z)$.*

Therefore, the transitive closure $\overline{R}$ of a reflexive and symmetric fuzzy relation is a FER. Moreover, it is straightforward to prove that $\overline{R}$ is a relation greater or equal than R ($\overline{R} \geq R$) and it can be shown [Valverde 1985] that if E is a FER greater or equal than R, then $E \geq \overline{R}$. In other words, the transitive closure $\overline{R}$ of R is the smallest FER that contains R and is therefore the best upper approximation of R by means of such relations.

In fact, the following proposition can be proved.

Proposition 5.3. *Given a reflexive and symmetric fuzzy relation R on a set X, let A be the set of FER on X greater than or equal to R and $\overline{R}$ its transitive closure. Then, for any $x, y \in X$,*

$$\overline{R}(x,y) = \inf_{E \in A} \{E(x,y)\}.$$

Next, the generation of FER will be studied from the point of view of the Representation Theorem 5.1, first stated for the product t-norm by S. Ovchinikov [see chapter 4] and then for arbitrary continuous t-norms by Valverde [Valverde 1985; Jacas 1988]. This theorem allows not only the construction of relations starting on a set of fuzzy subsets, but also the study of their structure as it will be seen in the next sections.

Lemma 5.1. *Given a fuzzy subset h of a set X, the fuzzy relation E_h on X defined by*

$$E_h(x,y) = Max\{h(x), h(y)\} \rightarrow Min\{h(x), h(y)\}$$

is a FER on X.

In particular, if X is the unit interval and h is the identity ($h(x) = x$, $x \in [0,1]$), we have the FER E_* defined by

$$E_*(x,y) = \text{Max}\{x,y\} \rightarrow \text{Min}\{x,y\} = (x \rightarrow y) \wedge (y \rightarrow x)$$

that we will be termed the **natural equivalence relation** associated to $*$.

Example 5.2.

5.2.1. If $* = \text{Ł}$ is the Lukasiewicz t-norm ($x * y = \text{Max}\{x + y - 1, 0\}$), then

$$E_h(x,y) = 1 - |h(x) - h(y)|.$$

5.2.2. If $* = \Pi$ is the product t-norm, then

$$E_h(x,y) = \begin{cases} \operatorname{Min}\left\{\frac{h(x)}{h(y)}, \frac{h(y)}{h(x)}\right\}, & \text{if } h(x) \neq h(y) \\ 1, & \text{otherwise.} \end{cases}$$

5.2.3. If $* =$Min, then

$$E_h(x,y) = \begin{cases} 1 & \text{if } h(x) = h(y) \\ \operatorname{Min}\{h(x), h(y)\} & \text{otherwise.} \end{cases}$$

Lemma 5.2. *If $\{E_j\}_{j\in J}$ is a family of FER on a set X, then $E = \inf_{j\in J} E_j$ is a FER on X.*

From Lemmas 5.1 and 5.2 it follows:

Theorem 5.1. (Representation Theorem). *A fuzzy relation E on a set X is a FER if and only if there exists a family $\{h_j\}_{j\in J}$ of fuzzy subsets of X such that*

$$E = \inf_{j\in J} E_{h_j} = \inf_{j\in J} E_* \circ (h_j \times h_j).$$

Example 5.3. Let $X = \{a_1, a_2, a_3, a_4\}$ and h_1, h_2 the two fuzzy subsets of X defined by

$h_1(a_1) = 0.2 \quad h_2(a_1) = 0.3$
$h_1(a_2) = 0.8 \quad h_2(a_2) = 0.2$
$h_1(a_3) = 0.9 \quad h_2(a_3) = 0.4$
$h_1(a_4) = 1 \quad h_2(a_4) = 0.1$

The FER $E_{Ł}, E_\Pi, E_{\text{Min}}$ generated by $\{h_1, h_2\}$ with respect to the t-norms of Lukasiewicz, product and minimum are

$$E_{Ł} = \begin{pmatrix} 1 & 0.4 & 0.3 & 0.2 \\ 0.4 & 1 & 0.8 & 0.8 \\ 0.3 & 0.8 & 1 & 0.7 \\ 0.2 & 0.8 & 0.7 & 1 \end{pmatrix} \qquad E_\Pi = \begin{pmatrix} 1 & 0.25 & 0.22 & 0.20 \\ 0.25 & 1 & 0.50 & 0.50 \\ 0.22 & 0.50 & 1 & 0.25 \\ 0.20 & 0.50 & 0.25 & 1 \end{pmatrix}$$

$$E_{\text{Min}} = \begin{pmatrix} 1 & 0.2 & 0.2 & 0.1 \\ 0.2 & 1 & 0.2 & 0.1 \\ 0.2 & 0.2 & 1 & 0.1 \\ 0.1 & 0.1 & 0.1 & 1 \end{pmatrix}$$

□

In the same way as the transitive closure of a reflexive and symmetric fuzzy relation R can be characterized as the smallest FER that is greater than or equal to R (Proposition 5.2.), it can be proved that the FER E generated by the columns of R (i.e. by the family $(h_x)_{x\in X}$ of fuzzy subsets of X defined by $h_x(y) = E(x,y)$) is the greatest of such relations smaller than or equal to R satisfying $R(x,z) * E(x,y) \leq R(y,z)$ for all x, y, z in X. E is then a kind of lower approximation of R which is related to complete linkage clustering methods in the same way as the transitive closure is related to single linkage. In this setting, although there are different ways of finding maximal reflexive and transitive lower approximations of a reflexive and symmetric relation R [Fodor and Roubens 1994; Wagenknecht 1995], it is still a challenging open problem to find maximal FER lower than or equal to R.

A third way to build FER, is by generating decomposable relations by means of a fuzzy set.

Proposition 5.4. *Given a t-norm $*$ and a fuzzy set h on a set X, the fuzzy relation on X defined by*

$$E(x,y) = \begin{cases} h(x) * h(y), & \textit{if } x \neq y \\ 1, & \textit{otherwise} \end{cases}$$

is a FER.

Definition 5.4. *A FER E on a set X is called decomposable if and only if there exists a fuzzy subset h on X such that E can be expressed as in the preceding proposition.*

It is interesting to connect decomposable relations with the Representation Theorem and the transitive closure.

If E_h^{RT} and E_h^d represent the FER generated by the fuzzy subset h via the Representation Theorem and as decomposable relation respectively, then $E_h^d \leq E_h^{RT}$ and both coincide when the t-norm is the minimum. In fact, E_h^d and E_h^{RT} can be viewed as the most pessimistic and the most optimistic way to generate a FER with a given fuzzy subset h.

The relation between the transitive closure procedure and decomposable relations can be stablished in the following way: Given a reflexive and symmetric fuzzy relation R on a set X, we can generate a decomposable FER with each column of R. Although the union of FER is not, in general, a transitive relation, the following result holds:

Proposition 5.5. *Let R be a reflexive and symmetric fuzzy relation on a set X, $(h_z)_{z\in X}$ the family of columns of R and $(E_{h_z})_{z\in X}$ the family of decomposable FER generated by them. Let S be the union of these relations. Then $S = R^2$.*

5.3. FUZZY EQUIVALENCE RELATIONS AND GENERALIZED METRICS

The original idea relating FER and distances comes from Menger [Menger 1951] (Probabilistic Theory of relations) when he shows that any probabilistic relation E is canonically attached to a distance d_E defined on X.

The key structure is the concept of pseudo-distance introduced by Fréchet (1928) as a mapping $d : X \times X \to R^+$ such that for any x, y, z of X, satisfies

1. $d(x, x) = 0$
2. $d(x, y) = d(y, x)$
3. $d(x, y) + d(y, z) \geq d(x, z)$ (Triangular inequality).

If $d(x, y) = 0$ implies $x = y$ then we have a distance.

Usually when classifying a set of objects we start from a dissimilarity index, a reflexive and symmetric relation in X, that is further transformed into a "distance" that is used to classify the objects of X. So a pseudodistance between elements is built in order to give an structural description of the data.

The ideal case is the one that has as associated pseudodistance, a discrete one d_0 that assigns to each pair of elements $(x, y) \in X \times X$ a value of the set $\{0, 1\}$. In this case the relation E associated with d_0 defined by: $x, y \in E$ iff $d_0(x, y) = 0$, is a classical equivalence relation that determines a partition in X. Nevertheless, this is not true in general. That is, given a pseudodistance d in X with values in $I = [0, 1]$ and $\alpha \in I$, the classical relation R_α defined by $(x, y) \in R_\alpha$ iff $d(x, y) \leq \alpha$, is reflexive and symmetric, but it is not transitive in general, in the classical sense. Therefore, if the $(x, y) \in R_\alpha$ is interpreted as "x and y are indistinguishable (at level α)" we have an "equality" that is not transitive. Only the case that d is an ultrametric gives such transitivity as it is shown in Chapter 4.

In Zadeh's seminal paper on similarity relations [Zadeh 1971] it is pointed out as a specially interesting feature of such relations, that reversing the "scale" in [0,1] by means of the function $\phi(x) = 1 - x$, a special type of metric (an ultrametric) is obtained.

J.C. Bezdek and J.O. Harris (1978) and later E. Ruspini (1982) indicate as one of the main reasons for introducing the likeness relations (fuzzy equivalences with respect to the Lukasiewicz t-norm $(x * y) = \text{Max}(x + y - 1, 0)$) is that the associated metric in [0,1] is the restriction to the unit interval of the Euclidean metric that refines the ultrametric given by similarity relations.

Finally, in Valverde (1985) the existing link between fuzzy equality relations and a special type of generalized metrics termed $\perp$-metrics is clearly shown. This duality allows us to construct any $\perp$-metric by means of the representation theorem. By means of the mentioned theorem it is possible to introduce the concept of a generator of a $\perp$-metric. The generators of a $\perp$-metric have a particulary appealing characterization and they form a fuzzy topology.

In general, if $*$ denotes a continuous t-norm $*$, $\perp$ a continuous t-conorm [Alsina et al. 1983; Schweizer and Sklar 1983] and I the unit interval [0,1], we have the following result: If E is a fuzzy equality relation in a non-empty set X, $*$ an archimedean t-norm and t is an additive generator, $d = t \circ E$ is a pseudometric on X and conversely, given a pseudometric d on X and decreasing and continuous bijection t from $[0, +\infty]$ into [0,1], $E = t \circ d$ is a FER where t^{-1} is an additive generator of $*$ (see chapter 2).

This procedure only applies when $*$ is Archimedean. In general, if we reverse the "scale" in [0,1] by means of continuous and order reversing bijection, we do not obtain pseudometrics but $\perp$-pseudometrics [Valverde 1985] defined as follows:

Definition 5.5. *A $\perp$-pseudometric on X is a map m from $X \times X$ into [0,1] such that*

i) $m(x,x) = 0.$

ii) $m(x,y) = m(y,x),$

iii) $m(x,y) \perp m(y,z) \geq m(x,z)$ *($\perp$-triangular inequality).*

where $\perp$ denotes a continuous t-conorm.

A $\perp$-metric is a $\perp$-pseudometric such that $m(x,y) = 0$ iff $x = y$.

The $\perp$-pseudometrics are a type of "generalized metrics" introduced in [Trillas and Alsina 1978].

$\perp$-pseudometrics and $*$-FER are dual concepts in the sense of the following theorem:

Theorem 5.2. *If E is a FER and ϕ a continuous and order reversing bijection from [0,1] into [0,1] then, $m = \phi \circ E$ is a $\perp$-pseudometric and conversely, where $\perp$ is the dual t-conorm associated with $*$ defined by $\perp(x,y) = \phi \circ (\phi^{-1}(x) * \phi^{-1}(y))$.*

Due to this duality, the FER representation theorem gives a similar one for $\perp$-pseudometrics.

That is, given a continuous t-conorm $\perp$, $\rightarrow_\perp$ will represent its residuation [Valverde 1985] i.e. $x \rightarrow_\perp y = \inf\{\alpha \in [0,1] \mid \alpha \perp x \geq y\}$. Then, it can be proved that $m_\perp(x,y) = \text{Min}\{x,y\} \rightarrow_\perp \text{Max}\{x,y\} = (x \rightarrow_\perp y) \vee (y \rightarrow_\perp x)$ is a $\perp$-metric in the unit interval

Example 5.4. If $x \perp y = \text{Max}\{x,y\}$, $x \rightarrow_\perp y = 0$ for $x \geq y$ and $x \rightarrow_\perp y = y$ otherwise, then

$$m_{\text{Max}}(x,y) = \begin{cases} \text{Max}\{x,y\}, & \text{if } x \neq y \\ 0, & \text{otherwise.} \end{cases}$$

If $x \perp y = xy - xy$ than $x \rightarrow_\perp y = \text{Max}\{0, (y-x)/(1-x)\}$, and then

$$m_{\text{prod}}(x,y) = \begin{cases} |x-y| / \text{Max}\{1-x, 1-y\}, & \text{if } x \neq y \\ 0, & \text{otherwise.} \end{cases}$$

If $x \perp y = \text{Min}\{x+y, 1\}$, $x \rightarrow_\perp y = \text{Max}\{y-x, 0\}$, then

$$m_{\text{Ł}} = |x-y|.$$

Furthermore, any $\perp$-metric in X is defined by a family of functions $\{h_j\}_{j\in J}$ that "value" the elements of X on [0,1] and by a natural $\perp$-metric from $[0,1]\times[0,1]$ into [0,1] that really establishes the metric relation.

Therefore, we are in a position to give a more compact version of the representation theorem [Valverde 1985]

Theorem 5.3. *($\perp$-metric representation theorem). [Jacas 1993] Let $\perp$ be a continuous t-conorm and $m_\perp$ its associated natural metric, a map m from $X \times X$ into [0,1] is a $\perp$-pseudometric if, and only if, there exists a family $\{h_j\}_{j\in J}$ of functions from X into [0,1] such that*

$$m = \sup_{j\in J}(m_\perp \circ (h_j \times h_j)).$$

5.4. THE GENERATORS SET: GRANULARITY, OBSERVABILITY AND APPROXIMATION

Given a FER E on a set X with respect to a continuous t-norm $*$, the Representation Theorem suggests the following defintions [Jacas 1988].

Definition 5.6. *$H \subseteq \mathcal{F}(X)$ is a generating family of E if $E(x,y) = \inf_{h\in H}\{E_*(h(x),h(y))\}$ for all $x,y \in X$.*

Definition 5.7. *A fuzzy set $h \in \mathcal{F}(X)$ is a generator of E if there exists a generating family H such that $h \in H$.*

Let us note by H_E the set of all generators of E. It can be immediately shown that $h \in H_E$ if and only if $E_h(x,y) \geq E(x,y)$ for all $x,y \in X$.

The generators play an important role in the field of FER. The study of generators can be arranged into three major groups of topics. A group is about the notion of generator, its meaning and its relation with granularity, observability and cluster analysis. A second group deals with the mathematical structure of H_E, and it includes topics such fuzzy topologies, lattices or morphisms. Finally, a third group is devoted to the problems about basis and dimensions, which are originated by the calculation of minimal generating families. This section is devoted to the first and second points. The problems related to bases and dimensions will be discussed later in this chapter.

Let us recall some basic definitions. As usual, we denote $\leq$ the pointwise order between fuzzy subsets (so, $h \leq h'$ if, and only if, $h(x) \leq h'(x)$, for any $x \in X$). It is a well known fact that $(\mathcal{F}(X), \leq)$ is a complete lattice, with meet ($\wedge$) and joint ($\vee$) defined in the natural way by $(h \vee h')(x) = \sup\{h(x), h'(x)\}$, and $(h \wedge h')(x) = \inf\{h(x), h'(x)\}$, for any $x \in X$.

The classical topological concepts can also be extended to the fuzzy framework [Lowen 1976] and will be useful tools to give a topological insight to the structure of H_E.

Definition 5.8. *A subset τ of $\mathcal{F}(X)$ is a fuzzy topology on X, if and only if, it satisfies the following properties:*

(1) If $A \subset \tau$, then $\sup\{h|h \in A\} \in \tau$.

(2) If $h, h' \in \tau$, then $h \wedge h' \in \tau$.

(3) If $h_k \in \mathcal{F}(X)$ is a constant fuzzy set (i.e., $h_k(x) = k\ \forall x \in X$), then $h_i \in \tau$.

Definition 5.9. *A map $\psi : \mathcal{F}(X) \to \mathcal{F}(X)$ is a fuzzy interior operator, if and only if, for all h, h' of $\mathcal{F}(X)$ it satisfies:*

(1) $h \geq \psi(h)$,

(2) $\psi(h \wedge h') = \psi(h) \wedge \psi(h')$,

(3) $\psi^2(h) = \psi(h)$,

(4) $\psi(h_k) = h_k$ for each constant fuzzy set h_k.

Definition 5.10. *A map $\phi : \mathcal{F}(X) \to \mathcal{F}(X)$ is a fuzzy closure operator, if and only if, for all h, h' of $\mathcal{F}(X)$ it satisfies:*

(1) $h \leq \phi(h)$,

(2) $\phi(h \vee h') = \phi(h) \vee \phi(h')$,

(3) $\phi^2(h) = \phi(h)$,

(4) $\phi(h_k) = h_k$ for each constant fuzzy set h_k.

The elements of a fuzzy topology τ are called *open fuzzy sets*, and it is easy to prove that a fuzzy interior operator ψ defines a fuzzy topology $\tau_\psi = \{h \in \mathcal{F}(X) \ / \ \psi(h) = h\}$ (so, the open fuzzy sets are the fixed points of ψ).

On the other hand, given a *strong negation* n (i.e., a continuous decreasing bijection $n : [0,1] \to [0,1]$ such that $n(n(x)) = x$ for all $x \in [0,1]$) [Trillas 1982; Trillas 1998] a fuzzy set is called *closed* if, and only if, $n \circ h$ is open.

For any closure operator ϕ, and any strong negation n, the fixed points of ϕ are the closed fuzzy sets of a fuzzy topology τ defined by $\tau = \{h \in \mathcal{F}(X) \ / \ \phi(n \circ h) = n \circ h\}$. A more extensive study of this topic can be found in [Jacas 1993].

The notion of extensionality is widely used in the literature about fuzzy equalities, e.g., see [Höhle 1988; Höhle 1990; Kruse et al. 1994] and plays an important role in a sound foundation of fuzzy sets theory. As we can see in the next results it is strongly related to the concept of a generator.

Definition 5.11. *$h \in \mathcal{F}(X)$ is extensional with respect to a FER E if $E(x, y) * h(x) \leq h(y)$, for all $x, y \in X$.*

Proposition 5.7. *$h \in H_E$ if and only if h is extensional with respect to E.*

So the generators are the extensional sets with respect to E.

As it will be shown later in this chapter, only those fuzzy sets that are extensional are compatible with the FER and they can be interpreted as the *observable sets* under this relation.

Proposition 5.8. *[Jacas 1988] $(H_E, \leq)$ is a complete lattice.*

Proof. Given a family $\{h_j\}_{j\in J} \subseteq H_E$, we have

$E(x,y) * h_j(x) \leq h_j(y)$ for all $x, y \in X$, $j \in J$, and then since $*$ is a continuous t-norm, $\sup_{j\in J}\{E(x,y) * h_j(x)\} = \sup_{j\in J}\{h_j(x) * E(x,y)\} \leq \inf_{j\in J}\{h_j(x)\}$, for all $x, y \in X$, so $\bigwedge_{j\in J}\{h_j\}$ and $\bigvee_{j\in J}\{h_j\}$ are extensional, and $\bigvee_{j\in J}\{h_j\}, \bigwedge_{j\in J}\{h_j\} \in H_E$. □

From the properties of the residuation of the t-norm $*$ [Boixader 1998] the next lemma follows

Lemma 5.3. *For any $h \in H_E$ and $\alpha \in [0,1]$, we have:*

(a) $(\alpha * h) \in H_E$

(b) $(\alpha \rightarrow h) \in H_E$

(c) $(h \rightarrow \alpha) \in H_E$

And due to the reflexivity of the relation E we also obtain

Lemma 5.4. *If $h \in H_E$ then $\sup_{x\in X}\{h(x) * E(x,y)\} = h(y)$.*

The next theorem gives an axiomatic characterization of H_E.

Theorem 5.4. *[Klawonn and Castro 1995] Given a family $H \subseteq \mathcal{F}(X)$, there exists a $*$-FER E such that $H = H_E$ if and only if it satisfies:*

(a) $(\alpha * h) \in H$ *for all* $\alpha \in [0,1]$, $h \in H$.

(b) $(\alpha \rightarrow h) \in H$ *for all* $\alpha \in [0,1]$, $h \in H$.

(c) $(h \rightarrow \alpha) \in H$, *for all* $\alpha \in [0,1]$, $h \in H$.

(d) $(H, \leq)$ *is a complete sublattice of* $(\mathcal{F}(X), \leq)$.

It is important to remark the existence of a special kind of generator: i.e. the columns of E.

Definition 5.12. *Given a FER E, the fuzzy set C_x defined by $C_x(y) = E(x,y)$ for all $, y \in X$ is the column of associated with the element x.*

Let us denote by $C_E = \{C_x\}_{x\in X}$ the set of all columns of E.

Note that the $*$-transitivity condition, $E(x,y) * E(y,z) \leq E(x,z)$ for all $x,y,z \in X$, ensures that C_x is extensional for all $x \in X$, and $C_E \subseteq H_E$. However, there are elements in H_E different from those of C_E (For example, $C_x \vee C_y \in H_E - C_E$, provided that E is a separating FER). As it is shown in many relevant works [Zadeh 1971; Dubois and Prade 1994; Jacas and Valverde 1990], the columns of E can be naturally thought as the associated fuzzy equivalence classes, the fuzzy clusters of E.

Going back to the Representation Theorem, it is worth noting that, given a generating family H of E, two elements $x, y \in X$ are equivalent by E at a certain level α, if it does not exist any fuzzy subset $h \in H$ such that $E_*(h(x), h(y)) < \alpha$.

This is the fuzzy version, scaled in [0,1] by means of E_*, of the classical equivalence between two elements induced by a set of crisp criteria (i.e. a set of crisp subsets of the universe of discourse X) in the sense that two elements are equivalent if, under any considered criterion, they both take the same truth value (zero or one). That is, if any a feature, measure, etc. that we can consider does not allow us to distinguish them.

So, taking into account that, in any context where we have a (fuzzy) equivalence relation, the information levels, or the granularity of knowledge in this context, become more degraded as the level of equivalence increases, a natural problem arises: how to fit in a coherent way, both the relation and the pieces of information expressed by means of fuzzy sets.

In presence of a new available piece of fuzzy information, namely a fuzzy set $h \notin H_E$ this problem may be addressed in two different ways:

a) to modify E in order to accomodate the incoming information, or

b) to modify h in order to make it compatible with E.

In the first case, to modify E in a suitable manner can be easily achieved by adding h to the set H_E of all generators of E and then generating a new FER accordingly to the Representation Theorem, i.e., $E'(x,y) = \inf_{h\in H'}\{E_*(h(x), h(y))\}$, where $H' = H \cup \{h\}$.

On the other hand, there are mainly two different ways to distort a fuzzy set h in order to make it compatible with a given FER E. Both ways consist in approximating h by a generator, but they differ depending on whether upper $(\phi_E(h))$ or lower $(\psi_E(h))$ approximations of h are chosen.

Definition 5.13. *Let E be a FER on a set X. The map $\phi_E : \mathcal{F}(X) \to \mathcal{F}(X)$ is defined by*

$$\phi_E(h)(x) = \sup_{y\in X}\{h(y) * E(x,y)\}, \text{ for all } x \in X.$$

Some relevant properties of ϕ_E are:

(a) $\phi_E(h) \geq h$

(b) If $h \leq h'$, then $\phi_E(h) \leq \phi_E(h')$

(c) $\phi_E(h \vee h') = \phi_E(h) \vee \phi_E(h')$, for all $h, h' \in \mathcal{F}(X)$.

Furthermore, ϕ_E sends each fuzzy set h to the smaller (i.e. less specific with respect the usual order in fuzzy sets) generator $\phi_E(h)$ that contains h (i.e. $h \leq \phi_E(h)$).

Proposition 5.9. $\phi_E(h) \in H_E$ for all $h \in \mathcal{F}(X)$.

Proof. $\phi_E(h) = \sup_{z \in X}\{h_z\}$, where

$$\begin{aligned} h_z : X &\longrightarrow [0,1] \\ x &\longrightarrow h_z(x) = (h(z) * E(z,x)) = (h(z) * C_z(x)) \end{aligned}$$

Since $C_z \in H_E$, we see that $\phi_E(h) \in H_E$. □

Putting together the previous results we can characterize the generators by the following theorem

Theorem 5.5. *[Jacas 1988] $\phi_E(h) = h$ if and only if $h \in H_E$.*

For this reason the generators have been also termed as eigenvectors of the FER [Jacas and Recasens1995] or fixed points of the operator ϕ_E [Jacas and Recasens 1994].

As immediate consequence of the previous result we have that

(a) $\phi_E(h) = \bigwedge_{h' \in H_E}\{h' \geq h\}$, for all $h \in \mathcal{F}(X)$.

(b) $\phi_E^2 = \phi_E$

(c) $\phi_E(h_k) = h_k$ for each constant fuzzy set h_k.

and therefore that ϕ_E is a fuzzy closure operator and the set H_E is the set of closed sets of a fuzzy topology.

The fuzzy closure operator ϕ_E is also closely related to the set C_E.

Proposition 5.10. *$\phi_E(\{x\}) = C_x$, for all $x \in X$.*

Proof. Let us recall that the characteristic function of $\{x\}$ is δ_x defined by $\delta_x(y) = 1$ if $y = x$, and $\delta_x(y) = 0$ otherwise.

So $\phi_E(\{x\})(u) = \phi_E(\delta_x)(u) = \sup_{y \in X}\{\delta_x(y) * E(y, \delta)\} = E(x,u) = C_x(u)$ for all $x, u \in X$.

In other words, the columns are the images of the singletons of X and they are also termed the *singletons* associated with E [Höhle 1990] and the set C_x can be interpreted as the fuzzified version of $\{x\}$ via the fuzzy equality E.

Furthermore, if A is a crisp subset of X, we have:

(a) $\phi_E(A) = \bigvee_{x \in A}\{C_x\}$, for all $A \subseteq X$.

(b) Let $\mathcal{P}(X)$ be the power set of X. $\phi_E(\mathcal{P}(X) - \{\phi\})$ is the semilattice H_{C_E} of H_E generated by subfamilies of $\{E(x, \cdot)\}_{x \in X}$ via suprema.

So H_E can be interpreted as the set of fuzzy sets of the quotient set X/E (i.e. $H_E = \mathcal{F}(X/E)$) and $\phi_E : \mathcal{F}(X) \to \mathcal{F}(X/E)$ is the canonical map. Note that if the equivalence relation is a crisp one, then $\phi_{E|X}$ is the crisp canonical map $\pi : X \to X/E$.

Remark. Let us observe that the operator ϕ_E gives a natural way for fuzzyfying crisp concepts (crisp fuzzy sets) of an universe of discourse X when a FER is defined in it. This point of view gives a clear semantical meaning to fuzzify-defuzzify process involved in most of applications specially in fuzzy control [Boixader et al. 1997]. In the same line, triangular and trapezoidal shaped fuzzy sets are the most commonly used ones. A similarity-based approximation to this fact leads to a clear semantical interpretation for these fuzzy sets and gives a logical justification for its use [Jacas and Recasens 1993; Klawonn and Kruse 1993; Kruse et al. 1994].

Now, let us turn our attention to the map ψ_E that sends each fuzzy set h to the greater (i.e. more specific) generator $\psi_E(h)$ contained in it (i.e. $h \geq \psi_E(h)$) and defined as follows

Definition 5.14. *Let E be a FER on a set X. The map $\psi_E : \mathcal{F}(X) \to \mathcal{F}(X)$ is defined by $\psi_E(h)(x) = \inf_{y \in X}\{E(x, y) \to h(y)\}$, for all $x \in X$.*

The results concerning ψ_E are dual of those obtained for ϕ_E.

Proposition 5.11. *For all $h, h' \in \mathcal{F}(X)$, we have:*

(a) $\psi_E(h) \leq h$,

(b) If $h \leq h'$ then $\psi_E(h) \leq \psi_E(h')$

(c) $\psi_E(h \wedge h') = \psi_E(h) \wedge \psi_E(h')$

Proposition 5.12. *$\psi_E(h) \in H_E$ for all $h \in \mathcal{F}(X)$.*

Theorem 5.6. *$\psi_E(h) = h$ if, and only if, $h \in H_E$.*

From this theorem it follows immediately that

(a) $\psi_E(h) = \bigvee_{h' \in H_E}\{h' \leq h\}$ for all $h \in \mathcal{F}(X)$.

(b) $\phi_E^2 = \phi_E$.

(c) $\phi_E(h_k) = h_k$ for each constant fuzzy set h_k.

and therefore ψ_E is a fuzzy interior operator and H_E is the set of open sets of a fuzzy topology.

So the generators are the open sets of a fuzzy topology τ_{ψ_E} ($H_E = \tau_{\psi_E} = \{h \in \mathcal{F}(X) \;/\psi_E(h) = h\}$) and the closed sets associated to a fuzzy topology τ_{ϕ_E} ($\tau_{\phi_E} = \{h \in \mathcal{F}(X) \mid \phi_E(n \circ h) = n \circ h\}$, where n stands for a strong negation).

In general, $\tau_\phi \neq \tau_\psi$, and it turns of interest to study the conditions under which both topologies coincide.

Obviously, $\psi_E = n \circ \phi_E \circ n$ is a sufficient condition, since

$$\begin{aligned} H_E &= \tau_{\psi_E} = \{h \in \mathcal{F}(X) \mid \psi_E(h)\} = \{h \in \mathcal{F}(X) \mid n \circ \phi_E \circ n(h) = h\} = \\ &= \{h \in \mathcal{F}(X) \mid \phi_E \circ n(h) = n(h)\} = \tau_{\phi_E}. \end{aligned}$$

Conversely, from $\tau_\phi = \tau_\psi$ it follows that for any $h \in \mathcal{F}(X)$, $\phi_E(n \circ h) = n \circ h$ and $\psi_E(h) = h$ are equivalent, and this condition easily leads us to $n \circ \phi_E \circ n = \psi_E$.

So $\tau_{\phi_E} = \tau_{\psi_E}$ if, and only if, $n \circ \phi_E \circ n = \psi_E$.

In many contexts such as Rough Sets theory and modal logic, the lower bound of h is defined by $n \circ \phi_E \circ n(h)$, where n is a strong negation. Nevertheless, the obtained fuzzy subset is not a generator in general.

In order to study the cases in which given a set X, the relation $\psi_E = n \circ \phi_E \circ n$ is fullfilled for any FER, let us recall the characterization theorem for strong negations [Trillas 1982; Trillas 1998].

Theorem 5.7. *A map $n : [0,1] \longrightarrow [0,1]$ is a strong negation if and only if there exists a continuous and decreasing map $t : [0,1] \longrightarrow R^+$ with $t(1) = 0$ such that*

$$n(x) = t^{-1}(t(0) - t(x)) \quad \forall x \in [0,1].$$

A map t as defined in the preceding theorem is termed a *generator* of the strong negation n.

Theorem 5.7 has the following interpretation in the setting of MV-algebras [Chang 1958]:

Proposition 5.13. *n is a strong negation if and only if there exists a non strict archimedean t-norm $*$ such that $n(x) = x \rightarrow_* 0$.*

Proposition 5.14. *Let X be a set, $*$ a continuous t-norm, n a strong negation and E a fuzzy equality relation on X. If $*$ is a non strict archimedean t-norm with the same generator as n or E it is a classical equivalence relation, then $n \circ \phi_E \circ n(h) = \psi_E(h)$ for all fuzzy subsets h of X.*

Finally, the following reciprocal result also holds.

Proposition 5.15. *Let X be a set, $*$ a continuous t-norm, n a strong negation. If $n \circ \phi_E \circ n(h) = \psi_E(h)$ for all fuzzy sets h of X then $*$ is a non-strict archimedean t-norm and $n_{|D} = n_{*|D}$ where $D = \{E(x,y) \mid x, y \in X\}$ and n_* the strong negation associated with E ($n_*(x) = x \rightarrow_* 0$).*

Another interesting concept closely related to the approximations $\phi_E(h)$ and $\psi_E(h)$ is the *observability* [Boixader et al. 1997]. Given a crisp equivalence realtion on a set X, it is not possible to deal with the singletons $\{x\}$ for any $x \in X$, but only with their equivalence classes $\overline{x}$ (of course, in a extreme case, it may happen that $\overline{x} = \{x\}$ for all $x \in X$). The same occurs with the subsets $A \subseteq X$ which are not the union of equivalence classes (i.e. $A \neq \bigcup_{x \in A} \overline{x}$). This is due to the fact that it is impossible to split any equivalence class without overcoming the information level of the given equivalence relation.

We refer a subset $A \subset X$ such that $A = \bigcup_{x \in A} \overline{x}$ as an *observable set* under the equivalence relation E, and there are mainly two different ways to fuzzify this concept.

First, we can rewrite $A = \bigcup_{x \in A} \overline{x}$ as $x \in A \Leftrightarrow \exists u \in X : u \in A \cap E(x,u)$, and then to extend it to the fuzzy context as

$$A(x) = \sup_{y \in X} (A(y) * E(x,y))$$

so that $A \in \mathcal{F}(X)$ is observable if, and only if, $A = \phi_E(A)$. The other way is obtained when crisp observability is rewriten as $x \in A \Leftrightarrow \forall y \in X$: $E(x,y) = 1$ implies $y \in A$.

In this case, $A(x) = \inf_{y \in X}(E(x,y) \rightarrow A(y))$ and $A = \psi_E(A)$ is obtained.

Note that, in both cases, the observable fuzzy sets are exactly the generators of E, as a consequence of Theorem 5.5 and 5.6.

In the last part of this section, we are going to present two results that allow us to decide whether a given generator is a column also termed singleton [Höhle 1990; Klawonn and Kruse 1993]. As it has been commented before the set of columns can be interpreted in the fuzzy framework, as the set of clusters or fuzzy classes associated to the FER [Zadeh 1971].

Unlike the crisp case, in which any column C_x (i.e. any cluster) is made up from prototypical elements (i.e. $x_0 \in X$ such that $C(x_0) = 1$), in the fuzzy case we have to accept the existence of clusters with their prototypes lying out of X.

This is due to the fact that, given a FER $\overline{E}$ on a set $\overline{X}$, any subset $X \subseteq \overline{X}$ inherites the structure of $\overline{E}$, and is also a FER with $E : X \times X \rightarrow [0,1]$ defined by $E(x,y) = \overline{E}(x,y)$ for all $x, y \in X$. So it is possible to have a generator of E, h coming from a column C_y of $\overline{E}$, (i.e. $h = C_{y|X}$), and thus having its prototypical element y_0 out of X.

Definition 5.15. Given two FER (X, E) and $(\overline{X}, \overline{E})$, we say that $(\overline{X}, \overline{E})$ is an extension of (X, E) if:

(1) $X \subseteq \overline{X}$

(2) $\overline{E}(x,y) = E(x,y)\ \forall x,y \in X$

Let us note $C_E = \{h \in \mathcal{F}(X)\ /\ \exists(\overline{X},\overline{E})$ extension of (X,E) s.t. $h \in C_{\overline{E}}\}$.

Theorem 5.8. *Let (X,E) be a FER. Given $h \in H_E$, the following statements are equivalent:*

(a) $h(x_1) * h(x_2) \leq E(x_1,x_2),\ \forall x_1,x_2 \in X.$

(b) There exists an extension $(\overline{X},\overline{E})$ of (X,E) such that $h = h_y|_X,\ y \in \overline{X}$ (i.e. $h(x) = \overline{E}(y,x)\ \forall x \in X$).

Note. If h is normal, then $(\overline{X},\overline{E}) = (X,E)$, and $h = h_x$, $x \in X$. [Kruse et al. 1994].

Definition 5.16. *Let E be a FER on a set X. The map $\Lambda_E : \mathcal{F}(X) \to \mathcal{F}(X)$ is defined by*

$$\Lambda_E(h)(x) = \inf_{u \in X} \{h(u) \to E(u,x)\}, \quad \forall x \in X.$$

Note that this definition fuzzifies the notion of (crisp) cluster, when it is expressed by

$$x \in h \Leftrightarrow \forall u \in X : u \in h \Rightarrow E(u,x) = 1.$$

Proposition 5.16. *For any $h, h' \in \mathcal{F}(X)$, we have.*

(a) If $h \leq h'$ then $\Lambda_E(h) \geq \Lambda_E(h')$.

(b) $\Lambda_E(h \vee h') = \Lambda_E(h) \wedge \Lambda_E(h')$

(c) $\Lambda_E(h \wedge h') \geq \Lambda_E(h) \vee \Lambda_E(h')$.

(d) $\Lambda_E(h) \in H_E$, for all $h \in \mathcal{F}(X)$.

Note. In general, $\mathrm{Im}\ \Lambda_E \neq H_E$. Take, for example, $X = \{x_1,x_2,x_3\}$, $E(x_i,x_j) = 0$ if $i \neq j$, and $h \in H_E$ defined by $h(x_1) = 0.5$, $h(x_2) = h(x_3) = 0$.

Proposition 5.17. *Let E be a FER, and $h \in \mathcal{F}(X)$. If $\Lambda_E(h) = h$, then $h \in C_E$.*

The next theorem characterizes columns of E as fixed points of Λ_E.

Theorem 5.9. *Let E be a FER, and $h \in \mathcal{F}(X)$ a normal fuzzy subset (i.e. $\exists x_0 \in X$ such that $h(x_0) = 1$) $\Lambda_E(h) = h$ if, and only if, $h = h_x$, $(x \in X)$.*

Theorem 5.9 characterizes only the columns of E, and cannot be extended to the whole set C_E, as it is shown in next example.

Example 5.5. $X = \{x_1, x_2\}$, $E(x_1, x_2) = 0$, $*$ an arbitrary t-norm. We define an extension of (X, E): $\overline{X} = X \cup \{y\}$, $\overline{E}(x_1, y) = \overline{E}(x_2, y) = 0$.

The column of y is (restricted to X), the constant fuzzy set $h(x_1) = h(x_2) = 0$. So $\Lambda_E(h)$ is the constant fuzzy set $\Lambda_E(h)(x_1) = \Lambda_E(h)(x_2) = 1$.

Finally, Table 5.1 summarizes the relation between some of the concepts contained in this section for the crisp case and their fuzzy versions.

Table 5.1

Crisp equivalence relation	*FER*
observable set	generator
upper approximation	$h \mapsto \phi_E(h)$
lower approximation	$h \mapsto \psi_E(h)$
cluster	column $h_x(y) = E(x, y)$

5.5. DIMENSION AND BASIS. THEIR CALCULATION.

The Representation Theorem of FER states that every FER can be generated by a family of fuzzy subsets. Nevertheless, there is no uniqueness in the election of the family, different families, even having different cardinals, can generate the same relation. This point gives great interest to the theorem, since if we interpret the elements of the family as degrees of matching between the elements of the initial set X and a set of prototypes, we can choose different features in order to stablish this matching, giving therefore different semantical interpretations to the same FER .

Among the generating families of a relation, the ones with "low" cardinality are of special interest, since they have an easy semantical interpretation and also because the information contained in the matrix representing the relation can be "packed" in a few (sometimes even one) fuzzy sets.

This section studies such minimal sets and solves the problem of finding them for the minimum and for archimedean t-norms in finite universes.

The case of the minimum is solved in a very combinatorial way, whereas in the archimedean case, a geometric interpretation of the set of generators is exploded.

Definition 5.17. *Let E be a FER on a set X. The dimension of E is the minimal of the cardinalities of the generating families of E, in the sense of the Representation Theorem. A generating family with this cardinality is called a basis of E.*

In the rest of the section, it will be assumed that X is a finite set of cardinality n ($|X| = n$) and the fuzzy set h_k defined by $h_k(x) = E(k, x)$ will be termed as the *column associated* with $k \in X$.

The case of the minimum (similarities)

Let us first study the case of the minimum t-norm. A FER with respect to the Minimum t-norm is commonly termed a similarity [Zadeh 1971].

In this case, it will be assumed that, for any x, y in X such that $x \neq y$, $E(x, y) \neq 1$. Otherwise, if $\simeq$ is the equivalence relation in X, $x \simeq y$ if, and only if $E(x, y) = 1$, then the induced similarity relation $\overline{E}$ on the quotient set $X^{\sim} = X/\simeq$, satisfies the mentioned condition.

Moreover, h is a generator of E if and only if the induced fuzzy set $\overline{h}$ on $X^{\sim}$ is a generator of $\overline{E}$. So, we can restrict ourselves to such separating relations in order to solve the problem of finding dimensions and basis.

In [Jacas 1988], it is proved that if E is a similarity relation and there exists $k \in X$ such that h_k is injective, then dim $E = 1$.

In order to determine the dimension of a given similarity relation E, the following definitions will be needed:

Definition 5.18. *A column h_k is termed a maximal column if,*

$$|\{h_k(x)|x \in X\}| \geq |\{h_u(x)|x \in X\}| \text{ for any } u \in X.$$

That is, the column has a maximal number of different values. H_M will denote the set of the maximal columns of E.

If h_k is a maximal column and $\alpha \in \{h_k(x)|x \in X\}$, then X^α is the subset of X defined by $X^\alpha = \{x \in X | h_k(x) = \alpha\}$ and $|X^\alpha|$ will be termed the *frequency of α in h_k* (fr (α)).

Definition 5.19. *The order of a maximal column h_k ($\theta(h_k)$) is the maximum of the frequencies of the values in h_k, i.e.*

$$\theta(h_k) = Max\{fr(h_k(x))|x \in X\}$$

Definition 5.20. *The order of a similarity E ($\theta(E)$) is the minimum of the orders of its maximal columns:*

$$\theta(E) = Min\{\theta(h_k)|h_k \in H_M\}$$

Example 5.6. Let E be a similarity relation over a set X with $|X| = 10$ defined by the matrix in Table 5.2.

The maximal columns are printed boldface and their orders are all 7. Therefore, the similarity order is $\theta(E) = 7$.

Table 5.2 Similarity relation matrix of Example 5.6.

	x_0	x_1	x_2	x_3	x_4	x_5	x_6	x_7	x_8	x_9
x_0	**1.00**	0.10	**0.24**	0.10	**0.10**	**0.10**	0.10	0.10	0.12	0.10
x_1	**0.10**	1.00	**0.10**	0.10	**0.10**	**0.10**	0.10	0.10	0.10	0.10
x_2	**0.24**	0,10	**1.00**	0.10	**0.10**	**0.10**	0.10	0.10	0.12	0.10
x_3	**0.10**	0.10	**0.10**	1.00	**0.10**	**0.10**	0.33	0.10	0.10	0.10
x_4	**0.10**	0.10	**0.10**	0.10	**1.00**	**0.42**	0,10	0.10	0.10	0.17
x_5	**0.10**	0.10	**0.10**	0.10	**0.42**	**1.00**	0.10	0.10	0.10	0.17
x_6	**0.10**	0.10	**0.10**	0.33	**0.10**	**0.10**	1.00	0.10	0.10	0.10
x_7	**0.10**	0.10	**0.10**	0.10	**0.10**	**0.10**	0.10	1.00	0.10	0.10
x_8	**0.12**	0.10	**0.12**	0.10	**0.10**	**0.10**	0.10	0.10	1.00	0.10
x_9	**0.10**	0.10	**0.10**	0.10	**0.17**	**0.17**	0.10	0.10	0.10	1.00

The next lemma plays a basic role in determining the cardinality of $\mathrm{Im}(E)$.

Lemma 5.5. *Let E be a similarity relation, h_k a maximal column and α one of its values. If*

$$D^{\alpha} = \{E(x,y) \mid E(x,y) \neq \alpha; x, y \in X^{\alpha}\}$$

then $|D^{\alpha}| \leq |X^{\alpha}|$.

From this lemma it immediately follows that if E is a similarity relation over a finite set and $|X| = n$ then $|\mathrm{Im}(E)| \leq n$. That is, a similarity relation on a set X with $|X| = n$ takes, at most n different values.

Lemma 5.6. *If $\{h_j\}_{j\in J}$ is a generating family of a similarity relation E, there exists another family of generators $\{h'_j\}_{j\in J}$ with the same cardinality such that for any $j \in J$ and $x \in X$, $h'_j(x) \in Im(E)$.*

That is given a generating family, we can find another one with the same cardinality, whose elements take values within the set of values of E,

The next theorem gives an upper bound for the dimension of a similarity relation and allows us to build an algorithm for the automatic search of minimal generating systems.

The bound obtained is highly efficient because the algorithm finds a set of generators whose cardinality is dim E+1, at the most.

Theorem 5.10. *[Jacas 1990] If $\theta(E)$ is the order of a similarity E and σ is the smallest natural number such that $2^{\sigma} \geq \theta(E)$, then dim $E \leq \sigma + 1$.*

Based on the pevious theorem we present two algorithms that basically differ in their efficiency and complexity. The first one always selects a minimal generating system, but presents higher difficulties then the second one in its implementation.

The second algorithm, in some pathological cases, selects a generating system with one element more than the dimension of the similarity but it presents a more iterative structure and, therefore, enables an easier implementation.

Algorithm 5.1.

1. Compute the orders of the columns and select the maximal ones.
2. Compute the similarity order and the upper bound of the dimension ($\sigma +$ 1).
3. Select a maximal column h_k whose order is the similarity order.
4. Build $\sigma + 1$ column vectors (h_j) of dimension n and initialize them with 1's.
5. For any unrepeated value $h_k(x)$ in h_k, do $h_j(x) \leftarrow h_k(x)$ for j from 1 to $\sigma + 1$. Do this assignment for all unrepeated values in h_k.
6. Select the a repeated value α in h_k and find its associated set X^α.
7. For any pair x_i, x_j of elements of X^α, such that $E(x_i, x_j) = \alpha' \neq \alpha$, do $h_1(x_i) = h_1(x_j) \leftarrow \alpha$, $h_2(x_i) \leftarrow 1$, $h_2(x_j) \leftarrow \alpha'$.
8. For any element x_r of X^α not selected in the preceding step, do $h_j(x_r) \leftarrow \alpha$ for $j = 1, 2, ...,$ until for any pair of elements x_r, x_s included in this step there exists a j such that $h_j(x_r) \neq h_j(x_s)$.
9. If there exists in h_k another repeated value, select it and go to step 6.
10. For any pair of elements x', x'' with different repeated values α, α', such that $h_j(x') = h_j(x'')$ for all j, select p such that $h_p(x') = h_p(x'') = 1$ and do $h_p(x') \leftarrow E(x', x'')$.
11. Delete all constant columns.
12. End.

The next example shows an application of this algorithm.

Example 5.7. Like in Example 5.6. let S be a similarity relation defined over a set X with $|X| = 10$, defined by the matrix in Table 5.2.

Step 1. Compute the column orders and select the maximal ones. Maximal columns are the 3rd and 8th (printed boldface), both with order 2.

Step 2. Compute the similarity order and the upper bound of the dimension. Both are 2.

Step 3. Select a maximal column of order 2. We select for example the 8th column.

Steps 4 and 5. Build two initialized vectors and list the unrepeated values like in schema 1 of Table 5.4.

Table 5.3 Similarity relation matrix of Example 5.7.

	x_0	x_1	$\mathbf{x}_2$	x_3	x_4	x_5	x_6	$\mathbf{x}_7$	x_8	x_9
x_0	1.00	0.10	**0.10**	0.10	0.10	0.10	0.10	**0.10**	0.10	0.25
x_1	0.10	1.00	**0.27**	0.27	0.27	0.27	0.13	**0.27**	0.13	0.10
x_2	0.10	0.27	**1.00**	0.35	0.36	0.30	0.13	**0.40**	0.13	0.10
x_3	0.10	0.27	**0.35**	1.00	0.35	0.30	0.13	**0.35**	0.13	0.10
x_4	0.10	0.27	**0.36**	0.35	1.00	0.30	0.13	**0.36**	0.13	0.10
x_5	0.10	0.27	**0.30**	0.30	0.30	1.00	0.13	**0.30**	0.13	0.10
x_6	0.10	0.13	**0.13**	0.13	0.13	0.13	1.00	**0.13**	0.28	0.10
x_7	0.10	0.27	**0.40**	0.35	0.36	0.30	0.13	**1.00**	0.13	0.10
x_8	0.10	0.13	**0.13**	0.13	0.13	0.13	0.28	**0.13**	1.00	0.10
x_9	0.25	0.10	**0.10**	0.10	0.10	0.10	0.10	**0.10**	0.10	1.00

Step 6. Select the repeated value $\alpha_1 = 0.10$, $X^{\alpha_1} = \{x_0, x_9\}$. Do the same with the value $\alpha_2 = 0.13$, $X^{\alpha_2} = \{x_6, x_8\}$.

Steps 7 to 10. Since $S(x_0, x_9) = 0.25 \neq 0.10$, place them in the second column, and do the same with the value $\alpha_2 = 0.13$, because $S(x_6, x_8) = 0.28 \neq 0.13$ (schema 2 in Table 5.4).

The columns of schema 2 form a minimal generating system (a basis) of S.

Table 5.4 Results of Example 5.7.

Schema 1		*Schema 2*	
h_1	h_2	h_1	h_2
1.00	1.00	0.10	0.25
0.27	0.27	0.27	0.27
0.40	0.40	0.40	0.40
0.35	0.35	0.35	0.35
0.36	0.36	0.36	0.36
0.30	0.30	0.30	0.30
1.00	1.00	0.13	0,28
1.00	1.00	1.00	1.00
1.00	1.00	0.13	1.00
1.00	1.00	0.10	1.00

Algorithm 5.2.

0. Counter $\leftarrow 0$, $|X| = n \leftarrow N$.

Table 5.5 Similarity relation matrix of example 5.7.

1.00	0.10	**0.12**	**0.12**	0.11	0.12	0.12	0.10	0.12	0.12
0.10	1.00	**0.10**	**0.10**	0.10	0.10	0.10	0.22	0.10	0.10
0.12	0.10	**1.00**	**0.36**	0.11	0.13	0.23	0.10	0.16	0.19
0.12	0.10	**0.36**	**1.00**	0.11	0.13	0.23	0.10	0.16	0.19
0.11	0.10	**0.11**	**0.11**	1.00	0.11	0.11	0.10	0.11	0.11
0.12	0.10	**0.13**	**0.13**	0.11	1.00	0.13	0.10	0.13	0.13
0.12	0.10	**0.23**	**0.23**	0.11	0.13	1.00	0.10	0.16	0.19
0.10	0.22	**0.10**	**0.10**	0.10	0.10	0.10	1.00	0.10	0.10
0.12	0.10	**0.16**	**0.16**	0.11	0.13	0.16	0.10	1.00	0.16
0.12	0.10	**0.19**	**0.19**	0.11	0.13	0.19	0.10	0.16	1.00

1. Compute the columns orders and select the maximal ones.
2. Compute the similarity order and calculate σ.
3. Select a maximal column h_k whose order equals of similarity order.
4. Counter $\leftarrow$ Counter+1 (C$\leftarrow$C+1).
5. If C=1, build $\sigma + 1$ columns h_i of dimension N, initialize them with 1's and do $h_1(x) \leftarrow h_k(x)$ for any x in X. Go to 6, otherwise.
6. Select a repeated value α of h^k and find the set X^α associated to α.
7. If $\sigma = 1$, go to 13.
8. Select the similarity $E' = E|X^\alpha$. Do $E \leftarrow E'$, $X \leftarrow X^\alpha\, N \leftarrow |X^\alpha|$.
9. If order of E is $n-1$, go to 12.
10. Do $h_C(x) \leftarrow h_k(x)$ for any x in X such that $h_k(x) \neq \alpha$.
11. For any element x not being selected in 10, do $h_j(x) \leftarrow \alpha$ for $j = 1, \ldots$ until for any pair of elements x_r, x_s selected in this step there exists a j with $h_j(x_r) \neq h_j(x_s)$ and $C \leq j \leq \sigma + 1$.
12. Select another repeated value and go to 4.
13. End.

Example 5.8. We present an example where algorithm 5.2, already implemented, has been applied to a similarity over X with $|X| = 10$ shown in Table 5.5.

The maximal columns are the 3rd and 4th, both with order 2. The similarity order is 2 and the similarity dimension upper bound is also 2.

Table 5.6 Generating system of example 5.8.

0.12	0.12
0.10	0.22
0.36	0.36
1.00	1.00
0.11	0.11
0.13	0.13
0.23	0.23
0.10	1.00
0.16	0.16
0.19	0.19

In this case, the algorithm selects a minimal generating system, shown in Table 5.6.

Example 5.9. The generating set obtained exceeds in one element the dimension of the given similarity. See tables 5.7 and 5.8.

Table 5.7 Similarity relation matrix of example 5.9.

1.00	0.10	0.10	0.10	0,10	0.10	0.10	0.10	0.10	0.10
0.10	1.00	0.10	0.10	0.10	0.18	0.10	0.10	0.10	0.10
0.10	0.10	1.00	0.17	0.17	0.10	0.17	0.23	0.10	0.23
0.10	0.10	0.17	1.00	0.17	0.10	0.17	0.17	0.10	0.17
0.10	0.10	0.17	0.17	1.00	0.10	0.17	0.17	0.10	0.17
0.10	0.18	0.10	0.10	0.10	1.00	0.10	0.10	0.10	0.10
0.10	0,10	0.17	0.17	0.17	0.10	1.00	0.17	0.10	0.17
0.10	0.10	0.23	0.17	0.17	0.10	0.17	1.00	0.10	0.25
0.10	0.10	0.10	0.10	0.10	0.10	0.10	0.10	1.00	0.10
0.10	0.10	0.23	0.17	0.17	0.10	0.17	0.25	0.10	1.00

The case of archimedean t-norms.

A natural geometric representation of the set of generators of FER E defined on a set X is exploded in order to calculate explicitly a basis of E, provided that $*$ is an archimedean t-norm [Valverde 1985] and X is finite ($|X| = n$).

Let us order the elements of the finite set X (i.e., $X = \{a_1, a_2, \ldots, a_n\}$). Then, every fuzzy subset h of X can be identified with the point $(h(a_1), h(a_2), \ldots, h(a_n))$ of $[0, 1]^n$.

Table 5.8 Results of Algorithm 5.2 for example 5.9

Generators selected			*Minimal system*	
0.10	0.10	0.10	1.00	0.10
0.10	0.18	0.18	0.10	0.18
0.23	0.23	0.23	0.23	0.23
0.17	0.17	0.17	1.00	0.17
0.17	0.17	1.00	0.17	1.00
0.10	1.00	1.00	0.10	1.00
0.17	1.00	1.00	0.17	0.17
0.25	0.25	0.25	0.25	0.25
0.10	0.10	1.00	0.10	0.10
1.00	1.00	1.00	1.00	1.00

If h is a generator of a given FER E on X, then $E_h \geq E$ or, in a more explicit way

$$\text{Max}\{h(a_i), h(a_j)\} \rightarrow \text{Min}\{h(a_i), h(a_j)\} \geq E(a_i, a_j) \quad \forall i, j = 1, 2, \ldots, n.$$

Therefore, in general we have the following result

Theorem 5.11. *The set of generators of a FER E on X is the solution of the following system of inequalities:*

$$Max\{x_i, x_j\} \rightarrow Min\{x_i, x_j\} \geq E(a_i, a_j), \quad 1 \leq i < j \leq n,\ 0 \leq x_i \leq 1.$$

This system becomes specially simple if the t-norm is the product Π, the Minimum or the Lukasiewicz t-norm $Ł(x, z) = \text{Max}\{x + y - 1, 0\}$, that is

(a) If $* = \Pi$, then H_E is the polyhedron A solution of the system

$$x_i - E(a_i a_j) x_j \geq 0, \quad 0 \leq x_i \leq 1, \quad i, j = 1, 2, \ldots, n.$$

(b) If $* = Ł$, then H_E is the polyhedron solution of

$$x_i - x_j \leq 1 - E(a_i a_j), \quad 0 \leq x_i \leq 1, \quad i, j = 1, 2, \ldots, n.$$

(c) If $* = \text{Min}$, then H_E is the solution of

$$\text{Min}\{x_i, x_j\} \geq E(a_i, a_j) \ \text{ if } x_i \neq x_j, \quad 0 \leq x_i \leq 1, \quad i, j = 1, 2, \ldots, n.$$

For $* = \Pi$ and $* = Ł$, the location of the elements of a basis is relatively strict as it is shown in the following proposition.

Proposition 5.18. *If $* = \Pi$ or $* = Ł$, the elements of a basis of a FER E are located in the (hyper) faces of H_E.*

Theorem 5.12. *[Jacas and Recasens1995] In the case $* = \Pi$ or $* = Ł$, it is always possible to find a basis of E with all its elements on the edges of H_E.*

Algorithm 5.3. Calculation of a basis of a FER E on a finite set X ($|X| = n$) for $* = \Pi$ or $* = Ł$:

(1) Calculate the edges of the set H_E [Solodóvnikov 1980] (Theorems 5.11 and 5.12).

(2) *Count* = 1.

(3) Build a set A obtained selecting a generator from each edge of H_E.

(4) Define B(*Count*) = the set of subsets of A of *Count* elements.

(5) Select a set H of B(*Count*) and generate the relation E_H.

(6) If $E_H = E$ then end.

(7) Do step 5 and step 6 for all different elements of B(*Count*).

(8) *Count* = *Count* + 1. Go to 4.

Example 5.10.. Let us consider the relation E, transitive with respect to the t-norm product and represented by the matrix

$$E = \begin{pmatrix} 1.00 & 0.12 & 0.41 & 0.13 \\ 0.12 & 1.00 & 0.12 & 0.23 \\ 0.41 & 0.12 & 1.00 & 0.27 \\ 0.13 & 0.23 & 0.27 & 1.00 \end{pmatrix},$$

($|X| = 4$). The set of edges in this case is:

$$\begin{array}{ll}
\{(0.12\ \ 1.00\ \ 0.29\ \ 0.23) & (0.41\ \ 0.12\ \ 1.00\ \ 0.52) \\
(0.12\ \ 1.00\ \ 0.25\ \ 0.92) & (0.41\ \ 0.12\ \ 1.00\ \ 0.27) \\
(0.12\ \ 1.00\ \ 0.29\ \ 0.92) & (0.66\ \ 0.23\ \ 0.27\ \ 1.00) \\
(0.12\ \ 1.00\ \ 0.12\ \ 0.44) & (1.00\ \ 0.12\ \ 0.48\ \ 0.13) \\
(0.12\ \ 1.00\ \ 0.12\ \ 0.23) & (1.00\ \ 0.12\ \ 0.41\ \ 0.13) \\
(0.13\ \ 0.23\ \ 0.32\ \ 1.00) & (1.00\ \ 0.12\ \ 0.41\ \ 0.52) \\
(0.13\ \ 0.23\ \ 0.27\ \ 1.00) & (1.00\ \ 0.12\ \ 1.00\ \ 0.27) \\
(0.29\ \ 1.00\ \ 0.12\ \ 0.44) & (1.00\ \ 0.12\ \ 1.00\ \ 0.52) \\
(0.29\ \ 1.00\ \ 0.12\ \ 0.23) & (1.00\ \ 0.57\ \ 0.41\ \ 0.13) \\
(0.35\ \ 1.00\ \ 0.85\ \ 0.23) & (1.00\ \ 0.57\ \ 0.48\ \ 0.13)\}
\end{array}$$

and a basis is

$$\begin{pmatrix} 0.12 & 1.00 & 0.25 & 0.92 \\ 0.29 & 1.00 & 0.12 & 0.23 \end{pmatrix}$$

Let us recall that an isomorphism between two t-norms $*$, $*'$ is a bijective map f such that $f \circ * = *' \circ (f \times f)$. It is well known that every archimedean t-norm is isomorphic either to Π or to Ł.

On the other hand, we have the following theorem.

Theorem 5.13. *If f is an isomorphism between the t-norms $*$ and $*'$ and E is a $*$-FER on X, then $E' = f \circ E$ is a $*'$ FER on X. Moreover, if $\{h_i\}_{i \in I}$ is a basis of E, then $\{f(h_i)\}_{i \in I}$ is a basis of $f \circ E$. [Jacas and Recasens1995].*

Therefore, this result allows us the calculation of a basis for any archimedean t-norm.

Acknowledgments

Research partially supported by DGICYT project number PB94-1208.

References

Alsina, C., Trillas, E. and Valverde, L. (1983) On some logical connectives for Fuzzy Set Theory, *J. Math. Ann. Appl.* **93**, 15-26.

Bezdek, J.C. and Harris, J.O. (1978) Fuzzy partitions and relations. An axiomatic basis for clustering. *Fuzzy Sets and Systems* **1**, 112-127.

Boixader, D., Jacas, J. and Recasens, J. (1997) Similarity-based approach to deffuzification. *Proceedings FUZZ'IEEE-97.* Barcelona. 761-765.

Boixader, D., Jacas, J. and Recasens, J. (1998) Uppper and lower approximation of fuzzy sets. *International Journal of General Systems*, (to appear).

Chang, C.C. (1958) Algebraic Analysis of Many-Valued Logics. *Trans. Amer. Math. Soc.* **88**, 467-490.

Dubois, D. and Prade, H. (1994) Similarity-Based Approximate Reasoning, In: *Computational Intelligence Imitating Life.* Zurada, J.M., Marles, R.J., Robinson, C.J. Eds. IEEE Press, New York, 69-80.

Fodor J. and Roubens, M. (1994) *Structure of transitive valued binary relations.* Preprint n. 94.001, Institut de Mathématique. Univ. de Liège.

Garcia, P. and Valverde, L. (1989) Isomorphisms between De Morgan triplets, *Fuzzy Sets and Systems* **30**, 1, 27-36.

Godo, L., Jacas, J. and Valverde, L. (1991) Fuzzy values in Fuzzy Logic, *Int. Journal of Intelligent Systems* **6**, 199-212.

Höhle, U. (1988) Quotients with respect to similarity relations. *Fuzzy Sets and Systems* **27**, 1, 31-44.

Höhle, U. (1990) The Poincaré paradox and the cluster problem. In: A. Dress, A. von Haeseler (Eds.), *Trees and Hierarchical Structures*, Lectures Notes in Biomathematics, 84, 117-124, Springer-Verlag.

Höhle, U. (1992) M-valued sets and sheaves over integral commutative CL-monoids. In: S.E. Rodabaugh, E.P. Klement and U. Höhle (Eds.), *Applications of Category Theory to Fuzzy Subsets*, 33-72, Kluwer, Dordrecht.

Jacas, J. (1988) On the generators of T-indistinguishability operators. *Stochastica* **12**, 49-63.

Jacas, J. (1990) Similarity relations-The calculation of minimal generating families. *Fuzzy Sets and Systems* **35**, 151-162.

Jacas, J. (1993) Fuzzy topologies induced by S-metrics. *The Journal of Fuzzy Mathematics* **1**, 1.

Jacas, J. and Recasens, J. (1993) Fuzzy Numbers and Equality Relations. *Proceedings FUZZ'IEEE 93*, San Francisco, 1298-1301.

Jacas, J. and Recasens, J. (1994) Fixed points and generators of fuzzy relations. *J. Math. Anal. Appl.* **186**, 21-29.

Jacas, J. and Recasens, J. (1995) Fuzzy T-transitive relations: eigenvectors and generators. *Fuzzy Sets & Systems*, **72** 147-154.

Jacas, J. and Valverde, L. (1990) On fuzzy relations, metrics and cluster Analysis. En Verdegay, Delgado Eds. *Appproximate Reasoning Tools for A I* (Verlag TUV, Rheinland) 15R, 96.

Klawonn, F. and Castro, J.L. (1995) Similarity in Fuzzy Reasoning, *Mathware & Soft Computing* **2**, 197-228.

Klawonn, F. and Kruse, R. (1993) Equality Relations as the Basis for Fuzzy Control. *Fuzzy Sets and Systems* **54**, 147-156

Kruse, R., Gebhard, J. and Klawoon, F. (1994) *Foundations of Fuzzy Systems.* Wiley, Chichester.

Lowen, R. (1976) Fuzzy topological spaces and fuzzy compactness. *J. Math. Anal. Appl.* **56**, 621-633.

Menger, K. (1951) Probabilistic theory of relations. *Proc. Nat. Acad. Sci. USA* **37**, 178-180.

Ovchinnikov, S.V. (1984) Representation of transitive fuzzy relations. In H.J. Skala et alt. eds. *Aspects of vagueness.* (Reidel, Dordrecht) 105-118.

Poincaré, H. (1902) *La science et l'hypothèse.* Flamarion ed., Paris.

Ruspini, E. (1982) Recent developments in fuzzy clustering. En. Yager ed. *Fuzzy sets and possibility theory: Recent developments.* Pergamon Press. NY, 133-147.

Schweizer, B. and Sklar, A. (1983) *Probabilistic metric spaces*, North-Holland, Amsterdam.

Solodóvnikov, A.S. (1980) *Sistemas de desigualdades lineales* Ed. Mir, Moscow.

Trillas, E. (1982) Assaig sobre les relacions d'indistingibilitat. *Actes del Primer Congrés Català de Lògica Matemàtica.* Barcelona. 51-59.

Trillas, E. (1998) On negation functions in fuzzy set theory, In: *Advances in Fuzzy Logic*, S. Barro et. al. Eds. Universidade de Santiago de Compostela, 31-43.

Trillas, E. and Alsina, C. (1978) Introducción a los Espacios Métricos Generalizados . *Fund. J. March* Serie Universitaria 49, Madrid.

Trillas, E. and Valverde, L. (1984) Implication and indistinguishability in the setting of fuzzy logic, In: J. Kacpryzk and R.R. Yager (Eds.) *Management Decision support systems using Fuzzy Sets and Possibility Theory*, Verlag TUV, Rheinland.

Valverde, L. (1985) On the structure of F-indistinguishability operators. *Fuzzy Sets & Systems* **17**, 313-328.

Wagenknecht, M. (1995) On transitive solutions of fuzzy equations, inequalities and lower approximations of fuzzy relations. *Fuzzy Sets and Systems* **75**, 229-240.

Zadeh, L. A. (1971) Similarity relations and fuzzy orderings. *Inform. Sci.* **3**, 177-200.

6 ANALYTICAL SOLUTION METHODS FOR FUZZY RELATIONAL EQUATIONS

Bernard De Baets*

Abstract: Fuzzy relational equations are without doubt the most important inverse problems arising from fuzzy set theory, and in particular from fuzzy relational calculus. Indeed, the calculus of fuzzy relations is a powerful one, with applications in fuzzy control and fuzzy systems modelling in general, approximate reasoning, relational databases, clustering, etc. In this chapter, fuzzy relational equations are approached from an order-theoretical point of view. It is shown how all inverse problems can be reduced to systems of polynomial lattice equations. The exposition is limited to the description of exact solutions, and analytical ways are presented for obtaining the complete solution set when working in a broad and interesting class of distributive lattices. Ample literature pointers to approximate solution methods and application areas are provided.

6.1 INTRODUCTION

Mathematical relations, just as human relations, are invaluable to any scientific researcher. As Goguen (1967) writes, "Science is, in a sense, the discovery of relations between observables. Zadeh has shown the study of relations to be equivalent to the general study of systems (a system is a relation between an "input" and an "output" space)." However, the typical behaviour of a mathematician to describe relationships in a black-or-white manner is not suitable for every-day problems. Fuzzy relations, however, allow for a description of gradual relationships. Depending on the context, fuzzy relations can be interpreted

*Post-Doctoral Fellow of the Fund for Scientific Research - Flanders (Belgium)

in two ways: a disjunctive way, or a conjunctive way, see e.g. De Baets (1996) and Dubois and Prade (1992).

In the *disjunctive interpretation*, a fuzzy relation is considered as an elastic restriction on the more or less possible values of a couple of variables, each of which takes exactly one value in reality. In this interpretation, fuzzy relations are extensively used for modelling systems of which the input-output behaviour is known only through linguistic descriptions. The basic operations on fuzzy relations are the *direct image* and the *composition.* The generalized modus ponens and the generalized modus tollens, the key operations behind forward and backward chaining inference mechanisms in fuzzy rule-based systems, perfectly fit within this disjunctive interpretation and are immediate applications of the basic operations. The design and deployment of such systems requires the solution of inverse problems connected with the basic operations: from known input-output behaviour a system model is built through the solution of fuzzy relational equations. The importance of fuzzy relational equations is best described with the words of Zadeh from the foreword to the book of Di Nola, Sessa, Pedrycz and Sanchez (1989): "In the context of applications, the importance of the theory of fuzzy relational equations derives from the fact that human knowledge may be viewed as a collection of facts and rules, each of which may be represented as the assignment of a fuzzy relation to the unconditional or conditional possibility distribution of a variable. What this implies is that knowledge may be viewed as a system of fuzzy relational equations. In this perspective, then, inference from a body of knowledge reduces to the solution of a system of fuzzy relational equations."

The notion of *mutually exclusive values* is no longer present in the conjunctive interpretation. In a diagnostic problem, for instance, a single cause may lead to multiple defects. In a decision problem, alternatives are evaluated on a number of criteria, of which, of course, more than one may be satisfied. Although most attention has been directed towards the conjunctive interpretation, this disjunctive interpretation also has an incredible potential. The basic operations from the conjunctive interpretation are still applicable in the disjunctive interpretation. However, Bandler and Kohout (1980a, 1980b) realized that in the conjunctive interpretation additional operations play an important role. The most important such operations are their *triangular compositions.* The contributions of Bandler and Kohout are, without doubt, among the most fundamental in the study of fuzzy relations. However, the solution of inverse problems related to these 'new' operations has received only little attention.

In this chapter, we present a uniform framework for solving the inverse problems arising in both interpretations. Remarkably, particular solutions in the disjunctive interpretation, for instance, can be described by means of operations from the conjunctive one. This demonstrates that it is really opportune to treat these problems in one common framework. This chapter is organized as follows. In Section 2, we briefly introduce the reader to the basic operations of fuzzy relational calculus, followed by an identification of the corresponding inverse problems in Section 3. The most important fuzzy relational equations are related to systems of sup-$\mathcal{T}$ equations. Their solution procedures are explained in Section 4. Equally interesting and important, but less known, are the fuzzy relational equations that can be formulated as systems of left inf-$\mathcal{I}$ or systems of right inf-$\mathcal{I}$ equations. These problems are discussed in Sections 5 and 6. In Section 7, we briefly comment some approximate solution methods. We conclude with some literature pointers in Section 8.

6.2 IMAGES AND COMPOSITIONS

6.2.1 *Relational calculus and Boolean equations*

In essence, there exist two basic types of operations in relational calculus: *images* and *compositions*. Consider a relation R from X to Y. The *afterset* xR of $x \in X$ is the subset of Y defined as $xR = \{y \in Y \mid (x, y) \in R\}$. The *foreset* Ry of $y \in Y$ is the subset of X defined as $Ry = \{x \in X \mid (x, y) \in R\}$. Given a subset A of X, the *direct image* $R(A)$, *subdirect image* $R_{\triangleleft}(A)$ and *superdirect image* $R_{\triangleright}(A)$ of A under R are the subsets of Y defined as:

$$
\begin{aligned}
R(A) &= \{y \in Y \mid (\exists x \in X)(x \in A \wedge (x, y) \in R)\} \\
&= \{y \in Y \mid A \cap Ry \neq \emptyset\}
\end{aligned}
$$

$$
\begin{aligned}
R_{\triangleleft}(A) &= \{y \in Y \mid (\forall x \in X)(x \in A \Rightarrow (x, y) \in R)\} \\
&= \{y \in Y \mid A \subseteq Ry\}
\end{aligned}
$$

$$
\begin{aligned}
R_{\triangleright}(A) &= \{y \in Y \mid (\forall x \in X)((x, y) \in R \Rightarrow x \in A)\} \\
&= \{y \in Y \mid Ry \subseteq A\}.
\end{aligned}
$$

Similarly, given a second relation S from Y to Z, the *(round) composition* $R \circ S$, *subcomposition* $R \triangleleft S$ and *supercomposition* $R \triangleright S$ of R and S are the relations from X to Z defined as:

$$
\begin{aligned}
R \circ S &= \{(x, z) \in X \times Z \mid (\exists y \in Y)((x, y) \in R \wedge (y, z) \in S)\} \\
&= \{(x, z) \in X \times Z \mid xR \cap Sz \neq \emptyset\}
\end{aligned}
$$

$$
\begin{aligned}
R \triangleleft S &= \{(x, z) \in X \times Z \mid (\forall y \in Y)((x, y) \in R \Rightarrow (y, z) \in S)\} \\
&= \{(x, z) \in X \times Z \mid xR \subseteq Sz\}
\end{aligned}
$$

$$
\begin{aligned}
R \triangleright S &= \{(x, z) \in X \times Z \mid (\forall y \in Y)((y, z) \in S \Rightarrow (x, y) \in R)\} \\
&= \{(x, z) \in X \times Z \mid Sz \subseteq xR\}.
\end{aligned}
$$

Strictly speaking, for the sub- and superdirect images, and the sub- and supercompositions, one should also take into account some non-emptiness conditions, see De Baets and Kerre (1993a). In the context of relational equations, however, the above formulations are more suitable. Sub- and supercompositions of relations have been studied extensively by Bandler and Kohout (1980a, 1980b) under the name *triangle relational products*.

The above operations, simple yet powerful, are used in a variety of application areas. Their very nature is typical for many mathematical theories: the evaluation of non-empty intersections versus inclusions. We mention, for instance, possibility theory, rough set theory, mathematical morphology, set-valued analysis, Dempster–Shafer theory of evidence, topology, etc.

It is now not difficult to imagine what types of inverse problems one can consider:

(i) *image equations*, i.e. equations of the type $R(A) = B$, $R_{\triangleleft}(A) = B$ or $R_{\triangleright}(A) = B$ in the unknown set A or in the unknown relation R;

(ii) *composition equations*, i.e. equations of the type $R \circ S = T$, $R \triangleleft S = T$ or $R \triangleright S = T$ in the unknown relation R or in the unknown relation S.

Identifying sets and relations with their characteristic mapping, these inverse problems, called relational equations, can be translated into Boolean equations (see e.g. Rudeanu (1974)). Note that the characteristic mapping χ_A of a crisp subset A of X is defined by

$$\chi_A(x) = \begin{cases} 1 & , \text{ if } x \in A \\ 0 & , \text{ elsewhere} \end{cases} .$$

6.2.2 *Fuzzy relational calculus*

The above images and compositions are extended to fuzzy relations in a by now traditional manner: supremum and infimum are playing the role of the existential and universal quantifiers and a conjunctor and implicator are modelling pointwise intersection and inclusion. The most natural mathematical framework for dealing with these operations is without doubt the lattice-theoretic one (see e.g. Birkhoff (1967), Davey and Priestley (1990)). Hence, we will work with so-called L-fuzzy relations, not to artificially increase the level of mathematical abstractness, but because lattice theory provides the proper framework for formulating and solving fuzzy relational equations.

Consider a complete lattice $(L, \leq)$ with smallest element 0 and greatest element 1. Denote the meet and join operation by $\frown$ and $\smile$. An L-fuzzy set A in a universe X is simply an $X \to L$ mapping; an L-fuzzy relation R from X to Y is an $X \times Y \to L$ mapping, see Goguen (1967). From here on, we will omit the prefix L if no confusion can occur. The binary Boolean logical operations $\wedge$, $\vee$ and $\to$, for instance, can be extended in the following obvious way:

(i) a *conjunctor* $\mathcal{C}$ is a binary operation on L with order-preserving partial mappings such that $\mathcal{C}(0,1) = \mathcal{C}(1,0) = 0$ and $\mathcal{C}(1,1) = 1$;

(ii) a *disjunctor* $\mathcal{D}$ is a binary operation on L with order-preserving partial mappings such that $\mathcal{D}(0,1) = \mathcal{C}(1,0) = 1$ and $\mathcal{D}(0,0) = 0$;

(iii) an *implicator* $\mathcal{I}$ is a binary operation on L with order-reversing first partial mappings and order-preserving second partial mappings such that $\mathcal{I}(0,1) = \mathcal{I}(1,0) = 0$ and $\mathcal{I}(1,1) = 1$.

Obviously, additional properties can be imposed on these operators. For instance, if a conjunctor (resp. disjunctor) is also required to be commutative, associative and to have 1 (resp. 0) as neutral element, then we are in the setting of t-norms (resp. t-conorms), see De Cooman and Kerre (1994) and Schweizer and Sklar (1983). For the sake of simplicity and in view of their familiarity to the reader, we will mainly restrict to t-norms and t-conorms from here on. Most results, however, can be stated much more generally. The three most important continuous t-norms on the real unit interval are the minimum operator M, the (algebraic) product P and the Łukasiewicz t-norm W defined by $W(x,y) = \max(x+y-1,0)$.

The unary Boolean logical operator $\neg$ is extended as follows:

(i) a *negator* $\mathcal{N}$ is an order-reversing unary operation on L such that $\mathcal{N}(0) = 1$ and $\mathcal{N}(1) = 0$;

(ii) a *strict negator* $\mathcal{N}$ is a permutation of L that satisfies

$$(\forall(x,y) \in [0,1]^2)(x \leq y - \mathcal{N}(x) \geq \mathcal{N}(y));$$

(iii) an *involutive negator* $\mathcal{N}$ is an involutive order-reversing unary operation on L.

Clearly, any involutive negator is strict. Any implicator $\mathcal{I}$ induces a negator $\mathcal{N}_{\mathcal{I}} = \mathcal{I}(\cdot,0)$. To any conjunctor $\mathcal{C}$ and negator $\mathcal{N}$ corresponds an implicator $\mathcal{I}_{\mathcal{C},\mathcal{N}}$, defined by

$$\mathcal{I}_{\mathcal{C},\mathcal{N}}(x,y) = \mathcal{N}(\mathcal{C}(x,\mathcal{N}(y))).$$

In case of a t-norm $\mathcal{T}$, the implicator $\mathcal{I}_{\mathcal{T},\mathcal{N}}$ clearly induces the negator $\mathcal{N}$.

The implicators obtained from the three t-norms above and the standard negator $\mathcal{N}_s$, i.e. $\mathcal{N}_s(x) = 1-x$, are listed next:

(i) $\mathcal{I}_{M,\mathcal{N}_s}(x,y) = \max(1-x,y)$ (the Kleene-Dienes implicator);

(ii) $\mathcal{I}_{P,\mathcal{N}_s}(x,y) = 1-x+xy$;

(iii) $\mathcal{I}_{W,\mathcal{N}_s}(x,y) = \min(1-x+y,1)$ (the Łukasiewicz implicator).

We are now ready to introduce the basic operations of fuzzy relational calculus. Consider a fuzzy relation R from X to Y. The *afterset* xR of $x \in X$ is the fuzzy set in Y defined by $xR(y) = R(x,y)$. The *foreset* Ry of $y \in Y$ is the fuzzy set in X defined by $Ry(x) = R(x,y)$. Let $\mathcal{C}$ be a conjunctor and $\mathcal{I}$ be an implicator. Given a fuzzy set A in X, the *direct image* $R(A)$, *subdirect image* $R_{\triangleleft}(A)$ and *superdirect image* $R_{\triangleright}(A)$ of A under R are the fuzzy sets in Y defined by:

$$\begin{aligned} R(A)(y) &= \sup_{x\in X} \mathcal{C}(A(x),R(x,y)) \\ R_{\triangleleft}(A)(y) &= \inf_{x\in X} \mathcal{I}(A(x),R(x,y)) \\ R_{\triangleright}(A)(y) &= \inf_{x\in X} \mathcal{I}(R(x,y),A(x)). \end{aligned}$$

Given a second fuzzy relation S from Y to Z, the *(round) composition* $R \circ S$, *subcomposition* $R \triangleleft S$ and *supercomposition* $R \triangleright S$ of R and S are the fuzzy relations from X to Z defined by:

$$\begin{aligned} R \circ S(x,z) &= \sup_{y\in Y} \mathcal{C}(R(x,y), S(y,z)) \\ R \triangleleft S(x,z) &= \inf_{y\in Y} \mathcal{I}(R(x,y), S(y,z)) \\ R \triangleright S(x,z) &= \inf_{y\in Y} \mathcal{I}(S(y,z), R(x,y)). \end{aligned}$$

For an in-depth study of these operations and their applications, we refer to De Baets and Kerre (1995).

6.3 TYPES OF INVERSE PROBLEMS

As in the Boolean case, we can consider the following types of inverse problems:

(i) *image equations*, i.e. equations of the type $R(A) = B$, $R_{\triangleleft}(A) = B$ or $R_{\triangleright}(A) = B$ in the unknown fuzzy set A or in the unknown fuzzy relation R;

(ii) *composition equations*, i.e. equations of the type $R \circ S = T$, $R \triangleleft S = T$ or $R \triangleright S = T$ in the unknown fuzzy relation R or in the unknown fuzzy relation S.

These inverse problems are what one usually calls *fuzzy relation(al) equations.* They can all be formulated as systems of particular lattice equations. Consider a binary operator $\mathcal{O}$ on L with monotone partial mappings, then we can distinguish the following four basic types of equations, given a family $(a_i)_{i\in I}$ in L and b in L:

- left sup-$\mathcal{O}$ equation: $\sup\limits_{i\in I} \mathcal{O}(x_i, a_i) = b$;
- right sup-$\mathcal{O}$ equation: $\sup\limits_{i\in I} \mathcal{O}(a_i, x_i) = b$;
- left inf-$\mathcal{O}$ equation: $\inf\limits_{i\in I} \mathcal{O}(x_i, a_i) = b$;
- right inf-$\mathcal{O}$ equation: $\inf\limits_{i\in I} \mathcal{O}(a_i, x_i) = b$.

These equations can be considered as *polynomial lattice equations.* Finite index sets will be denoted I_n, containing the first n integers. A first observation is that, due to the monotonicity of the operator $\mathcal{O}$, the solution set of any of the above equations is *order convex*, i.e. if $(x_i)_{i\in I}$ and $(z_i)_{i\in I}$ are two solutions of one of the above equations, then so is $(y_i)_{i\in I}$, whenever $x_i \leq y_i \leq z_i$, for any $i \in I$. This allows us to focus on extremal solutions, i.e. maximal (greatest) and minimal (smallest) solutions.

In case of a commutative operator $\mathcal{O}$, there is, of course, no need to talk about left or right equations. In this chapter, we will be dealing with sup-$\mathcal{T}$

equations and with left and right inf-$\mathcal{I}$ equations, where $\mathcal{T}$ is a t-norm and $\mathcal{I}$ is a border implicator, i.e. an implicator satisfying the *neutrality principle*: $(\forall y \in L)(\mathcal{I}(1, y) = y)$. Note that the associativity of the t-norm $\mathcal{T}$ plays no role, while the commutativity is only imposed for reducing the number of residual operators associated with it (see Section 4). It would therefore be sufficient to consider a t-seminorm, see De Cooman and Kerre (1994).

Our treatment of fuzzy relational equations is built up in the following way: we will gradually impose conditions on the logical operators involved and restrict the class of lattices considered, in order to go from a description of greatest or smallest solutions, and necessary and sufficient solvability conditions, to a full description of the solution set.

Conditions imposed on the logical operators are typically continuity conditions, or in the lattice-theoretic framework, *morphism* conditions. Recall that an $L \to L$ mapping is called a *sup-morphism* if for any non-empty subset A of L it holds that $f(\sup A) = \sup f(A)$, and a *dual sup-morphism* if for any non-empty subset A of L it holds that $f(\sup A) = \inf f(A)$. The notions of an *inf-morphism* and a *dual inf-morphism* are defined similarly. A mapping is called a *(dual) homomorphism* if it is both a (dual) sup-morphism and a (dual) inf-morphism. For instance, in the more familiar case of the unit interval, sup-morphims, resp. inf-morphisms, are nothing else but increasing left-continuous, resp. right-continuous, mappings.

Let us consider the image equations $R(A) = B$ and composition equations $R \circ S = T$:

- the equation $R(A) = B$ in the unknown fuzzy relation R, called image equation of *type 1*, is equivalent to a family of independent sup-$\mathcal{C}$ equations in the foresets Ry, $y \in Y$:

$$\sup_{x \in X} \mathcal{C}(A(x), Ry(x)) = B(y);$$

- the equation $R(A) = B$ in the unknown fuzzy set A, called image equation of *type 2*, is equivalent to a system of sup-$\mathcal{C}$ equations, $y \in Y$:

$$\sup_{x \in X} \mathcal{C}(A(x), R(x, y)) = B(y);$$

- the composition equation $R \circ S = T$ in the unknown fuzzy relation R or S can be reformulated as a system of image equations of type 1 or as a family of independent image equations of type 2.

Of course, the same observations hold for the other images and compositions. We will next show how to solve these fuzzy relational equations.

6.4 SUP-$\mathcal{T}$ EQUATIONS

6.4.1 *The equation* $\mathcal{T}(a, x) = b$

The most basic equation we consider is the equation $\mathcal{T}(a, x) = b$ in the unknown x, with $\mathcal{T}$ a t-norm on a complete lattice $(L, \leq)$. In order to describe the solution set of this equation, we associate to a given t-norm $\mathcal{T}$ two binary operators $\mathcal{I}_\mathcal{T}$ and $\mathcal{L}_\mathcal{T}$ on L, called *residual operators*, defined by

$$\mathcal{I}_\mathcal{T}(x,y) = \sup\{z \in L \mid \mathcal{T}(x,z) \leq y\}$$

$$\mathcal{L}_\mathcal{T}(x,y) = \inf\{z \in L \mid \mathcal{T}(x,z) \geq y\}.$$

The first operator is called the *residual implicator* of $\mathcal{T}$ and satisfies $(\forall (x,y) \in L^2)(x \leq y \Rightarrow \mathcal{I}_\mathcal{T}(x,y) = 1)$, the second one has no particular interpretation. Note that the definition of the operator $\mathcal{L}_\mathcal{T}$ slightly differs from the one in Di Nola, Sessa, Pedrycz and Sanchez (1989): on $[0,1]$, the corresponding operator is defined 0 whenever $x < y$, while in this case $\mathcal{L}_\mathcal{T}(x,y) = 1$.

For the three most important continuous t-norms on the real unit interval, the corresponding residual operators are listed in Table 6.1. Note that $\mathcal{I}_W$ coincides with $\mathcal{I}_{W,\mathcal{N}_s}$.

Table 6.1 Residual operators of the three main t-norms.

$\mathcal{T}$	$\mathcal{I}_\mathcal{T}(x,y)$	$\mathcal{L}_\mathcal{T}(x,y)$
M	$\begin{cases} 1 & \text{, if } x \leq y \\ y & \text{, elsewhere} \end{cases}$	$\begin{cases} 1 & \text{, if } x < y \\ y & \text{, elsewhere} \end{cases}$
P	$\begin{cases} 1 & \text{, if } x \leq y \\ y/x & \text{, elsewhere} \end{cases}$	$\begin{cases} 1 & \text{, if } x < y \\ y/x & \text{, if } 0 < y \leq x \\ 0 & \text{, elsewhere} \end{cases}$
W	$\min(1-x+y, 1)$	$\begin{cases} 1 & \text{, if } x < y \\ 1-x+y & \text{, if } 0 < y \leq x \\ 0 & \text{, elsewhere} \end{cases}$

Since the solution set of the equation $\mathcal{T}(a,x) = b$ can be seen as the intersection of the solution sets of the inequalities $\mathcal{T}(a,x) \leq b$ and $\mathcal{T}(a,x) \geq b$, we start by investigating the latter.

On a complete lattice. The following observations hold true:

- If the partial mapping $\mathcal{T}(a,\cdot)$ is a sup-morphism, then the solution set of the inequality $\mathcal{T}(a,x) \leq b$ is given by $[0, \mathcal{I}_\mathcal{T}(a,b)]$.

- If the partial mapping $\mathcal{T}(a,\cdot)$ is an inf-morphism, then the solution set of the inequality $\mathcal{T}(a,x) \geq b$ is given by

$$\begin{cases} [\mathcal{L}_\mathcal{T}(a,b), 1] & \text{, if } b \leq a \\ \emptyset & \text{, elsewhere} \end{cases}.$$

- If the partial mapping $\mathcal{T}(a,\cdot)$ is a homomorphism, then the solution set of the equation $\mathcal{T}(a,x) = b$ is given by

$$\begin{cases} [\mathcal{L}_\mathcal{T}(a,b), \mathcal{I}_\mathcal{T}(a,b)] & \text{, if } b \in \text{rng}(\mathcal{T}(a,\cdot)) \\ \emptyset & \text{, elsewhere} \end{cases};$$

if the partial mapping $\mathcal{T}(a,\cdot)$ is a maximally surjective homomorphism (i.e. $\text{rng}(\mathcal{T}(a,\cdot)) = [0,a]$), then the solution set of the equation $\mathcal{T}(a,x) = b$ is given by

$$\begin{cases} [\mathcal{L}_{\mathcal{T}}(a,b), \mathcal{I}_{\mathcal{T}}(a,b)] & , \text{if } b \leq a \\ \emptyset & , \text{elsewhere} \end{cases} .$$

The partial mappings of the meet operation $\frown$ on L, the greatest t-norm on L, are always maximally surjective.

On the real unit interval. A t-norm $\mathcal{T}$ on $[0,1]$ is continuous if and only if all of its partial mappings are homomorphisms. Moreover, the partial mappings of a continuous t-norm $\mathcal{T}$ on $[0,1]$ are maximally surjective. Hence, if $\mathcal{T}(a,\cdot)$ is continuous, then the solution set of the equation $\mathcal{T}(a,x) = b$ is given by

$$\begin{cases} [\mathcal{L}_{\mathcal{T}}(a,b), \mathcal{I}_{\mathcal{T}}(a,b)] & , \text{if } b \leq a \\ \emptyset & , \text{elsewhere} \end{cases} .$$

Therefore, a necessary and sufficient solvability condition is given by $b \leq a$.

6.4.2 *Greatest solution – solvability conditions*

Let $(a_i)_{i\in I}$ be an arbitrary family in L and $b \in L$, then we want to determine the solution set of the equation

$$\sup_{i\in I} \mathcal{T}(a_i, x_i) = b$$

in the family of unknowns $(x_i)_{i\in I}$ in L. The families $(a_i)_{i\in I}$ and $(x_i)_{i\in I}$ can be seen as L-fuzzy sets in the index set I. The problem can therefore be reformulated as follows. Given $A \in \mathcal{F}_L(I)$ (with $\mathcal{F}_L(I)$ the set of L-fuzzy sets in I) and $b \in L$, determine the solution set of the equation

$$\sup_{i\in I} \mathcal{T}(A(i), X(i)) = b \tag{6.1}$$

in the unknown L-fuzzy set X in I.

Recall that $(\mathcal{F}_L(I), \subseteq)$ is a complete lattice, with order relation $\subseteq$ (inclusion) defined by

$$A \subseteq B - (\forall i \in I)(A(i) \leq B(i)),$$

and as infimum and supremum the intersection and union defined by

$$\left(\bigcap_{j\in J} A_j\right)(i) = \inf_{j\in J} A_j(i) \quad \text{and} \quad \left(\bigcup_{j\in J} A_j\right)(i) = \sup_{j\in J} A_j(i).$$

The solution set of equation (6.1) then is a subset of the complete lattice $(\mathcal{F}_L(I), \subseteq)$ and can be seen as the intersection of the solution sets of the inequalities

$$\sup_{i\in I} \mathcal{T}(A(i), X(i)) \leq b \tag{6.2}$$

and

$$\sup_{i \in I} \mathcal{T}(A(i), X(i)) \geq b. \tag{6.3}$$

Obviously, due to monotonicity considerations, inequality (6.2) always has a solution, namely $\chi_\emptyset$, while inequality (6.3) has a solution if and only if χ_I is a solution, i.e. if and only if

$$b \leq \sup_{i \in I} A(i). \tag{6.4}$$

In general, inequality (6.3) is a lot harder to solve than inequality (6.2). For the moment, we are only concerned with the greatest solution of equation (6.1).

On a complete lattice. The following statements can be verified:

- If the partial mappings of $\mathcal{T}$ are sup-morphisms, then the solution set of inequality (6.2) is given by $[\chi_\emptyset, G]$ with G defined by

  $$G(i) = \mathcal{I}_\mathcal{T}(A(i), b). \tag{6.5}$$

  In particular, this result shows us the solution set for equation (6.1) in case $b = 0$.

- If the partial mappings of $\mathcal{T}$ are sup-morphisms, then the solution set of equation (6.1) is not empty if and only if the fuzzy set G defined by (6.5) is a solution. Moreover, if G is a solution, then it is the greatest solution.

Necessary and sufficient solvability conditions. Notice that, in general, no simple necessary and sufficient solvability condition for equation (6.1) can be given: one has to construct the potential greatest solution and verify that it is indeed a solution. In some cases, however, such a simpler condition can be formulated:

- In any case, condition (6.4) is a necessary condition for the existence of a solution to equation (6.1). If L is a complete chain and the partial mappings of $\mathcal{T}$ are maximally surjective homomorphisms, then condition (6.4) becomes also a sufficient condition; if L is also dense, i.e.

  $$(\forall (a, b) \in L^2)(a < b \Rightarrow (\exists c \in L)(a < c < b)),$$

  then the condition of maximal surjectivity is always fulfilled. Moreover in case $\mathcal{T}$ is the meet operation, then L need not be totally ordered (in this case, we are working in a complete Brouwerian lattice), see Zhao (1987).

- If the partial mappings of $\mathcal{T}$ are maximally surjective homomorphisms then the condition

  $$(\exists k \in I)(b \leq A(k)), \tag{6.6}$$

  a stronger version of condition (6.4), is a sufficient condition for solvability.

On the real unit interval. For a left-continuous t-norm $\mathcal{T}$ on $[0,1]$, the solution set of equation (6.1) is not empty if and only if the fuzzy set G defined by (6.5) is the greatest solution. Moreover, if $\mathcal{T}$ is continuous, then condition (6.4) is a necessary and sufficient solvability condition.

6.4.3 Complete solution set

For most practical considerations, it is sufficient to know the greatest solution of equation (6.1). From a mathematical point of view, one is, of course, highly interested in the complete solution set. Before we can discuss this solution set, we need to develop an appropriate representation for it. Obviously, it would be desirable to find a representation that is closed under (arbitrary) intersections, since that would learn us how to solve systems of equations at the same time. Fortunately, such a representation exists.

Root systems. A subset R of an ordered set $(P, \leq)$ is called a *root system* if there exists an element σ in P and an antichain O in $\downarrow \sigma = \{x \in P \mid x \leq \sigma\}$ such that

$$R = \bigcup_{\omega \in O} [\omega, \sigma].$$

For a root system R, the corresponding element σ and antichain O are unique. The element σ is called the *stem* and the elements of the antichain O are called the *offshoots* of the root system. A root system is called *finitely generated* if the set of offshoots is finite. The stem is the greatest element and the offshoots are the minimal elements of the root system. Hence, a root system clearly is an order convex subset that is completely determined by its greatest element and its minimal elements.

The following results are of extreme importance in the present discussion, see De Baets (1995c, 1998b):

- Let $(R_i)_{i \in I}$ be a family of finitely generated root systems of a complete lattice $(L, \leq)$ with stem σ_i and set of offshoots O_i. If the intersection $\bigcap_{i \in I} R_i$ is not empty, then it is a root system with stem $\sigma = \inf_{i \in I} \sigma_i$ and as offshoots the minimal elements of the set

$$\{\sup_{i \in I} \omega_i \mid (\forall i \in I)(\omega_i \in O_i \wedge \omega_i \leq \sigma)\}. \tag{6.7}$$

- If the intersection of a finite family of finitely generated root systems of a complete lattice is not empty, then it is a finitely generated root system. These offshoots can be determined in an exhaustive way, or following an iterative algorithm, see De Baets (1995c, 1998b). We will call this method the pen-and-paper method.

If the root systems are not finitely generated, then, in general, nothing can be said about their intersection.

On a complete lattice. So far, we know the greatest solution of equation (6.1), if there exists a solution at all. Under very general conditions, we can always find a root system of $(\mathcal{F}_L(I), \subseteq)$ that is contained in the solution set, see De Baets (1995c). Indeed, if the partial mappings of $\mathcal{T}$ are maximally surjective homomorphisms and condition (6.6) holds, then the solution set of equation (6.1) contains the root system with stem G defined by (6.5) and as set of offshoots the set $\{M_k \mid b \leq A(k)\}$, where M_k is defined by

$$M_k(i) = \begin{cases} \mathcal{L}_{\mathcal{T}}(A(k), b) & , \text{if } i = k \\ 0 & , \text{elsewhere} \end{cases} . \tag{6.8}$$

Moreover, these offshoots are minimal solutions. If the index set I is finite, then this root system is, of course, finitely generated. This result shows how particular minimal solutions, that could be considered as Dirac-solutions or point solutions, can be constructed. In general, it is not known how other minimal solutions can be determined. As pointed out further on, in some cases, the above root system coincides with the solution set. For instance, under the same conditions on $\mathcal{T}$ and condition (6.6), but for a complete chain L and a finite index set I, the solution set coincides with the above root system, which is then finitely generated.

On a distributive, complete lattice with join-irreducible and join-decomposable elements. So far, the above discussion reveals only the complete solution set in case of a complete chain and under some additional conditions. However, there is an interesting class of lattices, in particular from a practical point of view, in which the complete solution set can be determined. Let us first recall some order-theoretic notions, see Birkhoff (1967) and Davey and Priestley (1990):

(i) A lattice $(L, \leq)$ is called *distributive* if the following property holds:

$$(\forall(a, b, c) \in L^3)(a \frown (b \smile c) = (a \frown b) \smile (a \frown c)).$$

For instance, any chain is distributive and also the Cartesian product of distributive lattices is distributive.

(ii) An element a of a lattice $(L, \leq)$ is called *join-irreducible* if

$$(\forall(b, c) \in L^2)(b \smile c = a \Rightarrow (b = a \vee c = a)),$$

and *join-decomposable* if there exists a set A of join-irreducible elements of L, with $\#A \geq 2$, such that $a = \sup A$.

For instance, all elements of a chain are join-irreducible, and all elements of a finite distributive lattice or a Cartesian product of a finite number of chains are join-irreducible or join-decomposable.

In order to describe the complete solution set of equation (6.1), we now direct our attention towards inequality (6.3), inequality (6.2) being solved already on a complete lattice. The following results hold, for a distributive, complete lattice $(L, \leq)$ and a finite index set I, see De Baets (1995b):

- **The case of a join-irreducible right-hand side.** If b is join-irreducible, the partial mappings of $\mathcal{T}$ are inf-morphisms and the solution set of inequality (6.3) is not empty (see condition (6.4)), then it is a finitely generated root system with stem χ_I and as set of offshoots the set $\{M_k \mid b \leq A(k)\}$, where M_k is defined by (6.8).

- **The case of a join-decomposable right-hand side.** If b has a join-decomposition $b = \sup_{j \in J} b_j$, the partial mappings of $\mathcal{T}$ are inf-morphisms and the solution set of inequality (6.3) is not empty (see condition (6.4)), then it is a root system with stem χ_I and as offshoots the *minimal* elements of the set

$$\{\bigcup_{j \in J} M_k^j \mid b_j \leq A(k)\},$$

  where M_k^j is defined by

$$M_k^j(i) = \begin{cases} \mathcal{L}_{\mathcal{T}}(A(k), b_j) & \text{, if } i = k \\ 0 & \text{, elsewhere} \end{cases} . \tag{6.9}$$

  If the join-decomposition is finite, then this root system is also finitely generated.

Combining this information with our knowledge of the solution set of inequality (6.2) leads to a full description of the solution set of equation (6.1), again for a distributive, complete lattice and a finite index set I, see De Baets (1995b):

- **The case of a join-irreducible right-hand side.** If b is join-irreducible, the partial mappings of $\mathcal{T}$ are homomorphisms and the solution set of equation (6.1) is not empty, then it is a finitely generated root system with stem G defined by (6.5) and as set of offshoots the set $\{M_k \mid b \leq A(k) \wedge M_k \subseteq G\}$, where M_k is defined by (6.8); moreover, if the partial mappings of $\mathcal{T}$ are maximally surjective, then the set of offshoots can be written as $\{M_k \mid b \leq A(k)\}$. In the case of a complete chain, the latter result was already mentioned above.

- **The case of a join-decomposable right-hand side.** If b has a join-decomposition $b = \sup_{j \in J} b_j$, the partial mappings of $\mathcal{T}$ are homomorphisms and the solution set of equation (6.1) is not empty, then it is a root system with stem G defined by (6.5) and as offshoots the *minimal* elements of the set

$$\{\bigcup_{j \in J} M_k^j \mid b_j \leq A(k) \wedge M_k^j \subseteq G\},$$

where M_k^j is defined by (6.9). If the join-decomposition is finite, then this root system is also finitely generated, and if $\mathcal{L}_\mathcal{T} \leq \mathcal{I}_\mathcal{T}$, then the offshoots are the *minimal* elements of the set

$$\{\bigcup_{j\in J} M_k^j \mid b_j \leq A(k)\}.$$

Note that if L is a complete chain and the partial mappings of $\mathcal{T}$ are maximally surjective, then the inequality $\mathcal{L}_\mathcal{T} \leq \mathcal{I}_\mathcal{T}$ always holds.

These results have been obtained by Zhao (1987) in the case of a complete Brouwerian lattice, i.e. when choosing as t-norm the meet operation, of which all elements have a finite irredundant decomposition in join-irreducible elements. They have been generalized and rephrased in the terminology of root systems by De Baets (1995b).

On the real unit interval. The real unit interval is a distributive, complete lattice of which all elements are join-irreducible. Hence, the above results can be applied. For a continuous t-norm $\mathcal{T}$ and a finite index set I, the solution set of equation (6.1) is not empty if and only if condition (6.6) holds. This also follows from the previous subsection, since for finite I conditions (6.4) and (6.6) are equivalent. If the solution set is not empty, then it is the finitely generated root system with stem G defined by (6.5) and as set of offshoots the set $\{M_k \mid b \leq A(k)\}$, where M_k is defined by (6.8).

Example 1 For example, consider equation (6.1) with $I = I_5$,

$$A = (\ 0.6\ \ 0.4\ \ 0.1\ \ 1\ \ 0\ )$$

and $b = 0.4$. Since $b < \max_{i\in I_5} A(i) = 1$, the solution set is not empty for any continuous t-norm $\mathcal{T}$ and has a greatest solution G and three minimal solutions M_1, M_2 and M_4. In case $\mathcal{T} = M$, these are given by:

$$\begin{aligned} G &= (\ 0.4\ \ 1\ \ 1\ \ 0.4\ \ 1\ ) \\ M_1 &= (\ 0.4\ \ 0\ \ 0\ \ 0\ \ 0\ ) \\ M_2 &= (\ 0\ \ 0.4\ \ 0\ \ 0\ \ 0\ ) \\ M_4 &= (\ 0\ \ 0\ \ 0\ \ 0.4\ \ 0\ ) \end{aligned}$$

On the real unit hypercube. A very attractive lattice satisfying the above requirements is the unit hypercube $([0,1]^m, \leq)$. From an implementational point of view, the real unit hypercube is the ultimate ordinal and numerical working environment; it allows for incomparability and offers at the same time the possibility to fall back on the underlying real unit interval. Moreover, it is a distributive, complete lattice and all elements of it are either join-irreducible or have a finite join-decomposition. Consider $a \in [0,1]^m$, $a \neq (0,\ldots,0) = 0_m$, then a can be decomposed as

$$a = \sup_{a(i)\neq 0} \underline{a}_i$$

with $\underline{a}_i = (0,\ldots,a(i),\ldots,0)$.

From the above discussion, it is clear that we should consider a t-norm $\mathcal{T}$ on $[0,1]^m$ with partial mappings that are homomorphisms. However, this implies that the t-norm $\mathcal{T}$ is the direct product of m t-norms $\mathcal{T}_i$, $i = 1, \ldots, m$, on $[0,1]$, i.e. for any x and y in $[0,1]^m$ it holds that

$$\mathcal{T}(x,y) = (\mathcal{T}_1(x(1), y(1)), \ldots, \mathcal{T}_m(x(m), y(m))).$$

The latter follows even under weaker conditions, see De Baets and Mesiar (1998). For the sake of simplicity, let us consider a t-norm $\mathcal{T}$ on $[0,1]$ and the t-norm $\mathcal{T}^{(m)}$ on $[0,1]^m$ defined by:

$$\mathcal{T}^{(m)}(x,y) = (\mathcal{T}(x(1), y(1)), \ldots, \mathcal{T}(x(m), y(m))).$$

The corresponding operators $\mathcal{I}_{\mathcal{T}^{(m)}}$ and $\mathcal{L}_{\mathcal{T}^{(m)}}$ are then given by, with $1_m = (1, \ldots, 1)$, see De Baets (1995b):

$$\mathcal{I}_{\mathcal{T}^{(m)}}(x,y) = (\mathcal{I}_{\mathcal{T}}(x(1), y(1)), \ldots, \mathcal{I}_{\mathcal{T}}(x(m), y(m)))$$

and

$$\mathcal{L}_{\mathcal{T}^{(m)}}(x,y) = (\mathcal{L}_{\mathcal{T}}(x(1), y(1)), \ldots, \mathcal{L}_{\mathcal{T}}(x(m), y(m)))$$

if $(\forall i \in I_m)(y(i) \leq x(i))$, and $\mathcal{L}_{\mathcal{T}^{(m)}}(x,y) = 1_m$ elsewhere.

Combining the above results on join-decomposable and join-irreducible right-hand sides, we can state the next result. The most important difference is that now the offshoots can be written down *immediately*, and do not have to be determined as minimal elements of an auxiliary set. If $\mathcal{T}$ is continuous, I is finite and the solution set of the equation

$$\sup_{i \in I} \mathcal{T}^{(m)}(A(i), X(i)) = b \tag{6.10}$$

is not empty, then it is a finitely generated root system with stem G defined by

$$G(i) = \mathcal{I}_{\mathcal{T}^{(m)}}(A(i), b)$$

and as set of offshoots the set

$$\{ \bigcup_{j \in J \,\wedge\, b(j) \neq 0} M_k^j \mid b(j) \leq A(k)(j)\}$$

with M_k^j defined by

$$M_k^j(i) = \begin{cases} \mathcal{L}_{\mathcal{T}^{(m)}}(A(k), \underline{b}_j) & \text{, if } i = k \\ 0_m & \text{, elsewhere} \end{cases}.$$

Note that $\mathcal{L}_{\mathcal{T}^{(m)}}(A(k), \underline{b}_j) = (0, \ldots, \mathcal{L}_{\mathcal{T}}(A(k)(j), b(j)), \ldots, 0)$.

Example 2 Consider equation (6.10) with $m = 2$, $I = I_3$, $\mathcal{T} = W$,

$$A = (\ (0.4, 0.6)\ \ (0.7, 0.4)\ \ (0.6, 0.5)\ )$$

and $b = (0.3, 0)$. We construct the potential greatest solution G:

$$\begin{aligned}
G(1) &= \mathcal{I}_{W^{(2)}}((0.4,0.6),(0.3,0)) = (\mathcal{I}_W(0.4,0.3),\mathcal{I}_W(0.6,0)) = (0.9,0.4)\\
G(2) &= \mathcal{I}_{W^{(2)}}((0.7,0.4),(0.3,0)) = (\mathcal{I}_W(0.7,0.3),\mathcal{I}_W(0.4,0)) = (0.6,0.6)\\
G(3) &= \mathcal{I}_{W^{(2)}}((0.6,0.5),(0.3,0)) = (\mathcal{I}_W(0.6,0.3),\mathcal{I}_W(0.5,0)) = (0.7,0.5),
\end{aligned}$$

and hence $G = (\ (0.9,0.4)\ \ (0.6,0.6)\ \ (0.7,0.5)\ )$. One easily verifies that G is a solution. Since $(\forall k \in I_3)(b(1) \leq A(k)(1))$, there are three minimal solutions M_1^1, M_2^1 and M_3^1, given by:

$$\begin{aligned}
M_1^1 &= (\ (0.9,0)\ \ (0,0)\ \ (0,0)\ )\\
M_2^1 &= (\ (0,0)\ \ (0.6,0)\ \ (0,0)\ )\\
M_3^1 &= (\ (0,0)\ \ (0,0)\ \ (0.7,0)\ )
\end{aligned}$$

For instance, we have that

$$M_1^1(1) = \mathcal{L}_{W^{(2)}}((0.4,0.6),(0.3,0)) = (\mathcal{L}_W(0.4,0.3),\mathcal{L}_W(0.6,0)) = (0.9,0).$$

Example 3 Consider equation (6.10) with $m = 3$, $I = I_3$, $\mathcal{T} = M$,

$$A = (\ (0.2,0.6,0.5)\ \ (0.7,0.5,0.9)\ \ (1,0.8,0.7)\ )$$

and $b = (0.7, 0.5, 0.7)$. The potential greatest solution G is given by:

$$G = (\ (1,0.5,1)\ \ (1,1,0.7)\ \ (0.7,0.5,1)\ )$$

One easily verifies that G is a solution. The minimal solutions are given by

$$\begin{aligned}
M_2^1 \cup M_1^2 \cup M_2^3 &= (\ (0,0.5,0)\ \ (0.7,0,0.7)\ \ (0,0,0)\ )\\
M_2^1 \cup M_1^2 \cup M_3^3 &= (\ (0,0.5,0)\ \ (0.7,0,0)\ \ (0,0,0.7)\ )\\
M_2^1 \cup M_2^2 \cup M_2^3 &= (\ (0,0,0)\ \ (0.7,0.5,0.7)\ \ (0,0,0)\ )\\
M_2^1 \cup M_2^2 \cup M_3^3 &= (\ (0,0,0)\ \ (0.7,0.5,0)\ \ (0,0,0.7)\ )\\
M_2^1 \cup M_3^2 \cup M_2^3 &= (\ (0,0,0)\ \ (0.7,0,0.7)\ \ (0,0.5,0)\ )\\
M_2^1 \cup M_3^2 \cup M_3^3 &= (\ (0,0,0)\ \ (0.7,0,0)\ \ (0,0.5,0.7)\ )\\
M_3^1 \cup M_1^2 \cup M_2^3 &= (\ (0,0.5,0)\ \ (0,0,0.7)\ \ (0.7,0,0)\ )\\
M_3^1 \cup M_1^2 \cup M_3^3 &= (\ (0,0.5,0)\ \ (0,0,0)\ \ (0.7,0,0.7)\ )\\
M_3^1 \cup M_2^2 \cup M_2^3 &= (\ (0,0,0)\ \ (0,0.5,0.7)\ \ (0.7,0,0)\ )\\
M_3^1 \cup M_2^2 \cup M_3^3 &= (\ (0,0,0)\ \ (0,0.5,0)\ \ (0.7,0,0.7)\ )\\
M_3^1 \cup M_3^2 \cup M_2^3 &= (\ (0,0,0)\ \ (0,0,0.7)\ \ (0.7,0.5,0)\ )\\
M_3^1 \cup M_3^2 \cup M_3^3 &= (\ (0,0,0)\ \ (0,0,0)\ \ (0.7,0.5,0.7)\ )
\end{aligned}$$

6.4.4 *Systems of sup-$\mathcal{T}$ equations*

It should be clear from the above that in many important cases the solution set of a sup-$\mathcal{T}$ equation, if not empty, is a finitely generated root system. Recalling that the intersection of a family of finitely generated root systems, if not empty, is again a root system, leads immediately to the description of solution procedures for systems of sup-$\mathcal{T}$ equations.

The problem we are concerned with is the following: given a complete lattice $(L, \leq)$ and a family $(A_s)_{s\in S}$ in $\mathcal{F}_L(I)$ and a family $(b_s)_{s\in S}$ in L, determine the solution set of the system of equations $(E_s)_{s\in S}$

$$\sup_{i\in I} \mathcal{T}(A_s(i), X(i)) = b_s \tag{6.11}$$

in the unknown L-fuzzy set X in I.

Greatest solution. If the partial mappings of $\mathcal{T}$ are sup-morphisms, then the solution set of system (6.11) is not empty if and only if the fuzzy set $G = \bigcap_{s\in S} G_s$, with G_s the potential greatest solution of equation E_s, i.e.

$$G(i) = \inf_{s\in S} \mathcal{I}_{\mathcal{T}}(A_s(i), b_s), \tag{6.12}$$

is a solution. Moreover, if G is a solution, then it is the greatest solution.

In general, no simpler necessary and sufficient solvability conditions are known. Recall that in the case of a single equation, under some very general conditions, we could always designate a root system contained in the solution set with offshoots that are minimal solutions. Of course, the intersection of these root systems will also be contained in the solution set of the systems of equations. Unfortunately, there is no guarantee that the offshoots of this intersection (if not empty) are again minimal solutions of the system of equations. Let us therefore turn our attention to the description of the complete solution set.

On a distributive, complete lattice with join-irreducible or finitely join-decomposable elements. Consider a finite index set I and assume that the right-hand sides of system (6.11) are join-irreducible or finitely join-decomposable. If the partial mappings of $\mathcal{T}$ are homomorphisms and the solution set of system (6.11) is not empty, then it is a root system; if the system is finite, then this root system is finitely generated. Suppose that the solution is not empty, and that the solution sets of the equations E_s are the root systems with stem G_s and set of offshoots O_s, then the solution set of the system $(E_s)_{s\in S}$ is the root system with stem $G = \bigcap_{s\in S} G_s$ and as set of offshoots the set

$$\{\bigcup_{s\in s} N_s \mid N_s \in O_s \wedge N_s \subseteq G\}.$$

A necessary and sufficient solvability condition therefore is that G is a solution. The case of a complete Brouwerian lattice was again discussed by Zhao (1987), the above more general case by De Baets (1998a).

A finite system, $S = I_m$, of sup-$\mathcal{T}$ equations, with finite index set $I = I_n$, can obviously be written in matrixform, as follows:

$$\begin{pmatrix} A_1(1) & \dots & A_1(n) \\ \vdots & \ddots & \vdots \\ A_m(1) & \dots & A_m(n) \end{pmatrix} \begin{pmatrix} X(1) \\ \vdots \\ X(n) \end{pmatrix} = \begin{pmatrix} b_1 \\ \vdots \\ b_m \end{pmatrix},$$

or in compact form, $AX = B$, with $A_{i,j} = A_i(j)$ and $B_i = b_i$. The formal analogy with the multiplication of matrices is clear: the multiplication is replaced by a t-norm and the addition by the supremum operation.

Example 4 Consider the system of sup-$\mathcal{T}$ equations (6.11) on the real unit interval, with $I = I_7$, $S = I_3$,

$$A = \begin{pmatrix} 0.69 & 0.75 & 0.96 & 1.00 & 0.56 & 0.48 & 0.36 \\ 0.48 & 0.78 & 0.69 & 0.53 & 0.42 & 0.99 & 0.47 \\ 0.29 & 0.84 & 0.56 & 0.71 & 0.80 & 0.46 & 0.88 \\ 0.68 & 0.48 & 0.52 & 0.47 & 0.55 & 0.78 & 0.95 \end{pmatrix} \qquad B = \begin{pmatrix} 0.48 \\ 0.48 \\ 0.48 \\ 0.48 \end{pmatrix}.$$

For $\mathcal{T} = W$, this system has a greatest solution G and six minimal solutions given by

$$\begin{array}{lcl} G & = & (\ 0.79\ 0.64\ 0.52\ 0.48\ 0.68\ 0.49\ 0.53\ ) \\ M_1^1 \cup M_6^2 \cup M_2^3 \cup M_7^4 & = & (\ 0.79\ 0.64\ 0.00\ 0.00\ 0.00\ 0.49\ 0.53\ ) \\ M_3^1 \cup M_6^2 \cup M_2^3 \cup M_7^4 & = & (\ 0.00\ 0.64\ 0.52\ 0.00\ 0.00\ 0.49\ 0.53\ ) \\ M_4^1 \cup M_6^2 \cup M_2^3 \cup M_7^4 & = & (\ 0.00\ 0.64\ 0.00\ 0.48\ 0.00\ 0.49\ 0.53\ ) \\ M_1^1 \cup M_6^2 \cup M_5^3 \cup M_7^4 & = & (\ 0.79\ 0.00\ 0.00\ 0.00\ 0.68\ 0.49\ 0.53\ ) \\ M_3^1 \cup M_6^2 \cup M_5^3 \cup M_7^4 & = & (\ 0.00\ 0.00\ 0.52\ 0.00\ 0.68\ 0.49\ 0.53\ ) \\ M_4^1 \cup M_6^2 \cup M_5^3 \cup M_7^4 & = & (\ 0.00\ 0.00\ 0.00\ 0.48\ 0.68\ 0.49\ 0.53\ ) \end{array}$$

6.4.5 Fuzzy relational equations

We are now in a position to tackle what are called fuzzy relational equations. For the sake of brevity, we will restrict ourselves to the real unit interval; the reader can easily generalize the results to the more general setting of distributive, complete lattices with join-irreducible or finitely join-decomposable elements.

Image equations of type 1. Consider a fuzzy set A in X and a fuzzy set B in Y. As mentioned before, the image equation $R(A) = B$ in the unknown fuzzy relation R is equivalent to a family of independent sup-$\mathcal{T}$ equations in the foresets Ry, $y \in Y$. This equation is an important identification problem in fuzzy rule modelling: determine a relational correspondence given the input A and the output B. Clearly, the solution set of a family of independent equations is isomorphic to the Cartesian product of the solution sets of the individual equations. Since the Cartesian product of root systems obviously also is a root system, our knowledge of sup-$\mathcal{T}$ equations is sufficient for solving this type of image equation. Note that the set of offshoots of a Cartesian product of root systems is nothing else but the Cartesian product of the corresponding sets of offshoots. For instance, the solution set is not empty if and only if the solution sets of all individual equations are not empty.

We then have that for a left-continuous t-norm $\mathcal{T}$, the solution set of the above image equation is not empty if and only if the fuzzy relation G defined by

$$G(x, y) = \mathcal{I}_{\mathcal{T}}(A(x), B(y)) \tag{6.13}$$

is the greatest solution. In case of a complete Brouwerian lattice and the meet operation, this result is known as one of the very first results on fuzzy relational equations, see Sanchez (1977). The case of the real unit interval was discussed by Di Nola, Sessa, Pedrycz and Sanchez (1989). For a continuous t-norm $\mathcal{T}$, a necessary and sufficient solvability condition is given by

$$\sup_{y \in Y} B(y) \leq \sup_{x \in X} A(x), \tag{6.14}$$

or, in words, the height of A is not smaller than the height of B. In the case of the real unit interval, this condition can also be found in the work of Gottwald and Pedrycz (1988) and Gottwald (1994).

Using the above framework of sup-$\mathcal{T}$ equations and root systems, the complete solution set can be written down immediately. If the universe X is finite, $\mathcal{T}$ is continuous and the solution set of the equation $R(A) = B$ is not empty, then it is a root system with stem G defined by (6.13) and as set of offshoots the set

$$O = \{M \mid (\forall y \in Y)(My \in O_y)\},$$

where

$$O_y = \{M_u^y \mid B(y) \leq A(u)\}$$

with M_u^y defined by

$$M_u^y(x) = \begin{cases} \mathcal{L}_{\mathcal{T}}(A(u), B(y)) & \text{, if } x = u \\ 0 & \text{, elsewhere} \end{cases} .$$

If the universe Y is also finite, then this root system is finitely generated.

The complete solution set was described first by Sanchez (1977) for the case of the real unit interval, finite underlying universes and the minimum operator as t-norm. The generalization to continuous t-norms has been done by Di Nola, Sessa, Pedrycz and Sanchez (1989). In both sources, explicit proofs are given.

Example 5 Consider the equation $R(A) = B$ in the unknown fuzzy relation R, $\#X = 4$, $\#Y = 3$, $A = \begin{pmatrix} 0.3 \\ 0 \\ 0.9 \\ 0.7 \end{pmatrix}$ and $B = \begin{pmatrix} 0.5 \\ 0.9 \\ 0.2 \end{pmatrix}$. For $\mathcal{T} = W$, the equation has a greatest solution $G = \begin{pmatrix} 1 & 1 & 0.9 \\ 1 & 1 & 1 \\ 0.6 & 1 & 0.3 \\ 0.8 & 1 & 0.5 \end{pmatrix}$ and 6 minimal solutions

$$M_1 = \begin{pmatrix} 0 & 0 & 0.9 \\ 0 & 0 & 0 \\ 0.6 & 1 & 0 \\ 0 & 0 & 0 \end{pmatrix} \quad M_2 = \begin{pmatrix} 0 & 0 & 0 \\ 0 & 0 & 0 \\ 0.6 & 1 & 0.3 \\ 0 & 0 & 0 \end{pmatrix} \quad M_3 = \begin{pmatrix} 0 & 0 & 0 \\ 0 & 0 & 0 \\ 0.6 & 1 & 0 \\ 0 & 0 & 0.5 \end{pmatrix}$$

$$M_4 = \begin{pmatrix} 0 & 0 & 0.9 \\ 0 & 0 & 0 \\ 0 & 1 & 0 \\ 0.8 & 0 & 0 \end{pmatrix} \quad M_5 = \begin{pmatrix} 0 & 0 & 0 \\ 0 & 0 & 0 \\ 0 & 1 & 0.3 \\ 0.8 & 0 & 0 \end{pmatrix} \quad M_6 = \begin{pmatrix} 0 & 0 & 0 \\ 0 & 0 & 0 \\ 0 & 1 & 0 \\ 0.8 & 0 & 0.5 \end{pmatrix}$$

Image equations of type 2. Consider a fuzzy relation R from X to Y and a fuzzy set B in Y. The image equation $R(A) = B$ in the unknown fuzzy set A is equivalent to a system of sup-$\mathcal{T}$ equations. This inverse problem plays a crucial role in backward chaining in fuzzy rule-based systems. Here, we can immediately apply the results from the previous subsection.

For a left-continuous t-norm $\mathcal{T}$, the solution set of this equation is not empty if and only if the fuzzy set G defined by

$$G(x) = \inf_{y \in Y} \mathcal{I}_{\mathcal{T}}(R(x, y), B(y)) \tag{6.15}$$

is the greatest solution. Denoting the converse (or transpose) of a fuzzy relation R by R^t, i.e. $R^t(y, x) = R(x, y)$, we can rewrite G as follows:

$$G = (R^t)_{\triangleright}(B),$$

using as implicator the residual implicator $\mathcal{I}_{\mathcal{T}}$. In view of the importance of this equation, we also mention the complete solution set. If the universe X is finite, $\mathcal{T}$ is continuous and the solution set of the equation $R(A) = B$ is not empty, then it is a root system with stem G defined by (6.15) and as offshoots the minimal elements of the set

$$\{\bigcup_{y \in Y} M_u^y \mid B(y) \leq R(u, y) \wedge M_u^y \subseteq G\}$$

where M_u^y is defined by

$$M_u^y(x) = \begin{cases} \mathcal{L}_{\mathcal{T}}(R(u, y), B(y)) & , \text{if } x = u \\ 0 & , \text{elsewhere} \end{cases} . \tag{6.16}$$

This type of image equation has received a lot of attention in the literature. Most attempts concern finite universes, the real unit interval (or a complete chain) and mostly the minimum operator. Prévot (1981) has written an algorithm for determining the minimal solutions; however, only in particular cases, this algorithm effectively leads to all minimal solutions. Czogała *et al.* (1982) have described the complete solution set, an algorithm for determining the minimal solutions and upper bounds for the number of minimal solutions. Their reasoning is rather complicated and not always completely correct, as was noticed by Higashi and Klir (1984), who revised the algorithm of Czogała *et al.* and proposed also an alternative algorithm. Yet another algorithm for determining the minimal solutions has been designed by Xu *et al.* (1982). The same problem has been solved once again by Pappis and Sugeno (1985), with more transparant proofs. The minimal solutions were also discussed by Miyakoshi and Shimbo (1986).

Other approaches are based on some type of quasi-characteristic matrix. Peeva (1985, 1992) proposed an algorithm that builds the minimal solutions from this matrix, improved later on by Li (1994). A similar approach is followed by Wang and Hsu (1992), and results in a more efficient algorithm than that of Higashi and Klir (1984). Recently, rule-based approaches for finding the minimal solutions were presented, see Arnould and Tano (1994a, 1994b).

All the results mentioned are based on the explicit expressions of the operators $\mathcal{I}_M$ and $\mathcal{L}_M$, and in particular on the fact that they only differ on the diagonal. The case of a (left-)continuous t-norm is also discussed in Pedrycz (1985), Di Nola, Pedrycz and Sessa (1987) and Di Nola, Sessa, Pedrycz and Sanchez (1989). However, using our knowledge about the intersection of finitely generated root systems, the above results can be been derived easily and more generally.

Cheng and Peng (1988) considered the equation $R(A) = B$ in the unknown fuzzy set A, using as conjunctor a binary operation V on $[0, 1]$ with first partial mappings that are sup-morphisms and satisfy $V(x, 0) = 0$, and that fulfills the 'existence condition of the least solution': if the equation $V(a, x) = b$ has a solution, then it also has a smallest solution. The latter is, of course, not a very elegant way of circumventing the problem.

Concerning the algorithms for determining the minimal solutions, it is instructive to mention that the algorithm for determining the offshoots of a finite family of finitely generated root systems, as described in De Baets (1995c, 1998b), can always be applied, independent of the t-norm or the complete lattice involved. Of course, this does not mean that for the specific case of the minimum operator, none of the algorithms mentioned could be more efficient.

Composition equations. Consider a fuzzy relation S from Y to Z and a fuzzy relation T from X to Z. The composition equation $R \circ S = T$ in the unknown fuzzy relation R is equivalent to a family of independent image equations $S(xR) = xT$ in the unknown aftersets xR, for all $x \in X$. The solution set therefore is isomorphic to the Cartesian product of the solution sets of these image equations, each of which is nothing else but a system of sup-$\mathcal{T}$ equations. Of course, the equation $R \circ S = T$ in the unknown fuzzy relation S is solved at the same time, since it can be reformulated as follows: $S^t \circ R^t = T^t$.

Therefore, for a left-continuous t-norm $\mathcal{T}$, the solution set of this equation is not empty if and only if the fuzzy relation G defined by

$$G(x, y) = \inf_{z \in Z} \mathcal{I}_{\mathcal{T}}(S(y, z), T(x, z)) \tag{6.17}$$

is the greatest solution. We can rewrite G as follows:

$$G = T \triangleright S^t,$$

using as implicator the residual implicator $\mathcal{I}_{\mathcal{T}}$. In the case of a complete Brouwerian lattice and the meet operation as t-norm, this is again one the first results of Sanchez (1977); in this case, there also exists an alternative necessary and sufficient solvability condition by Di Nola (1990). In the case of finite universes, a completely distributive complete lattice $(L, \leq)$ and the meet operation as t-norm, the solution set is also a root system. Indeed, Di Nola (1987, 1990) has shown that for any solution at least one underlying minimal solution can be constructed; however, not all minimal solutions can be constructed explicitly. This result complements the results obtained in our general framework. It is clear that for specific operators and in specific situations, additional results can be obtained. Of course, in case of join-irreducibility and join-decomposability, we are again in our framework.

However, there is another way of dealing with the equation $R \circ S = T$. Indeed, it can be seen as a system of image equations $S(xR) = xT$ in the unknown fuzzy relation S, for all $x \in X$. At first sight, the latter approach seems to be the most practical one, in particular in view of determining the minimal solutions. Indeed, since each of the image equations $S(xR) = xT$ in the unknown fuzzy relation S stands for a set of independent sup-$\mathcal{T}$ equations, their minimal solutions can be written down immediately and the pen-and-paper method has to be applied only once to find the minimal solutions of the composition equation $R \circ S = T$. In the first approach, the minimal solutions of each of the independent image equations $S(xR) = xT$ in the unknown aftersets xR have to be found using the pen-and-paper method. The minimal solutions of the composition equation $R \circ S = T$ are then found immediately. However, in practice it is difficult to say which method is the most efficient one. In the second approach, the pen-and-paper method has to be applied only once, but then possibly on a huge number of *fuzzy relations*, while for each of the independent image equations the minimal solutions have to be found among a smaller number of *fuzzy sets*. The second approach is the one usually followed in the literature, see Di Nola, Sessa, Pedrycz and Sanchez (1989), while we favour the first one, in particular from an implementational point of view.

We conclude our discussion with the study of the equation $Q \circ R \circ S = T$ in the unknown fuzzy relation R, given a fuzzy relation Q from X to Y, a fuzzy relation S from Z to U and a fuzzy relation T from X to U. Of course, the above equation only makes sense when the composition of fuzzy relations is associative. The latter is, for instance, fulfilled when using as conjunctor a left-continuous t-norm. This equation has been studied by Ohsato and Sekiguchi (1982) for finite universes and the minimum operator. They have shown that this equation can be transformed into an image equation of type 2. De Baets and Kerre (1994b) have generalized these results for an arbitrary t-norm and arbitrary universes. We briefly explain the transformation. Define the following universes: $V = Y \times Z$ and $W = X \times U$. The *direct product* of the fuzzy relations Q^t and S is the fuzzy relation $Q^t \odot S$ from V to W, defined by, for $(v,w) = ((y,z),(x,u)) \in V \times W$,

$$Q^t \odot S(v,w) = \mathcal{T}(Q^t(y,x), S(z,u)).$$

The composition equation $Q \circ R \circ S = T$ in the unknown R is then equivalent to the image equation

$$(Q^t \odot S)(R) = T,$$

where the unknown fuzzy relation R is treated as a fuzzy set in V and the fuzzy relation T as a fuzzy set in W. It is easy to see that the other composition equations are special cases of this more complicated equation.

Example 6 Consider the equation $Q \circ R \circ S = T$ in the unknown fuzzy relation R, with $\#X = \#Y = 3$, $\#Z = 4$, $\#U = 2$ and

$$Q = \begin{pmatrix} 0.3 & 0.85 & 0.24 \\ 0.4 & 0.44 & 0.48 \\ 0.7 & 0.76 & 1 \end{pmatrix} \qquad S = \begin{pmatrix} 0.66 & 0.5 \\ 0.22 & 1 \\ 0.34 & 0.27 \\ 0.68 & 0.14 \end{pmatrix}$$

and

$$T = \begin{pmatrix} 0.66 & 0.5 \\ 0.48 & 0.48 \\ 0.66 & 0.5 \end{pmatrix}$$

For $\mathcal{T} = P$ and $\mathcal{T} = W$ this equation has no solution. For $\mathcal{T} = M$, it has a greatest solution and six minimal solutions. The direct product $Q^t \odot S$ is given by:

$$Q^t \odot S = \begin{pmatrix} 0.3 & 0.3 & 0.4 & 0.4 & 0.66 & 0.5 \\ 0.22 & 0.3 & 0.22 & 0.4 & 0.22 & 0.7 \\ 0.3 & 0.27 & 0.34 & 0.27 & 0.34 & 0.27 \\ 0.3 & 0.14 & 0.4 & 0.14 & 0.68 & 0.14 \\ 0.66 & 0.5 & 0.44 & 0.44 & 0.66 & 0.5 \\ 0.22 & 0.85 & 0.22 & 0.44 & 0.22 & 0.76 \\ 0.34 & 0.27 & 0.34 & 0.27 & 0.34 & 0.27 \\ 0.68 & 0.14 & 0.44 & 0.14 & 0.68 & 0.14 \\ 0.24 & 0.24 & 0.48 & 0.48 & 0.66 & 0.5 \\ 0.22 & 0.24 & 0.22 & 0.48 & 0.22 & 1 \\ 0.24 & 0.24 & 0.34 & 0.27 & 0.34 & 0.27 \\ 0.24 & 0.14 & 0.48 & 0.14 & 0.68 & 0.14 \end{pmatrix}$$

Hence, this equation is reduced to a system of 6 sup-M equations in 12 unknowns. The greatest solution G and the minimal solutions $M_1, \ldots, M_6$ are given by:

$$G = \begin{pmatrix} 1 & 0.5 & 1 & 0.66 \\ 1 & 0.5 & 1 & 0.66 \\ 1 & 0.5 & 1 & 0.66 \end{pmatrix}$$

$$M_1 = \begin{pmatrix} 0 & 0 & 0 & 0 \\ 0.66 & 0 & 0 & 0 \\ 0.48 & 0 & 0 & 0 \end{pmatrix} \qquad M_2 = \begin{pmatrix} 0 & 0 & 0 & 0 \\ 0.66 & 0 & 0 & 0 \\ 0 & 0.48 & 0 & 0.48 \end{pmatrix}$$

$$M_3 = \begin{pmatrix} 0 & 0 & 0 & 0 \\ 0.5 & 0 & 0 & 0.66 \\ 0 & 0.48 & 0 & 0.48 \end{pmatrix} \qquad M_4 = \begin{pmatrix} 0 & 0 & 0 & 0 \\ 0.5 & 0 & 0 & 0.66 \\ 0.48 & 0 & 0 & 0 \end{pmatrix}$$

$$M_5 = \begin{pmatrix} 0 & 0 & 0 & 0 \\ 0 & 0.5 & 0 & 0.66 \\ 0.48 & 0 & 0 & 0 \end{pmatrix} \qquad M_6 = \begin{pmatrix} 0 & 0 & 0 & 0 \\ 0 & 0.5 & 0 & 0.66 \\ 0 & 0.48 & 0 & 0.48 \end{pmatrix}$$

6.5 LEFT INF-$\mathcal{I}$ EQUATIONS

6.5.1 The equation $\mathcal{I}(x, b) = a$

The second basic equation we consider is the equation $\mathcal{I}(x, b) = a$ in the unknown x, with $\mathcal{I}$ a *border implicator* on a complete lattice $(L, \leq)$. In order to describe the solution set of this equation, we associate to $\mathcal{I}$ two binary operators $\mathcal{I}_\mathcal{I}$ and $\mathcal{L}_\mathcal{I}$ on L defined by, see De Baets (1995b, 1995d, 1998c):

$$\mathcal{I}_\mathcal{I}(x, y) = \sup\{z \in L \mid \mathcal{I}(z, y) \geq x\}$$

$$\mathcal{L}_\mathcal{I}(x, y) = \inf\{z \in L \mid \mathcal{I}(z, y) \leq x\}.$$

In general, the operator $\mathcal{I}_\mathcal{I}$ satisfies all conditions of an implicator except for the behaviour in the point $(1, 0)$. In many important cases, however, the behaviour $\mathcal{I}_\mathcal{I}(1, 0) = 0$ can be guaranteed:

- if the negator $\mathcal{N}_\mathcal{I}$ induced by $\mathcal{I}$ is strong;
- if $\mathcal{I} = \mathcal{I}_\mathcal{T}$, with $\mathcal{T}$ a positive t-norm, i.e.

  $$(\forall(x, y) \in L^2)(\mathcal{T}(x, y) = 0 \Rightarrow (x = 0 \vee y = 0)),$$

  or with $\mathcal{T}$ a t-norm with partial mappings that are sup-morphisms.

The second operator $\mathcal{L}_\mathcal{I}$ has no particular interpretation.

Some important closure theorems can be stated for the operator $\mathcal{I}_\mathcal{I}$, see De Baets (1995d, 1998c):

- If $\mathcal{I} = \mathcal{I}_{\mathcal{T},\mathcal{N}}$, with $\mathcal{T}$ a t-norm and $\mathcal{N}$ an involutive negator, then the corresponding implicator $\mathcal{I}_\mathcal{I}$ is given by

  $$\mathcal{I}_\mathcal{I}(x, y) = \mathcal{I}_\mathcal{T}(\mathcal{N}(y), \mathcal{N}(x)).$$

  For instance, for $\mathcal{I} = \mathcal{I}_{M,\mathcal{N}_s}$ it holds that

  $$\mathcal{I}_\mathcal{I}(x, y) = \begin{cases} 1 & \text{, if } x \leq y \\ 1 - x & \text{, elsewhere} \end{cases}.$$

- If $\mathcal{I} = \mathcal{I}_{\mathcal{T}}$, with $\mathcal{T}$ a t-norm with partial mappings that are sup-morphisms, then the following equality holds:

$$\mathcal{I}_{\mathcal{I}_{\mathcal{T}}} = \mathcal{I}_{\mathcal{T}}.$$

  This result obviously applies to the residual implicators listed in Table 6.1.

The operators $\mathcal{L}_{\mathcal{I}}$ corresponding to the implicators $\mathcal{I}_{\mathcal{T},\mathcal{N}_s}$ and $\mathcal{I}_{\mathcal{T}}$ are listed in Table 6.2.

Table 6.2 Operators of the type $\mathcal{L}_{\mathcal{I}}$.

$\mathcal{I}$	$\mathcal{L}_{\mathcal{I}}(x, y)$
$\mathcal{I}_{M,\mathcal{N}_s}$	$\begin{cases} 1 & \text{, if } x < y \\ 1-x & \text{, elsewhere} \end{cases}$
$\mathcal{I}_{P,\mathcal{N}_s}$	$\begin{cases} 1 & \text{, if } x < y \\ \dfrac{1-x}{1-y} & \text{, if } y \leq x \text{ and } y \neq 1 \\ 0 & \text{, elsewhere} \end{cases}$
$\mathcal{I}_{W,\mathcal{N}_s} = \mathcal{I}_W$	$\begin{cases} \min(1-x+y, 1) & \text{, if } x \neq 1 \\ 0 & \text{, elsewhere} \end{cases}$
$\mathcal{I}_M$	$\begin{cases} 1 & \text{, if } x < y \\ y & \text{, if } y \leq x < 1 \\ 0 & \text{, elsewhere} \end{cases}$
$\mathcal{I}_P$	$\begin{cases} 1 & \text{, if } x < y \\ y/x & \text{, if } 0 < y \leq x < 1 \\ 0 & \text{, elsewhere} \end{cases}$

The solution set of the equation $\mathcal{I}(x, b) = a$ can again be seen as the intersection of the solution sets of the inequalities $\mathcal{I}(x, b) \geq a$ and $\mathcal{I}(x, b) \leq a$.

On a complete lattice. The following observations hold true:

- If the partial mapping $\mathcal{I}(\cdot, b)$ is a dual sup-morphism, then the solution set of the inequality $\mathcal{I}(x, b) \geq a$ is given by $[0, \mathcal{I}_{\mathcal{I}}(a, b)]$.

- If the partial mapping $\mathcal{I}(\cdot, b)$ is a dual inf-morphism, then the solution set of the inequality $\mathcal{I}(x, b) \leq a$ is given by

$$\begin{cases} [\mathcal{L}_{\mathcal{I}}(a, b), 1] & \text{, if } b \leq a \\ \emptyset & \text{, elsewhere} \end{cases}.$$

- If the partial mapping $\mathcal{I}(\cdot, b)$ is a dual homomorphism, then the solution set of the equation $\mathcal{I}(x, b) = a$ is given by

$$\begin{cases} [\mathcal{L}_{\mathcal{I}}(a, b), \mathcal{I}_{\mathcal{I}}(a, b)] & \text{, if } a \in \text{rng}(\mathcal{I}(\cdot, b)) \\ \emptyset & \text{, elsewhere} \end{cases};$$

if the partial mapping $\mathcal{I}(\cdot, b)$ is a maximally surjective dual homomorphism (i.e. $\text{rng}(\mathcal{I}(\cdot, b)) = [b, 1]$), then the solution set of the equation $\mathcal{I}(x, b) = a$ is given by

$$\begin{cases} [\mathcal{L}_\mathcal{I}(a, b), \mathcal{I}_\mathcal{I}(a, b)] & , \text{ if } b \leq a \\ \emptyset & , \text{ elsewhere} \end{cases}.$$

On the real unit interval. If the partial mapping $\mathcal{I}(\cdot, b)$ is continuous, then the solution set of the equation $\mathcal{I}(x, b) = a$ is given by

$$\begin{cases} [\mathcal{L}_\mathcal{I}(a, b), \mathcal{I}_\mathcal{I}(a, b)] & , \text{ if } b \leq a \\ \emptyset & , \text{ elsewhere} \end{cases}.$$

A necessary and sufficient solvability condition therefore is $b \leq a$.

6.5.2 *Greatest solution – solvability conditions*

Given $B \in \mathcal{F}_L(I)$ and $a \in L$, we want to determine the solution set of the equation

$$\inf_{i \in I} \mathcal{I}(X(i), B(i)) = a \tag{6.18}$$

in the unknown L-fuzzy set X in I. The solution set of this equation can be seen as the intersection of the solution sets of the inequalities

$$\inf_{i \in I} \mathcal{I}(X(i), B(i)) \geq a \tag{6.19}$$

and

$$\inf_{i \in I} \mathcal{I}(X(i), B(i)) \leq a. \tag{6.20}$$

Obviously, due to monotonicity considerations, inequality (6.19) always has a solution, namely $\chi_\emptyset$, while inequality (6.20) has a solution if and only if χ_I is a solution, i.e. if and only if

$$\inf_{i \in I} B(i) \leq a. \tag{6.21}$$

Note that in the literature, this equation has only been studied using as implicator the residual implicator of a t-norm. The framework offered here is much wider.

On a complete lattice. The following statements can be verified:

- If the first partial mappings of $\mathcal{I}$ are dual sup-morphisms, then the solution set of inequality (6.19) is given by $[\chi_\emptyset, G]$ with G defined by

  $$G(i) = \mathcal{I}_\mathcal{I}(a, B(i)). \tag{6.22}$$

  In particular, this result shows us the solution set for equation (6.18) in case $a = 1$.

- If the first partial mappings of $\mathcal{I}$ are dual sup-morphisms, then the solution set of equation (6.18) is not empty if and only if the fuzzy set G

defined by (6.22) is a solution. Moreover, if G is a solution, then it is the greatest solution.

Necessary and sufficient solvability conditions. As for sup-$\mathcal{T}$ equations, in general, no simple necessary and sufficient solvability condition can be given.

- In any case, condition (6.21) is a necessary condition for the existence of a solution to equation (6.18). If L is a complete chain and the first partial mappings of $\mathcal{I}$ are maximally surjective dual homomorphisms, then condition (6.21) becomes also a sufficient condition; if L is also dense, then the condition of maximal surjectivity can be omitted.

- If the first partial mappings of $\mathcal{I}$ are maximally surjective dual homomorphisms, then the condition

$$(\exists k \in I)(B(k) \leq a), \tag{6.23}$$

  a stronger version of condition (6.21), is a sufficient condition for solvability.

On the real unit interval. For a border implicator with continuous first partial mappings, the solution set of equation (6.18) is not empty if and only condition (6.21) holds. If the solution set is not empty, then the fuzzy set G defined by (6.22) is the greatest solution.

6.5.3 Complete solution set

In the description of the complete solution set of equation (6.18), root systems again play a prominent role. Let us immediately turn to the most general framework within which we can describe this solution set. The following results are obtained by studying the solution sets of both inequalities (6.19) and (6.20) and combining the results.

On a distributive, complete lattice with meet-irreducible and meet-decomposable elements. Similarly as before, an element a of a lattice $(L, \leq)$ is called *meet-irreducible* if

$$(\forall (b, c) \in L^2)(b \frown c = a \Rightarrow (b = a \vee c = a)),$$

and *meet-decomposable* if there exists a set A of meet-irreducible elements of L, with $\#A \geq 2$, such that $a = \inf A$. Of course, all elements of a chain are meet-irreducible.

We consider again a distributive, complete lattice $(L, \leq)$. For a finite index set I, it holds that:

- **The case of a meet-irreducible right-hand side.** If a is meet-irreducible, the first partial mappings of $\mathcal{I}$ are dual homomorphisms and the solution set of equation (6.18) is not empty, then it is a finitely generated root system with stem G defined by (6.22) and as set of offshoots the set $\{M_k \mid B(k) \leq a \wedge M_k \subseteq G\}$, where M_k is defined by

$$M_k(i) = \begin{cases} \mathcal{L}_\mathcal{I}(a, B(k)) & \text{, if } i = k \\ 0 & \text{, elsewhere} \end{cases} ; \tag{6.24}$$

moreover, if the first partial mappings of $\mathcal{I}$ are maximally surjective, then the set of offshoots can be written as $\{M_k \mid B(k) \leq a\}$.

- **The case of a meet-decomposable right-hand side.** If a has a meet-decomposition $a = \inf_{j \in J} a_j$, the first partial mappings of $\mathcal{I}$ are dual homomorphisms and the solution set of equation (6.18) is not empty, then it is a root system with stem G defined by (6.22) and as offshoots the *minimal* elements of the set

$$\{\bigcup_{j \in J} M_k^j \mid B(k) \leq a_j \wedge M_k^j \subseteq G\},$$

where M_k^j is defined by

$$M_k^j(i) = \begin{cases} \mathcal{L}_\mathcal{I}(a_j, B(k)) & \text{, if } i = k \\ 0 & \text{, elsewhere} \end{cases} ; \tag{6.25}$$

if the meet-decomposition is finite, then this root system is also finitely generated and if $\mathcal{L}_\mathcal{I} \leq \mathcal{I}_\mathcal{I}$, then the offshoots are the *minimal* elements of the set

$$\{\bigcup_{j \in J} M_k^j \mid B(k) \leq a_j\}.$$

On the real unit interval. All elements of the real unit interval are meet-irreducible. Hence, the above results can be applied. For a border implicator $\mathcal{I}$ with continuous first partial mappings and a finite index set I, the solution set of equation (6.18) is not empty if and only if condition (6.23) holds. If the solution set is not empty, then it is the finitely generated root system with stem G defined by (6.22) and as set of offshoots the set $\{M_k \mid B(k) \leq a\}$, where M_k is defined by (6.24).

Example 7 For example, consider equation (6.18) with $I = I_4$,

$$B = \begin{pmatrix} 0.3 & 0.6 & 1 & 0.2 \end{pmatrix}$$

and $a = 0.5$. Since $a > \min_{i \in I_4} B(i) = 0.2$, the solution set is not empty for any border implicator $\mathcal{I}$ with continuous first partial mappings, and has a greatest solution G and two minimal solutions M_1 and M_4. In case $\mathcal{I} = \mathcal{I}_{M,\mathcal{N}_s}$, these are given by:

$$\begin{aligned} G &= \begin{pmatrix} 0.5 & 1 & 1 & 0.5 \end{pmatrix} \\ M_1 &= \begin{pmatrix} 0.5 & 0 & 0 & 0 \end{pmatrix} \\ M_4 &= \begin{pmatrix} 0 & 0 & 0 & 0.5 \end{pmatrix} \end{aligned}$$

On the real unit hypercube. Since all elements of the real unit hypercube are meet-irreducible or meet-decomposable, once again the above results can be applied. This is left as an exercise to the reader.

6.5.4 Systems of left inf-$\mathcal{I}$ equations

The problem we are concerned with is the following: given a complete lattice $(L, \leq)$ and a family $(B_s)_{s\in S}$ in $\mathcal{F}_L(I)$ and a family $(a_s)_{s\in S}$ in L, determine the solution set of the system of equations $(E_s)_{s\in S}$

$$\inf_{i\in I} \mathcal{I}(X(i), B_s(i)) = a_s \tag{6.26}$$

in the unknown L-fuzzy set X in I.

Greatest solution. If the first partial mappings of $\mathcal{I}$ are dual sup-morphisms, then the solution set of system (6.18) is not empty if and only if the fuzzy set $G = \bigcap_{s\in S} G_s$, with G_s the potential greatest solution of equation E_s, i.e.

$$G(i) = \inf_{s\in S} \mathcal{I}_{\mathcal{I}}(a_s, B_s(i)), \tag{6.27}$$

is a solution. Moreover, if G is a solution, then it is the greatest solution. Again, in general, no simpler necessary and sufficient solvability conditions are known.

On a distributive, complete lattice with meet-irreducible or finitely meet-decomposable elements. Consider a finite index set I and assume that the right-hand sides of system (6.26) are meet-irreducible or finitely meet-decomposable. If the first partial mappings of $\mathcal{I}$ are dual homomorphisms and the solution set of system (6.26) is not empty, then it is a root system; if the system is finite, then this root system is finitely generated. Suppose that the solution is not empty, and that the solution sets of the equations E_s are the root systems with stem G_s and set of offshoots O_s, then the solution set of the system $(E_s)_{s\in S}$ is the root system with stem $G = \bigcap_{s\in S} G_s$ and as offshoots the minimal elements of the set

$$\{\bigcup_{s\in S} N_s \mid N_s \in O_s \wedge N_s \subseteq G\}.$$

A necessary and sufficient solvability condition therefore is that G is a solution.

Example 8 Consider the system of left inf-$\mathcal{I}$ equations (6.26) on the real unit interval, with $I = I_9$, $S = I_3$,

$$B = \begin{pmatrix} 0.50 & 0.60 & 0.70 & 0.80 & 0.50 & 1.00 & 0.60 & 0.58 & 0.69 \\ 0.36 & 0.48 & 0.49 & 0.86 & 0.40 & 0.86 & 0.96 & 0.25 & 0.74 \\ 0.39 & 0.86 & 0.47 & 0.85 & 0.52 & 0.81 & 0.69 & 0.36 & 0.48 \end{pmatrix}$$

and $A = \begin{pmatrix} 0.7 \\ 0.37 \\ 0.48 \end{pmatrix}$. For $\mathcal{I} = \mathcal{I}_{M,\mathcal{N}_s}$ and $\mathcal{I} = \mathcal{I}_{P,\mathcal{N}_s}$, this system has no solution. For $\mathcal{I} = \mathcal{I}_{W,\mathcal{N}_s}$, the system has a greatest solution G and a smallest solution S given by

$$\begin{aligned} G &= (\ 0.8\ \ 0.9\ \ 0.99\ \ 1\ \ 0.8\ \ 1\ \ 0.9\ \ 0.88\ \ 0.99\ ) \\ S &= (\ 0\ \ 0\ \ 0\ \ 0\ \ 0\ \ 0\ \ 0\ \ 0.88\ \ 0\ ) \end{aligned}$$

6.5.5 *Fuzzy relational equations*

We will again restrict ourselves to the real unit interval. A number of interesting fuzzy relational equations can be solved based upon the knowledge of (systems of) left inf-$\mathcal{I}$ equations. Also, we will comment only on the greatest solution and necessary and sufficient solvability conditions; the discussion of the complete solution set can again be derived from the above.

Image equations of type 1. Consider a fuzzy set A in X and a fuzzy set B in Y. The image equation $R_{\triangleright}(A) = B$ in the unknown fuzzy relation R is equivalent to a family of independent left inf-$\mathcal{I}$ equations in the foresets Ry, $y \in Y$. We then have that for a border implicator $\mathcal{I}$ with left-continuous first partial mappings, the solution set of the above image equation is not empty if and only if the fuzzy relation G defined by

$$G(x, y) = \mathcal{I}_{\mathcal{I}}(B(y), A(x)) \tag{6.28}$$

is the greatest solution. If the first partial mappings are continuous, then a necessary and sufficient solvability condition is given by

$$\inf_{x \in X} A(x) \leq \inf_{y \in Y} B(y). \tag{6.29}$$

For an implicator of the type $\mathcal{I}_{\mathcal{T}}$, with $\mathcal{T}$ a left-continuous t-norm, this result was mentioned by Miyakoshi and Shimbo (1985).

Image equations of type 2. Consider a fuzzy relation R from X to Y and a fuzzy set B in Y. The image equation $R_{\triangleleft}(A) = B$ in the unknown fuzzy set A is equivalent to a system of left inf-$\mathcal{I}$ equations. For a border implicator $\mathcal{I}$ with left-continuous first partial mappings, the solution set of this equation is not empty if and only if the fuzzy set G defined by

$$G(x) = \inf_{y \in Y} \mathcal{I}_{\mathcal{I}}(B(y), R(x, y)) \tag{6.30}$$

is the greatest solution, i.e. $G = (R^t)_{\triangleleft}(B)$, using as 'implicator' the operator $\mathcal{I}_{\mathcal{I}}$.

Composition equations. Consider a fuzzy relation S from Y to Z and a fuzzy relation T from X to Z. The composition equation $R \triangleleft S = T$ in the unknown fuzzy relation R is equivalent to a family of independent image equations $S_{\triangleleft}(xR) = xT$ in the unknown aftersets xR, for all $x \in X$. The equation $R \triangleright S = T$ in the unknown fuzzy relation S can be reformulated as follows: $S^t \triangleleft R^t = T^t$.

For a border implicator $\mathcal{I}$ with left-continuous first partial mappings, the solution set of this equation is not empty if and only if the fuzzy relation G defined by

$$G(x, y) = \inf_{z \in Z} \mathcal{I}_{\mathcal{I}}(T(x, z), S(y, z)) \tag{6.31}$$

is the greatest solution. We can rewrite G as follows:

$$G = T \triangleleft S^t,$$

again using as 'implicator' the operator $\mathcal{I}_{\mathcal{I}}$.

6.6 RIGHT INF-$\mathcal{I}$ EQUATIONS

6.6.1 *The equation* $\mathcal{I}(a, x) = b$

The last basic equation we consider is the equation $\mathcal{I}(a, x) = b$ in the unknown x, with $\mathcal{I}$ a *border implicator* on a complete lattice $(L, \leq)$. We associate to $\mathcal{I}$ two binary operators $\mathcal{C}_\mathcal{I}$ and $\mathcal{R}_\mathcal{I}$ on L defined by, see De Baets (1995b, 1995d, 1998c):

$$\mathcal{C}_\mathcal{I}(x, y) = \inf\{z \in L \mid \mathcal{I}(x, z) \geq y\}$$

$$\mathcal{R}_\mathcal{I}(x, y) = \sup\{z \in L \mid \mathcal{I}(x, z) \leq y\}.$$

The operator $\mathcal{C}_\mathcal{I}$ always is a conjunctor, while the operator $\mathcal{R}_\mathcal{I}$ again has no interesting interpretation.

Note that a border implicator $\mathcal{I}$ can only be contrapositive w.r.t. the negator $\mathcal{N}_\mathcal{I}$ it induces, see De Baets (1995a), i.e.

$$\mathcal{I}(x, y) = \mathcal{I}(\mathcal{N}_\mathcal{I}(y), \mathcal{N}_\mathcal{I}(x)).$$

If the latter holds, then we have the following relationships, see De Baets (1995d, 1998c):

$$\begin{aligned} \mathcal{L}_\mathcal{I}(x, y) &= \mathcal{N}_\mathcal{I}(\mathcal{R}_\mathcal{I}(\mathcal{N}_\mathcal{I}(y), x)) \\ \mathcal{R}_\mathcal{I}(x, y) &= \mathcal{N}_\mathcal{I}(\mathcal{L}_\mathcal{I}(y, \mathcal{N}_\mathcal{I}(x))) \\ \mathcal{I}_\mathcal{I}(x, y) &= \mathcal{N}_\mathcal{I}(\mathcal{C}_\mathcal{I}(\mathcal{N}_\mathcal{I}(y), x)) \\ \mathcal{C}_\mathcal{I}(x, y) &= \mathcal{N}_\mathcal{I}(\mathcal{I}_\mathcal{I}(y, \mathcal{N}_\mathcal{I}(x))). \end{aligned} \tag{6.32}$$

Some important closure theorems can be stated for the operator $\mathcal{C}_\mathcal{I}$:

- if $\mathcal{I} = \mathcal{I}_{\mathcal{T},\mathcal{N}}$, with $\mathcal{T}$ a t-norm and $\mathcal{N}$ an involutive negator, then the corresponding conjunctor $\mathcal{C}_\mathcal{I}$ is given by

  $$\mathcal{C}_\mathcal{I}(x, y) = \mathcal{N}(\mathcal{I}_\mathcal{T}(x, \mathcal{N}(y))).$$

  For instance, for $\mathcal{I} = \mathcal{I}_{M,\mathcal{N}_s}$ it holds that

  $$\mathcal{C}_\mathcal{I}(x, y) = \begin{cases} 0 & \text{, if } x + y \leq 1 \\ y & \text{, elsewhere} \end{cases} .$$

- if $\mathcal{I} = \mathcal{I}_\mathcal{T}$, with $\mathcal{T}$ a t-norm with partial mappings that are sup-morphisms, then the following equality holds:

  $$\mathcal{C}_{\mathcal{I}_\mathcal{T}} = \mathcal{T}.$$

The operators $\mathcal{R}_\mathcal{I}$ corresponding to the implicators $\mathcal{I}_{\mathcal{T},\mathcal{N}_s}$ and $\mathcal{I}_\mathcal{T}$ are listed in Table 6.3. The first ones follow from Table 6.2 by applying relationships (6.32).

Table 6.3 Operators of the type $\mathcal{R}_{\mathcal{I}}$.

$\mathcal{I}$	$\mathcal{R}_{\mathcal{I}}(x,y)$
$\mathcal{I}_{M,\mathcal{N}_s}$	$\begin{cases} 0 & \text{, if } x+y<1 \\ y & \text{, elsewhere} \end{cases}$
$\mathcal{I}_{P,\mathcal{N}_s}$	$\begin{cases} 0 & \text{, if } x+y<1 \\ \dfrac{x+y-1}{x} & \text{, if } 1 \leq x+y \text{ and } x \neq 0 \\ 1 & \text{, elsewhere} \end{cases}$
$\mathcal{I}_{W,\mathcal{N}_s} = \mathcal{I}_W$	$\begin{cases} \max(x+y-1,0) & \text{, if } y \neq 1 \\ 1 & \text{, elsewhere} \end{cases}$
$\mathcal{I}_M$	$\begin{cases} \min(x,y) & \text{, if } y \neq 1 \\ 1 & \text{, elsewhere} \end{cases}$
$\mathcal{I}_P$	$\begin{cases} xy & \text{, if } y \neq 1 \\ 1 & \text{, elsewhere} \end{cases}$

The solution set of the equation $\mathcal{I}(a,x) = b$ can again be seen as the intersection of the solution sets of the inequalities $\mathcal{I}(a,x) \geq b$ and $\mathcal{I}(a,x) \leq b$.

On a complete lattice. The following observations hold true:

- If the partial mapping $\mathcal{I}(a,\cdot)$ is an inf-morphism, then the solution set of the inequality $\mathcal{I}(a,x) \geq b$ is given by $[\mathcal{C}_{\mathcal{I}}(a,b),1]$.

- If the partial mapping $\mathcal{I}(a,\cdot)$ is a sup-morphism, then the solution set of the inequality $\mathcal{I}(a,x) \leq b$ is given by

$$\begin{cases} [0,\mathcal{R}_{\mathcal{I}}(a,b)] & \text{, if } \mathcal{N}_{\mathcal{I}}(a) \leq b \\ \emptyset & \text{, elsewhere} \end{cases} .$$

  Note that the condition for solvability involves the negator $\mathcal{N}_{\mathcal{I}}$ induced by $\mathcal{I}$.

- If the partial mapping $\mathcal{I}(a,\cdot)$ is a homomorphism, then the solution set of the equation $\mathcal{I}(a,x) = b$ is given by

$$\begin{cases} [\mathcal{C}_{\mathcal{I}}(a,b),\mathcal{R}_{\mathcal{I}}(a,b)] & \text{, if } b \in \text{rng}(\mathcal{I}(a,\cdot)) \\ \emptyset & \text{, elsewhere} \end{cases} ;$$

  if the partial mapping $\mathcal{I}(a,\cdot)$ is a maximally surjective homomorphism (i.e. $\text{rng}(\mathcal{I}(a,\cdot)) = [\mathcal{N}_{\mathcal{I}}(a),1]$), then the solution set of the equation $\mathcal{I}(a,x) = b$ is given by

$$\begin{cases} [\mathcal{C}_{\mathcal{I}}(a,b),\mathcal{R}_{\mathcal{I}}(a,b)] & \text{, if } \mathcal{N}_{\mathcal{I}}(a) \leq b \\ \emptyset & \text{, elsewhere} \end{cases} .$$

On the real unit interval. If the partial mapping $\mathcal{I}(a,\cdot)$ is continuous, then the solution set of the equation $\mathcal{I}(a,x) = b$ is given by

$$\begin{cases} [\mathcal{C}_\mathcal{I}(a,b), \mathcal{R}_\mathcal{I}(a,b)] & \text{, if } \mathcal{N}_\mathcal{I}(a) \leq b \\ \emptyset & \text{, elsewhere} \end{cases}.$$

A necessary and sufficient solvability condition therefore is $\mathcal{N}_\mathcal{I}(a) \leq b$. For instance, the second partial mappings of the residual implicator $\mathcal{I}_P$ are continuous and the corresponding negator $\mathcal{N}_{\mathcal{I}_P}$ is given by

$$\mathcal{N}_{\mathcal{I}_P}(x) = \begin{cases} 1 & \text{, if } x = 0 \\ 0 & \text{, elsewhere} \end{cases}.$$

This negator is the smallest one possible and is denoted $\mathcal{N}_l$. Hence, the condition $\mathcal{N}_{\mathcal{I}_P}(x) \leq y$ is fulfilled when $x \neq 0$ or $(x,y) = (0,1)$. For the residual implicator $\mathcal{I}_M$, only the partial mapping $\mathcal{I}_M(0,\cdot)$ is continuous.

6.6.2 Smallest solution – solvability conditions

Given $A \in \mathcal{F}_L(I)$ and $b \in L$, we want to determine the solution set of the equation

$$\inf_{i\in I} \mathcal{I}(A(i), X(i)) = b \tag{6.33}$$

in the unknown L-fuzzy set X in I. The solution set of this equation can be seen as the intersection of the solution sets of the inequalities

$$\inf_{i\in I} \mathcal{I}(A(i), X(i)) \geq b \tag{6.34}$$

and

$$\inf_{i\in I} \mathcal{I}(A(i), X(i)) \leq b. \tag{6.35}$$

Due to monotonicity considerations, inequality (6.34) always has a solution, namely χ_I, while inequality (6.35) has a solution if and only if $\chi_\emptyset$ is a solution, i.e. if and only if

$$\inf_{i\in I} \mathcal{N}_\mathcal{I}(A(i)) \leq b. \tag{6.36}$$

Compared to conditions (6.4) and (6.21), the above condition is not operator-independent, as it still involves the negator $\mathcal{N}_\mathcal{I}$ induced by $\mathcal{I}$.

On a complete lattice. The following statements can be verified:

- If the second partial mappings of $\mathcal{I}$ are inf-morphisms, then the solution set of inequality (6.34) is given by $[K, \chi_I]$ with K defined by

  $$K(i) = \mathcal{C}_\mathcal{I}(A(i), b). \tag{6.37}$$

  In particular, this result shows us the solution set for equation (6.33) in case $b = 0$.

- If the second partial mappings of $\mathcal{I}$ are inf-morphisms, then the solution set of equation (6.33) is not empty if and only if the fuzzy set K defined by (6.37) is the smallest solution.

Necessary and sufficient solvability conditions. Again, no simple necessary and sufficient solvability condition can be given.

- In any case, condition (6.36) is a necessary condition for the existence of a solution to equation (6.33). If L is a complete chain and the second partial mappings of $\mathcal{I}$ are maximally surjective homomorphisms, then condition (6.36) becomes also a sufficient condition; if L is also dense, then the condition of maximal surjectivity can be omitted.

- If the second partial mappings of $\mathcal{I}$ are maximally surjective homomorphisms, then the condition

$$(\exists k \in I)(\mathcal{N}_{\mathcal{I}}(A(k)) \leq b), \tag{6.38}$$

  a stronger version of condition (6.36), is a sufficient condition for solvability.

On the real unit interval. For a border implicator with continuous second partial mappings, the solution set of equation (6.33) is not empty if and only if condition (6.36) holds. If the solution set is not empty, then the fuzzy set K defined by (6.37) is the smallest solution.

6.6.3 Complete solution set

In contrast to sup-$\mathcal{T}$ equations and left inf-$\mathcal{I}$ equations, right inf-$\mathcal{I}$ equations do not have a greatest solution, but a smallest one. For the representation of the complete solution set, we therefore need the dual structure of a root system called *crown*.

Crowns. A subset C of an ordered set $(P, \leq)$ is called a *crown* if there exists an element σ in P and an antichain O in $\uparrow \sigma = \{x \in P \mid x \geq \sigma\}$ such that

$$R = \bigcup_{\omega \in O} [\sigma, \omega].$$

The corresponding element σ, called *stem* and antichain O, called the set of *offshoots*, are unique. A crown is called *finitely generated* if the set of offshoots is finite. The stem is the smallest element and the offshoots are the maximal elements of the crown. In the framework of complete lattices, the following results hold, see De Baets (1995c, 1998b):

- Let $(C_i)_{i\in I}$ be a family of finitely generated crowns of a complete lattice $(L, \leq)$ with stem σ_i and set of offshoots O_i. If the intersection $\bigcap_{i\in I} C_i$ is not empty, then it is a crown with stem $\sigma = \sup_{i\in I} \sigma_i$ and as offshoots the maximal elements of the set

$$\{\inf_{i\in I} \omega_i \mid (\forall i \in I)(\omega_i \in O_i \wedge \omega_i \geq \sigma)\}. \tag{6.39}$$

- If the intersection of a finite family of finitely generated crowns of a complete lattice is not empty, then it is a finitely generated crown.

The following results are obtained by studying the solution sets of both inequalities (6.34) and (6.35) and combining the results.

On a distributive, complete lattice with meet-irreducible and meet-decomposable elements. We consider again a distributive, complete lattice $(L, \leq)$. For a finite index set I, it holds that:

- **The case of a meet-irreducible right-hand side.** If b is meet-irreducible, the second partial mappings of $\mathcal{I}$ are homomorphisms and the solution set of equation (6.33) is not empty, then it is a finitely generated crown with stem K defined by (6.37) and as set of offshoots the set $\{M_k \mid \mathcal{N}_\mathcal{I}(A(k)) \leq b \wedge M_k \supseteq K\}$, where M_k is defined by

$$M_k(i) = \begin{cases} \mathcal{R}_\mathcal{I}(A(k), b) & \text{, if } i = k \\ 1 & \text{, elsewhere} \end{cases} ; \tag{6.40}$$

  moreover, if the second partial mappings of $\mathcal{I}$ are maximally surjective, then the set of offshoots can be written as $\{M_k \mid \mathcal{N}_\mathcal{I}(A(k)) \leq b\}$.

- **The case of a meet-decomposable right-hand side.** If b has a meet-decomposition $b = \inf_{j\in J} b_j$, the second partial mappings of $\mathcal{I}$ are homomorphisms and the solution set of equation (6.33) is not empty, then it is a crown with stem K defined by (6.37) and as offshoots the *maximal* elements of the set

$$\{\bigcap_{j\in J} M_k^j \mid \mathcal{N}_\mathcal{I}(A(k)) \leq b_j \wedge M_k^j \supseteq K\},$$

  where M_k^j is defined by

$$M_k^j(i) = \begin{cases} \mathcal{R}_\mathcal{I}(A(k), b_j) & \text{, if } i = k \\ 1 & \text{, elsewhere} \end{cases} ; \tag{6.41}$$

  if the meet-decomposition is finite, then this crown is also finitely generated and if $\mathcal{C}_\mathcal{I} \leq \mathcal{R}_\mathcal{I}$, then the offshoots are the *maximal* elements of the set

$$\{\bigcap_{j\in J} M_k^j \mid \mathcal{N}_\mathcal{I}(A(k)) \leq b_j\}.$$

On the real unit interval. For a border implicator $\mathcal{I}$ with continuous second partial mappings and a finite index set I, the solution set of equation (6.33) is not empty if and only if condition (6.38) holds. If the solution set is not empty, then it is the finitely generated crown with stem K defined by (6.37) and as set of offshoots the set $\{M_k \mid \mathcal{N}_\mathcal{I}(A(k)) \leq b\}$, where M_k is defined by (6.40). The case of the real unit hypercube can be treated similarly.

Example 9 Consider equation (6.33) with $I = I_6$,

$$A = (\ 0.8\ 0.3\ 0.4\ 0.3\ 0.2\ 1\ )$$

and $b = 0.6$. Since $\min_{i \in I_6} \mathcal{N}_s(A(i)) = 0 < b$, the solution set is not empty for any border implicator with continuous second partial mappings that induces the standard negator $\mathcal{N}_s$, and has a smallest solution K and three maximal solutions M_1, M_3 and M_6. For $\mathcal{I} = \mathcal{I}_{P,\mathcal{N}_s}$, these are given by

$$\begin{aligned} K &= (\ 0.5\ 0\ 0\ 0\ 0\ 0.6\ ) \\ M_1 &= (\ 0.5\ 1\ 1\ 1\ 1\ 1\ ) \\ M_3 &= (\ 1\ 1\ 0\ 1\ 1\ 1\ ) \\ M_6 &= (\ 1\ 1\ 1\ 1\ 1\ 0.6\ ) \end{aligned}$$

Since $\mathcal{N}_l \leq \mathcal{N}_s$, it also holds that $\min_{i \in I_6} \mathcal{N}_l(A(i)) = 0 < b$. Hence, for the border implicator $\mathcal{I} = \mathcal{I}_P$, the equation has a smallest solution K and 6 maximal solutions:

$$\begin{aligned} K &= (\ 0.48\ 0.18\ 0.24\ 0.18\ 0.12\ 0.6\ ) \\ M_1 &= (\ 0.48\ 1\ 1\ 1\ 1\ 1\ ) \\ M_2 &= (\ 1\ 0.18\ 1\ 1\ 1\ 1\ ) \\ M_3 &= (\ 1\ 1\ 0.24\ 1\ 1\ 1\ ) \\ M_4 &= (\ 1\ 1\ 1\ 0.18\ 1\ 1\ ) \\ M_5 &= (\ 1\ 1\ 1\ 1\ 0.12\ 1\ ) \\ M_6 &= (\ 1\ 1\ 1\ 1\ 1\ 0.6\ ) \end{aligned}$$

6.6.4 *Systems of right inf-$\mathcal{I}$ equations*

Given a complete lattice $(L, \leq)$ and a family $(A_s)_{s \in S}$ in $\mathcal{F}_L(I)$ and a family $(b_s)_{s \in S}$ in L, we want to determine the solution set of the system of equations $(E_s)_{s \in S}$

$$\inf_{i \in I} \mathcal{I}(A_s(i), X(i)) = b_s \tag{6.42}$$

in the unknown L-fuzzy set X in I.

Smallest solution. If the second partial mappings of $\mathcal{I}$ are inf-morphisms, then the solution set of system (6.33) is not empty if and only if the fuzzy set $K = \bigcup_{s \in S} K_s$, with K_s the potential smallest solution of equation E_s, i.e.

$$K(i) = \sup_{s \in S} \mathcal{C}_{\mathcal{I}}(A_s(i), b_s), \tag{6.43}$$

is the smallest solution. No simpler necessary and sufficient solvability conditions are known.

On a distributive, complete lattice with meet-irreducible or finitely meet-decomposable elements. Consider a finite index set I and assume that the right-hand sides of system (6.42) are meet-irreducible or finitely meet-decomposable. If the second partial mappings of $\mathcal{I}$ are homomorphisms and the solution set of system (6.42) is not empty, then it is a crown; if the system is finite, then this crown is finitely generated. Suppose that the solution set is not empty, and that the solution sets of the equations E_s are the crowns with stem K_s and set of offshoots O_s, then the solution set of the system $(E_s)_{s\in S}$ is the crown with stem $K = \bigcup_{s\in S} K_s$ and as offshoots the maximal elements of the set

$$\{\bigcap_{s\in S} N_s \mid N_s \in O_s \wedge N_s \supseteq K\}.$$

A necessary and sufficient solvability condition therefore is that K is a solution.

Example 10 Consider the system of right inf-$\mathcal{I}$ equations (6.42) on the real unit interval, with $I = I_7$, $S = I_5$,

$$A = \begin{pmatrix} 0.26 & 0.24 & 0.34 & 0.51 & 0.22 & 0.16 & 0.92 \\ 0.44 & 0.89 & 0.68 & 0.66 & 0.46 & 0.39 & 0.42 \\ 0.56 & 0.34 & 0.18 & 0.74 & 1.00 & 0.44 & 0.48 \\ 0.48 & 1.00 & 0.23 & 0.37 & 0.83 & 0.36 & 0.16 \\ 0.62 & 0.56 & 0.49 & 0.52 & 0.14 & 0.38 & 0.29 \end{pmatrix} \qquad B = \begin{pmatrix} 0.38 \\ 0.38 \\ 0.38 \\ 0.52 \\ 0.38 \end{pmatrix}.$$

For $\mathcal{I} = \mathcal{I}_P$, this system has no solution. For $\mathcal{I} = \mathcal{I}_{W,\mathcal{N}_s}$, the system has a greatest solution G and two maximal solutions M_1 and M_2 given by

$$\begin{aligned} K &= (\ 0 \ \ 0.52 \ \ 0.06 \ \ 0.12 \ \ 0.38 \ \ 0 \ \ 0.3\ ) \\ M_1 &= (\ 0 \ \ 1 \ \ 0.06 \ \ 1 \ \ 0.38 \ \ 1 \ \ 0.3\ ) \\ M_2 &= (\ 0 \ \ 1 \ \ 0.06 \ \ 0.12 \ \ 1 \ \ 1 \ \ 0.3\ ) \end{aligned}$$

6.6.5 *Fuzzy relational equations*

Several types of fuzzy relational equations are related to (systems of) right inf-$\mathcal{I}$ equations. Hereafter, we list the most important ones.

Image equations of type 1. Consider a fuzzy set A in X and a fuzzy set B in Y. The image equation $R_{\triangleleft}(A) = B$ in the unknown fuzzy relation R is equivalent to a family of independent right inf-$\mathcal{I}$ equations in the foresets Ry, $y \in Y$. We then have that for a border implicator $\mathcal{I}$ with right-continuous second partial mappings, the solution set of the above image equation is not empty if and only if the fuzzy relation K defined by

$$K(x,y) = \mathcal{C}_{\mathcal{I}}(A(x), B(y)) \tag{6.44}$$

is the smallest solution. Note that K could also be written as

$$K = A \times B,$$

modelling the Cartesian product of fuzzy sets by means of the conjunctor $\mathcal{C}_\mathcal{I}$. In the case of the real unit interval and using as implicator the residual implicator of a left-continuous t-norm $\mathcal{T}$, this result has been obtained by Miyakoshi and Shimbo (1985). The same situation has also been studied by Gottwald and Pedrycz (1988, 1994). However, they claim that this equation always has a solution provided that the t-norm $\mathcal{T}$ is strictly increasing, which is false.

If the second partial mappings are continuous, then a necessary and sufficient solvability condition is given by

$$\inf_{x\in X} \mathcal{N}_\mathcal{I}(A(x)) \leq \inf_{y\in Y} B(y). \tag{6.45}$$

For a border implicator with continuous second partial mappings, also the complete solution set can be decribed. This has also been done in Di Nola, Sessa, Pedrycz and Sanchez (1989) in the case of finite universes, a complete Brouwerian lattice and using as implicator the residual implicator $\mathcal{I}_\frown$ of the meet operation $\frown$. Note that the second partial mappings of $\mathcal{I}_\frown$ are not necessarily sup-morphisms, and can impossibly all be maximally surjective (unless we are working on a chain with two elements). These results are among the few that do not follow from the general framework described here; its proof is based on the explicit expression of $\mathcal{I}_\frown$. Of course, it is again possible to describe the solution of a single right inf-$\mathcal{I}_\frown$ equation and apply the Cartesian product of crowns (in that case, the finiteness of the second universe is even not required).

Image equations of type 2. Consider a fuzzy relation R from X to Y and a fuzzy set B in Y. The image equation $R_\triangleright(A) = B$ in the unknown fuzzy set A is equivalent to a system of right inf-$\mathcal{I}$ equations. For a border implicator $\mathcal{I}$ with right-continuous second partial mappings, the solution set of this equation is not empty if and only if the fuzzy set K defined by

$$K(x) = \sup_{y\in Y} \mathcal{C}_\mathcal{I}(R(x,y), B(y)) \tag{6.46}$$

is the smallest solution, i.e. $K = R^t(B)$, using $\mathcal{C}_\mathcal{I}$ as conjunctor. In the case of the real unit interval and the residual implicator of a left-continuous t-norm, this result was mentioned by Miyakoshi and Shimbo (1985); the case a complete Brouwerian lattice and the implicator $\mathcal{I}_\frown$ was discussed by Di Nola, Sessa, Pedrycz and Sanchez (1989).

Composition equations. Consider a fuzzy relation S from Y to Z and a fuzzy relation T from X to Z. The composition equation $R \triangleright S = T$ in the unknown fuzzy relation R is equivalent to a family of independent image equations $S_\triangleright(xR) = xT$ in the unknown aftersets xR, for all $x \in X$. The equation $R \triangleleft S = T$ in the unknown fuzzy relation S is again solved at the same time.

For a border implicator $\mathcal{I}$ with right-continuous second partial mappings, the solution set of this equation is not empty if and only if the fuzzy relation K defined by

$$K(x,y) = \sup_{z\in Z} \mathcal{C}_\mathcal{I}(S(y,z), T(x,z)) \tag{6.47}$$

is the smallest solution. We can rewrite K as follows:

$$K = T \circ S^t,$$

using $\mathcal{C}_\mathcal{I}$ as conjunctor.

6.7 APPROXIMATE SOLUTION METHODS

The preceding sections were devoted solely to the description of the solution sets of various types of fuzzy relational equations. It should be clear that also systems of fuzzy relational equations can be treated within the same framework. However, as the complexity of these inverse problems grows, it becomes harder to find exact solutions. From a practical point of view, one might not be satisfied with a negative answer when no solutions exist. Therefore, various authors have proposed alternative methods to come up with approximate solutions: "solutions" maximizing some solution index or minimizing some distance function, "solutions" obtained by neural networks or by means of genetic algorithms. We briefly comment on these approaches and give appropriate references.

Solvability indices and structural modifications. In various papers, solvability indices of fuzzy relational equations are discussed, see e.g. Gottwald and Pedrycz (1986, 1988) and Gottwald (1994, 1995). Typically, if a fuzzy relational equation is solvable then its solvability index is equal to 1. In the case of the image equation $R(A) = B$ of type 1, for instance, structural changes on A and B have been proposed when the solvability index is smaller than one in order to obtain an equation which is effectively solvable. One way consists of replacing A and B by A^γ and B^γ, where $A^\gamma(x) = \max(\gamma, A(x))$. In this way, any equation can be transformed into a solvable one (which can be meaningless, e.g. if $\gamma = 1$). The above way of modifying the inputs and outputs was criticized by Yuan and Klir (1993) who proposed alternative structural modifications. They also used the notion of an approximate solution as introduced by Wu (1986). Recent results on solvability indices are due to Perfilieva and Tonis (1997).

Numerical and probabilistic methods. Numerical methods for solving fuzzy relational equations have been studied by Pedrycz (1983). Various techniques, for instance Newton's and quasi Newton's, have been investigated for minimizing the Euclidean distance between $R(A)$ and B, where either R or A are unknown. Another reference is Pedrycz, Czogała and Hirota (1984).

In case a system of fuzzy relational equations has no solution, a way out could be the removal of one or more equations. Statistical methods that can offer a way out have been investigated by Pedrycz (1990a). Probabilistic descent methods were discussed by Ikoma *et al.* (1993).

Neural networks. Saito and Mukaidono (1991) proposed a max-min style single layer neural network whose learning algorithm can run properly only in such a case that there exist exact solutions. Pedrycz (1991a) proposed a single layer neural network and derived a learning algorithm optimizing an index for the given set of input and output fuzzy data. Further work along this line was done by Blanco *et al.* (1994, 1995). A multilayer neural network with a back-propagation learning algorithm for solving yet another type of fuzzy relational equations (see Ohsato and Sekiguchi (1983)) has been developed by Chun and Bien (1993). Both conventional and adaptive gradient based learning methods were proposed by Valente de Oliveira (1993), generalizing Pedrycz pioneering work. Recent work has been done by Li and Ruan (1997).

Genetic algorithms. The use of genetic algorithms for solving fuzzy relational equations was suggested by Sanchez (1993). Results in this direction have been obtained by Negoiță *et al.* (1994). More extensive work on this topic has been done by Rondeau, Dubois, Levrat and Lamotte (1994).

6.8 FURTHER READING

6.8.1 Various generalizations

In the literature, various other types of fuzzy relational equations are considered, corresponding to binary operators $\mathcal{O}$ other than t-norms and implicators. For instance, Di Nola, Pedrycz and Sessa (1988) have considered also sup-$\mathcal{O}$ equations, where $\mathcal{O}$ is an equality operator, and inf-$\mathcal{O}$ equations, where $\mathcal{O}$ is a difference operator. Moreover, Pedrycz (1993) has studied s–t equations, replacing the supremum by a general t-conorm, see also Levrat *et al.* (1993) and Di Nola, Pedrycz and Sessa (1995).

Other studies concern the search for necessary and sufficient conditions for having a unique solution, see e.g. Lettieri and Liguori (1984, 1985), Di Nola and Sessa (1988), Sessa (1989) and Cechlárová (1995), or exactly one minimal solution (e.g. in the case of an image equation $R(A) = B$ of type 2), see e.g. Guo *et al.* (1988), Li (1990) and Imai *et al.* (1997).

6.8.2 Miscellaneous problems

Over the past decades, various other types of inverse problems in fuzzy relational calculus have been considered. They usually concern the search for specific solutions, such as solutions with minimal fuzziness, Boolean solutions, etc., of the equations discussed above. Also other types of equations have been considered, such as fuzzy eigen equations and fuzzy arithmetical equations. We will restrict ourselves to giving some literature pointers. If not stated otherwise, all results here concern the real unit interval and the minimum operator.

Fuzziness of solutions. Sanchez (1977) suggested that the study of the composition equation $R \circ S = T$, for instance, could be enlarged by studying the fuzziness of the solutions. Di Nola and Pedrycz (1982) have established a necessary and sufficient condition for the existence of a Boolean (i.e. crisp) solution with entropy measure equal to zero. Another paper dealing with Boolean solutions is due to Di Nola, Pedrycz and Sessa (1987) Following the suggestion of Sanchez, Di Nola and Sessa (1983a) have developed algorithms which minimize the fuzziness measure of a given solution. These algorithms are used in

Di Nola and Sessa (1983b) for determining solutions with a minimal fuzziness measure. Di Nola and Sessa (1983c) discussed the existence of a solution with the greatest energy measure or with the smallest entropy measure. Further, Di Nola (1984) has characterized all solutions having a minimal or maximal measure of fuzziness measured by means of an assigned functional. Di Nola, Pedrycz and Sessa (1985a) have continued this work for more general types of logical operators.

Eigen fuzzy sets and fuzzy relations. Also the study of eigen fuzzy sets equations was initiated by Sanchez (1978). He presented algorithms for determining the greatest eigen fuzzy set associated with a given binary fuzzy relation R on a finite universe X, i.e. the greatest fuzzy set A for which it holds that $R(A) = A$. The need for this study comes from the field of medical diagnosis, e.g. for finding invariants in therapeutic recommendations. In Sanchez (1981) it is shown that this greatest eigen fuzzy set is the same as the one corresponding to the transitive closure of R. Also, all binary fuzzy relations R keeping a given fuzzy set A invariant, i.e. satisfying $R(A) = A$, are obtained. Another variant of this problem has been studied by Wagenknecht and Hartmann (1986) who were looking for eigen fuzzy sets in a-priori given regions. It is related to their study of fuzzy relational equations with tolerances, i.e. inverse problems of the type $B_1 \subseteq R(A) \subseteq B_2$ in the unknown fuzzy set A, see Wagenknecht and Hartmann (1987, 1988). Another invariance study carried out by Miyakoshi and Shimbo (1987) concerns the determination of all fuzzy relations R for which the solution set of the image equation $R(A) = B$, with fixed B, is the same. Eigen fuzzy sets were discussed in the context of the cut technique by Kitainik (1995). An application of fuzzy eigen vectors was discussed by Kickert (1979).

The decomposition problem and transitive solutions. The decomposition problem of fuzzy relations was first tackled by Di Nola, Sessa and Pedrycz (1985) for a binary L-fuzzy relation on a finite universe X, where L is a finite chain, and can be formulated as follows: given a binary L-fuzzy relation R on X, find a (all) binary L-fuzzy relation(s) Z on X such that $Z \circ Z = R$, using as t-norm the meet operation. They have proposed an algorithm which produces the decomposition or indicates that the given relation is non-decomposable, within a finite sequence of steps. Minimal and maximal solutions of this decomposition problem were discussed by Di Nola, Sessa, Pedrycz and Higashi (1985).

Solutions of fuzzy relational equations on a finite space with different types of transitivity (max-min transitivity, matrices with Schein rank equal to 1) have been characterized by Di Nola, Kołodziejczyk and Sessa (1990). Further work in this direction has been done by Wagenknecht (1995). In connection with the decomposition problem, a very general concept of transitivity was defined in Di Nola, Pedrycz, Sessa and Sanchez (1990). Related to this, the canonical forms and properties of convergence of powers of fuzzy relations were pointed out in connection with the theory of fuzzy relational equations by Di Nola, Martini and Sessa (1992). More work in this direction has been done by Kołodziejczyk (1990). Kołodziejczyk (1988) has also studied relationships between decomposable and transitive fuzzy relations and has dealt with the decomposition problem of transitive fuzzy relations. Various other types of solutions, such as (ir-)reflexive, (anti-)symmetric, complete, etc., solutions have been studied by Drewniak (1995).

Arithmetical equations. Sanchez (1984) has shown that fuzzy arithmetical equations, where the arithmetical operations are defined via Zadeh's extension principle, can be treated within the framework of fuzzy relational equations. Applied to the addition of fuzzy numbers, for instance, this leads to a non-standard subtraction. See also Gottwald (1986), Kawaguchi and Da-Te (1993) and Kawaguchi, Da-Te and Nonaka (1994).

Other problems. Yet other inverse problems have been considered. Tang (1988) presented an algorithm for finding the greatest solution of a fuzzy bilinear equation, i.e. an equation of the type $R_1(A) = R_2(A)$ in the unknown fuzzy set A. Interval-valued fuzzy relational equations have been studied by Wang and Chang (1991).

6.8.3 Implementations

Only few implementations have been reported in the literature. Pappis and Adomopoulos (1991) presented ZBasic code for solving an image equation $R(A) = B$ of type 2. Similarly, composition equations were discussed in Pappis and Adomopoulos (1992). A Pascal library called FRESH (an acronym for Fuzzy Relational Equations Solving Help) for solving systems of sup-$\mathcal{T}$ equations and systems of inf-$\mathcal{I}$ equations and the corresponding fuzzy relational equations on the real unit interval has been developed by De Baets (1995c); an earlier version was reported on in De Baets and Kerre (1993d). The library has been rewritten in MATLAB and extended for equations on the real unit hypercube by Janssens (1996). Other Pascal source code was published by Peeva (1992) (and implemented by Radev) for the image equation $R(A) = B$ of type 2, based on the specific expressions of the residual operators of the minimum operator.

6.8.4 Applications

In many methods of describing a system it is assumed that there exists a functional relation between the input and the output variables of the system. Statistical methods are then widely used for determining an acceptable mathematical model of the process. In many ill-defined processes, decision algorithms may be set up based on numerical and non-numerical (linguistic) kinds of information. One way of modelling such fuzzy systems is by means of fuzzy relational equations. In fact, the idea of fuzzy relational equations used as a model of a system could be useful for a wide class of problems where we are faced with fuzzy information, i.e. economical, social or industrial. Excellent overviews have been written by Pedrycz (1990b, 1991b, 1991c) and Di Nola, Pedrycz, Sessa and Sanchez (1991).

Fuzzy control. In the context of control theory, various aspects of the above interpretation have been explored by Czogała and Pedrycz (1982), such as simultaneous identification and control, stabilization control, control with constraints and generation of control rules of a fuzzy controller. The correspondence of the underlying fuzzy relational equations with difference equations was presented by Czogała and Pedrycz (1984). Further applicational aspects of the identification of fuzzy systems can be found in Pedrycz (1984). Another reference is Yi and Chung (1993).

Approximate reasoning and expert systems. Pedrycz (1985) and Di Nola, Pedrycz and Sessa (1985b) have dealt extensively with algorithms for inference mechanisms in expert systems for which a knowledge base consists of conditional statements. The background of these algorithms is formed by the use of the theory of fuzzy relational equations. Further investigations on the modus ponens connected with fuzzy relational equations with equality operators have been done by Di Nola, Sessa and Pedrcyz (1990). Further investigations are due to Di Nola, Sessa and Pedrycz (1992). The connections with causal reasoning were pointed out by Dubois and Prade (1995).

Clustering. The use of approximate solutions of fuzzy relational equations in the problem of discovering structure in a data set has been exploited by Pedrycz (1988), in particular in the context of hierarchical clustering.

References

Arnould T. and Tano S. (1994a). A rule-based method to calculate the widest solution sets of a max-min fuzzy relational equation, *Int. J. Uncertainty, Fuzziness and Knowledge-Based Systems*, **2**, 247–256.

Arnould T. and Tano S. (1994b). A rule-based method to calculate the widest solution sets of a max-min fuzzy relational inequality, *Fuzzy Sets and Systems*, **64**, 39–58.

Bandler W. and Kohout L. (1980a). Fuzzy relational products as a tool for analysis and synthesis of the behaviour of complex natural and artificial systems, *Fuzzy Sets: Theory and Application to Policy Analysis and Information Systems* (Wang P. and Chang S., eds.), Plenum Press, 341–367.

Bandler W. and Kohout L. (1980b). Semantics of implication operators and fuzzy relational products, *Int. J. Man-Machine Studies*, **12**, 89–116.

Birkhoff G. (1967). *Lattice Theory,* AMS Colloquium Publications, Volume XXV, American Mathematical Society.

Blanco A., Delgado M. and Requena I. (1994). Max-min fuzzy neural networks for solving relational equations, *Mathware & Soft Computing*, **3**, 335–345.

Blanco A., Delgado M. and Requena I. (1995a). Identification of fuzzy relational equations by fuzzy neural networks, *Fuzzy Sets and Systems*, **71**, 215–226.

Blanco A., Delgado M. and Requena I. (1995b). Improved fuzzy neural networks for solving relational equations, *Fuzzy Sets and Systems*, **72**, 311–322.

Cao Z. (1982). The eigen fuzzy sets of a fuzzy matrix, *Approximate Reasoning in Decision Analysis* (Gupta M. and Sanchez E., eds.), North-Holland, 61–62.

Cechlárová K. (1995). Unique solvability of max-min fuzzy equations and strong regularity of matrices over fuzzy algebra, *Fuzzy Sets and Systems*, **75**, 165–177.

Cheng L. and Peng B. (1988). The fuzzy relation equation with union or intersection preserving operator, *Fuzzy Sets and Systems*, **25**, 191–204.

Chun M.-G. and Bien Z. (1993). Neurocomputational approach to solve a convexly combined fuzzy relational equation with generalized connectives, *Fuzzy Sets and Systems*, **57**, 321–333.

Czogała E., Drewniak J. and Pedrycz W. (1982). Fuzzy relation equations on a finite set, *Fuzzy Sets and Systems*, **7**, 89–101.

Czogała E. and Pedrycz W. (1982). Control problems in fuzzy systems, *Fuzzy Sets and Systems*, **7**, 257–273.

Czogała E. and Pedrycz W. (1984). Identificaton and control problems in fuzzy systems, *TIMS/Studies in the Management Sciences*, **20**, 447–466.

Davey B. and Priestley H. (1990). *Introduction to Lattices and Order*, Cambridge University Press.

De Baets B. (1995a). Model implicators and their characterization, *Proc First ICSC Internat. Symposium on Fuzzy Logic* (Zürich, Switzerland) (Steele N., ed.), ICSC Academic Press, A42–A49.

De Baets B. (1995b). An order-theoretic approach to solving sup-$\mathcal{T}$ equations, *Fuzzy Set Theory and Advanced Mathematical Applications* (Ruan D., ed.), Kluwer Academic Publishers, 67–87.

De Baets B. (1995c). *Oplossen van vaagrelationele vergelijkingen: een ordetheoretische benadering*, Ph.D. dissertation, University of Gent, 389 p.

De Baets B. (1995d). Residual operators of implicators, *Proc Third European Congress on Intelligent Techniques and Soft Computing* (Aachen, Germany) (Zimmermann H.-J., ed.), Vol. 1, ELITE, 136–140.

De Baets B. (1996). Disjunctive and conjunctive fuzzy modelling, *Proc Symposium on Qualitative System Modelling, Qualitative Fault Diagnosis and Fuzzy Logic and Control* (Budapest and Balatonfüred, Hungary), 63–70.

De Baets B. (1998a). Sup-$\mathcal{T}$ equations: state of the art, *Computational Intelligence: Soft Computing and Fuzzy-Neural Integration with Applications* (Kaynak O., Zadeh L., Turksen B. and Rudas I., eds.), NATO ASI Series F: Computer and Systems Sciences, Vol. 162, Springer Verlag, 1998, pp. 80-93.

De Baets B. (1998b), Crowns and root systems, submitted.

De Baets B. (1998c), Closure theorems for residual operators of implicators, submitted.

De Baets B. and Kerre E. (1993a). A revision of Bandler–Kohout compositions of relations, *Math. Pannon.*, **4**, 59–78.

De Baets B. and Kerre E. (1993b). Fuzzy relational compositions, *Fuzzy Sets and Systems*, **60**, 109–120.

De Baets B. and Kerre E. (1993c). Triangular fuzzy relational compositions revisited, *Intelligent Systems with Uncertainty* (Bouchon–Meunier B., Yager R. and Valverde L., eds.), North-Holland, 257–267.

De Baets B. and Kerre E. (1993d). FRESH: A Pascal library for solving fuzzy relational equations on the unit interval, *Proc First European Congress on Fuzzy and Intelligent Technologies* (Aachen, Germany) (Zimmermann H.-J., ed.), Vol. 1, ELITE, 537–541.

De Baets B. and Kerre E. (1994a). A representation of solution sets of fuzzy relational equations, *Cybernetics and Systems '94, Proc Twelfth European Meeting on Cybernetics and Systems Research* (Vienna, Austria) (Trappl R., ed.), Vol. I, World Scientific Publishing, 1994, 287–294.

De Baets B. and Kerre E. (1994b). A primer on solving fuzzy relational equations on the unit interval, *Int. J. Uncertainty, Fuzziness and Knowledge-Based Systems*, **2**, 205–225.

De Baets B. and Kerre E. (1995). Fuzzy relations and applications, *Advances in Electronics and Electron Physics* (Hawkes P., ed.), **89**, Academic Press, 1995, 255–324.

De Baets B. and Mesiar R. (1998). Triangular norms on product lattices, *Fuzzy Sets and Systems*, Special Issue "Triangular Norms" (B. De Baets and R. Mesiar, eds.), **104** (1999), 61-75.

De Cooman G. and Kerre E. (1994). Order norms on bounded partially ordered sets, *J. Fuzzy Math.*, **2**, 281–310.

Di Nola A. (1984). On functionals measuring the fuzziness of solutions in relational equations, *Fuzzy Sets and Systems*, **14**, 249–258.

Di Nola A. (1987). Fuzzy equations in infinitely distributive lattices, *Proc Second IFSA World Congress* (Tokyo, Japan), 533–534.

Di Nola A. (1990). On solving relational equations in Brouwerian lattices, *Fuzzy Sets and Systems*, **34**, 365–376.

Di Nola A., Kołodziejczyk W. and Sessa S. (1990). Transitive solutions of relational equations on finite sets and linear lattices. *Proc Third Internat. Conference on Information Processing and Management of Uncertainty in Knowledge-Based Systems* (Paris, France), 119–121.

Di Nola A., Martini G. and Sessa S. (1992). Convergence of powers of a class of transitive matrix and related matrix equations, *Simon Stevin*, **66**, 214–256.

Di Nola A. and Pedrycz W. (1982). Entropy and energy measure characterization of resolutions of some fuzzy relational equations, *BUSEFAL*, **10**.

Di Nola A., Pedrycz W. and Sessa S. (1985a). On measures of fuzziness of fuzzy relation equations with generalized connectives, *J. Math. Anal. Appl.*, **106**, 443–453.

Di Nola A., Pedrycz W. and Sessa S. (1985b). Fuzzy relation equations and algorithms of inference mechanisms in expert systems, *Approximate Reasoning in Expert Systems* (Gupta M., Kandel A., Bandler W. and Kiszka J., eds.), North-Holland, 355–367.

Di Nola A., Pedrycz W. and Sessa S. (1987). Fuzzy relation equations under lsc and usc t-norms and their Boolean solutions, *Stochastica*, **11**, 151–183.

Di Nola A., Pedrycz W. and Sessa S. (1988). Fuzzy relation equations with equality and difference composition operators, *Fuzzy Sets and Systems*, **25**, 205–215.

Di Nola A., Pedrycz W. and Sessa S. (1995). Fuzzy relational structures: The state-of-art, *Fuzzy Sets and Systems*, **75**, 241–262.

Di Nola A., Pedrycz W., Sessa S. and Sanchez E. (1991). Fuzzy relation equations theory as a basis of fuzzy modelling: An overview, *Fuzzy Sets and Systems*, **40**, 415–429.

Di Nola A., Pedrycz W., Sessa S. and Wang P. (1984). Fuzzy relation equation under a class of triangular norms: a survey and new results, *Stochastica*, **8**, 99–145.

Di Nola A. and Sessa S. (1983a). On the set of solutions of composite fuzzy relation equations, *Fuzzy Sets and Systems*, **9**, 275–285.

Di Nola A. and Sessa S. (1983b). On the fuzziness of solutions of σ-fuzzy relation equations on finite spaces, *Fuzzy Sets and Systems*, **11**, 65–77.

Di Nola A. and Sessa S. (1983c). On measures of fuzziness of solutions of composite fuzzy relation equations, *Proc IFAC Fuzzy Information* (Marseille, France), 277–281.

Di Nola A. and Sessa S. (1988). Finite fuzzy relation equations with a unique solution in linear lattices, *J. Math. Anal. Appl.*, **132**, 39–49.

Di Nola A., Sessa S. and Pedrycz W. (1985). Decomposition problem of fuzzy relations, *Int. J. General Systems*, **10**, 123–133.

Di Nola A., Sessa S. and Pedrycz W. (1990). Modus ponens for fuzzy data realized via equations with equality operators, *Int. J. Intell. Sys.*, **5**, 1–14.

Di Nola A., Sessa S. and Pedrycz W. (1992). A study on approximate reasoning mechanisms via fuzzy relation equations, *Int. J. Approx. Reason.*, **6**, 33–44.

Di Nola A., Sessa S., Pedrycz W. and Higashi M. (1985). Minimal and maximal solutions of a decomposition problem of fuzzy relations, *Int. J. General Systems*, **11**, 103–116.

Di Nola A., Sessa S., Pedrycz W. and Sanchez E. (1989). *Fuzzy Relation Equations and their Applications to Knowledge Engineering,* Theory and Decision Library. Series D. System Theory, Knowledge Engineering and Problem Solving, Kluwer Academic Publishers.

Drewniak J. (1984). Fuzzy relation equations and inequalities, *Fuzzy Sets and Systems*, **14**, 237–247.

Drewniak J. (1995). Equations in classes of fuzzy relations, *Fuzzy Sets and Systems*, **75**, 215–228.

Dubois D. and Prade H. (1992). Upper and lower images of a fuzzy set induced by a fuzzy relation: applications to fuzzy inference and diagnosis, *Inform. Sc.*, **64**, 203–232.

Dubois D. and Prade H. (1995). Fuzzy relation equations and causal reasoning, *Fuzzy Sets and Systems*, **75**, 119–134.

Goguen J. (1967). L-fuzzy sets, *J. Math. Anal. Appl.*, **18**, 145–174.

Gottwald S. (1986). Characterization of solvability of fuzzy equations, *Elektron. Inf. verarb. Kybern.*, **22**, 67–91.

Gottwald S. (1994). Approximately solving fuzzy relational equations: some mathematical results and some heuristic proposals, *Fuzzy Sets and Systems*, **66**, 175–193.

Gottwald S. (1995). Approximate solutions of fuzzy relational equations and a characterization of t-norms that define metrics for fuzzy sets, *Fuzzy Sets and Systems*, **75**, 189–201.

Gottwald S. and Pedrycz W. (1986). Solvability of fuzzy relational equations and manipulation of fuzzy data, *Fuzzy Sets and Systems*, **18**, 45–65.

Gottwald S. and Pedrycz W. (1988). On the methodology of solving fuzzy relational equations and its impact on fuzzy modelling, *Fuzzy Logic in Knowledge-Based Systems, Decision and Control* (Gupta M. and Yamakawa T., eds.), North-Holland, 197–210.

Guo S.-Z., Wang P.-Z., Di Nola A. and Sessa S. (1988). Further contributions to the study of finite fuzzy relation equations, *Fuzzy Sets and Systems*, **26**, 93–104.

Higashi M. and Klir G. (1984). Resolution of finite fuzzy relation equations, *Fuzzy Sets and Systems*, **13**, 65–82.

Ikoma N., Pedrycz W. and Hirota K. (1993). Estimation of fuzzy relational matrix by using probabilistic descent method, *Fuzzy Sets and Systems*, **57**, 335–349.

Imai H., Miyakoshi M. and Da-te T. (1997). Some properties of minimal solutions for a fuzzy relation equation, *Fuzzy Sets and Systems*, **90**, 335–340.

Janssens S. (1996). *Een implementatie van oplossingsmethoden van vaagrelationele vergelijkingen over* $[0,1]^n$, M.Sc. thesis, University of Gent, 131 p.

Kawaguchi M. and Da-Te T. (1986). A calculation method for solving fuzzy arithmetic equations with triangular norms, *Proc Second IEEE Conference on Fuzzy Systems* (San Francisco, USA), Vol. I, 470–476.

Kawaguchi M. and Da-Te T. (1994). A necessary condition for solvability of fuzzy arithmetic equations with triangular norms, *Proc. Third IEEE Conference on Fuzzy Systems* (Orlando, USA), Vol. II, 1148–1152.

Kickert W. (1979). Towards an analysis of linguistic modelling, *Fuzzy Sets and Systems*, **2**, 293–307.

Kitainik L. (1995). Cut technique in valued relational systems: mainsprings and applications, *Fuzzy Sets and Systems*, **75**, 143–164.

Kołodziejczyk W. (1988). Decomposition problem of fuzzy relations – further results, *Int. J. General Systems*, **14**, 307–315.

Kołodziejczyk W. (1990). On transitive solutions of σ-fuzzy relation equations describing fuzzy systems, *Int. J. General Systems*, **17**, 277–288.

Lettieri A. and Liguori F. (1984). Characterization of some fuzzy relation equations provided with one solution on a finite set, *Fuzzy Sets and Systems*, **13**, 83–94.

Lettieri A. and Liguori F. (1985). Some results on fuzzy relation equations provided with one solution, *Fuzzy Sets and Systems*, **17**, 199–209.

Levrat E., Dubois G., Bombardier V. and Lamotte M. (1993). Generalisation of the resolution of a fuzzy relational equation, *Proc Second IEEE Conference on Fuzzy Systems* (San Francisco, USA), Vol. II, 1414–1418.

Li J. (1990). The smallest solution of max-min fuzzy equations, *Fuzzy Sets and Systems*, **41**, 317–327.

Li J. (1994). On an algorithm for solving fuzzy linear systems, *Fuzzy Sets and Systems*, **61**, 369–371.

Li X. and Ruan D. (1997). Novel neural algorithms based on fuzzy δ-rules for solving fuzzy relational equations. Part I, *Fuzzy Sets and Systems*, **90**, 11-23.

Miyakoshi M. and Shimbo M. (1985). Solutions of composite fuzzy relational equations with triangular norms, *Fuzzy Sets and Systems*, **16**, 53–63.

Miyakoshi M. and Shimbo M. (1987). Sets of solution-set-invariant coefficient matrices of simple fuzzy relation equations, *Fuzzy Sets and Systems*, **21**, 59–83.

Negoiță M., Fagarasan F. and Agapie A. (1994). Applications of genetic algorithms in solving fuzzy relational equations, *Proc Second European Congress on Intelligent Techniques and Soft Computing* (Aachen, Germany) (Zimmermann H.-J., ed.), Vol. 2, ELITE, 1126–1129.

Ohsato A. and Sekiguchi T. (1982). Method of solving the polynomial form of composite fuzzy relation equations and its application to a group decision problem, *Approximate Reasoning in Decision Analysis* (Gupta M. and Sanchez E., eds.), North-Holland, 33–44.

Ohsato A. and Sekiguchi T. (1983). Convexly combined form of fuzzy relational equations and its applications to knowledge representation, *Proc IEEE Conference on Systems, Man and Cybernetics* (Bombay and New Delhi, India), Vol. 1, 294–299.

Pappis C. and Adomopoulos G. (1991). A computer algorithm for the solution of the inverse problem of fuzzy systems, *Fuzzy Sets and Systems*, **39**, 279–290.

Pappis C. and Adomopoulos G. (1992). A software routine to solve the generalized inverse problem of fuzzy systems, *Fuzzy Sets and Systems*, **47**, 319–322.

Pappis C. and Sugeno M. (1985). Fuzzy relational equations and the inverse problem, *Fuzzy Sets and Systems*, **15**, 79–90.

Pedrycz W. (1983). Numerical and applicational aspects of relational equations, *Fuzzy Sets and Systems*, **11**, 1–18.

Pedrycz W. (1984). An identification algorithm in fuzzy relational systems, *Fuzzy Sets and Systems*, **13**, 153–167.

Pedrycz W. (1985). Applications of fuzzy relational equations for methods of reasoning in presence of fuzzy data, *Fuzzy Sets and Systems*, **16**, 163–175.

Pedrycz W. (1988). Approximate solutions of fuzzy relational equations, *Fuzzy Sets and Systems*, **28**, 183–202.

Pedrycz W. (1990a). Algorithms for solving fuzzy relational equations in a probabilistic setting, *Fuzzy Sets and Systems*, **38**, 313–327.

Pedrycz W. (1990b). Fuzzy systems: analysis and synthesis, from theory to applications, *Int. J. General Systems*, **17**, 139–156.

Pedrycz W. (1991a). Neurocomputations in relational systems, *IEEE Trans. Pattern Analysis and Machine Intelligence*, **13**, 289–297.

Pedrycz W. (1991b). Processing in relational structures: Fuzzy relational equations, *Fuzzy Sets and Systems*, **40**, 77–106.

Pedrycz W. (1991c). Fuzzy modelling: Fundamentals, construction and evaluation, *Fuzzy Sets and Systems*, **41**, 1–15.

Pedrycz W. (1993). *s–t* Fuzzy relational equations, *Fuzzy Sets and Systems*, **59**, 189–195.

Pedrycz W., Czogała E. and Hirota K. (1984). Some remarks on the identification problem in fuzzy systems, **12**, 185–189.

Peeva K. (1985). Systems of linear equations over a bounded chain, *Acta Cybernetica*, **7**, 195–202.

Peeva K. (1992). Fuzzy linear systems, *Fuzzy Sets and Systems*, **49**, 339–355.

Perfilieva I. and Tonis A. (1997). Criterion of solvability of fuzzy relational equations system, *Proc Seventh IFSA World Congress* (Prague, Czech Republic), Vol. I, Academia, 90–95.

Prévot M. (1981). Algorithm for the solution of fuzzy relations, *Fuzzy Sets and Systems*, **5**, 319–322.

Rondeau L., Dubois G., Levrat E. and Lamotte M. (1994). Résolution d'un système d'équations de relation floue par les algorithmes génétiques, *Proc Fifth International Conference on Information Processing and Management of Uncertainty in Knowledge-Based Systems* (Paris, France), 1093–1098.

Rudeanu S. (1974). *Boolean Functions and Equations*, North-Holland, Amsterdam.

Saito T. and Mukaidono M. (1991). A learning algorithm for max-min network and its application to solve fuzzy relation equations, *Proc Fourth IFSA World Congress*, 184–187.

Sanchez E. (1974). *Equations de relations floues*, Thèse Biologie Humaine, Faculté de Médecine de Marseille.

Sanchez E. (1976). Resolution of composite fuzzy relation equations, *Information and Control*, **30**, 38–48.

Sanchez E. (1977). Solutions in composite fuzzy relation equations: application to medical diagnosis in Brouwerian logic, *Fuzzy Automata and Decision Processes* (Gupta M., Saridis G. and Gaines B., eds.), North-Holland, New York, 221–234.

Sanchez E. (1978). Resolution of eigen fuzzy sets equations, *Fuzzy Sets and Systems*, **1**, 69–74.

Sanchez E. (1981). Eigen fuzzy sets and fuzzy relations, *J. Math. Anal. Appl.*, **81**, 399–421.

Sanchez E. (1984). Solution of fuzzy equations with extended operations, *Fuzzy Sets and Systems*, **12**, 237–248.

Sanchez E. (1993). Fuzzy genetic algorithms in soft computing environment, *Proc Fifth IFSA World Congress*, XLIV–L.

Sessa S. (1989). Finite fuzzy relation equations with unique solution in complete Brouwerian lattices, *Fuzzy Sets and Systems*, **29**, 103–113.

Schweizer B. and Sklar A. (1983). *Probabilistic Metric Spaces*, Elsevier.

Tang F. (1988). Fuzzy bilinear equations, *Fuzzy Sets and Systems*, **28**, 217–226.

Valente de Oliveira J. (1993). Neuron inspired learning rules for fuzzy relational structures, *Fuzzy Sets and Systems*, **57**, 41–53.

Wagenknecht M. and Hartmann K. (1986). On the construction of fuzzy eigen solutions in given regions, *Fuzzy Sets and Systems*, **20**, 55–65.

Wagenknecht M. and Hartmann K. (1987). On direct and inverse problems for fuzzy equation systems with tolerances, *Fuzzy Sets and Systems*, **24**, 93–102.

Wagenknecht M. and Hartmann K. (1988). Application of fuzzy sets of type 2 to the solution of fuzzy equations systems, *Fuzzy Sets and Systems*, **25**, 183–190.

Wagenknecht M. (1995). On transitive solutions of fuzzy equations, inequalities and lower approximations of fuzzy relations, *Fuzzy Sets and Systems*, **75**, 229–240.

Wang H. and Chang Y.-C. (1991). Resolution of composite interval-valued fuzzy relation equations, *Fuzzy Sets and Systems*, **44**, 227–240.

Wang H. and Hsu H. (1992). An alternative approach to the resolution of fuzzy relation equations, *Fuzzy Sets and Systems*, **45**, 203–213.

Wu W. (1986). Fuzzy reasoning and fuzzy relational equations, *Fuzzy Sets and Systems*, **86**, 67–78.

Xu W., Wu C. and Cheng W. (1982). An algorithm to solve the max-min fuzzy relational equations, *Approximate Reasoning in Decision Analysis* (Gupta M. and Sanchez E., eds.), North-Holland, 47–49.

Yi S. and Chung M. (1993). Identification of fuzzy relational model and its application to control, *Fuzzy Sets and Systems*, **59**, 25–33.

Yuan B. and Klir G. (1993). On approximate solutions of fuzzy relation equations, *Proc Twelfth NAFIPS Workshop* (Mc Allister M., ed.), 237–241.

Zhao C. (1987). On matrix equations in a class of complete and completely distributive lattices, *Fuzzy Sets and Systems*, **22**, 303–320.

III Uncertainty

7 POSSIBILITY THEORY, PROBABILITY AND FUZZY SETS

MISUNDERSTANDINGS, BRIDGES AND GAPS

Didier Dubois
Hung T. Nguyen
Henri Prade

Abstract: Possibility theory was coined by L.A. Zadeh in the late seventies as an approach to model flexible restrictions constructed from vague pieces of information, described by means of fuzzy sets. Possibility theory is also a basic non-classical theory of uncertainty, different from but related to probability theory. This chapter discusses the basic elements of the theory: possibility and necessity measures (as well as two other set functions associated with a possibility distribution), the minimal specificity principle which underlies the whole theory, the notions of possibilitic conditioning and possibilistic independence, the combination, and projection of joint possibility distributions, as well as the possibilistic counterparts to integration. The relations and differences between this approach and other uncertainty frameworks, and especially probability theory, are pointed out. The difference between probability theory and fuzzy set theory is thus tentatively clarified. Lastly, decision-theoretic justifications of possibility theory are given.

7.1 INTRODUCTION

One of the most controversial issues in uncertainty modelling and information sciences is the relationship between probability theory and fuzzy sets. One of the reasons for controversies is the claim that fuzzy sets have something to say with regard to uncertainty modelling, thus creating a competition between fuzzy and probabilistic approaches. Sometimes the debate has been very harsh because of claims made by fuzzy set advocates that "probability deals with objective uncertainty, while fuzzy sets deal with subjective uncertainty", or yet "probability is a special case of fuzzy sets". The first statement only shows that fuzzy set theory tends to ignore the subjective school of Bayesian probability. The second statement, taken stricto sensu, is wrong, but may it contain some truth if fuzzy sets are cast in the setting of random sets or upper and lower probability. Besides, because fuzzy set models were motivated by reasoning tasks, and Bayesianism oriented toward rational decision-making, they came up with different understandings of what is uncertainty. Indeed, as advocated by Smets (1990) (see also Dubois, Prade and Smets, 1996), it is not clear that uncertainty should be represented in the same way for inference purposes and for decision purposes. This difference in background, along with the presence of dogmatic or overenthusiastic point of views, has led to the big debate between probability and fuzzy sets that seems far from extinct. We claim that, to some extent, controversies exist because of mutual ignorance of each field. It seems that a lot of controversies might have been avoided if protagonists had been patient enough to build a common language and to share their scientific backgrounds.

A key notion that has the potential to solve part of the misunderstanding is possibility theory, which stands at the cross-roads between fuzzy sets and probability theory. The work of Zadeh, started with (Zadeh, 1978a), is an effort to put the intuitive and primitive concept of "possibility" on a firm mathematical ground, suitable for a quantitative analysis of this type of uncertainty, especially in artificial intelligence problems. This was possible with the development of fuzzy logic and fuzzy integral. This is quite similar to two cases in uncertainty analysis, namely:

i) the primitive concept of "chance" was mathematized by A. Kolmogorov. The mathematical theory of probability is based upon the theory of functions and measure theory,

ii) the concept of "evidence" was mathematized by Shafer (1976). The mathematical background is the theory of capacities (or equivalently, the theory of random sets, see appendix).

Possibility theory equips fuzzy sets with the full setting required for a sound comparison with probability: set-functions, conditioning tools, notions of independence, and even decision-theoretic foundations. Possibility distributions are modelled by fuzzy set membership functions and play for possibility theory the role of probability densities for probability theory. A major difference between possibility and probability is that probability is a self-dual notion while possibility is not. Namely the negation of a proposition is improbable if and only if the proposition is probable. On the contrary, the negation of a proposition is impossible if and only if this proposition is necessarily true. Hence in possibility theory we work with pairs of possibility and necessity functions that, as mathematical models of partial belief capture dual notions of plausibility and certainty. The specificity of uncertainty theories that use such pairs is to capture ignorance, or partial lack of information. This is impossible with a single probability function, in general. A second major difference between possibility and probability theories, is that the former can be purely qualitative, non numerical, and yet remain very simple. Comparative probabilities do exist but they are much more complex than numerical probabilities. Moreover it is very difficult to exploit a finite scale of probability degrees because the additivity axiom makes no sense. On the contrary, possibility theory has a simple, although non trivial, purely ordinal version.

Possibility theory drastically enhances the modelling capabilities of fuzzy sets with regards to uncertainty modelling. Fuzzy set theory, as presented in the previous chapters, is a consistent body of mathematical tools that, as such, cannot be fully compared with the setting of probability theory. Although fuzzy sets and probability measures are distinct mathematical objects, several bridges relating them have been proposed, and noticeably the so-called random sets. Besides, mathematical objects that behave like fuzzy sets exist in probability theory (for instance likelihood functions). It does not mean that fuzziness is reducible to randomness. Indeed there are ways of approaching fuzzy sets and possibility theory that owe nothing to probability theory. Interpretations of probability theory are multiple, especially frequentist versus subjectivist views (Fine, 1973); several interpretations of fuzzy sets also exist. Some interpretations of fuzzy sets are in agreement with probability calculus and some are not.

This paper is meant i) to survey the literature pertaining to the "probability versus fuzzy set" debate; ii) to try and overcome misunderstandings that have pervaded the literature; iii) to present essentials of possibility theory, the uncertainty approach that is based on fuzzy sets, emphasizing the existence of several theories of possibility; iv) to display the fruitful and various links existing between probability and possibility theories, suggesting their complementarity rather than their antagonism.

Section 2 addresses some classical misunderstandings between fuzzy sets and probabilities. They must be solved before any discussion can take place. Then we present the basic notions of possibility theory in Section 3. Especially we distinguish between comparative possibility theory, ordinal possibility theory and quantitative possibility theory. In the first case possibility relations on a set of events are considered, whereby some event is considered at least as possible as another. Ordinal possibility measures take value in a finite totally ordered scale. They share with quantitative possibility theory the basic maxitivity axiom, but they differ from each other when it comes to conditioning. Quantitative possibility functions are related to probability theory while ordinal possibility functions are not. Section 4 scans the various probabilistic semantics of quantitative possibility, reviews possibility-probability transformations and tries to clarify some lurking controversies on that topic. Then we discuss operational semantics of possibility distributions in Section 5, ranging from frequentist to subjectivist view of probability bounds to similarity and preference. The last section considers possibilistic decision-making that, in the quantitative setting, is based on the use of Choquet integrals, and in the qualitative setting on Sugeno integrals. The latter justifies the formulation of possibility and necessity of fuzzy events. In conclusion, we briefly mention several subfields of fuzzy set research where fuzzy sets, possibility and and probability are conjointly used.

7.2 SOME MISUNDERSTANDINGS BETWEEN FUZZY SETS AND PROBABILITY

Criticizing fuzzy sets as being nothing but misunderstood, fake or ill-conceived probabilities has been sort of popular in some circles, and still is, although opponents seem to be less numerous as of to-date. Examples of such critical texts coming from probability theory are Stallings (1977), Tribus (1978), French (1984), Cheeseman (1988), Laviolette and Seaman (1994), Laviolette et al (1995), with replies (Bezdek, 1994; Zadeh, 1995). Fuzzy sets have suffered the fate of new-born theories when they do not have solid foundations yet. Criticisms come from long since well-founded theories whose tenants forget that their forerunners had to face similar doubts from the scientific circles of their times. Sometimes ill-founded objections come from scholars that are just ignorant of current developments. It also occurs that some claims made by scientists from the fuzzy camp are unwarranted (like for instance the strange statements that fuzzy sets capture subjective uncertainty while probability represents only objective phenomena, or the erroneous idea that fuzzy sets generalize probabilities). More often than not, critical comments addressed to fuzzy set theory raise interesting questions. Addressing such questions generally leads to improving our undesrstanding of what fuzzy sets are and are useful for, and they lead to better anchoring them in the scientific tradition. This section tries to clarifies various issues pertaining the the ever-lasting debate between fuzzy sets and probabilities.

7.2.1 Membership Function and Probability Measure

The problem of quantifying partial belief is different from the one of modelling vagueness or predicates that come by degrees (see the discussion in Dubois Ostasiewicz and Prade, 1999, Chapter 1 of this volume). Partial belief is basically due to incomplete or conflicting information. If you do not have a complete and precise description of the state of the world, you cannot reply with certainty to all queries pertaining to this world. This problem is one of partial belief only, in our terminology. Fuzzy sets, in their preliminary version, have little to offer for the modelling of partial belief. It is only since the emergence of possibility theory in 1978 that fuzzy sets can be clearly related to this problem. See Section 7.3.

The modelling of vagueness is a problem of representing what is sometimes called lexical imprecision of linguistic terms. It tries to capture the fact that some predicates underlie some gradualness, or some notion of intensity in their relevance to qualify a given situation, and, as such, have no clearcut extensions. In that context, membership functions are only preference profiles induced by linguistic habits.

At the formal level, the relative situation of fuzzy sets and probability is clear. After Zadeh (1965) a fuzzy set F on a universe U is identified to a membership function $F(\cdot) : U \to [0,1]$ and F(u) is the grade of membership of *element u* in F. Here let us restrict to finite universes for simplicity. In contrast, a probability measure P is a mapping $2^U \to [0,1]$ that assigns a number P(A) to each *subset* of U, and satisfies the Kolmogorov axioms (in the finite case):

$$P(U) = 1;\ P(\emptyset) = 0;$$
$$\text{if } A \cap B = \emptyset,\ P(A \cup B) = P(A) + P(B). \tag{7.1}$$

However some risk for confusion is already present in the way the definitions are understood. P(A) is the probability that an ill-known single-valued variable x ranging on U hits the fixed well-known set A. A first misunderstanding is to confuse P(A) with a membership grade in A. However, when we consider a membership grade F(u), the element u is fixed and known and the set is ill-defined, while when we consider P(A), the set A is well defined, but the value of the underlying variable x, to which P is attached, is unknown (and sometimes randomly moves).

F(u) has more similarity with the distribution P({u}) = p(u) (or a probability density in the continuous case), except for two things. First, the normalization condition $\Sigma_{u \in U}$ p(u) = 1 must hold while it is not so for F(·). A set-theoretic calculus for probability distributions has nevertheless been built up under the name of Lebesgue logic (Bennett *et al.*,1992); however the normalization condition make set-theoretic operations difficult to handle in a simple way, and the obtained algebraic structure is not so nice. A second difference lies in the underyling assumption regarding a probability distribution. Namely all values of positive probability are mutually exclusive. They represent what is known about the unique value of a variable. This assumption is not at all made in fuzzy set theory. A fuzzy set is often a conjunct of elements. For instance, in image processing, imprecise regions are often modelled by fuzzy sets. However the pixels in the region are not mutually exclusive (although they do not overlap). Namely the region contains several pixels, not a single unkown one. When the assumption of mutual exclusion of elements of a fuzzy set is explicitly made, then, and only then, the membership function is interpreted as a possibility distribution; this is the case of fuzzy numbers describing the ill-located unique value of a parameter.

Consequently Laviolette and Seaman (1994) are right to question the idea, sometimes found in fuzzy set papers, that fuzzy sets generalize probability. Especially the fact that a probability density can be obtained by suitably renormalizing a membership function is not a valid argument in favor of this thesis, since it does not consider the mutual exclusiveness assumption for values of positive probability not does it account for the remainder of the probability construction (set-functions, conditioning, etc.).

Indeed at the mathematical level the domain of the mapping P is the Boolean algebra 2^U. P is a set-function, while a fuzzy set is not. Moreover, the set of fuzzy sets is $[0,1]^U$, i.e., can never be a Boolean algebra. It has necessarily a weaker structure, depending on the choice of fuzzy set operations. One should not identify (as done for instance by Cooke, 1990) the definition of a fuzzy set connective, e.g., union

$$F \cup G(u) = S(F(u), G(u)) \tag{7.2}$$

with an hypothetic decomposability property

$$P(A \cup B) = S'(P(A), P(B)) \tag{7.3}$$

of the probability measure with respect to two given events A and B. First (7.3) cannot hold for all A and B when P is a probability measure while (7.2) can be assumed for all $u \in U$. The risk of getting confused lies in the fact (due to the Boolean algebra structure of 2^U) that S and S' can both be chosen among a family of semi-groups of the unit interval (i.e., triangular co-norms such as max(a,b), a + b – ab, or min(1, a + b), see Chapter 2 by Fodor and Yager, in this volume). This is because the properties usually required for membership grade combinations (e.g., the associativity of union and the empty set acting as an identity leads to choose S among triangular co-norms), and the properties induced by the Boolean structure of 2^U on degrees of probability are the same (e.g., Dubois and Prade, 1982). Concludingly, using min and max, or triangular norms and co-norms, in a problem is not enough to claim that fuzzy sets are at work ; similarly the fact that sometimes $P(A \cup B) = \max(P(A), P(B))$ (e.g., if $A \subseteq B$) is not enough to claim that fuzzy sets are probabilities. This type of confusion between degrees of probability and membership grades pervades the literature of uncertainty in expert systems (see Duda *et al.* (1981), Horvitz *et al.* (1986) for instance). However fuzzy sets can be related to probability in the random set or likelihood function views. Or they can be viewed as standing fully apart from probability theory.

A dual point of view is to start with a poorly structured set of sentences $\mathcal{S}$ and consider various ways of attaching numbers to sentences $\phi \in \mathcal{S}$, say $v(\phi) \in [0,1]$. The question is then to analyze how the structure of $\mathcal{S}$ under the logical connectives AND, OR, NOT is mapped on the unit interval through the valuation mapping v. Gaines (1978) has shown that according to the axiomatic system satisfied by the connectives in $\mathcal{S}$ and the valuation v, one may find a particular representation of $\mathcal{S}$ as 2^U and $v(\phi) = P([\phi])$ where $[\phi]$ is the subset of worlds in U where ϕ is true, and another representation where $\mathcal{S} = [0,1]^U$ and $v(\phi) = [\phi](u)$ where u is fixed in U, and ϕ is a fuzzy statement. Both fuzzy sets and probability can thus be cast in a more general framework of weighted logics. Only in such an abstract setting does it make sense to compare, say Bellman and Giertz (1973) fuzzy set axioms and Kolmogorov axioms. However this framework is too general to be used in practice. It covers both the case where the set of sentences is made of Boolean propositions and the weights are probabilities and the case where it is made of fuzzy propositions whose degrees of truth are evaluated in some fixed possible world. Beyond its mathematical and philosophical interest, Gaines' setting should not however lead us to confuse degrees of probability and membership grades. This confusion is still latent in subsequent works e.g., Blockley (1985), Kovalerchuk and Shapiro (1988)).

7.2.2. Fuzzy Relative Cardinality and Conditional Probability

It F is a fuzzy set on U, its cardinality is $|F| = \Sigma_u F(u)$. A well-known index of inclusion of F in another fuzzy set G is (e.g., Dubois and Prade, 1982c)

$$I(F,G) = \frac{|F \cap G|}{|F|} \tag{7.4}$$

where intersection is defined as minimum. Kosko (1990) has pointed out the analogy between I(F,G) and a conditional probability P(B | A) where B and G (resp. A and F) play the same role. He even noticed that the analogy carries over to Bayesian inference, by noticing that

$$I(F,G) \geq \frac{I(G,F) \cdot I(U,G)}{I(G,F) \cdot I(U,G) + I(\bar{G},F) \cdot I(U,\bar{G})} \tag{7.5}$$

where I(U,G) is nothing but the relative cardinality $\frac{|G|}{|U|}$. Indeed, (7.5) reminds us of Bayes theorem

$$P(B \mid A) = \frac{P(A \mid B) \cdot P(B)}{P(A \mid B) \cdot P(B) + P(A \mid \bar{B}) \cdot P(\bar{B})}. \tag{7.6}$$

where A and B are Boolean sets. The inequality in (7.5) is related to the following one: $I(F,G) + I(F,\bar{G}) \geq 1$, due to the lack of excluded middle law and the use of min for $\cap$ in (7.4). The equality holds in (7.5) when G is a crisp set. It would hold as well with $\cap$ expressed as a product in (7.4). Kosko (1990) claims that (7.5) implies Bayes theorem when both G and F are crisp. This claim is debatable because in that case (7.4) is only a special kind of conditional probability, namely $P(B \mid A) = \frac{|A \cap B|}{|A|}$ provided that P is uniformly distributed on U.

A natural way to generalize both (7.5) and (7.6) is to consider a probability measure P on U, and $[0,1]^U$ as a set of fuzzy events (Zadeh, 1968). Then

$$P(F) = \Sigma p(u)F(u), \tag{7.7}$$

is the probability of the fuzzy event F. The natural extension of (7.4) is

$$P(G \mid F) = \frac{P(F \cap G)}{P(F)} \tag{7.8}$$

and (7.5) can be generalized changing I(X,Y) into P(Y | X). (7.8) generalizes both a conditional probability and the fuzzy inclusion index advocated by Kosko (1990). Note that a whole literature on conditional probabilities of fuzzy events exists, starting with Okuda *et al.* (1978). See Dubois, Prade and Yager (1993), and especially Gebhardt *et al.* (1998) in the companion volume of the Handbook of Fuzzy Sets for a list of basic references. See Young and Wang (1998) for more general, Choquet-integral based methods for conditioning non-additive measures using fuzzy events.

Concludingly, the comparison between fuzzy relative cardinality and conditional probability should not again lead us to conclude with Kosko(1990) that «probability is a special case of fuzziness», but hints on probability of fuzzy events, a situation where, as Pearl (1988) himself admits, probability and fuzzy sets are orthogonal.

7.2.3 Fuzzy Sets Can Be Cast in Random Set Theory

Several scholars such as Fortet and Kambouzia (1976), Orlov (1978), Kampé de Fériet (1982), Sales (1982), Goodman (1982), Wang and Sanchez (1982), Dubois and Prade (1982c), Wang (1983), Nguyen (1984), have pointed out that given a random set $\mathcal{R}$ on U, i.e., a set of subsets $A_1, \ldots, A_n \subseteq U$, and a mass assignment m such that $m(A_i) = m_i \in (0,1]$ and $\Sigma_{i=1,n}\, m_i = 1$, then a membership function could be defined as

$$F(u) = \Sigma_{u \in A_i}\, m_i. \tag{7.9}$$

Viewing $\{(A_i,m_i),\ i = 1,n\}$ as a random set $\mathcal{F}$ with realizations A_i, the membership grade F(u) is then interpreted as the probability that $u \in \mathcal{F}$. Note that again the element u is perfectly known, but the crisp representative of $\mathcal{F}$ is unknown. This kind of probabilistic interpretation of a membership function seems to go back to E. Borel, as pointed out by Godal and Goodman (1980). We emphasize the term "interpretation": this interpretation should not mean that the two mathematical concepts are equivalent! For example, the concept of (Choquet) capacities has nothing to do with probability, or random elements; the Choquet's theorem stating that an infinitely alternating capacity characterizes a random set (see, e.g., Mathéron, 1975), does not mean that there is something random within the theory of capacities in potential theory! See also, the interpretation of Shafer (1976) belief functions in terms of random sets (Nguyen, 1978a).

In the infinite setting the random set interpretation of a fuzzy set F goes as follows. Let $F: U \to [0,1]$ be a fuzzy set. Let $(\Omega, \mathcal{A}, P)$ be a probability space and $\alpha: \Omega \to [0,1]$ be a random variable, uniformly distributed on [0,1]. Consider the random set $F_\alpha: \Omega \to 2^U$ (power set of U) defined by $F_{\alpha(\omega)} = \{u: F(u) \geq \alpha(\omega)\}$ (that is, we randomly select the α-cuts of F), then, after (Höhle, 1976):

$$F(u) = P\{\omega: u \in F_{\alpha(\omega)}\} = \int_0^1 F_\alpha(u)\, d\alpha., \tag{7.10}$$

where $F_\alpha(u) = 0$ or 1 according to whether $F(u) < \alpha$ or not. Li (1997) has studied a generalized version of this equation, when $\alpha: \Omega \to [0,1]$ is a general random variable.

Thus, for a given $F(\cdot)$, there exists some random set $\mathcal{R}$ (not unique), such that the *one-point-coverage function* of $\mathcal{R}$, namely $u \to P(u \in \mathcal{R})$ is precisely F. In general, without additional assumptions on a membership function F, we know, through Goodman (1982)'s work, that F corresponds to the one-point-coverage function of more than one random set. In other words, F only specifies *a class* of random sets. The knowledge of F is not sufficient to determine completely a distribution of a random set. This weaker form of randomness is similar to the well-known situation in probability theory, namely the problem of moments: the knowledge of moments (mean, second moment, etc.) is not sufficient to determine the probability law of the random variable (or vector) involved. Note that while the set of level-cuts is nested, the other random sets whose one-point coverage function coincide with F are generally not nested. Liang and Song (1996) show the coherence between the random set view of fuzzy sets and the idea of average membership function over a population. They consider an experiment where for a given membership grade λ, each individual ω in the population would supply a level cut $F_\lambda{}^\omega$. Using (7.10) one obtains a fuzzy set synthetizing the view of the level-cut F_λ for the population. Integrating over λ yields the average membership function one would obtain via pointwise averaging if each individual in the population supplied a genuine membership function instead of a level-cut.

7.2.4 Membership Functions as Likelihood Functions

Another point of view is to interpret the membership function of a fuzzy set as a likelihood function. This idea is actually quite old in fuzzy set theory (e.g., an early Russian paper by Loginov (1966), whose English translation talks about "hussy" sets!). It has been the basis of experimental methods for constructing membership functions (See also chapter 3, by Bilgic and Turksen, of this volume). Given a population of individuals and a fuzzy concept F, each individual is asked whether a given element $u \in U$ can be called an F or not. The likelihood function P('F' | u) is then obtained and represents the proportion of individuals that answered yes to the question. 'F' is then the corresponding non-fuzzy event. Then it is natural to let

$$F(u) = P('F' \mid u),\ \forall\, u \in U. \tag{7.11}$$

This view is more convincing than the identification between a membership functon and a probability distribution. Indeed no normalization constraint applies to the likelihood values $\{P('F' \mid u), \forall\, u \in U\}$. This type of yes-no experiments have been used by Hersh and Caramazza (1976), and Hisdal (1991). Cheeseman (1988) claims that on such grounds fuzzy sets are nothing new. See also Stallings (1977) for an older, but similar critique. A systematic analysis of fuzzy sets and their connectives, in the framework of conditional probabilities is carried out by Mabuchi (1992, 1997) with the view to preserve all structural properties of sets.

Here, we suggest that this link leads to a cross-fertilisation of fuzzy set and likelihood theories, provided that one does not stick to a dogmatic Bayesian position (see Manton *et al.*, 1994).

Thomas (1979, 1995) indicates that the Bayesian inference procedure that exploits an a priori probability p on U and the probability of observation O when the value of the state variable is u, say

$$p(u \mid O) = \frac{P(O \mid u) \cdot p(u)}{P(O)} \tag{7.12}$$

can be reinterpreted in terms of fuzzy observations. For instance it p represents the base rate of the size of some population, and one learns the concerned individual is "tall", where "tall" is defined by membership function μ_{tall}, then the a posteriori probability can be computed in accordance with (7.8) as follows

$$p(u \mid tall) = \frac{\mu_{tall}(u) \cdot p(u)}{P(tall)} \tag{7.13}$$

where $\mu_{tall}(u)$ is interpreted as a likelihood function, and $P(tall) = \int \mu_{tall}(u)\, dP(u)$ is Zadeh (1968)'s probability of a fuzzy event (see also Dubois and Prade, 1988b).

The two probability-oriented views (random sets and likelihood functions) of fuzzy sets and possibility distributions are not antagonistic and can be reconciled. For instance, the random set view (7.9) corresponds to another experiment for constructing membership functions whereby individuals are asked to point out a single crisp subset $A_i \subseteq U$ that represents some fuzzy concept F best. The weight m_i represents the proportion of individuals for which F is best described by A_i. It makes sense to relate this experiment with the likelihood Yes-No experiment provided that, when each individual chooses A_i as representing F, it means that he would answer yes in the other experiment to the question "is u an F ?" if and only if $u \in A_i$. Then as pointed out in Dubois and Prade (1989):

$$P('F' \mid u) = \Sigma_{u \in A_i}\, m_i.$$

The above results suggest how to get likelihood functions from random sets and confirms the interest of relating fuzzy sets to likelihood functions. Another way of relating the two approaches is to notice that the use of Dempster rule (Shafer, 1976) to intersect two random sets, one of which being a probability function, leads to a probability function that is computed via (7.12) or (7.13) exactly. Namely let $\mathcal{R}_1$ be defined by $\{(A_i,m_i), i = 1,n\}$ and $\mathcal{R}_2 = P$, then the probability function P' obtained by computing the combination of $\mathcal{R}_1$ and $\mathcal{R}_2$ by Dempster rule.

$$\forall\ u \in U,\ P'(u) = \frac{\Sigma_{i:u\in A_i}\, m_i \cdot p(u)}{\Sigma_{u,i:u\in A_i}\, m_i \cdot p(u)} = \frac{F(u) \cdot p(u)}{\sum_{u'\in U} F(u') \cdot p(u')}$$

where F is defined by (7.9). Only the plausibility of singletons (i.e., the possibility distribution) is useful for the combination, and it plays the role of a likelihood function.

7.3 POSSIBILITY THEORY

In 1978, Zadeh published a paper proposing a theory of possibility based on fuzzy sets. The aim was to propose an approach to model flexible restrictions constructed from vague pieces of information, described by means of fuzzy sets. Possibility theory can also be viewed as a non-classical theory of uncertainty, different from probability theory. The comparison between possibility and probability theories is made easy by the parallelism of the constructs, which is not the case for fuzzy sets and probability. There are not so many extensive works on possibility theory. The first book on this topic (Dubois and Prade, 1988) emphasizes the close links between possibility theory and fuzzy sets, and mainly deals with numerical possibility and necessity measures. Klir and Folger (1988) emphasize the links between possibility theory and belief functions, with again a numerical flavor. The recent detailed survey by the authors (Dubois and Prade, 1998) clearly distinguishes between quantitative and qualitative sides of the theory. Basic mathematical aspects of quantitative possibility theory are studied at length by De Cooman (1997). The edited volume by De Cooman *et al.* (1995) reports on both theoretical and applied sides. This section discusses the basic elements of the theory: possibility and necessity measures (as well as two other set functions associated with a possibility distribution), the minimal specificity principle which underlies the whole theory, the combination, projection, and conditioning of possibility distributions. There are three possible settings for possibility theory: the comparative setting where a complete preoder of elements on the set U generates possibility and necessity relations between events; the finite ordinal framework where possibility degrees take values on a finite symbolic totally ordered set, and the numerical setting, where possibility degrees are valued on the unit interval, or a subset thereof. However before presenting these three settings, let us comments on the various meanings of the term "possibility".

7.3.1 *The Meaning of Possibility*

In the same way as probabilities can be interpreted in different ways (e.g., frequentist view vs. subjective view), possibility theory can support various interpretations. Basically there are four ideas each of which can be conveyed by the word 'possibility'. First is the idea of *feasibility*, such as ease of achievement, also referring to the solution to a problem, satisfying some constraints. At the linguistic level this meaning is at work in expressions such as "it is possible *to* solve this problem". Another notion of possibility is that of *plausibility*, referring to the propensity of events to occur. At the grammatical level this semantics is expressed by means of sentences such as "it is possible *that* the train arrives on time". Another view of possibility is logical and it refers to *consistency* with available information. Namely, stating that a proposition is possible means that it does not contradict this information. The last semantics of possibility is deontic, whereby possible means *allowed*, permitted by the law. In this chapter we shall not be concerned with this latter aspect.

The Logical View. Possibility as logical consistency has been put forward by Yager (1980), Dubois and Prade (1980), and Prade (1982). Namely, let a piece of incomplete information be given as a set E such that "$x \in E$" is known for sure where x is some ill-known object. Let set E refer to the information possessed by an agent. It is incomplete insofar as E contains more than one element, i.e., the value of x can be any one (but only one) of the elements in E. Given such a piece of information, a set-function Π_E is built from E by the following procedure

$$\begin{aligned} \Pi_E(A) &= 1 \text{ if } A \cap E \neq \emptyset \ (x \in A \text{ and } x \in E \text{ are consistent}) \\ &= 0 \text{ otherwise } (A \text{ and } E \text{ are mutually exclusive}). \end{aligned}$$

Clearly, $\Pi_E(A) = 1$ means that given that $x \in E$, $x \in A$ is possible because the intersection between set A and set E is not empty, while $\Pi_E(A) = 0$ means that $x \in A$ is impossible knowing that $x \in E$. It is easy to verify that Π_E satisfies the axiom, sometimes called "maxitivity" (Shilkret, 1971), or "f-additivity" (Sugeno, 1974), namely

$$\Pi_E(A \text{ OR } B) = \max(\Pi_E(A), \Pi_E(B)). \tag{7.14}$$

Objective vs. Subjective Possibility. Hacking (1975) also pointed out that possibility can be understood either as an objective notion (referring to properties of the physical world) or as an epistemic one (referring to the state of knowledge of an agent). Feasibility and plausibility can be envisaged from both points of view. Possibility, as objective feasibility, means *physically easy to achieve*, as in the sentence "it is possible *for* Hans *to* eat six eggs for breakfast". Physical possibility has been advocated by Zadeh (1978) so as to justify the axiomatic rule of possibility measures (7.14). The degree of ease of some action that produces A OR B ($A \cup B$, formally speaking), is given by the easiest of two actions, one that produces A and one that produces B. The idea of ease of attainment often comes along with the idea of preference: considering mutually exclusive alternatives, the most feasible(s) one(s) (in some sense) is/are usually preferred. Hence, subjective feasibility can be interpreted as the willingness of an agent to make a decision. Here possibility refers to *preference*. As shown later the whole calculus of possibility theory can be interpreted in the light of preference modeling and optimization theory as much as in terms of partial belief.

The epistemic notion of plausibility, as another side of possibility, has been put forward by Shackle (1961) who proposed a calculus of degrees of "potential surprize" that matches the possibilistic framework. Following Shackle, the potential surprise attached to an event by an agent reflects the level of contradiction between the occurence of this event and the agent's knowledge. Plausibility here means lack of surprize and embodies the above mentioned logical view of consistency with available knowledge. Shackle claimed that decisions are guided at least as much by expectations that look possible as by those that look probable, because probabilities are often not available. This is the subjectivist side of possibility in relation to uncertainty modelling. Actually, this kind of possibility, like probability, has objectivist sides as well. Rather early works (Wang, 1983, Dubois and Prade, 1986) have developed a frequentist view of possibility, which bridges the gap between possibility theory and the mathematics of probability theory. Possibility degrees then offer a simple approach to imprecise (set-valued) statistics, in terms of upper bounds of frequency. However possibility degrees were introduced in terms of ease of attainment and flexible constraints by Zadeh and as an epistemic uncertainty notion by Shackle, without any reference to probability and/or statistics.

Necessity. Plausibility is dually related to certainty, in the sense that the certainty of an event reflects a lack of plausibility of its opposite. This is a striking difference with probability which is self-dual. The expression *It is not probable that "not A"* means that *It is probable that A,* while *It is not possible that "not A"* does not mean *It is possible that A*. It has a stronger meaning: *It is necessary that A*. Conversely, *it is possible that A* does not entail anything about the possibility nor the impossibility of "not A". Here we are dealing with a dual pair possibility/ necessity.

A semantic analysis of *necessity* can be done paralleling the one of possibility. There exists physical necessity as well as epistemic necessity (the latter coincides with the notion of *certainty*). Objective necessity refers to laws of nature: if we drop an object it is necessary that it falls. Subjective necessity viewed as the dual of preference means *priority,* as shown later. Objective necessity refers to lower bounds of frequency measurements, while subjective certainty means belief, acceptance, probable provability. Namely believing a proposition means accepting to reason using this proposition as if it were true, until some new information disconfirms the proposition. A proposition is certain if and only if it logically derives from the available knowledge. Hence necessity also conveys a logical meaning in terms of deduction, just as possibility is a matter of logical consistency (Yager, 1980). The certainty of the event "$x \in A$", knowing that "$x \in E$", is then evaluated by the following index, called necessity measure:

$$N_E(A) = 1 \text{ if } E \subseteq A \\ = 0 \text{ otherwise.} \tag{7.15}$$

Clearly the information $x \in E$ logically entails $x \in A$ when E is contained in A, so that certainty applies to events that are *logically* entailed by the available evidence. It can be easily seen that $N_E(A) = 1 - \Pi_E(\text{not } A)$, i.e., A is necessarily true if and only if 'not A' is impossible.

Qualitative vs. Quantitative. An important question connected to the semantics of possibility is the choice of a formal setting for expressing levels of possibility and necessity. This setting is not unique contrary to probability theory. There are two families of possibility theory: qualitative versus quantitative possibility theories. In quantitative theories, degrees of possibility are numbers and one must indicate where they come from, in order to understand their meaning. In qualitative theories, we can describe possibility by means of a suitable relation between events following the pioneering work of Lewis (1973, 1979); one may also use a totally ordered scale, or a complete lattice of possibility levels attached to events. All theories agree on the maxitivity axiom $\Pi(A \cup B) = \max(\Pi(A), \Pi(B))$, but they diverge on the conditioning and combination operations.

Cognitive studies for the empirical evaluation of possibility theory have recently appeared (Raufaste, Da Silva Neves, 1998). Their experiments suggest "that human experts might behave in a way that is closer to possibilistic predictions than probabilistic ones". The cognitive validation of possibility theory is clearly an important issue for a better understanding of when possibility theory is most appropriate.

7.3.2 Possibility Distributions

A possibility distribution π on a set of possible situations U is a mapping from U to a bounded, linearly ordered valuation set (L, <). This ordered set can be anything, but it may be exemplified by the unit interval [0,1], equipped by the maximum and the minimum operation, and an order-reversing map that is denoted n, such that $n(0) = 1$ and $n(1) = 0$. So, a possibility distribution is not necessarily numerical. In the numerical setting, when L = [0, 1], function n is generally taken as $1 - (\cdot)$. When L is finite, we continue to denote 1 and 0 the top and bottom of L. The referential set U represents what Zadeh calls "universe of discourse", what Shafer (1976) calls a "frame of discernment", and is a set of "states of affairs" (Shackle, 1961), each state being an unambiguous description of a cluster of situations, at a certain level of granularity. In this section, possibility is understood in terms of plausibility. However the mathematical tools presented make sense under the preference view as well. A possibility distribution is often attached to a variable x ranging on U and then denoted π_x. Such a variable x stands for the unknown unique value taken by one or several attributes used to describe states of affairs. Possible values of x are mutually exclusive. For instance, this variable may refer to the age of a man, the size of a building, the temperature of a room, etc. A possibility distribution represents a state *of knowledge* (about the state of affairs) distinguishing what is plausible from what is less plausible, what is the normal course of thing from what is not, what is surprizing from what is expected. The function π_x represents a flexible restriction on the value of x with the following conventions:

$\pi_x(u) = 0$ means that x = u is rejected as impossible;
$\pi_x(u) = 1$ means that x = u is totally possible (= plausible).

Distinct values may simultaneously have a degree of possibility equal to 1. Flexibility in this description is modelled by letting $\pi_x(u)$ between 0 and 1 for some values u. The quantity $\pi_x(u)$ thus represents the degree of possibility of the assignment x = u, some values u being more possible than others. Clearly, if U is the complete range of x, at least one of the elements of U should be fully possible as a value of x, so that

$$\exists u, \pi_x(u) = 1 \text{ (normalization)}.$$

In this paper we consider only normalized possibility distributions. A possibility distribution such that $\pi(u) > 0\ \forall u$, is called non-dogmatic as it does not definitely excludes any value of x.

Strictly speaking a possibility distribution can be viewed as the generalized characteristic function of a fuzzy set (Zadeh, 1965). The fundamental point made by Zadeh (1978) is the following: as set-characteristic functions can be used to express equipossibility, fuzzy set membership functions are the basis of gradual possibility. However this does not mean that the two concepts of a fuzzy set and of a possibility distribution are the same thing. Zadeh's equation $\pi_x(u) = F(u)$ is rather similar to equating a probability function to a frequency (which does not mean that frequency and probability are the same). Indeed $\pi_x(u)$ estimates the possibility that the variable x is equal to u, given the incomplete state of knowledge "x is F", and F(u) estimates the degree of compatibility of the precise information x = u with the statement "x is F".

The above possibility distributions are called absolute, as they refer to a possibility scale, symbolic or numeric. To each possibility distribution π, we can associate its comparative counterpart, a complete preorder denoted by $\geq_\pi$, defined by $u \geq_\pi u'$ if and only if $\pi(u) \geq \pi(u')$, which induces the well-ordered partition (Spohn, 1988) $\{E_1, \ldots, E_n\}$ of Ω, that is, $\{E_1, \ldots, E_n\}$ is a partition of U such that:

$$\forall u \in E_i, \forall u' \in E_j, \pi(u) \geq \pi(u') \text{ iff } i \leq j \text{ (for } n \geq j \geq i \geq 1).$$

The strict part of the complete preorder is denoted $>_\pi$. And $u =_\pi u'$ iff $u \geq_\pi u'$ and $u' \geq_\pi u$. By convention E_1 represents the most normal states of affairs. Thus, a possibility distribution partitions U into classes of equally possible states. Note that each possibility distribution has exactly one comparative counterpart, but a given comparative possibility distribution $>_\pi$ admits of an infinite number of representations on the unit interval. In particular the comparative setting does not lend itself as easily to the expression of impossible values as the absolute setting involving a possibility scale authorizes, since from the only knowledge of the plausibility ordering, it is not clear whether the possibility distribution is dogmatic or not.

7.3.3 Information Content of a Possibility Distribution

A possibility distribution π is said to be *at least as specific* as another π' if and only if for each state of affairs u: $\pi(u) \leq \pi'(u)$ (Yager, 1983). Then, π is at least as restrictive, hence, informative as π'. In the possibilistic framework, extreme forms of partial knowledge can be captured, namely

- *complete knowledge*: for some u_0, $\pi_x(u_0) = 1$ and $\pi_x(u) = 0$, $\forall\ u \neq u_0$ (only $x = u_0$ is possible)
- *complete ignorance*: $\pi_x(u) = 1$, $\forall u$ (all values in U are possible).

Indices measuring specificity are based on the evaluation of the cardinality of the fuzzy set with membership function π when the possibility scale is numerical. See Yager (1992), Sudkamp (1993), Klir and Bo Yuan (1995) and the next chapter, by Klir, for variants of such specificity or non-specificity index. The non-specificity index having the best properties seems to be the so-called U-uncertainty function of Higashi and Klir (1982)

$$U(\pi) = \sum_{i=1}^{n} (\tilde{\pi}(u_{i+1}) - \tilde{\pi}(u_i))\cdot\log i \tag{7.16}$$

where $\tilde{\pi}$ operates a decreasing rearrangement of the possibility distribution π.

Notions of informational distances between possibility distributions can also be defined. Higashi and Klir (1983) defined a basic function $g(\pi, \pi') = U(\pi) - U(\pi')$, and two companion informational distances:

$$G(\pi, \pi') = g(\pi, \max(\pi, \pi')) + g(\pi', \max(\pi, \pi')) = 2U(\max(\pi, \pi')) - U(\pi) - U(\pi')$$
$$H(\pi, \pi') = g(\min(\pi, \pi'), \pi) + g(\min(\pi, \pi'), \pi') = U(\pi) + U(\pi') - 2U(\min(\pi, \pi')).$$

G is a metric while H is not. Properties of these distances were studied and their characterization achieved by Ramer (1990a, b).

Possibility theory is driven by the *principle of minimal specificity*. Given a set of constraints restricting a feasible subset of possibility distributions, the best representative is the least specific feasible possibility distribution, which assigns the highest degree of possibility to each state, since it is also the least committed one. Several examples of this kind of reasoning will be given later. The idea is to assume that whenever some state of affairs is not provably impossible, then admit it is possible. A dual principle may be at work in some situations, namely the one of maximal specificity, when the available knowledge is too weak and minimal specificity destroys the information. Then it is assumed that anything that has not been observably possible is considered impossible. This principle is also at work when a possibility distribution is used as a weak approximation to richer pieces of information: the aim is to waste as little information as possible, as in the case of possibility-probability transformations (see Section 7.4.4)

The concept of specificity can be applied to comparative possibility. A comparative possibility distribution $\geq_{\pi'} = \{E'_1, \ldots, E'_n\}$ is less specific than $\geq_\pi = \{E_1, \ldots, E_n\}$ in the wide sense iff $\forall$ j=1, max(n,n'), $\bigcup_{i=1,j} E_i \subseteq \bigcup_{i=1,j} E'_i$, where for $j > \min(n,n')$ we use $E_j = \emptyset$ when $n < n'$. It is clear that the less specific $\geq_\pi$ is, the more numerous are the elements in the classes E_j of low rank j (high possibility). Hence minimizing specificity comes down to minimizing the number of equivalence classes E_j so as to assign as many worlds as possible to classes of lower ranks. It is obvious that if $\pi \geq \pi'$ for absolute possibility distributions then $\geq_\pi$ is less specific than $\geq_{\pi'}$ in the wide sense. Conversely, if $\geq_\pi$ is less specific than $\geq_{\pi'}$ in the wide sense, then there exists absolute representations π and π' of there orderings such that $\pi \geq \pi'$. But not all absolute representations obey the specificity inequality. An approach to measure the specificity of possibility distributions where possibility degrees belong to an ordinal scale is described by Yager (1990). It evaluates to what extent the fuzzy set representing the possibility distribution contains "at least one element but not much more than one", using fuzzy quantifiers and ordered weighted maximum (see chapter 2 by Fodor and Yager, in this volume).

There exists another natural ordering between comparative possibility distributions, introduced by Yager (1993a, b):

$\geq_\pi$ is said to refine $\geq_{\pi'}$ iff $\forall\, u_1, u_2$ if $u_1 \geq_{\pi'} u_2$ then $u_1 \geq_\pi u_2$.

It means that the relation $\geq_\pi$ is more discriminant than $\geq_{\pi'}$. A maximally discriminant comparative possibility distribution is a linear ordering of U. In the non-dogmatic case, if $\geq_\pi$ refines $\geq_{\pi'}$ then $\geq_{\pi'}$ is less specific than $\geq_\pi$. The converse is false. Maximally discriminant comparative possibility distributions are maximally specific when non-dogmatic (Benferhat *et al.* 1997b). However these results are not valid for absolute possibility distributions in general. For instance, maximally specific possibility distributions are singletons. They are dogmatic and share the referential into only two sets: a singleton and its complement. Conversely, a non-dogmatic, fully discriminant numerical possibility distribution such that $\pi(u_0) = 1$, that generates a linear ordering on U is less specific (in Yager's sense) than the fully informative possibility distribution π' such that $\pi(u_0) = 1$ and 0 otherwise. Numerical evaluations of how discriminant a comparative possibility distribution is are studied by Yager (1993a, b) under the name "buoyancy functions".

These results point out that comparative possibility distributions and absolute ones do not have the same power of expressiveness even if absolute possibility degrees are symbolic.

7.3.4 Possibility and Necessity of Events

In the following, subsets of the set U of states are denoted A, B, C... A^c denotes the complement of A. Given a simple query of the form "does x belong to A?" where A is a prescribed subset of situations, the response to the query can be obtained by computing the partial belief induced on A by the knowledge encoded by the possibility distribution π, noticeably to what extent:

– A is consistent with π, with degree of possibility

$$\Pi(A) = \sup_{u \in A} \pi(u) \tag{7.17}$$

– A is certainly implied by π, with degree of necessity

$$N(A) = n(\Pi(A^c)) = \inf_{u \notin A} n(\pi(u)) \tag{7.18}$$

where n is an order-reversing map on L. In the following we take L = [0, 1] and n(·) is 1 - (·).

These definitions hold for any absolute possibility distributions. In the numerical case where L = [0, 1], $N(A) = 1 - \Pi(A^c)$. Note that $\Pi(U) = 1$ (for normalized possibility measures), and $\Pi(\emptyset) = 0$ by convention. Necessity measures do not appear in the original paper by Zadeh (1978). They were introduced in (Dubois and Prade, 1980; see also Prade, 1979), and mentioned as being certainty functions in Zadeh (1979b). Conversely the possibility distribution can be retrieved as:

$$\pi(u) = \Pi(\{u\}) \tag{7.19}$$

The equations (7.17) and (7.19) spell out the fact that there is one-to-one correspondence between possibility distributions and possibility measures, a fact different from what happens in probability theory. Possibility theory appears as a more direct challenger of probability theory than fuzzy sets, because it also proposes a set-function that quantifies the uncertainty of events. A possibility measure *on a finite set* U is a mapping from 2^U to [0,1] such that

$$\Pi(\emptyset) = 0$$
$$\Pi(A \cup B) = \max(\Pi(A), \Pi(B)). \tag{7.20}$$

This property is called f-additivity by Sugeno (1974) in his general theory of monotonic measures. It is called here maxitivity, following Shilkret(1973). Necessity measures satisfy an axiom dual of the one of possibility measures, namely (Dubois and Prade 1988):

$$N(A \cap B) = \min(N(A), N(B)) \tag{7.21}$$

These finite decomposability axioms presuppose a finite setting in order to be characteristic, and allow to recover the possibility distribution from the possibility measure as $\pi(u) = \Pi(\{u\})$. Otherwise the axiom must hold for infinite families of sets. For instance if $U = \mathbb{N}$, the set of integers, then the set function $\Pi(A) = 1$ if A is infinite and 0 otherwise, is finitely maxitive, but it has no possibility distribution it derives from, since $\Pi(\{n\}) = 0$ for all integers n. In the following, for non numerical possibility distributions, the referential set U is finite unless otherwise stated. The infinite version of the maxitivity axiom for possibility measures (called complete maxitivity) is as follows (Nguyen, 1978b): for any index set I, and any B_i, $i \in I$, $B_i \subseteq U$,

$$\Pi(\bigcup_I B_i) = \sup_{i \in I} \Pi(B_i). \tag{7.22}$$

Then the possibility distribution can be recovered from the possibility measure as $\pi(u) = \Pi(\{u\})$, and $\Pi(A) = \sup_{u \in A} \pi(u)$ (e.g., Wang, 1982). Note that countable maxitivity is not as strong as complete maxitivity.

The degree of possibility $\Pi(A)$ is defined by assuming that, if it is only known that A occurs, then the most plausible situation compatible with A is the one that takes place. It expresses a level of unsurprizingness. For instance, the degree of possibility of the event "the light is off" is computed after the degree of possibility of "someone switched off the light" rather than "the bulb is broken", insofar as the latter is very unusual compared to the former. The basic maxitivity axiom of possibility measures (7.20 or 7.22) is justified by the assumption of jumping to the most plausible situation. A positive degree of necessity $N(A) \geq \alpha > 0$ means that the most plausible situation where A is false is rather impossible, i.e., not possible to a level greater than $n(\alpha)$. These set-functions have been given other names in the past. For instance Nahmias (1978) calls possibility measures *scales*, and Agbeko (1991) calls them optimal measures because of the use of the maximum in the definition. Wang (1982) calls a possibility measure "contactability" of the fuzzy subset induced by the possibility distribution.

Possibility and necessity measures Π and N are set-functions that can provide simple ordinal representations of graded belief. Their particular character lies in their ordinal nature, i.e., the valuation set L may be used only to rank-order the various possible situations in U, in terms of their compatibility with the normal course of things as encoded by the possibility distribution π. As said above, a systematic assumption in possibility theory is that the actual situation is normal, i.e., it is any u such that $\pi(u)$ is maximal given other known constraints. It justifies the evaluation $\Pi(A)$, and contrasts with the probabilistic evaluation of the likelihood of events. Here, "possible" means "unsurprizing", "plausible". Moreover $N(A) > 0$ means that A holds in all the most normal situations. Since the assumption of normality is always made, $N(A) > 0$ thus means that A is an accepted belief, i.e., one may act as if A were true. This assumption is always a default one and can be revised if further pieces of evidence contradict it. As a mathematical confirmation of the acceptance view of necessity measures, Biacino and Gerla (1992) have pointed out that a necessity measure is closed with respect to modus ponens ($N(B) \geq \min(N(A), N(A^c \cup B))$, and this property is characteristic when $N(U) = 1$. Also characteristic is the fact that for any threshold $0 \leq \alpha < 1$ the family of sets $\{A: N(A) > \alpha\}$ is deductively closed (= forms a filter, in the algebraic sense).

Possibility is not Compositional. The controversy between possibility and probability measures is related to the one about fuzzy sets. Cheeseman (1988) criticizes possibility theory because he thinks $\Pi(A \cap B) = \min(\Pi(A), \Pi(B))$ holds. But in fact only the inequality $\Pi(A \cap B) \leq \min(\Pi(A), \Pi(B))$ universally holds, and part of Cheeseman's objections thus make no sense. But his mistake may be due to the dangerous temptation to consider that possibility measures satisfy a companion axiom to (7.20) for conjunction as well (since possibility theory is based on fuzzy sets). This is based on a misunderstanding of the kind pointed cut in Section 7.2.1, i.e., confusing membership functions and set functions.

In fact possibility theory is NOT compositional (contrary to what Pearl (1990 p. 679) supposes). Possibility measures are decomposable with respect to union only, and the degree of certainty or necessity $N(A) = 1 - \Pi(A^c)$, where A^c is the complement of A, is compositional with respect to intersection only, i.e., $N(A \cap B) = \min(N(A),N(B))$. For instance when one is totally ignorant about event A we have $\Pi(A) = \Pi(A^c) = 1$ (and $N(A) = N(A^c) = 0$) while $\Pi(A \cap A^c) = 0$ and $N(A \cup A^c) = 1$. This ability to model ignorance in a non-biased way is typical of possibility theory, while probability theory in the Bayesian approach, cannot account for ignorance, since it forbids interval-valued probabilities. In particular, one might expect that, in a state of total ignorance, all non trivial events should have the same a priori pausibility. This is true for possibility measures deriving from uniform possibility distributions. However this is generally not true for probability measures induced from uniform distributions. Hence the latter are much more informed than some scholars claim. Actually, possibility theory and probability theory, since modelling different kinds of states of knowledge, are complementary theories of uncertainty.

Comparative Possibility. It is not always meaningful to relate uncertainty to frequency. Some types of events can be rare, unrepeatable, or statistical data may simply be unavailable. It does not prevent people from thinking that some events are more possible, probable or certain than others. And indeed the most primitive way of comparing events is to define a complete pre-ordering on 2^U, that expresses comparative possibility, probability or belief. Comparative counterparts of probability have been searched for since the thirties (see Fishburn (1986) for an overview). However no fully satisfying justification of additive probability has been proposed so far, in the finite case. In the case of possibility, the comparative and absolute notions exactly match because of the ordinal nature of the maxitivity axiom. Any comparative possibility distribution $\geq_\pi$ on a finite set U also induces a comparative possibility relation $\geq_\Pi$ defined for any pair of events A and B as follows :

$A \geq_\Pi B$ (resp. $A >_\Pi B$) iff there exists $u \in A$ such that for each $u' \in B$, $u \geq_\pi u'$ (resp. $u >_\pi u'$) and $\forall A \neq \emptyset$, $A >_\Pi \emptyset$.

In other words, $A \geq_\Pi B$ means that there is a situation realizing A which is at least as plausible as any situation realizing B. The relation $\geq_\Pi$ is a complete preordering on the sets of events. Qualitative necessity relations are defined by duality, i.e., $A \geq_N B \Leftrightarrow B^c \geq_\Pi A^c$. Comparative possibility relations were actually introduced by Lewis (1973) in the setting of modal logics of counterfactuals, and independently rediscovered by Dubois (1986) as a counterpart to De Finetti-Savage comparative probalilities. They can be equivalently defined by the following properties proposed by Lewis:

i) completeness ($A \geq_\Pi B$ or $B \geq_\Pi A$);
ii) transitivity;
iii) $U >_\Pi \emptyset$, where $>_\Pi$ is the strict part of the ordering $\geq_\Pi$;
iv) $A \geq_\Pi \emptyset$;
(Π) $A \geq_\Pi B \Rightarrow C \cup A \geq_\Pi C \cup B$ (characteristic property).

For comparative necessities, axiom iv) becomes $U \geq_N A$ and axiom (Π) is changed into :

(N) $A \geq_N B \Rightarrow C \cap A \geq_N C \cap B$ (characteristic property).

In the finite case, Dubois (1986) has shown that the only numerical counterparts to comparative possibility (resp. necessity relations) are possibilisty (resp. necessity) measures: $A \geq B \Leftrightarrow \Pi(A) \geq \Pi(B)$ (resp. $A \geq B \Leftrightarrow N(A) \geq N(B)$) for some absolute possibility measure, if and only if axioms (i-iv) and (Π) (resp., i-iii, $U \geq_N A$, (N)) are satisfied by an ordering of events $\geq$. Qualitative necessity relations are closely related to the epistemic entrenchment relation underlying any revision of a belief set in the sense of Gärdenfors (1988), and comparative possibility relations are used to represent epistemic states by Grove (1988) who represents comparative possibility distributions as systems of spheres, in the tradition of Lewis.

N.B. The problem of extending a preference relation defined on a set to its power set has been extensively investigated by Halpern (1996). He particularly discusses the case where the preference relation $\geq_\pi$ is only partially defined (i.e., there may exist u and u', where neither $u \geq_\pi u'$ nor $u' \geq_\pi u$ hold). He first defines $\geq_\Pi$ on formulas from $\geq_\pi$ in a natural but slightly different, way: $A \geq_\Pi B$ iff $\forall \omega' \in B$, $\exists \omega \in A$ such that $u \geq_\pi u'$. He points out two non-equivalent definitions for the strict part:

- $A >_\Pi B$ iff $\forall u' \in B$, $\exists u \in A$ such that $u >_\pi u'$
- $A >_\Pi B$ iff $A \geq_\Pi B$ but not $B \geq_\Pi A$.

However, both definitions coincide in the case of a complete pre-order.

Guaranteed Possibility. Apart from Π and N, two other quantities can be defined using 'sup' or 'inf', namely (Dubois and Prade, 1992a, Prade and Yager, 1994):

– a measure of 'guaranteed possibility': $\Delta(A) = \inf_{u \in A} \pi_x(u)$ which estimates to what extent *all* the values in A are actually possible for x according to what is known, i.e., each value in A is at least possible for x at the degree Δ(A). Clearly Δ is a stronger measure than Π, i.e., $\Delta \leq \Pi$, since Π only estimates the existence of at least one value in A compatible with the available knowledge, while the evaluation provided by Δ concerns *all* the values in A. Note also that Δ and N are unrelated.

– a dual measure of 'potential certainty': $\nabla(A) = 1 - \Delta(A^c) = \sup_{u \notin A} 1 - \pi_x(u)$ which estimates to what extent there exists at least one value in the complement of A which has a low degree of possibility; this is a necessary condition for having '$x \in A$' somewhat certain (but in general far from being sufficient, except if A^c has only one element). Obviously we have $N \leq \nabla$.

Δ and ∇ are monotonically *decreasing* set-functions (in the wide sense) with respect to set inclusion, e.g., $\Delta(A \cup B) = \min(\Delta(A), \Delta(B))$; it contrasts with Π and N which are monotonically increasing.

Set-functions intermediary between Δ and Π can be defined if instead of computing the degree to which at least one element in A is possible (Π(A)), or all elements of A are possible (Δ(A)) the degree to which *most* elements in A are possible is calculated, where *most* is a fuzzy quantifier. This fuzzy quantifier can be expressed by means of an ordered weighted average (See Chapter 2 by Fodor and Yager in this volume), applied to combine the values $\pi(u)$, $u \in A$. It can also be expressed by means of an order-statistics applied to select the kth greatest in the same set (Prade and Yager, 1994), or an ordered weighed minimum or maximum. See (Dubois, Prade and Yager, 1999) in a companion volume of the Handbooks of Fuzzy sets.

7.3.5 *Joint Possibility, Separability and Non-Interactive Variables*

A joint possibility distribution is a fuzzy restriction on the simultaneous values of several variables $x_1, \ldots, x_m$ (where x_i ranges on domain U_i). Joint possibility distributions on $U_1 \times \ldots \times U_m$ may derive from the statement of links between variables, expressed as fuzzy relations R, i.e., $\pi_{x_1,\ldots,x_m}(u_1, \ldots, u_m) = \mu_R(u_1, \ldots, u_m)$. When several variables $x_1, \ldots, x_m$ are linked together through a restriction encoded by the possibility distribution $\pi_{x_1,\ldots,x_m}$, the possibility distribution π_{x_i} representing the induced restriction on the possible values of x_i, can be calculated as the *projection* of $\pi_{x_1,\ldots,x_m}$ on U_i, defined by (Zadeh, 1975a,b, 1979a):

$$\begin{aligned}\pi_{x_i}(u_i) &= \Pi(U_1 \times \ldots \times U_{i-1} \times \{u_i\} \times U_{i+1} \times \ldots \times U_m) \\ &= \sup_{u_j,\, j\neq i} \pi_{x_1,\ldots,x_m}(u_1, \ldots, u_m) \end{aligned} \tag{7.23}$$

This projection is nothing but the possibility of the event $x_i = u_i$, the other variables being unrestricted. Generally, $\pi_{x_1,\ldots,x_m} \leq \min(\pi_{x_1}, \ldots, \pi_{x_m})$. When equality holds, $\pi_{x,y}$ is then said to be separable, and the variables are said to be *non-interactive* (Zadeh, 1975a,b). Note that projection is also in accordance with the principle of minimal specificity since in (7.23), $\pi_{x_i}(u_i)$ is calculated from the highest possibility value of m-tuples $(x_1, \ldots, x_m)$ where $x_i = u$. Yager (1985a) has generalized this type of projections with fuzzy quantifiers in the finite case. For instance, instead of computing the best value in (7.23), one may compute the kth best.

Consider next the case of two variables x and y with possibility distributions π_x and π_y respectively. The principle of minimal specificity leads us to define the joint possibility distribution $\pi_{x,y}$ as

$$\pi_{x,y} = \min(\pi_x,\pi_y) \tag{7.24}$$

as the greatest solution of the inequalities $\pi_{x,y}(u,v) \leq \pi_x(u), \forall v$ and $\pi_{x,y}(u,v) \leq \pi_y(v), \forall u$. These inequalities express that the possibility of x = u (or y = v) cannot be less than the possibility of both x = u *and* y = v. These inequalities do not account by themselves for relational links between the variables x and y (since they hold $\forall v$ and $\forall u$ respectively). The principle of minimal specificity does not lead to assume any link between x and y, as long as no relation is explicitly stated between x and y. Note that if there is a link between x and y, then (7.24) provides upper bounds on degrees of possibility, i.e., conclusions deduced from (7.24) are always correct but might be less informative than those derived from the actual joint possiility distribution. Indeed some pairs of values (u,v) may be considered as possible using (7.24) while if $\pi_{x,y} < \min(\pi_x,\pi_y)$, they are not so. This upper bounding effect is in agreement with the entailment principle (Zadeh, 1979a) which more generally states that if "x lies in A" then "x lies in B", provided that $A \subseteq B$, or $A(\cdot) \leq B(\cdot)$. Note that while in probability theory, independence plays the same role as non-interactivity in possibility theory (variables can be assumed to be stochastically independent by virtue of the principle of maximum entropy), stochastic independence does not lead to bounding properties as non-interactivity does. This is because stochastic independence assumes an actual absence of correlation while non-interactivity expresses a lack of knowledge about the links between x and y and a lack of commitment as to whether x and y are linked or not. It contrasts with the assumption of independence between sources of knowledge providing π_x, π_y, which, in the numerical case, leads to use a product (instead of the minimum) for combining π_x and π_y, so as to model a reinforcement effect.

The non-interactivity of variables, i.e., $\pi_{x,y} = \min(\pi_x, \pi_y)$ makes possibility and necessity measures decomposable with respect to conjunction and disjunction of logically independent events respectively

$$\Pi(A \times B) = \min(\Pi_x(A), \Pi_y(B))$$
$$N(A + B) = \max(N_x(A), N_y(B))$$

where $\times$ and $+$ denote the Cartesian product and the associated co-product, the latter being $A + B = (A^c \times B^c)^c$, A and B are subsets of the universes U and V (that is the ranges of x and y), Π_x, Π_y, and N_x, N_y are possibility and necessity measures based on a normalized possibility distribution π_x, and π_y respectively.

The most general combination setting is when several possibility distributions $\pi^1, \ldots, \pi^n$, where π^j is now a fuzzy restriction on *a set* of variables $\{x_i \mid i \in I_j\}$, encoding pieces of information that are available. The principle of minimal specificity leads us again to use the min operation

$$\pi = \min_{j=1,n} \pi^j. \tag{7.25}$$

Clearly, (7.24) is a particular case of (7.25) for n = 2, when the possibility distributions pertain to disjoint sets of variables. When combining a joint possibility distribution $\pi_{x_1,\ldots,x_m}$, with other distributions involving other variables say $x_{m+1}, \ldots, x_s$, $\pi_{x_1,\ldots,x_m}$ is extended to the universe $U_1 \times \ldots \times U_m \times U_{m+1} \times \ldots \times U_s$ by means of a cylindrical extension, i.e., for all values $u_{m+1}, \ldots, u_s$,

$$\pi_{x_1,\ldots x_m, x_{m+1},\ldots x_s}(u_1, \ldots, u_m, u_{m+1}, \ldots, u_s) = \pi_{x_1,\ldots,x_m}(u_1, \ldots, u_m).$$

This cylindrical extension is also in accordance with the principle of minimal specificity since it produces the greatest possibility distribution on $U_1 \times \ldots \times U_s$ whose projection is $\pi_{x_1,\ldots,x_m}$. Note that the complete reliability of the sources presupposes that $\sup_{\bar{u}} \min_j \pi_j(\bar{u}) = 1$, where $\bar{u}$ denotes a tuple of instances.

Otherwise, the sources of information providing the π_j's are conflicting as soon as $\pi(\bar{u}) < 1$, $\forall \bar{u}$. Indeed subnormalization corresponds to partial inconsistency since no assignment of the variables is fully compatible with all the information. See Spott (1999) for a reasoned derivation of combination and projection as the basis for possibilistic reasoning. Combination and projection are used in graphical representations of multi-dimensional possibility distributions, using Markov tree-structures (Kruse *et al.*, 1994). Shenoy (1992) develops possibilistic reasoning with the normalized product as a combination rule.

The notions of joint possibility and projection enable properties of indices of non-specificity to be studied. In particular, the U-uncertainty (7.16) of Higashi and Klir (1982) is additive, namely $U(\min(\pi_x, \pi_y)) = U(\pi_x) + U(\pi_y)$. Moreover the informational distance H also introduced in Section 7.3.3 is also additive, that is $H(\min(\pi_x, \pi_y), \min(\pi'_x, \pi'_y)) = H(\pi_x, \pi'_x) + H(\pi_y, \pi'_y)$.

7.3.6 Certainty and Possibility Qualification and the Extension Problem

A possibility distribution can be implicitly specified through the qualification of ordinary subsets of the referential U, either in terms of certainty and possibility. Let A be an *ordinary* subset of U. The statement "it is certain at least at degree α that the value of x is in A", can be interpreted as "any value outside A is at most possible to the complementary degree, namely $1 - \alpha$", in other words, $\forall u \notin A$, $\pi_x(u) \leq 1 - \alpha$, which leads to translate a proposition like "A is α-certain for x" by:

$$\forall u \in U, \pi_x(u) \leq \max(A(u), 1 - \alpha). \tag{7.26}$$

It can be checked that this is equivalent to $N(A) \geq \alpha$. Note that for $\alpha = 1$, the inequality $\pi_x \leq A(\cdot)$ is recovered. When α decreases from 1 to 0, our knowledge evolves from complete certainty in A to acknowledged ignorance about x.

When several sources, that can be considered as reliable, supply pieces of information of the form $N(A_i) \geq \alpha_i$, the possibility distribution that accounts for the whole information is induced by the principle of minimal specificity. Let $\pi_i = \max(A_i(u), 1 - \alpha_i)$ be the possibility distribution supplied by source i. All π's such that $\pi \leq \pi_i$ are compatible with source i, since they are more specific. The least specific distribution induced by all sources is

$$\pi^* = \min_{i = 1,\dots,n} \max(A_i(u), 1 - \alpha_i)$$

and is a faithful representation of the available information. This result is at the basis of the development of the semantics of possibilistic logic; see (Dubois, Lang and Prade, 1994) and the companion volume on Approximate Reasoning and Information Systems edited by Bezdek *et al.* (1999). Biacino and Gerla (1992) provide an algebraic analysis of possibility and necessity measures generated in this way; see also Wang and Klir (1992).

Of course, if the sets A_i have an empty intersection then π^* may fail to be normalized. Even if the sets A_i have at least one common element it many turn out that $N^*(A_i) > \alpha_i$, where N^* is induced by π^*, if other implicit information is present that enforces it. The equality $N^*(A_i) = \alpha_i$ is warranted only if the weights α_i are coherent with the characteristic properties of the necessity measure in the sense of the following property for any subset I of $\{1,\dots n\}$ and any index $j \in \{1,\dots, n\}$:

$$\text{if } \cap_{i \in I} A_i \subseteq A_j \text{ then } \min_{i \in I} \alpha_i \leq \alpha_j.$$

De Baets *et al.* (1998) consider the more general case when the A_i's are fuzzy sets of given possibility levels and solve the corresponding representation problem using fuzzy relational equations.

The statement "A is a possible range for x at least at the degree α" will be understood as $\forall u \in A, \pi_x(u) \geq \alpha$, which leads to translate a proposition like "A is α-possible for x" by:

$$\forall u \in U, \min(A(u), \alpha) \leq \pi_x(u). \tag{7.27}$$

It can be checked that this is equivalent to $\Delta(A) \geq \alpha$. Note that for $\alpha = 1$, (7.27) reduces to $\pi_x \geq A(\cdot)$. The latter is a natural translation of the statement "A is possible" understood as "everything in A is possible". When α decreases from 1 to 0, our knowledge evolves from the certainty that A is the minimal range of our ignorance to a total lack of information whatsoever. This kind of possibility-qualification goes back to Zadeh (1978b) and Sanchez (1978).

The principle of minimal specificity cannot be applied to possibility-qualification. Indeed the constraint $\Delta(A) \geq \alpha$ is satisfied by the possibility distribution modelling total ignorance. This should not be surprizing. The principle of minimal specificity leads to restrictions on variables, which are not more restrictive than what is required by upper bounding constraints. Possibility qualification does not express such a restriction strictly speaking, but a lower bound on possibility degrees. In order to extract non-trivial information we need to apply a principle of *maximal informativeness*, selecting the least possibility distribution that satisfies (7.27):

$$\pi_* = \max_{i = 1,\ldots,n} \min(A_i(u), \alpha_i)$$

As a consequence several possibility-qualified statements will be combined disjunctively by max-operation (and not conjunctively as in (7.25). In case this distribution is subnormalized, any normal possibility distribution that pointwisely dominates it is a possible representation of the information. Note that possibility qualification involves Δ, not Π. Indeed the constraint $\Pi(A) \geq \alpha$ is too weak and even a unique *most* specific solution does not exist since $\forall u_0 \in A$, any possibility distribution such that $\pi_x(u_0) = \alpha$, $\pi_x(u) = 0$ if $u \neq u_0$ is a least solution.

7.3.7 Conditional Possibility and Possibilistic Independence

Conditioning in possibility theory has been studied as a counterpart to probabilistic conditioning. However there is no longer a unique meaningful notion.

Ordinal Conditioning. Let us consider the case of a finite possibility scale. The qualitative notion of conditional possibility measure goes back to Hisdal (1978) who introduced the set function $\Pi(\cdot \mid A)$ through the equality

$$\forall B,\ B \cap A \neq \emptyset,\ \Pi(A \cap B) = \min(\Pi(B \mid A), \Pi(A)). \tag{7.28}$$

In order to overcome the existence of several solutions to this equation, the qualitative conditional possibility measure can be defined, as proposed by Dubois and Prade (1988a), as the least specific solution to this equation, that is, when $\Pi(A) > 0$,

$$\begin{aligned} \Pi(B \mid A) &= 1 \text{ if } \Pi(A \cap B) = \Pi(A) \\ &= \Pi\,(A \cap B) \text{ otherwise.} \end{aligned} \tag{7.29}$$

The only difference with conditional probability is that the renormalisation via division cannot be carried out and is changed into a simple move of the most plausible elements in A to 1. The conditioning equation agrees with a purely ordinal view of possibility theory and makes sense in a finite setting only. Indeed this form of conditioning applied to infinite numerical settings creates discontinuities, and the infinite maxitivity axiom is then not preserved by conditioning (see De Baets *et al.* (1999). Especially $\Pi(B \mid A) = \sup_{u \in B} \pi(u \mid A)$ may fail to hold for *non-compact* events B (De Cooman, 1997). The possibility distribution associated with (7.29) is given by

$$\begin{aligned} \pi(u \mid A) &= 1 \text{ if } \pi(u) = \Pi(A),\ u \in A \\ &= \pi(u) \text{ if } \pi(u) < \Pi(A),\ u \in A \\ &= 0 \text{ if } u \notin A. \end{aligned} \tag{7.30}$$

Note that if $\Pi(A) = 0$ then $\pi(\cdot \mid A)$ is still a solution to (7.28) and is equal to the characteristic function of A, which is simply substituted to π. Ramer (1989) noticed that only one element with possibility $\Pi(A)$ needs to receive possibility 1 after conditioning by A, in order to minimize the change in the possibility values.

The conditional necessity function is defined as $N(B \mid A) = n(\Pi(B^c \mid A))$ by duality. This notion of conditional necessity is akin to the concept of "would counterfactual" following Lewis (1973), while the counterpart of Lewis' "might conditional" is of course $\Pi(B \mid A)$. The careful reader may wonder why conditioning is not directly defined in terms of necessity rather than possibility measures. Namely the equation $N(A \cap B) = \min(x, N(A))$ sounds as natural a starting point as the other one (7.28). The above conditional necessity verifies this equation. However the natural solution to be extracted by the principle of minimal specificity leads to minimize x. The least committed solution is given by $N(B \parallel A) = N(A \cap B)$, and corresponds to a definition of conditional possibility measure $\Pi(B \parallel A) = \Pi(A^c \cup B)$ based on the material conditional. Hence this approach fails to define conditioning as a distinguished concept.

Note that $N(B \mid A) > 0 \Leftrightarrow \Pi(A \cap B) > \Pi(A \cap B^c)$, which expresses that B is provisionally accepted in the context A if and only if B is more possible than B^c

when A is true. Moreover one may have $N(B \mid A) > 0$ and $N(B^c \mid A \cap C) > 0$, a non-monotonicity phenomenon. This remark has led to model default rules with exceptions by means of constraints of the form $N(B \mid A) > 0$ in a purely ordinal setting (see Benferhat et al, 1997a, b; 1998). The idea is to let a generic rule "if A then normally B" mean "B holds in the most normal situations where A is true" (Shoham, 1988). The extent to which a situation u is normal is evaluated by $\pi(u)$, and the most normal situations in A have possibility $\Pi(A)$. Then the constraint $\Pi(A \cap B) > \Pi(A \cap B^c)$ means that all situations in A where B is false are less normal than $\Pi(A)$, so that B is expected to be true in A-worlds.

Modal accounts of qualitative possibility theory involving conditional statements are proposed in Lewis (1973) (this is called the VN conditional logic), Farinas del Cerro and Herzig (1991), Boutilier (1994), Fariñas del Cerro *et al.* (1994), Klir and Harmanek (1994), Hajek *et al.* (1994), Hajek *et al.* (1995).

Qualitative Bayes Theorem. A counterpart of Bayes theorem holds for possibilistic conditioning, namely

$$\min(\Pi(B \mid A), \Pi(A)) = \min (\Pi(A \mid B), \Pi(B)) \tag{7.31}$$

where $(\Pi(B), \Pi(B^c))$ is the a priori knowledge about x on U, $\Pi(A \mid B)$ the possibilistic likelihood of observing A when B is present, and $\Pi(A) = \max(\min(\Pi(A \mid B), \Pi(B)), \min(\Pi(A \mid B^c), \Pi(B^c)))$. It is then possible to compute the posterior possibilities $\Pi(B \mid A)$ and $\Pi(B^c \mid A)$ when A has been observed. The flexibility of the possibilistic approach lies in the convenient unbiased handling of the case of no a priori knowledge about B, that is, $\Pi(B) = \Pi(B^c) = 1$. Then (7.31) becomes

$$\Pi(A \mid B) = \min(\Pi(B \mid A), \max (\min(\Pi(A \mid B), \Pi(B)), \min(\Pi(A \mid B^c), \Pi(B^c)))).$$

Finally, one may let $\Pi(B \mid A) = \Pi(A \mid B)$ in accordance with the use of likelihood functions for abductive reasoning; a qualitative normalization is possible which consists in moving to 1 the greatest possibility degrees and leaving the other degrees unchanged.

More generally, when prior knowledge is available, the confirmation problem in the qualitative setting (checking if $\Pi(B \mid A) > \Pi(B)$ for instance) has been studied by Garbolino (1989), Zwirn and Zwirn (1995) in the scope of epistemology, Dubois and Prade (1992c) in the scope of diagnosis, and Dubois, Fariñas, Herzig and Prade (1997) in the scope of belief revision. In the latter case, the approach seems to be intuitively appealing since it unifies Gärdenfors (1988) revision theory (see Dubois and Prade, 1992d), non-monotonic reasoning and possibility theory (Benferhat *et al.*, 1997a). But the use of this qualitative approach as a substitute to the Bayesian approach for learning purposes meets some difficulties pointed out by the above authors. This is because some of them view ordinal possibilities as if they were numerical. Let us describe confirmation in terms of necessity functions (since possibility values smaller than 1 have a negative interpretation as "not fully possible"). Recall N(B) >0 means that B is a priori accepted, no matter how large N(B) is. However if the value N(B) in the necessity scale is viewed as a measure of the intensity of acceptance of B, the posterior necessity seems to have a counterintuitive behavior, namely it can be proved that either $N(B \mid A) = 0$ (then the assumption B is cancelled), or $N(B \mid A) \geq N(B)$ (confirmation). The case of attenuation ($0 < N(B \mid A) < N(B)$), whereby B is still accepted after observing A, but to a lesser extent, is mathematically impossible. This observation suggests that ordinal possibility theory should be taken for what it is: a qualitative calculus of acceptance. The phenomenon of graceful degradation, useful in learning processes, cannot be captured, but by numerical possibility theories.

Numerical Possibilistic Conditioning. The main difference between numerical and qualitative possibility theories lies in the conditioning process. In case of a numerical valuation set, the use of the product instead of min in the conditioning equation equation (2.28) avoids discontinuity problems. In close agreement with probablity theory, it leads to

$$\forall B, B \cap A \neq \emptyset, \Pi(B \mid A) = \frac{\Pi(A \cap B)}{\Pi(A)} \tag{7.32}$$

provided that $\Pi(A) \neq 0$. Then $N(B \mid A) = 1 - \Pi(B^c \mid A)$. This is formally like Dempster rule of conditioning, specialized to possibility measures, i.e., consonant plausibility measures of Shafer (1976). The corresponding conditional possibility distribution is

$$\pi(u \mid A) = \frac{\pi(u)}{\Pi(A)}, \forall u \in A \tag{7.33}$$
$$= 0 \text{ otherwise.}$$

The possibilistic counterpart of Bayes theorem in the quantitative case looks formally the same as in probability theory:

$$\Pi(B \mid A) \cdot \Pi(A) = \Pi(A \mid B) \cdot \Pi(B).$$

Considering the problem of testing hypothesis A against its complement, upon observing B, the maxitivity axiom leads to an expression of $\Pi(B \mid A)$ in terms of $\Pi(A \mid B)$, $\Pi(A \mid B^c)$, $\Pi(B)$, $\Pi(B^c)$ that differs from the probabilistic form. Namely

$$\Pi(B \mid A) = \frac{\Pi(A \mid B) \cdot \Pi(B)}{\max(\Pi(A \mid B) \cdot \Pi(B), \Pi(A \mid B^c) \cdot \Pi(B^c))}$$

The uninformed case, with uniform possibility $\Pi(B) = \Pi(B^c) = 1$. leads to compute $\Pi(B \mid A) = \dfrac{\Pi(A \mid B)}{\max(\Pi(A \mid B),\ \Pi(A \mid B^c))}$ and compare it to $\Pi(B^c \mid A)$. It corresponds to some existing practice in statistics, called likelihood ratio tests, whereby the likelihood function is renormalized via a proportional rescaling (Edwards, 1972; Barnett, 1973). This updating method does not have the limitation of the qualitative one: the case of attenuation $0 < N(B \mid A) < N(B)$ is now possible. This approach has been recently successfully developed for use in practical applications by Lapointe (1999) (see also Lapointe and Bobée, 1999).

The dual definition of conditioning stems from the necessity-based conditioning equation $N(A \cap B) = N(B \mid A) \cdot N(A)$. But then the conditional necessity and the dual conditional possibility are undefined as soon as $N(A) = 0$, which encompasses the state of total ignorance. This property is very counterintuitive and contrasts with Dempster conditioning which only presupposes $\Pi(A) \neq 0$. Another deficient proposal is based on substituting product with the Lukasiewicz conjunction $\max(0, a + b - 1)$ in the conditioning equation. Then the conditioning possibility takes the form $\Pi(B \mid A) = \min(1, 1 - \Pi(A) + \Pi(A \cap B))$. However it is easy to see that $\Pi(\emptyset \mid A) \neq 0$, in general and this is counterintuitive. See De Baets *et al.* (1999) for a complete mathematical study of this question, leading to the unicity of the product-based possibilistic conditioning, in the infinite setting. The problem of reasoning with conditional possibilities based on minimum or product, namely that of deriving conditional possibilities from constraints on other conditional possibilities based on minimum or product is studied in Maung (1995). He gives an axiomatic justification of the selection of the least informative possibility distribution in the sense of non-specificity.

Yet another definition of numerical possibilistic conditioning can be obtained if we notice that in probability theory, $P(B \mid A)$ is indeed an increasing function of both $P(B \cap A)$ and $P(A^c \cup B)$ and this function is exactly $f(x,y) = \frac{x}{x + 1 - y}$. Then a natural counterpart of Bayesian conditioning in quantitative possibility theory is

$$\Pi(B \mid A) = \frac{\Pi(A \cap B)}{\Pi(A \cap B) + 1 - \Pi(A^c \cup B)} = \frac{\Pi(A \cap B)}{\Pi(A \cap B) + N(A \cap B^c)}. \qquad (7.34)$$

Then, the dual conditional necessity is such that

$$N(B \mid A) = 1 - \Pi(B^c \mid A) = \frac{N(A \cap B)}{N(A \cap B) + \Pi(A \cap B^c)}.$$

This view of conditioning, that we shall call Bayesian possibilistic conditioning (Dubois and Prade, 1996, 1997a; Walley, 1996) is in accordance with De Campos *et al.* (1990) and Fagin and Halpern (1989) proposals for Sugeno measures and belief functions respectively. Bayesian conditioning preserves consonance of possibility measure and the corresponding conditional possibility distribution has support B ($\pi(u \mid A) = 0$ if $u \notin A$) and, if $u \in A$:

$$\pi(u \mid A) = \max\left(\pi(u), \frac{\pi(u)}{\pi(u) + N(A)}\right) \qquad (7.35)$$

which indicates that the result is less specific that π on A and coincides with the characteristic function of A if N(A)= 0. It contrasts with Dempster conditioning which always supplies more specific results than the above. Note that minimum information reasoning is more difficult with the Bayesian possibilistic conditioning than with other ones (See Dubois and Prade, 1997a). This is because the set of possibility measures such that $\Pi(B \mid A) = \alpha$, in the sense of (7.34) has no unique least specific element.

Another type of possibilistic conditioning was proposed very early by Nguyen (1978b), whereby the conditional possibility distribution $\pi_{x|y}(u|v)$ on a Cartesian product UxV is defined from the joint possibility distribution $\pi_{(x, y)}$ via its projections π_x and π_y as:

$$\pi_{x|y}(u|v) = \pi_{(x, y)}(u, v) \cdot \min\left(1, \frac{\pi_x(u)}{\pi_y(v)}\right).$$

The main motivation for this definition seems to be the satisfaction of the equality $\pi_{(x, y)} = \min(\pi_{x|y}, \pi_{y|x})$. This notion has been more recently investigated by Gwét (1997) who applied it to statistics for the description of possibilistic correspondences. See De Cooman (1997, Part 2) for an extensive discussion of various proposals of conditional possibilities.

Conditioning with Uncertain Events. An extension of conditionalization to uncertain events of the form $N(A) \geq \alpha > 0$ has been defined (Dubois and Prade, 1991b, 1992d). The uncertain event is interpreted as a more or less informed piece of fuzzy information F such that F(u) = 1 if $u \in A$, and $F(u) = (1 - \alpha)$ if $u \in A^c$, that may be useful or not for refining the current information according to the value of α; in particular the input information may be discarded if α is too low. It is defined for $\alpha > 0$ as

$$\begin{aligned} \pi(u \mid^+ (A\ \alpha)) &= \pi(u \mid A) \text{ if } u \in A \\ &= \min(\pi(u), (1 - \alpha)) \text{ if } u \in A^c. \end{aligned} \tag{7.36}$$

If $\alpha = 0$, $\pi(u \mid^+(A\ \alpha)) = \pi(u)$ since the F = U, i.e., the possibility distribution is unchanged if the input event is completely uncertain. Of course, $\pi(u \mid^+ (A\ \alpha)) = \pi(u \mid A)$ if $\alpha = 1$. This form of extended conditioning corresponds to a process of knowledge acquisition that discards uninformed inputs.

Another view of conditioning by an uncertain event $N(A) \geq \alpha$ is when the uncertain input acts as a constraint that the resulting possibility distribution must satisfy. It is defined as follows:

$$\begin{aligned} \pi(u \mid (A\ \alpha)) &= \pi(u \mid A) \text{ if } u \in A \\ &= \min(\pi(u \mid A^c), (1 - \alpha)) \text{ if } u \in A^c. \end{aligned} \tag{7.37}$$

Clearly $N(A \mid (A\ \alpha)) = \alpha$. This is a formal qualitative counterpart of Jeffrey's rule of generalized conditioning, that corresponds to a non-symmetric revision process generalizing the conditioning. See (Dubois and Prade, 1997b) for more details and discussions.

This form of conditioning can be applied to quantitative possibility measures, changing minimum into product and using Dempster product-based conditioning instead of the ordinal conditioning. For instance the possibilistic counterpart of Jeffrey's rule is $\pi(u \mid (A\ \alpha)) = \max(\pi(u \mid A), (1 - \alpha) \cdot \pi(u \mid A^c))$.

Independence. Possibilistic independence is a difficult and composite topic whose full elucidation is a big challenge for future research. First, one must distinguish between independence between variables and independence between events. Moreover the situation of independence in quantitative possibility theory is significantly different from the one in qualitative possibility theory.

Consider independence between events. Nahmias (1978) made the following proposal under the name "unrelatedness": A and B are unrelated propositions in Nahmias's sense iff $\Pi(A \cap B) = \min(\Pi(A), \Pi(B))$. This definition is clearly the possibilistic counterpart of the multiplication rule of probability theory. This situation holds if and only if $\Pi(A \cap B) \geq \min(\Pi(A \cap B^c), \Pi(A^c \cap B))$. So, A and B are related means that $A \cap B$ is an implausible event (since it then holds that $\Pi(A \cap B) < \min(\Pi(A), \Pi(B))$), i.e., A and B are (more or less) mutually exclusive. So, independence in the sense of the minimum rule is only a weakening of the logical notion of consistency between A and B, and it can be called "epistemic consistency", although the case when $\Pi(A) = 0$ or $\Pi(B) = 0$ is encompassed. This notion is not very demanding. Moreover it is local in the sense that it is sensitive to negation, contrary to the product-based probabilistic independence: if A and B are epistemically consistent, it does not say anything about the other events A^c and B, A and B^c, B^c and A^c. Enforcing epistemic consistency between all these pairs (non-interactivity between the corresponding propositional variables) corresponds to a generalization of logical independence. De Cooman (1997, Part III) defines the independence between events in this very way, so as to solve the paradox of Nahmias definition whereby A is unrelated with itself.

A notion of qualitative independence has been recently introduced (Dubois, Fariñas, Herzig and Prade, 1994, 1997) based on ordinal conditioning. It is not equivalent, but stronger than the minimum rule, in possibility theory. Namely to define B as independent of A when the conditional measure of B given A is equal to the unconditional measure of B. Here we have two uncertainty functions Π and N. Hence we can define independence either as $\Pi(B|A) = \Pi(B)$ or as $N(B|A) = N(B)$. But since $N(B|A) = N(B)$ is equivalent to $\Pi(B^c|A) = \Pi(B^c)$ one of them is enough. Let us consider the equality $N(B|A) = N(B)$ in order to express the positive statement that A is irrelevant to (the level of acceptance of) B. It turns out that this notion of irrelevance is rather heterogeneous because it expresses the disjunction of two distinct forms of independence according to whether B is apriori believed or not:

(i) $N(B|A) = N(B) = 0$ if and only if
$$1 = \max(\Pi(A^c \cap B^c), \Pi(A \cap B^c)) \text{ and } \Pi(A \cap B^c) \geq \Pi(A \cap B)$$

(ii) $N(B|A) = N(B) > 0$ if and only if
$$\Pi(A \cap B) > \Pi(A \cap B^c) \geq \Pi(A^c \cap B^c).$$

The last requirement may be seen as too strong. Recall that in the ordinal setting the preservation of absolute values does not mean much. Especially the situation where $N(B) > N(B|A) > 0$ is provably impossible. A weaker statement, that is purely ordinal is the following:

B is ordinally independent of A if and only if $N(B) > 0$ and $N(B \mid A) > 0$ hold simultaneously.

It expresses that B is an accepted belief independently of A. It is non-symmetric and sensitive to negation. It can be checked that this notion implies but is stronger than the condition $\Pi(A \cap B) = \min(\Pi(A), \Pi(B))$. These notions have been studied in detail and axiomatized by Dubois, Fariñas *et al.* (1997).

When conditioning is defined in the numerical setting via the product rule, then the product-based symmetric independence notion $\Pi(A \cap B) = \Pi(A) \cdot \Pi(B)$ is trivially equivalent to $\Pi(B|A) = \Pi(B)$ for positive values and is thus much closer to probabilistic independence.

The study of conditional independence between variables is driven by the attempt to represent joint possibility distributions by means of graphical structures similar to Bayesian networks. It enables local computations to be performed, provided that the conditional independence relations satisfy so-called graphoid and semi-graphoid properties (De Campos *et al.*, 1995). Fonck (1997) has proposed several symmetrized definitions of possibilistic independence between variables, in the numerical setting based on the invariance of possibility levels under conditioning and the use of the three basic triangular norms. She studies the use of the three obtained conditional independence and the axioms they satisfy in the scope of devising tools for local computations in inference networks. De Cooman (1997) defines the possibilistic independence between variables using the product rule of decomposition, generalized to triangular norms.

Three definitions of qualitative possibilistic independence between variables have been also studied by De Campos and Huete (1999, Part 1). The first notion means that information is invariant under conditioning, that is $\pi(x \mid y, z) = \pi(x \mid z)$. It is proved to be too restrictive because it forces the information about one of the variables to be null. The second idea is not gaining information under conditioning : $\pi(x \mid y, z) \geq \pi(x \mid z)$. It turns out to be equivalent to the decomposability property $\pi(x, y, z) = \min(\pi(x, y), \pi(x, z))$ for the joint possibility distribution, and the same for the conditional possibility $\pi(x, y \mid z) = \min(\pi(x \mid z), \pi(y \mid z))$. This definition leads to equip a set of conditional independence assumptions with a semi-graphoid structure. Yet another non-symmetric proposal of the authors is to keep the information invariance condition on a modified conditioning notion whereby the prior information is kept if conditioning leads to losing information. This notion, weaker than the first and stronger than the second one has the merit of preserving the graphoid axiom that is lost. The second part of the paper by De Campos and Huete (1999) carries out the same investigation for the product-based conditioning.

Finally, conditional independence in possibility theory can be defined in terms of similarity between $\pi(x \mid y, z)$ and $\pi(x \mid z)$. De Campos and Huete (1999) advocate the use of equivalence relations on the set of possibility distributions for the expression of similarity. The authors envisage three forms of similarity: preservation of ordering of possibility levels, partitioning of the possibility scale, equality $\pi(x \mid y, z) = \pi(x \mid z)$ above a possibility threshold. Sanguessa, Cabos and Cortes (1998) envisage degrees of similarity between possibility distributions with applications to learning from data. Indeed conditional independence notions can be exploited for possibility distributions as much as for probability, in order to encode them by means of directed acyclic graphs, like Bayesian networks. See De Campos (1995), Gebhardt and Kruse (1996), and Sanguessa, Cortes and Gisolfi (1998) for techniques that learn possibilistic networks from data.

7.3. 8 Combination Rules in Possibility Theory

The advantage of founding possibility theory on fuzzy set theory becomes very patent when it comes to methods for aggregating uncertain information represented by possibility distributions. See Dubois, Prade and Yager (1999) in the companion volume on Approximate Reasoning and Information Systems edited by Bezdek *et al.* (1999). Let us briefly survey this interesting topic.

Qualitative possibility theory can be used for the problem of information combination. Combination of possibility distributions pertaining to the same problem (e.g., the same variable) imperatively requires the existence of a possibility scale common to all possibility distributions, namely a totally ordered set L, that may only be a complete lattice in the most general case. Combining several comparative possibility distributions is impossible as such due to Arrow's theorem on the impossibility of combining ordering relations. The basic tools for information combination in qualitative possibility theory are set-theoretic (Dubois and Prade, 1988c, 1994). The main modes are the conjunctive and the disjunctive ones that respectively translate into an intersection and a union of fuzzy sets: given two possibility distributions π and π' ranging on the scale L, we thus get

$$\pi^{\cap} = \min(\pi, \pi') \text{ (conjunctive mode)}$$
$$\pi^{\cup} = \max(\pi, \pi') \text{ (disjunctive mode).}$$

The conjunctive mode corresponds to the pooling of fully reliable information, that deletes elements not considered possible by one source. It copes with possible dependence between sources due to the idempotency of min. However $\pi^{\cap}$ may be subnormalized if sources are conflicting. A qualitative normalization exists which consists in moving to 1 the level of possibility of the values having the greatest possibility degree already and leaving the other degrees unchanged, thus generalizing the qualitative conditioning (7.30). The apparent asymmetry of (7.30) is then interpreted by the fact that one information is binary while the other is not. On the contrary, in the Jeffrey-like rule of the previous section, the conditioning is viewed as basically non-symmetrical.

When one source only is reliable while the other is not, the disjunctive rule provides a safe, but potentially very imprecise, result where the possibly faulty information is hidden. The disjunctive rule is the only one that agrees with the axioms of possibility theory (Dubois and Prade, 1990c); namely $\max(\Pi, \Pi')$ is a possibility measure again, with distribution $\max(\pi, \pi')$, while $\min(\Pi, \Pi')$ is not a possibility measure, in general. This remark is the basis of a qualitative counterpart of the concept of probabilistic mixtures (Dubois Fodor *et al.*, 1996) that is the building block of qualitative decision theory (Section 7.6, below).

The quantitative setting is richer than the qualitative setting for what concerns the problem of information combination. The full gamut of fuzzy-set-theoretic operations is a priori available for combining numerical possibility distributions. Here we discuss only the basic combination modes mentioned above, namely the conjunctive and the disjunctive ones. The minimum rule makes full sense in the numerical setting for possibly dependent reliable sources. However the normalization is carried out in a more standard way. If π and π' are two possibility distributions to be combined, the normalized idempotent conjunctive rule reads

$$\pi^{\cap} = \frac{\min(\pi, \pi')}{h(\pi, \pi')} \tag{7.38}$$

where $h(\pi, \pi') = \sup_{u \in U} \min(\pi(u), \pi'(u))$ that evaluates the mutual degree of consistency of π and π'. It extends Dempster conditioning (7.33) which is recovered with $\pi' = A(\cdot)$. However this rule is not associative (Dubois and Prade, 1988c). Sudkamp (1993) showed that there is no reasonable normalization technique that, applied to $\min(\pi, \pi')$, could preserve associativity. Note that the renormalization of the results of the combination does away with the partial inconsistency between the two sources. If we want to keep track of it, we can define the result of the combination (which remains conjunctive) as

$$\forall u, \pi^{\cap}(u) = \min(\pi(u), \pi'(u)) + 1 - h(\pi, \pi') \tag{7.39}$$

as first suggested by Yager (1985b); it expresses that due to the partial inconsistency, we are unsure that the value of x lies in the intersection of the supports of π and π'.

In numerical settings combination operations other than min in (7.38) make sense; for instance, a reinforcement effect is obtained using the product instead of the minimum:

$$\pi_{\bullet}^{\cap} = \frac{\pi \cdot \pi'}{h^{\bullet}(\pi, \pi')} \tag{7.40}$$

where $h^{\bullet}(\pi, \pi') = \sup_{u \in U} \pi(u) \cdot \pi'(u)$. The combination is then associative. This non-idempotent combination rule presupposes independent sources.

It is possible to use Lukasiewicz triangular norm max(a + b – 1, 0) instead of product in (7.40) (with a suitable adaptation of the normalization coefficient); this is a very drastic combination rule that discards elements that both sources consider as little possible. It has been proposed by Mundici (1992) with a different terminology. The drastic reinforcement effect is explained in the latter reference by assuming that all possibility distributions come from the same unreliable source that lies only a given number of times which allow for the fact that some values become impossible although none of the sources discard them.

Disjunctive rules dual to conjunctive rules (i.e., operations of the form 1 – (1 – a) * (1 – b) where * is a conjunctive operation) can be used (without normalization) for merging information coming from unreliable sources (see Dubois Prade and Yager (1999) for more details).

7.4 QUANTITATIVE POSSIBILITY THEORY AS A BRIDGE BETWEEN PROBABILITY AND FUZZY SETS

Instead of pursuing sterile polemics between fuzzy sets and probability, another attitude is to try to relate fuzzy sets, possibility distributions, and probability at the mathematical and interpretive levels. Again, such bridges make full sense when fuzzy sets are viewed as numerical possibility distributions. The history of such relationships is already quite long and various types of link exist between quantitative possibility and upper/lower probabilities, belief functions, likelihood functions and infinitesimal probabilities, respectively. These links also suggest natural transformations between probabilistic and possibilistic representations. Moreover possibility distributions offer a proposal for addressing the problems of uninformed priors in Bayesian statistics. This section is based on a previous paper (Dubois and Prade, 1993) which it extends significantly.

7.4.1 Possibility Theory and Bayesian Statistics

Consider the simplest case in Bayesian statistics. The probability distribution of a random variable x is known except for some parameter θ_0. Random samples from x will be used to estimate θ_0. In the Bayesian spirit, besides "objective" data (random samples), one should use also any (reliable) additional information about θ_0 in order to improve the estimation. This type of information is referred to as "prior information", and is based upon the nature of θ_0, commonsense or "knowledge" about the problem. Thus, the prior information is "subjective" in nature. Let Θ denote the space in which θ_0 lies. Obviously, if we know that, in fact, θ_0 lies in some proper subset A of Θ, we can get a more reliable estimate for θ_0 (based on random samples). Thus, this kind of prior information is a type of all-or nothing possibilistic "information of localization" about θ_0. Without having such a precise information, one usually knows that θ_0 is "more likely" in one region than in another. In other words, verbally, a prior information specifies different "degrees of possibility" of where θ_0 might be. Now, since random samples will be used mainly to estimate θ_0, we are going to face the problem of "data fusion" when we take into account of prior information. By "data fusion", we mean a procedure that combines *different* types of data: probabilistic data (random samples) and possibilistic data (prior subjective knowledge).

The Bayesian methodology avoids this problem by modeling prior information within the context of probability theory (so that both types of information about θ_0 are of the same nature, namely probabilistic). Technically speaking, by doing so, Bayesian estimation can be carried entirely by using the calculus of probability. Specifically, the subjective, prior information of location of θ_0 is given in the form of a probability distribution, say, a prior probability density function $p(\theta)$ on Θ. This requires to view the unknown (but constant!) θ_0 as a *random variable* taking values in Θ. In particular, when Θ is a bounded interval of the real line $\mathbb{R}$,

$$p(\theta) = 1 / |\Theta|, \forall \theta \in \Theta,$$

where $|\Theta|$ denotes the length of Θ, expressing a uniform density function, corresponding to the case of no "preference" (or total ignorance) about location of θ_0 on Θ. Note that, however, except for the case where Θ is discrete (finite or infinitely countable), the value $\pi(\theta)$ is *not* the probability that the random variable θ_0 might take θ as value. On the other hand, if Θ is an*unbounded* interval, such as $\Theta = \mathbb{R}$, it is not possible to define a uniform density on it! Since, as a formal probability density, one need to have

$$\int_0^{+\infty} p(\theta)\, d\theta = 1$$

This will be violated for any positive constant value assigned to $p(\cdot)$. In the literature of Bayesian statistics, when Θ is unbounded, one asssumes $\pi(\cdot)$ to be constant, but then refer to it as an "improper prior"! Now, it is patent that, as far as information of localization is concerned, a function like $p(\cdot)$ has also the flavor of "possibility". The notion of possibility does reflect localization information. The problem of "improper uniform priors" can be avoided technically, if we view $p(\cdot)$ as a *possibility distribution* instead as a probability density function. Let $\pi(\theta)$ represent the *degree of possibility* that θ_0 is θ, even when Θ is infinitely uncountable; it can be obtained (subjectively) by eliciting experts' opinion. Setting $\pi(\theta) = 1, \forall \theta \in \Theta$ to express total ignorance about the location of θ_0 on, say, $\Theta = \mathbb{R}$, is natural, since in *possibility theory*, there is *no* requirement such that

$$\int_{\mathbb{R}} \pi(\theta)d\theta = 1.$$

And yet, even in this unbounded case, the intuitive fact that the possibility that $\theta_0 \in \mathbb{R}$ is 1 is preserved since:

$$\Pi(\theta_0 \in \mathbb{R}) = \sup_{\theta \in \mathbb{R}} \pi(\theta) = 1.$$

When viewing $\Pi(\theta)$ as a possibility distribution, θ_0 is referred to as a possibilistic variable. The above discussion tries and sheds some light on the motivation of the concept of possibility distributions in uncertainty analysis, as well as on the differences with probability density functions.

7.4.2. Upper and Lower Probabilities

A first point of view is to acknowledge a possibility measure as an upper probability envelope. Namely if we consider a set of probabilities $\mathcal{P}(\Pi) = \{P \mid P(A) \leq \Pi(A), \forall A \subseteq U\}$, then the induced upper probabilities $P^*(A) = \sup\{P(A), P \in \mathcal{P}(\Pi)\}$ coincide with possibility degrees $\Pi(A)$. The possibility distribution π is then defined by :

$$\pi(u) = P^*(\{u\}), \forall u \in U. \tag{7.41}$$

The set $\mathcal{P}(\Pi)$ is never empty due to the normalisation of the possibility distribution. This view is adopted in Dubois and Prade (1988b, 1992b, 1997a) as well as De Cooman and Aeyels (1996), De Cooman and Walley (1999). A characteristic condition under which a set of probability measures induced by lower bounds of probabilities of specified events lead to possibility measures can be given. Namely $\mathcal{P} = \{P \mid P(A_i) \geq a_i, i = 1,n\} = \mathcal{P}(\Pi)$ for some possibility measure Π if and only if the A_i's are nested (Dubois and Prade, 1992). The extension of these results to infinite settings is carried out by De Cooman and Aeyels (1996), who cast possibility theory in the subjectivist framework for imprecise probabilities due to Walley (1991). Upper probabilities are closely related to Choquet (1953) capacities. Hence so are possibility measures. See the mathematical appendix.

From the point of view of conditioning, viewing possibility degrees as upper bounds of probabilities leads to what we have called, after Jaffray (1992), Bayesian conditionalization of possibility measures. The justification of this form of conditionalization is due to the following equality (Dubois and Prade, 1996b, 1997; Walley, 1996):

$$\forall B, \Pi(B \mid A) = \frac{\Pi(A \cap B)}{\Pi(A \cap B) + N(B^c \cap A)} = \sup\{P(\cdot \mid A) \mid P \in \mathcal{P}(\Pi)\}$$

such that $\mathcal{P}(\Pi) = \{P \mid P(A) \leq \Pi(A), \forall A \subseteq U\}$, and provided that $N(A) > 0$ (which is required for the safe definition of $P(\cdot \mid A)$). However $\{P(\cdot \mid A) \mid P \in \mathcal{P}(\Pi)\}$ is generally strictly included in $\mathcal{P}(\Pi(\cdot \mid A))$ (see Jaffray, 1992). Dempster's rule of conditioning corresponds to *maximal likelihood* conditioning.

Banon (1995), via algebraic considerations based on Galois connections, has noticed that any fuzzy measure σ (monotonic with respect to inclusion) dominates a possibility measure: there is Π such that $\sigma \geq \Pi$, and any fuzzy measure is the intersection of a family of possibility measures: $\sigma(A) = \inf\{\Pi(A): \sigma \geq \Pi\}$. More specifically, using results in mathematical morphology (Banon and Barbera, 1993), they prove that this result holds when restricting the possibility measures to simple ones based on distributions of the form $\pi_A{}^\lambda(u) = 1$ if $u \in A$ and λ otherwise. This result is closely connected to the work of Biacino and Gerla (1992) and to possibilistic logic. In particular, this result leads to consider, dually to what this subsection suggests, a probability measure as induced by a family of possibility measures that dominate it.

7.4.3 Possibility Distributions as Special Cases of Random Sets and Belief Functions

The same probabilistic interpretation of fuzzy sets as given by Goodmann (1982) provides an interpretation for possibility distributions. See also Nguyen (1984). This view of possibility measures in terms of random sets and belief function theory (Shafer 1976) has been adopted in the late seventies. It is a more restricted framework than the one of possibility measures as upper bounds of a family of probability measures. Namely let $\mathcal{F} \subseteq 2^U$ be a family of subsets of U called focal elements, and let m_i be the positive mass of $A_i \in \mathcal{F}$ such that $\Sigma_{Ai \in \mathcal{F}}\, m_i = 1$; $m(\emptyset) = 0$. A random set is thus defined. Neverthteless, the A_i's are interpreted as imprecise observations, hence disjunctive sets. The two following set functions are defined in belief function theory (Shafer 1976)

$$Bel(A) = \Sigma_{A_i \subseteq A}\, m_i;\ Pl(A) = \Sigma_{A_i \cap A \neq \emptyset}\, m_i \tag{7.42}$$

and respectively called belief and plausibility functions. Pl(A) is the expected possibility of A with respect to the random set, in the sense of logical possibility. Formally Pl(A) is mathematically identical to a special type of upper probability and Bel(A) a special type of lower probability. Moreover Shafer (1976) showed in his book that $Pl(A \cup B) = \max(Pl(A), Pl(B))$ if and only if the A_i's are nested. Then we may define $\pi(u) = Pl(\{u\}) = \Sigma_{u \in A_i}\, m_i$ in the finite setting. Hence a possibility measure is a special case of a plausibility function, itself a special case of upper probability. Similarly the necessity function is a special case of belief function, and the guaranteed possibility function Δ (Section 7.3.4) is a special case of the Shafer commonality function, defined by reversing the inclusion sign in the above equation of Bel(A).

Conversely, consider a numerical possibility distribution π on a finite set U such that $\pi(u_i) = \pi_i$ and $\pi_1 = 1 \geq \pi_2 \geq \ldots \geq \pi_n$. The belief function equivalent to π has focal elements which are the level cuts of π, i.e., $A_j = \{u_i, \pi(u_i) \geq \pi_j\}$ for j = 1,n and $m(A_j) = \pi_j - \pi_{j+1}$ for j = 1,n, with $\pi_{n+1} = 0$ by convention (Dubois and Prade, 1982a; Klir and Bo Yuan, 1995). For a possibility distribution over the reals (fuzzy numbers), a consonant random set can be built as pointed out in Section 7.2.3.

Possibility measures can also be generated by mapping an underlying probability space (Ω, $\mathcal{B}$, P) to a set U via a consonant multivalued mapping Γ, particularizing the construction of Dempster (1967) for belief and plausibility functions (See Section 7.2.3). Iterating this process comes down to studying images of a possibility distribution via such a consonant multivalued mapping, thus leading to systems of upper and lower possibilities and necessities (Dubois and Prade, 1985). In particular, upper possibilities are possibility measures, and lower necessities are necessity measures. However lower possibilities and upper necessities are different set-functions. Tsiporkova and De Baets (1998) have investigated this construction in a systematic way.

More generally, even if $\mathcal{F}$ is not nested, necessity (and possibility) measures provide inner and outer approximations of belief (and plausibility) functions; for instance, a possibility measure Π_* is derived, based on $\pi_*(u) = \sum_{u \in A_i} m_i$ as the best inner approximation of ($\mathcal{F}$, m) in the sense of a random set inclusion that relates to the inequality $\Pi_*(A) \leq Pl(A)$, ∀A. Π_* is optimal in the sense that for any approximation Π of Pl, $\Pi \leq Pl \Rightarrow \Pi \leq \Pi_*$ (best inner approximation); see (Dubois and Prade, 1990a), where the more difficult problem of outer approximations $\Pi^*(A) \geq Pl$ is discussed.

The idea that a possibility measure is a compressed approximate representation of more elaborated uncertainty functions accounting for set-valued statistical data has been advocated by Gebhardt and Kruse (1993, 1994b) in the framework of the so-called "context model". In particular, they have introduced the notion of α-correctness so as to justify the representation of possibility in terms of level-cuts. Suppose that the underlying set Ω forms a finite set of contexts, each having subjective probability p(ω). Recall that for each context ω, the image Γ(ω) is the most specific set that is known to contain the value of a variable x (in accordance with the meaning of focal elements in belief function theory). So in this context, it is safe to state "x ∈ Γ(ω)" but not x ∈ A for a subset A of Γ(ω). The α-correctness asumption presupposes the existence of a fixed confidence level α, and a subset of contexts C such that P(C) ≥ α and ∀ω ∈ C, x ∈ Γ(ω). In this case the α-cut of the possibility distribution induced by the multiple-valued mapping Γ on

U is the most specific correct representation of the knowledge on the value of x, given the representation Γ of the knowledge and the α-correctness assumption. In particular, the extension principle of fuzzy set theory, commonly used in fuzzy interval analysis (see Chapter 10 by Dubois, Kerre et al, 1999, in this volume) can be derived as a theorem (Gebhardt and Kruse, 1994a, b; Kruse *et al.* 1994).

It is worth noticing that while possibility measures are special cases of upper and lower probabilities (e.g., of plausibility measures), possibility theory does not completely agree with these theories for combination schemes. For instance, Dempster rule of combination (without normalization) is like a random set intersection that, applied to possibility measures, does not preserve the consonant nature of these set-functions (although the product of possibility distributions is a good approximation of the result). See (Dubois and Prade, 1990a).

Shafer (1987) has criticized possibility theory on the basis of such discrepancies. However when a theory is developed, combination operations are imagined so that their result remains within the scope of the theory. Possibilistic combination rules are of this kind: consonance is preserved by assuming that the fuzzy sets underlying the possibility measures are pointwisely combined so as to result in a new fuzzy set. Similarly a natural combination scheme for families of probability measures is the intersection of these families. It only partially agrees with the minimum rule applied to possibility distributions (Dubois and Prade, 1992b).

From the point of view of conditioning, adopting Dempster conditioning for possibility measures comes down to considering numerical possibility theory as a subchapter of belief function theory, where necessity measures are interpreted as belief functionsin the Transferable Belief Model (Smets and Kennes, 1994).

7.4.4 Possibility -Probability Transformations

The problem of transforming a possibility distribution into a probability distribution and conversely has received more and more attention in the past years. This question is meaningful in the scope of uncertainty combination with heterogeneous sources (some supplying statistical data, other linguistic data, for instance).

Possibility-Probability Consistency. The starting point is that some consistency exists between possibilistic and probabilistic representations of uncertainty. Zadeh (1978) defined the degree of consistency between a possibility distribution π and a probability distribution p as follows: $p(\pi) = \sum_{i=1,n} \pi_i p_i$. It is the probability of the fuzzy event whose membership function is π. However Zadeh also described the consistency principle between possibility and probability in an informal way, whereby what is probable should be possible. Dubois and Prade (1980) have translated this requirement via the inequality $\Pi(A) \geq P(A)$ that founds the interpretation of possibility measures as upper probability bounds. P and Π are then said to be consistent. In the finite case, if we let $\pi(u_i) = \pi_i$ and $P(\{u_i\}) = p_i$ and assume $\pi_1 = 1 \geq \ldots \geq \pi_n \geq \pi_{n+1} = 0$, then P and Π are consistent if and only if $\pi_i \geq \Sigma_{j=i,n} p_j \; \forall i = 1, \ldots, n$ (Delgado and Moral, 1987). More conditions for consistency could be added. For instance one may also require that the orderings induced by the probability and the possibility distributions on U be the same. However we cannot require the same condition for events since the possibility ordering is generally coarser than the probability ordering on finite sets. See Drakopoulos (1995) for a compared analysis of the relative expressive power of probability and possibility theories.

Possibility and probability theories do not have the same descriptive power. Although it is clear that there are some states of information that probability can describe while possibility cannot (e.g., total randomness) and conversely (a single probability distribution cannot express ignorance, as advocated et length by Dubois and Prade (1990b) and Dubois, Prade and Smets (1996)), it can be advocated that the possibilistic representation of ignorance is weaker than the probabilistic representation, in the sense that the first is additive and supplies precise degrees of confidence and the other relies on an ordinal structure induced by the consonance assumption and only provides bounds on degrees of probability. It does not mean that precise probabilistic representations subsume possibilistic representations, since possibilistic representations can capture weaker states of information that probability distributions cannot.

Transformations Based on Optimizing Information Content. If we accept that possibility distributions are weaker representations of uncertainty than probability distributions, the transformation problem can be stated in a clear way. Namely going from possibility to probability leads to increase the informational content of the considered representation, while going the other way around means a loss of information. However the adopted transformations must be as little arbitrzary as possible. Hence, the principles behind the two following transformations are different and are not the converse of each other (Dubois *et al.* 1993) :

- *From possibility to probability* : a generalized Laplacean indifference principle is adopted: the weights m_i bearing on the nested family of levels cuts of π are uniformly distributed on the elements of these level-cuts. This transformation, already proposed by Dubois and Prade (1982a) comes down to selecting the gravity center of the set $\mathcal{P} = \{P \mid \forall A, P(A) \leq \Pi(A)\}$ of probability distributions dominated by Π. This transformation also coincides with the so-called pignistic transformation of belief functions (Smets 1990) and the Shapley value in game theory, where a game can be viewed as a non additive set function (Shapley, 1954). The rationale behind this transformation is to minimize arbitrariness by preserving the symmetry properties of the representation.
- *From probability to possibility* : in this case the rationale is to preserve as much information as possible ; hence we select as the result of the transformation of P, the most specific element of the set $\mathcal{F}(P)$ of possibility measures dominating P. This most specific element is unique if P induces a linear ordering on U.

Otherwise if there are equiprobable elements, unicity is preserved if equipossibility of the corresponding elements is enforced. In particular uniform probability is then transformed into uniform possibility. See Dubois and Prade (1982c), and Delgado and Moral (1987) for two different characterizations of this transformation.

In the finite case, if we let $\pi(u_i) = \pi_i$ and $p(u_i) = p_i$ and assume $\pi_1 = 1 > \ldots > \pi_n > \pi_{n+1} = 0$ and $p_1 > \ldots > p_n > p_{n+1} = 0$ the two transformations (as expected they are not the converse of each other) take the following forms

$$p_i = \Sigma_{j=i,n} \frac{\pi_j - \pi_{j+1}}{j} \tag{7.43}$$

$$\pi_i = \Sigma_{j=i,n}\, p_j. \tag{7.44}$$

However if $p_1 = \ldots = p_n = 1/n$ then selecting any linear ordering of elements and applying the above formula gives a most specific possibility distribution consistent with p. More generally if $E_1, \ldots, E_k$ is the well-ordered partition of U induced by p (elements in E_i have the same probability which is greater than the one of elements in E_{i+1}), then the most specific possibility distributions consistent with p are given by (b) applied to any linear ordering of U coherent with $E_1, \ldots, E_k$ (arbitrarily reordering elements within each E_i).

Confidence Intervals. Applied to continuous universes, the transformation of a unimodal probability density p with strictly monotonic sides into a possibility distribution is closely related to the notion of confidence interval. First it is a known fact that the confidence interval of length L, i. e., the interval with maximal probability is $I_L = [a_L, a_L + L]$ such that $p(a_L) = p(a_L + L)$. This interval has degree of confidence $P(I_L)$ (often taken as 95%). The most specific possibility distribution consistent with p is π such that $\forall L > 0$, $\pi(a_L) = \pi(a_L + L) = 1 - P(I_L)$. Hence the α-level cut of the optimal (most specific) π is the 1 – α- confidence interval of p (Dubois *et al.*, 1993). Note that these confidence intervals are around the mode of p, viewed as the "most frequent value". Going from probability to possibility means adopting a representation of uncertainty in terms of confidence intervals.

More recently Lasserre *et al.* (1998) has found more results along this line: she proved that each side of the optimal possibilistic transform is convex and there is no derivative for the mode of π. Hence given a probability density on a bounded interval [a, b], the triangular fuzzy number whose core is the mode of p and the support is [a, b] is an upper approximation of p regardless of its shape. In the case of a uniform distribution on [a, b], any triangular fuzzy number with support [a, b] provides a most specific upper approximation. Considerations of symmetry then suggests to choose the mode of the possibility distribution in the middle of the interval. These result justify the use of triangular fuzzy numbers as fuzzy counterparts of uniform probability distributions. Lasserre *et al.* (1998) also views well-known inequalities of probability theory, such as Chebyshev and Camp-Meidel ones, as yielding possibilistic approximations of probability functions, since they provide families of (loose) intervals. However they provide shapes of fuzzy numbers that can act as generic possibilistic approximations of probability functions, regardless of their shapes.

Berstein *et al.* (1996) have provided a general framework for representing a probability space by a family of nested sets. Namely if (U, $\mathcal{B}$, P) is a probability space ($\mathcal{B}$ being a σ-algebra of measurable subsets of U). Assume the size or volume of sets A $\in$ $\mathcal{B}$ makes sense, and let V(A) denote this volume. Then let $\mathcal{A}(\lambda) = \{A \in \mathcal{B} : P(A) = \lambda\}$, and E($\lambda$) be a set of minimal volume in $\mathcal{A}(\lambda)$. E(λ) is called a minimal event. The authors prove that, under mild conditions, a minimal event is uniquely determined by level λ. Minimal events are nested and can be represented by a possibility distribution. This construct clearly generalizes the probability-possibility transformation based on confidence intervals of greatest probability. The authors point out that this representation of probability spaces by means of possibility distributions enable them to be classified using fuzzy set inclusion as the basis for ordering them.

Transformations Based on Commensurate Information Indices. Klir (see Klir, 1990; Geer and Klir, 1992) has suggested a measurement-theoretic view of possibility/probability transformations. For him, the transformation should be based on three assumptions :

– A scaling assumption that forces each value π_i to be a function of p_i/p_1 (where $p_1 \geq p_2 \geq \ldots \geq p_n$, that can be ratio-scale, interval scale, Log-interval scale transformations, etc. ;

– An uncertainty invariance assumption according to which the entropy H(p) should be numerically equal to the measure of information E(π) contained in the transform π of p. E(π) can be the logarithmic imprecision index (7.16) of Higashi and Klir (1982) or the measure of total uncertainty, namely, the sum of two heterogeneous terms estimating imprecision and discord respectively (after Klir and Ramer, 1990).

– Transformations should satisfy the consistency condition $\pi(u) \geq p(u)$, $\forall u$, stating that what is probable must be possible.

The uncertainty invariance equation E(π) = H(p), along with a scaling transformation assumption (e.g., $\pi(x) = \alpha p(x) + \beta$, $\forall x$), reduces the problem of computing π from p to that of solving an algebraic equation with one or two unknowns.

Klir's approach relies on two prerequisites that may be discussed. First the scaling assumption leads to assume that $\pi(u)$ is a function of p(u) only. This is not the case with the transformation based on the most informative possibilistic upper approximation of p. This pointwiseness assumption may conflict with the probability/possibility consistency principle that requires $\Pi \geq P$ (see e.g., Dubois and Prade, 1980, pp. 258-259 where the simple transformation $\pi_i = p_i/p_1$ is proved to violate consistency ; see also Klir (1990). A second prerequisite is to admit that possibilistic and probabilistic information measures are commensurate. The basic idea is that the choice between possibility and probability is a mere matter of translation between languages "neither of which is weaker or stronger than the other" (quoting Klir and Parviz, 1992). It means that entropy and imprecision capture the same facet of uncertainty, albeit in different guises. Clearly the previous approach, based on (7.43) and (7.44), disagrees with this postulate. The last point of divergence is that Klir does not try to respect the full probability/possibility consistency principle (for events, not only singletons), which enables a nice link between possibility and probability to be maintained, casting possibility measures in the setting of upper and lower probabilities.

The most questionable assumption with this approach lies in the assumed equality $H(p) = E(\pi)$ between quantities whose commensurability is not obvious, especially if $E(\pi)$ is an index of non-specificity. Changing it into proportionality is not enough since it would claim that the entropy ordering of probability functions exactly reflects the specificity ordering of possibility, again a point hard to admit.

However more recently Harmanec and Klir (1997) considered yet another measure of information content, denoted $AU(\pi)$, of a possibility distribution π as the Shannon entropy of the least informative probability distribution consistent with π (maximizing $H(p)$ over all probability distributions p such that $\Pi(A) \geq P(A)$). Then, the commensurateness of $AU(\pi)$ and $H(p)$ sounds much more natural.

The computation of $AU(\pi)$ is achieved by an algorithm originally proposed by Maeda *et al.* (1993). It has some connection with transformation (7.43). Indeed starting with $\pi_1 = 1 \geq \ldots \geq \pi_n \geq \pi_{n+1} = 0$, the first step consists in finding i such that $\frac{\pi_1 - \pi_{i+1}}{i}$ is maximal. Then equal probability $p_j = \frac{\pi_1 - \pi_{i+1}}{i}$ is granted to each of $u_1, \ldots, u_j, \ldots, u_i$. The algorithm is then applied again to $\{\pi_{i+1}, \ldots, \pi_n\}$, and so on.

Even more interesting, the authors prove that, given p, the possibility distributions consistent with p and satisfying $AU(\pi) = H(p)$ are precisely the ones given as follows: let $E_1, \ldots, E_k$ is the well-ordered partition of U induced by p. Fix an arbitrary linear ordering of U, coherent with $E_1, \ldots, E_k$. Any possibility distribution is a solution to the above problem if and only if $\pi_{|E_i|+1} = \sum_{j=|E_i|+1}^{n} p_j$ and $\pi_{j-1} \geq \pi_j \geq \sum_{\ell=j}^{n} p_\ell$ for other indices j. Then the minimally specific among those possibility distributions is unique and enforces equal possibility within each E_i. The maximally specific such possibility distributions are precisely given by the optimal transformation (7.44) applied to any linear ordering of U coherent with E_1, …, E_k. Hence the two approaches to probability-possibility transformations, respectively on optimizing information content and on commensurate uncertainty indices, are here fully coherent.

Other Transformations. Other approaches have been proposed for the possibility-probability transformation problem. For instance, Dubois and Prade (1983) have tried to justify the following transformation of a probability into a possibility distribution:

$$\pi_i = \Sigma_{j=1,n} \min(p_j, p_i). \tag{7.45}$$

It gives results that are less specific than (7.44). Nevertheless, it is precisely the transformation converse to (7.43). Wonneberger(1994) has extended (7.45) to the continuous case. Civanlar and Trussell (1986) have considered a transformation T from probability to possibility by requiring that the probability of the fuzzy event described by $\pi = T(p)$ should be above a given threshold, and $\Sigma\pi_i^2$ should be minimum. The first requirement writes $p(\pi) = \Sigma_{i=1,n} \pi_i p_i \geq c$. Note that $p(\pi)$ was introduced by Zadeh as a probability/possibility consistency index. The drawbacks of this method are that its result is threshold-dependent, and may violate the probability/possibility consistency principle. Moreover minimizing $\Sigma\pi_i^k$ for $k \neq 2$ would give another solution. Lastly it was proved in Dubois and Prade (1990b) that $p(\pi)$ is constant for all π that are minimum elements (in the sense of inclusion) among possibility distributions that dominate p. Chanas and Nowakowski (1988) solve the problem of deriving a Gaussian probability distribution from a triangular possibility distribution on the real line, so as to maximize $p(\pi)$. However by construction, the obtained probability distribution is not consistent with π. See Gupta (1993) for further discussions on probability-possibility consistency.

The Multidimensional Case. An interesting question pertaining to possibility-probability transformations is their consistency with respect to second order properties such as marginalization, conditionalisation, independence and the like. Sudkamp (1992) proved general negative results along this line regarding exact coherence. For instance, the projection of a transform is generally different from the transform of the projection. Also, the transform of a product measure is generally not the joint possibility of the transforms of the marginal probabilities. These results should not be surprizing because the probability-possibility transformations cannot be structural homomorphisms due to the strikingly different properties of the basic operations in each calculus. If some consistency is to be looked for, it must be a weak form of consistency like the following:

If $P \in \mathcal{P}(\Pi)$ then $Proj_1(P) \in \mathcal{P}(Proj_1(\Pi))$ where P and Π are defined on $U_1 \times U_2$, and $Proj_1$ is the marginalization operator on U_1.

The reason why this property holds is that the marginalization process in belief function theory subsumes those of possibility and probability theory (Shafer 1987). Anyway the underlying difficulty is the one of the partial lack of consistency between confidence domains of multidimensional random variables and the confidence intervals built from their projections.

7.4.5 Possibility Theory and the Calculus of Likelihoods

Several authors have suggested to see likelihood functions (Edwards, 1972) of the form $\{P(O \mid u), u \in U\}$, as possibility distributions $\pi(u) = P(O \mid u)$. Direct relationships between possibility distributions and likelihood functions have been pointed out by Smets (1982) and Thomas (1979, 1995). Indeed the likelihood function is treated as a possibility distribution in classical statistics for so-called likelihood ratio tests, as already pointed out in Section 7.3.7. Namely if some hypothesis of the form $u \in A$ is to be tested against the opposite hypothesis $u \notin A$ on the basis of observation O alone, and the knowledge of the elementary likelihood functions $P(O \mid u), u \in U$, then the likelihood ratio test methodology suggests the comparison between $\max_{u \in A} P(O \mid u)$ and $\max_{u \notin A} P(O \mid u)$, i.e., $\Pi(A)$ and $\Pi(A^c)$ letting $\pi(u) = P(O \mid u)$ (see e.g., Barnett (1973, p. 150)). Stallings (1977) points out the similarity between maximum likelihood pattern classification and fuzzy pattern classification. Natvig (1983) interprets likelihood functions as possibility distributions in the law of total probabilities. The use of possibility theory in classification and diagnosis problems, as discussed in (Dubois and Prade, 1999), makes sense under the likelihood setting especially. Note that in continuous frameworks one cannot directly assume that $\pi(u) = p(O \mid u)$ if O is a pointwise obverved value, and p is the conditional density, since $p(O \mid u) > 1$ may occur. One way out of this problem is to turn each density $p(\cdot \mid u)$ into a possibility distribution.

The likelihood interpretation of quantitative possibility theory takes advantage of the probabilistic inequalities $\inf\{P(A \mid b), b \in B\} \leq P(A \mid B) \leq \sup\{P(A \mid b), b \in B\}$. Putting together the upper probability approach and the likelihood approach to possibility theory leads to consider a family of likelihood functions and to define a possibility measure from this family (see Dubois, Moral and Prade, 1997). Namely, it is possible to see a possibility measure Π (resp.: a guaranteed possibility function Δ of Section 7.3.4) as the upper (resp.: lower) envelope of a family of likelihood functions. These authors also investigate combination rules of fuzzy set theory in the light of likelihoods. It is interesting to notice that while the upper and lower probability view of possibility theory captures possibility and necessity measures, the likelihood view advocates the Π and Δ functions while the necessity functions are more difficult to interpret. In the belief function setting, however all three possibilistic functions are captured (since Δ corresponds to the commonality function).

As recalled in Dubois, Moral and Prade (1997), there is a very close proposal by Shafer (1976b). In the latter paper, it is suggested that an observation O defines a consonant belief function (whose associated plausibility function is a possibility measure) with mass assignment given by

$$m_O(A_i) = P(O \mid u_i) - P(O \mid u_{i+1})$$

$$m_O(U) = P(O \mid u_n)$$
$$m_O(A) = 0 \text{ otherwise}$$

where $U = \{u_1, \ldots, u_n\}$ with $P(O \mid u_1) \geq \ldots \geq P(O \mid u_n)$ and $A_i = \{u_1, \ldots, u_i\}$. This is a non-normalized mass assignment. However if $P(O \mid u_1) = 1$, and letting $\pi(u) = P(O \mid u)$ then this procedure is exactly the converse of equation (7.9) stated for possibility distributions.

Rather than interpreting possibility distributions in terms of conditional probability, one may think of using other conditional uncertainty measures as well, for instance a possibility measure, i.e.

$$F(u) = \Pi(\text{'F'} \mid u), \forall u \in U.$$

This equality presupposes the identity between the membership function describing the fuzzy class F and a possibilistic likelihood function expressing the possibility that an element u is classified in F. See Section 7.3.7.

7.4.6 Probabilistic Interpretations of Fuzzy Set Operations

Viewing the membership function F as a likelihood function leads to interpret fuzzy set-theoretic connectives in the framework of probability theory. For instance the minimum rule of fuzzy sets is then seen as computing

$$P(\text{'F'} \cap \text{'G'} \mid u) = \min(P(\text{'F'} \mid u), P(\text{'G'} \mid u)) \tag{7.46}$$

where 'F' and 'G' are the crisp events consisting of assigning fuzzy labels F on G to u respectively. People have then claimed that (7.46) really means the total dependence between the fuzzy sets F and G, thus pointing out a limitation in fuzzy set theory. This claim is debatable for two reasons: first the crisp events 'F' and 'G' are related to the fuzzy sets F and G only through a random experiment that is built as a model for constructing the membership function. So the dependence is between the crisp events, not the fuzzy sets. This remark also points out that we do not succumb here to the fallacy described in Section 7.2.1. A second reason is that if we remember the role of possibility distributions as upper bounds the equality in (7.46) should be turned into an inequality, the right-hand side being an upper bound of the left-hand side. In that perspective the minimum rule expresses a lack of knowledge about dependence (this is the meaning of the non-interactivity assumption in possibility theory) rather than an assumption of strong dependence (see, e.g., Dubois, Moral and Prade, 1997). The likelihood interpretation of membership functions cannot capture all fuzzy set connectives anyway. Only some connectives below minimum and above maximum are amenable to a likelihood interpretation, because due to inclusion monotonicity

$$P(A \cap B \mid x) \leq \min(P(A \mid x), P(B \mid x))$$
$$\text{and } P(A \cup B \mid x) \geq \max(P(A \mid x), P(B \mid x)).$$

Hence averaging fuzzy connectives and symmetric sums (see Chapter 2 of this volume by Fodor and Yager) cannot be captured in this way. Paris (1997) has proposed similar probabilistic semantics for fuzzy set connectives (especially min and product). Membership grades are calculated in terms of relative counts of independent arguments for or against membership.

The (consonant) random set (resp. belief function) view of membership functions (resp. possibility distributions) also provides a semantics to the main possible choices for defining fuzzy set union and intersection Orlov (1978), Goodman (1982) (see Dubois and Prade, 1989 for an overview). In particular the product conjunction of fuzzy sets is closely related to Dempster rule of combination of belief functions and to the intersection of independent random sets. Namely the product of membership functions is the one-point coverage function of the result of this random set intersection. The same remark applies for fuzzy set union with the probabilistic sum connective ($a + b - ab$) and the disjunction of independent random sets (see also Sudkamp, 1992). For other fuzzy conjunctions such as minimum and $\max(0, a+b-1)$, see also Dubois and Yager (1992). This framework seems to be much more flexible than the one of likelihoods for describing fuzzy set connectives: while fuzzy set intersections or unions correspond to forms of random set intersections or unions, other connectives such as averages continue to make sense with random sets.

Lastly fuzzy set connectives can also be interpreted in the setting of upper and lower probabilities. In that case again the minimum rule should not be understood as induced by a strong dependence assumption but as a minimal commitment rule when nothing is known about dependence (see Dubois and Prade, 1992b).

7.4.7 Possibility Degrees as Infinitesimal Probabilities

Spohn (1988) has proposed a theory for the representation of epistemic states that bear strong similarities with possibility and necessity measures, and tries to model the idea of ordering between possible worlds. What he calls an ordinal conditional function comes down to a mapping from a finite set of events to the set of positive integers, denoted κ and such that $\kappa(A \cup B) = \min(\kappa(A), \kappa(B))$; moreover there is an elementary event $\{u\}$ such that $\kappa(\{u\}) = 0$. $\kappa(A)$ expresses a degree of disbelief in A (Shenoy, 1991), and grows as A becomes less plausible. It is easy to check that for any real $k > 1$, $1 - k^{-\kappa(A^c)}$ is a degree of necessity (see Dubois and Prade, 1991a). The quantity $\kappa(A^c)$ represents the belief in A if positive (Shenoy, 1991).

A probabilistic interpretation of $\kappa(A)$ has been suggested by Spohn (1990). Namely, $\kappa(A) = n$ is interpreted as a small probability of the form ε^n, where ε is positive and close to 0, that is the probability of a rare event. Indeed, if A has a small probability with order of magnitude ε^n and B has also a small probability of the form ε^m, then $P(A \cup B)$ is of order of magnitude $\varepsilon^{\min(m,n)}$. These remarks may lead to an interpretation of possibility and necessity measures in terms of probabilities of rare events. Clearly, this relationship between possibility and probability theory is completely unrelated to the ones in the preceding paragraphs. However it explains the strong similarities between Pearl (1988)'s approach to non-monotonic reasoning based on infinitesimal probabilities and the possibilistic approach to the same problem (Benferhat *et al.* 1997a).

Spohn (1988) also introduces conditioning concepts for an OCF κ:

– the A-part of κ such that $\forall$ u $\in$ A, $\kappa(u \mid A) = \kappa(u) - \kappa(A)$. This is induced by probabilistic conditioning via infinitesimals. Translating this notion into the possibilistic setting, it is easy to recognize the product-based conditionalization of a possibility measure. Conditional independence in this setting has been investigated by Hunter (1991) and Studeny (1995) for natural conditional functions. They prove that under mild conditions, conditional independence relations form a graphoid structure, very close to the one of probabilistic conditional independence.

– the (A,n) conditionalization of κ, say $\kappa(u \mid (A,n))$ defined by

$$\begin{aligned}\kappa(u \mid (A,n)) &= \kappa(u \mid A) \text{ if } u \in A \\ &= n + \kappa(u \mid A^c) \text{ if } u \in A^c.\end{aligned}$$

This conditioning is the infinitesimal form of Jeffrey's rule for updating probabilities upon uncertain events: $P(B \mid (A,\alpha)) = \alpha \cdot P(B \mid A) + (1 - \alpha) \cdot P(B \mid A^c)$. Translated in to the possibilistic setting, this is the numerical counterpart of the possibilistic Jeffrey's rule of Section 7.3.7. While this possibilistic revision rule can be justified in terms of infinitesimal probabilities, there is not yet such a justification in terms of upper and lower probabilities, for instance.

Shenoy (1991) has also proposed a combination rule for Spohn functions:

$$\kappa_1 \oplus \kappa_2(u) = \kappa_1(u) + \kappa_2(u) - K,$$

where $K = \min_{u \in U} \kappa_1(u) + \kappa_2(u)$. It corresponds to the normalized product combination rule for possibility distributions.

More recently, and independently of the above research trends, Maslov (1989) has started an analytical study of idempotent set-functions taking their values on the interval $[0, +\infty)$, and obeying the corresponding counterpart of the maxitivity axiom of possibility measures, where 0 means complete possibility and $+\infty$ total impossibility, in the style of Spohn theory. The aim of this study is the description of discrete event systems. More works along this line are found in (Maslov and Samborski, 1992), (Akian *et al.*, 1994, 1996).

NB. Kramosil (1996) has presented yet another non-standard approach to possibility measures, taking values in the Cantor subset of the unit interval. He proves that the distribution of any continuous real-valued random variable defined over a general abstract probability space can be defined by an appropriate nonstandard possibility measure.

7.5 TOWARDS OPERATIONAL SEMANTICS OF POSSIBILITY DISTRIBUTIONS AND FUZZY SETS

In this section, several possible settings for assessing possibility measures are surveyed. Clearly, the links with probability theory enable possibility theory to inherit probabilistic paradigms provided that suitable assumptions are adopted. However, instead of interpreting fuzzy sets and possibility theory in a probabilistic framework, one might on the contrary look for interpretive settings in which the additive notion of probability is no longer needed. Two such settings are reviewed, that respectively relate possibility to preference and to similarity.

7.5.1 Frequentist Possibility

It is noteworthy that possibility measures can be given a purely frequentist interpretation (Wang and Sanchez, 1982; Dubois and Prade, 1986, 1989). Indeed a probability degree is sometimes understood as the idealization of a frequency, and possibility degrees, viewed as probability bounds inherit this interpetation as well. Frequentist views of possibility measures can adopt the following election poll example as a metaphor. Let U be a set of candidates to an election, and consider an opinion poll where individuals have not made up their mind completely yet, and are allowed to express it by proposing a subset of candidates containing their future choice. Let $\mathcal{F} \subseteq 2^U$ be the set of observed responses, and m(E), $E \in \mathcal{F}$, be the proportion of responses having the form of set E ($\Sigma_{E\in \mathcal{F}}$ m(E) = 1; m(Ø) = 0). Response E is a commitment to vote for one candidate in E only. More generally we can consider any given statistical experiment with outcomes in U, where a precise observation of outcomes is out of reach for some reason. The possibility distribution drawn from a body of evidence ($\mathcal{F}$,m) corresponds to what Shafer called "contour function", i.e.,

$$\pi(u) = \Sigma_{E:\, u\in E}\, m(E) = P^*(\{u\}).$$

However the election example is likely to produce conflicting information (the subsets E_i have no reason to be nested in this case), so that it is a better metaphor for general 'frequentist' belief functions. But the contour function formula can serve as the basis for the elicitation of frequentist possibility distributions. Such possibility distributions may represent membership functions for vague concepts whose meaning for a group of people in similar conditions generally meets some consensus (Dubois and Prade, 1986; Kampé de Fériet, 1982; Wang and Sanchez, 1982). Each individual in the group then provides a set that to him matches best the meaning of the vague concept F. m(E) is then the proportion of individuals whose response is E, and F(u)= π(u) is the proportion of individuals for whom u is consistent with their view of F. While it is not sure, even then, that $\mathcal{F}$ will be a nested family, one may expect the set in $\mathcal{F}$ to be minimally consistent, i.e., $\exists$ u $\in$ U, $\forall$ E $\in \mathcal{F}$, u $\in$ E. Namely, there is a consensus value being a prototype of F. In that case, a possibilistic approximation of the random set by the contour function works at its best (Dubois and Prade, 1990a). In particular, the minimal consistency of $\mathcal{F}$ is equivalent to the normalization of π. However only the nestedness of $\mathcal{F}$

ensures that the information contained in π is equivalent to that contained in $\mathcal{F}$. What is obtained is a "consensual" rather than a personal membership function. See Kruse *et al.* (1994), and Joslyn (1997) for an application of this view to the measurement of possibility distributions, using possibilistic histograms. There are also techniques that derive possibilistic networks from incomplete statistical data, and that exploit the random set view of possibility distributions, possibilistic independence and conditioning to decompose the joint distribution, and non-specificity indices for optimizing the structure of the network and the local possibility distributions (see De Campos (1995), Gebhardt and Kruse (1996), and Sanguessa, Cortes and Gisolfi (1998)).

7.5.2 Uncertainty Measures and Scoring Rules

In knowledge representation problems, it seems obvious that there are various types of uncertainty involved. With the success of probability theory in handling randomness in various branches of science, attempts to use set-functions to capture and model other types of uncertainty have started in the mid 70's. Basically, it was found that, since these uncertainties are different from randomness, their mathematical models differ from probability measures in a crucial point, namely they are non-additive. Although non-additive set-functions are abundant in mathematical analysis (such as Lebesgue outer measures, Choquet capacities, Hausdorff dimension, etc.), their use in uncertainty modeling raised concerns from a rational (mathematical) viewpoint, exemplified by the paper of Lindley (1982).

The problem is this. You are free to use any kind of set-functions to model uncertainty, but unless they are "admissible" within some reasonable framework, their use remains doubtful as far as rational decision-making is concerned. This is a decision viewpoint which is referred to as the coherence principle. To be rigorous, Lindley extended De Finetti (1972)'s decision framework for subjective probabilities and investigated the admissibility of various proposed set-functions. Namely the probability of A reflects the amount of money ss(A) an individual would accept to pay to a banker in order to assert the occurrence of an event A, with a commitment to receive \$1 if A does occur. So the buyer's gain is A – ss(A) where A = 1 if it occurs and 0 otherwise. Moreover the banker exchanges his role with the buyer if he finds the buyer's proposed price is too low. Suppose the buyer buys ticket for the two complementary events A and A^c. So the buyer's gain is $1 - ss(A) - ss(A^c)$ and the banker's gain is $ss(A) + ss(A^c) - 1$. Clearly the only way for the buyer to avoid losing money is to enforce $ss(A) + ss(A^c) = 1$, that is to use probability theory. A scoring rule is a numerical function $f(\alpha, A)$ of the number $\alpha = ss(A)$ that is the assessment of the likelihood occurence of A by a person. This function gives a penalty $f(\alpha, A)$ to the person giving assessment α. The scoring rule of a conditional bet on "A if B" is $f(\alpha, A)\cdot B(\cdot)$, and for several conditional bets the scoring rule is the sum of scores for each bet. An admissible assessment of likelihood values is one such that the person cannot in any circumstance reduce his penalty without increasing it in some other.

Lindley's main mathematical result is this. If an uncertainty measure is admissible, then necessarly it has to be a *function* of probability. From this result, the message is "you cannot avoid probabilities", in other words, probabilities suffice and are the only ones which can represent uncertainty; all non-additive uncertainty measures are inadmissible. While the mathematical result is correct, its interpretation, leading to the above message, was misleading. Indeed, a *function* of a probability need not be an additive set-function! For example, if P is a probability measure on a finite set Θ, then P^2 is non-additive, and in fact, it is a belief function (in the sense of Shafer (1976)). In the response to Lindley's message, Goodman *et al.* (1991) re-examined carefully Lindley's framework. Their mathematical findings are as follows.

(i) Even under the basic assumption of Lindley's decision framework, namely "additive score functions", there exist large classes of non-additive set-functions which are "admissible" in Lindley's sense.

(ii) Examples of such admissible non-additive set-functions are belief functions, pseudo-additive (decomposable) measures (see Chapter 1 of this volume, by Dubois *et al.*), and Zadeh's possibility measures are uniform limits of admissible measures.

(iii) When the additivity of score functions used in Lindley's decision framework is removed, the door is wide open for admissibility of other set-functions. In fact, in this case, admissible uncertainty measures need not even be transforms of probability measures. For details see (Goodman *et al.*, 1991). Hence Lindley's argument is sort of circular: additivity of the uncertainty measure derives from the additivity of scoring rules.

7.5.3 Betting Possibilities

Possibility measures can indeed partially capture a betting behavior paradigm in accordance with the sujectivist view of upper and lower probabilities (Smith, 1961; Giles, 1982; Walley 1991). It differs from De Finetti's bet setting by the rejection of exchangeable bets. Namely $P^*(A)$ reflects the minimal amount of money an individual would require in order to deny the occurrence of A, with a commitment to pay \$1 if A does occur. The companion amount $P^*(A^c)$ is defined likewise, independently of $P^*(A)$. The first bet defines an upper bound for an unknown probability P(A); the second bet defines the upper bound of $P(A^c)$, i.e., a lower bound $1 - P^*(A^c) = P_*(A)$ for P(A). Clearly if $A \subseteq B$, it is safer to deny A rather than B. So, less money is required to deny A, and $P^*(B) \geq P^*(A)$. Moreover if $P^*(A) \leq a$, $P^*(B) \leq b$, then the individual will accept to deny both A and B for a \$(a + b) fee, and a fortiori he must agree to deny A or B for such a premium. Hence the rationality assumptions:

$$\max(P^*(A), P^*(B)) \leq P^*(A \cup B) \leq P^*(A) + P^*(B)$$

and particularly $P^*(A) + P^*(A^c) \geq 1$, i.e., $P_*(A) \leq P^*(A)$. The family of confidence measures thus captured is that of upper and lower envelopes, as first proposed by

C.A.B. Smith (1961), i.e., P* is the upper limit function of a set $\mathcal{P}$ of probability measures. This is the view extensively developed in Walley (1991)'s book.

When $P^*(A \cup B) = \max(P^*(A), P^*(B))$ is forced, it clearly corresponds to individuals that always ask for the maximal amount of money for one of A or A^c, its complement, according to whether A is considered safer than A^c or not. This accounts for the property of possibility measures: $\max(\Pi(A), \Pi(A^c)) = 1$. Moreover the individual is allowed to be completely risk-averse by requesting the maximal safety amount for both A and A^c, when he feels ignorant about their occurrence. This is clearly an attitude of extremely cautious gambler; it clearly stands in total contrast with Bayesian gamblers who ask for a maximal safety amount on A *only* in case of complete certainty about A^c.

So, possibility measures can be motivated in the same framework as subjective probabilities; possibilistic gamblers cannot be claimed "irrational", although they do not accept standard gambling rules due to their cautiousness. Bayesians accept the idea of a forced bet while in the upper/lower probability setting the player can give up the bet. Note that the betting behavior approach to possibility as outlined by Giles (1982) leads to the interpretation of the membership grade F(u) in a fuzzy set as the least amount of money requested for a safe enough claim that u will lose in some chance game, where a wrong statement leads to a \$1 fine and where it is allowed to simultaneously request \$1 safety fee for all u's. See De Cooman and Aeyels (1996) for a systematic study of the subjectivist view of numerical possibility measures in the style of Walley, and Walley (1997) for an application of this view to representing membership functions of vague terms.

Operational interpretations of possibility theory as orders of magnitude of the frequency of occurence of rare events, as suggested by Spohn's framework, have not been really considered yet, although they have been extensively used in Artificial Intelligence for the purpose of belief revision modeling (Darwiche and Pearl, 1997; Dubois Moral and Prade, 1998) in the language of kappa-functions.

7.5.4 Possibility as Similarity

The frequentist and gamble settings shed some light on possiblity measures viewed as evaluations of the plausibility of occurence of events. However the fuzzy connection hints on a purely metrical view of possibility, namely an occurence is all the more plausible as it is similar to a maximally plausible occurence. Indeed membership grades in fuzzy sets are often construed as a function of a distance to a prototype of the fuzzy concept. There is a whole trend in fuzzy set theory according to which the degree of membership F(u) reflects the similarity between u and an ideal prototype u_F of F (for which $F(u_F) = 1$). This interpretation of partial membership is clearly related to the relation of distance, and not to probability. Then if a variable x is attached a possibility distribution $\pi = F(\cdot)$, x = u is all the more possible as u looks like u_F, is close to u_F.

This trend is well represented by the so-called "type B membership function" described by Zimmermann (1987) and for which direct scaling methods can be applied. Clearly, the elicitation of numerical possibility distributions is a measurement problem which depends both on the mathematical setting and the intuitive interpretation of possibility degrees. This line of thought has been investigated by Norwich and Türksen (1982) (see also Chapter 3 of this volume, by Bilgic and Türksen) but many problems remain unsolved on this issue.

The fuzzy linear regression approach of Tanaka *et al.* (1993) is a technique that derives joint possibility distributions on $\mathbb{R}^n$, purely based on metric considerations, from a data set. What is obtained is a quadratic possibility distribution, solution to a fuzzy linear program. See Diamond and Tanaka (1998) for a review of fuzzy regression techniques. Tanaka and Ishibuchi (1993) have also proposed a parameterized representation of normal joint possibility distributions, closely related to the quadratic representation. Operations such as projection, product combination and conditioning can be more easily performed by simple operations on the parameters. See also Tanaka and Guo (1999) and Wolkenhauer, O. (1998) for possibility theory applications to data analysis.

Ruspini (1991) has pushed the metric view further by exploring how the similarity relation R that exists between elements of the referential U affects the inference process between propositions. For Ruspini, given two classical propositions A and B, the extent to which A implies B is the minimal amount of "stretching"of B that is required to include the set A in the neighborhood of the element of set B. More formally the following metric counterpart of Carnap's confirmation degree is obtained

$$I(B \mid A) = \inf_{u \in A} \sup_{u' \in B} \mu_R(u,u') \qquad (7.47)$$

where R is a fuzzy proximity relation, that expresses indiscernibility on U. In order to clarify the meaning of $I(B \mid A)$, let us consider the case when R is a standard equivalence relation. Let B^* be the upper approximation of B in the sense of rough set theory (Pawlak 1991), i.e., $B^* = \{u: \exists u', u \mathrel{R} u' \text{ and } u' \in B\}$; then $I(B \mid A) = 1$ corresponds to the classical entailment $A \subseteq B^*$. More generally this view leads to considering a fuzzy set F as the max-min combination of a set A (the core of F) and the fuzzy relation R. Note that the view of possibility as plausibility leads to another non-probabilistic degree of entailment, namely the qualitative conditional necessity N(B|A) (e.g., Benferhat *et al.* 1997a).

7.5.5 Possibility as Preference and Graded Feasibility

Besides, the existence of a preference driven view of possibility functions mentioned in Section 7.3.1 leads to completely different interpretive and operational settings since a possibility distribution is then viewed as a kind of utility function describing a soft constraint. Many paths remain unexplored on the problem of devising thought experiments for the elicitation of such possibility degrees.

Fuzzy Constraints. A classical hard constraint C is represented by a classical set of solutions to a problem, i.e., only using degrees of membership 0 or 1, and the characteristic function that is attached to the constraint is a binary possibility distribution π_x, where x is a vector of *decision* variables. Introducing intermediary levels of feasibility, C is called a fuzzy, or soft constraint: C(u) = 1 means that a solution u totally satisfies C while C(u) = 0 means that it totally violates C (u is unfeasible). If $0 < C(u) < 1$, u satisfies C only partially; $C(u) > C(u')$ indicates that C is more satisfied by u than by u' (C prefers u to u'). Hence, like an objective function, a fuzzy constraint rank-orders the feasible decisions and C(u) denotes the degree of possibility of the decision assignment x = u, where possibility stands for feasibility. But, contrary to an objective function, a fuzzy constraint also models a threshold (represented by the bottom level 0) beyond which a solution will be rejected, regardless of other considerations; this is like a veto threshold in multiple criteria decision-making methods (Vincke, 1982). In fact, a fuzzy constraint can be viewed as the association of a constraint (defining the support of C) and a non-compensatory criterion which rank-orders the solutions satisfying the constraints. A soft constraint C' will be at least as loose as another one C if and only if $C(\cdot) \leq C'(\cdot)$, that is, if any solution to C is at least as feasible for C'.

In this interpretive framework, a membership function, construed as a possibility distribution, is similar to a utility function, or, better, a value function, but it may range on a qualitative valuation set. Contrary to the previous sections, here, possibility distributions encode feasibility profiles of soft constraints relating decision variables. In the presence of very loose constraints it is clear that the use of soft constraints can help breaking ties among feasible solutions, just as objective functions do. On the contrary when hard constraints are tight, there may be no feasible solution and it may be time-consuming to discover it. One way of coping with the lack of solutions is to relax the constraints, but automating this process is not so easy. Again, the use of soft constraints is one way out of the difficulty. Very often, the cores of the fuzzy constraints lead to an unfeasible problem, their supports lead to an underconstrained problem, and fuzzy constraint satisfaction will achieve a trade-off between antagonistic soft requirements.

Prioritized Constraints. What is usually done for problem relaxation, is to assign priorities to constraints. Solutions that satisfy all constraints are selected if any. Otherwise the less prioritary constraints are dropped, and the chosen solution satisfies only higher priority constraints. It can be proved that this decision strategy can be captured in the setting of possibility theory. A priority level ρ of constraint C can be modelled by a degree of necessity estimating how imperative C is. The pair (C, ρ), where C is a crisp constraint, can be modelled as a special kind of fuzzy constraint C' (Dubois, Fargier and Prade, 1994):

$$C'(u) = 1 \text{ if u satisfies C}$$
$$C'(u) = n(\rho) \text{ if u violates C.}$$

where $n(\rho) = 1 - \rho$ if the scale [0, 1] is used; function n is more generally an order-reversing map of the possibility scale. Since ρ represents to what extent it is necessary to satisfy C, n(ρ) indicates to what extent it is possible to violate it. In other words, any potential solution u that violates C satisfies (C, ρ) to a degree equal to n(ρ). While possibility degrees are interpreted in terms of feasibility, necessity degrees are levels of priority. If C is itself a soft constraint with priority ρ, it can be modeled by the fuzzy constraint C' with membership function: C'(u) = max(n(ρ), C(u)).

There is a different understanding of priorities, namely by considering that the constraint C is sufficiently satisfied by u if the level of satisfaction for u reaches some threshold θ, i.e., $C(u) \geq \theta$. Then C(u) will be changed into $C^*(u) = 1$ and C^* will no longer affect the conjunctive aggregation process between levels of satisfaction of solution u. If $C(u) < \theta$ we may either consider that the requirement is satisfied at the level which is reached, i.e., $C^*(u) = C(u)$, or in order to avoid discontinuity in a numerical setting, make $C^*(u)$ equal to the *relative* level of satisfaction C(u)/θ. Then the degree of satisfaction of the requirement (C, θ) by a solution u is $\theta \rightarrow C(u)$ where implication → is Gödel implication in the first case ($a \rightarrow b = 1$ if $a \leq b$, $a \rightarrow b = b$ if $a > b$), and Goguen's in the second one, namely $a \rightarrow b = \min(1, b/a)$ if $a \neq 0$ and $a \rightarrow b = 1$ if $a = 0$. For $\theta = 1$, $C^* = C$ in any case. It contrasts with the use of Dienes implication $a \rightarrow b = \max(n(a), b)$ in the other priority handling method. All the implications we considered above are such that $a \rightarrow b \geq b$, and thus the prioritizing procedures relax C. It would not be the case with Rescher-Gaines implication ($a \rightarrow b = 1$ if $a \leq b$, $a \rightarrow b = 0$ if $a > b$), with which C would be made crisp.

Fuzzy constraint satisfaction. The notion of fuzzy set as viewed by Bellman and Zadeh (1970) meant to represent constraints as well as objective functions by fuzzy subsets C of possible decisions. If $C_1, \ldots, C_m$ denote m fuzzy sets, the fuzzy decision set was defined by the joint possibility distribution

$$D(u) = \min_{i=1,m} C_i(u). \tag{7.48}$$

However this definition of a decision set is not really in accordance with the usual paradigm of multiple-criteria decision-making since an optimal solution in the sense of Bellman and Zadeh does not achieve compensation between the membership values $C_i(u)$. On the contrary, an optimal solution to (7.48) is one that least violates the most violated constraint. Hence, Bellman and Zadeh's proposal is a constraint-directed view of problem-solving, where constraints are flexible: if there is a solution u that completely satisfies all constraints, this solution is optimal. Otherwise $\lambda^* = \sup_u D(u) < 1$, and this indicates that the constraints are partially contradictory. To accept a solution such that $D(u) = \lambda^*$ means to partially relax some of the constraints C_i. In that sense, constraints C_i are flexible. When the constraints are prioritized, the minimum in (7.48) becomes a thresholded minimum of one of the following forms:

$$D(u) = \min_{i=1,m} \max(n(\rho_i), C_i(u)),$$

$$D(u) = \min_{i=1,m} C^*_i(u) = \min_{i=1,m} \theta_i \to C_i(u),$$

where the priorities modify the minimum aggregation like weights modify an arithmetic mean. But here the priorities are thresholds, not weights proper. The first method was first proposed by Yager (1981). The other one dates back to Yager (1984).

The notion of maximin optimality is not new in the literature of decision making. It is well-known in game theory as the most cautious strategy for players. It is also well-known in decision-making under uncertainty, where again it is a cautious strategy. Despite these formal analogies, the problem tackled by flexible constraint satisfaction is very different, since index i pertains neither to an opposing player nor to a state of the world. The paradigm most akin to Bellman and Zadeh's is to be found in social choice theory (Moulin, 1988) and $\min_{i=1,n}C_i(u)$ is an egalitarist social welfare function, $C_i(u)$ being the welfare index of an individual i under decuision u. Maximizing $\min_{i=1,n} C_i(u)$ tends to select solutions which equally satisfy all constraints. An axiomatic decision-theoretic justification of the minimum is supplied by Pirlot (1995), and for prioritized minimum by Fortemps (1997).

The Most Feasible Solutions. The maxmin ordering of solutions is coarse. Moreover, not all maximin optimal solutions are interesting because some of them fail to be Pareto-optimal. Subclasses of Pareto-optimal solutions have been laid bare (Dubois, Fargier and Prade, 1996a), namely letting $v_j = C_j(u)$ and $w_j = C_j(u')$ for solutions u and u':

– *discrimin optimal solutions*: they are maximal solutions to the so-called *discrimin relation* between vectors $\bar{v}$ and $\bar{w}$, defined as follows:

$$\bar{w} >_{discrimin} \bar{v} \text{ iff } \min_{j\in\mathcal{D}(\bar{w},\bar{v})} w_j > \min_{j\in\mathcal{D}(\bar{w},\bar{v})} v_j$$

where $\mathcal{D}(\bar{w},\bar{v})$ is the set $\{j, w_j \neq v_j\}$ of discriminating components. $>_{disc}$ is a partial ordering;

– *leximin optimal solutions*: they are maximal solutions to the so-called leximin ordering between vectors, defined as follows:

$$\bar{w} \geq_{lm} \bar{v} \text{ iff } \exists k, w_{[i]} = v_{[i]} \text{ for } i < k \text{ and } w_{[k]} \geq v_{[k]}$$

where $w_{[1]} \leq w_{[2]} \leq \ldots \leq w_{[m]}$ and $v_{[1]} \leq v_{[2]} \leq \ldots \leq v_{[m]}$. It corresponds to a lexicographic ordering of increasingly rearranged vectors. $\geq_{lm}$ is a complete preordering that refines the discrimin relation. The leximin ordering is found as the limit of other types of multiple criteria rankings based on triangular norms and ordered weighted averages (Dubois *et al.* 1997). Note that the discrimination power of the leximin ordering is as good as the one of any symmetrical numerical aggregation function. These refinements of the min-based orderings thus responses to critiques of the minimum operation on the basis of lack of discriminating power.

An important issue is how to compute discrimin and leximin optimal solutions. This has been investigated by Dubois and Fortemps (1999).

7.5.6 *Refinements of Qualitative Possibility Theory*

It is tempting to exploit discrimin and leximin orderings within possibility theory itself, for achieving a finer-grained representation of qualitative uncertainty (see Dubois, Fargier and Prade, 1998). It enables possibility and necessity orderings of events to be refined on the basis of the only knowledge of the underlying possibility ordering, and without necessarily resorting to the use of numbers.

Two events A and B cannot be discriminated using the dual pair (N, Π) as soon as $\Pi(A) = \Pi(B)$ and $N(A) = N(B)$, which occurs quite often, namely, as soon as $\max_{u \in A} \pi(u) = \max_{u \in B} \pi(u) < 1$ or $\max_{u \notin A} \pi(u) = \max_{u \notin B} \pi(u) < 1$.

The equality $\Pi(A) = \Pi(B)$ may be due to a high value of $\Pi(B \cap A)$. However in the spirit of the discrimin relation, this weight should not affect the comparison between A and B, since it is common to both. The same reasoning also applies to $\Pi(B^c \cap A^c)$ with respect to the equality $N(A) = N(B)$. Only $B^c \cap A$ and $B \cap A^c$ should matter in telling A from B. Let us introduce the relation $>_{\Pi L}$ on 2^U, different from although closely related to the original necessity and possibility orderings modelled by N and its dual Π:

$$A >_{\Pi L} B \text{ iff } \Pi(B^c \cap A) > \Pi(B \cap A^c).$$
$$A \geq_{\Pi L} B \text{ iff not } (B >_{\Pi L} A).$$

The relation $>_{\Pi L}$ is a refinement of both possibility and necessity orderings. Now if π denotes the possibility distribution associated to Π (i.e., such that $\Pi(A) = \max_{u \in A} \pi(u)$), and let $\vec{a}$ be the vector $(a_1, \ldots, a_n)$ such that, denoting $U = \{u_1, u_2, \ldots, u_n\}$in the finite case:

$$a_i = \pi(u_i) \text{ if } u_i \in A$$
$$a_i = 0 \text{ otherwise}$$
$$\text{then } A >_{\Pi L} B \Leftrightarrow \vec{a} >_{\text{discrimax}} \vec{b},$$

where the discrimax partial ordering is like the discrimin partial ordering, changing min into max. This partial ordering is purely ordinal, yet being almost as much discriminant as comparative probability orderings (Fishburn, 1986) with which they share many properties.

$$A \subseteq B \Rightarrow B \geq_{\Pi L} A \text{ (monotony);}$$
$$A \cap (B \cup C) = \emptyset \Rightarrow (B \geq_{\Pi L} C \Leftrightarrow A \cup B \geq_{\Pi L} A \cup C) \text{ (additivity);}$$
$$A \geq_{\Pi L} B \Leftrightarrow B^c \geq_{\Pi L} A^c \text{ (self duality).}$$

If A is a proper subset of B then B $>_{\Pi L}$ A always hold for positive possibility distributions. The possibilistic relation $\geq_{\Pi L}$ is called *discrimax likelihood* and A $\geq_{\Pi L}$ B reads "A is at least as likely as B", so as to emphasize the self-duality. However, because of the lack of transitivity of the indifference relation, we cannot generally represent such relations by means of probability measures, nor can we assume properties that are usually derived from comparative probabilities. However the *leximax* extension of a possibility measure (defined similarly to the leximin relation above, but reordering components in a decreasing order) is a genuine comparative probability relation on finite sets and can be represented by a special class of numerical probabilities (see Dubois, Fargier and Prade, 1998).

7.6 POSSIBILITY AND NECESSITY OF FUZZY EVENTS: A TOOL FOR DECISION UNDER UNCERTAINTY

This section is devoted to the counterparts of probabilistic expectations in possibility theory, and the underlying decision-theoretical foundations. The decision-theoretic construct of Savage is one of the most convincing justifications of probabilistic representations of uncertainty. Such an approach makes sense in the scope of repeated decisions whose results compensate. In the setting of possibility theory, two counterparts of expectation can be used: Sugeno integral which only requires a totally ordered scale, and Choquet integral, a genuine extension of expectation to nonadditive numerical set-functions including possibility measures. These expectations have been used in fuzzy set theory in order to evaluate the uncertainty of a fuzzy event. However they can also be exploited as tools for non-classical approaches to decision under uncertainty, thus suggesting behavioral, observable justifications of possibility theory.

7.6.1 Possibility and Necessity of Fuzzy Events

Events, which we would like to evaluate the uncertainty of, are often ill-defined as well: e.g., we would like to estimate whether somebody is "rich" without being able to describe in *precise* quantitative terms what it means. Possibility and fuzzy set theory jointly offer a simple tool for modelling both the partial lack of available information and the ill-definition of relevant events. Fuzzy sets enable all set-functions representing partial belief to be generalized by extending the usual concept of "event" and the algebra of events so as to accommodate ill-defined events. Algebras of fuzzy events are no longer Boolean.

Possibility and necessity measures can be easily extended to fuzzy events provided that possibility degrees and membership grades range on the same scale L. Using the ideas of consistency and entailment as the basis for possibility and necessity, extensions of possibility measures and necessity measures are obtained. Namely, if F is a L-fuzzy set on U, with membership function ranging on L, the possibility of a fuzzy event is defined by:

$$\Pi(F) = \sup_u F(u) * \pi_x(u) \qquad (7.49)$$

where $*$ is a monotonic conjunctive-like operation on L (such that $1 * t = t$ and $0 * t = 0$ in order to recover (7.17) when F is non-fuzzy), and the necessity of a fuzzy event is

$$N(F) = \inf_u I(\pi_x(u), F(u)) \tag{7.50}$$

where I is an implication operation of the form $I(a,b) = 1 - [a * (1 - b)]$, in order to preserve the identity $N(F) = 1 - \Pi(F^c)$ for the usual fuzzy set complementation $F^c(\cdot) = 1 - F(\cdot)$. $(1 - (\cdot))$ denotes an order-reversing function on L). (7.49) has been proposed by Zadeh (1978) with $* = \min$. If operation $*$ is increasing in the wide sense in both places, the characteristic properties (7.20-7.21) of possibility and necessity measures with respect to fuzzy disjunction and conjunction still hold (respectively defined by the pointwise maximum and the minimum of membership functions). Moreover, if $a * b \le \min(a,b)$, then $\Pi(F) \ge N(F)$, $\forall$ F also holds (if π_x is normalized). See Prade (1982) for a level-cut based derivation of the necessity of a fuzzy event of the form (7.50), with $I(a,b) = \max(1 - a, b)$, from the comparison of level-cuts. Inuiguchi and Kume (1994) provide a systematic study of possibility and necessity measures of fuzzy events with various implications. They point out three kinds of fuzzy necessity measures: strong ones such that $N(F) = 1$ if and only if the support of π_x is contained in the core of F (this is when $I(a,b) = \max(1 - a, b)$, for instance); medium ones such that $N(F) = 1$ if an only if π_x defines a fuzzy subset included in F in the usual sense ($\pi_x \le F(\cdot)$; this is when $I(a,b) = 1$ as soon as $a \le b$, especially residuated implications, see Bouchon-Meunier *et al.*, 1999); and weak ones π_x such that $N(F) = 1$ if and only if the support of π_x is contained in the support of F.

Here, the measurement of possibility and necessity is viewed as a semantic matching process between the representation of the available information (π_x) and the description of some request in the form of a fuzzy event F. Pattern matching is one of the basic ideas in Artificial Intelligence; when knowledge is expressed in a symbolic and declarative way, access to pieces of knowledge proceeds by pattern matching between symbol strings, rather than addressing. Hence possibility theory enriches the pattern matching paradigm of Artificial Intelligence, allowing for graded semantic matching (Cayrol *et al.*, 1982; Dubois, Prade and Testemale, 1988). Given a collection of objects for which we have some (possibly incomplete) knowledge about the value of an attribute x, represented by means of a possibility distribution for each object, the set of objects such that "x is F" is an ill-known set described by a pair of fuzzy sets, namely the fuzzy set of objects which more or less possibly satisfy the requirement "x is F" and the fuzzy set of objects which more or less necessarily (i.e., certainly) satisfy it. Because generally, $\Pi(F) \ge N(F)$ for each object, the latter fuzzy set is included in the former. Such a pair of fuzzy sets is called a twofold fuzzy set (Dubois and Prade, 1987). The possibilistic framework thus enables us to capture the idea of sub-definite sets, viewed as pairs of upper and lower approximations of sets, considered by various authors (e.g., Narin'yani, 1980).

More generally, fuzzy pattern matching takes advantage of the following decomposition formulas for dealing with multiple attribute patterns. Given two normalized fuzzy sets F and G defined on U and V respectively (i.e., $\exists u$, F(u) = 1, $\exists v$, G(v) = 1), and a *separable* possibility distribution $\pi_{(x,y)} = \min(\pi_x(u), \pi_y(v))$, defined on U × V, using * = min, we have the decomposability properties

$$\Pi(F \times G) = \min(\Pi_x(F), \Pi_y(G));$$
$$N(F \times G) = \min(N_x(F), N_y(G));$$
$$\Pi(F + G) = \max(\Pi_x(F), \Pi_y(G));$$
$$N(F + G) = \max(N_x(F), N_y(G));$$

where Π_x and N_x (resp. Π_y and N_y) are the possibility and necessity measures associated with π_x on U (resp. π_y on V). The Cartesian product F× G is defined by $(F \times G)(u, v) = \min(F(u), G(v))$, and the Cartesian coproduct is of the form $F + G = (F^c \times G^c)^c$. Identities expressing N(F× G) and (Π(F + G) hold even if $\pi_{(x,y)}$ is not separable, but the two others will not in the general case.

Certainty and possibility qualification (7.26) and (7.27) can be generalized to the case where F is a *fuzzy* set. Namely, consider statements of the form 'x is F is α–certain' and 'x is F is α–possible'. The semantic translation of these statements are supposed to be respectively $\pi_x(u) \le \max(F(u), 1 - \alpha)$ and $\min(F(u), \alpha) \le \pi_x(u)$ as in Section 7.3.6. These inequalities are still equivalent to $N(F) \ge \alpha$ and $\Delta(F) \ge \alpha$ provided we use the following appropriate extensions of N and Δ when F is fuzzy

$$N(F) = \inf_{u \in U} (1 - F(u)) \to (1 - \pi_x(u)) \tag{7.51}$$
$$\Delta(F) = \inf_{u \in U} F(u) \to \pi_x(u) \tag{7.52}$$

where → is Gödel's implication $r = s \to t = \begin{cases} 1 & \text{if } s \le t \\ t & s > t \end{cases}$. Indeed, in the unit interval we have the equivalences $\max(f, 1 - \alpha) \ge \pi \Leftrightarrow \alpha \le (1 - f) \to (1 - \pi)$ and $\min(f,\alpha) \le \pi \Leftrightarrow \alpha \le f \to \pi$ (where we recognize $\max(F(u), 1 - \alpha) \ge \pi_x(u) \Leftrightarrow \alpha \le N(F)$ and $\min(\mu_F(u), \alpha) \le \pi_x(u) \Leftrightarrow \alpha \le \Delta(F)$). Although N(F) in (7.51) reduces to the standard necessity degree (7.18) when F is non-fuzzy, it differs from (7.50) when * = max and I(a, b) = 1 - (a*(1 - b)). This should not be surprizing. In (7.50), N(F) = 1 if and only if $\{u, \pi_x(u) > 0\} \subseteq \{u, F(u) = 1\}$, i.e., F is completely certain whatever the crisp interpretation we may make of the fuzzy set F, since the event corresponding to the core of F has certainty 1 (i.e., this event is true for any non-impossible value of x). In (7.51), N(F) = 1 if and only if $\pi_x \le F(\cdot)$, since we would hold "x is F" for certain (with α = 1) in this case, in agreement with the entailment principle. This also agrees with the interpretation of the certainty qualification "F is α–certain" as the family of level cut-based statements "$F_{\overline{1-\lambda}}$ s min(α,λ)–certain", i.e., $\forall \lambda \in [0,1]$, $N(F_{\overline{1-\lambda}}) \ge \min(\alpha,\lambda)$ with $F_{\overline{\alpha}} = \{u, \mu_F(u) > \alpha\}$; it expresses that the larger the crisp interpretation of the fuzzy event F, the greater the certainty should be. Similarly $\Delta(F)$ coincides with the guaranteed possibility index of Section 7.3.4 when F is not fuzzy.

7.6.2 Sugeno Integrals

Possibility and fuzzy set theory enable fuzzy statements to be evaluated in terms of partial belief, on the basis of fuzzy evidence. When evidence is of a different nature (e.g., probabilistic), the evaluation of fuzzy statements is still meaningful and achievable, (viz. works by Zadeh (1968) on the probability of a fuzzy event). More generally, fuzzy statements can be evaluated by means of monotone set functions called fuzzy measures.

Roughly speaking, fuzzy measures are monotone increasing (with respect to set inclusion) set-functions. Namely If A is a subset of B it is assumed that $\mathcal{M}(A) \leq \mathcal{M}(B)$, where $\mathcal{M}(A)$ is a degree of confidence of A. In his thesis, Sugeno (1974) was inspired by fuzzy logic operations in formalizing the concept of integral of a function F valued on a totally ordered scale L, with respect to a fuzzy measure $\mathcal{M}$. Replacing addition and multiplication of numbers by max and min, respectively, he arrived at the following definition where $F_\alpha = \{u \in U: F(u) \geq \alpha\}$ is the level-cut of the fuzzy set with membership function $F(\cdot)$ (Sugeno, 1974, 1977):

$$\int F d\mathcal{M} = \sup_{\alpha \geq 0} \min(\alpha, \mathcal{M}(F_\alpha)). \tag{7.53}$$

For a careful comparison of Sugeno integral and Lebesgue integral, see Ralescu and Adams (1980). Note that if α varies on a numerical, continuous scale, this definition can also be written as

$$\int F d\mathcal{M} = \inf_{\alpha \geq 0} \max(\alpha, \mathcal{M}(F_\alpha)).$$

Consider the case of a finite set $U = \{u_1, \ldots, u_m\}$. The Sugeno integral reduces to the following expression:

$$\begin{aligned} \int F d\mathcal{M} &= \max_{i=1,m} \min(F(u_{\sigma(i)}), \mathcal{M}(\{u_{\sigma(i)}, \ldots, u_{\sigma(m)}\}) \\ &= \min_{i=1,m} \max(F(u_{\sigma(i)}), \mathcal{M}(\{u_{\sigma(i+1)}, \ldots, u_{\sigma(m)}\}) \end{aligned}$$

using an increasing reordering of vector $(F(u_1, \ldots, F(u_m))$ according to a permutation σ. S(u) is actually the median of $2m - 1$ terms $\{F(u_1, \ldots, F(u_m), \mathcal{M}(\{u_{\sigma(2)}, \ldots, u_{\sigma(m)}\}), \ldots, \mathcal{M}(\{u_{\sigma(m)}\})\}$ (Dubois and Prade, 1980).

Possibility and necessity of fuzzy events are particular cases of Sugeno integrals, where $\mathcal{M}$ is a possibility measure and a necessity measure respectively. Namely $\int F d\Pi = \sup_{\alpha \in [0,1]} \{\alpha \wedge \Pi(u: F(u) \geq \alpha)\} = \Pi(F)$ and $\int F dN = \sup_{\alpha \in [0,1]} \{\alpha \wedge$

$N(u: F(u) \geq \alpha))\} = N(F)$ as given in (7.49) and (7.50) respectively, and $* = \min$, $I(a,b) = \max(1 - a, b)$; see for instance Dubois and Prade (1980), Inuiguichi *et al.* (1989), Grabisch *et al.*, (1992).

Variants of Sugeno integrals exist such as Shilkret integral

$$\int_{prod} F d\mathcal{M} = \sup_{\alpha \geq 0} \alpha \cdot \mathcal{M}(F_\alpha)$$

where the minimum is replaced by a product in (7.53). Shilkret (1971) only studies the case when the fuzzy measure is maxitive, i.e., a possibility measure. In that case $\int_{prod} F d\Pi = \Pi(F)$ with $*$ = product in (7.49) a notion studied by Kaufmann (1978) under the name "admissibility function" and Agbeko (1991) under the name "optimal average". Later on, particularizations of the extension of Sugeno integral with t-norms and conorms in the finite setting have been considered by Suarez Garcia and Gil Alvarez (1986). T-norm- and conorm-based possibility integrals have been studied by De Cooman *et al.* (1992), Yan (1993, 1994), De Cooman and Kerre (1996). They correspond to the following Sugeno integral-like schemes, called normed and conormed possibility integrals, respectively :

$$\int_{*} F d\Pi = \sup_{\alpha \geq 0} \alpha * \Pi(F_\alpha);$$
$$\int_{\perp} F d\Pi = \inf_{\alpha \geq 0} \alpha \perp N(F_\alpha).$$

These expressions can be simplified into a form of $\Pi(F)$ as in (7.49) and $N(F)$ in (7.50) (see for instance De Cooman and Kerre (1996)). On such a basis, De Cooman (1997, Part I) outlines a theory of possibilistic integration, where possibility levels are taken in a complete lattice, and the operation $*$ is a so-called semi-norm (a fuzzy conjunction that is not necessarily commutative or associative, but is increasing in the wide sense, and such as $1 * a = a * 1 = a$). He proves possibilistic counterparts of Radon-Nikodym and Fubini theorems. In a series of papers, Mesiar (1992, 1994, 1997) studies the representation of possibility of fuzzy events using Shilkret and Sugeno integrals via Markov kernels.

7.6.3 Quantitative Possibility and Choquet Integrals

For numerical possibility and necessity measures, another kind of evaluation of a function F with respect to a fuzzy measure $\mathcal{M}$ makes sense. It is based on a generalization of Lebesgue integral, called Choquet integral, defined as follows, when F is the membership function of a fuzzy set:

$$\mathcal{M}_c(F) = \int_0^1 \mathcal{M}(F_\alpha)\, d\alpha \qquad (7.54)$$

See the appendix for a mathematical introduction. When $\mathcal{M} = P$, a probabibility measure, it reduces to a Lebesgue integral. In particular, Zadeh (1968)'s probability of a fuzzy event takes this form, as noticed by Höhle (1976):

$$P(F) = \int_U F(u)dP(u) = \int_0^1 P(F_\alpha)d\alpha.$$

For $\mathcal{M} = \Pi$ or N we get another definition of the possibility and necessity of fuzzy events. However the maxitivity of $\Pi_c(F)$ is not preserved. In particular the interval $[\int_0^1 N(F_\alpha)\, d\alpha, \int_0^1 \Pi(F_\alpha)\, d\alpha]$ is the range of the expectations of F with respect to probability measures P such that $\Pi \geq P$. Moreover

$$N_c(F) = \int_0^1 N(F_\alpha)\, d\alpha = \int_{\mathbb{R}} F\, d\Phi_* = \int_0^1 \inf\{F(u): \pi(f(u)) \geq \alpha\}\, d\alpha,$$

$$\Pi_c(F) = \int_0^1 \Pi(F_\alpha)\, d\alpha = \int_{\mathbb{R}} F\, d\Phi^* = \int_0^1 \sup\{F(u): \pi(f(u)) \geq \alpha\}\, d\alpha,$$

where $\Phi^*(\alpha) = \Pi(\{u, F(u) \leq \alpha\})$ and $\Phi_*(\alpha) = N(\{u, F(u) \leq \alpha\})$. When, $U = \mathbb{R}$, and the possibility distribution is a fuzzy interval M, and F is the identity function, it leads to compute the upper and lower expected values of a fuzzy interval (see Chapter 10 by Dubois, Kerre, Mesiar and Prade in this volume).

In the finite setting, these expressions read (Dubois and Prade, 1985b; Grabisch *et al.*, 1995; Marichal, 1998) :

$$\begin{aligned} Nc(F) &= \Sigma_{i = 1,\dots n} F(u_{\sigma(i)}) (\max_{k \leq i} \pi(u_{\sigma(k)}) - \max_{k < i} \pi(u_{\sigma(k)})) \\ &= \Sigma_{i = 1,\dots n} (F(u_{\sigma(i)}) - F(u_{\sigma(i-1)})) \min_{k < i} 1 - \pi(u_{\sigma(k)}) \\ &= \Sigma_{i = 1,\dots n} (\pi_i - \pi_{i+1}) \min\{F(u): u \in A_i\} \end{aligned}$$

$$\begin{aligned} \Pi_c(F) &= \Sigma_{i = 1,\dots n} F(u_{\sigma(i)}) (\max_{k \geq i} \pi(u_{\sigma(k)}) - \max_{k > i} \pi(u_{\sigma(k)})) \\ &= \Sigma_{i = 1,\dots n} (F(u_{\sigma(i)}) - F(u_{\sigma(i-1)})) \max_{k \geq i} \pi(u_{\sigma(k)}) \\ &= \Sigma_{i = 1,\dots n} (\pi_i - \pi_{i+1}) \max\{F(u): u \in A_i\} \end{aligned}$$

where, without loss of generality, $\pi_1 = 1 > \dots > \pi_n > \pi_{n+1} = 0$, and A_i is the π_i–level-cut of F, and $F(u_{\sigma(1)}) \leq \dots \leq F(u_{\sigma(n)})$.

7.6.4 Decision- Theoretic Foundations of Possibility Theory

In the Savage (1954) framework, the consequence of a decision depends on the state of the world in which it takes place. For instance, in a restaurant one must choose our dishes without being totally sure that the food is good. If U is a set of states and X a set of possible consequences, a decision is a mapping from U to X. The decision-maker has some knowledge of the actual state and some preference on the consequences of his decision. The usual approach to decision-making under uncertainty is based on representing uncertainty on states by a unique probability distribution and the decision-maker by a utility function on consequences. Then decisions are selected on the basis of expected utility.

When uncertainty is just due to a lack of information and only the result of the present decision matters (as in the restaurant), the idea of expectation is less attractive. For this reason a purely possibilistic approach to decision under uncertainty that addresses this kind of situation has been developed.

Possibilistic Decision Criteria. It makes sense, if information is qualitative, to represent the incomplete knowledge on the state by a possibility distribution π on U with values in a plausibility scale L and the decision-maker's preference on X by means of another possibility distribution μ with values on a preference scale T. The utility of a decision d whose consequence in state u is $x = d(u) \in X$ for all states u, can be evaluated by combining the plausibilities $\pi(u)$ and the utilities $\mu(x)$ is a suitable way. Two quatitative criteria that evaluate the worth of decision d have been put forward in the literature, provided that a commensurability assumption between plausibility and preference is made:

– *a pessimistic criterion*

$\mathcal{U}_*(d) = \inf_{u \in U} \max(n(\pi(u)), \mu(d(u)))$

which generalizes the max-min Wald criterion in the absence of probabilistic knowledge. Mapping n is order-reversing from L to T.

– *an optimistic criterion*

$\mathcal{U}^*(d) = \sup_{u \in U} \min(m(\pi(u)), \mu(d(u)))$

which generalizes the maximax optimistic criterion. Mapping m is order-preserving from L to T.

These criteria are clearly based on possibility and necessity of fuzzy events, and are special cases of Sugeno integrals. The optimistic criterion has been first proposed by Yager (1979) and the pessimistic criterion, under the form of a maximin disutility, by Whalen (1984), and also used in Inuiguchi *et al.* (1989).

Maximizing $\mathcal{U}_*(d)$ means to find a decision d, all the highly plausible consequences of which are also highly preferred. The definition of "highly plausible" is decision-dependent and reflects the compromise between high plausibility and low utility expressed by the order-reversing map between the plausibility valuation set L and the utility valuation set S. The pessimistic utility $\mathcal{U}_*(d)$ is small as soon as there exists a possible consequence of d which is both highly plausible and bad with respect to preferences. This is clearly a risk-averse and thus a pessimistic attitude. It generalizes the max-min Wald criterion in the absence of probabilistic knowledge since if π is the characteristic function of a subset A of states, $\mathcal{U}_*(d)$ is the utility of the worst consequence of states in A, however unlikely they are. But the possibilistic criterion is less pessimistic. It focuses on the idea of usuality and relies on the worst *plausible* consequences induced by the decision. Some unlikely states are neglected by a variable thresholding and the threshold is determined by comparing the possibility distributions valued on L and S via the mapping η. On the contrary, $\mathcal{U}^*(d)$ corresponds to an optimistic attitude since it is high as soon as there exists a possible consequence of d which is both highly plausible and highly prized. It generalizes the maximax optimistic criterion.

For binary acts the expression of the utilities is rather simple: let xAy denote a decision that gives consequence x if A occurs and y otherwise, and assume x is preferred to y ($\mu(x) > \mu(y)$). Observe that $N(A) = n(\Pi(A^c))$ so that:

$$\begin{aligned}\mathcal{U}_*(xAy) &= \max(\mu(y), \min(N(A), \mu(x))) \\ &= \min(\mu(x), \max(N(A), \mu(y)))\end{aligned}$$

This form of the pessimistic utility is easy to understand: if the agent is sure enough that A occurs ($N(A) > \mu(x)$) then the utility of the act xAy is $\mu(x)$. If the agent has too little knowledge ($\max(N(A), N(A^c)) < \mu(y)$)), then cautiousness prevails and the utility is $\mu(y)$, the worst case. Of course the same happens if the agent is at least somewhat certain that A^c occurs. If the agent's certainty that A occurs is positive but not extreme, the utility reflects the certainty level and is equal to N(A). Note that the utility of the binary decision is the median of $\{\mu(x), N(A), \mu(y)\}$, thus contrasting with expected utility, which is a mean.

Similarly, the optimistic utility of the binary act takes the simplified form (if function m = identity):

$$\mathcal{U}^*(xAy) = \max(\min(\Pi(A), \mu(x)), \mu(y))$$

and can be interpreted similarly as the median of $\{\mu(x), \Pi(A), \mu(y)\}$, but here the utility is $\mu(x)$ as soon as the agent believes that obtaining x is possible enough ($\Pi(A) > \mu(x)$).

Axiomatics for Possibilistic Lotteries. The pessimistic criterion has been axiomatically justified by Dubois and Prade (1995) in the style of Von Neumann and Morgenstern utility theory; see (Dubois, Godo, Prade and Zapico, 1998) for the simplified set of axioms given below. The idea is that if the uncertainty on the state is represented by π, each decision induces, on the set of consequences X, a possibility distribution such that $\pi_d(x) = \Pi(d^{-1}(x))$. So ranking decisions comes down to ranking possibility distributions on X. Assume the decision-maker supplies an ordering between possibility distributions on X, thus expressing his attitude in front of risk, that is, in front of various possibilities of happy and unhappy consequences in X. Let $\pi_d(x)$ be the plausibility of getting x under decision d. The question is to know what kind of axioms on the ordering between possibility distributions on X make it representable by the ranking of decisions according to the above pessimistic or optimistic criteria. Let $(\lambda/\pi, \mu/\pi')$ denote the "qualitative lottery" yielding π with plausibility λ and π' with plausibility μ. Of course, $\max(\lambda, \mu) = 1$. The following are the pessimistic axioms:

1: The set of possibility distribution is equipped with a structure ($\succcurlyeq, \sim, \succ$) of *complete preordering*, where $\succcurlyeq, \sim, \succ$ are the weak preference, indifference and strict preference.

2: *Independence*: $\pi_1 \sim \pi_2 \Rightarrow (\lambda/\pi_1, \mu/\pi) \sim (\lambda/\pi_2, \mu/\pi)$

3: *Continuity*: if $\pi \succ \pi'$ then $\exists\ \lambda \in T$, $\pi' \sim (1/\pi, \lambda/X)$

4: *Reduction of lotteries*: $(\lambda/(\alpha/\pi, \beta/\pi'), \mu/\pi') \sim (\gamma/\pi, \delta/\pi')$ where $\gamma = \min(\alpha,\lambda)$ and $\delta = \max(\mu, \min(\lambda,\beta))$

5: *Risk aversion*: $\pi \geq \pi' \Rightarrow \pi' \succcurlyeq \pi$.

The risk aversion axiom states that the less informative is π, the more risky is the situation: The worst epistemic state is total ignorance. Continuity says that the utility of π goes down if the uncertainty about π raises. It can be proved that if the knowledge is represented by a subset A of possible states, then $\exists\ x \in A$, $x \sim A$. This property, violated by expected utility, suggests that contrary to it, the pessimistic utility is not based on the idea of average and repeated decisions, but makes sense for one-shot decisions. The reduction of lotteries suggests that if lottery $(\alpha/\pi, \beta/\pi')$ is embedded in the lottery$(\lambda/x, \mu/\pi')$ at place x, π obtains in two sequential steps with possibility $\min(\alpha,\lambda)$ and π' obtains either directly with possibility μ or in two steps with possibility $\min(\lambda,\beta)$.

The reduction of lotteries property is due to the fact that the whole setting relies on a qualitative counterpart of the structure of probabilistic mixtures that underlies the Von Neumann and Morgenstern approach (Herstein and Milnor, 1953). This is due to the closure of the set of possibility measures under weighted maxima: if Π and Π' are possibility measures then, namely $\max(f(\Pi), f'(\Pi'))$ is a possibility measure again (Dubois and Prade, 1990c) where f and f' are monotonic transformations of the possibility scale, with $f(0) = f'(0) = 0$, and $\max(f(1), f'(1)) = 1$. This is the only aggregation that preserves possibility measures.

Let H be the set of ordered pairs (α, β) such that $\max(\alpha, \beta) = 1$. A possibilistic mixture set is a pair (G, M) where G is a set of possibilistic lotteries, M is a function from G×G×H to G that computes for each pair of possibilistic lotteries π

and π' and each pair of weights (α, β) in H another element of G denoted M(π, π'; α, β) such that (Dubois, Fodor, Prade and Roubens,1996):

M1: $M(\pi, \pi'; 1, 0) = \pi$
M2: $M(\pi, \pi'; \alpha, \beta) = M(\pi, \pi'; \beta, \alpha)$
M3: $M(M(\pi, \pi'; \alpha, \beta), \pi'; \lambda, \mu) = M(\pi, \pi'; \min(\alpha, \lambda), \max(\min(\beta,\lambda), \mu))$

M1 says that lottery (1/π, 0/π') is π; M2 says that lotteries (λ/π, μ/π') and (μ/π', λ/π') are the same; M3 is axiom 4 (which can be dropped if M3 is adopted as a computation rule). M1-M2-M3 imply the following:

M4 $M(\pi, \pi; \alpha, \beta) = \pi$
M5: $M(M(\pi, \pi'; \alpha, \beta), M(\pi, \pi'; \gamma, \delta); \lambda, \mu)$
$= M(\pi, \pi'; \max(\min(\alpha, \lambda) \min(\gamma, \mu)), \max(\min(\beta, \lambda) \min(\delta, \mu))$

The last expression explains the form of the optimistic criterion. The form of the other criterion is obtained by working with degrees of necessity (ordered pairs (α, β) such that min(α, β) = 0).

Act-Based Justification of Possibility Theory. An alternative axiomatic justification has been developed in the style of Savage (Dubois, Prade and Sabbadin, 1998). The problem is one of representing the preference relation of an agent over the set $D = X^U$ of potential decisions d by means of a utility function whose form is dictated by the properties of this preference relation $\succcurlyeq$. Let dAd' denotes a decision that coincides with d if A occurs and d' if not. An example of such a set of axiom for the pessimistic utility is as follows:

S1. (D, $\succcurlyeq$) is a complete preorder.

S2. There exist two decisions d and d' where d is strictly preferred to d' (d $\succ$ d').

S3. If **x** and **y** denote *constant acts* that respectively yield consequence x and y for any state and for any decision d and any event A: **x** $\succcurlyeq$ **y** implies **x**Ad $\succcurlyeq$ **y**Ad.

S4. If d $\succ$ d' and d $\succ$ d'' then d $\succ$ d'$\vee$ d'', where d' is a constant act.

S5. If d' $\succ$ d and d'' $\succ$ d then d'$\wedge$ d''$\succ$ d.

If the preference relation over D satisfies the above axioms then there exists a possibility distribution on U, a preference function on X and an order-reversing function between plausibility and preference scales such that d $\succ$ d' if and only if $\mathcal{U}_*(d) > \mathcal{U}_*(d')$ using the pessimistic possibilistic utility. The optimistic criterion can be justified likewise, moving the constant act condition from axiom S4 to S5. In fact axioms S1, S2, S3 are enough to show that the uncertainty on states is captured by a monotonic (Sugeno) set-function. We have also proved that S1, S2, S3, S4, and S5 restricted to constant acts d', are enough to show that the preference relation $\succ$ is representable by a Sugeno integral (Dubois, Prade and Sabbadin, 1998a). S5 ensures the min-decomposition of the uncertainty measure (which is then a necessity function). The pessimistic nature of S5 can be guessed from its equivalent form: d'Ad $\succ$ d implies d $\succcurlyeq$ dAd'. If decision d is improved by d' when A occurs it means that A is sure enough to occur and there is no way of improving decision d by changing it in case A does not occur (the agent neglects this possibility of non-occurence).

This decision-theoretic approach sounds realistic in settings where information about plausible states and preferred consequences is poor and linguistically expressed, and where decisions will not be repeated.

Preference vs. Uncertainty in Fuzzy Sets: Why Bother? In possibilistic decision theory, both the preference profile of decision makers and their uncertainty can be modelled by possibility distributions. This situation is a bit dangerous because there is a great temptation to confuse uncertainty and preference in a fuzzy approach, as sometimes found in the literature. Not making the distinction may lead to the wrong approach to modelling fuzzy decision problems as shown now.

Example (Dubois, Fargier and Prade, 1995). Suppose Tom wants to attend a meeting in the morning at 8 a.m. He wants to know when to get up, so as to arrive on time. He has to take the bus to reach the venue of the meeting, and the bus ride takes about 1 hour. Besides he does not want to leave his house too early, say before 7, and in any case not earlier than 6:30 h. Clearly the time when he leaves his house is his choice. He can express preferences on this decision variable by means of a linear fuzzy interval R with a nondecreasing membership function, lower modal value 7 and support [6.5, 12]. As for the travel time, Tom has no control on it. It depends on the waiting time (not more than 15 mn) and the traffic situation, which never causes more than a 15 mn delay. His experience suggests that the usual overall travel time (including the wait at the bus stop) is 1 hour. But if Tom is lucky (no wait, no traffic jam) it can take only 45 mn. and, if unlucky, 1 hour and 15mn. This information builds up a triangular fuzzy number T that models the uncertainty about the travel time; it does not reflect preference. Tom has preference on his arrival time: he would like to arrive at 8 or before, but certainly not after 8:15 a.m. This makes up the fuzzy arrival time D. How can Tom choose his actual starting time? Treating all these pieces of information in a symmetric way (without separating uncertainty and preference the problem becomes one of finding both s and t which maximize the membership of the starting time, the driving time, and the arrival hour, i.e., $\max_{s,t} \min(\mu_R(s), \mu_T(t), \mu_D(s+t)))$. This problem is trivially solved by letting s = 7 and t = 1 with membership 1. But nobody can believe it is a sound solution to the problem because in many cases the travel will take more than one hour and Tom will be in trouble.

Assuming that the specifications for the starting time R, the travel time T, and arrival time D are intervals, what Tom wants to find is a starting time s that does not force him to get up too early (hence $s \in R$) such that whatever the travel time (hence for any $t \in T$) he arrives on time at the meeting ($s+t \in D$). Mathematically this means:

$$\exists\, s, \text{ s.t. } s \in R \text{ and } \forall t, \text{ if } t \in T \text{ then } s+t \in D.$$

When R, T, D are fuzzy sets it comes down to finding s that maximizes

$$\min(\mu_R(s), \inf_t \max(1- \mu_T(t), \mu_D(s+t))).$$

The maximization and minimization are the fuzzy counterpart of the universal ($\forall$) and existential ($\exists$) quantifiers; and max(1-a, b) is a multiple-valued implication. Clearly, the term $\inf_t \max(1.- \mu_T(t), \mu_D(s+t))$ represents a pessimistic utility function and is equal to the degree of necessity of the fuzzy event $s+t \in D$ given the uncertain information $t \in T$. The calculation is detailed in (Dubois *et al.* 1995). The reader can check that the optimal satisfaction level is .75, and that the optimal time for Tom to start safely enough is $6.875 = \mu_{R^{-1}}(.75)$ (= 6: 52' 30" a.m.). This result presupposes that the travel time does not exceed 1.1875 (= 1h 11' 20"); note that this is very likely. Moreover taking these conditions for granted, Tom will arrive at his meeting not later than $\mu_{D^{-1}}(.75) = 8.0625$ (= 8: 03' 45" a.m.). Note that this solution sounds quite reasonable in practice. The previous solution assumed a very optimistic attitude.

These qualitative counterparts of the expected utility theory nicely fit the setting of flexible constraint propagation (Dubois, Fargier and Prade, 1996b) as suggested by the previous example, illustrating the difference between a fuzzy set modelling preference and a fuzzy set modelling uncertainty, for making decisions. Of course the previous example also suggests that the possibilistic criteria are useful in quantitative settings as well. However we may then think of alternative decision criteria:

$$\mathcal{U}_*(d) = \inf_{u \in U} n(\pi(u)) \perp \mu(d(u))),$$
$$\mathcal{U}^*(d) = \sup_{u \in U} m(\pi(u)) * \mu(d(u))),$$

where $\perp$ and $*$ are triangular conorms and norms. Interestingly the structure of possibilistic mixtures is preserved in this extension (Dubois, Fodor, Prade and Roubens, 1996). Alternatively one may use Choquet integrals for representing pessimistic and optimistic criteria. This is in accordance with the view of possibility measures as families of probabilities, since Choquet integral representations of utility functions have been obtained by Schmeidler (1986) in that setting. However Schmeidler theorem does not justify the use of possibility measures proper for representing uncertainty.

7.7 CONCLUSION

This chapter has investigated the relationships between fuzzy set, possibility and probability. The concept of possibility distribution plays a central role in the bridge between probability theory and fuzzy sets. Possibility theory is a theory of uncertainty dual to probability: while the latter synthetizes scattered pieces of precise information (as produced by precisely observed random phenomena, or by the behavior of gamblers sharing a stake among mutually exclusive alternatives), possibility theory can represent a body of imprecise but coherent evidence. Possibility theory and the revealed links between probability and fuzzy sets lead to new uncertainty and decision modeling tools. They also shed new light on old practice. For instance fuzzy set-theoretic operations can be justified from several probabilistic points of view according to whether the membership function is viewed as an upper probability, the one-point coverage of a random set, or a likelihood function. Conversely, the possibilistic nature of likelihood functions seems to be in accordance with the way statisticians have used them, it leads to use fuzzy set theory as a basis for deriving likelihood functions, and fuzzy events can be considered as relevant evidence in statistics. Possibility theory is also a very simple theory of uncertainty that lends itself either to a purely ordinal and efficient treatment of uncertainty, or in its numerical versions to a versatile, computationally attractive tool for approximating families of probability functions.

There are already several domains of application where fuzzy sets and probability are conjointly used. A typical example is fuzzy random variables (e.g., Kruse and Meyer, 1987; Negoita and Ralescu, 1987) that can handle opinion polls where people can give fuzzy (linguistic) answers. Statistics with fuzzy events has received extensive treatment in the past from Okuda *et al.* (1987) to Gil (1988), and Viertl and Hule (1991), Viertl (1996). The use of fuzzy sets in statistics is surveyed in a chapter of a companion volume of the Handbook of Fuzzy Sets (Gil and Kruse, 1998). Lastly fuzzy sets can model linguistic probabilities as verified experimentally (Wallsten *et al.* 1986). Linguistic probabilities can be handled in inference processes, as done by Zadeh (1985), Dubois and Prade (1988d), Jain and Agogino (1990), or in decision analysis as done by Watson *et al.* (1979), and Nau (1992) for instance. These examples should convince us that instead of considering probability and fuzzy sets as conflicting rivals, it sounds more promising to build bridges and take advantage of the enlarged framework for modeling uncertainty and vagueness they conjointly offer.

MATHEMATICAL APPENDIX

The following notions are briefly recalled: first random sets, which generalize fuzzy sets to a non-nested family of sets, upper and lower probabilities which particularize into belief functions which can be viewed as stemming from disjunctive random sets (elements in the sets are mutually exclusive), and lastly fuzzy measures which can be viewed as monotonic set functions. The corresponding notions of integrals are also introduced.

Random sets.

For concreteness, consider the Euclidean space $\mathbb{R}^d$. Let $\mathcal{F}$, $\mathcal{G}$, $\mathcal{K}$ denote the classes of closed, open and compact sets of $\mathbb{R}^d$, respectively. A (closed) random set S is a measurable mapping from some measurable space $(\Omega,\mathcal{A})$ to $\mathcal{F}$. To specify this measurability, we need to specify a σ-field $\sigma_{\mathcal{F}}$ on $\mathcal{F}$. For this purpose, for each $A \subseteq \mathbb{R}^d$, let $\mathcal{F}^A = \{F \in \mathcal{F}: F \cap A = \emptyset\}$, $\mathcal{F}_A = \{F \in \mathcal{F}: F \cap A \neq \emptyset\}$.

The space F is endowed with the hit-or-miss topology generated by

$$\mathcal{F}^K_{G_1,G_2,\ldots,G_n} = \mathcal{F}^K \cap \mathcal{F}_{G_1} \cap \ldots \cap \mathcal{F}_{G_n}$$

for $n \geq 0$, $K \in \mathcal{K}$, $G_i \in \mathcal{G}$. The σ-field $\sigma_{\mathcal{F}}$ is taken to be the Borel σ-field on this topological space $\mathcal{F}$.

If P is a probability measure on $(\Omega,\mathcal{A})$, then the probability law of S is the probability measure image P_S on $(\mathcal{F},\sigma_{\mathcal{F}})$, that is $\forall A \in \sigma_{\mathcal{F}}$, $P_S(A) = P[S^{-1}(A)]$. By abuse of language, any probability measure on $(\mathcal{F},\sigma_{\mathcal{F}})$ is called a (closed) random set (the probabilistic "behavior" of S is completely characterized by such a measure). It turns out that P_S is characterized by the following set-function T, defined only on $\mathcal{K}$:

$$T(K) = P_S(\mathcal{F}_K),\ K \in \mathcal{K}.$$

Specifically, let T be defined on $\mathcal{K}$. Then, there is a unique probability measure Q on $(\mathcal{F},\sigma_{\mathcal{F}})$ with $Q(\mathcal{F}_K) = T(K)$, $\forall K \in \mathcal{K}$ if and only if T is an alternating Choquet capacity of infinite order such that $0 \leq T(K) \leq 1$, $\forall K \in \mathcal{K}$, and $T(\emptyset) = 0$ (this is Choquet's theorem). This kind of capacity in Choquet's theorem is a function $T: \mathcal{K} \to [0,1]$ such that

a) $T(\emptyset) = 0$
b) if $K_n, K \in \mathcal{K}$ and $K_n \searrow K$ then $T(K_n) \searrow T(K)$
c) define $S_1(K_0;K) = T(K_0 \cup K) - T(K_0)$, and more generally, $S_n(K_0; K_1, K_2, \ldots, K_n) = S_{n-1}(K_0; K_1, \ldots, K_{n-1}) - S_{n-1}(K_0 \cup K_n; K_1, K_2, \ldots, K_{n-1})$, for $n \geq 2$, then $S_n \geq 0$, $\forall n \geq 1$. (Note that since $S_1 \geq 0$, T is monotone increasing).

Since the properties of T resemble those of cumulative distribution functions of real-valued random variables, we call T the distribution of a random set S.

In the theory of extremal processes (see, e.g., Norberg, 1986), we encounter capacities T such that $T(K_1 \cup K_2) = \max(T(K_1), T(K_2))$. Such capacities are called *maxitive capacities*. Here is an example showing that possibility measures are maxitive capacities when the associated possibility distributions π are upper semi-continuous (u.s.c.); let $\Pi: \mathbb{R}^d \to [0,1]$ be u.s.c. Define $\hat{\Pi}(K) = \sup_{x \in K} \pi(x)$, $K \in \mathcal{K}$. Then, $\hat{\Pi}$ is the distribution of the random set S defined by

$$S(\omega) = \{x \in \mathbb{R}^d: \pi(x) \geq \alpha(\omega)\}$$

where α: $(\Pi, \mathcal{A}, P) \rightarrow [0,1]$ is a random variable, uniformly distributed. See also (Molchanov, 1993), and Nguyen (1984). Note that the unicity of the random set S comes from a stronger condition than the one-point-coverage function of Goodman (1982), namely:

$$\forall K \in \mathcal{K},\ \hat{\Pi}(K) = P\{\omega: S(\omega) \cap K \neq \emptyset\}.$$

Clearly, this is even stronger than the many-point coverage function $P\{\omega: \{x_1, x_2, \ldots, x_n\} \subseteq S(\omega)\}$.

As a matter of fact, Robbin's formula (Robbins, 1944) for the mathematical expectation says it all: let S be a (closed) random set on $\mathbb{R}^d$ as before. Let $\mathcal{M}(S)$ denote the volume of S. Then $\mathcal{M}(S)$ is a non-negative random variable, its expectation is given by

$$E\mathcal{M}(S) = \int_{\mathbb{R}^d} \pi(x) d\mathcal{M}(x)$$

where $d\mathcal{M}(x)$ is the Lebesgue measure on $\mathbb{R}^d$, and $\pi(x) = P\{\omega: x \in S(\omega)\}$. Thus, knowing $\pi(\cdot)$, we know the first moment of the random variable $\mathcal{M}(S)$, but not its distribution. It is interesting to consider the following simple example: let X be a non-negative random variable. Define the (closed) random set S by $S(\omega) = [0, X(\omega)]$ then the length of S is $|S(\omega)| = X(\omega)$. The one-point-coverage function of S is $\pi(x) = P(\omega: x \in [0, X(\omega)]) = P(\omega: 0 \leq x \leq X(\omega))$. So that Robbins' formula becomes:

$$E\ |S(\omega)| = EX = \int_0^{+\infty} P(X \geq x)dx$$

(note that, for $x < 0$, $P\{\omega: x \in [0, X(\omega)]\} = 0$) which is a well-known alternative formula to compute EX when $X \geq 0$.

This well-known formula is in fact a form of *Choquet integral* of the measurable function X (random variable) with respect to the set-function P. When P is replaced by a capacity T (which is not necessarily an additive set-function), or more generally, by a monotone increasing set-function (called a *fuzzy measure*), the expression $\int_0^{+\infty} T(f \geq x)dx$ where f is any measurable, non-negative function, is still well-defined, since:

(i) f being measurable, the sets $(f \geq x)$ belong to the domain of T,
(ii) T being monotone increasing, the function $x \rightarrow T(f \geq x)$ is monotone, and hence measurable.

When T is a fuzzy measure, the above expression is termed the *fuzzy integral* of f with respect to T.

Non-Additive Set-Functions

Possibility measures are examples of set-functions which are non-additive (as opposed to ordinary measures such as probability measures). In this part, we recall other non-additive set-functions, used specifically for measuring uncertainty, from a mathematical standpoint. Reasoning with incomplete information is an every-day activity of humans, and witnesses aspects of their "intelligence". Thus it seems reasonable to assume that they use some form of "general degrees of belief" in various tasks, such as making decisions, assessments of risk, etc.

Probabilistic Envelopes. Consider first the case of "incomplete *probabilistic* knowledge". Let X be a random quantity of interest. The "true" probability measure P_0 on Ω of X is the complete knowledge about X. Saying that the knowledge of P_0 is incomplete amounts simply to say that we only know that P_0 belongs to some (known) class $\mathcal{P}$ of probability measures, but it is not known which one. This is a form of "imprecise" information.

As in robust Bayesian statistics, see, e.g., Walley (1991), in a conservative manner, one could use "safe" bounds on P_0, namely

$$P_*(\cdot) \leq P_0(\cdot) \leq P^*(\cdot) \text{where } P_*(\cdot) = \inf_{\mathcal{P}} P(\cdot),\ P^*(\cdot) = \sup_{\mathcal{P}} P(\cdot).$$

They are termed *lower* and *upper probabilities*, respectively. Since, $\forall$ event A, we have $P^*(A) = 1 - P_*(A^c)$, where A^c denotes the set-complement of A, one needs only to focus on one of the bounds, for example on the lower envelope P_*. Thus, without any other additional information, the incomplete probabilistic knowledge is translated into the set-function P_*. Without the precise knowledge of P_0, we are forced to use P_* in all inferential purposes. In using P_*, the first thing to observe is that, unlike P_0, P_* is no longer *additive*, and hence is *not* a probability measure (of course when the class $\mathcal{P}$ is different from the singleton $\{P_0\}$). In general, $P_*(\emptyset) = 0$, $P_*(\Omega) = 1$, and $A \subseteq B$ implies $P_*(A) \leq P_*(B)$.

As our example of inference using P_*, suppose, in a simplest decision-framework, we wish to use expected value $E_{P_0}(X)$ to make decisions. There are two obvious courses of action: use, place of the unknown $E_{P_0}(X)$, either $\inf_{\mathcal{P}} E_P(X)$, or $\mathcal{E}_{P_*}(X)$. The latter symbol stands for the *Choquet integral* of X with respect to the monotone-increasing set-function P^*, namely

$$\mathcal{E}_{P*}(X) = \int_0^{+\infty} P*(\omega: X(u) > t)dt + \int_{+\infty}^{0} [P*(\omega: X(\omega) > t) - 1]dt$$

(see the next section on fuzzy integrals). Obviously, the computation of $\mathcal{E}_{P*}(X)$ is much simpler than that of $\inf_{\mathcal{P}} E_P(X)$. There are *cases* where we do not need to choose among the above two alternatives, since they coincide! Here is a simple example. See also Wasserman (1987). Let $X: \Omega \to \Theta = \{1, 2, 3, 4\}$. The true density f_0 of X on Θ is not known completely. Suppose partial information about f_0 is given as

$$f_0(1) \geq 0.4,\ f_0(2) \geq 0.2,\ f_0(3) \geq 0.2,\ f_0(4) \geq 0.1. \qquad (*)$$

For a density function f on Θ, we write P_f for its associated probability measure on Θ, i.e.,

$$A \subseteq \Theta,\ P_f(A) = \Sigma_{\theta \in A}\, f(\theta).$$

Since f_0 is only partially specified, all we can say about the "true" law P_{f_0} of X is that $P_{f_0} \in \mathcal{P}$, where $\mathcal{P} = \{P_f\text{: f satisfies } (*)\}$. Then, it can be verified that $\mathcal{E}_F(X) = \inf_{\mathcal{P}} E_{P_f}(X)$ (see (Nguyen and Walker, 1994) for details), where F is the lower envelope of the P_f over $\mathcal{P}$. General conditions for this equivalence are provided in Denneberg (1994).

Belief Functions. Let's examine this example a little further. The incomplete probabilistic knowledge is captured by the set-function F, defined on subsets of Θ, which can be completely specified. Besides properties mentioned above in the general case, we can verify also that, using the explicit values of F(A), $A \subseteq \Theta$, F has the additional property:

$$\forall\, n \geq 1,\ F(\bigcup_I A_i) \geq \Sigma_{\emptyset \neq J \subseteq I} (-1)^{|J|+1} F(\bigcap_J A_j)$$

where $I = \{1, 2, \ldots, n\}$, |J| is the cardinality of J. With this additional property, F is referred to as an *infinitely monotone capacity*. Now consider the *Möbius inverse* of F, namely

$$m(A) = \Sigma_{B \subseteq A} (-1)^{|A-B|} F(B),\ \forall\, A \subseteq \Theta$$

where $A - B = \{\theta: \theta \in A, \theta \notin B\}$. We see that $m \geq 0$, and in fact $m: 2^{\Theta} \to [0,1]$. It can be verified that m{1} = 0.4, m{2} = 0.2, m{3} = 0.2, m{4} = 0.1, m{1, 2, 3, 4} = 0.1 and m(A) = 0 for all other subsets A of $\Theta = \{1, 2, 3, 4\}$. But then

$$\forall\, A \subseteq \Theta,\ F(A) = \Sigma_{B \subseteq A}\, m(B),$$

so that $\Sigma_{A \subseteq \Theta}$ m(A) = F(Θ) = 1. Thus, m(·) is a *bona fide probability density*, not on Θ, but on 2^{Θ}.

If we view, *formally*, m(·) as the probability density function of some random set S which takes subsets of Θ as values, then infinitely monotone capacities (called *belief functions* by Shafer (1976)) are distributions of random sets. See for instance Nguyen (1978a). Here the situation is clear in vertu of the one-to-one correspondence between F and m. So, in incomplete probabilistic information cases, such as the above, uncertainty is handled by the capacity F which can be *argued* to model "general degrees of belief". Note that in Shafer's work, belief functions are defined *axiomatically*, that is without referring to any underlying probabilistic structures. However, his work was inspired from that of Dempster (1967) in which lower and upper probabilities are of concern.

The above example is in fact a special case of a more general setting in the continuous case (i.e., Θ needs not be finite). Specifically, $\mathscr{E}_{P_*}(X) = \inf_{\mathscr{P}} E_P(X)$ is true (although the inf might not be attained as the finite case), provided that P_* is a belief function (see Wasserman, 1987). For an exposition of evidential reasoning in the general case, see Hestir *et al.* (1991) and Smets and Kennes (1994). Possibility theory corresponds to the study of consonant belief functions, that is, when the sets $\{A, m(A) > 0\}$ are nested.

Fuzzy Measures and Integrals

Roughly speaking, fuzzy measures are monotone increasing (with respect to set inclusion) set-functions. We have seen that upper and lower probabilities are such set-functions. However, fuzzy measures are more general, since it is easy to find examples such that if F is an arbitrary increasing set-function, say on Ω, then the class $\mathscr{P}$ of probability measures P such that F(·) ≤ P(·) might be empty, so that an arbitrary fuzzy measure needs not be a lower probability.

Fuzzy measures were first introduced by Sugeno (1974) in an effort to model subjective evaluation S. His motivation is this. Suppose the "true state of nature" $\theta_0 \in \Theta$ is unknown. For each subset $A \subseteq \Theta$, you are asked to guess whether $\theta_0 \in A$. First, this reminds us right away of the situation in statistical confidence interval estimation. For example, let θ_0 be the unknown mean of a random variable X (population). From a random sample $X_1, X_2, \ldots, X_n$ drawn from X, we form a random set $S(X_1, X_2, \ldots, X_n)$ by some procedure. Before collecting data, S is random, so that we can talk about the probability that S will contain θ_0, that is the probability of coverage $P(\theta_0 \in S)$. Suppose we set $P(\theta_0 \in S) = \alpha$ in order to construct S. Then, after making observations $X_1 = x_1, X_2 = x_2, \ldots, X_n = x_n$, we get the set $S(x_1, x_2, \ldots, x_n)$. But then, either $\theta_0 \in S(x_1, \ldots, x_n)$ or not. The value α has to be then referred to as something else than a probability value, and we call it "degree of confidence" in statistics.

Now, it can be argued that, without a random mechanism, humans still can express their "degrees of trust" or "grades of fuzziness" subjectively in a similar situation. Thus, to each $A \subseteq \Theta$, a value $\mathcal{M}(A) \in [0,1]$ is assigned. Obviously $\mathcal{M}(\emptyset) = 0$, $\mathcal{M}(\Theta) = 1$, and $\mathcal{M}$ is increasing. What is fuzzy about $\mathcal{M}$? The domain of $\mathcal{M}$ are crisp sets, and $\mathcal{M}$ is an ordinary map $2^{\Theta} \to [0,1]$. The word "fuzzy" reflects the degrees of uncertainty in the meaning of values $\mathcal{M}(A)$. Thus the term "fuzzy" is used in a somewhat different meaning from degrees of membership of fuzzy concepts in the theory of fuzzy sets of Zadeh. However the set of values $\mathcal{M}(\{u\})$, $u \in U$ forms a fuzzy set.

Measure theory is in fact a theory of integration. Recall that, in standard measure theory, the σ-additivity property of a measure is essential, since it implies the monotone continuity which, in turn, in crucial for defining Lebesgue integral. In his thesis, Sugeno (1974), while relaxing additivity, kept monotone continuity as an axiom for fuzzy measure. Later, Murofushi and Sugeno (1989) realized that in their constructions of fuzzy integrals, whether original Sugeno (1974)'s integral, or Choquet integral, this property of monotone continuity is not used! So, they drop it! This story is very interesting, and should be brought to light.

With the monotone continuity as an axiom, possibility measures are not fuzzy measures (as pointed out by (e.g., Puri and Ralescu, 1982) since possibility measures are only continuous from below; without the continuity axiom, they are! The new view of fuzzy measures (i.e., without the axiom of monotone continuity), fuzzy measures form the largest class of uncertainty measures, including upper and lower probabilities, belief functions and possibility measures. Perhaps without being aware of this new definition of fuzzy measures, several researchers still work with Sugeno's old definition! More recently, Murofushi and Sugeno (1993) characterize possibility measures that are continuous from above, thus solving the problem of when possibility measures are fuzzy measures completely. Continuous from above possibility measures are such that the support $S(\Pi) = \{u: \Pi(\{u\}) > 0\}$ is countable at most, and there is a sequence of values $\pi_n = \Pi(\{u_n\})$ for $u_n \in S(\Pi)$, such that $\lim_{n \to \infty} \pi_n = 0$, and $\Pi(A) = \sup\{\pi_n : u_n \in A\}$. On separable metric spaces, continuous from above possibility measures must be completely maxitive in the sense of (7.22).

In his thesis, Sugeno was inspired by fuzzy logic operations in formalizing the concept of integral of a function f with respect to a fuzzy measure $\mathcal{M}$. Replacing addition and multiplication of numbers by max and min, respectively, he arrived as

$$\int f d\mathcal{M} = \sup_{\alpha \geq 0} \{\alpha \wedge \mathcal{M}(f > \alpha)\}.$$

Recognizing much later that this integral is *not* a generalization of Lebesgue integral (in the sense that, if $\mathcal{M}$ becomes additive, the above integral does not reduce to ordinary integral). So, with Murofushi (Murofushi and Sugeno, 1989), they even made a *big change*: from now on, there are two types of integrals with respect to fuzzy measures: Sugeno's integral and Choquet integral. The natural approach now in the numerical setting is to use Choquet integral as fuzzy integral. This is quite reasonable, since Choquet integral does reduce to Lebesgue integral when the fuzzy measure becomes additive. However Sugeno integral proves interesting and natural in non-numerical ordinal structures. See Wang and Klir (1992) for mathematical results concerning Sugeno integral.

Below, we sketch the fundamentals of Choquet integral.

Definition: A fuzzy measure M on Ω, with domain are arbitrary class A of subsets of Ω, is a map:

$$\mathcal{M}: \mathcal{A} \to \overline{\mathbb{R}}^{+} = [0, +\infty] \text{ such that}$$

(i) $\mathcal{M}(\emptyset) = 0$
(ii) $A, B \in \mathcal{A}$ with $A \subseteq B \Rightarrow \mathcal{M}(A) \leq \mathcal{M}(B)$.

Very often, the class $\mathcal{A}$ can be taken to be a σ-field. As in analysis, when defining integral for real-valued functions, it is *convenient* to allow $\pm\infty$, so that expressions like $\sup_{\omega \in A} f(\omega)$ always exist. Of course, arithmetic operations on $\mathbb{R} = (-\infty, +\infty)$ are extended to $\overline{\mathbb{R}} = [-\infty, +\infty]$ in a usual way. Let $\mathcal{M}$ be a fuzzy measure on $(\Omega, \mathcal{A})$ and $f: \Omega \to \overline{\mathbb{R}}$. Consider the function φ: $\mathbb{R} \to \overline{[0, +\infty]}$ defined by $\varphi(t) = \mathcal{M}(\omega: f(\omega) > t)$. This function is well-defined if, $\forall\, t \in \mathbb{R}$, the set $\{\omega: f(\omega) > t\} \in \mathcal{A}$. Thus, if we task $\mathcal{A}$ to be a σ-field, then the above condition holds when f is $\mathcal{A} - \mathcal{B}(\overline{\mathbb{R}})$ measurable, where $\mathcal{B}(\overline{\mathbb{R}})$ is the Borel σ-field of $\overline{\mathbb{R}}$ defined as follows. $\mathcal{B}(\overline{\mathbb{R}})$ is the σ-field generated by the Borel σ-field $\mathcal{B}(\mathbb{R})$ of $\mathbb{R}$ (i.e., generated by open intervals of $\mathbb{R}$) and $\{-\infty, +\infty\}$. Note that the Borel σ-field $\mathcal{B}(\overline{\mathbb{R}}^{+})$ is $\{A \cap \overline{\mathbb{R}}^{+}: A \in \mathcal{B}(\overline{\mathbb{R}})\} = \{A: A \subseteq \overline{\mathbb{R}}^{+} \cap \mathcal{B}(\overline{\mathbb{R}})\}$.

As in ordinary integration theory, to avoid meaningless expressions like $-\infty +\infty$, we start out by considering functions with constant sign (e.g., we consider non-negative functions). It is well-known that, *if* $\mathcal{M}$ is a σ-additive measure, then the Lebesgue integral of a non-negative function f can be writen as:

$$\int_{\Omega} f(\omega) d\,\mathcal{M}(\omega) = \int_{0}^{+\infty} \mathcal{M}(f > t) dt$$

where dt is the Lebesgue measure on $\overline{\mathbb{R}}^{+}$. Note that, f being measurable, the set $(f > t) \in \mathcal{A}$, so $\mathcal{M}(f > t)$ makes sense. Moreover, $\mathcal{M}$ being increasing, the integrand $t \to \mathcal{M}(f > t)$ is monotone, and hence measurable, thus

$$\int_{0}^{+\infty} \mathcal{M}(f > t) dt \text{ exists, which can be} = +\infty.$$

This second way of writting the abstract integral $\int_\Omega f(\omega)d\mathcal{M}(\omega)$ is suitable for generalization. Indeed, in $\int_0^{+\infty} \mathcal{M}(f > t)dt$, we simply replace $\mathcal{M}$ by a fuzzy measure (that is, dropping σ-additivity). $\mathcal{M}$ enters only in the integrand $\varphi(t) = \mathcal{M}(f > t)$, the measure relative to which this integral is defined remains the Lebesgue measure dt on $\overline{\mathbb{R}}^+$. Now, for f: $\Omega \to \overline{\mathbb{R}}$, we write $f = f^+ - f^-$ where

$$f^+(\omega) = f(\omega)I_{(f\geq 0)}(\omega)$$
$$f^-(\omega) = -f(\omega)I_{(f<0)}(\omega).$$

When $\mathcal{M}$ is σ-additive, $\int_\Omega fd\mathcal{M} = \int_\Omega f^+d\mathcal{M} + \int_\Omega f^-d\mathcal{M}$ holds, provided that not both terms in the right are $+\infty$. And when $\mathcal{M}(\Omega) < +\infty$, we have

$$\int_\Omega fd\mathcal{M} = \int_0^{+\infty} \mathcal{M}(f > t)dt + \int_{-\infty}^0 [\mathcal{M}(f \geq t) - \mathcal{M}(\Omega)]dt.$$

Thus, the *fuzzy integral* of a measurable function f with respect to a monotone set function $\mathcal{M}$ is defined by :

$$⨎_\Omega fd\mathcal{M} = \int_0^{+\infty} \mathcal{M}(f > t)dt + \int_{-\infty}^0 [\mathcal{M}(f \geq t) - \mathcal{M}(\Omega)]dt$$

(for $\mathcal{M}(\Omega) < +\infty$). While the Choquet integral (as a fuzzy integral) is monotone and positively homogeneous of degree one, that is, if $f \leq g$ then $⨎ fd\mathcal{M} \leq ⨎ gd\mathcal{M}$, and for $r > 0$, $⨎ rfd\mathcal{M} = r ⨎ fd\mathcal{M}$, its additivity fails. For example, $f = (1/4)A(\cdot)$, $g = (1/2)B(\cdot)$ with $A \cap B = \emptyset$ ($A(\cdot)$ denotes the indicator function of the set A) then

$$⨎_\Omega (f + g)d\mathcal{M} \neq ⨎_\Omega fd\mathcal{M} + ⨎_\Omega gd\mathcal{M}.$$

However, let $f = aA(\cdot) + bB(\cdot)$ with $A \cap B = \emptyset$, and $0 \leq a \leq b$, $g = \alpha A(\cdot) + \beta B(\cdot)$ with $0 \leq \alpha \leq \beta$, we do have

$$⨎_\Omega (f + g)d\mathcal{M} \neq ⨎_\Omega fd\mathcal{M} + ⨎_\Omega gd\mathcal{M}.$$

In fact, additivity of the fuzzy integral holds for pairs of functions f, g satisfying the condition:

$$\forall \omega, \omega' \in \Omega, (f(\omega) - f(\omega'))(g(\omega) - g(\omega')) \geq 0.$$

This condition is called *co-monotonicity*. See (Hardy *et al.*, 1934; Dellacherie, 1971; Anger, 1977; Schmeidler, 1986). Since the fuzzy integral is additive for co-monotonic functions, we call it a *co-monotonic additive functional* (since it operates on functions). It turns out that co-monotonic additivity properties characterizes fuzzy integral. See, e.g., Schmeider, 1986; Denneberg, 1994.

REFERENCES

Agbeko K. (1991). Optimal average, *Ann. Univ. Sci. Budapest, Sect. Comp.*, **16**, 5-10 (Full version: Report 73/1990, Mathematical Institute, Hungarian Academy of Sciences, Budapest).

Akian M., Quadrat J.P. and Viot M. (1994). Bellman processes, *Proc. 11th Conf. on Analysis and Optimization of Systems*, L.N. in Control and Inf. Sci. n°199, Springer Verlag.

Akian M., Quadrat J.P. and Viot M. (1996). Duality between probability and optimization, *Idempotency*, (Gunawardena J., ed.), Cambridge University Press.

Anger B. (1977). Representation of capacities, *Math. Annalen*, **229**, 245-258.

Banon G. (1995). Constructive decomposition of fuzzy measures in terms of possibility and necessity measures, *Proc. 6th IFSA World Congress*, Sao Paulo, 217-220.

Banon G. and Barrera J. (1993). Decomposition of mappings between complete lattices by mathematical morphology, *Signal Processing*, **30**, 299-327.

Barnett V. (1973). *Comparative Statistical Inference*, J. Wiley, New York.

Bellman R.E. and Giertz M. (1973). On the analytic formalism of the theory of fuzzy sets, *Information Sciences*, **5**, 149-157.

Bellman R.E. and Zadeh L.A. (1970). Decision making in a fuzzy environment, *Management Science,* **17**, B141-B164.

Benferhat S., Dubois D. and Prade H. (1997a). Nonmonotonic reasoning, conditional objects and possibility theory, *Artificial Intelligence*, **92**, 259-276.

Benferhat S., Dubois D. and Prade H. (1997b). Possibilistic and standard probabilistic semantics of conditional knowledge, *Proc. of the 14th National Conf. on Artificial Intelligence* (AAAI'97), 70-75. Extended version to appear in *J. Logic & Comput.*

Benferhat S., Dubois D. and Prade H. (1998). Practical handling of exception-tainted rules and independence information in possibilistic logic, *Applied Intelligence*, **9**, 101-127.

Bennett B.M., Hoffman D.D. and Murthy P. (1992). Lebesgue order on probabilities and some applications to perception, *J. of Mathematical Psychology*.

Berstein L.S., Bronevitch A.G., Karkishchenko A.N. and Zakharevitch V.G. (1996). Statistical classes and possibilistic models of classifying probability distributions, *BUSEFAL*, **65**, 19-26.

Bezdek J.C. (Ed.) (1994). Special issue: Fuzziness v. probability - the n-th round, *IEEE Trans. on Fuzzy Systems*, **2**, 1-42.

Bezdek J.C., Dubois D. and Prade H. (Eds.) (1999). *Fuzzy Sets in Approximate Reasoning and Information Systems.* The Handbooks of Fuzzy Sets Series (Dubois D. and Prade H., eds.), Kluwer Acad. Publ., Boston.

Biacino L. and Gerla G. (1992). Generated necessities and possibilities, *Int. J. Intelligent Systems*, **7**, 445-454.

Bilgic T. and Türksen I.B. (1999). Measurement of membership functions: Theoretical and empirical work, *Fundamentals of Fuzzy Sets* (Dubois D. and Prade H. eds.), Kluwer Acad. Publ., 1999. This volume.

Blockley D.I. (1985). Fuzziness and probability: A discussion of Gaines' axioms, *Civ. Engng Syst.*, **2**, 195-200.

Bouchon-Meunier B., Dubois D., Godo L. and Prade H. (1999). Fuzzy sets and possibility theory in approximate and plausible reasoning, *Fuzzy Sets in Approximate Reasoning and Information Systems* (Bezdek J.C., Dubois D. and Prade H., eds.), The Handbooks of Fuzzy Sets Series, Kluwer Academic Publ., Boston, 15-190.

Boutilier C. (1994). Modal logics for qualitative possibility theory, *Int. J. Approximate Reasoning*, **10**, 173-201.

Cayrol M., Farreny H. and Prade H. (1982). Fuzzy pattern matching, *Kybernetes*, **11**, 103-116.

Chanas S. and Nowakowski M. (1988). Single value simulation of fuzzy variable, *Fuzzy Sets and Systems*, **25**, 43-57.

Cheeseman P. (1988). Probabilistic versus fuzzy reasoning, *Uncertainty in Artificial Intelligence 1* (Kanal L. and Lemmer J., Eds.), North-Holland, Amsterdam, 85-102.

Choquet G. (1953). Theory of capacities, *Ann. Inst. Fourier* (Grenoble), **5**, 131-295.

Civanlar M.R. and Trussell H.J. (1986). Constructing membership functions using statistical data, *Fuzzy Sets and Systems*, **18**, 1-13.

Cooke R. (1990). *Experts in Uncertainty*, Oxford University Press, Oxford.

Darwiche A. and Pearl J. (1997). On the logic of iterated revision, *Artificial Intelligence*, **89**, 1-29.

De Baets B., De Cooman G. and Kerre E. (1998). The construction of possibility measures from samples on T-semi-partitions, *Information Sciences,* **106**, 3-24.

De Baets B., Tsiporkova E. and Mesiar R. (1999). Conditioning in possibility with strict order norms, *Fuzzy Sets and Systems*, **106**, 221-229.

De Campos L.M. (1995). Independence relationships in possibility theory and their application to learning belief networks, *Mathematical end Statistical Methods in Artificial Intelligence* (Della Riccia R., Kruse R. and Viertl R., Eds.), CISM Courses and Lectures Vol. 363, Springer Verlag, Berlin, 119-130.

De Campos L.M., Gebhardt J. and Kruse R. (1995). Axiomatic treatment of possibilistic independence, *Symbolic and Quantitative Approaches to Reasoning and Uncertainty* (Froidevaux C. and Kohlas J., Eds.), Lecture Notes in AI Vol. 946, Springer Verlag, 77-88.

De Campos L.M. and Huete J.F. (1999). Independence concepts in possibility theory, Part I: *Fuzzy Sets and Systems*, **103**, 127-152; Part II: *Fuzzy Sets and Systems*.

De Campos L.M., Lamata M.T. and Moral S. (1990). The concept of conditional fuzzy measure, *Int. J. of Intelligent Systems*, **5**, 237-246.

De Cooman G. (1997). Possibility theory — Part I: Measure- and integral-theoretics groundwork; Part II: Conditional possibility; Part III: Possibilistic independence, *Int. J. of General Systems*, **25**(4), 291-371.

De Cooman G. and Aeyels D. (1996). On the coherence of supremum preserving upper previsions, *Proc. of the 6th Inter. Conf. Information Processing and Management of Uncertainty in Knowledge-Based Systems (IPMU'96)*, Granada, 1405-1410. Extended version to appear in *Information Sciences.*

De Cooman G. and Kerre E. (1996). Possibility and necessity integrals, *Fuzzy Sets and Systems*, **77**, 207-229.

De Cooman G., Kerre E. and Vanmassenhove F.R. (1992). Possibility theory and integral-theoretic approach, *Fuzzy Sets and Systems*, **46**, 287-300.

De Cooman G., Ruan D. and Kerre E.E. (Eds.) (1995). *Foundations & Applications of Possibility Theory* (Proceedings of the FAPT'95, Ghent, Belgium, Dec. 13-15, 1995), World Scientific, Singapore.

De Finetti B. (1974). Theory of Probability — A Critical Introductory Treatment — Vol. 1, John Wiley & Sons, Chichester, UK.

De Finetti B. (1975). Theory of Probability — A Critical Introductory Treatment — Vol. 2, John Wiley & Sons, Chichester, UK.

Delgado M. and Moral S. (1987). On the concept of possibility-probability consistency, *Fuzzy Sets and Systems*, **21**, 311-318.

Dellacherie C. (1971). Quelques commentaires sur les prolongements de capacités, *Séminaire de Probabilité* (1969/70), Lecture Notes in Mathematics Vol. 191, Springer Verlag, Berlin, 77-81.

Dempster A.P. (1967). Upper and lower probabilities induced by a multivalued-mapping, *Ann. Math. Statist.*, **38**, 325-339.

Denneberg D. (1994). *Non-Additive Measure and Integral*, Kluwer Academic Publ., Dordrecht.

Diamond P. and Tanaka H. (1998). Fuzzy regression analysis, *Fuzzy Sets in Decision Analysis, Operations Research and Statistics* (Slowinski R., ed.), The Handbooks of Fuzzy Sets Series, Kluwer Academic Publ., Boston, 349-387.

Drakopoulos J.A. (1995). Probabilities, possibilities and fuzzy sets, *Fuzzy Sets and Systems,* **75**, 1-15.

Dubois D. (1986). Belief structures, possibility theory and decomposable confidence measures on finite sets, *Computers and Artificial Intelligence* (Bratislava), **5**, 403-416.

Dubois D., Fargier H., Fortemps P. and Prade H. (1997). Leximin optimality and fuzzy set-theoretic operations, *Proc. of the 7th World Congress of the Inter. Fuzzy Systems Assoc. (IFSA'97)*, Prague.

Dubois D., Fargier H. and Prade H. (1994). Propagation and satisfaction of flexible constraints, *Fuzzy Sets, Neural Networks and Soft Computing* (Yager R.R. and Zadeh L.A., Eds.), Van Nostrand Reinhold, New York, 166-187.

Dubois D., Fargier H. and Prade H. (1995). Fuzzy constraints in job-shop scheduling, *J. of Intelligent Manufacturing*, **6**, 215-234.

Dubois D., Fargier H. and Prade H. (1996a). Refinements of the maximin approach to decision-making in fuzzy environment, *Fuzzy Sets and Systems,* **81**,103-122.

Dubois D., Fargier H; and Prade H. (1996b). Possibility theory in constraint satisfaction problems: Handling priority, preference and uncertainty, *Applied Intelligence*, **6**, 287-309.

Dubois D., Fargier H. and Prade H. (1998). Possibilistic likelihood relations, *Proc. of the 7th Inter. Conf. on Information Processing and Management of Uncertainty in Knowledge-Based Systems (IPMU'98)*, Paris, 1196-1203.

Dubois D., Fariñas del Cerro L., Herzig A. and Prade H. (1994). An ordinal view of independence with application to plausible reasoning, *Proc. of the 10th Conf. on Uncertainty in Artificial Intelligence* (Lopez de Mantaras R. and Poole D., eds.), Seattle, WA, July 29-31, 195-203.

Dubois D., Fariñas del Cerro L., Herzig A. and Prade H. (1997). Qualitative relevance and independence: A roadmap, *Proc. of the 15h Inter. Joint Conf. on Artificial Intelligence (IJCAI'97)*, Nagoya, Japan, Aug. 23-29, 62-67.

Dubois D., Fodor J.C., Prade H. and Roubens M. (1996). Aggregation of decomposable measures with application to utility theory, *Theory and Decision*, **41**, 59-95.

Dubois D. and Fortemps P. (1999). Computing improved optimal solutions to fuzzy constraint satisfaction problems, *Eur. J. Operational Research.*, **118**, 95-126.

Dubois D., Godo L., Prade H. and Zapico A. (1998). Making decision in a qualitative setting: From decision under uncertainty to case-based decision, *Proc. of the 6th Inter. Conf. on Principles of Knowledge Representation and Reasoning (KR'98)* (Cohn A.G., Schubert L. and Shapiro S.C., eds.), Trento, Italy, 594-605.

Dubois D., Lang J. and Prade H. (1994). Possibilistic logic, *Handbook of Logic in Artificial Intelligence and Logic Programming — Vol. 3: Nonmonotonic Reasoning and Uncertain Reasoning* (Gabbay D.M., Hogger C.J., Robinson J.A. and Nute D., eds.), Clarendon Press, Oxford, 439-513.

Dubois D., Kerre E., Mesiar R. and Prade H. (1999). Fuzzy interval analysis, *Fundamentals of Fuzzy Sets* (Dubois D. and Prade H., eds.), Kluwer Acad. Publ., 1999. This volume.

Dubois D., Moral S. and Prade H. (1997). A semantics for possibility theory based on likelihoods, *J. of Mathematical Analysis and Applications*, **205**, 359-380.

Dubois D. and Prade H. (1980). *Fuzzy Sets and Systems: Theory and Applications*, Academic Press, New York.

Dubois D. and Prade H. (1982a). A class of fuzzy measures based on triangular norms, *Int. J. of General Systems*, 8, 43-61.

Dubois D. and Prade H. (1982b). A unifying view of comparison indices in a fuzzy set-theoretic framework, *Fuzzy Sets and Possibility Theory — Recent Developments* (Yager R.R., Ed.), Pergamon Press, pp. 1-13.

Dubois and Prade H. (1982c). 0n several representations of an uncertain body of evidence, *Fuzzy Information and Decision Processes* (Gupta M.M. and Sanchez E., eds.), North-Holland, Amsterdam, 167-181.

Dubois D. and Prade H. (1983). Unfair coins and necessity measures: towards a possibilistic interpretation of histograms, *Fuzzy Sets and Systems*, **10**, 15-20.

Dubois D. and Prade H. (1985a). Evidence measures based on fuzzy information, *Automatica*, **21**, 547-562.

Dubois D. and Prade H. (1985b). The mean value of a fuzzy number, *Fuzzy Sets and Systems*, **24**, 279-300.

Dubois D. and Prade H. (1986). Fuzzy sets and statistical data, *Europ. J. Operations Research*, **25**, 345-356.

Dubois D. and Prade H. (1987). Twofold fuzzy sets and rough sets — Some issues in knowledge representation, *Fuzzy Sets and Systems*, **23**, 3-18.

Dubois D. and Prade H. (1988a). Modelling uncertainty and inductive inference, *Acta Psychologica*, **68**, 53-78.

Dubois D. and Prade H. (1988b). *Possibility Theory*, Plenum Press, New York.

Dubois D. and Prade H. (1988c). Representation and combination of uncertainty with belief functions and possibility measures, *Computational Intelligence*, **4**, 244-264.

Dubois D. and Prade H. (1988d). Weighted fuzzy pattern matching, *Fuzzy Sets and Systems*, **28**, 313-331.

Dubois D. and Prade H. (1989). Fuzzy sets, probability and measurement, *Europ. J. of Operations Research*, **40**, 135-154.

Dubois D. and Prade H. (1990a). Consonant approximations of belief functions, *Int. J. Approximate Reasoning*, **4**, 419-449.

Dubois D. and Prade H. (1990b). Modelling uncertain and vague knowledge in possibility and evidence theories, *Uncertainty in Artificial Intelligence 4* (Shachter R., *et al.*, Eds.), North-Holland, Amsterdam, 303-318.

Dubois D. and Prade H. (1990c). Aggregation of possibility measures, *Multiperson Decision Making Using Fuzzy Sets and Possibility Theory* (Kacprzyk J. and Fedrizzi M., eds.), Kluwer Academic Publ., Dordrecht, 55-63.

Dubois D. and Prade H. (1991a). Epistemic entrenchment and possibilistic logic, *Artificial Intelligence*, **50**, 223-239.

Dubois D. and Prade H. (1991b). Updating with belief functions, ordinal conditional functions and possibility measures, *Uncertainty in Artificial Intelligence* **6** (Bonissone P.P., Henrion M., Kanal L.N. Lemmer J.F., eds.), North-Holland, Amsterdam, 311-329.

Dubois D. and Prade H. (1992a). Fuzzy rules in knowledge-based systems — Modelling gradedness, uncertainty and preference, *An Introduction to Fuzzy Logic Applications in Intelligent Systems* (Yager R.R. and Zadeh L.A., Eds.), Kluwer Academic Publ., Dordrecht, The Netherlands, 45-68.

Dubois D. and Prade H. (1992b). When upper probabilities are possibility measures, *Fuzzy Sets and Systems*, **49**, 65-74.

Dubois D. and Prade H. (1992c). Possibilistic abduction, *IPMU'92 - Advanced Methods in Artificial Intelligence* (Bouchon-Meunier B. *et al.*, eds), Lecture Notes in Computer Science, Vol. 682, Springer Verlag, Berlin, 3-12.

Dubois D. and Prade H. (1992d). Belief change and possibility theory, *Belief Revision* (Gärdenfors P., ed.), Cambridge University Press, 1992, 142-182.

Dubois D. and Prade H. (1993). Fuzzy sets and probability: Misunderstandings, bridges and gaps, *Proc. of the 2nd IEEE Inter. Conf. on Fuzzy Systems (FUZZ-IEEE'93)*, San Francisco, CA, March 28-April 1st, 1059-1068.

Dubois D. and Prade H. (1994). Possibility theory and data fusion in poorly informed environments, *Control Engineering Practice*, **2**(5), 811-823.

Dubois D. and Prade H. (1995). Possibility theory as a basis for qualitative decision theory, *Proc. of the 14th Inter. Joint Conf. on Artificial Intelligence (IJCAI'95)*, Montréal, Canada, 1924-1930.

Dubois D. and Prade H. (1996). Focusing vs. revision in possibility theory. In *Proc. of the 5th IEEE Inter. Conf. on Fuzzy Systems (FUZZ-IEEE'96)*, New Orleans, LO, Sept. 8-11, 1700-1705.

Dubois D. and Prade H. (1997a). Bayesian conditioning in possibility theory, *Fuzzy Sets and Systems*, **92**, 223-240.

Dubois D. and Prade H. (1997b). A synthetic view of belief revision with uncertain inputs in the framework of possibility theory, *Int. J. of Approximate Reasoning*, **17**(2/3), 295-324.

Dubois D. and Prade H. (1998). Possibility theory: Qualitative and quantitative aspects, *Handbook of Defeasible Reasoning and Uncertainty Management Systems — Vol. 1* (Gabbay D.M. and Smets P., eds.), Kluwer Academic Publ., Dordrecht, 169-226.

Dubois D. and Prade H. (1998c). Belief change rules in ordinal and numerical uncertainty theories, *Handbook of Defeasible Reasoning and Uncertainty Management Systems — Vol. 3: Belief Change* (Dubois D. and Prade H., eds.), Kluwer Academic Publ., Dordrecht, 311-392.

Dubois D. and Prade H. (1999). An overview of ordinal and numerical approaches to causal diagnostic problem solving, *Handbook of Defeasible Reasoning and Uncertainty Management Systems — Vol. 4: Abduction and Diagnosis* (Gebhardt J. and Kruse R., eds.), Kluwer Academic Publ., Dordrecht, to appear.

Dubois D., Prade H. and Sabbadin R. (1998). Qualitative decision theory with Sugeno integrals, *Proc. of the 14th Conf. on Uncertainty in Artificial Intelligence* (Cooper G.F. and Moral S., eds.), Morgan Kaufmann, San Mateo, CA, 121-128.

Dubois D., Prade H. and Sandri S.A. (1993). On possibility/probability transformations, *Fuzzy Logic: State of the Art* (Lowen R., Ed.), Kluwer Academic Publ.

Dubois D., Prade H. and Smets P. (1996). Representing partial ignorance, *IEEE Trans. on Systems, Man and Cybernetics*, **26**, 361-377.

Dubois D., Prade H. and Yager R.R. (Eds.) (1993). *Readings in Fuzzy Sets and Intelligent Systems*, Morgan Kaufmann, San Mateo, CA.

Dubois D., Prade H. and Yager R.R. (1999). Merging fuzzy information, *Fuzzy Sets in Approximate Reasoning and Information Systems* (Bezdek J.C., Dubois D. and Prade H., eds.), The Handbooks of Fuzzy Sets Series, Kluwer Academic Publ., Boston, 335-401.

Dubois D. and Yager R.R. (1992). Fuzzy set connectives as combinations of belief structures, *Information Sciences*, **66**, 245-275.

Duda R., Gaschnig J. and Hart P. (1981). Model design in the Prospector consultant system for mineral exploration, *Expert Systems in the Micro-Electronics Age* (Michie D., ed.), Edinburgh University Press, 153-167.

Edwards W. F. (1972). *Likelihood*, Cambridge University Press, Cambridge, U.K.

Fagin R. and Halpern J.Y. (1989). *A new approach to updating beliefs*, Research Report RJ 7222, IBM, Research Division, San Jose, CA.

Fariñas del Cerro L. and Herzig A. (1991). A modal analysis of possibility theory, *Fundamentals of Artificial Intelligence Research* (FAIR'91) (Jorrand Ph. and Kelemen J., Eds.), Lecture Notes in Computer Sciences, Vol. 535, Springer Verlag, Berlin, 11-18.

Fariñas del Cerro L., Herzig A. and Lang J. (1994). From ordering-based nonmonotonic reasoning to conditional logics, *Artificial Intelligence*, **66**, 375-393.

Fine T. L. (1973). *Theories of Probability*, Academic Press, New York.

Fishburn P.C. (1986). The axioms of subjective probability, *Statistical Science*, **1**, 335-358.

Fodor J. and Yager R.R. (1999). Fuzzy set-theoretic operators and quantifiers, *Fundamentals of Fuzzy Sets* (Dubois D. and Prade H., eds.), Kluwer Acad. Publ., 1999. This volume.

Fonck P. (1997). A comparative study of possibilistic conditional independence and lack of interaction, *Int. J. Approximate Reasoning*, **16**, 149-172.

Fortemps P. (1997). *Fuzzy sets for Modelling and Handling Imprecision and Flexibility,* Ph. D. Thesis, Faculté Polytechnique de Mons, Belgium.

Fortet R. and Kambouzia M. (1976). Ensembles aléatoires et ensembles flous, *Publications Econométriques*, **IX**, 1-23.

French S. (1984). Fuzzy decision analysis: some criticisms, *TIMS Studies in the Management Sciences*, **20**, 29-44

Garbolino P. (1989). On the plausibility of inverse possibility, Discussion paper 8, *Ann. Univ. Ferrara,* III (Philosophy).

Gaines B.R. (1978). Fuzzy and probability uncertainty logics, *Information and Control*, **38**, 154-169.

Gärdenfors P. (1988). *Knowledge in Flux — Modeling the Dynamics of Epistemic States*, The MIT Press, Cambridge, MA.

Gebhardt J., Gil M.A. and Kruse R. (1998). Fuzzy set-theoretic methods in statistics, *Fuzzy Sets in Decision Analysis, Operations Research and Statistics* (Slowinski R., ed.), The Handbooks of Fuzzy Sets Series, Kluwer Academic Publ., Boston, 311-347.

Gebhardt J. and Kruse R. (1993). The context model – an intergating view of vagueness and uncertainty. *Int. J. Approximate Reasoning*, **9**, 283-314.

Gebhardt J. and Kruse R. (1994a). A new approach to semantic aspects of possibilistic reasoning, *Symbolic and Quantitative Approaches to Reasoning and Uncertainty* (Clarke M. *et al.,* Eds.), Lecture Notes in Computer Sciences Vol. 747, Springer Verlag, 151-160.

Gebhardt J. and Kruse R. (1994b). On an information compression view of possibility theory, *Proc 3rd IEEE Int. Conference on Fuzzy Systems.* Orlando,Fl., 1285-1288.

Gebhardt J. and Kruse R. (1996). Automated construction of possibilistic networks from data, *Applied Mathematics and Computer Science*, **6**, 529-564.

Geer J.F. and Klir G.J. (1992). A mathematical analysis of information-preserving transformations between probabilistic and possibilistic formulations of uncertainty, *Int. J. of General Systems*, **20**, 143-176.

Gil M.A. (1988). On the loss of information due to fuzziness in experimental observations, *Ann. Inst. Statist. Math.*, **4**, 627-639.

Giles R. (1982). Foundations for a theory of possibility, *Fuzzy Information and Decision Processes* (Gupta M.M. and Sanchez E., eds.), North-Holland, 183-195.

Godal R.C. and Goodman T.J. (1980). Fuzzy sets and Borel, *IEEE Trans. on Systems, Man and Cybernetics*, **10**, 637.

Goodman I.R. (1982). Fuzzy sets as equivalence classes of random sets, *Fuzzy Sets and Possibility Theory* (Yager R.R., Ed.), Pergamon Press, Oxford, 327-342.

Goodman I.R., Nguyen H.T. and Rogers G.S. (1991). On the scoring approach to admissibility of uncertainty measures in expert systems, *J. Math. Anal. and Appl.*, **159**, 550-594.

Grabisch M., Murofushi T. and Sugeno M. (1992). Fuzzy measure of fuzzy events defined by fuzzy integrals, *Fuzzy Sets and Systems*, **50**, 293-313.

Grabisch M., Nguyen H.T. and Walker E.A. (1995). *Fundamentals of Uncertainty Calculi with Applications to Fuzzy Inference*, Kluwer Academic Publ., Dordrecht.

Grove A. (1988). Two modellings for theory change, *J. Philos. Logic*, **17**, 157-170.

Gupta C.P. (1993). A note on the transformation of possibilistic information into probabilistic information for investment decisions, *Fuzzy Sets and Systems,* **56**, 175, 182.

Gwét H. (1997). Normalized conditional possibility distributions and informational connection between fuzzy variables, *Int. J. Uncertainty, Fuzziness and Knowledge-based Systems*, **5**, 177-198.

Hacking I. (1975). All kinds of possibility, *Philosophical Review*, **84**, 321-347.

Hajek P. (1994). A qualitative fuzzy possibilistic logic, *Int. J. of Approximate Reasoning*, **12**, 1-19.

Hajek P., Harmancova D., Esteva F., Garcia P. and Godo L. (1994). On modal logics for qualitative possibility in a fuzzy setting, *Proc. of the 11th Conf. on Uncertainty in Artificial Intelligence* (Lopez de Mantaras R. and Poole D., eds.), Morgan Kaufmann, San Francisco, CA, 278-285.

Halpern J. (1996). Defining relative likelihood in partially-ordered preferential structures, *Proc. of the 12th Conf. on Uncertainty in Artificial Intelligence* (Horvitz E. and Jensen J., eds.), Morgan Kaufmann,, San Francisco, 299-306.

Hardy G.H., Littlewood J.E. and Polya G. (1934). *Inequalities*, Cambridge University Press. 2nd edition, 1952.

Harmanec D. and Klir G.J. (1997). On information-preserving transformations, *Int. J. of General Systems*, **26**(3), 265-290.

Hersh H.M. and Caramazza A. (1976). A fuzzy set approach to modifiers and vagueness in natural language, *J. of Exp. Psycho., General*, **105**, 254-276.

Herstein I.N. and Milnor J. (19..). An axiomatic approach to measurable utility, *Econometrica*, **21**, 291-297.

Hestir K., Nguyen H.T. and Rogers G.S. (1991). A random set formalism for evidential reasoning, *Conditional Logic in Expert Systems* (Goodman I.R. *et al.*, Eds.), North-Holland, Amsterdam, 309-344.

Higashi and Klir G. (1982). Measures of uncertainty and information based on possibility distributions, *Int. J. General Systems,* **8**, 43-58.

Higashi and Klir G. (1983). On the notion of distance representing information closeness, *Int. J. General Systems,* **9**, 103-115.

Hisdal E. (1978). Conditional possibilities independence and noninteraction, *Fuzzy Sets and Systems*, **1**, 283-297.

Hisdal E. (1991). Naturalized logic and chain sets, *Information Sciences*, **57-58**, 31-77.

Höhle U. (1976). MaBe auf unscharfen Mengen, *Zeitschrift für Wahrscheinlichkeitstheorie and verwandte Gebiete*, **36**, 179-188.

Horvitz E.J., Heckerman D.E. and Langlotz C.P. (1986). A framework for comparing alternative formalisms for plausible reasoning, *Proc. of the 5th National Conf. on Artificial Intelligence (AAAI'86)*, Philadelphia, PA, 210-214.

Huber P.J. (1981). *Robust Statistics*, John Wiley & Sons, New York.

Hunter D. (1991). Graphoids and natural conditional functions, *Int. J. Approximate Reasoning*, **5**, 489-504.

Inuiguchi M. and Kume Y. (1994). Necessity measures defined by level set inclusions: Nine kinds of necessity measures and their properties, *Int. J. of General Systems*, **22**, 245-275.

Inuiguichi M., Ichihashi H. and Tanaka H. (1989). Possibilistic linear programming with measurable multiattribute value functions, *ORSA J. on Computing*, **1**, 146-158.

Jaffray J.Y. (1992). Bayesian updating and belief functions, *IEEE Trans. on Systems, Man and Cybernetics*, **22**, 1144-1152.

Jain P. and Agogino A.M. (1990). Stochastic sensitivity analysis using fuzzy influence diagrams, *Uncertainty in Artificial Intelligence 4* (Shachter R.D., Levitt T.S., Kanal L.N. and Lemmer J.F., Eds.), North-Holland, Amsterdam, 79-92.

Joslyn C. (1997). Measurement of possibilistic histograms from interval data, *Int. J. of General Systems*, **26**(1-2), 9-33.

Kampé de Fériet J. (1982). Interpretation of membership functions of fuzzy sets in terms of plausibility and belief, *Fuzzy Information and Decision Processes* (Gupta M.M. and Sanchez E., Eds.), North-Holland, Amsterdam, 93-98.

Kaufmann A. (1978). Le calcul des admissibilités. Private communication.

Klir G.J. (1990). A principle of uncertainty and information invariance, *Int. J. of General Systems*, **17**, 249-275.

Klir G.J. and Yuan B. (1995). *Fuzzy Sets and Fuzzy Logic — Theory and Applications*, Prentice Hall, Upper Saddle River, NJ.

Klir G.J. and Folger T. (1988). *Fuzzy Sets, Uncertainty and Information*, Prentice Hall, Englewood Cliffs, NJ.

Klir G.J. and Harmanek D. (1994). On modal logic interpretations of possibility theory, *Int. J. of Uncertainty, Fuzziness, and Knowledge-Based Syst.*, **2**, 237-245.

Klir G.J. and Parviz B. (1992). Probability-possibility transformations: A comparison, *Int. J. of General Systems*, **21**, 291-310.

Klir G. J. (1999). Measures of uncertainty and information, *Fundamentals of Fuzzy Sets* (Dubois D. and Prade H., eds.), Kluwer Acad. Publ., 1999. This volume.

Kosko B. (1990). Fuzziness vs. probability, *Int. J. of General Systems*, **17**, 211-240.

Kovalerchuk B.Y. and Shapiro D.I. (1988). On the relation of the probability theory and the fuzzy setz theory foundations, *Computers and Artificial Intelligence*, **7**, 385-396.

Kramosil I (1996). Nonstandard approach to possibility theory, *Int. J. of Uncertainty, Fuzziness, and Knowledge-Based Syst.*, **4**, 275-301.

Kruse R. and Meyer K.D. (1987). *Statistics with Vague Data*, D. Reidel, Dordrecht.

Kruse R., Gebhardt J., Klawonn F. (1994). *Foundations of Fuzzy Systems*, Wiley, New York.

Lapointe S. (1999). *Traitement Subjectif de l'Incertitude par la Théorie des Possibilités: Application à la Prévision et au Suivi des Crues*, Ph. D. Thesis, INRS-Eau, Québec University.

Lapointe S. Bobée B. (1999). Revision of possibility distributions: A Bayesian inference pattern, to appear in *Fuzzy Sets and Systems.*

Lasserre V., Mauris G. and Foulloy L. (1998). A simple modelisation of measurement uncertainty: The truncated triangular possibility distribution, *Proc 7th Int. Conf. on Information Processing and Management of Uncertainty in Knowledge-based Systems (IPMU'98)*, Paris, Editions Medicales et Scientifiques, 10-17.

Laviolette M. and Seaman Jr. J. W. (1994). The efficacy of fuzzy representations of uncertainty. *IEEE Trans. on Fuzzy Systems,* **2**, 4-15.

Laviolette M., Seaman Jr. J. W., Barrett J.D. and Woodall W.H. (1995). Fuzziness vs. probability, *Technometrics*, **37**, 249-259.

Lewis D. (1973). *Counterfactuals*, Basil Blackwell, Oxford. 2nd edition, Billing and Sons Ltd., Worcester, UK, 1986.

Lewis D. L. (1979). Counterfactuals and comparative possibility, *Ifs* (Harper W. L., Stalnaker R. and Pearce G., eds.), D. Reidel, Dordrecht, 57-86.

Li Q.D. (1997). The random set and the cutting of random fuzzy sets, *Fuzzy Sets and Systems*, **86**, 223-234.

Liang P. and Song F. (1996). What does a probabilistic interpretation of fuzzy sets mean?, *IEEE trans. on Fuzzy Systems,* **4,** 200-205.

Lindley D.V. (1982). Scoring rules and the inevitability of probability, *Int. Statist. Rev.*, **50**, 1-26.

Loginov V.I. (1966). Probability treatment of Zadeh membership functions and their use in pattern recognition, *Eng. Cyber.*, 68-69.

Mabuchi, S. (1992). An interpretation of membership functions and the properties of general probabilistic operators as fuzzy set operators–Part I: Case of type 1 fuzzy sets. *Fuzzy Sets and Systems*, **49**, 271-283.

Mabuchi, S. (1997). An interpretation of membership functions and the properties of general probabilistic operators as fuzzy set operators–Part II. Extensions to three-valued and interval-valued fuzzy sets. *Fuzzy Sets and Systems*, **92**, 31-50.

Maeda, Y. Nguyen, H.T. and Hichihashi H. (1993). Maximum entropy algorithms for uncertainty measures, *Int. J. of Uncertainty, Fuzziness, and Knowledge-Based Syst.*, **1**, 69-93.

Manton K.G., Woodbury M.A. and Tolley H.D. *Statistical Applications Using Fuzzy Sets.*, Wiley, New York.

Marichal J.-L. (1998). On Sugeno integral as an aggregation function, Tech. Rep. 9710, GEMME, Fac. d'Ecomomie, Université de Liège, Belgium. *Proceedings of EUFIT'98 — Vol. I*, Aachen, Germany, 540-544.

Maslov V. (1987). *Méthodes Opératorielles*, Editions MIR, Moscow.

Maslov V. and Samborski S.N. (1992)*Idempotent Analysis*. Advances in Soviet Mathematics **13**, Amer. Math. Soc. , Providence.

Mathéron G. (1975). *Random Sets and Integral Geometry*, John Wiley & Sons, New York.

Maung I. (1995). Two characterizations of a minimum information principle for possibilistic reasoning, *Int. J. Approximate Reasoning*, **12**, 133-156.

Mesiar R. (1992). Characterisation of possibility measures of fuzzy events using Markov kernels, *Fuzzy Sets and Systems*, **46**, 301-303.

Mesiar R. (1994). On the integral representation of fuzzy possibility measures, *Int. J. of General Systems*, **21**, 109-121.

Mesiar R. (1997). Possibility measures, integration, and fuzzy possibility measures, *Fuzzy Sets and Systems*, **92**, 191-196.

Molchanov I.S. (1993). *Limit Theorems for Unions of Random Closed Sets*, Lecture Notes in Mathematics, Vol. 1561, Springer Verlag, Berlin.

Moulin H. (1988). *Axioms of Cooperative Decision Making*, Cambridge University Press, Cambridge, MA.

Mundici D. (1992). The logic of Ulam games with lies, *Knowledge, Belief and Strategic Interaction* (Bicchieri C. and Dalla Chiara M., eds.), Cambridge University Press, 275-284.

Murofushi T. and Sugeno M. (1989). In interpretation of fuzzy measure and the Choquet integral as an integral with respect to a fuzzy measure, *Fuzzy Sets and Systems*, **29**, 201-227.

Murofushi T. and Sugeno M. (1993). Continuous from above possibility measures and f-additive fuzzy measures on separable metic spaces: Characterization and regularity, *Fuzzy Sets and Systems,* **54**, 351-354.

Nahmias S. (1978). Fuzzy variables, *Fuzzy Sets and Systems,* **1**, 97-110.

Narin'yani A.S. (1980). Sub-definite set — New data-type for knowledge representation, Memo n° 4-232, Computing Center, Novosibirsk, URSS.

Natvig B. (1983). Possibility versus probability, *Fuzzy Sets and Systems*, **10**, 31-36.

Nau R.F. (1992). Decision analysis with indeterminate or incoherent probabilities, *Annals of Operations Research — Vol : Choice Under Uncertainty* (Fishburn P.C. and LaValle I.H., Eds.), to appear.

Negoita C.V. and Ralescu D. (1987). *Simulation, Knowledge-Based Computing, and Fuzzy Statistics*, Van Nostrand Reinhold Comp., New York.

Nguyen H.T. (1978a). On random sets and belief functions, *J. Math. Anal. and Appl.*, **65**, 531-542.

Nguyen H.T. (1978b). On conditional possibility distributions, *Fuzzy Sets and Systems*, **1**, 299-309.

Nguyen H.T. (1984). On modeling of linguistic information using random sets, *Information Science*, **34**, 265-274.

Nguyen H.T. and Walker E.A. (1994). On decision making using belief functions, *Advances in the Dempster-Shafer Theory of Evidence* (Yager R.R. *et al.*, Eds.), John Wiley & Sons, New York, 311-330.

Norberg T. (1986). Random capacities and their distributions, *Proba. Th. Rel. Fields*, **73**, 281-297.

Okuda T., Tanaka H. and Asai K. (1978). A formulation of fuzzy decision problems with fuzzy information, using probability measures of fuzzy events, *Information and Control*, **38**, 135-147.

Orlov A.I. (1978). Fuzzy and random sets, *Prikladnoï Mnogomiernii Statisticheskii Analyz* (Nauka, Moscow), 262-280 (in Russian).

Pan Y. and Yuan B. (1997). Bayesian inference of fuzzy probabilities, *Int. J. of General Systems*, **26**(1-2), 73-90.

Paris J.B. (1997). A semantics for fuzzy logic, *Soft Computing*, **1**, 143-147.

Pawlak Z. (1991). *Rough sets — Theoretical Aspects of Reasoning about Data*, Kluwer Academic Publ., Dordrecht.

Pearl J. (1988). *Probabilistic Reasoning Intelligent Systems: Networks of Plausible Inference*, Morgan Kaufmann, San Mateo, CA.

Pearl J. (1990). Integrating probability and logic, *Readings in Uncertain Reasoning* (Shafer G. and Pearl J., Eds.), Morgan Kaufmann, San Mateo, CA, 677-679.

Pirlot M. (1995). A characterization of 'min' as a procedure for exploiting valued preference relations and related results, *J. of Multi-Criteria Decision Analysis*, **4**, 37-56.

Prade H. (1979). Nomenclature of fuzzy measures, *Proc. 1st Int. Seminar on Theory of Fuzzy Sets* , Linz, Austria, Sept. 24-29, 9-25.

Prade H. (1982). Modal semantics and fuzzy set theory. In: Fuzzy Set and Possibility Theory. Recent Developments. (R.R. Yager, ed.), Pergamon Press, New York, 232-246.

Prade H. and Yager R.R. (1994). Estimations of expectedness and potential surprize in possibility theory, *Int. J. Uncertainty, Fuzziness and Knowledge-based Systems*, **2**, 417-428.

Puri M. L. and Ralescu D. (1982). A possibility measure is not a fuzzy measure, *Fuzzy Sets and Systems*, **7**, 311-313.

Ralescu D. and Adams G. (1980). The fuzzy integral, *J. Math. Anal. and Appl.*, **75**, 562-570.

Ramer A. (1989). Conditional possibility measures, *Cybernetics and Systems*, **20**, 233-247.

Ramer A. (1990a). Structure of possibilistic information metrics and distances: Properties, *Int. J. of General Systems*, **17**, 21-32.

Ramer A. (1990b). Possibilistic information metrics and distances: Characterizations of structure, *Int. J. of General Systems*, **18**, 1-10.

Ramer A. (1990c). Uncertainty in the Dempster-Shafer theory: A critical re-examination, *Int. J. of General Systems*, **18**, 155-166.

Raufaste E. and Da Silva Neves R. (1998). Empirical evaluation of possibility theory in human radiological diagnosis, *Proc. of the 13th Europ. Conf. on Artificial Intelligence (ECAI'98)* (Prade H., ed.), John Wiley & Sons, 124-128.

Robbins H.E. (1944). On the measure of a random set, *Ann. Math. Statist.*, **15**, 70-74.

Ruspini E.H. (1991). Approximate reasoning: Past, present, future, *Information Sciences*, **57-58**, 297-317.

Sales T. (1982). Fuzzy sets as set classes, *Stochastica*, **6**, 249-264.

Sanchez E. (1978). On possibility-qualification in natural languages, *Information Sciences*, **15**, 45-76.

Sanguessa R., Cabos J. and Cortes U. (1998). Possibilistic conditional independence: A similarity-based measure and its application to causal network learning, *Int. J. Approximate Reasoning*, **18**, 145-167.

Sanguessa R., Cortes U. and Gisolfi A. (1998). Possibilistic conditional independence: a similarity-based measure and its application to causal network learning, *Int. J. Approximate Reasoning*, **18**, 251-270.

Schmeidler D. (1986). Integral representation without additivity, *Proc. Amer. Math. Soc.*, **97**(2), 255-261.

Shackle G. L.S. (1961). *Decision, Order and Time in Human Affairs*, (2nd edition), Cambridge University Press, UK.

Shafer G. (1976a). *A Mathematical Theory of Evidence*, Princeton University Press, Princeton, NJ.

Shafer G. (1976b). A theory of statistical evidence, *Foundations of Probability Theory, Statistical Inference, and Statistical Theories of Sciences — Vol. II*, (Harper and Hooker, eds.), D. Reidel, Dordrecht, 365-436.

Shafer G. (1987). Belief functions and possibility measures, *Analysis of Fuzzy Information — Vol. I: Mathematics and Logic* (Bezdek J.C., Ed.), CRC Press, Boca Raton, FL, 51-84.

Shapley S. (1971). Cores of convex games, *Int. J. of Game Theory*, **1**, 12-26.

Shenoy P. (1991). On Spohn's rule for revision of beliefs, *Int. J. Approximate Reasoning*, **5**, 149-181.

Shenoy P. (1992). Using possibility theory in expert systems, *Fuzzy Sets and Systems*, **52**, 129-142.

Shilkret N. (1971). Maxitive measure and integration, *Indag. Math.*, **33**, 109-116.

Shoham Y. (1988). *Reasoning About Change — Time and Causation from the Standpoint of Artificial Intelligence*, The MIT Press, Cambridge, MA.

Slowinski R. (Eds.) (1998). *Fuzzy Sets in Decision Analysis, Operations Research and Statistics*, The Handbooks of Fuzzy Sets Series (Dubois D. and Prade H., eds.), Kluwer Academic Publ., Boston.

Smets P. (1982). Possibilistic inference from statistical data, *Proc. of the 2nd World Conf. on Mathematics at the Service of Man*, Las Palmas (Canary Island), Spain, 611-613.

Smets P. (1990). Constructing the pignistic probability function in a context of uncertainty, *Uncertainty in Artificial Intelligence 5* (Henrion M. *et al.*, Eds.), North-Holland, Amsterdam, 29-39.

Smets P. and Kennes R. (1994). The transferable belief model, *Artificial Intelligence*, **66**, 191-234.

Smith C.A.B. (1961). Consistency in statistical inference and decision, *J. Royal Statist. Soc.*, **B-23**, 1-37.

Spohn W. (1988). Ordinal conditional functions: A dynamic theory of epistemic states, *Causation in Decision, Belief Change and Statistics* (Harper W. and Skyrms B., Eds.), 105-134.

Spohn W. (1990). A general non-probabilistic theory of inductive reasoning, *Uncertainty in Artificial Intelligence 4* (Shachter R.D., Levitt T.S., Kanal L.N. and Lemmer J.F., Eds.), North-Holland, Amsterdam, 149-158.

Spott M. (1999). A theory of possibility distributions, *Fuzzy Sets and Systems*, **102**, 135-155.

Stallings W. (1977). Fuzzy set theory versus Bayesian statistics, *IEEE Trans. on Systems, Man and Cybernetics*, 216-219.

Studeny M. (1995). Conditional independence and natural conditional functions, *Int. J. Approximate Reasoning*, **12**, 43-68.

Suarez Garcia F. and Gil Alvarez P. (1986). Two families of fuzzy integrals, *Fuzzy Sets and Systems*, **18**, 67-81

Sudkamp T. (1992a). On probability-possibility transformations, *Fuzzy Sets and Systems*, **51**, 73-81.

Sudkamp T. (1992b). The semantics of plausibility and possibility, *Int. J. of General Systems*, **21**, 273-289.

Sudkamp T. (1993). Geometric measures of possibilistic uncertainty, *Int. J. of General Systems*, **22**, 7-23.

Sugeno M. (1974). *Theory of fuzzy integrals and its applications*, Ph. D. Thesis, Tokyo Institute of Technology, Japan.

Sugeno M. (1977). Fuzzy measures and fuzzy integrals — A survey, *Fuzzy Automata and Decision Processes* (Gupta M.M., Saridis G.N. and Gaines B.R., eds.), North-Holland, Amsterdam, 89-102.

Tanaka H. and Guo P.J. (1999). *Possibilistic Data Analysis for Operations Research*, Physica-Verlag, Heidelberg.

Tanaka H. and Ishibuchi H. (1993). Evidence theory of exponential possibility distributions. *Int. J. Approximate Reasoning*, **8**, 123-140.

Tanaka H., Ishibuchi H., Hayashi I. (1993). Identification method of possibility distributions and its application to discriminant analysis, *Fuzzy Sets and Systems*, **58**, 41-50.

Thomas S.F. (1979). *A theory of semantics and possible inference with application to decision analysis*, PhD Thesis, University of Toronto, Canada.

Thomas S.F. (1981). Possibilistic uncertainty and statistical inference, *ORSA/TIMS Meeting*, Houston, Texas.

Thomas S.F. (1995). *Fuzziness and Probability,* ACG Press, Wichita, Kansas.

Tribus M. (1978). Comments on fuzzy sets, fuzzy algebras, and fuzzy statistics, *Proc. IEEE*, **67**, 1168-1169.

Tsiporkova E. and De Baets B. (1998). A general framework for upper and lower possibilities and necessities, *Int. J. Uncertainty, Fuzziness and Knowledge-based Systems*, **6**, 1-33.

Viertl R. (1996). *Statistical Methods for Non-Precise Data*, CRC Press, Boca Raton, FL.

Viertl R. and Hule H. (1991). On Bayes' theorems for fuzzy data, *Statistical Papers*, 32, 115-122.

Vincke P. (1982). Aggregation of preferences: A review, *Europ. J. Operational res.*, **9**, 17-22.

Walley P. (1991). *Statistical Reasoning with Imprecise Probabilities*, Chapman and Hall.

Walley P. (1996). Measures of uncertainty in expert systems, *Artificial Intelligence*, **83**, 1-58.

Walley P. (1997). Statisticaél inferences based on a second-order possibility distribution, *Int. J. of General Systems*, **26**, 337-383.

Walley P. and de Cooman G.(1999) A behavioural model for linguistic uncertainty, *Computing With Words* (Wang P.P., ed.), Wiley, to appear.

Wallsten T.S., Budescu D.V., Rapoport A., Zwick R. and Forsyth B. (1986). Measuring the vague meanings of probability terms, *J. of Experimental Psychology: General*, **115**, 348-365.

Wang P.Z. (1983). From the fuzzy statistics to the falling random subsets, *Advances in Fuzzy Sets, Possibility Theory and Applications* (Wang P.P., Eds.), Plenum Press, New York, 81-96.

Wang P.Z. (1982). Fuzzy contactability and fuzzy variables, *Fuzzy Sets and Systems*, **8**, 81-92.

Wang P.Z. and Sanchez E. (1982). Treating a fuzzy subset as a projectable random set, *Fuzzy Information and Decision Processes* (Gupta M.M. and Sanchez E., Eds.), North-Holland, Amsterdam, 213-220.

Wang Z. and Klir G. (1992). *Fuzzy measure theory*, Plenum Press, New-York.

Wasserman L.A. (1987). *Some applications of belief functions to statistical inference*, Thesis, Univ. of Toronto, Canada.

Watson S.R., Weiss J.J. and Donnell M. (1979). Fuzzy decision analysis, *IEEE Trans. on Systems, Man and Cybernetics*, **9**, 1-9.

Whalen T. (1984). Decision making under uncertainty with various assumptions about available information, *IEEE Trans. on Systems, Man and Cybernetics*, **14**, 888-900.

Wolkenhauer O. (1988). *Possibility Theory with Applications to Data Analysis*, Research Studies Press, Hertfordshire, UK.

Wonneberger S. (1994). Generalization of an invertible mapping between probability and possibility, *Fuzzy Sets and Systems*, **64**, 229-240.

Yager R.R. (1979). Possibilistic decision making, *IEEE Trans. on Systems, Man and Cybernetics*, **9**, 388-392.

Yager R.R. (1980). A foundation for a theory of possibility, *J. Cybernetics*, **10**, 177-204.

Yager R.R. (1981). A new methodology for ordinal multiobjective decisions based on fuzzy set, *Decision Sciences*, **12**, 589-600.

Yager R.R. (1983). An introduction to applications of possibility theory, *Human Systems Management*, **3**, 246-269.

Yager R.R. (1984). General multiple objective decision making and linguistically quantified statements, *Int. J. of Man-Machine Studies*, **21**, 389-400.

Yager R.R. (1985a). Q-projections of possibility distributions, *IEEE trans. on Systems Man & Cybern.*, **15**, 775-777.

Yager R.R. (1985b). On the relationships of methods of aggregation evidence in expert systems, *Cybernetics and Systems*, **16**, 1-21.

Yager R.R. (1990). Ordinal measures of specificity, *Int. J. Gener. Syst.*, **17**, 57-72.

Yager R.R. (1992). On the specificity of a possibility distribution, *Fuzzy Sets and Systems*, **50**, 279-292.

Yager R.R. (1993a). On the completion of qualitative possibility measures, *IEEE Trans. on Fuzzy Systems*, **1**, 184-195.

Yager R.R. (1993b). On the completion of priority orderings in non-monotonic reasoning systems. *Int. J. of Uncertainty, Fuzziness, and Knowledge-Based Syst.*, **1**, 139-165.

Yan B. (1993). Semiconormed possibility integrals for application-oriented modelling, *Fuzzy Sets and Systems*, **57**, 239-248.

Yan B. (1994). Seminormed possibility integrals for application-oriented modelling, *Fuzzy Sets and Systems*, **61**, 189-198.

Young V.R. and Wang S.S. Upating non-additive measures with fuzzy information. *Fuzzy Sets and Systems*, **94**, 355-366.

Zadeh L.A. (1965). Fuzzy sets, *Information and Control*, **8**, 338-353.

Zadeh L.A. (1968). Probability measures of fuzzy events, *J. Math. Anal. Appl.*, **23**, 421-427.

Zadeh L. A. (1975a). The concept of a linguistic variable and its application to approximate reasoning, *Information Sciences,* Part I: **8**, 199-249; Part II: **8**, 301-357; Part III: **9**, 43-80.

Zadeh L. A. (1975b). Calculus of fuzzy restrictions, *Fuzzy Sets and Their Applications to Cognitive and Decision Processes* (Zadeh L. A., Fu K. S., Shimura M. and Tanuka K., eds.), Academic Press, New York, 1-39.

Zadeh L.A. (1978). Fuzzy sets as a basis for a theory of possibility, *Fuzzy Sets and Systems*, **1**, 3-28. Reprinted in *Fuzzy Sets and Applications : Selected Papers* (Zadeh L.A., Yager R.R., Ovchinnikov S., Tong R.M. and Nguyen H.T., Eds.), Wiley, New York, 1987, 193-218.

Zadeh L.A. (1979a).A theory of approximate reasoning, *Machine Intelligence*, Vol. 9 (Hayes J. E., Michie D. and Mikulich L. I., eds.), John Wiley & Sons, New York, 149-194.

Zadeh L.A. (1979b). Fuzzy sets and information granularity. In: *Advances in Fuzzy Set Theory and Applications*, (M.M. Gupta, R.K. Ragade, R.R. Yager, eds.), North-Holland, Amsterdam, 3-18.

Zadeh L.A. (1985). Syllogistic reasoning in fuzzy logic and its application to usuality and reasoning with dispositions, *IEEE Trans. on Systems, Man and Cybernetics*, **15**, 754-763.

Zadeh L.A. (1995). Discussion: Probability theory and fuzzy logic are complementary rather than competitive, *Technometrics*, **37**, 271-276.

Zimmermann H.J. (1987). *Fuzzy Sets, Decision Making, and Expert Systems*, Kluwer Academic Publ., Boston.

Zwirn D. and Zwirn H. (1995). Confirmation non-probabiliste, *Méthodes Logiques pour les Sciences Cognitives* (Dubucs J. and Lepage F., eds.). Hermès, Paris., 77-98.

8 MEASURES OF UNCERTAINTY AND INFORMATION

George J. Klir

Abstract: This chapter is an overview of current research regarding measures of uncertainty and information in five mathematical theories: classical set theory, fuzzy set theory, probability theory, possibility theory, and the Dempster-Shafer theory of evidence. Three types of uncertainty are involved: nonspecificity, entropy-like uncertainty, and fuzziness.

8.1 INTRODUCTION

The purpose of this chapter is to introduce the main developments and results regarding the measurement of uncertainty. Although the focus is on the measurement of uncertainty in fuzzy set theory and possibility theory, also covered are measures of uncertainty in classical set theory, probability theory, and the Dempster-Shafer theory of evidence. These are theories within which the issue of measuring uncertainty has been thoroughly investigated.

The nature of uncertainty depends on the mathematical theory within which uncertainty pertaining to various problem-solving situations is formalized. Each formalization is a mathematical model of the situation. A more general theory is capable of capturing uncertainty of the situation more faithfully than its less general competitors. However, it usually involves greater computational demands.

It is generally recognized that uncertainty is a multidimensional concept. Which of its dimensions is actually manifested is determined by the mathematical theory employed. Three types of uncertainty have thus far been recognized. The first, and perhaps the most fundamental type of uncertainty, which originates in classical set theory, is usually referred to as *nonspecificity* (or *imprecision*).

The second type of uncertainty, which originates in probability theory, has been discussed in the literature under different names. In this chapter, we use the term *entropy-like uncertainty*; the term is borrowed from probability theory, where uncertainty is measured by the well-known entropy measure introduced by Shannon [1948].

The third type of uncertainty, referred to as *fuzziness*, is uniquely connected with fuzzy sets. It is thus applicable to fuzzy set theory and to any of the other uncertainty theories that are fuzzified. Contrary to the other two types of uncertainty, which result solely from information deficiency, fuzziness is connected with information only indirectly; it results primarily from linguistic vagueness.

Uncertainty is connected with information in the following way. Assume that we can measure the amount of uncertainty involved in a problem-solving situation formalized in a particular mathematical theory. Assume further that the amount of uncertainty can be reduced by obtaining relevant information as a result of some action (finding a relevant new fact, performing a relevant experiment and observing the experimental outcome, etc.). Then, the amount of information obtained by the action may be measured by the reduction of uncertainty that results from the action.

Measures of uncertainty based on information deficiency (nonspecificity and entropy-like uncertainty) are thus applicable for measuring information as well. However, information measured solely by the reduction of uncertainty does not capture the rich notion of information in human communication and cognition. The former is not explicitly concerned with the semantic and pragmatic aspects of information viewed in the latter sense. Although these aspects are not ignored when uncertainty is viewed in terms of uncertainty reduction, they are assumed to be fixed in each particular application. To distinguish information conceived solely in terms of uncertainty reduction from other conceptions of information, the former is sometimes referred to as *uncertainty-based information.*

There are several overview papers in the literature that cover the various developments regarding measurement of uncertainty in more details than their summaries in this chapter. They were written by Bouchon [1988], Dubois and Prade [1987], Klir [1987, 1991, 1993] and Pal et al. [1992, 1993]. The most comprehensive coverage of the material summarized in this chapter is presented in a book by Klir and Wierman [1998].

8.2 MEASURES OF NONSPECIFICITY

8.2.1 Classical Set Theory

To measure uncertainty and the associated information was first conceived in terms of classical set theory. It was shown by Hartley [1928] that using a function from the class of functions

$$U(A) = c \cdot \log_b |A|,$$

where $|A|$ denotes the cardinality of a finite set A and b,c are positive constants ($b > 1$, $c > 0$), is the only sensible way to measure the amount of uncertainty associated with a finite set of possible alternatives. Each choice of values of the constants b and c determines the unit in which the amount of uncertainty is measured. When $b = 2$ and $c = 1$, which is the most common choice, uncertainty is measured in *bits* and we obtain

$$U(A) = \log_2 |A|. \tag{8.1}$$

One bit of uncertainty is equivalent to the total uncertainty regarding the truth or falsity of one elementary proposition.

Function U defined by (8.1) is usually called a Hartley function. Its uniqueness as a measure of uncertainty (in bits) associated with finite sets of alternatives was also proven on axiomatic grounds by Rényi [1970]; see also Klir and Folger [1988]. The type of uncertainty measured by this function is usually called *nonspecificity*.

When the Hartley function U is applied to subsets of a given finite universal set X, it has the form

$$U: \mathcal{P}(X) \rightarrow [0, \infty)$$

where $\mathcal{P}(X)$ denotes the power set of X. In this case, the range of U is

$$0 \leq U(A) \leq \log_2 |X|.$$

Consider now that a situation characterized by a set A of possible alternatives is reduced to its subset B by some action. Then, the amount of relevant information, I(A, B), produced by the action is equal to the amount of reduced uncertainty (i.e., nonspecificity), which is given by the difference U(A) - U(B). That is,

$$I(A, B) = \log_2 \frac{|A|}{|B|}, \tag{8.2}$$

where $B \subset A$. When the action eliminates all alternatives except one (i.e., when $|B| = 1$), we obtain $I(A, B) = \log_2 |A| = U(A)$. This means that U(A) may also be viewed as the amount of information needed to characterize one alternative in set A.

Consider now two universal sets, X and Y, and assume that a relation $R \subset X \times Y$ describes a set of possible alternatives in some situation of interest. Assume further that the domain and range of R are sets $R_X \subset X$ and $R_Y \subset Y$, respectively. Then three distinct Hartley functions are applicable, defined on the power sets of X, Y, and $X \times Y$. To identify clearly which universal set is involved in each case, it is useful (and a common practice) to write U(X), U(Y), U(X, Y) instead of $U(R_X)$, $U(R_Y)$, U(R), respectively.

Functions

$$U(X) = \log_2 |R_X|, \tag{8.3}$$

$$U(Y) = \log_2 |R_Y| \tag{8.4}$$

are called *simple uncertainties*, while function

$$U(X,Y) = \log_2 |R| \tag{8.5}$$

is called a *joint uncertainty*. Two additional Hartley functions are defined,

$$U(X \mid Y) = \log_2 \frac{|R|}{|R_Y|}, \tag{8.6}$$

$$U(Y \mid X) = \log_2 \frac{|R|}{|R_X|}, \tag{8.7}$$

which are called *conditional uncertainties*.

Observe that the ratio $|R| / |R_Y|$ in U(X|Y) represents the average number of elements of X that are possible alternatives under the condition that an element Y has already been selected. This means that U(X|Y) measures the average nonspecificity regarding alternative choices from X for all particular choices from Y. Function U(Y|X) has clearly a similar meaning with the roles of sets X and Y exchanged. Observe also that the conditional uncertainties can be expressed in terms of the joint uncertainty and the two simple uncertainties:

$$U(X|Y) = U(X,Y) - U(Y), \tag{8.8}$$
$$U(Y|X) = U(X,Y) - U(X). \tag{8.9}$$

Furthermore,

$$U(X) - U(Y) = U(X|Y) - U(Y|X), \tag{8.10}$$

which follows immediately from Eqs. (8.8) and (8.9).

If possible alternatives from X do not depend on selections from Y, and vice versa, then $R = X \times Y$ and the sets X and Y are called *noninteractive*. Then, clearly,

$$U(X|Y) = U(X), \tag{8.11}$$
$$U(Y|X) = U(Y), \tag{8.12}$$
$$U(X,Y) = U(X) + U(Y). \tag{8.13}$$

In all other cases, when sets X and Y are *interactive*, these equations become the inequalities

$$U(X|Y) < U(X), \tag{8.14}$$
$$U(Y|X) < U(Y), \tag{8.15}$$
$$U(X,Y) < U(X) + U(Y). \tag{8.16}$$

The last inequality is usually referred to as *subadditivity*, while the equality (8.13) (based on the assumption that $R = X \times Y$) is called *additivity*.

The following symmetric function, which is usually referred to as *information transmission*, is a useful indicator of the strength of constraint between sets X and Y:

$$T(X,Y) = U(X) + U(Y) - U(X,Y). \tag{8.17}$$

When the sets are noninteractive, $T(X,Y) = 0$; otherwise, $T(X,Y) > 0$. Using Eqs. (8.8) and (8.9), T(X,Y) can also be expressed in terms of the conditional uncertainties:

$$T(X,Y) = U(X) - U(X|Y), \tag{8.18}$$
$$T(X,Y) = U(Y) - U(Y|X). \tag{8.19}$$

The Hartley function is applicable only to finite sets. Its counterpart, ${}^cU$, for convex subsets of the n-dimensional Euclidean space was proposed by Klir and Yuan [1995]. Given a universal set X that is convex and such that $X \subseteq \mathbb{R}^n$ for some finite $n \geq 1$, function ${}^cU$ is defined for each convex subset A of X by the formula

$${}^cU(A) = \min_{t \in T} \ln\left[\prod_{i=1}^{n}[1 + \mu(A_{i_t})] + \mu(A) - \prod_{i=1}^{n}\mu(A_{i_t})\right], \tag{8.20}$$

where μ denotes the Lebesgue measure, T denotes the set of all transformations from one orthogonal coordinate system to another, and A_{i_t} denotes the i-th projection of A within the coordinate system t. Klir and Yuan [1995] proved that function ${}^cU$ is monotonic with respect to set inclusion, subadditive, continuous, invariant with respect to isometric transformations of the coordinate system, and ${}^cU(A) = 0$ iff $A = \{x\}$ for some $x \in X$. They also proved its additivity for sets with equal edge lengths and for all convex sets when $n \leq 3$; moreover, they tested the additivity of the function on the computer for thousands of randomly generated examples for each $n = 4, 5, \cdots, 10$. A general proof of additivity of ${}^cU$ was completed by Ramer [1987].

8.2.2 Fuzzy Set Theory

A natural generalization of the Hartley function from classical set theory to fuzzy set theory was proposed by Higashi and Klir [1983a] under the name *U-uncertainty*. For any nonempty fuzzy set A defined on a finite universal set X, the U-uncertainty is defined by the formula

$$U(A) = \frac{1}{h(A)} \int_0^{h(A)} \log_2 |A_\alpha| \, d\alpha, \tag{8.21}$$

where $|A_\alpha|$ denotes the cardinality of the α-cut of A and h(A) is the height of A. It is easy to see that the U-uncertainty is a weighted average of the Hartley function for all distinct α-cuts of the normalized counterpart of A, defined by A(x)/h(A) for all x $\in$ X. Each weight is a difference between the values of α of a given α-cut and the immediately preceding α-cut. For any nonempty fuzzy sets A, B on X, if A(x)/h(A) = B(x)/h(B) for all x $\in$ X, then U(A) = U(B). This means that sets that are equal when normalized have the same nonspecificity according to the U-uncertainty.

The advantage of the U-uncertainty is that it extends the well-justified Hartley function from crisp sets to fuzzy sets via their α-cuts. As a consequence, it satisfies fuzzy counterparts of the properties that make the Hartley function significant: monotonicity, subadditivity, additivity, and the range $[0, \log_2 |X|]$. Moreover, Eqs. (8.8) - (8.19) holds for the U-uncertainty as well.

An alternative measure of nonspecificity for nonempty finite fuzzy sets was proposed by Yager [1982]. First, he introduces a *measure of specificity*, Sp, by the formula

$$Sp(A) = \int_0^{h(A)} \frac{d\alpha}{|A_\alpha|}, \tag{8.22}$$

where all symbols have the same meaning as in (8.21). Since

$$|X|^{-1} \leq Sp(A) \leq 1,$$

the associated measure of nonspecificity, Nsp, is given by the formula

$$NSp(A) = 1 - Sp(A) \tag{8.23}$$

for each nonempty finite fuzzy set A. This measure satisfies only some of the requirements that are generally considered essential for a measure of uncertainty. It is monotonic and attains its minimum and maximum under the same conditions as the U-uncertainty, but it is neither additive nor subadditive, and it does not coincide with the Hartley function for crisp sets. Hence, its justification as a measure of uncertainty and uncertainty-based information is wanting. Nevertheless, the functions Sp and NSp have some utility in problems that do not depend on the missing properties; this utility has been explored by Yager [1990, 1991].

Function ${}^cU$ defined (8.20) can be extended to fuzzy sets via the formula

$${}^cU(A) = \frac{1}{h(A)} \int_0^{h(A)} {}^cU(A_\alpha) d\alpha, \tag{8.24}$$

where A denotes a convex fuzzy subset of the n-dimensional Euclidean space for some finite $n \geq 1$.

8.2.3 Possibility Theory

The U-uncertainty was investigated more thoroughly within possibility theory, utilizing ordered and normal possibility distributions. In this domain, function U has the form

$$U : R \rightarrow \mathbb{R}^+,$$

where R denotes the set of all finite, ordered, and normal possibility distributions. Given a possibility distribution

$$\mathbf{r} = (r_1, r_2, \cdots, r_n) \text{ such that } 1 = r_1 \geq r_2 \geq \cdots \geq r_n,$$

the U-uncertainty of $\mathbf{r}$, $U(\mathbf{r})$, can be expressed by a convenient form

$$U(\mathbf{r}) = \sum_{i=2}^{n} (r_i - r_{i+1}) \log_2 i = \sum_{i=2}^{n} r_i \log_2 \frac{i}{i-1}, \tag{8.25}$$

where $r_{n+1} = 0$ by convention [Klir and Folger, 1988].

Among properties that the U-uncertainty expressed by (8.25) possesses are *additivity*, *subadditivity*, and the following properties [Higashi and Klir, 1983a]:

- *Expansibility* —when components of zero are added to a given possibility distribution, the value of the U-uncertainty does not change.
- *Symmetry*—$U(\mathbf{r}) = U(\mathbf{q})$, where $\mathbf{q}$ is any permutation of $\mathbf{r}$.
- *Continuity* — the U-uncertainty is a continuous function.
- *Monotonicity* — for any pair ${}^1\mathbf{r}$, ${}^2\mathbf{r}$ of possibility distributions of the same length, if ${}^1\mathbf{r} \leq {}^2\mathbf{r}$, then $U({}^1\mathbf{r}) \leq U({}^2\mathbf{r})$.
- *Measurement unit* — since $U(1,1) = 1$, the U-uncertainty measures non-specificity in bits.
- *Range* — within a given universal set X, $U(r) \in [0, \log_2 |X|]$, where the minimum and maximum are obtained for $\mathbf{r} = (1, 0, \cdots, 0)$ and $\mathbf{r} = (1,1, \cdots, 1)$, respectively.
- *Branching* — for every possibility distribution $\mathbf{r} = (r_1, r_2, \cdots, r_n)$ of any length n,

$$U(r_1,r_2,\cdots,r_n) = U(r_1,r_2,\cdots,r_{k-2}, r_k, r_k, r_{k+1},\cdots,r_n) \tag{8.26}$$
$$+ (r_{k-2} - r_k)U\left(\mathbf{1}_{k-2}, \frac{r_{k-1} - r_k}{r_{k-2} - r_k}, \mathbf{0}_{n-k+1}\right) - (r_{k-2} - r_k)\, U(\mathbf{1}_{k-2}, \mathbf{0}_{n-k+2})$$

for each $k = 3,4, \cdots, n$, where, for any given integer i, $\mathbf{1}_i$, denotes a sequence of i ones and $\mathbf{0}_i$ denotes a sequence of i zeroes.

The branching property, which can also be formulated in other forms [Klir and Mariano, 1987; Klir and Folger, 1988], needs some explanation. The property basically shows that the U-uncertainty has the capability of measuring possibilistic nonspecificity in two ways. It can be measured either directly for the given possibility distribution or indirectly by adding U-uncertainties associated with a combination of possibility distributions that reflect a two-stage measuring process. In the first stage of measurement, the distinction between two neighboring components, r_{k-1} and r_k, is ignored (r_{k-1} is replaced with r_k) and the U-uncertainty of the resulting, less refined possibility distribution is calculated. In the second stage, the U-uncertainty is calculated in a local frame of reference, which is defined by a possibility distribution that distinguishes only between the two neighboring possibility values that are not distinguished in the first stage of measurement. The U-uncertainty calculated in the local frame must be scaled back to the original frame by a suitable weighting factor. The sum of the two U-uncertainties obtained by the two stages of measurement is equal to the total U-uncertainty of the given possibility distribution.

The first term on the right-hand side of (8.26) represents the U-uncertainty obtained in the first stage of measurement. The remaining two terms represent the second stage, associated with the local U-uncertainty. The first of these two terms expresses the loss of uncertainty caused by ignoring the component r_{k-1} in the given possibility distribution, but it introduces some additional U-uncertainty equivalent to the uncertainty of a crisp possibility distribution with k-2 components. This additional U-uncertainty is excluded by the last term in (8.26). That is, the local uncertainty is expressed by the last two terms in (8.26).

It was proven by Klir and Mariano [1987] that the U-uncertainty is a unique measure of possibilistic nonspecificity when expansibility, monotonicity, additivity, branching, and appropriate normalization (designating bits as measurement units) are chosen as axioms. Although other axioms could possibly be used for proving the uniqueness of function U, Ramer and Lander [1987] showed that the branching property is necessary for any such proof.

The notion of distance between possibility measures based on information closeness was investigated in terms of possibility distributions by Higashi and Klir [1993b]. For any pair of possibility distributions ${}^1\mathbf{r}$, ${}^2\mathbf{r}$ on X such that ${}^1\mathbf{r} \leq {}^2\mathbf{r}$, they define the gain of information, $g({}^1\mathbf{r}, {}^2\mathbf{r})$, when ${}^2\mathbf{r}$ is replaced with ${}^1\mathbf{r}$ as

$$g({}^1\mathbf{r}, {}^2\mathbf{r}) = U({}^2\mathbf{r}) - U({}^1\mathbf{r}).$$

They also define another function, G, for any pair of possibility distributions ${}^1\mathbf{r}$, ${}^2\mathbf{r}$ on X by the formula

$$G({}^1\mathbf{r}, {}^2\mathbf{r}) = g({}^1\mathbf{r}, {}^1\mathbf{r} \vee {}^2\mathbf{r}) + g({}^2\mathbf{r}, {}^1\mathbf{r} \vee {}^2\mathbf{r}). \tag{8.27}$$

They prove that function G is a metric distance.

The question of how to measure uncertainty and information for continuous possibility distributions was investigated by Ramer [1990]. He showed that the U-uncertainty is not directly applicable to continuous possibility distributions, but it can be employed in a relative form

$$W(\mathbf{r}, \mathbf{s}) = \sum_{i=2}^{n} (s_i - r_i) \log_2 \frac{i}{i-1}, \tag{8.28}$$

which involves two possibility distributions $\mathbf{r} = (r_1, r_2, \cdots, r_n)$ and $\mathbf{s} = (s_1, s_2, \cdots, s_n)$ such that $r_i \leq s_i$ for all $i = 1,2, \cdots, n$. When W is restricted to $\mathbf{s} = (1,1, \cdots, 1) = \mathbf{1}$, we obtain

$$W(\mathbf{r}) = \log_2 n - U(\mathbf{r}), \tag{8.29}$$

where $W(\mathbf{r})$ is a shorthand symbol for $W(\mathbf{r},\mathbf{1})$. This function has properties similar to the U-uncertainty, except that all inequality signs are reversed.

Ramer showed that

$$W(\mathbf{r}) = \int_0^1 \frac{1 - \tilde{r}(x)}{x} dx \tag{8.30}$$

is the continuous counterpart of Eq. (8.29), where $\tilde{r}$ denotes the decreasing rearrangement [Hardy et al., 1934] of the possibility distribution function r.

8.2.4 Evidence Theory

Once the U-uncertainty was well established as a possibilistic generalization of the Hartley function, Dubois and Prade [1985] showed that it can be further generalized to measure nonspecificity in the Dempster-Shafer evidence theory [Shafer, 1976]. To distinguish the U-uncertainty from its more general counterpart in evidence theory, let the latter function be denoted by V. It is defined by the formula

$$V(m) = \sum_{A \in \mathcal{F}} m(A) \log_2 |A|, \tag{8.31}$$

where $(\mathcal{F},m)$ is an arbitrary body of evidence.

It was proven by Ramer [1987] that function V (or V-uncertainty) defined by Eq. (8.31) is the only function that satisfies additivity for noninteractive marginal bodies of evidence, subadditivity for interactive ones, normalization based on choosing bits as measurement units ($V(m) = 1$ when $m(A) = 1$ and $|A| = 2$), and the following two properties:

- *Symmetry* — V is invariant with respect to permutations of values of the basic probability assignment within each group of subsets of X that have equal cardinalities
- *Branching* — $V(m) = V(m_1) + V(m_2)$ for any three bodies of evidence $(\mathcal{F}, m)$, $(\mathcal{F}_1, m_1)$, and $(\mathcal{F}_2, m_2)$ such that

  $F = \{A, B, C, \cdots\}$,
  $F_1 = \{A_1, B, C, \cdots\}$,
  $F_2 = \{A, B_1, C_1, \cdots\}$,

  where $A_1 \subset A$, $B_1 \subset B$, $C_1 \subset C$, $\cdots$, $|A_1| = |B_1| = |C_1| = \cdots = 1$, and

  $m(A) = m_1(A_1) = m_2(A)$,
  $m(B) = m_1(B) = m_2(B_1)$
  $m(C) = m_1(C) = m_2(C_1)$, etc.

Function V is clearly a weighted average of the Hartley function for all focal elements. The weights are values of the basic probability assignment. For each focal element A, m(A) indicates the degree of evidence focusing on A, while $\log_2|A|$ indicates the lack of specificity of this evidential claim. The larger the value of m(A), the stronger the evidence; the larger the set A (and $\log_2|A|$), the less specific the evidence. Consequently, it is reasonable to view function V as a *measure of nonspecificity*.

The range of function V is, as expected, $[0, \log_2 |X|]$. The minimum, $V(m) = 0$, is obtained when $m(\{x\}) = 1$ for some $x \in X$ (no uncertainty); the maximum, $V(m) = \log_2 |X|$, is reached for $m(X) = 1$ (total ignorance). It can easily be shown that Eqs. (8.8) - (8.19) remain valid when the U-uncertainty is generalized to the V-uncertainty.

Since focal elements in probability measures are singletons, $|A| = 1$ and $\log_2 |A| = 0$ for each focal element. Consequently, $V(m) = 0$ for every probability measure.

That is, probability theory is not capable of incorporating nonspecificity, one of the basic types of uncertainty.

In addition to the mentioned properties of the V-uncertainty, which make it well justified as a measure of nonspecificity in evidence theory, the function has some other, surprisingly strong properties. As shown by Klir and Wierman [1987], for example, the V-uncertainty of a joint body of evidence obtained from nested marginal bodies of evidence by the rules of possibility theory (applying the minimum operator to marginal possibility distributions) is exactly the same as the V-uncertainty of a totally different joint body of evidence obtained from the same marginal by the rules of evidence theory (applying the product operator to marginal basic probability assignments). The V-uncertainty is also additive not only for noninteractive bodies of evidence, but also for any joint body of evidence on $X \times Y$ whose only focal elements are Cartesian products of some subsets of X and some subsets of Y.

8.3 ENTROPY-LIKE MEASURES

8.3.1 Probability Theory

As indicated in the previous section, the measure of nonspecificity V does not discriminate among probability measures. All probability measures are fully specific: when $(\mathcal{F},m)$ is a probabilistic body of evidence ($\mathcal{F}$ consists of singletons), then V(m) = 0). What, then, is actually measured by the Shannon entropy, which is a well-established measure of uncertainty in probability theory? Before attempting to answer this question, let's overview key properties of the Shannon entropy.

The *Shannon entropy*, H, which is applicable only to finite probability measures, assumes in evidence theory the form

$$H(m) = -\sum_{x \in X} m(\{x\}) \log_2 m(\{x\}), \tag{8.32}$$

where X denotes a finite universal set and m denotes a given basic probability assignment function. Function H, which forms the basis of *classical information theory* (see, e.g., Blahut [1987] or Guiasu [1977], measures the average uncertainty (in bits) associated with the prediction of outcomes in a random experiment; its range is $[0, \log_2 |X|]$. Clearly, H(m) = 0 when $m(\{x\}) = 1$ for some $x \in X$; $H(m) = \log_2|X|$ when m defines the uniform probability distribution of X (i.e., $m(\{x\}) = 1/|X|$ for all $x \in X$).

As the name suggests, function H was proposed by Shannon [1948]. It was proven in numerous ways, from several well-justified axiomatic characterizations, that this function is the only sensible measure of uncertainty in probability theory. A good overview was prepared by Aczél and Daróczy [1975]. It is also well known that Eqs. (8.8) - (8.19) are valid when function U is replaced with function H.

Since values $m(\{x\})$ are required to add to 1 for all $x \in X$, (8.32) can be rewritten as

$$H(m) = -\sum_{x \in X} m(\{x\}) \log_2 \left[1 - \sum_{y \neq x} m(\{y\}) \right]. \tag{8.33}$$

The term

$$\text{Con}(\{x\}) = \sum_{y \neq x} m(\{y\})$$

in (8.33) represents the total evidential claim pertaining to focal elements that are different from the focal element $\{x\}$. That is, $\text{Con}(\{x\})$ expresses the sum of all evidential claims that fully conflict with the one focusing on $\{x\}$. Clearly, $\text{Con}(\{x\}) \in [0,1]$ for each $x \in X$. The function $-\log_2 [1 - \text{Con}(\{x\})]$, which is employed in (8.33), is monotonic increasing with $\text{Con}(\{x\})$ and extends its range from $[0,1]$ to $[0,\infty)$. The choice of the logarithmic function is a result of the requirement that function H be additive.

It follows from these facts and from the form of (8.33) that the Shannon entropy may be viewed as the mean (expected) value of the *conflict* among evidential claims within a given probabilistic body of evidence.

When a probability measure is defined on a real interval [a,b] by a probability density function f, the Shannon entropy is not directly applicable as a measure of uncertainty (Guiasu [1977], Klir and Folger[1988]). It can be employed only in a modified form

$$D[f(x), g(x) \mid x \in [a,b]] = \int_a^b f(x) \log_2 \frac{f(x)}{g(x)} dx, \tag{8.34}$$

which involves two probability density functions, f(x) and g(x), defined on [a,b]. Its finite counterpart is the function

$$D[p(x), q(x) \mid x \in X] = \sum_{x \in X} p(x) \log_2 \frac{p(x)}{q(x)}, \tag{8.35}$$

which is known in information theory as the *Shannon cross-entropy* or *directed divergence.* Function D measures uncertainty in relative rather than absolute terms.

When f(x) in (8.34) is replaced with a density function, f(x,y), of a joint probability distribution on $X \times Y$, and g(x) is replaced with the product of density functions of marginal distributions on X and Y, $f(x) \cdot f(y)$, D becomes equivalent to the information transmission given by (8.17). This means that the continuous counterpart of the information transmission can be expressed as

$$D(f(x,y), f(x) \cdot f(y) | x \in [a,b], y \in [\alpha,\beta])$$

$$= \int_a^b \int_\alpha^\beta f(x,y) \log_2 \frac{f(x,y)}{f(x) \cdot f(y)} dx dy. \tag{8.36}$$

8.3.2 Evidence Theory

What form should the generalized counterpart of the Shannon entropy in evidence theory take? The answer is by no means obvious, as exhibited by several candidates proposed in the literature.

Two of the candidates were proposed in the early 1980s. One of them is function E defined by the formula

$$E(m) = -\sum_{A \in \mathcal{F}} m(A) \log_2 Pl(A), \tag{8.37}$$

which is usually called a measure of *dissonance*; this function was proposed by Yager [1983]. The other one is function C defined by the formula

$$C(m) = -\sum_{A \in \mathcal{F}} m(A) \log_2 Bel(A), \tag{8.38}$$

which is referred to as a measure of *confusion*; it was proposed by Höhle [1982]. It is obvious that both of these functions collapse into the Shannon entropy when m defines a probability measure.

To examine the meaning of function E, we may rewrite it in the form

$$E(m) = -\sum_{A \in \mathcal{F}} m(A) \log_2 \left[1 - \sum_{A \cap B = \emptyset} m(B) \right]. \tag{8.39}$$

The term

$$K(A) = \sum_{A \cap B = \emptyset} m(B)$$

in (8.39) represents the total evidential claim pertaining to focal elements that are disjoint with the set A. That is, K(A) expresses the sum of all evidential claims that fully conflict with the one focusing on the set A. Clearly, $K(A) \in [0,1]$. The function

$$-\log_2[1 - K(A)],$$

which is employed in (8.39), is monotonic increasing with K(A) and extends its range from [0,1] to $[0,\infty)$. The logarithmic function is chosen to conform to the Shannon entropy when m defines a probability measure.

It follows from these facts and from the form of Eq. (8.39) that E(m) is the mean (expected) value of the conflict among evidential claims within a given body of evidence $(\mathcal{F},m)$; it measures the conflict in bits and its range is $[0, \log_2|X|]$.

Function E is not fully satisfactory since it is not subadditive and, moreover, we feel intuitively that m(B) conflicts with m(A) not only when $B \cap A = \emptyset$. A broader view of conflict is expressed by the measure of confusion C given by (8.38), which may also be written in the form

$$C(m) = -\sum_{A \in \mathcal{F}} m(A) \log_2 \left[1 - \sum_{B \not\subset A} m(B) \right]. \tag{8.40}$$

The term

$$L(A) = \sum_{B \not\subset A} m(B)$$

in (8.40) expresses the sum of all evidential claims that conflict with the one focusing on the set A according to the following broader view of conflict: m(B) conflicts with m(A) whenever $B \not\subset A$. The reason for using the function

$$-\log_2[1 - L(A)]$$

instead of L(A) in Eq. (8.40) is the same as already explained in the context of function E. The conclusion is that C(m) is the mean (expected) value of the conflict, viewed in the broader sense, among evidential claims within a given body of evidence ($\mathcal{F}$,m).

Function C is also not fully satisfactory as a measure of conflicting evidential claims within a body of evidence, but for a different reason than function E. Although it employs the broader, and more satisfactory, view of conflict, it does not properly scale each particular conflict of m(B) with respect to m(A) according to the degree of violation of the subsethood relation $B \subset A$. It is clear that the more this subsethood relation is violated the greater the conflict. In addition, function C also has some undesirable mathematical properties. For example, its maximum is greater than $\log_2|X|$ and, similarly as function E, it is not subadditive.

To overcome the deficiencies of functions E and C as adequate measures of conflict in evidence theory, two additional functions, D and S, were proposed in which the individual conflicts in evidential claims among focal elements are properly scaled in two distinct ways. These functions are defined by the formulas

$$D(m) = -\sum_{A\in\mathcal{F}} m(A)\log_2 \sum_{B\in\mathcal{F}} m(B)\frac{|A\cap B|}{|B|} \tag{8.41}$$

and

$$S(m) = -\sum_{A\in\mathcal{F}} m(A)\log_2 \sum_{B\in\mathcal{F}} m(B)\frac{|A\cap B|}{|A|}. \tag{8.42}$$

These functions are referred to as *discord* (Klir and Ramer, 1990] and *strife* (Vejnarova and Klir, 1993], respectively. As shown by Klir and Yuan [1993], the distinction between discord and strife reflects the distinction between conjunctive and disjunctive set-valued propositions.

Although functions D and S are better justified on both intuitive and mathematical grounds than their predecessors (Ramer and Klir, 1993], they do not satisfy the essential requirement of subadditivity. To overcome the lack of subadditivity of all the proposed entropy-like measures, a novel approach to measuring uncertainty in evidence theory was proposed by Harmanec and Klir [1994]. In this approach, the two components of uncertainty in evidence theory, nonspecificity and conflict, are not distinguished. Instead, a measure of total uncertainty is defined by using the principle of maximum entropy. This measure is defined by a function, AU, that assigns a nonnegative real number to each given finite belief measure in the following way.

Given a finite set X and a belief measure, Bel, on $\mathcal{P}(X)$, the amount of uncertainty contained in Bel, denoted as AU(Bel), if defined by

$$AU(Bel) = \max[-\sum p_x \log_2 p_x],$$

where the maximum is taken over all probability distributions on X that satisfy, in addition to axioms of probability theory, the inequalities:

$$\text{Bel}(A) \leq \sum_{x \in A} p_x \leq 1 - \text{Bel}(\overline{A})$$

for all A ∈ $\mathcal{P}$(X).

It is significant that function AU, which was also conceived by Maeda and Ichihashi [1993] and Chokr ad Kreinovich [1994], satisfies all requirements that are considered essential for a measure of uncertainty in evidence theory (Harmanec and Klir, 1994]. Although the uniqueness of function AU for measuring uncertainty in evidence theory has not been proven as yet, it was proven by Harmanec [1995] that AU is the smallest measure among all possible aggregate measures of uncertainty in evidence theory. Moreover, efficient algorithms for computing function AU were developed by Meyerowitz et al. [1994] and Harmanec et al.[1996].

Various other measures of total uncertainty in evidence theory have been proposed in the literature. For example, Lamata and Moral [1988] proposed the sum of functions N and E, Klir and Ramer [1993] proposed the sum of functions V and D, Vejnarova and Klir [1993] proposed the sum of functions V and S, and Pal et al. [1993] proposed the sum of function V and a new function, G, defined by the formula

$$G(m) = -\sum_{A \in \mathcal{F}} m(A) \log_2 m(A). \tag{8.43}$$

Unfortunately, none of these measures is subadditive. Moreover, the last measure does not satisfy the proper range [0, $\log_2|X|$].

A measure of uncertainty-based information for evidence theory, which is fundamentally different from the described measures, was proposed by Smets [1993]. He formulated requirements for such a measure without considering bodies of evidence on Cartesian products. Hence, his formulation does not contain the usual requirements of additivity and subadditivity. Instead, he requires that the information content of two non-contradictory bodies of evidence be the same as the information content of the combined body of evidence obtained by the Dempster rule of combination (Shafer, 1976], in addition to monotonicity and appropriate boundary conditions. Under these requirements, Smets shows that the only meaningful measure of information in bits is a function I defined by the formula

$$I(\text{Bel}) = -\sum_{A \in \mathcal{P}(X)} \log_2 Q(A) \tag{8.44}$$

for each given non-dogmatic belief function Bel on $\mathcal{P}$(X), (i.e., m(X) > 0 and, hence, Q(X) > 0), where Q denotes the commonality function [Shafer, 1976].

8.3.3 Possibility Theory

Since the value of function E, which was the entropy-like measure proposed first for evidence theory, if zero for any nested body of evidence, possibilistic evidence was often considered as conflict free. This belief was revised with the emergence of the other entropy-like measure, particularly the functions D (discord) and S(strife). These functions were investigated within possibility theory by Geer and Klir [1991]. They found that mathematical properties of these functions in possibility theory are quite similar. In particular, their maxima depend on n = |X| in exactly the same way: they increase with n and converge to a constant, estimated as

0.892 as $n \to \infty$. Hence, according to these measures, possibility theory is not conflict free, but almost conflict free. For large bodies of evidence, these measures are negligible when compared with nonspecificity.

As already mentioned, neither discord nor strife are subadditive. Hence, their justification is questionable. The only fully justified measure of uncertainty in evidence theory is the function AU. When applying this measure to possibilistic bodies of evidence, the entropy-like portion of AU (obtained by subtracting the U-uncertainty from the total uncertainty AU) is not bounded from above as functions D and S are. The computation of function AU in possibility theory is very simple, almost trivial.

8.4 MEASURES OF FUZZINESS

8.4.1 Fuzzy Set Theory

The question of how to measure fuzziness is one of the fundamental issues of fuzzy set theory. As already mentioned, fuzziness does not result from information deficiency, but rather from linguistic imprecision (or vagueness). In general, a *measure of fuzziness* is a function

$$f: \mathcal{F}(X) \to \mathbb{R}^+,$$

where $\mathcal{F}(X)$ denotes the set of all fuzzy subsets of X (fuzzy power set). For each fuzzy set A, this function assigns a nonnegative real number f(A) that expresses the degree to which the boundary of A is not sharp.

In order to qualify as a sensible measure of fuzziness, function f must satisfy some requirements that adequately capture our intuitive comprehension of the degree of fuzziness. The following three requirements are essential:

1. $f(A) = 0$ if and only if A is a crisp set.
2. f(A) attains its maximum if and only if $A(x) = 0.5$ for all $x \in X$, which is intuitively conceived as the highest fuzziness.
3. $f(A) \le f(B)$ when set A is undoubtedly sharper than set B, which, according to our intuition, means that for all $x \in X$, $A(x) \le B(x)$ when $B(x) \le 0.5$, while $A(x) \ge B(x)$ when $B(x) \ge 0.5$.

There are different ways of measuring fuzziness that all satisfy the three essential requirements. One way is to measure fuzziness of any set A by a metric distance between its membership grade function and the membership grade function (or characteristic function) of the nearest crisp set [Kaufmann, 1975]. Even when committing to this conception of measuring fuzziness, the measurement is not unique. To make it unique, we have to choose one of many possible distance functions.

Another way of measuring fuzziness, which seems more practical as well as more general, is to view fuzziness of a set in terms of the lack of distinction between the set and its complement [Yager, 1979, 1980]. Indeed, it is precisely the lack of distinction between sets and their complements that distinguishes fuzzy sets from crisp sets. The less a set differs from its complement, the fuzzier it is. Measuring fuzziness in this way is dependent on the definition of a fuzzy complement (Klir and Folger, 1988]. This issue, which was thoroughly investigated by Higashi and Klir [1982], is not discussed here. For the sake of simplicity, only the standard fuzzy complement is considered. If other types of fuzzy complements were also considered, the second and third of the three requirements for function f would have to be properly generalized [Higashi and Klir, 1982].

Employing the standard fuzzy complement, we can still choose different distance functions to express the lack of distinction of a set and its complement. Choosing, for example, the Hamming distance, the local distinction (one for each $x \in X$) of a given set A and its complement is measured by

$$|A(x) - (1 - A(x))| = |2A(x) - 1|,$$

and the lack of each local distinction is measured by

$$1 - |2A(x) - 1|.$$

When X is finite, the measure of fuzziness, f(A), is then obtained by adding all these local measurements:

$$f(A) = \sum_{x \in X} [1 - |2A(x) - 1|]. \tag{8.45}$$

This formula can readily be modified to fuzzy sets defined on infinite but bounded subsets of $\mathbb{R}$ (or $\mathbb{R}^n$ for some finite $n \geq 1$). Consider, for example, that $X = [a,b]$, which is perhaps the most practical case. Then, (8.45) can be modified by replacing the summation with integration:

$$f(A) = \int_a^b (1 - |2A(x) - 1|)dx. \tag{8.46}$$

An alternative measure of fuzziness,

$${}^e f(A) = -\sum_{x \in X} [A(x) \log_2 A(x) + (1 - A(x)) \log_2 (1 - A(x))], \tag{8.47}$$

based on a different distance function within the same general view of fuzziness, was proposed and investigated by DeLuca and Termini [1972, 1974, 1977]. This measure, which is frequently cited in the literature, is usually called an *entropy of fuzzy sets*. The name *entropy* was apparently chosen due to the similarity of the product terms in (8.47) with the product terms in the Shannon entropy. This choice is unfortunate since the two functions measure fundamentally different types of uncertainty. Function ${}^e f$ satisfies, in addition to the three required properties, also the valuation property:

$${}^e f(A \cup B) + {}^e f(A \cap B) = {}^e f(A) + {}^e f(B) \tag{8.48}$$

for all $A, B \in \mathcal{F}(X)$, where $\cup$ and $\cap$ are the standard operations of fuzzy union and intersection, respectively.

Numerous other measures of fuzziness has been proposed in the literature. An excellent summary of these measures was prepared by Pal and Bezdek [1994]. This paper is recommended for further details on this subject.

8.4.2 Fuzzified Evidence Theory

Any measure of fuzziness is applicable not only to fuzzy set theory, but also to any other theory that is fuzzified. When focal elements in evidence theory are fuzzy sets, it is reasonable to express the degree of fuzziness of a body of evidence $(\mathcal{F},m)$ by the function

$$\tilde{f}(m) = \sum_{A \in \mathcal{F}} m(A) f(A), \qquad (8.50)$$

which represents the weighted average of the degree of fuzziness of focal elements.

8.5 CONCLUSIONS

Measures of uncertainty of relevant types are now well established in classical set theory, fuzzy set theory, probability theory, possibility theory, and evidence theory. Among these measures, the best understood and justified is the Shannon entropy in probability theory and the measures of nonspecificity in all these five theories of uncertainty. Although the measures of nonspecificity are known to possess a surprising number of important properties, their further investigation still seems worthwhile. As discovered by Ramer [1987], for example, function V can be interpreted as the value of the first derivative of the logarithm of the Dirichlet function at zero. Exploring this and other possible avenues is likely to advance further our understanding of this delicate function.

The entropy-like measures are less understood. Although the measurement of total uncertainty in evidence theory and possibility theory by the function AU is well justified, the uniqueness of this function AU is still an open question. Moreover, function AU is rather insensitive to changes in evidence that seems significant on intuitive grounds. The relationship among the total uncertainty, nonspecificity, and entropy-like uncertainty is also not understood as yet. A comparative study of the many measures of fuzziness based on sensible requirements is needed to make convincing choices in various classes of applications. The question of uniqueness should eventually be addressed as well.

A host of unresolved problems are connected with the various principles of uncertainty, such as the principles of maximum uncertainty, minimum uncertainty, and uncertainty invariance [Klir and Wierman, 1998]. Ideally, these principles should be derived from appropriate consistency axioms, similarly as Shore and Johnson [1980] did for the maximum entropy principle. To make these principles operational, efficient algorithms need to be developed to deal with problems resulting from the various applications of these principles.

References

Aczél, J. and Daróczy, Z. [1975], *On Measures of Information and Their Characterizations.* Academic Press, New York.

Blahut, R. E. [1987], *Principles and Practice of Information Theory.* Addison-Wesley, Reading, Mass.

Bouchon, B. [1988], "Entropic models: A general framework for measures of uncertainty and information." In: Gupta, M. M. and Yamakawa, T., eds., *Fuzzy Logic in Knowledge-Based Systems, Decision and Control.* North-Holland, New York, pp. 93-105.

Chokr, B. A. and V. Y. Kreinovich [1994], "How far are we from the complete knowledge? Complexity of knowledge acquisition in the Dempster-Shafer approach." Yager, R. R. et al. eds, *Advances in the Dempster-Shafer Theory of Evidence*. John Wiley, New York, pp. 555-576.

De Luca, A. and Termini, S. [1972], "A definition of a nonprobabilistic entropy in the setting of fuzzy sets theory." *Information and Control*, 20(4), pp. 301-312.

De Luca, A. and Termini, S. [1974], "Entropy of L-fuzzy sets." *Information and Control*, 24, pp. 55-73.

De Luca, A. and Termini, S. [1977], "Measures of ambiguity in the analysis of complex systems." In: *Mathematical Foundations of Computer Science*. Springer-Verlag, New York, pp. 382-389.

Dubois, D. and Prade, H. [1985], "A note on measures of specificity for fuzzy sets." *Intern. J. of General Systems*, 10(4), pp. 279-283.

Dubois, D. and Prade, H. [1987], "Properties of measures of information in evidence and possibility theories." *Fuzzy Sets and Systems*, 24(2), pp. 161-182.

Geer, J. F. and Klir, G. J. [1991], "Discord in possibility theory." *Intern. J. of General Systems*, 19(2), pp. 119-132.

Guiasu, S. [1977], *Information Theory with Applications*. McGraw-Hill, New York.

Hardy, G. H., Littlewood, J. E. and Polya, G. [1934], *Inequalities*. Cambridge University Press, Cambridge.

Harmanec, D. [1995], "Toward a characterization of uncertainty measure for the Dempster-Shafer theory." In: *Proc. of the Eleventh Intern. Conf. on Uncertainty in Artificial Intelligence*, Montreal.

Harmanec, D. and Klir, G. J. [1994], "Measuring total uncertainty in Dempster-Shafer theory: A novel approach." *Intern. J. of General Systems*, 22(4), pp. 405-419.

Harmanec D., G.Resconi, G.J.Klir and Y.Pan [1996], "On the computation of uncertainty measure in Dempster-Shafer theory." *Intern. J. of General Systems*, **25**(2), pp. 153-163.

Hartley, R. V. L. [1928], "Transmission of information." *The Bell System Technical J.*, 7, pp. 535-563.

Higashi, M. and Klir, G. J. [1982], "On measure of fuzziness and fuzzy complements." *Intern. J. of General Systems*, 8(3), pp. 169-180.

Higashi, M. and Klir, G. J. [1983a], "Measures of uncertainty and information based on possibility distributions." *Intern. J. of General Systems*, 9(1), pp. 43-58.

Higashi, M. and Klir, G. J. [1983b], "On the notion of distance representing information closeness: possibility and probability distributions." *Intern. J. of General Systems*, 9(2), pp. 103-115.

Höhle, U. [1982], "Entropy with respect to plausibility measures." In: *Proc. 12th IEEE Intern. Symp. on Multiple-Valued Logic*, Paris, pp. 167-169.

Kaufmann, A. [1975], *Introduction to the Theory of Fuzzy Subsets*. Academic Press, New York.

Klir, G. J. [1987], "Where do we stand on measures of uncertainty, ambiguity, fuzziness, and the like?" *Fuzzy Sets and Systems*, 24(2), pp. 141-160.

Klir, G. J. [1991], "Generalized information theory." *Fuzzy Sets and Systems*, 40(1), pp. 127-142.

Klir, G. J. [1993], "Developments in uncertainty-based information." In: Yovits, M. C., ed., *Advances in Computers*. Academic Press, San Diego, pp. 255-332.

Klir, G. J. and Folger, T. [1988], *Fuzzy Sets, Uncertainty, and Information*. Prentice Hall, Englewood Cliffs, NJ.

Klir, G. J. and Mariano, M. [1987], "On the uniqueness of possibilistic measure of uncertainty and information." *Fuzzy Sets and Systems*, 24(2), pp. 197-219.

Klir, G. J. and Ramer, A. [1990], "Uncertainty in the Dempster-Shafer theory : A critical re-examination." *Intern. J. of General Systems*, 18((2)), pp. 155-166.

Klir, G. J. and Wierman, M. [1987], "On properties of the V-uncertainty." In: *Proc. NAFIPS Meeting*, pp. 96-106.

Klir, G. J. and Wierman M.J. [1998], *Uncertainty-Based Information: Elements of Generalized Information Theory.* Physica-Verlag/Springer-Verlag, Heidelberg and New York.

Klir, G. J. and Yuan, B. [1993], "On measures of conflict among set-valued statements." *Proc. 1993 World Congress on Neural Networks*, Portland, Oregon.

Klir, G. J. and Yuan, B. [1995], "On nonspecificity of fuzzy sets with continuous membership functions." In: *Proc. 1995 Intern. Conf. on Systems, Man, and Cybernetics*, Vancouver.

Lamata, M. T. and Moral, S. [1988], "Measures of entropy in the theory of evidence." *Intern. J. of General Systems*, 14(4), pp. 297-305.

Maeda, Y. and Ichihashi, H. [1993], "An uncertainty measure with monotonicity under the random set inclusion." *Intern. J. of General Systems*, 21(4), pp. 379-392.

Meyerowitz, A., Richman, F. and Walker, E. A. [1994], "Calculating maximum-entropy probability densities for belief functions." *Intern. J. of Uncertainty, Fuzziness, and Knowledge-Based Systems*, 2(4), pp. 377-389.

Pal, N. R. and Bezdek, J. C. [1994], "Measuring fuzzy uncertainty." *IEEE Transactions on Fuzzy Systems*, 2(2), pp. 107-118.

Pal, N. R., Bezdek, J. C. and Hemashina, R. [1992], "Uncertainty measures for evidential reasoning I: A review." *Intern. J. of Approximate Reasoning*, 7(3,4), pp. 165-183.

Pal, N. R., Bezdek, J. C. and Hemashina, R. [1993], "Uncertainty measures for evidential reasoning II: A new measure of total uncertainty." *Intern. J. of Approximate Reasoning*, 8(1), pp. 1-16.

Ramer, A. [1987], "Uniqueness of information measure in the theory of evidence." *Fuzzy Sets and Systems*, 24(2), pp. 183-196.

Ramer, A. [1990], "Concepts of fuzzy information measures on continuous domains." *Intern. J. of General Systems*, 17(2-3), pp. 241-248.

Ramer, A. [1997], "Euclidean specificity: two solutions and few problems." *Proc. Seventh IFSA Congress,* Vol. IV, *Prague*, pp. 268-271.

Ramer, A. and Klir, G. J. [1993], "Measures of discord in the Dempster-Shafer theory." *Information Sciences*, 67(1&2), pp. 35-50.

Ramer, A. and Lander, L. [1987], "Classification of possibilistic uncertainty and information functions." *Fuzzy Sets and Systems*, 24(2), pp. 221-230.

Rènyi, A. [1970], *Probability Theory*. North-Holland, Amsterdam (Chapter IX, Introduction to Information Theory , pp.540-616).

Shafer G. [1976], *A Mathematical Theory of Evidence*. Princeton Univ. Press, Princeton, NJ.

Shannon, C. E. [1948], "The mathematical theory of communication." *The Bell System Technical J.*, 27, pp. 379-423, 623-656.

Shore, J. E. and Johnson, R. W. [1980], "Axiomatic derivation of the principle of maximum entropy and the principle of minimum cross -entropy." *IEEE Trans. on Information Theory*, IT-26, pp. 26-37.

Smets, P. [1983], "Information content of an evidence." *Intern. J. of Man-Machine Studies*, 19, pp. 33-43.

Vejnarova, J. and Klir, G. J. [1993], "Measure of strife in Dempster-Shafer theory." *Intern. J. of General Systems*, 22(1), pp.25-42.

Yager, R. R. [1979], "On the measure of fuzziness and negation. Part I: Membership in the unit interval." *Intern. J. of General Systems*, **5**(4), pp. 189-200.

Yager, R. R. [1980], "On the measure of fuzziness and negation. Part II: Lattices." *Information and Control*, 44, pp. 236-260.

Yager, R. R. [1982], "Measuring tranquility and anxiety in decision making: an application of fuzzy sets." *Intern. J. of General Systems*, 8(3), pp. 139-146.

Yager, R. R. [1983], "Entropy and specificity in a mathematical theory of evidence." *Intern. J. of General Systems*, 9(4), pp. 249-260.

Yager, R. R. [1990], "Ordinal measures of specificity." *Intern. J. of General Systems*, 17(1), pp. 57-72.

Yager, R. R. [1991], "Similarity based specificity measures." *Intern. J. of General Systems*, 19(2), pp. 91-105.

9 QUANTIFYING DIFFERENT FACETS OF FUZZY UNCERTAINTY

Nikhil R. Pal
and James C. Bezdek

Abstract : In this chapter we discuss two facets of uncertainty, namely, *fuzziness* and *non-specificity*, that may be associated with fuzzy sets. After presenting a set of desirable axioms that a measure of fuzziness should satisfy, various attempts to quantify fuzziness are discussed. In this regard, the need for and quantification of higher order measures of fuzziness and weighted fuzziness are also presented. Finally, we discuss a few attempts to quantify non-specificity.

9.1 INTRODUCTION

Fuzzy sets are a powerful tool for modeling non-statistical uncertainty and play a significant role in many deployed systems (Sugeno,1985). Therefore, characterization and quantification of uncertainties associated with fuzzy sets are important issues that affect the management of uncertainty in system modeling and design. This chapter deals with two quantitative measures of fuzzy uncertainty, namely, fuzziness and non-specificity. Suppose a die is thrown and you are asked to guess the top face. Your uncertainty about the outcome is attributed to randomness. The best way to approach this question might be to describe the status of the die in terms of a probability distribution on the six faces. Uncertainty that arises due to chance is called *Probabilistic Uncertainty* (PU). There are many measures of probabilistic uncertainty, of which Shannon's entropy is the most celebrated.

To make the situation more complex, suppose an artificial vision system analyzes a digital image of the top face. Based on the evidence gathered, the system might suggest that the top face is either a 5 or 6, but cannot be more specific. This kind of uncertainty arises from limitations (for example, sensor resolution) of the evidence gathering system. Uncertainty in this second situation reflects ambiguity in specifying the exact solution, and is known as *Non-Specificity* (NS). If we are certain that the top face is either a 5 or 6, this case involves only non-specificity.

More generally, the vision system might also supply a certainty factor with its information. For example, the system might suggest that the top face is either a 5 or a 6 with a belief of 0.8 and it is any other face with a belief of 0.2. In this case uncertainty due to chance is also present, because any one of the two hypotheses, top face is $\{ 5,6 \}$ or top face is $\{1, 2, 3, 4 \}$, could be true. So the system contains both PU and NS. There have been several attempts to quantify uncertainty associated with such systems (Klir and Ramer,1990; Pal et al.,1992; Pal et al.,1993; George and Pal,1996). Chapter 8 by Klir in this handbook provides a good exposition of this.

Finally, suppose the segmented version of the top face produced by the vision system is so ambiguous that the system expresses it by a fuzzy set, say, HIGH, characterized by a membership function. Here two types of uncertainties appear, *Fuzzy Uncertainty* (FU) due to linguistic imprecision or vagueness; and a type of non-specificity which depends on the possible solutions (top face) on which the fuzzy concept HIGH is defined. Fuzzy uncertainty differs from PU because it deals with situations where set boundaries are not sharply defined. Probabilistic uncertainty is not due to ambiguity about set-boundaries, but rather, about the belonging of elements or events to crisp sets. Non-specificity can be associated with both crisp and fuzzy sets.

Uncertainty due to fuzziness is sometimes related to probabilistic uncertainty. For example, in the die experiment the occurrence of a 6 supports the fuzzy event HIGH more than a 3 does, but there is still an element of chance about the outcome of a throw, so the system contains PU and FU. Moreover, NS also appears in this third case, so the most complex systems may exhibit all three types. Many authors have defined measures of the total uncertainty in a system (Lamata and Moral,1987; Klir and Ramer,1990; Pal et al.,1993). This article is devoted primarily to measures of FU and NS in the context of fuzzy sets.

9.2 DIFFERENT FACETS OF FUZZY UNCERTAINTY

Fuzzy sets can model vagueness due to linguistic imprecision, for example, a set of TALL persons. Here there is no precise boundary between the sets TALL and NOT TALL. Intuitively we all feel that there is a gradual transition from NOT TALL to TALL. Fuzziness can also arise due to lack of precision in sensors. For example, an image of a real scene, which has colors, when taken by a camera loaded with a black and white film shows many intermediate gray shades instead of the actual colors in the scene. These gray shades may arise even if the scene (say, a printed page from a book) appears black and white.

Let X be any set. A fuzzy set (FS) A on X is characterized by a membership function $\mu_A : X \to [0, 1]$. The value of $\mu_A(x)$ represents the grade or degree of membership of x in A. When μ_A is valued in {0,1}, it is the characteristic function of a crisp or nonfuzzy set. Subjective characterization of the membership function of a fuzzy set is never unique. Different individuals may define different μ_A's for the same fuzzy concept. For example, the membership function of a fuzzy set TALL defined by an American for the Americans would usually be quite different from the TALL membership function defined by an Oriental for the Orientals. Thus, as rightly pointed out by Jumerie (1986), the definition of fuzzy sets refers neither to the observed sets nor to the observation process. This led Jumerie to define *relativistic* fuzzy sets. The relativistic fuzzy set describes the fuzziness which is involved in a set A with respect (relative) to an observer. Fuzziness in the relativistic framework is beyond the scope of this chapter; interested readers referred to (Jumarie,1986). In this chapter we concentrate on fuzziness as defined by Zadeh(1965).

There are actually two facets of the uncertainty that are associated with a fuzzy set. Let A=TALL; and let μ_A be the membership function defined on a set of discrete heights $X = \{4, 4.5, 5, 5.5, 6, 6.5\}$ where each $x_i \in X$ is in feet. There are two questions that we can ask : how specific is the concept (set) TALL as defined by μ_A; and how ambiguous is the concept (set) TALL - i.e., how ill defined is the boundary of the set TALL? Note that these are different questions and refer to two different aspects of uncertainty. Consider two definitions of μ_A, say as μ_{A1} and μ_{A2}; $\mu_{A1} = \{0.3, 0.45, 0.45, 0.5, 1.0, 0.5\}$ and $\mu_{A2} = \{0.8, 0.8, 1.0, 1.0, 1.0, 1.0\}$. The values of μ_{A1} imply that we know very precisely what TALL is; $\mu_{A1}(6) = 1$ identifies 6 feet as TALL, and as we go away from 6 feet the membership falls drastically. For μ_{A2}, the values are not very specific about what height should represent the concept TALL; rather, we say that a person having the TALL property to the maximum extent can take any of the four heights {5, 5.5, 6, 6.5}.

Hence μ_{A1} is more specific than μ_{A2}. This facet of uncertainty is known as the lack of specificity or *non-specificity.* On the other hand, fuzziness is more with μ_{A1} because four of the heights have memberships which are nearly 0.5. For μ_{A2} the lack of precision is high, but the fuzziness is quite low because for every height we are almost sure whether it belongs to the set or not.

9.3 MEASURING FUZZINESS

Before discussing measures of fuzziness we introduce some of the basic definitions. Let $X = \{x_1, x_2, ..., x_n\}$ be any finite set. $\mu_A(x) : X \to [0, 1]$ characterizes a fuzzy set A on X. We let $\mu_i = \mu_A(x_i)\ \forall i$ when no confusion is possible. Let $\mathcal{P}_n(X)$ denote the set of all fuzzy subsets of X. Also let x be a discrete random variable that takes values in X with probabilities $P = \{p_1, p_2, ..., p_n\}$; we call (X, P) a discrete probabilistic framework.

Of the various definitions of compliments, we will frequently use Zadeh's (1965) complement, A^c_Z of A, defined as $\mu_{A^c_Z}(x) = 1 - \mu_A(x)$. Similarly, unless stated otherwise, the union and intersection operators are Zadeh's (1965) definitions :

$$\mu_{A \cup B}(x) = \max\{\mu_A(x), \mu_B(x)\} \ \ \forall x \in X$$

and

$$\mu_{A \cap B}(x) = \min\{\mu_A(x), \mu_B(x)\} \ \ \forall x \in X.$$

Most measures of fuzziness estimate the *average* ambiguity in a fuzzy set in some well defined sense : that is, the average ambiguity in making a decision as to whether an element belongs to the set or not. Yager (1979, 1980), on the other hand, viewed fuzziness as the lack of distinction between a fuzzy set and its Zadeh-complement. From Yager's view point any measure of the lack of distinction between A and A^c can also be taken as a measure of fuzziness.

9.3.1 POSTULATES OF MEASURES OF FUZZINESS

We now discuss some properties that seem plausible for a measure of fuzziness. The fuzziness of a crisp set using any measure should be zero, as there is no ambiguity about whether an element belongs to the set or not. If a set is maximally ambiguous ($\mu_A(x) = 0.5\ \forall\ x$), then its fuzziness should be maximum. A fuzzy set A^* is called a *sharpened* version of A if the following conditions are satisfied :

$$\begin{aligned} \mu_{A^*}(x) \le \mu_A(x) \quad & if \quad \mu_A(x) \le 0.5; \\ & and \\ \mu_{A^*}(x) \ge \mu_A(x) \quad & if \quad \mu_A(x) \ge 0.5. \end{aligned} \tag{9.1}$$

When a membership value approaches either 0 or 1, the ambiguity about belongingness of the concerned element decreases. Therefore, for a sharpened version A^* of a fuzzy set A the measure of fuzziness should decrease as sharpening reduces ambiguity. A fourth intuitively desirable property is that the fuzziness of a set and its Zadeh complement be equal. For example, the ambiguity present in TALL and NOT TALL (note that NOT TALL is not necessarily SHORT) should be the same.

Formally, a *measure of fuzziness* H is a mapping $H : \mathcal{P}_n(X) \to \mathcal{R}^+$ that quantifies the degree of fuzziness present in A. Ebanks (1983) suggested that a measure of fuzziness should satisfy the following properties for $A, B \in \mathcal{P}_n(X)$:

Minimality P1 : $H(A) = 0 \Leftrightarrow \mu_A(x) = 0$ *or* $1 \; \forall \; x \in X$;
Maximality P2 : $H(A)$ is maximum $\Leftrightarrow \mu_A(x) = 0.5 \; \forall \; x \in X$;
Resolution P3 : $H(A) \ge H(A^*)$ where A^* is a sharpened version of A satisfying (9.1).
Symmetry P4 : $H(A) = H(A^c)$ where $\mu_{A^c}(x) = 1 - \mu_A(x) \;\; \forall x \in X$;
Valuation P5 : $H(A \cup B) + H(A \cap B) = H(A) + H(B)$.

Ebanks also proposed a sixth requirement called generalized additivity:

Generalized additivity P6 : There exist mappings $\sigma, \tau : \mathcal{R}^+ \to \mathcal{R}^+$ such that $H(A \times B) = H(A)\tau(S(B)) + \sigma(S(A))H(B)$, where $\mu_{A\times B}(x, y) = \mu_A(x)\mu_B(y) \;\; \forall x \in X, y \in Y$; S, the sigma count (Ebanks call it *power*) of A is $S(A) = \sum_{i=1}^{n} \mu_A(x_i)$.

The objective of P6 is to establish a connection between the fuzziness of $A \times B$ on one hand and the fuzziness of A and B, on the other. Ebanks (1983) remarked that the exact form of this connection may be debated. P6, in the suggested form, states that to get $H(A \times B)$, $H(A)$ and $H(B)$ should be multiplied by some factors reflecting the "sizes" of A and B.

We believe that P1-P5 are the most desirable properties of a measure of fuzziness. Hence, our discussion focuses on P1-P5; however, as and when needed, we will also refer to P6. Ebanks gave the following necessary and sufficient conditions, on functions that satisfy requirements $P1 - P5$ for discrete fuzzy sets :

Theorem E (1983, p. 28) Let $H : \mathcal{P}_n(X) \to \mathcal{R}^+$ and $\mu_i = \mu_A(x_i) \; \forall i$. Then H satisfies $P1 - P5$ if and only if H has the from $H_E(A) = \sum_{i=1}^{n} g(\mu_i)$ for some function $g : [0, 1] \to \mathcal{R}^+$ that satisfies :

$G1$: $g(0) = g(1) = 0; g(t) > 0 \; \forall \; t \in (0, 1)$;
$G2$: $g(t) < g(0.5) \; \forall \; t \in [0, 1] - \{0.5\}$;
$G3$: g is nondecreasing on [0.0,0.5) and nonincreasing on (0.5,1];
$G4$: $g(t) = g(1 - t) \; \forall \; t \in [0, 1]$.

In the sequel we call any function satisfying H_E an *E-function.* Theorem E provides a way to test measures of fuzziness against $P1 - P5$.

Emptoz (1981), in fact, in addition to $P1 - P5$, proposed a set of five properties for a measure of fuzziness (Emptoz did not call the proposed function a measure of fuzziness, but a non-probabilistic entropy measure of fuzzy sets). However, Emptoz (1981) did not suggest any axiom for computation of $H(A \times B)$ from $H(A)$ and $H(B)$ as advocated by Ebanks (1983) or Sander (1989).

9.3.2 VARIOUS MEASURES OF FUZZINESS

We group various measures of fuzziness based on their characterization : (i) Entropy like measures, (ii) Distance like measures, and (iii) General classes of measures. The first category deals with measures which are motivated by entropy measures of a probability distribution, while the second category includes measures which are based on some form of distance from crisp sets. Finally, the third category discusses various attempts to general characterization of fuzziness measures.

Entropy Like Measures. Possibly the first attempt to quantify the uncertainty associated with a fuzzy set in the context of a discrete probability framework was made by Zadeh (1968), who defined the [weighted] entropy of a fuzzy set, $A \in \mathcal{P}_n(X)$ as

$$H_{ZE}(A, P) = -\sum_{i=1}^{n} \mu_i p_i \log p_i \tag{9.2}$$

H_{ZE} incorporates both fuzzy and probabilistic uncertainties and it does not satisfy $P1 - P5$, nor was it meant to. Note that H_{ZE} of a fuzzy event with respect to a probability distribution P is less than Shannon's entropy $H_S(P) = -\sum_{i=1}^{n} p_i \log p_i$, of P alone.

The idea of measuring FU based on concepts borrowed from classical measures of probabilistic uncertainty began with Deluca and Termini (1972), who defined the entropy of $A \in \mathcal{P}_n$ using the Shannon's functional form,

$$H_{DTE}(A) = -K \sum^{n} \{\mu_i \log \mu_i + (1 - \mu_i) \log(1 - \mu_i)\} \tag{9.3}$$

where K is a normalizing constant. Although (9.3) is a fuzzy analog of Fermi's entropy, its meaning is quite different. Equation (9.3) satisfies properties $P1-P5$. Jumarie (1990) remarked that the second term $(1-\mu_i)\log(1-\mu_i)$ in (9.3) is "theoretically" unnecessary; however, dropping this term causes H_{DTE} to fail $P2$ and $P4$ and it no longer remains a measure of fuzziness.

The sum in (9.3) can be factored into terms involving A and A^c_Z, and this form has an alternate interpretation as the entropy of a fuzzy 2-partition of any n points. Bezdek generalized this in 1973 by defining the *classification entropy* of a constrained fuzzy c-partition U of $X = \{\mathbf{x}_1, \mathbf{x}_2, ..., \mathbf{x}_n\} \subset \mathcal{R}^p$ as $H_{BE} = -1/n \sum_{k=1}^{n} \sum_{i=1}^{c} \mu_{ik} \log \mu_{ik}$, where $U = [\mu_{ik}]_{c\times n}$ is a $c \times n$ matrix whose columns sum to 1, and whose rows are the membership functions of the c fuzzy clusters in X (Bezdek,1973). Bezdek also proposed the use of the Euclidean norm, called the *partition coefficient* in (Bezdek,1973), $H_B = 1 - 1/n \sum_{k=1}^{n} \sum_{i=1}^{c} \mu_{ik}^2$, as a measure of fuzziness of a constrained c partition U of X. Both of these measures apply to a single fuzzy set $A \in \mathcal{P}_n(X)$ by taking c=1 in these formulae.

In the same year (1973), Capocelli and DeLuca (1973) and subsequently DeLuca and Termini (1974) studied fuzziness measures through the postulation of axioms such as properties P1-P5, and widened the specific form (9.3) to the first class of functions that satisfied intuitively desirable properties.

The fuzzy entropy of Deluca and Termini, H_{DT}, can also be interpreted as Shannon's entropy. Let $X = \{x_1, x_2, ..., x_n\}$ and $\mu : X \to [0, 1]$ be a fuzzy set. Let us interpret $\mu(x)$ as the probability that $x \in X$ possesses a property P. Also assume that the process of deciding whether x_i possesses P or not as a probabilistic experiment. If we repeat the experiment n times, once for each x_i, then the outcome of these experiments would be one of the possible 2^n crisp subsets P_i of X, where $\{P_1, P_2, ..., P_{2^n}\}$ is the collection of all possible subsets of X. Thus, whether $x \in X$ is in P_i or not is decided by making a probabilistic trial. Let π_i be the probability of observing P_i, then assuming $\mu(x)$ is independent of $\mu(y)$, $x, y \in X, x \neq y$ we get

$$\pi_i = \prod_{x\in P_i} \mu(x) \prod_{x\in X-P_i} (1-\mu(x)). \tag{9.4}$$

Note that π_i is not the probability of occurrence of $x_i \in X$; and here x_i does not have to be the outcome of a probabilistic experiment. Now Shannon's entropy associated with the probability distribution defined on $\{P_1, P_2, ..., P_{2^n}\}$ is given by

$$H_S(\{P_1, ..., P_{2^n}\}) = -\sum_{i=1}^{2^n} \pi_i \log \pi_i. \tag{9.5}$$

Equation (9.5) simplifies to H_{DTE} in (9.3). Ebanks attributed this result to a personal communication of Forte (Ebanks,1983).

Sander (1989) interpreted the entropy of a fuzzy set as "an average intrinsic information arising from a decision about the degree or intensity of belonging or not belonging to a given fuzzy set". If we replace "intrinsic information" in this quote by "difficulty" we get our definition - so Sander's view is consistent with ours. Sander (1989) emphasized P1, P5 and P6 to characterize fuzzy entropy and suggested the following function, having the exactly the same form as Shannon's probabilistic entropy, as a measure for fuzzy entropy :

$$D_S(A) = -c\sum_{i=1}^{n} \{\mu_i \log \mu_i\}. \tag{9.6}$$

We already mentioned that Jumarie (1990) favors equation (9.6). Thus, $H_{DT}(A) = D_S(A) + D_S(A^c_Z)$. Here a very important observation of Sander (1989) is worth mentioning. Although P6 was introduced by Ebanks (1983), analysis of the proof of one of Ebanks' lemmas reveals that Ebanks (1983) used not *P6*, but $P6'$: $H(A \times B) = H(A)S(\tau(B)) + S(\sigma(A))H(B)$ where $\tau, \sigma : [0,1]^n \to \mathcal{R}^n$.

Kaufmann also used the functional form of Shannon's probabilistic entropy to define the entropy of a discrete fuzzy set. Letting $\psi(\mu_i) = \mu_i / \sum_{j=1}^{n} \mu_j; i = 1, 2, ..., n$, Kaufmann's entropy of $A \in \mathcal{P}_n(X)$ is :

$$H_{KaE}(A) = -\frac{1}{\log(n)} \sum_{i=1}^{n} \psi(\mu_i) \log \psi(\mu_i). \tag{9.7}$$

Equation (9.7) does not depend directly on the membership values ; rather it is a function of their values relative to normalization by ψ. This can cause an unexpected effect. For example, consider two fuzzy sets A and B with membership values $A = \{0.2, 0.2, ..., : n\ times\}$ and $B = \{0.7, 0.7, ..., : n\ times\}$. In this case $H_{KaE}(A) = H_{KaE}(B)$. Moreover (9.7) does not satisfy properties P2- P6.

Pal and Pal (1989) introduced a definition for probabilistic entropy based on an exponential gain function and used it as a basis for defining the following measures of fuzziness of $A \in \mathcal{P}_n(X)$:

$$H_{PPE}(A) = K\sum_{i=1}^{n} \left\{\mu_i e^{(1-\mu_i)} + (1-\mu_i)e^{\mu_i}\right\}. \tag{9.8}$$

In (9.8) $e = 2.718\cdots$ and K is a positive constant. Equation (9.8) can be normalized so that it satisfies properties P1 through P5.

Bhandari and Pal (1993) use Renyi's entropy to define a nonprobabilistic entropy (a measure of fuzziness) for fuzzy sets as

$$H_{BPE}(\alpha; A) = \frac{K}{1-\alpha}\sum_{i=1}^{n}\log(\mu_i^\alpha + (1-\mu_i)^\alpha), \tag{9.9}$$

where $\alpha > 0, \alpha \neq 1$ and K is a positive constant. Equation (9.9) satisfies properties P1-P5.

For differentiable membership functions Jumarie (1990) suggested the following definition. Let $\mu' = \frac{d\mu}{dx}$, then the entropy of the fuzzy set $A \in \mathcal{P}_n(X)$ is defined as :

$$H_{JE}(A) = \int_{-\infty}^{\infty} |\mu'(x)| \log |\mu'(x)| dx. \tag{9.10}$$

Jumarie considered $|\mu'(x)|$ as the counterpart of a probability density and $\log |\mu'(x)|$ as the density of "uncertainty due to fuzziness". Regarding (9.10) we make the following observation: any probability density function $f(x)$ satisfies $\int_{-\infty}^{+\infty} f(x)dx = 1$, but this constraint is usually not satisfied by the fuzzy density function $|\mu'(x)|$. As a result of this the integral in (9.10) can take any value! One solution to this problem would be to divide (9.10) by $\int_{-\infty}^{+\infty} |\mu'(x)|dx$, but even with this modification it is difficult to interpret H_{JE} as a measure of fuzziness.

Distance Like Measures. Kaufmann (1975) introduced an index of fuzziness for $A \in \mathcal{P}_n(X)$,

$$H_{Ka}(A) = \frac{2}{n^k} d(A, A^{near}), \tag{9.11}$$

where $k \in \mathcal{R}^+$, d is a metric on $\mathcal{P}_n(X) \times \mathcal{P}_n(X)$, A^{near} is the crisp set nearest to A :

$$\mu_{A^{near}}(x) = \left\{ \begin{array}{ll} 1 & if\ \mu_A(x) \geq 0.5 \\ 0 & otherwise \end{array} \right\}.$$

The value of k in (9.11) depends on d. Using the (Minkowski) q-norms to induce a metric on $\mathcal{P}_n(X) \times \mathcal{P}_n(X)$, $d(A, A^{near})$ and $H_{Ka}(A)$ take the following forms:

$$d^q(A, A^{near}) = [\sum_{i=1}^{n} |\mu_i - \mu_{A^{near},i}|^q]^{1/q} = ||\bar{\mu}_A - \bar{\mu}_{A^{near}}||_q \tag{9.12}$$

where $\bar{\mu}_A, \bar{\mu}_{A^{near}} \in \mathcal{R}^n$ are vectors of memberships in A and A^{near}. Substituting (9.12) into (9.11) yields

$$H_{Ka,q}(A) = \frac{2}{n^{1/q}}[\sum_{i=1}^{n} |\mu_i - \mu_{A^{near},i}|^q]^{1/q} \tag{9.13}$$

where $q \in [1, \infty]$. $H_{Ka,q}$ is called the *linear* or *quadratic index of fuzziness* when $q = 1$ or $q = 2$, respectively.

Yager (1979, 1980) associated fuzziness with the lack of distinction between a proposition and its negation. This is intuitively quite appealing because as a concept (fuzzy set) becomes more vague, the contrast between it and its complement decreases. Yager used Zadeh's definition (Zadeh,1965) of complement to define his fuzziness measure :

$$H_Y(q, A) = (d^q(Y, Y^c) - d^q(A, A^c_Z))/d^q(Y, Y^c), \tag{9.14}$$

where $A \in \mathcal{P}_n(X)$, d^q is the q-norm having form (9.12), and Y is an arbitrary crisp subset of X with complement Y^c. Note that $d^q(Y, Y^c)$ is the maximum distance between any pair of sets in $\mathcal{P}_n(X) \times \mathcal{P}_n(X)$.

Higashi and Klir (1982) extended Yager's concept to a very general class of fuzzy complements. Let A^c be a general fuzzy complement function. Higashi and Klir introduced a class of measures based on aggregation of differences between the membership values in A and A^c. Let d aggregate the absolute differences $|\mu_i - \mu_i^c|$ in membership grades between A and A^c. Higashi and Klir's measure of fuzziness is

$$H_{HK}(c, d; A) = \{d(Y, Y^c) - d(A, A^c)\}/d(Y, Y^c), \tag{9.15}$$

where Y is an arbitrary crisp subset of X. $d(Y, Y^c)$ is again the maximum distance in $\mathcal{P}_n(X) \times \mathcal{P}_n(X)$, and does not depend on the choice of crisp set Y. The difference between Yager's measure (9.14) and the measure of Higashi and Klir (9.15) is in the complement and aggregation functions; (9.14) is a special case of (9.15) where c is Zadeh's complement and d is the q-norm, d^q.

Kosko (1986) defined a fuzziness measure as the ratio of the distance between the fuzzy set A and A^{near} to the distance between A and its farthest nonfuzzy set $A^{far} = (A^{near})^c$:

$$H_{KoE}(q, A) = d^q(A, A^{near})/d^q(A, A^{far}), \tag{9.16}$$

where d^q is specified in (9.12). H_{KoE} satisfies properties P1 through P4, but may not satisfy P5.

Bhandari and Pal (1993) measured fuzziness as the ratio of fuzzy divergence between A and A^{near} to that between A and A^{far}. This is close in spirit to Kosko's definition in (9.16). The fuzzy divergence is an information measure for discrimination between two fuzzy sets. The definition of Bhandari and Pal satisfies properties P1-P4. To study the average behavior of a fuzzy set, and to evaluate the extent to which μ separates the elements of X (compared to the ideal discrete case), a concept of separation and of the separating power of a fuzzy set was discussed by Dujet (1988). Dujet suggested an entropy measure based on the concept of separation.

In order to characterize the information possessed by a fuzzy event Pardo (1985) defined the information energy associated with a fuzzy event, and derived many interesting properties of this function. Pardo's information energy is not related to fuzziness or non-specificity of fuzzy sets, so we do not discuss it further.

General classes of measures. Knopfmacher (1975) and Loo (1977) have shown that (9.3) and (9.13) are special cases of a larger class of measures defined as follows:

$$H_{KL}(A) = h\left\{\sum_{i=1}^{n} c_x g_x(\mu_i)\right\} \tag{9.17}$$

where,

1. $c_x \in \mathcal{R}^+$ is a constant;
2. $g_x : [0,1] \to \mathcal{R}^+$ is monotone increasing on [0,0.5] and monotone decreasing on [0.5,1];
3. $g_x(0) = g_x(1) = 0$;
4. g_x has a unique maximum at 0.5; and
5. $h : \mathcal{R}^+ \to \mathcal{R}^+$ is monotone increasing.

Equation (9.17) yields (9.3) (up to a constant) by defining g_x as $g_x(\mu_A(x)) = -\{\mu_A(x)\log\mu_A(x) + (1-\mu_A(x))\log(1-\mu_A(x))\}$ for all x, and h the identity function on $\mathcal{R}^+$.

Trillas and Riera (1978) proposed a general class of fuzziness measures called $\otimes - *$ entropies which were defined as

$$H_{TR,\otimes-*}(A) = \otimes_{i=1}^{n}\alpha_i * N(\mu_i), \tag{9.18}$$

where $\alpha_i > 0 \ \ \forall i$ and $N : [0,1] \to \mathcal{R}^+$ is an N-function that satisfies three properties:

N1. N(0) = N(1) = 0 ; otherwise, $N(t) > 0$;

N2. N is non-decreasing on [0, 0.5] and non-increasing on [0.5, 1];and
N3. N attains a maximum at 0.5.

In (9.18) $*$ is an inner operator that combines pairs $(\alpha_i, N(\mu_i))$; and $\otimes$ is an outer operator that aggregates these combinations. This class of entropies includes the important inner operators $* =$ minimum and $* =$ product; and outer operators $\otimes =$ sum and $\otimes =$ maximum. In particular, $H_{TR,sum-prod}(A) = \sum_{i=1}^{n} \alpha_i N(\mu_i)$ has the form of an E-function in Theorem E with $g(\mu_i) = \alpha_i N(\mu_i)$. For this choice of operators, properties N1-N3 of Trillas and Riera are respectively equivalent to properties G1, G3 and G2 of Ebanks. Recalling G4, we see that $H_{TR,sum-prod}(A)$ satisfies P1-P5 if and only if the N-function in N1-N3 satisfies the additional property $N(t) = N(1-t)$ for all t in [0,1].

Two wide classes of measures of fuzziness, namely the *multiplicative* class and the *additive* class were proposed by Pal and Bezdek (1994). These two classes satisfy the five properties $P1 - P5$. To define the *multiplicative class* Pal and Bezdek (1994) proceed as follows : Let $f : [0,1] \to \mathcal{R}^+$ be a concave increasing function on [0,1]; that is, $f'(t) > 0$ and $f''(t) < 0 \ \forall t \in [0,1]$. Now define

$$\hat{g}(t) = f(t)f(1-t); \tag{9.19}$$

and with $\hat{g}$, define

$$g(t) = \hat{g}(t) - \min_{0 \le t \le 1} \{\hat{g}(t)\}. \tag{9.20}$$

Now, for $A \in \mathcal{P}_n(X)$, define

$$H_*(A) = K \sum_{i=1}^{n} g(\mu_i), K \in \mathcal{R}^+, \tag{9.21}$$

where K is a positive constant. Since $H_*(A)$ has the form of an E-function as shown in Theorem E, the multiplicative class satisfies P1-P5 if and only if g in (9.20) satisfies G1-G4. It can be shown either through G1-G4 or directly that (9.21) satisfies $P1 - P5$. Theorem 1 in Pal and Bezdek (1994) gives this result.

As an example of the multiplicative class consider

$$f(t) = te^{(1-t)}. \tag{9.22}$$

It can be shown that f in (9.22) satisfies the requirements for the multiplicative class, and hence, using (9.21) we get the following measure of fuzziness :

$$H_{*QE}(A) = K \sum_{i=1}^{n} \mu_i(1-\mu_i). \tag{9.23}$$

The constant $e = 2.718\cdots$ has here been absorbed in the constant K. It is easy to show directly that properties P1-P5 are satisfied by (9.23). Pal and Bezdek (1994) call H_{*QE} the *quadratic entropy* of the fuzzy set A because of its similarity to Vajda's probabilistic quadratic entropy (Vajda,1970) for a discrete probabilistic framework (X,P),

$$H_{VQE}(P) = \sum_{i=1}^{n} p_i(1 - p_i). \tag{9.24}$$

Just as H_{DTE} at (9.3) is the fuzzy analog of Shannon's probabilistic entropy, H_{*QE} is the fuzzy analog of Vajda's probabilistic entropy. Quadratic entropy was probably first used in theoretical physics by Fermi. It has also been used for risk evaluation of the nearest neighbor classification rule (Cover and Hart,1967). The choice of f in (9.22) was motivated by the expression for exponential entropy given in (Pal and Pal,1989). In Chapter 8 you have seen that most measures of conflict/dissonance/ confusion associated with a basic probability assignment (BPA) reduce to Shannon's probabilistic entropy for a Bayesian belief (or probability) measure. It is interesting observe that the conflict measure derived by George and Pal (1996) for a BPA reduces to Vajda's quadratic entropy for a Bayesian belief measure.

Another interesting family of fuzziness measures can be obtained from the multiplicative class by considering the function

$$f(t) = t^{\alpha}, 0 < \alpha < 1. \tag{9.25}$$

Equation (9.25) satisfies the requirements of the multiplicative class. Hence, we get a measure of fuzziness satisfying P1-P5 from (9.25) as

$$H_{\alpha QE}(\alpha, A) = K \sum_{i=1}^{n} \mu_i^{\alpha}(1 - \mu_i)^{\alpha}. \tag{9.26}$$

Pal and Bezdek (1994) proved that $H_{\alpha QE} \rightarrow H_{*QE}$ as $\alpha \rightarrow 1$; they also analyzed the effect of α on the characteristics of $H_{\alpha QE}$.

For the *additive class* of fuzziness measures, f is less restricted than for the multiplicative case. Pal and Bezdek (1994) require only that $f : [0, 1] \rightarrow \mathcal{R}^{+}$ be concave, i.e., $f''(t) < 0 \; \forall \; t \in [0, 1]$. For the additive class the functions g and H_{+} are defined as follows:

$$\hat{g}(t) = f(t) + f(1 - t); \tag{9.27}$$

and with $\hat{g}$, define, as in (9.20),

$$g(t) = \hat{g}(t) - \min_{0 \leq t \leq 1} \{\hat{g}(t)\}. \tag{9.28}$$

A measure of fuzziness for $A \in \mathcal{P}_n(X)$ is then defined as

$$H_+(A) = K \sum_{i=1}^{n} g(\mu_i), K \in \mathcal{R}^+, \tag{9.29}$$

where K is a positive constant. Since H_+ in (9.29) also has the form shown in Theorem E, the additive class satisfies P1-P5 if g at (9.28) satisfies G1-G4. Pal and Bezdek (1994) proved that H_+ in (9.29) is a measure of fuzziness satisfying $P1 - P5$.

To illustrate H_+, we consider the same f as in (9.22); i.e.,

$$f(t) = te^{(1-t)}. \tag{9.30}$$

Since f in (9.30) satisfies the requirements for the multiplicative class, it satisfies the requirement for the additive class. Using (9.30) in (9.29), we get :

$$H_+(A) = K \sum_{i=1}^{n} \mu_i e^{(1-\mu_i)} + (1 - \mu_i)e^{\mu_i} - n. \tag{9.31}$$

Comparing (9.31) to H_{PPE} at (9.8) for this f shows that $H_+(A) = H_{PPE}(A) - n$, up to adjustment of K. Similarly, H_{DTE} at (9.3) can be realized as a member of the additive class by choosing $f = -t \log t$.

The multiplicative and additive classes are easily extendible to fuzzy sets on any real interval. Let μ_A be a fuzzy subset of $X \subset \mathcal{R}$. Then, to extend the multiplicative class, define

$$H_{*c}(A) = K \int_X g(\mu_A(x))dx, \tag{9.32}$$

where K is a positive constant and g is a continuous function having the form in (9.20).

Similarly, to extend the additive class define

$$H_{+c}(A) = K \int_X g(\mu_A(x))dx, \tag{9.33}$$

where K is a positive constant and g is a continuous function whose form is defined in (9.28). Pal and Bezdek (1994) showed that (9.32) and (9.33) are measures of fuzziness satisfying $P1 - P4$.

The additive and multiplicative classes are consistent with Yager's view, since the complement of any argument of g is $\mu_{A^c_Z}(x) = 1 - \mu_A(x)$. Now $g(\mu_A(x))$ increases as $\mu_A(x)$ and $\mu_{A^c_Z}(x)$ become closer to each other. In other words, as the distinction between $\mu_A(x)$ and $\mu_{A^c_Z}(x)$ decreases, $g(\mu_A(x))$ increases monotonically. Therefore, $g(\mu_A(x))$ can be viewed as a measure of lack of distinction between the values $\mu_A(x)$ and $\mu_{A_Z^c}(x)$. Hence, H_*, H_+, H_{+c}, and H_{*c}, each measures the overall lack of distinction between a fuzzy set and its Zadeh complement.

For a comparative analysis of computational cost and sensitivity of different measures with respect to μ readers are referred to Pal and Bezdek (1994).

9.4 GENERALIZED MEASURE OF FUZZINESS

All measures of fuzziness that we discussed so far attempt to quantify how fuzzy a fuzzy set is. In this section we discuss a few other measures of fuzziness which address issues different from just the fuzziness of a set.

9.4.1 HIGHER ORDER MEASURES OF FUZZINESS

The measures of fuzziness discussed earlier can be used to quantify our difficulty in making a decision as to whether an *element* belongs to the concerned fuzzy set or not. But these measures do not assess the ambiguity associated with a *collection* of elements of X. For this Pal and Pal (1992) introduced the concept of $r^{th}(1 \leq r \leq n)$ order (higher order) measures of fuzziness. Let X be a set (team) of players with say, n = 20 members, and A be the set of GOOD players on the team (X). Suppose we form a team of size $r = 11 \leq n$. On average, to what extent is this team good? This raises two issues : the degree to which a collection possesses the property of "GOODness"; and how to measure the average ambiguity associated with such collections. An answer to the first question depends on the problem at hand. Given a property characterized by a fuzzy set A in X, how can we extend the same property to a collection of elements from X? Let us denote the ith subset of X with r members (among all subsets of X with cardinality r) as S_i^r. For example, let X be a set of acrobats, and suppose that a subset S_i^r of X are standing in such a manner that if one of the members of S_i^r falls the entire team falls. In this case the extent to which S_i^r is GOOD may plausibly be defined as $\min_{x \in S_i^r}\{\mu_A(x)\}$. On the other hand, if X is a team of quiz takers, then the degree to which the team S_i^r is GOOD be better computed as $\max_{x \in S_i^r}\{\mu_A(x)\}$, because in a quiz team, if one member succeeds, the team succeeds. If X is a group of football players or members of a tug-of-war team, possibly some other aggregation operator (may be average) will be a better choice.

Let $\mathcal{O}$ be an aggregation operator. Given a fuzzy set A on X, the operator $\mathcal{O}$ can be used to define a fuzzy set *A-team* (e.g., GOOD-

team) on the universe containing all possible subsets (of size r) of X. Let $\mu(S_i^r) = \mathcal{O}\{\mu_A(x), x \in S_i^r\}$ be the degree to which S_i^r possesses the property *A-team*.

Now to answer the second question, i.e., how to get a measure of fuzziness of the fuzzy set *A-team*, we may use any one of the functions that quantifies fuzziness in A. For example, beginning with the quadratic fuzzy entropy in (9.23), *the r-th order fuzzy entropy* (Pal and Pal,1992) of A can be defined as $H_{*QE}^r = K\sum_i \mu(S_i^r)(1-\mu(S_i^r))$. This reduces to (9.23) when r=1; and otherwise, H_{*QE}^r provides an assessment of the overall ambiguity induced by μ_A in the set of all subsets of X with r distinct elements. Don't confuse this with the idea of measuring fuzziness in a higher-order fuzzy set. Properties of higher order measures of fuzziness can be found in (Pal and Pal,1992).

9.4.2 WEIGHTED FUZZINESS

We noted earlier that Zadeh's (1968) entropy of a fuzzy set shown at (9.2) can be viewed as Shannon's entropy with each term weighted by the corresponding fuzzy membership. In the same year Belis and Guiasu (1968) introduced a slightly more general weighted Shannon entropy,

$$H_{BGE}(P,W) = -\sum_{i=1}^{n} w_i p_i \log p_i. \tag{9.34}$$

In (9.34) $W = \{w_i \geq 0; i = 1, ..., n\}$ is a set of non-negative coefficients denoting the subjective importance of different outcomes of the discrete probabilistic framework (X,P). If all the weights are valued in [0,1], (9.34) takes the same form as (9.2). Belis and Guiasu suggested that the occurrence of an event removes two types of uncertainty : the quantitative type related to its probability of occurrence, and the qualitative type relative to its utility (importance) for the fulfillment of some goal set by the experimenter. For example, depending on the goal, a low probability event might have high utility; while in a different context the same event might have low utility. If the weights satisfy the constraint $\sum_{i=1}^{n} w_i = 1$, their interpretation is more plausible. Usually W is determined by the subjective goal set by the experimenter.

Weighted entropy allows subjective quantification of the information content associated with each event in (X,P). In this same spirit Pal and Bezdek (1994) introduced a similar concept for any discrete fuzzy set $A \in \mathcal{P}_n(X)$, by defining its weighted fuzziness as

$$H_{WFE}(A, W) = \sum_{i=1}^{n} w_i g(\mu_i), \tag{9.35}$$

where $\forall i, w_i \geq 0; \sum_{i=1}^{n} w_i = 1$, and the $\hat{g}$ defining g can have either the multiplicative (9.19) or additive (9.27) form. Sometimes the weights in (9.35) can be determined objectively. For example, when (X, P) is a discrete probability framework and $A \in \mathcal{P}_n(X)$, the expected fuzziness of outcomes in the system is found by taking $w_i = p_i \ \forall \ i$ in (9.35). In this case H_{WFE} does not measure the fuzziness associated with A as defined by P1-P5, nor it is equivalent to (9.2) or (9.34). In general, H_{WFE} incorporates the subjectivity (discussed earlier) associated with the supporting elements of A into the measures of fuzziness.

Expressions like (9.35) have been used for many applications (Pal and Dutta Majumder,1986). Next we focus on non-specificity, another facet of uncertainty associated with a fuzzy set.

9.5 MEASURING NON-SPECIFICITY

The concept of *specificity* or precision of a fuzzy set was introduced by Yager (Yager,1982). Specificity measures the extent to which a fuzzy set restricts a variable to a small number of values. According to Yager, a measure of specificity, Sp, of a fuzzy set should satisfy the following axioms: $\forall \ A, B \in \mathcal{P}_n(X)$,

A1. $Sp(A) \in [0, 1]$,

A2. $Sp(A) = 1$ if and only if $\exists \ x \in X, A = \{x\}$, i.e., A is a singleton of X,

A3. $A \subseteq B$(i.e., $\mu_A \leq \mu_B$) $\Rightarrow Sp(A) \geq Sp(B)$.

A2 suggests that only the singletons in $\mathcal{P}_n(X)$ are specific or precise because assigning a single value to a variable is the most specific case.

Based on axioms $A1 - A3$ Yager (1982) suggested the following expression as a measure of specificity of a *nonempty* finite fuzzy set A:

$$Sp(A) = \int_0^{h(A)} \frac{1}{|A_\alpha|} d\alpha, \tag{9.36}$$

where $A_\alpha = \{x : \mu_A(x) \geq \alpha\}$ = the α-cut of A, $|.|$ is the cardinality and $h(A)$ is the height of A. For a *normal* fuzzy set $h(A) = 1$ so (9.36) takes the form

$$Sp(A) = \int_0^1 \frac{1}{|A_\alpha|} d\alpha. \tag{9.37}$$

The value of $Sp(A)$ in (9.37) can be used to define a measure of *non-specificity* or *anxiety*, NSp, as

$$NSp(A) = 1 - Sp(A). \tag{9.38}$$

To illustrate how specificity is computed using (9.36) we use an example from Yager (1982). Let $X = \{x_1, x_2, x_3, x_4\}$ and $A = \{0.7, 0.5, 0.2, 0.1\}$. The level sets of A are :

$$\begin{array}{lll} A_\alpha = \{x_1, x_2, x_3, x_4\} & \text{for} & \alpha \le 0.1, \\ A_\alpha = \{x_1, x_2, x_3\} & \text{for} & 0 < \alpha \le 0.2, \\ A_\alpha = \{x_1, x_2\} & \text{for} & 0.2 \le \alpha \le 0.5, \\ A_\alpha = \{x_1\} & \text{for} & 0.5 \le \alpha \le 0.7. \end{array}$$

Hence $Sp(A) = \int_{0.0}^{0.7} \frac{1}{|A_\alpha|} d\alpha = \int_{0.0}^{0.1} \frac{1}{4} d\alpha + \int_{0.1}^{0.2} \frac{1}{3} d\alpha + \int_{0.2}^{0.5} \frac{1}{2} d\alpha + \int_{0.5}^{0.7} d\alpha =$ 0.408.

For *normal* fuzzy sets the specificity takes the minimum value of $1/|X|$, i.e., non-specificity takes the maximum value of $1 - 1/|X|$, when $\mu_A(x) = 1 \ \forall \ x \in X$. This is a desirable property as in this case all alternatives satisfy the criterion completely and the fuzzy proposition "x *is* A" cannot specifically express any bias towards a specific value of X. In general, of all fuzzy sets on X with height $\alpha, 0 \le \alpha \le 1$, specificity, $Sp(A)$ takes the minimum value when every element assumes a membership value equal to the height of the fuzzy set, i.e., $\{\mu_A(x_i) = \alpha \ \forall \ i\}$, and the minimum value is $\frac{\alpha}{|X|}$, which goes to zero as $\alpha \to 0$; i.e., $A \to \Phi$. The limiting case says that when none of the choices satisfies the criterion at any positive level, the specificity is zero. In other words, the anxiety or non-specificity is maximal (globally) when we have to make a choice but no alternative is good. On the other hand, by A2 NSp takes the minimum value 0 (i.e., specificity takes the maximum value of 1) when $\mu_A(x_k) = 1, x_k \in X; \mu_A(x) = 0, \forall \ x \in X - \{x_k\}$. Since in this case we are certain about exactly one specific value, the specificity is maximum (non-specificity is minimum).

Observe also that both extrema of specificity occur for crisp sets; i.e. when fuzziness is zero. Consider the case when fuzziness is maximum, i.e., when $\mu_A(x) = 0.5 \ \forall \ x \in X$. In this case the specificity, $Sp = \frac{1}{2|X|}$. So when fuzziness is maximum, specificity (and also non-specificity) has a positive value which is neither the maximum nor the minimum of specificity. We emphasize that a measure of non-specificity does not express the difficulty in deciding whether an element belongs to a set or not but represents how particular (specific) we are about the fuzzy property. SP and NSp have many other interesting properties (Yager,1982).

For a probabilistic interpretation of non-specificity readers are referred to Chapter 8 of this handbook and also (Dubois and Prade,1985).

Higashi and Klir (1983) also suggested a measure of imprecision, called U-uncertainty, satisfying a set of axioms similar to $A1 - A3$: $\forall\ A, B \in \mathcal{P}_n(X)$,

U1. $U(A) \in [0, +\infty)$,

U2. $U(A) = 0$ if and only if $\exists\ x \in X, A = \{x\}$, i.e., A is a singleton of X,

U3. $A \subseteq B$(i.e., $\mu_A \leq \mu_B$) $\Rightarrow U(A) \leq U(B)$.

A measure of imprecision for a normal fuzzy set satisfying $U1 - U3$ as given by Higashi and Klir (1983) is :

$$U(A) = \int_0^1 \log_2 |A_\alpha| d\alpha. \tag{9.39}$$

Unlike (9.38), $U(A)$ in (9.39) satisfies the usual additivity requirement of measures of uncertainty.

If A is not normal, sometimes $U(A)$ is defined as $U_N(A)$ (see Chapter 8 of this volume) :

$$U_N(A) = \frac{1}{h(A)} \int_0^{h(A)} \log_2 |A_\alpha| d\alpha. \tag{9.40}$$

The numerical value of $U_N(A)$ should be interpreted with caution. Consider two fuzzy sets A and B on X as : $A = \{\mu_A(x_i) = \alpha,\ \forall\ x_i \in X; 0 < \alpha < 1$ and $B = \{\mu_B(x_i) = 1,\ \forall\ x_i \in X\}$. Using (9.40), $U_N(A) = U_N(B)$, which may not be desirable. But for Yager's measure $NSp(A) = 1 - \frac{\alpha}{|X|} > 1 - \frac{1}{|X|} = NSp(B)$. Thus, if we view imprecision as non-specificity in the sense of Yager, then NSp seems more reasonable than U_N because, intuitively, the anxiety in selecting an alternative which partially satisfies the desired criterion is more than that for the one which meets the criterion.

9.6 CONCLUSIONS

After an introduction to different types of uncertainty, we focused on two facets of uncertainties; namely, fuzziness and non-specificity, that are associated with fuzzy concepts. We then reviewed some well known measures of fuzziness, including a few general classes of measures for discrete fuzzy sets. We have discussed different axioms proposed by various authors and justified a set of five axioms that we feel are minimally necessary for a measure of fuzziness. We explained the need for higher order measures of fuzziness and related issues, and then discussed how they can be computed. We also discussed the concept of weighted fuzzy entropy and its power in incorporating subjectiveness in the measure of fuzziness. Finally, we discussed Yager's concept of specificity (tranquility) / non-specificity (anxiety) in decision-making and then presented two measures of this type of imprecision.

We have *not* discussed how the *total uncertainty* in a system can be estimated when fuzziness and/or non-specificity appear along with uncertainty of a probabilistic nature. There have been several attempts to combine fuzzy and probabilistic uncertainties. Attempts have also been made to combine probabilistic uncertainty with non-specificity. For detailed discussions on these issues readers are referred to Deluca and Termini (1972), Pal et al. (1992, 1993), Pal and Pal (1992), Xie and Bedrosian (1984), George and Pal (1996), and Klir and Ramer (1990). In general, of course, a complex system may involve all three : probabilistic uncertainty, fuzziness and non-specificity. Note that under Dempster-Shafer (Shafer, 1976) modeling, non-specificity may arise even without fuzziness. Various attempts have been made to axiomatically characterize each facet of uncertainty separately, and some attempts have also been made to axiomatically characterize total uncertainty when the system has only two facets of the uncertainty. However, as far as we know, no attempt has been made to axiomatically quantify total uncertainty when the system has all three types. This is a hard but important problem which needs further investigation.

References

M. Belis and S. Guiasu (1968), A qualitative - quantitative measure of information in cybernetic system, *IEEE Trans. Info. Theory*, IT 14, 593-594.

J.C. Bezdek (1973), *Fuzzy Mathematics for Pattern Classification*, PhD Thesis, Cornell University, Ithaca, NY, 1973.

D. Bhandari and N. R. Pal (1993), Some new information measures for fuzzy sets, *Information Sciences*, 67, 3, 209-228.

R. M. Capocelli and A. Deluca (1973), Fuzzy sets and decision theory, *Information and Control*, 23, 446-473.

T. M. Cover and P. E. Hart (1967), Nearest neighbor pattern classification, *IEEE Trans Info. Theory*, IT 13, 21-27.

A. Deluca and S. Termini (1972), A definition of nonprobabilistic entropy in the setting of fuzzy sets theory, *Information and Control*, 20, 301-312.

A. Deluca and S. Termini(1974), Entropy of L-fuzzy sets, *Information and Control*, 24, 55-73.

D. Dubois and H. Prade, A note on measures of specificity for fuzzy sets, *Int. Jour. Gen. Syst.*, 10, 279-283, 1985.

C. Dujet (1988), Separation and measures of fuzziness, *Fuzzy Sets and Syst.*, 28, 245-262.

B. R. Ebanks (1983), On measures of fuzziness and their representations, *Jour. Math Anal. and Appl.*, 94, 24-37.

H. Emptoz (1981), Nonprobabilistic entropies and indetermination measures in the setting of fuzzy sets theory, *Fuzzy Sets and Syst.*, 5, 307-317.

T. George and N. R. Pal (1996), Quantification of Conflict in Dempster-Shafer framework : a new approach, *Int. Jour. Gen. Systems*, 24, 4, 407-423.

M. Higashi and G. J. Klir (1982), On measures of fuzziness and fuzzy complements, *Int. Jour. Gen. Syst.*, 8, 169-180.

M. Higashi and G. J. Klir (1983), Measures of uncertainty and information based on possibility distributions, *Int. Jour. Gen. Syst.*, 9, 43-58.

G. M. Jumarie (1986), *Subjectivity, information, systems - introduction to a theory of relativistic cybernetics*, Gordon and Breach Science Publishers, New York.

G. M. Jumarie (1990), *Relative information- theories and applications*, Springer Verlag, New York.

A. Kaufmann (1975), *Introduction to the theory of fuzzy subsets: Volume 1*, Academic Press, New York.

G. J. Klir and A Ramer (1990), Uncertainty in Dempster Shafer theory: A critical re-examination, *Int. Jour. Gen. Syst.*, 18, 155-166.

J. Knopfmacher (1975), On measures of fuzziness, *Jour. Math. Anal. Appl.*, 49, 529-534.

B. Kosko (1986), Fuzzy entropy and conditioning, *Information Sciences*, 40, 165-174.

M. T. Lamata and S. Moral (1987), Measures of entropy in the theory of evidence, *Int. Jour. Gen. Syst.*, 14, 297-305.

S. G. Loo (1977), Measures of fuzziness, *Cybernetica*, 20, 201-210.

N. R. Pal and J. C. Bezdek (1994), Measuring fuzzy uncertainty, *IEEE Trans. Fuzzy Syst.*, 2, 2, 107-118.

S. K. Pal and D. Dutta Majumder (1986), *Fuzzy mathematical approach to pattern recognition*, John Wiley and Sons, New York.

N. R. Pal and S. K. Pal (1989), Object-background segmentation using new definitions of entropy, *IEE Proc.*, 136, Pt. E, 284-295.

N. R. Pal, J. C. Bezdek and R. Hemasinha (1992), Uncertainty measures for evidential reasoning I: a review, *Int. Jour. Approximate Reasoning*, 7 (3/4), 165-183.

N. R. Pal, J. C. Bezdek and R. Hemasinha (1993), Uncertainty measures for evidential reasoning II : a new measure, *Int. Jour. Approximate Reasoning*, 8(1).

N. R. Pal and S. K. Pal (1992), Higher order fuzzy entropy and hybrid entropy of a set, *Information Sciences*, 61, 3, 211-231.

L. Pardo (1985), Information energy of fuzzy event and a partition of fuzzy event, *IEEE Trans. Syst. Man and Cybern.*, 15, 1, 139-144.

W. Sander (1989), On measures of fuzziness, *Fuzzy Sets and Systs.*, 29, 49-55.

G. Shafer (1976), *A mathematical theory of evidence*, Princeton University Press, Princeton.

M. Sugeno (1985), *Industrial Applications of Fuzzy Control*, Elsevier Science,Amsterdam.

E. Trillas and T. Riera (1987), Entropies in finite fuzzy sets, *Information Sciences*, 15, 159-168.

I. Vajda (1970), A note on discrimination information and variation, *IEEE Trans. Info. Theory*, 16, 771-772.

R. R. Yager (1979), On measures of fuzziness and negation, Part I : Membership in the unit interval, *Int. Jour. Gen. Syst.*, 5, 221-229.

R. R. Yager (1980), On measures of fuzziness and negation, Part II : Lattices, *Information and Control*, 44, 236-260.

R. R. Yager (1982), Measuring tranquility and anxiety in decision making : an application of fuzzy sets, *Int. Jour. Gen. Syst.*, 8(3), 139-144.

R. R. Yager (1983), Entropy and specificity in a mathematical theory of evidence, *Int. Jour. Gen. Syst.*, 9, 249-260.

W. X. Xie and S. D. Bedrosian (1984), An information measure for fuzzy sets, *IEEE Trans. Syst. Man and Cybern.*, SMC-14(1), 151-156.

L. A. Zadeh (1965), Fuzzy Sets, *Information and Control*, 8, 338-353.

L. A. Zadeh (1968), Probability measures of fuzzy events, *Jour. Math. Anal. and Appl.*, 23, 421-427.

IV Fuzzy Sets on the Real Line

10 FUZZY INTERVAL ANALYSIS

Didier Dubois
Etienne Kerre
Radko Mesiar
Henri Prade

Abstract: This chapter is an overview of past and present works dealing with fuzzy intervals and their operations. A fuzzy interval is a fuzzy set in the real line whose level-cuts are intervals. Particular cases include usual real numbers and intervals. Usual operations on the real line canonically extend to operations between fuzzy quantities, thus extending the usual interval (or error) analysis to membership functions. What is obtained is a counterpart of random variable calculus, but where, contrary to the latter case, there is no compensation between variables. Many results pertaining to basic properties of fuzzy interval analysis are summed up in the chapter. Computational methods are presented, exact or approximate ones, based on parametric representations, or level-cut approximations. The generalized fuzzy variable calculus involving interactive variables is also discussed with emphasis on triangular-norm based fuzzy additions. Dual 'optimistic' operations on fuzzy intervals, i.e., with maximal error compensation are also presented; its interest lies in providing tools for solving fuzzy interval equations. This chapter also contains a reasoned survey of methods for comparing and ranking fuzzy intervals. The chapter includes some historical background, as well as pointers to applications in mathematics and engineering.

10.1 INTRODUCTION

The calculus of fuzzy quantities derives from the introduction of the notion of a fuzzy set (Zadeh, 1965) into the older area of interval analysis (Moore, 1966). Interval analysis defines rules for the propagation of errors (described by intervals) in calculation processes. Specifically, the influence of errors pervading data on the values of algebraic or arithmetic expressions combining these data is analysed. Commonly performed by physicists (Parratt, 1971), the topic of interval mathematics has become even more important with the advent of computers: the motivation is 'the quest for rigor in numerical computations on machines', (Moore,

1979), i.e., the propagation of rounding errors which stem from the finite precision encoding of real numbers.

The introduction of fuzzy sets in interval mathematics is also linked to the advent of the information technology era, and especially the trend towards encoding and processing forms of information that are more sophisticated than numbers or symbols, namely pieces of information provided by human beings. The concept of a fuzzy quantity is intended to be a mathematical model of a vaguely perceived or imprecisely defined quantitative piece of information, viewed as an ill-bounded set of possible values. Typical examples are found in natural language phenomena, where orders of magnitude are expressed in a linguistic way, such as 'high speed', 'low temperature', 'much more larger than 6'. The necessity of a mathematical modelling of vague quantitative information, commonly used in human communication and decision making, arises when we wish to build intelligent computer systems that can communicate with humans. Fuzzy quantities can also describe flexible requirements in design processes (Otto and Antonsson, 1991, 1994) or tuning strategies. For instance, when tuning a machine-tool for a manufacturing operation, the operator has a clear idea of the ideal speed and feed of the cutting tool, along with tolerance areas. Ultimately, he has to select precise values for these parameters, making a trade-off with other constraints such as a requested production rate (Dubois, 1987a). In such situations, fuzzy quantities represent preference profiles rather than uncertainty distributions.

The idea that fuzzy quantities could be arithmetically combined according to the laws of fuzzy set theory is due to Zadeh (1975a). Soon after, several researchers worked independently along these lines, such as Jain (1976), Mizumoto and Tanaka (1976a, 1976b), Nahmias (1978), Nguyen (1978), Dubois and Prade (1978a). It was only further on recognised that the mathematics of fuzzy quantities are an application of possibility theory, an extension of interval analysis as well as of the algebra of many-values quantities (Young, 1931). The latter, proposed by Young in 1931, is itself an extension of interval analysis to calculations with sets of numbers. Today there is an extensive body of available results about fuzzy quantities, and computational methods have been developed in noticeable classes of applications.

The practical importance of fuzzy interval analysis lies in the necessity, in decision or design problems or in risk analysis, of computing functions of uncertain variables, or of propagating flexible constraints so as to compute uncertainty distributions of parameters of interest, or preference profiles on decision variables. For instance, the body mass index (BMI), used to measure the corpulence of people, is defined as the quotient of the weight and the square of the height. In order to compute the BMI of a tall and heavy man, we need an extension of multiplication and division to fuzzy operands that represent 'heavy' and 'tall'. This is one aspect of what Zadeh (1996) calls 'computing with words'.

The main fields of application of fuzzy intervals seem to be operations research, design, decision analysis, and statistics. In the field of operations research, fuzzy quantities have been used for the modelling of optimisation problems where some data is only imprecisely known, and fuzzy constraints express local preference, as in production or project scheduling (Dubois *et al.*, 1995). In design, it is very important to introduce imprecision and tolerance in early specifications of a product, and to compute the consequences of these specifications or requirements on the potential decisions. Fuzzy interval analysis is useful to achieve a trade-off between antagonistic requirements in a flexible way, and to compute forecasts of performance measures of the objects at early stages of the design process (Otto and Antonsson, 1991, 1994). In decision analysis under imperfect knowledge one is seldom in the situation where probabilities of states and utility of consequences of potential acts can be precisely assessed. Again fuzzy intervals are useful to express this lack of precision faithfully to what a human decision maker can perceive (Dubois and Prade, 1982a). In multiple-criteria analysis the relative weight of criteria may be imprecisely assessed and fuzzy intervals are again useful (van Laarhoven and Pedrycz, 1983). In statistics, fuzzy intervals can also more faithfully describe the outcome of random experiments such as opinion polls, which yield linguistic hence imprecise data (Kruse and Meyer, 1987). Moreover, several attempts at founding a theory of fuzzy random variables (Kwakernaak, 1978, 1979; Puri and Ralescu, 1986) are based on fuzzy intervals and the extension principle, a principle which extends a function to fuzzy set arguments .

This chapter is a survey of the present state of the art about fuzzy quantities, and especially fuzzy intervals. It extends and updates the overview of Dubois and Prade (1987a). Several theoretical details and applications can be found e.g., in monographs of Kaufmann and Gupta (1985, 1988), and Mares (1994). In 1987, there was a special issue of Fuzzy Sets and Systems (Dubois and Prade, 1987c), devoted to the fuzzy intervals domain, and more recently another one has appeared (Fullér and Mesiar, 1997).

The chapter is organised as follows: Section 10.2 introduces fuzzy quantities, with emphasis on fuzzy intervals and their characteristics in terms of order of magnitude, levels of specificity and fuzziness. Section 10.3 presents the extension principle and the basics of fuzzy interval analysis, including the four operations of arithmetics. Section 10.4 is devoted to computational methods for standard fuzzy interval analysis with unrelated variables. Alternative fuzzy interval calculi are discussed in Section 10.5, namely fuzzy interval analysis with interactive variables, also called constrained fuzzy arithmetics, the addition of fuzzy intervals based on triangular norms, including the least t-norm and the continuous t-norms, and the error compensating fuzzy arithmetics, useful for discussing fuzzy equations. Fuzzy complex quantities and fuzzy numbers of higher dimension are briefly discussed in Section 10.5.3. Current results pertaining to the comparison of fuzzy intervals for ranking purposes are presented in Section 10.6, where existing methods are presented in a reasoned survey. Lastly, pointers to on-going areas of applications of fuzzy interval mathematics are provided in Section 10.7.

10.2 FUZZY QUANTITIES AND INTERVALS

The simplest definition of a fuzzy quantity consists in stating that it is a fuzzy set of numbers, of real numbers generally, but sometimes integers as well. On the real line, not all subsets of numbers are interesting. The most useful ones are intervals, hence the almost universal focus on fuzzy intervals in the literature. In order to make computation with fuzzy quantities, basic principles must be defined as to how to extend functions with real or integer arguments to fuzzy arguments. The so-called extension principle proposed by Zadeh (1965, 1975a) offers the necessary tool for computations with fuzzy quantities. A necessary intermediary step is to extend such functions to set-valued arguments, which is well-known in mathematics.

Possibility theory offers a natural framework for interpreting the extension principle (Dubois and Prade, 1988). In that framework, fuzzy quantities are interpreted as possibility distributions describing the ill-known values of variables, and the rules for computing fuzzy quantities are those of possibility theory. The extension principle is similar to the basic equation which enables functions of random variables to be calculated. This view has been first pointed out by Nahmias (1978), although with a different terminology. Since the most common functions to be calculated have several arguments, it is necessary to combine fuzzy arguments into a single multidimensional fuzzy set, and the technique to do so presupposes assumptions of how fuzzy variables interact. This section provides a typology of fuzzy quantities, discusses various understandings of them, and introduces the basic principles that drive fuzzy computations.

10.2.1 Definitions

As just said, a *fuzzy quantity*, Q, is any normal fuzzy set of the real line $\mathbb{R}$. Like any fuzzy set, a fuzzy quantity can be represented either by its membership function (that we write $Q(\cdot)$ so as to distinguish it from the name of the fuzzy set), that associates to any real value r the membership grade Q(r), or by its level-cuts $Q_\alpha = \{a \in \mathbb{R} : Q(a) \geq \alpha\}$ for $\alpha \in (0, 1]$. A fuzzy quantity is often interpreted as a possibility distribution on the values which a variable x may assume. Then x is called a *fuzzy variable* to which Q is attached. Sometimes this attachment is meant to describe the available information about the value of x, sometimes $Q(\cdot)$ is rather a preference profile depicting what are the best values for x. In the first case, computing with fuzzy quantities comes down to a form of uncertainty propagation. In the second case the aim of fuzzy computation is to make appropriate decisions by propagating preference in a network of constraints. Unless otherwise specified, Q is supposed to be normal, that is, Q(r) = 1 for some value r, and any element of its core (the set of values with membership 1) is called a *modal value* of Q. In the scope of possibility theory, a lack of normalisation means that the variable x may take values outside the universe of discourse (x is not a real number) or may take no value at all (if, for instance, x is the duration of a trip which might not be undertaken). More generally, the height of a fuzzy quantity can be viewed as a measure of its internal conflict.

Fuzzy Intervals. There is a special kind of fuzzy quantity which extends the concept of an interval: a *fuzzy interval* M is a fuzzy quantity with a quasi-concave membership function, i.e., a convex fuzzy subset of the real line $\mathbb{R}$ (Zadeh, 1965):

$$\forall a, b, \forall c \in [a, b], M(c) \geq \min((M(a), (M(b)). \tag{10.1}$$

Note that a fuzzy quantity is convex if and only if its level-cuts are convex, i.e., are intervals (bounded or not). Closed intervals are generalised by fuzzy intervals with upper semicontinuous (u.s.c.) membership functions, i.e., by definition, whose level-cuts are closed intervals. Note that M is an u.s.c. fuzzy interval if and only if the three following conditions are met (Dubois and Prade, 1980a):

i) The core $\{x: M(x) = 1\}$is either a closed interval $[\underline{m}, \overline{m}]$ or an interval of the form $[\underline{m}, +\infty)$, $(-\infty, \overline{m}]$, $(-\infty, +\infty)$.

ii) The restriction $M^-(\cdot)$ of $M(\cdot)$ to $(-\infty, \underline{m})$ (when appropriate) is a right-continuous non-decreasing function

iii) The restriction $M^+(\cdot)$ of $M(\cdot)$ to $(\overline{m}, +\infty)$ (when appropriate) is a left-continuous non-increasing function.

Note that an u.s.c. fuzzy interval may have a bounded support (the support of Q is the set $\text{supp}(Q) = \{a \in \mathbb{R}: Q(a) > 0\}$) or not. Compact subsets of the real line (i.e., closed and bounded ones) are extended by fuzzy u.s.c. intervals with a bounded support. A fuzzy number M is defined as a bounded support u.s.c. fuzzy interval with a unique modal value m; it may be a representation of the imprecisely specified quantity "approximately m". An extensive analysis of the relationships between the concepts of upper (lower) semi-continuity, as well as the links between monotony and convexity is supplied in Kerre and Van Schooten (1988).

N.B. Alternatively, an u.s.c. fuzzy interval can be represented by its set of level-cuts namely, $\{M_\lambda = [\underline{m}_\lambda, \overline{m}_\lambda], \lambda \in (0, 1]\}$ when they are bounded. Goetschel (1997) pushed this representation further by means of the two functions μ^+ and μ^- from $[0, 1]$ to $\mathbb{R}$ such that $\mu^-(\lambda) = \underline{m}_\lambda$ and $\mu^+(\lambda) = \overline{m}_\lambda$. Then it is tempting to represent a fuzzy interval by the pair (μ^-, μ^+) of functions. However in order to recover M from (μ^-, μ^+), the pair of functions must satisfy properties such as $\mu^-(\alpha) \leq \mu^-(\beta) \leq \mu^+(\beta) \leq \mu^+(\alpha) \forall\ \alpha \leq \beta$. Goetschel (1997) considered cases when these conditions do not hold, thus giving birth to what he called "antifuzzy numbers". He especially considered piecewise representations of such functions that are parameterized in terms of elementary linear components called *fuzzy darts*. An earlier paper by Tamura and Horiuchi (1993) pushed this view even further by considering any pair of functions (μ^-, μ^+) as a generalized fuzzy interval. They studied arithmetic operations on such representations.

Other Fuzzy Quantities. A fuzzy quantity Q is said to be *multimodal* if Q is the finite union (in the sense of standard fuzzy set union) of unimodal, not necessarily all normal, fuzzy quantities M_i with distinct modal values:

$$Q(x) = \max_{i = 1,\ldots,n} M_i(x). \tag{10.2}$$

A fuzzy integer is a fuzzy quantity whose support contains only integers: $\mathrm{supp}(Q) \subset \mathbb{N}$ = the set of natural numbers. It is an interesting case of a multimodal fuzzy quantity (even if they are unimodal on $\mathbb{N}$). Fuzzy integers are encountered when defining the fuzzy cardinality of fuzzy sets (Zadeh, 1979, Dubois and Prade, 1985; see also Chapter 1 in this volume by Dubois, Ostasiewicz and Prade). A fuzzy quantity Q can be made convex if we calculate its convex hull $\hat{Q}$ (Lowen, 1980) such that

$$\forall c \in \mathbb{R}, \hat{Q}(c) = \sup\{\min((Q(a), Q(b)): a \leq c \leq b\}. \tag{10.3}$$

The level-cuts of $\hat{Q}$ are the convex hulls of the level-cuts of Q.

Interpretation. A fuzzy interval is a convenient representation of imprecisely specified ranges of variables, conveying more information than a usual crisp interval. For instance, when one has to subjectively estimate the value of an uncontrollable unknown parameter, the interval representation tends to be a poor modelling tool: should pessimistic bounds be provided ? (and further computations will yield far too imprecise results to be of any use); or optimistic bounds ? (then more precise results will be obtained, but are they reliable ?). A fuzzy interval carries at the same time optimistic and pessimistic bounds: its support is built on reliable grounds while its core contains only the most plausible values. Similarly, when specifying requirements about a decision variable, a goal, a constraint to be respected, the all-or-nothing nature of intervals masks the presence of more or less preferred values. Fuzzy quantities then also offer more expressive models of flexible requirements.

The concept of a variable attached to a possibility distribution called "fuzzy variables" by Nahmias (1978) is similar to the concept of a random variable. Nahmias (1978) did introduce possibility measures (he calls them*scales*) and he considered fuzzy variables as real-valued mappings like random variables. In this chapter a variable is only the symbolic name of an unknown value whose range is fuzzy.

Thus, fuzzy quantities can be interpreted in two radically different ways: they are either uncertainty distributions that are possibilistic counterparts of probability densities; or preference profiles, which then make them akin to utility functions. Techniques for membership function elicitation are particularly adapted to the estimation of fuzzy intervals, whether based upon psychometry-oriented methods (see Chapter 3 by Bilgic and Türksen in this volume), or based on histograms and probability-possibility transformations, or yet set-valued statistics (Wang, 1983; see Dubois, Nguyen and Prade, Chapter 7 in this volume).

A fuzzy interval is a more cautious representation of an uncertain quantity than a probability distribution when the only source of information is a human being (hence subjective). One reason may be that the rules of possibility theory that will govern the calculus of fuzzy quantities involve only comparisons between degrees of possibility, i.e., involve a poorer structure (hence fewer assumptions) than the rules of probability theory where some justification of the additive law must be put forward (e.g., ideal gambling situation). This is especially true if a fuzzy quantity is viewed as a bunch of nested intervals, membership grades being taken from a qualitative scale.

Probabilistic Connections. Possibility measures usually refer to non-statistical types of imprecision, such as the one pervading natural language and subjective knowledge. However, a statistical interpretation of possibility measures has been proposed (Wang, 1983; Dubois and Prade, 1986) in the framework of random experiments yielding imprecise outcomes. It suggests a purely numerical view of u.s.c. fuzzy intervals related to probability theory. The cautiousness of the representation is also patent in this framework of imprecise statistics. The idea is based on the fact that fuzzy sets can be viewed as deriving from random sets, or similarly, that possibility measures are consonant plausibility functions in the sense of Shafer (1976). Namely, following the view of Dempster (1967), a fuzzy interval M can be obtained by carrying the Lebesgue measure from the unit interval to the real line via the multiple-valued mapping that assigns to each level $\alpha \in (0,1]$ the level-cut M_α by the formula

$$M(a) = \int_0^1 M_\alpha(a)d\alpha \tag{10.4}$$

where $M_\alpha(\cdot)$ denotes the indicator function of the α–level-cut of M ($M_\alpha(a) = 1$ if $a \in M_\alpha(\cdot)$ and 0 otherwise). The fuzzy interval M is viewed as a random interval $\mathcal{I}_M$ and $M(a) = P(a \in \mathcal{I}_M)$ the probability that a belongs to a realization of $\mathcal{I}_M$ (Dubois and Prade 1987b; Heilpern, 1992). Possibility and necessity measures Π_M and N_M induced by the possibility distribution $\pi(\cdot) = M(\cdot)$ are thus upper and lower bounds on degrees of probability P(A). A fuzzy interval M can thus be viewed as encoding a family of probabilities, a set of probability measures $\mathcal{P}_M$ defined by

$$\mathcal{P}_M = \{ P \mid \Pi_M(A) = \sup_{a \in A} M(a) \geq P(A), \text{A measurable}\}. \tag{10.5}$$

This view presupposes that a probability measure P could have been obtained from the available statistics if outcomes had been precisely observed. Let $[\underline{m}, \overline{m}]$ be the core of M. The upper bound in $\mathcal{P}_M$ is attained by the probability measure $\overline{P}$ with distribution function F^* such that

$$\forall a,\ F^*(a) = \overline{P}(\,(-\infty, a]\,) = \Pi_M(\,(-\infty, a]\,) = \sup\{M(x) \colon x \le a\}$$

That is (see Figure 1)

$$F^*(a) = M(a) \text{ if } a \le \underline{m} \qquad = 1 \text{ otherwise.} \tag{10.6}$$

Similarly, the lower bound of $\mathcal{P}_M$ is attained by the probability measure $\underline{P}$ with distribution function F_* such that:

$$\forall a,\ F_*(a) = \underline{P}(\,(-\infty, a]\,) = N_M(\,(-\infty, a]\,) = 1 - \Pi_M(\,(a, +\infty)\,) = \inf\{1 - M(x) \colon x > a\}.$$

That is (see Figure 1)

$$F_*(a) = 0 \text{ if } a < \overline{m} \qquad = 1 - \lim_{x \downarrow a} M(x) \text{ otherwise.} \tag{10.7}$$

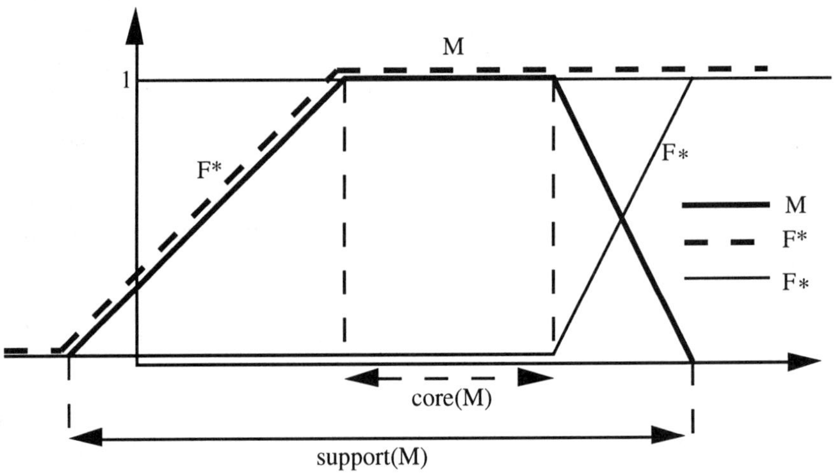

Figure 10.1

At this point it is important to notice that there are actually three probabilistic views of a fuzzy interval:

(i) The *imprecise probability view* whereby M encodes a set of probability measures, as shown above.

(ii) The *random set view* whereby M encodes the one point coverage function of a random interval, defined by a probability measure on the unit interval (for instance the uniformly distributed one) and a family of nested intervals, via a multivalued mapping from (0, 1] to ℝ, following Dempster (1967) (for instance the level-cuts) (see Heilpern, 1992, for instance).

(iii) The *pair of PDFs view*, whereby M is defined by two random variables x^- and x^+ with distributions F* and F_* and M stands for the random interval $[x^-, x^+]$ with possibly independent end-points (See Heilpern, 1992, 1997; Gil, 1992). Namely let $M_\lambda = [\underline{m}_\lambda, \overline{m}_\lambda]$ be the level-cuts of M. One may have x^- and x^+ depend on a single parameter λ such that $[x^-(\lambda), x^+(\lambda)] = [\underline{m}_\lambda, \overline{m}_\lambda]$, or the generated intervals may appear in an asymmetric way: e.g., $[x^-(\lambda), x^+(\lambda)] = [\underline{m}_\lambda, \overline{m}_{1-\lambda}]$. Or one may have x^- and x^+ depending on two independent parameters λ and λ' such that $[x^-(\lambda), x^+(\lambda')] = [\underline{m}_\lambda, \overline{m}_{\lambda'}]$. Note that the intersection of all such generated intervals is not empty so as to ensure the normalization of M.

The first view is more general than the second one, and can be understood as the specification of an imprecise probability function; and the third view is also more general, since the second view corresponds to a random process that delivers nested intervals as a whole (hence a clear dependence between endpoints) while the third view is one where end-points of the interval are separately generated, possibly in an independent way. However a clear development of each point of view is a matter of further research.

McCain (1983) regards a fuzzy interval M as a weighted set of nested confidence intervals $\mathcal{L}_s(M)$, which are its strong level-cuts $M_{>\lambda} = \{a\colon M(a) > \lambda\}$. And indeed, in statistics the set of confidence intervals attached to the value of an estimated parameter, and obtained by varying the level of confidence is nested. Under this view the membership function $1 - M(\cdot)$ of the complement M^c of M makes sense since the higher the level λ, the less confident we are that the fuzzy variable x fuzzily restricted by M has value in the strong level-cut $M_{>\lambda}$. Indeed if $S = M_{>\lambda}$, $\lambda = \Pi(S^c) = \sup\{M(a)\colon a \notin S\}$ is the degree of possibility that x lies outside the strong level-cut at this level. Hence our confidence that x lies inside the strong level-cut $M_{>\lambda}$ is $1 - \lambda$. Transforming a probability density into a possibility distribution of maximal specificity is also a technique to derive fuzzy intervals whose level-cuts coincide with the confidence intervals of the probability density (Dubois, Prade and Sandri, 1993). See Chapter 7 by Dubois, Nguyen and Prade in this volume.

NB.1. Fuzzy numbers as studied by Höhle (1981, 1987), are actually fuzzy sets of the real line whose membership functions have the shape of probability distribution functions (such as F* or F_*). This concept is thus different from that introduced here, where a fuzzy interval is equivalent to a pair of probability distribution functions. Höhle (1981) extends the idea of a number, while we are extending the notion of interval. The arithmetics of this type of fuzzy numbers was studied by Klement (1985, 1987). Note that the random variables with values in the Höhle's fuzzy real line coincide with the fuzzy observables introduced by Mesiar (1993), see also Kolesarova and Riecan (1993). Rodabaugh (1982) and Lowen (1983a, 1996) consider the same concept as Höhle, where the interval [0, 1] is changed into a lattice. See Delgado *et al.* (1994) on the comparative merits of each kind of fuzzy numbers. It seems that fuzzy numbers viewed as extensions of intervals are more useful in practice than the other ones which are mainly motivated by mathematical considerations. Fuzzy numbers with membership function range in more general settings were introduced and extensively studied by Sinha (1990), Eklund and Gähler (1988), Gähler (1992), Gähler and Gähler (1994), mainly concentrating on the corresponding topological properties.

NB.2. Kaufmann (1981) has considered hybrid numbers viewed as a pair composed of a fuzzy interval and a random variable whose role is to blur the level-cuts of the fuzzy interval, see also Kaufmann and Gupta (1985). This composite notion seems to be related to the one of a probabilistic set, i.e., a fuzzy set with a random membership function; see Czogala and Hirota (1986).

10.2.2 Characteristics of a Fuzzy Interval

A fuzzy interval can be summarized by several characteristics: a representative scalar value describing its order of magnitude, a measure of imprecision describing how little specific it is, and a measure of fuzziness describing how clear-cut are its boundaries.

Mean Interval. The simplest order of magnitude value is the core of the fuzzy interval, or its mode when it is a fuzzy number. In Dubois and Prade (1978a, 1979a, 1980a), what is called mean value of a fuzzy interval is actually its modal value. Here the term mean value of a fuzzy number is used as follows: Under the random set interpretation of a fuzzy interval, upper and lower mean values of M in the sense of Dempster (1967), can be defined, i.e., $E^*(M)$ and $E_*(M)$, respectively, such that (Dubois and Prade, 1987b):

$$E^*(M) = \int_0^1 (\sup M_\alpha) d\alpha; \qquad E_*(M) = \int_0^1 (\inf M_\alpha) d\alpha . \tag{10.8}$$

Note that these expressions are Choquet integrals (Denneberg, 1994) with respect to the possibility and the necessity measures induced by M. Namely

$$E^*(M) = \int_{-\infty}^{+\infty} x dF_*(x) = \sup \left\{ \int_{-\infty}^{+\infty} x dP((-\infty, x]) : P \in \mathcal{P}_M \right\};$$

$$E_*(M) = \int_{-\infty}^{+\infty} x dF^*(x) = \inf \left\{ \int_{-\infty}^{+\infty} x dP((-\infty, x]) : P \in \mathcal{P}_M \right\}. \tag{10.9}$$

The mean value of a fuzzy interval M is defined as the interval E(M) = [E_*(M), E^*(M)] obtained by separately taking the mean values of the lower and of the upper bounds of the nested random interval equivalent to M. We shall call E(M) the *mean interval* of M. It is also the interval containing the mean values of all random variables compatible with M(i.e., $P \in \mathcal{P}_M$). That the mean value of a fuzzy interval is an interval seems to be intuitively satisfactory. Particularly the mean interval of a (regular) interval [a, b] is this interval itself, namely, $E_*([a, b]) = a$; and $E^*([a, b]) = b$. The same mean interval obtains in the three probabilistic settings for fuzzy intervals.

Gonzalez (1990) has generalized the mean interval of a fuzzy interval. He has introduced a strictly increasing continuous function s such that s(0) = 0 and s(1) = 1 (a linguistic modifier form) and he defines the so-called s-mean value E^s(M)with bounds:

$$E^s{}_*(M) = \int_0^1 (\inf M_\alpha) ds(\alpha); \quad E^{s*}(M) = \int_0^1 (\sup M_\alpha) ds(\alpha).$$

He proves that $E^s(M) = E(s(M))$, the mean interval of the modified fuzzy interval. Further generalisations on mean intervals of fuzzy intervals can be found in (Heilpern, 1992).

Defuzzification. Finding a scalar representative value of a fuzzy interval is often called defuzzification in the literature of fuzzy control (See Yager and Filev, 1993, 1994, and Van Leekwijk and Kerre, 1998 for an extensive overview). Various proposals exist:

-) the *mean of maxima* (MOM), which is the middle point in the core of the fuzzy interval M,

-) the *center of gravity* $GR(M) = \dfrac{\int_{-\infty}^{+\infty} xM(x)dx}{\int_{-\infty}^{+\infty} M(x)dx}$ This is the center of gravity of the support of M, weighted by the membership grade (Braae and Rutherford, 1978).

-) the *center of area (median):* the value COA(M) = m such that $\int_{-\infty}^{m} M(x)dx = \int_{m}^{+\infty} M(x)dx$. This is the point of the support of M that equally divides the area under the membership function.

-) the *middle point of the mean interval* (Yager, 1981)

$$\bar{E}(M) = \int_0^1 \frac{(\inf M_\alpha + \sup M_\alpha)}{2} d\alpha = \frac{E_*(M) + E^*(M)}{2}. \quad (10.10)$$

The MOM sounds natural as a representative of a fuzzy interval M in the scope of possibility theory where values of highest possibility are considered as default plausible values. This is in the particular case when the maximum is unique.. However it makes sense especially if a revision process is in order, that may affect the MOM, by deleting the most plausible values upon arrival of new information. In a static perspective, the MOM clearly does not exploit all the information contained in M since it neglects the membership function.

Yager and Filev (1994) generalize both the MOM and the center of gravity defuzzification, using a power of the membership function $(M)^k$ in GR(M). The MOM is the limit when k is high enough. Yager and Filev (1993) present a general methodology for extracting characteristic values from fuzzy intervals. They show that all methods come down to a possibility-probability transformation followed by the extraction of characteristic value such as a mean value. Note that the MOM, the center of gravity and the center of area come down to renormalizing the fuzzy interval as a probability distribution and computing its mode, expected value or its median, respectively. For instance, Yoon (1996) does the explicit conversion of membership functions into probability densities and concentrates on practical methods for finding the center of gravity GR(M) for complex membership functions M(·) using the Mellin transform. These approaches are debatable, because the renormalization technique is itself arbitrary since the obtained probablility may not belong to $\mathcal{P}_M$, the set of probabilities dominated by the possibility measure attached to M (Dubois and Prade, 1980a).

Only the mean interval as defined in (10.10) accounts for the specific possibilistic nature of the fuzzy interval. The choice of the middle point expresses a neutral attitude of the user and extends the mean of maxima to an average mean of level-cuts. This characteristic value has a natural interpretation in terms of simulation of a fuzzy variable, an idea originally due to Kaufmann (1980) and Yager (1982). Chanas and Nowakowski (1988) investigate this problem in greater detail. Namely, consider the two step random generator which selects a level-cut at random (by choosing $\alpha \in (0, 1]$), and a number in the level-cut. The corresponding random quantity is $x(\alpha, \lambda) = \lambda \cdot \inf M_\alpha + (1 - \lambda) \cdot \sup M_\alpha$. The mean value of this random variable is $\bar{E}(M)$. It corresponds to the mean value $\int_0^1 x.p_M(x)\, dx$ of the probability density $p_M \in \mathcal{P}_M$ (in fact, the center of gravity of $\mathcal{P}_M$; see Dubois, Prade and Sandri, 1993) obtained by considering level-cuts as uniformly distributed probabilities:

$$p_M(x) = \int_0^1 \frac{M_\lambda(x)}{(\sup M_\lambda - \inf M_\lambda)} d\lambda.$$

When the center of area is in the core of M, Zhao and Govind (1991c) called M regular. They noticed that in that case the center of area coincides with $\bar{E}(M)$. However this identity does not hold more generally, for non regular fuzzy intervals.. The middle point of the mean interval has been generalized by introducing a function for weighting level-cuts inside the integral by Delgado *et al.* (1998a). Moreover all these characteristic values (except the MOM) presuppose that the involved integrals are bounded, and make sense for unimodal fuzzy quantities only.

Level of Specificity. A fuzzy interval is also characterized by its specificity or precision. The lack of specificity of an interval is reflected by its width. Hence the specificity of a fuzzy interval can be defined as done by Yager (1982)

$$sp(M) = \int_0^1 \frac{d\alpha}{(\sup M_\alpha - \inf M_\alpha)} . \tag{10.11}$$

More simply a measure of imprecision or ambiguity is the surface of the area under the membership function,

$$am(M) = \int_0^1 (\sup M_\alpha - \inf M_\alpha) d\alpha = E^*(M) - E_*(M)$$

which can also been weighted as proposed by Delgado *et al.* (1998a).

Level of Fuzziness. Lastly, a fuzzy interval is also characterized by its level of fuzziness. See Chapters 8 and 9 by Klir anby Pal and Bezdek respectively in this volume. An interval is not fuzzy at all. So the fuzziness of an interval roughly depends on the size of the parts of the support that lie outside the core. Such an index is for instance (Delgado *et al.*, 1998b)

$$Fuz(M) = \int_0^{0.5} (\sup M_\alpha - \inf M_\alpha) d\alpha + \int_{0.5}^1 (\inf M_\alpha - \sup M_\alpha) d\alpha \tag{10.12}$$

$$= \int_{-\infty}^{+\infty} |M(x) - M_{0.5}(x)| \, dx$$

which corresponds to Kaufmann (1975) index of fuzziness based on the Hamming distance between M and its 0.5-level-cut. More on this point in Chapter 9 by Pal and Bezdek in this volume.

Distances Between Fuzzy Numbers. An interesting question is the embedding of fuzzy quantities in a topological or metrical structure. Metrical aspects of $[0, 1]^{\mathbb{R}}$ are dealt with by Goetschel and Voxman (1983), who introduce a Hausdorff metric between fuzzy quantities, see also (Puri and Ralescu, 1983). Among the several recent sources on the topic, see e.g., Heilpern (1997) and the next Chapter 11 by Diamond and Kloeden. Heilpern points out three approaches to the distance beween fuzzy intervals M and N:

i) Using characteristic values of the fuzzy intervals. If c_M and c_N are characteristic values of M and N (chosen among one of the above proposals), then take $\delta(M,N) = |c_M - c_N|$. For instance the mean distance, when $c_M = E(M)$, see for instance Campos and Gonzalez (1994).

ii) Using a generalized distance between intervals. If A = [a, b], B = [c, d] then let $\phi_p(x) = x^p$ and

$$\begin{aligned}\delta_p(A, B) &= \phi_p^{-1}(0.5(\phi_p(|a - c|) + \phi_p(|b - d|))), \ p \geq 1 \\ &= \max(|a - c|, |b - d|) \text{ if } p = +\infty.\end{aligned}$$

Then for two fuzzy intervals M and N, it is possible to define the following metric: $\delta_p(M, N) = \int_0^1 \delta_p(M_\lambda, N_\lambda)\, d\lambda$. For p = 1, the expression is simpler and sometimes coincides with the above mean distance. For trapezoidal fuzzy intervals it may also coincide with the geometrical distance between trapezoids (the average of the distance between their vertices); see Heilpern (1997). A variant of this expression is to replace the integral by a supremum over level-cuts. Another variant with a weighted sum is in Bertoluzza *et al.* (1995). A mathematical study of such distances is the book by Diamond and Kloeden (1994).

iii) Using a parametric representation of the fuzzy intervals and considering a standard distance in the parameter space. This was first suggested by Bonissone for the purpose of linguistic approximation (Tong and Bonissone, 1980, for instance). It applies to triangular fuzzy numbers represented by the abcissae of the support and the core.

There is in fact a fourth approach: fuzzy-valued distances, that picture the distribution of distances between level-cuts. For instance fuzzy Hausdorff distances are defined in Dubois and Prade (1983d). Related to the notion of distance is the associated notion of convergence. Dubois and Prade (1983a), Kaleva (1985), Nanda (1989) suggest several notions of convergence of sequences of fuzzy intervals. Fuzzy fixed-point problems are studied by Badard (1984). Topological issues and convergence problem in $[0, 1]^{\mathbb{R}}$ are considered in Lowen (1983b) at a theoretical level. The book by Diamond and Kloeden (1994) also covers this topic. More about convergence can also be found in the literature on series of fuzzy intervals, reviewed in Section 10.5.2. More generally, see the Chapter 11 by Diamond and Kloeden in this volume.

10.2.3 Noninteractive Fuzzy Variables

When the universe of discourse U is a Cartesian product, $U_1 x U_2$, fuzzy subsets of U are called fuzzy relations, denoted R, S, T ..., and fuzzily restrict the ranges of pairs (x, y) of variables. R can be viewed as a joint possibility distribution. The projections of a fuzzy relation R on U_1, and U_2 are denoted $P_1[R]$ and $P_2[R]$, respectively and are defined by (Zadeh, 1975a):

$$P_1[R](u_1) = \Pi_R(\{u_1\} x U_2) = \sup\{R(u_1, u_2) : u_2 \in U_2\}; \qquad (10.13)$$
$$P_2[R](u_2) = \Pi_R(U_1 x \{u_2\}) = \sup\{R(u_1, u_2) : u_1 \in U_1\}. \qquad (10.14)$$

If E_1 and E_2 are fuzzy subsets of U_1 and U_2, respectively, the cylindrical extensions $C[E_1]$ and $C[E_2]$ of E_1 and E_2 on U are defined by (Zadeh, 1975a):

$$C[E_1](u_1, u_2) = E_1(u_1) \text{ and } C[E_2](u_1, u_2) = E_2(u_2). \qquad (10.15)$$

The Cartesian product $E_1 x E_2$ is then defined as the min-based intersection $C[E_1] \cap C[E_2]$. Given a fuzzy relation R, it can be checked that

$$R \subseteq P_1[R] \times P_2[R]. \qquad (10.16)$$

When this inclusion holds with equality, R is said to be *separable* (Zadeh, 1975a), and the underlying variables x and y are said to be *noninteractive*, i.e., assigning a value to one of them has no impact on the assignment of a value to the other. This is exactly what happens when the range of (x, y) is a (usual) Cartesian product. If x and y are noninteractive variables then the joint possibility distribution $\pi_{(x, y)} = \min(\pi_1, \pi_2)$ decomposes by the minimum. Noninteraction plays the same simplification role in possibility theory as independence in probability theory, but they do not convey the same meaning at all: noninteraction simply means a lack of functional link between the variables, while, in probability, an event A is said to be independent of B if its probability (understood as frequency) is the same whether B occurs or not. Non-interactivity is a cautious assumption that ignores possible links between variables, at the expense of precision in computations.

Like in the probabilistic setting, noninteraction of fuzzy variables induces what Nahmias (1978) called unrelatedness of events, defined by $\Pi(A \cap B) = \min(\Pi(A), \Pi(B))$. This property indicates that A and B are not mutually exclusive. Nahmias (1978) did not provide such an interpretation of this notion, in terms of statistical experiments. The study of statistical independence under imprecise measurements in connection with fuzzy intervals has yet to be carried out. See (Rao and Rashed, 1981) for further discussions along Nahmias intuitions.

The above notion of joint possibility distributions can be justified in the setting of random sets (Dubois and Prade, 1991a). It turns out to be a very strong dependence assumption that given two nested random sets with associated one-point coverage functions π_1 and π_2 produces a nested random set of one-point coverage function $\min(\pi_1, \pi_2)$. It presupposes that if the first random process delivers a level-cut $(A_1)_\lambda$ of π_1 then the other one delivers $(A_2)_\lambda$ for the same value of λ. On the contrary, the assumption that level-cuts $(A_1)_\lambda$ and $(A_2)_\nu$ may be jointly observed will not lead to a nested random set, hence is not equivalent to a joint possibility distribution.

10.3 BASIC PRINCIPLES OF FUZZY INTERVAL ANALYSIS

The problem which must be dealt with in fuzzy interval mathematics can be stated as follows: given several noninteractive real variables x, y, z, ... whose ranges are fuzzily restricted by fuzzy quantities, given a function $f(x, y, z, \ldots)$, how can the fuzzy range of the variable $f(x, y, z, \ldots)$ be calculated?

10.3.1 The Extension Principle

The purpose of the extension principle proposed by Zadeh (1965, 1975a) is to extend functions having pointwise arguments to functions with fuzzy set arguments. The main application is fuzzy interval analysis. Zadeh's methodology can be cast in the more general setting of carrying a set function g from a set to another set via a mapping. Let us consider two sets U and V and a mapping f from U into V. Let g be a set function on U. Then, by means of the reciprocal images of subsets A of V via f, namely $f^{-1}(A) = \{u: f(u) \in A\}$, there is a canonical construction of a set function g^f on V, such that:

$$\forall A \subseteq V, g^f(A) = g(f^{-1}(A)). \tag{10.17}$$

It is easy to check that if g is a possibility measure Π, then so is Π^f. Similarly, if g is a probability measure and choosing A such that $f^{-1}(A)$ is measurable, the standard theory of functions of random variables is retrieved (e.g., Papoulis, 1965). Nahmias (1978) advocates this analogy but he calls a possibility measure a scale.

The possibility distribution π^f underlying the result of carrying the fuzzy set E from U to V is obtained as

$$\forall v \in V, \pi^f(v) = \Pi^f(\{v\}) \begin{array}{l} = \sup\{E(u): f(u) = v\} \text{ if } f^{-1}(\{v\}) \neq \emptyset \\ = 0 \text{ otherwise.} \end{array} \tag{10.18}$$

This equation is referred to as the extension principle in fuzzy set theory (Zadeh, 1965, 1975a). The general principle could be equivalently applied to the necessity measure N which is dual to Π, so that $N(A) = 1 - \Pi(A^c)$ without bringing more information. When Π is a crisp {0,1}-valued possibility measure, then Π^f is the crisp possibility measure whose possibility distribution is the characteristic function of the set $f(E)$, that is, $f(E) = \{f(u): u \in E\}$. Hence we are more generally allowed to write $\pi^f = f(E)$. If x is the variable with fuzzy range E, then $f(x)$ has a fuzzy range described by the extension principle (10.18).

When U is a Cartesian product $U_1 x U_2$, and E is separable, i.e., $E = \min(E_1, E_2)$ where E_1 and E_2 are fuzzy subsets of U_1 and U_2, respectively, then the membership function π^f of $f(E_1, E_2)$ becomes, $\forall v \in V$,

$$\begin{aligned}\pi^f(\{v\}) &= \sup\{\min(E_1(u_1), E_2(u_2)): f(u_1, u_2) = v\} \text{ if } f^{-1}(\{v\}) \neq \emptyset \\ &= 0 \text{ otherwise.}\end{aligned} \tag{10.19}$$

The variables x_1 and x_2, fuzzily restricted by E_1 and E_2, are noninteractive and $f(E_1, E_2)$ is the fuzzy range of $f(x_1, x_2)$. Note that when the addition of fuzzy intervals is performed, i.e., when $f(x, y) = x + y$, $x, y \in \mathbb{R}$, then (10.19) turns to be a Sugeno integral based convolution (Wang and Klir, 1992) with respect to a possibility measure (see also Mizumoto and Tanaka, 1976c).

The fuzzy set $f(E_1, E_2)$ can be defined in terms of the level-cuts of E_1 and E_2, since, as shown by Nguyen (1978):

$$\forall v \in V,\ f(E_1, E_2)(v) = \sup\{\lambda \in (0, 1]: v \in f((E_1)_\lambda, (E_2)_\lambda)\}$$

but we only have the following inclusion:

$$f((E_1)_\lambda, (E_2)_\lambda) \subseteq (f(E_1, E_2))_\lambda \tag{10.20}$$

and not the equality. The equality in this expression was proved by Nguyen (1978) to be equivalent to the existence of a supremum attained in (10.19) for all $v \in V$ (see Proposition 1 below). For further generalisations of Nguyen's result see, e.g., Fullér and Keresztfalvi (1991), Bertoluzza and Bodini (1998). However, if strong level-cuts are considered instead of level-cuts, then the equality holds in (10.20).

More generally, fuzzy variables may be interactive and the pair (x_1, x_2) of variables is then attached to a fuzzy relation playing the role of a joint possibility. This case is considered in Section 10.5. A special situation is when the fuzzy Cartesian product is generalized using a triangular norm instead of the minimum in (10.19) (Dubois and Prade, 1981). This weak form of interaction and the calculus of fuzzy intervals it leads to is considered in Section 10.5.2. Otto *et al.* (1993) change the minimum into idempotent monotonic combinations (other than triangular norms) in the extension principle.

NB. Other types of extension principles have been considered for carrying fuzzy sets by functions. Kaufmann (1975) applied the probabilistic extension principle (convolution) to define the addition of fuzzy quantities; the works of Mares (1977a,b, 1993a,b, 1994) are also along this line. Another exotic approach was introduced by Harman (1992a,b) who studies properties of the product of fuzzy intervals using again a probabilistic-like approach. Jain (1976) uses the extension principle (10.19) where the supremum is changed into the probabilistic sum a + b – ab. See Dubois and Prade (1978b) for criticism of this approach. The aforementioned attempts at combining fuzzy quantities clearly stand outside of possibility theory. Dijkman *et al.* (1983) consider nine different extension principles, seven of which also stand outside of this framework. Ramik 1986) has considered the fuzzy extension of set-to-set mappings.

The extension principle in the case of several non-interactive variables is similar to the calculus of independent random variables. However they cannot be directly compared in general since they are based on different assumptions. The calculus of fuzzy intervals does not sanction any compensation of error while this is taken for granted in probability theory due to the independence assumption. However it is possible to embed both calculi in a single setting: the one of random intervals. Rules of computation with random intervals do subsume both the calculus of independent random variables and the fuzzy extension principle. However the latter is incompatible with the calculus of independent random intervals; it rather corresponds to a strong dependence between nested random intervals (see Dubois and Prade, 1991a). If $x(\alpha, \lambda)$ and $y(\beta, \nu)$ are random variables obtained via a simulation process of fuzzy intervals M and N described in Section 10.2.2, then $f(M, N)$ is the fuzzy set attached to the random variable $f(x(\alpha, \lambda), y(\beta, \nu))$ provided that $\alpha = \beta$ and $\lambda = \nu$ (Chanas and Nowakowski, 1988).

There is another view of the extension principle (Yager, 1986) which is related to fuzzy relational calculus (Zadeh, 1975b). Namely $f(M, N)$ can be viewed as the result of a constraint propagation process. This is particularly natural if M and N are viewed as preference profiles of decision variables. Indeed consider the set of decision variables {x, y, z} such that x is fuzzily restricted by a preference profile M, y is fuzzily restricted by N and (x, y, z) should belong to the relation $R = \{(u, v, w) \mid w = f(u, v)\}$. Let us achieve arc-consistency, in the sense of fuzzy constraint propagation (Dubois *et al.* 1995), for variable z. It means that we project over the domain of z all constraints that relate z to other variables and may affect its admissibility domain. The fuzzy set of feasible 3-tuples (u, v, w) is $M \cap N \cap R$ and its projection $(M \cap N \cap R)^{\downarrow\{z\}}$ has membership function $\sup_{u,v} \min((\mu_A(u), \mu_B(v), \mu_R(u, v, w)) = \pi_{f(x,y)}(w)$, precisely the generalized convolution operation that founds the calculus of fuzzy intervals. In many applications of fuzzy interval analysis (e.g., in design) fuzzy intervals do represent preference profiles, and preference, not uncertainty, is propagated by the computations. Solutions of design problems obtained via fuzzy interval calculations then achieve egalitarist trade-offs between local flexible requirements (Otto and Antonsson, 1991, 1994; Antonsson and Otto, 1995).

10.3.2 Functions of Non-Interactive Fuzzy Variables: Basic Results

The calculus of fuzzy quantities is an application of the extension principle to fuzzy quantities. A first question is whether properties of fuzzy intervals are preserved through the extension principle. The following conditions for the preservation of properties have been established (for instance Biacino and Lettieri, 1989):

- If Q_1 and Q_2 are normal, then so is $f(Q_1, Q_2)$.
- If Q_1 and Q_2 are convex and f is continuous, then the convexity of $f(Q_1, Q_2)$ follows.

The main other result of this section is to provide conditions so that the level-cuts of a fuzzy real-valued expression can be expressed in terms of the level-cuts of the involved fuzzy operands. This result enables main properties of operations on fuzzy intervals to be studied, and lays bare the fact that fuzzy interval mathematics encompass the regular interval mathematics (Moore, 1966, 1979). Moreover, theoretical grounds for practical computation techniques (described in Section 10.4) are provided since these techniques very often come down to computing level-cuts via interval analysis.

Several sufficient conditions have been proposed, in order to ensure that any function commutes with level-cutting, which is expressed by the following equality:

$$f((Q_1)_\lambda, (Q_2)_\lambda) = (f(Q_1, Q_2))_\lambda, \ \forall \lambda \in (0, 1] \tag{10.21}$$

where f is a real-based real-valued function, and M, N are fuzzy quantities.

Proposition 1: *(Nguyen, 1978) If f is continuous and M and N are u.s.c. fuzzy quantities whose supports have compact closures then function f commutes with level-cutting.*

In Proposition 1, the condition of compact support closure can be relaxed to compact λ-level-cuts for all $\lambda > 0$ (see Dubois and Prade, 1987a). Note that if we change level-cuts into strong level-cuts $M_{>\lambda} = \{a \in \mathbb{R};\ M(a) > \lambda\}$ then the commutation property holds without any restriction. For a detailed contribution on the problems related to the level-cut representation of fuzzy quantities and the level-cutting approach to the extension principle, see Herencia (1997) who considers so-called "graded numbers" as any collection of nested intervals. Stojakovic and Stojakovic (1996) study the commutation with level-cutting of infinite sums of fuzzy intervals (for converging series).

Now, assume M and N are u.s.c. fuzzy intervals and the level-cut $M_\lambda = [\underline{m}_\lambda, \overline{m}_\lambda]$ when bounded.

Proposition 2: *(Dubois and Prade, 1987a) Let M and N be two u.s.c. fuzzy intervals, such that $\forall\lambda \in (0, 1]$, $M_\lambda \neq \mathbb{R}$, $N_\lambda \neq \mathbb{R}$. Any continuous non-decreasing function f, namely, such that $\forall u \geq u'$, $\forall v \geq v'$, $f(u, v) \geq f(u', v')$ commutes with level-cutting.*

When Proposition 1 or 2 applies, $f(M, N)$ is clearly u.s.c., and in particular, when Proposition 2 applies, $f(M, N)$ is a fuzzy interval. If M_λ and N_λ are bounded, then

$$(f(M, N))_\lambda = [f(\underline{m}_\lambda, \underline{n}_\lambda), f(\overline{m}_\lambda, \overline{n}_\lambda)]. \quad (10.22)$$

Proposition 2 is still valid under the following alternative assumptions (the continuity of f is taken for granted):

i) Function f is defined only on a Cartesian product of closed intervals (then consider the restrictions of M x N to this set).

ii) f is non-increasing, that is, $\forall u \geq u'$, $\forall v \geq v'$, $f(u, v) \leq f(u', v')$. Then, when M_λ and N_λ are compact intervals,

$$(f(M, N))_\lambda = [f(\overline{m}_\lambda, \overline{n}_\lambda), f(\underline{m}_\lambda, \underline{n}_\lambda)].$$

iii) f is hybrid, e.g., $\forall u \geq u'$, $\forall v \leq v'$, $f(u, v) \geq f(u', v')$. Then, when M_λ and N_λ are compact intervals,

$$(f(M, N))_\lambda = [f(\underline{m}_\lambda, \overline{n}_\lambda), f(\overline{m}_\lambda, \underline{n}_\lambda)].$$

iv) f has a finite number of arguments and is monotonic with respect to each of them.

Proposition 2 and the above extensions apply to a large set of operations between real numbers. A noticeable consequence of it is that the result of these operations applied to u.s.c. fuzzy intervals M and N can be carried out by combining the increasing parts of their membership functions and their decreasing parts separately (if f is non-decreasing, otherwise one must combine the increasing parts of some with the decreasing parts of others and conversely). Let $[\underline{m}, \overline{m}]$ and $[\underline{n}, \overline{n}]$ be cores of M and N, respectively and M^- (resp. M^+) denote the restriction of M to $(-\infty, \underline{m}]$ (resp. $[\overline{m}, +\infty)$); N^- and N^+ are similarly defined. The procedure that calculates $f(M, N)$ via level-cuts is pictured in Figure 2 for non-decreasing functions and expressed by the following equalities when M^- and N^- are strictly increasing, and M^+ and N^+ are strictly decreasing:

$\forall c \leq f(\underline{m}, \underline{n})$, $f(M, N)(c) = \sup\{\min(M^-(\underline{m}_\lambda), N^-(\underline{n}_\lambda)): f(\underline{m}_\lambda, \underline{n}_\lambda) = c\}$.

$\forall c \in [f(\underline{m}, \underline{n}), f(\overline{m}, \overline{n})]$, $f(M,N)(c) = 1$.

$\forall c \geq f(\overline{m}, \overline{n})$, $f(M, N)(c) = \sup\{\min(M^+(\overline{m}_\lambda), N^+(\overline{n}_\lambda)): f(\overline{m}_\lambda, \overline{n}_\lambda) = c\}$.

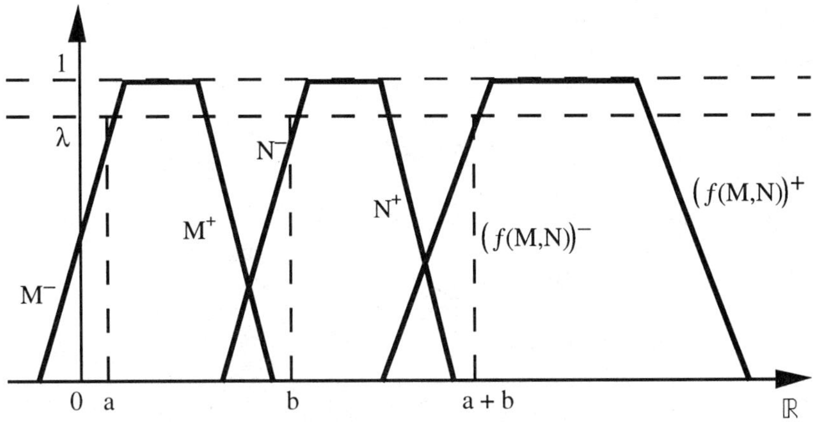

Figure 10.2

If moreover, function f is jointly strictly increasing ($\forall$ u > u', v > v', $f(u, v) > f(u', v')$) then for any c the pair $(\underline{m}_\lambda, \underline{n}_\lambda)$ (resp. $(\overline{m}_\lambda, \overline{n}_\lambda)$) such that $f(\underline{m}_\lambda, \underline{n}_\lambda) = c$ (resp. $f(\overline{m}_\lambda, \overline{n}_\lambda) = c$) is unique (Dubois and Prade, 1979a) and the inverse functions of $f(M, N)^+$, and $f(M, N)^-$ are expressed in closed form:

$$\forall \varepsilon \in \{+, -\}, \left(f(M, N)^{\varepsilon}\right)^{-1} = f\left((M^{\varepsilon})^{-1}, (N^{\varepsilon})^{-1}\right). \qquad (10.23)$$

Identity (10.23) is still valid when M^ε or N^ε are constant on intervals, outside of their cores, by extending the concept of reciprocal function so that an interval where M^ε is a constant becomes a jump for $(M^\varepsilon)^{-1}$ following the ideas of Klement (1985, 1987), see also Mesiar (1995, 1996a). Let M be a fuzzy interval. For $\lambda \in (0, 1]$, we define mappings $(M^+)^*$, $(M^-)_*$: $(0, 1] \to \mathbb{R}$ by:

$$(M^+)^*(\lambda) = \sup\{x; M(x) \geq \lambda\} = \overline{m}_\lambda;$$
$$(M^-)_*(\lambda) = \inf\{x; M(x) \geq \lambda\} = \underline{m}_\lambda.$$

It is immediate that both $(M^+)^*$ and $(M^-)_*$ are monotone. $(M^+)^*$ is non-increasing, while $(M^-)_*$ is non-decreasing. M is uniquely determined by the corresponding $(M^+)^*$ and $(M^-)_*$. Indeed, for all $x \in \mathbb{R}$, we have:

$$M(x) = \min(\sup \{\lambda: (M^-)_* (\lambda) \le x\}, \sup \{\lambda: (M^+)^*(\lambda) \le x\}).$$

Identity (10.23) makes it very clear that computing with fuzzy intervals may lead to simple calculations for a wide range of operations. Namely, finding $f(M, N)(c) \in (0, 1)$ when (10.23) is valid comes down to solving the pair of equations with unknown λ on their respective domains:

$$f((M^+)^*(\lambda), (N^+)^*(\lambda)) = c;$$
$$f((M^-)_*(\lambda), (N^-)_*(\lambda)) = c. \tag{10.24}$$

If $(M^-)_*$ and $(M^+)^*$ are injective, this backward determination of M can be performed by means of inverse functions. When this equation is tricky, one may resort to using the commutation property (10.22) for a selected numbers of level-cuts. We get an exact sampling of $f(M, N)$ and a smoothing procedure, if needed, can be used to complete the estimated membership function.

Turning to multimodal fuzzy quantities, two results given in (Dubois and Prade, 1979a) enable the computation of operations on such quantities to be turned into computations with fuzzy intervals.

Proposition 3: *Let Q_1 and Q_2 be two possibly non-normal fuzzy quantities such that $hgt(Q_1) \le hgt(Q_2)$. If Q_2^t is obtained by truncation of Q_2 at height $hgt(Q_1)$, i.e., $Q_2^t(x) = \min(hgt(Q_1), Q_2(x))$, then $f(Q_1, Q_2) = f(Q_1, Q_2^t)$. This is called the truncation effect.*

Proposition 4: *Let $Q_1 = \cup_{i=1,m_1} M_i^1$ and $Q_2 = \cup_{j=1,m_2} M_j^2$, where M_i^1 and M_j^2 are fuzzy subsets of the real line and $\cup$ is the fuzzy union (based on max). Then*

$$f(Q_1, Q_2) = \cup_{i=1,m_1, j=1,m_2} f(M_i^1, M_j^2).$$

Using Proposition 4, the calculation of $f(Q_1, Q_2)$ where Q_1, Q_2 are multimodal reduces to calculations on (non-normal) fuzzy intervals: each of these calculations may take advantage of the truncation effect to bring the fuzzy intervals down to identical heights.

The calculus of fuzzy intervals is closely related to interval analysis and set-valued mathematics. If Q_1 and Q_2 are two imprecise quantities modeled by subsets of the real line, $f(Q_1, Q_2)$, as given by the extension principle is equivalently expressed as (Young, 1931):

$$f(Q_1, Q_2) = \{f(a, b): a \in Q_1, b \in Q_2\}. \tag{10.25}$$

When $Q_1 = M$ and $Q_2 = N$ are closed intervals, this equation is an error-propagation computation, i.e., we recover interval analysis (Moore, 1966, 1979; Markov, 1995, 1996). The commutation with level-cuts property of (10.21) points out that when M and N are fuzzy intervals, $f(M, N)$ can be obtained by first applying the rules of interval analysis (i.e., Equation 10.25) to level-cuts or strong level-cuts, and then synthesizing the result via the representation theorem of fuzzy sets by level-cuts (see,e.g., Chapter 1 by Dubois, Ostasiewicz and Prade in this volume).

There is a full consistency between the properties of fuzzy and nonfuzzy interval analysis. For instance, set-inclusive monotonicity, as given by Moore (1979), holds for fuzzy quantities (where fuzzy set inclusion is defined by $Q \subseteq Q' \Leftrightarrow Q(\cdot) \leq Q'(\cdot)$)

$$Q \subseteq Q' \text{ implies } f(Q) \subseteq f(Q'). \tag{10.26}$$

This result just reminds us that when the operands become more imprecise, the precision of the result cannot but diminish. Due to its close relationship to interval analysis, the calculus of fuzzy quantities is clearly pessimistic about precision, since $f(Q_1, Q_2)$ is the largest fuzzy set (in the sense of fuzzy inclusion) such that the ranges of the arguments of f are Q_1 and Q_2, respectively.

An important problem is whether the cancellativity properties of operations on numbers remain when extended to fuzzy quantities. Cancellativity of a two-place function f means that $f((a,b) = f(c, b)$ implies $a = c$. Bouchon–Meunier *et al.*, (1997) have recently proved that addition is cancellative (the authors use the term 'invertible') on the class of compact convex fuzzy quantities.

10.3.3 Application to Usual Operations

The general results summarised below point out interesting properties of fuzzy intervals when the usual operations, especially the four operations of arithmetic and the minimum and maximum operations, are applied to them (Dubois and Prade 1979a; Mizumoto and Tanaka, 1979). First we consider unary operations in the real line.

Functions of One Variable. f a real-valued function f has only one argument and is injective, and Q is a fuzzy quantity, $f(Q)$ has a simple expression derived from (10.18):

$$\begin{aligned} \forall a \in \mathbb{R}, f(Q)(a) &= Q(f^{-1}(a)) \text{ if } f^{-1}(\{a\}) \neq \emptyset \\ &= 0 \text{ otherwise.} \end{aligned} \tag{10.27}$$

Applying this formula to the usual functions, Table 1 may be built. In this table all functions map fuzzy intervals to fuzzy intervals, except for the inverse function.

A positive fuzzy quantity Q is defined as having its support in the non-negative real line $\mathbb{R}^+$: inf supp(Q) ≥ 0. A negative fuzzy quantity Q is such that sup supp(Q) ≤ 0. The inverse of a positive (resp. negative) fuzzy interval is a positive (resp. negative) fuzzy interval. The exponential of a fuzzy interval is positive. The absolute value of Q, |Q| as shown in Table 1, is a positive fuzzy quantity and is not obtained by (10.27) (remember the non-injectivity of |x|). For more technical details see Kerre (1993), for instance.

Function	*f(x)*	*f(Q)*	*membership function*
Opposite	$-x$	$-Q$	$Q(-x)$
Product by a scalar a ≠ 0	ax	aQ	$Q(x/a)$
Inverse	$1/x$	$1/Q$	$Q(1/x)\ x{\neq}0$
Power (p≠0)	x^p	$(Q)^p$	$Q(x^{1/p})$
Exponential	e^x	e^Q	$Q(\ln x);\ x>0$
Absolute value	$\lvert x\rvert$	$\lvert Q\rvert$	$(Q\cup -Q)\cap\mathbb{R}^+$

Table 10.1

The Four Operations of Arithmetics. First it is useful to know that any commutative (respectively, associative) operation is still commutative (respectively, associative) when it has fuzzy arguments. The sum $Q_1 \oplus Q_2$ of the fuzzy quantities is defined by

$$[Q_1 \oplus Q_2](c) = \sup\{\min(Q_1(c-a), Q_2(a)) : a \in \mathbb{R}\}. \tag{10.28}$$

This is a sup-min counterpart of the convolution operation as said earlier. The set $\mathcal{I}(\mathbb{R})$ of u.s.c. fuzzy intervals, equipped with operation ⊕, is a semi-group with identity 0. The operation ⊕ coincides with the usual addition when operands are real numbers. It is important to notice that – Q is *not* the inverse of Q for operation ⊕, since $-Q \oplus Q \neq 0$, and is only a fuzzy quantity with modal value 0. $(\mathcal{I}(\mathbb{R}), \oplus)$ is a sub-semi-group of $([0,1]^{\mathbb{R}}, \oplus)$.

The upper and lower mean values (Equations (10.8) or (10.9)) are additive with respect to the fuzzy addition, since they satisfy, for u.s.c. fuzzy intervals (Dubois and Prade, 1987b; Heilpern, 1992):

$$E^*(M \oplus N) = E^*(M) + E^*(N);$$
$$E_*(M \oplus N) = E_*(M) + E_*(N).$$

This property is a consequence of the additivity of Choquet integral for the sum of comonotonic functions (Denneberg, 1994), that is functions g and h such that $g(a) > g(b)$ implies $h(a) \geq h(b)$ for all $a, b \in \mathbb{R}$. Indeed the fuzzy addition comes down to the pointwise addition of the inverses of the increasing parts and the decreasing parts of $M(\cdot)$ and $N(\cdot)$, as shown from (10.23). As a consequence, the middle point of the mean interval also satisfies

$$\bar{E}(M \oplus N) = \bar{E}(M) \oplus \bar{E}(N),$$

contrary to other characteristic values such as center-of-area, MOM, etc. (see Section 10.2.2). This linearity property (we also have $\bar{E}(aM) = a\bar{E}(M)$ for a scalar value a) recommends $\bar{E}(M)$ as a proper defuzzified value for a fuzzy interval M. Zhao and Govind (1991a,c) notice this linearity property for the center of area of so-called regular fuzzy intervals whose core contain this center of area. This is because the latter then coincides with $\bar{E}(M)$. The authors conclude on the interest of regular fuzzy intervals, while we rather advocate the worth of $\bar{E}(M)$. Another type of fuzzy intervals is considered by Fullér (1990): Lipschitzian fuzzy intervals M such that there is positive constant k such that $|M(a) - M(a')| \leq k|a - a'|$. He proves that the class of Lipschitzian fuzzy intervals is closed under fuzzy addition and scalar multiplication. This tool is useful to study the stability of fuzzy operations.

The *subtraction* $Q_1(-)Q_2$ of fuzzy quantities is defined by changing $c - a$ into $c + a$ in (28). Of course, $Q_1(-)Q_2 = Q_1 \oplus (-Q_2)$.

The *product* $Q_1 \otimes Q_2$ of fuzzy quantities is defined by:

$$[Q_1 \otimes Q_2](c) = \sup \{\min((Q_1(c/a), Q_2(a))): a \in \mathbb{R} \setminus \{0\}\} \text{ if } c \neq 0$$
$$[Q_1 \otimes Q_2](0) = \max(Q_1(0), Q_2(0)). \quad (10.29)$$

The set $\mathcal{I}(\mathbb{R}^+)$ of positive u.s.c. fuzzy intervals, equipped with the fuzzy product $\otimes$ is a semi-group with identity 1. On real numbers, $\otimes$ coincides with the usual product. The fuzzy inverse $1/Q$ is not a group-inverse since $Q \otimes (1/Q)$ is only a fuzzy quantity with modal value 1. Further, 0 is an absorbing element ($0 \otimes Q = 0$). Lastly,

$$Q_1 \otimes Q_2 = (-Q_1) \otimes (-Q_2); \ (-Q_1) \otimes Q_2 = Q_1 \otimes (-Q_2) = -(Q_1 \otimes Q_2).$$

The distributivity of $\otimes$ over $\oplus$ is not always valid; only a weak distributivity remains:

$$Q_1 \otimes (Q_2 \oplus Q_3) \subseteq ((Q_1 \otimes Q_2) \oplus (Q_1 \otimes Q_3)) \quad (10.30)$$

Distributivity is valid in the following special situations:

i) Q_1 is a real number;

ii) Q_1, Q_2, Q_3 are u.s.c. fuzzy intervals, Q_2 and Q_3 are both positive or negative;

iii) Q_1, Q_2, Q_3 are u.s.c. fuzzy intervals, Q_2 and Q_3 are symmetric intervals ($Q_2 = -Q_2$, $Q_3 = -Q_3$).

These results are based on the commutativity of operations with level-cutting (10.21) and have counterparts in interval analysis (Moore, 1979).

The quotient $Q_1(/)Q_2$ is defined by changing c/a into c·a in the first part of (10.29), and $[Q_1 (/) Q_2](0) = Q_1(0)$. It can be checked that $Q_1(/)Q_2 = Q_1 \otimes (1/Q_2)$. When Q_1 and Q_2 are both positive or negative, then if Q_1 and Q_2 are fuzzy intervals, so is $Q_1(/)Q_2$.

Addition and product of fuzzy intervals can be instrumental for computing the composition of a fuzzy set and a fuzzy proximity relation expressing tolerance. A proximity relation on $\mathbb{R}$ is a fuzzy subset P of $\mathbb{R} \times \mathbb{R}$ such that $P(x, x) = 1$ (reflexivity) and $P(x,y) = P(y, x)$ (symmetry). A quite general type of proximity relation on $\mathbb{R}$ can be defined by $P(x, y) = M(|x - y|)$, where M is a symmetric fuzzy interval such that $M(0) = 1$. The fuzzy set of numbers close to a fuzzy quantity Q in the sense of P, is defined by $Q \circ P$ such that

$$(Q \circ P)(z) = \Pi_P(\{z\} \times Q) = \sup \{\min(P(z, a), M(a)): a \in \mathbb{R}\}.$$

It is easy to see that $Q \circ P = M \oplus Q$ (Dubois and Prade, 1988a). Similarly, if we consider a fuzzy proximity relation R expressing relative neighborhood, that is defined from a fuzzy interval M as $R(x, y) = M(\min(x/y, y/x))$ where $M(1) = 1$ and $M(0) = 0$, then for any fuzzy quantity Q, $Q \circ R = M \otimes Q$ (See Dubois and Prade, 1989b).

The Maximum and Minimum Operations. The two functions max(x, y) and min(x, y) are non-decreasing; for interval arguments the interval analysis equation (10.25) yields

$$\max([a, a'], [b, b']) = [\max(a, b), \max(a', b')]$$
$$\min([a, a'], [b, b']) = [\min(a, b), \min(a', b')].$$

Hence the obvious construction of the fuzzy extensions of the maximum and minimum operations using level-cuts, when M, N are u.s.c. fuzzy intervals: intervals $\widetilde{\max}(M, N)$ and $\widetilde{\min}(M, N)$ are pictured on Fig. 3 where $\widetilde{\max}$ and $\widetilde{\min}$ denote the extended maximum and minimum operations, respectively. See

(Mizumoto and Tanaka, 1976 c), (Dubois and Prade, 1979a), (Ambrosio and Martini, 1984).

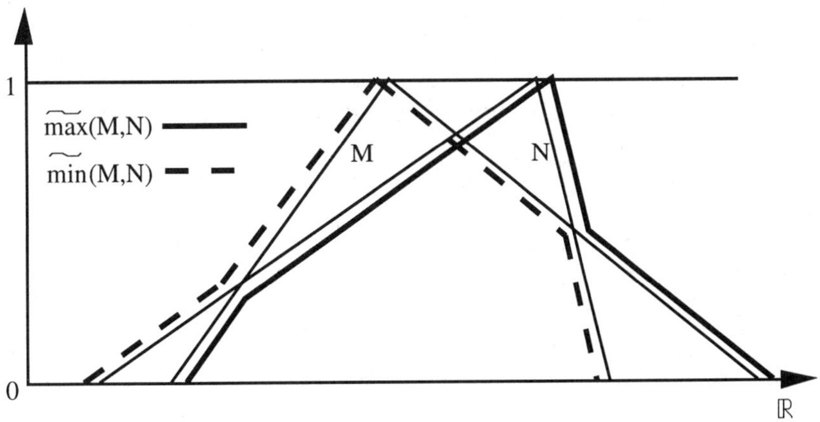

Figure 10.3

Operations $\widetilde{\max}$ and $\widetilde{\min}$ are commutative and associative, and

$$\widetilde{\max}(-M, -N) = -\widetilde{\min}(M, N).$$

On the set of u.s.c. fuzzy intervals, $\widetilde{\max}$ and $\widetilde{\min}$ are mutually distributive, idempotent, and satisfy the following properties:

$$\widetilde{\max}(M, N) \oplus \widetilde{\min}(M, N) = M \oplus N;$$

$$M \oplus \widetilde{\min}(N_1, N_2) = \widetilde{\min}(M \oplus N_1, M \oplus N_2);$$

$$M \oplus \widetilde{\max}(N_1, N_2) = \widetilde{\max}(M \oplus N_1, M \oplus N_2);$$

$$\widetilde{\max}(M, N) = M \text{ if and only if } \widetilde{\min}(M, N) = N.$$

Operations $\widetilde{\max}$ and $\widetilde{\min}$ enable fuzzy set-theoretic operations for type-2-fuzzy sets (i.e., with fuzzy membership grades) to be defined (Mizumoto and Tanaka, 1976c, Dubois and Prade, 1979b) and play a role in the comparison of fuzzy intervals.

10.3.4 Proper and Improper Representations of Functions

There always exist more than one possible mathematical representation of a real-based real-valued function. Let f and g be two such functions, with f = g on ℝ. It is easy to see that generally, their extensions with fuzzy arguments may be distinct fuzzy quantities. This fact is exemplified by the lack of distributivity of the fuzzy product over fuzzy addition (30). Moore (1979) is aware of this difficulty, which also appears with interval mathematics.

Let $f(x, y)$ be a mathematical expression of a two-place function from $\mathbb{R}^2$ to $\mathbb{R}$, where the variables x and y appear only once. $f(x, y)$ is then said to be a proper representation of this function. Let $g(x_1, x_2, \ldots x_p, y_1, y_2, \ldots y_q)$ be the mathematical expression of a function involving p + q variables, such that

$$\text{if } \forall i,\ x_i = x,\ \forall j,\ y_j = y, \text{ then } f = g$$

then, g is said to be an improper representation of f. For instance $g(x_1, x_2, x_3, y_1) = f(x_1, y_1) - x_2 + x_3$. Let $g(Q_1, Q_2)$ be the fuzzy set obtained by substituting Q_1 for $x_1 \ldots x_p$ and Q_2 for $y_1 \ldots y_q$. Then the following inclusion generally holds:

$$f(Q_1, Q_2) \subseteq g(Q_1, Q_2).$$

In the case of $f(x_1, y_1) + x_2 - x_3$, the inclusion is strict because the fuzzy quantity $Q_1(-)Q_1$ is imprecise and has modal value 0. The inclusion means that using improper representations of functions in calculations involving fuzzy quantities, may yield arbitrarily imprecise albeit valid results. The main reason is that in the expression $g(Q_1, Q_2)$, several distinct variables (here x_2, x_3) happen to have the same range. In other words, the same fuzzy quantity appearing several times in a mathematical expression implicitly pertains to distinct variables, but what we really mean is that these variables are the same (not only the ranges).

There are cases, however, when improper representations yield the least imprecise results. It has been proved, for instance, that equality holds in the above inclusion for u.s.c. fuzzy intervals as soon as $f(x, y)$ and $g(x_1 \ldots x_p, y_1 \ldots y_q)$ are both non-decreasing (Dubois and Prade, 1980a, 1987a). In contrast, $f(x_1, y_1) - x_2 + x_3$ is decreasing with some variables and increasing with others even when $f(x_1, y_1)$ is non decreasing. On the other hand, $f(x, y, z) = x(y + z)$ and $f(x, y, z, t) = xy + zt$ are both non-decreasing when their variables range over $[0, +\infty)$. This explains why, with positive fuzzy intervals, the distributivity property (30) holds with equality.

Along the same lines, the identity $M \oplus M = 2M$ is valid when M is a fuzzy interval but not a fuzzy quantity in general. The calculation of expressions of the form

$$s_n(Q) = p_1 \cdot Q \oplus \ldots \oplus p_n \cdot Q,$$

where $\sum p_i = 1$, has drawn the attention of several researchers as a model of 'fuzzy expectation' (Nahmias, 1979; Rao and Rashed, 1981), or in the scope of finding some possibilistic counterpart of the law of large numbers (Stein and Talati, 1981,

Dubois, 1982, Stein, 1983). When Q is a fuzzy interval, then $\forall n$, $s_n(Q) = Q$ (on this question see Hong and Kim, 1998); however, when Q is not convex then there are many instances when $\lim_{n \to +\infty} s_n(Q) = \widehat{Q}$ (the convex hull).

Similarly, the identity $M \otimes M = M^2$ requires M to be a positive fuzzy interval. Note that proper representations of functions in the above sense do not always exist: for instance $f(x, y, z) = xy + yz + zy$ or $\frac{\sum_i a_i x_i}{\sum_i x_i}$. The latter expression was first studied by Baas and Kwakernaak (1977) and several authors since then, and is generally not monotonic at all. Under a fully rigorous approach, such expressions involving fuzzy quantities must be calculated by applying the extension principle to the whole expression; the identity between variables appearing more than once is accounted for under the form of interaction constraints, see Section 10.5. Such mathematical expressions cannot be calculated using elementary operations between fuzzy operands unless the monotonicity property applies. The difficulty of representing functions with more than two fuzzy intervals operands by means of unary and binary operations is extensively studied by Nguyen *et al.* (1997): they give several impossibility theorems for smooth real functions.

10.4 PRACTICAL COMPUTING WITH NON-INTERACTIVE FUZZY INTERVALS

Computing with fuzzy intervals is generally easier than computing with random variables. Moreover such computations make sense either for propagation of incomplete information for purposes of risk assessment for instance, or for preference propagation for design tasks, for instance. Hence the study of practical methods for fuzzy interval analysis is very important for the applicability of fuzzy set theory. Such methods are of two kinds: exact or approximate calculations on the basis of parameterized representations, and approximate methods based on classical interval calculations and level-cutting.

10.4.1 Parameterized Representations of a Fuzzy Interval

A general parameterized representation of fuzzy intervals has been first proposed by Dubois and Prade (1978 a, 1980 a). Let L be any u.s.c. mapping from $[0, +\infty)$ to [0, 1] satisfying the following requirements:

$\forall x > 0$, $L(x) < 1$;
$\forall x < 1$, $L(x) > 0$;
$L(0) = 1$;
Either $L(1) = 0$ or ($L(x) > 0$, $\forall x$, and $\lim_{x \to +\infty} L(x) = 0$).

Under these requirements, L is said to be a *shape* function. Examples of shape-functions are $L(x) = \max(1 - x^p, 0)$, $\max(1 - x, 0)^p$, for $p > 0$, e^{-x}, e^{-x^2}, $\frac{1}{x+1}$. We consider the class of u.s.c. fuzzy intervals whose membership function $M(\cdot)$ can be described by means of two shape functions L and R and four parameters: $\underline{m}$, $\overline{m} \in \mathbb{R} \cup \{-\infty, +\infty\}$, $s, t \in [0, +\infty)$ having the form

$$M(x) = L\left(\frac{\underline{m} - x}{s}\right), \forall x < \underline{m}$$
$$M(x) = 1 \text{ if } x \in [\underline{m}, \overline{m}] \tag{10.31}$$
$$M(x) = R\left(\frac{x - \overline{m}}{t}\right), \forall x > \overline{m}.$$

For $s = 0$ (resp. $t = 0$), $M(x) = 0$ when $x < \underline{m}$ (resp. $x > \overline{m}$), by convention. A fuzzy interval M described by (31) is called an LR fuzzy interval and the notation $M = (\underline{m}, \overline{m}, s, t)_{LR}$ is adopted. L and R are called the left shape and the right shape functions, respectively. The parameters s and t are called the left spread and the right spread, respectively.

This class is very general since it encompasses all u.s.c. fuzzy intervals whose membership functions are such that: $\lim_{x \to +\infty} M(x) \in \{0,1\}$, $\lim_{x \to -\infty} M(x) \in \{0,1\}$, and $\text{core}(M) \neq \emptyset$. It also encompasses bounded-support fuzzy intervals.

The core of M is the interval $[\underline{m}, \overline{m}]$. The value $\underline{m}$ (resp. $\overline{m}$) is called the lower (respectively upper) modal value of M. If M has bounded support, then due to the representation, the closure of the support is $[\underline{m} - s, \overline{m} + t]$. The following conventions are adopted:

If $\underline{m} = \overline{m} = m$, M is called an LR fuzzy number with the notation $M = (m, s, t)_{LR}$.

A real number $m \in \mathbb{R}$ is also denoted $M = (m, m, 0, 0)_{LR} = (m, 0, 0)_{LR}$ (the choice of L and R is irrelevant).

$M = [a, b]$ is denoted $M = (a, b, 0, 0)_{LR}$ (for any L, R).

If M has a trapezoidal or triangular membership function, then $L(x) = R(x) = \max(0, 1-x)$.

If $\forall x > \overline{m}$, $M(x) = 1$, then $M = (\underline{m}, +\infty, s, t)_{LR}$ ($\forall$ R and t).

If $\forall x < \underline{m}$, $M(x) = 1$, then $M = (-\infty, \overline{m}, s, t)_{LR}$ ($\forall$ L and s).

There are basically three kinds of membership functions (notwithstanding that of the real line itself): bell-shaped ones, non-decreasing functions, and non-increasing functions (see Figure 4). The two last types can be viewed as fuzzy thresholds, and correspond to Höhle's (1981) idea of fuzzy numbers. The simplest fuzzy interval (resp. number) is the trapezoidal (resp. triangular) one. Several papers have been devoted to justification of some particular shapes. Pedrycz (1994) advocates the use of triangular membership functions in fuzzy modeling processes. Triangular membership functions can be also justified in terms of a canonic probability/possibility transformation as upper envelopes of the transforms of unimodal probability distributions with bounded supports; see Chapter 7 by Dubois, Nguyen and Prade in this volume. Puncochar *et al.* (1996) prove that a triangular or trapezoidal fuzzy interval is a natural approximation of an infinite sequence of intervals covering a point in the style of Cantor continuum. The idea is that when approximating a point-value by an interval containing it, one must in turn approximate the end points by (smaller) intervals, and so on. Klawonn and Kruse (1993) point out that when equality is replaced by a (fuzzy) similarity relation, singletons of the real line become fuzzy numbers. Triangular fuzzy numbers are precisely recovered when the fuzzy similarity is the dual of a standard metric. Delgado *et al.* (1998a) propose the use of trapezoidal fuzzy intervals as canonical and study the problem of finding a trapezoidal representation of any fuzzy interval having the same characteristic value (after defuzzification), the same specificity and the same degree of fuzziness. Jacas and Recasens (1993), Mares and Mesiar (1996), De Baets *et al.* (1997) considered the idea of generating fuzzy numbers as equivalence classes of particular fuzzy equivalences on the real line. For a given fuzzy number M with core {0}, and a function g monotone: $\mathbb{R} \to \mathbb{R}$, called "source of vagueness", the class of fuzzy intervals M_a, $a \in \mathbb{R}$ is defined by $M_a(b) = M(g(b) - g(a))$.

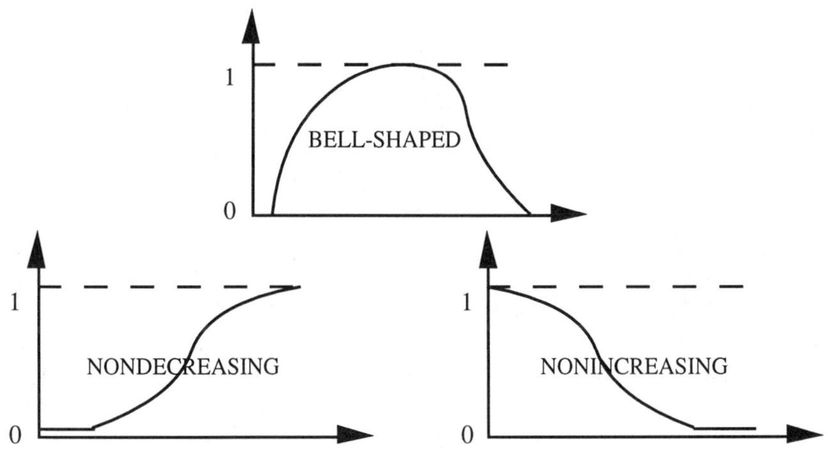

Figure 10.4

NB. A fuzzy interval M with non-empty core cannot be represented as an LR fuzzy interval (for any shapes L and R) when $\lim_{x \to +\infty} M(x) = c_M \in (0,1)$, or $\lim_{x \to -\infty} M(x) = d_M \in (0,1)$. However, for this type of fuzzy intervals similar calculations as with the LR fuzzy intervals can be performed. So, e.g., for the sum

$K = M \oplus N$, the level– cuts equality $K_\lambda = M_\lambda \oplus N_\lambda$ holds for all $\lambda \in (0, 1]$. It is evident that if $c_M = d_M \in (0, 1)$, then $M_\lambda = \mathbb{R}$ for all $\lambda \leq c_M$. Consequently, for any fuzzy interval N, $(M \oplus N)_\lambda = \mathbb{R}$ whenever $\lambda \leq c_M$.

10.4.2 Exact Calculation of the Four Arithmetic Operations

The basic results of this section date back to (Dubois and Prade, 1978a, 1979a, 1980a). Let $M = (\underline{m}, \overline{m}, s, t)_{LR}$, $N = (\underline{n}, \overline{n}, u, v)_{LR}$. For any LR-fuzzy interval, $M = (\underline{m}, \overline{m}, s, t)_{LR}$, it is easy to see that $(M^+)^* = \overline{m} + t\, R^*$ and $(M^-)_* = \underline{m} - t\, L_*$. Note that if L, R are injective (on their support), then $R^* = R^{-1}$ and $L_* = L^{-1}$ are the inverse functions to R and L restricted to the corresponding support, respectively. This is assumed below for simplicity. When a function f is non-decreasing then, the closed form equation (10.23), that enables $f(M, N)$ to be computed, becomes:

$$f(\underline{m} - s \cdot L^{-1}(\lambda), \underline{n} - u \cdot L^{-1}(\lambda)) = c \text{ if } c \leq f(\underline{m}, \underline{n}) \qquad (10.32)$$

$$f(\overline{m} + t \cdot R^{-1}(\lambda), \overline{n} + v \cdot R^{-1}(\lambda)) = c \text{ if } c \geq f(\overline{m}, \overline{n}) \qquad (10.33)$$

i.e., it reduces to two equations with respective unknowns $L^{-1}(\lambda)$ and $R^{-1}(\lambda)$, whose complexity is directly linked to that of f. These equations are of the form:

$$z = f(a + bx, c + dx) \qquad (10.34)$$

where the unknown is x. If $f(x, y) = x + y$, (10.34) is a first-order equation and (10.32–10.33) straightforwardly yield the following important result (Dubois and Prade, 1978 a):

$$M \oplus N = (\underline{m} + \underline{n}, \overline{m} + \overline{n}, s + u, t + v)_{LR}. \qquad (10.35)$$

More generally, for any fuzzy intervals M, N it holds that

$$((M \oplus N)^+)^* = (M^+)^* + (N^+)^* \text{ and } ((M \oplus N)^-)_* = (M^-)_* + (N^-)_*$$

of which equation (10.35) is an immediate consequence. Further $((-M)^+)^* = -(M^-)_*$ and similarly $((-M)^-)_* = -(M^+)^*$ for any fuzzy interval.

If P is a fuzzy proximity relation defined from a symmetric fuzzy interval $M = (-\varepsilon, \varepsilon, \eta, \eta)_{LL}$ (i.e., L = R) as in Section 10. 3.3, and $N = (\underline{n}, \overline{n}, u, v)_{LL}$ then $N \circ P = (\underline{n} - \varepsilon, \overline{n} + \varepsilon, u + \eta, v + \eta)_{LL}$.

Similarly, the subtraction $M(-)N$ is easily calculated if M and N are of opposite types LR and RL, respectively, i.e., $N = (\underline{n}, \overline{n}, u, v)_{RL}$:

$$M(-)\,N = (\underline{m} - \overline{n},\ \overline{m} - \underline{n},\ s + v,\ t + u)_{LR}$$

simply because $-N = (-\overline{n}, -\underline{n}, v, u)_{LR}$. A very strong invariance theorem is thus obtained for the sum of fuzzy intervals. Namely, let $\mathcal{I}_{LR}(\mathbb{R})$ be the set of fuzzy intervals of type LR then:

Proposition 5: *If M, N $\in \mathcal{I}_{LR}(\mathbb{R})$ then $M \oplus N \in \mathcal{I}_{LR}(\mathbb{R})$.*
If $M \in \mathcal{I}_{LR}(\mathbb{R})$, $N \in \mathcal{I}_{RL}(\mathbb{R})$ then $M(-)N \in \mathcal{I}_{LR}(\mathbb{R})$.

This invariance property is far stronger than the one which holds for the sum of random variables where only special distributions, such as Gaussian, are invariant. At the same time when Dubois and Prade (1978a) first proved this result, Nahmias (1978) obtained it for specific shape functions (linear, Gaussian, ...). In probability theory, the sum of two independent Gaussian random variables with mean values m and n and standard deviations σ_m and σ_n is a Gaussian variable with mean value m + n and standard deviation $\sqrt{\sigma_m^2 + \sigma_n^2}$. In contrast, the sum of two Gaussian fuzzy numbers (where $L(x) = R(x) = e^{-x^2}$) with modal values m and n and respective spreads $s = t = \sigma_m$ and $u = v = \sigma_n$ is a Gaussian fuzzy number with modal value m + n, and both spreads are $\sigma_m + \sigma_n > \sqrt{\sigma_m^2 + \sigma_n^2}$.

Of course, Proposition 5 does not hold for product nor quotient. However, equation (10.34) remains simple, for the product of two fuzzy intervals M > 0, N > 0: it is a second-order equation if M and N are both of the LR type (the product is non-decreasing over $[0, +\infty)$). Solving equations (10.32) and (10.33) then yields:

$$[M \otimes N](x) = L\left(\frac{\underline{n}s + \underline{m}u - \sqrt{(\underline{n}s + \underline{m}u)^2 + 4su(x - \underline{mn})}}{2su}\right) \qquad \text{if } x \le \underline{m} \cdot \underline{n}$$

$$[M \otimes N](x) = R\left(\frac{-\overline{n}t - \overline{m}v + \sqrt{(\overline{n}t + \overline{m}v)^2 + 4tv(x - \overline{mn})}}{2tv}\right) \qquad \text{if } x \ge \overline{m} \cdot \overline{n}$$

This is because for any nonnegative fuzzy intervals M, N:

$$((M \otimes N)^+)^* = (M^+)^* \cdot (N^+)^* \text{ and } ((M \otimes N)^-)_* = (M^-)_* \cdot (N^-)_* .$$

For $f(x, y) = x/y$ (a hybrid operation over $[0, +\infty)^2$, if M > 0, N > 0 are of type LR and RL, respectively, then equation (10.34) is once again a first-order equation. Solving equations (10.32-10.33) yield:

$$[M(/)N](x) = L\left(\frac{\underline{m} - \overline{n}x}{s + vx}\right) \qquad \text{if } x \le \underline{m} / \overline{n}$$

$$[M(/)N](x) = R\left(\frac{\underline{n}x - \overline{m}}{t + ux}\right) \qquad \text{if } x \geq \overline{m} / \underline{n}\,.$$

Note that equation (10.34) is also a second-order equation for sums of products of the form $\Sigma_i\, a_i{\cdot}M_i{\otimes}N_i$, where M_i and N_i are all positive fuzzy intervals of type LR and the a_i's are positive real numbers. On the other hand, the calculation of products of more than two fuzzy intervals requires the analytical solution of polynomial equations of order higher than 2. Nevertheless, for non-negative fuzzy intervals, $((1/M)^+)^* = 1/(M^-)_*$ and $((1/M)^-)_* = 1/(M^+)^*$. The division of fuzzy intervals also derives from this remark.

N.B.: Several properties of classical real functions also apply to fuzzy intervals. So, e.g., for all x, y $\in$ [0, +∞)

$$\ln x + \ln y = \ln x{\cdot}y; \;\; x{\cdot}y = \exp(\ln x + \ln y).$$

For positive fuzzy intervals M, N, it also holds:

$$M{\otimes}N = \exp(\ln M \oplus \ln N).$$

Consequently, the multiplication of fuzzy intervals can be derived from their addition.

10.4.3 Approximate Parametric Calculation of Functions of Fuzzy Intervals

Most of the time, if M and N are fuzzy intervals of type LR, f(M, N) is not. In order to preserve the representation, it is possible to resort to approximations of type LR. First, if f is differentiable, and L, R are continuous, then a first-order approximation of equations (10.32-10.33) can be made when λ is close to 1, i.e., $L^{-1}(\lambda)$ or $R^{-1}(\lambda)$ are close to zero; when f is non-decreasing, and $c \leq f(\underline{m},\underline{n})$; the following linear equation is obtained, with unknown $\omega = L^{-1}(\lambda)$:

$$c \approx f(\underline{m},\underline{n}) - (s{\cdot}f'_x(\underline{m},\underline{n}) + u{\cdot}f'_y(\underline{m},\underline{n})){\cdot}\omega$$

where f'_x and f'_y denote the partial derivatives of f(x, y).

In this case an approximation of type LR, for $f^{\#}$(M, N) is finally built and proves to be (Dubois and Prade 1987a):

$$f^{\#}(M, N) = (f(\underline{m},\underline{n}),\, f(\overline{m},\overline{n}),\, s{\cdot}f'_x(\underline{m},\underline{n}) + u{\cdot}f'_y(\underline{m},\underline{n}),\, t{\cdot}f'_x(\overline{m},\overline{n}) + v{\cdot}f'_y(\overline{m},\overline{n}))_{LR} \quad (10.36)$$

Equation 36 generalizes approximate formulae provided by Dubois and Prade (1978a) for the product and the quotient, and points out the link between fuzzy and non-fuzzy interval analysis. For instance

$$(\underline{m}, \overline{m}, s, t)_{LR} \otimes (\underline{n}, \overline{n}, u, v)_{LR} \approx (\underline{m}\cdot\underline{n}, \overline{m}\cdot\overline{n}), s\cdot\underline{n} + u\cdot\underline{m}, t\cdot\overline{n} + v\cdot\overline{m})_{LR}$$

which holds if spreads are small compared to modal values. It is easy to see that the membership function of $f^{\#}(M, N)$ is tangent to the one of $f(M,N)$ at the points $f(\underline{m},\underline{n})$, $f(\overline{m},\overline{n})$. More generally, it is possible to write first-order approximations of equations (10.32-10.33) in the neighborhood of the boundaries of any level-cut of $f(M, N)$, i.e., at points $f(\underline{m}_\lambda, \underline{n}_\lambda)$, $f(\overline{m}_\lambda, \overline{n}_\lambda)$, $\lambda \in (0,1]$. We may thus derive approximations denoted $f^{\#}_\lambda(M, N)$. Especially interesting are the expressions of $f^{\#}_\lambda(M, N)$ when λ is close to zero, i.e., in the neighborhood of the support boundaries, when the support is bounded.

$f^{\#}_\lambda(M, N)$ is a local approximation. A global LR-type approximation, denoted $f^{\approx}(M, N)$, can be obtained as the fuzzy interval of the LR type whose core is $[f(\underline{m},\underline{n}), f(\overline{m},\overline{n})]$ and whose support is $[f(\underline{m} - s, \underline{n} - u), f(\overline{m} + t, \overline{n} + v)]$ (when bounded); this approach, contrary to the local one, can be applied to $f(x, y) = \min(x, y)$ or $\max(x, y)$ since it does not require differentiability. For some types of operations f, the membership functions of $f^{\#}(M, N)$, $f^{\approx}(M, N)$, $f^{\#}_0(M, N)$ define an area which contains $f(M, N)$; this remark makes it possible to quantify the error involved in the proposed approximations. See Jimenez and Rivas (1998) for the estimation of error induced by linear approximations of fuzzy intervals.

An improved approximation scheme has recently been proposed by Giachetti (1996), see also (Giachetti and Young, 1997a, b). A positive fuzzy interval M is represented as a six-tuple, M = (a, b, c, λ, ρ, n), where b is the core of M, (a, c) = supp(M) is the support of M, λ and ρ are called spread ratios and are respectively λ = b/a, and ρ = c/a, and n is the order of the polynomial expression for the membership function. For a positive triangular fuzzy number M, n = 1. When computing the product or quotient of triangular fuzzy numbers, the idea of Giachetti is to improve the approximation by adding to it a correcting polynomial term of suitable order. More details, including the error analysis for the simple approximations, can be found in (Giachetti and Young, 1997a, b).

A more general representation consisting of piecewise approximations of fuzzy quantities can be efficiently exploited in computations, as proposed by Kerre *et al.* (1988). This type of fuzzy quantity is simple enough to be represented in a compact way while allowing for fast execution of operations. Moreover it remains under a stable form across the calculations.

A piecewise linear fuzzy quantity may be represented by means of a finite list

$$Q = ((x_1\ l_1\ f_1\ r_1)\ (x_2\ l_2\ f_2\ r_2)\ \dots\ (x_n\ l_n\ f_n\ r_n)),$$

where $\forall i \in \{1,2, \dots n\}\ x_i \in \mathbb{R}$, and $\forall i \in \{1,2, \dots n-1\}\ x_i < x_{i+1}$ and

$$Q(x) = l_1 \qquad \text{if } x \in (-\infty, x_1)$$

$$Q(x) = f_i \qquad \text{if } x = x_i\ \forall i \in \{1,2, \dots n\}$$

$$Q(x) = r_i + \frac{(x - x_i)(l_{i+1} - r_i)}{x_{i+1} - x_i} \qquad \text{if } x \in (x_i, x_{i+1})$$

$$Q(x) = r_n \qquad \text{if } x \in (x_n, +\infty).$$

Moreover $\lim_{x \uparrow x_i} Q(x) = l_i$ and $\lim_{x \downarrow x_i} Q(x) = r_i$. The real numbers x_1, x_2, ... x_n are called the breakpoints, f_i is the value of $Q(\cdot)$ at x_i, l_i is the left limit of $Q(\cdot)$ at x_i, r_i is the right limit of $Q(\cdot)$ at x_i, and $Q(\cdot)$ is linear between two breakpoints. The list is supposed to be minimal, that is, the line segment joining the points (x_{i-1}, r_{i-1}) and (x_{i+1}, r_{i+1}) does not contain (x_i, f_i) such that $f_i = r_i = l_i$. Particular instances of piecewise linear fuzzy quantities are:

- a real number 'a' has list representation ((a 0 1 0));
- a triangular fuzzy number with support (b, c) and core {a} has list representation ((b 0 0 0) (a 1 1 1) (c 0 0 0));
- a trapezoidal fuzzy interval with support (c, d) and core [a, b] has list representation ((c 0 0 0) (a 1 1 1) (b 1 1 1) (d 0 0 0)).

Triangular and trapezoidal fuzzy intervals are not closed for absolute value, complementation, union, intersection, fuzzy maximum and minimum, but piecewise representations are so. Namely if M and N are trapezoidal fuzzy intervals then $|M|$, M^c, $M \cap N$, $M \cup N$, $\widetilde{\max}(M, N)$, $\underset{\sim}{\min}(M, N)$ may fail to be trapezoidal fuzzy intervals. However if M and N are piecewise linear fuzzy quantities then the fuzzy quantities aM, $|M|$, M^c, $M \cap N$, $M \cup N$, $\widetilde{\max}(M, N)$, $\underset{\sim}{\min}(M, N)$ $M \oplus N$, $M(-)N$ are piecewise linear. Clearly this feature is not preserved by other non-linear operations such as division, product, inverse, logarithm, etc. However the results of these fuzzy operations can be conveniently approximated by piecewise linear quantities. See Kerre and Van Schooten (1988) for more details and an algorithm for the implementation of binary operations on this kind of representation. Steyaert *et al.* (1995) propose a practical implementation of piecewise linear quantities in C++ for unary and binary operations, that can be used on a PC. This implementation is based on continuity preservation properties proved in Baekeland and Kerre (1988).

10.4.4 *Approximate Calculation of Functions of Fuzzy Intervals Using Level-Cuts*

When $f(M, N)$ cannot be explicitly obtained in an analytic way, one may obtain a sampling of its membership function using level-cuts and applying interval analysis, taking advantage of the level-cut commutation property (10.21). The level-cut representation clearly comes down to a piecewise linear approximation. However in general, functions to which fuzzy arithmetic apply are not simply monotonic as assumed in the previous sections. Hence analytical approaches presented so far have limited applicability in practice. Dong and Wong (1987), and also Dong and Shah (1987), Dong *et al.* (1985) have proposed the vertex method, that apply to functions that are continuous in the Cartesian product of the supports of the fuzzy intervals involved.

Namely, consider a continuous function $f(x_1,\ldots,x_n)$ where for all $i = 1,\ldots, n$, x_i is assigned the fuzzy interval M_i. Consider a membership grade λ and let the level-cut $(M_i)_\lambda$ be the interval $[a_i, b_i]$. The vertices of the hypercube H defined as Cartesian product $\times_{i=1,\ldots,n} [a_i, b_i]$ are denoted v_j, for $j = 1,\ldots, 2^n$ and form the set $V_n(\lambda)$ of ridge points. Each vertex in $V_n(\lambda)$ is a n-tuple of values whose ith component is a_i or b_i. Assume that neither the infimum of f nor its supremum lie inside H. It means that they are attained at one of the vertices. Hence the remarkable property that generalizes (10.22) follows:

$$(f(M_1,\ldots, M_n))_\lambda = [\min_j f(v_j), \max_j f(v_j)].$$

This method has been successfully applied to weighted average functions of the form $f(y_1 \ldots y_n) = \dfrac{\sum_{i=1}^{n} c_i y_i}{\sum_{i=1}^{n} y_i}$ where the y_i's are restricted by fuzzy intervals W_i which act as fuzzy weights. Baas and Kwakernaak (1977) had provided a numerical technique for the computation of $f(W_1 \ldots W_n)$, based on comparing derivative signs for f and the membership functions. The vertex method is much simpler although of exponential complexity with the number of arguments of f; this complexity is cut down if the monotonicity of f with respect to some of its arguments is known (as for the c_i's in the weighted average if they become fuzzy variables). Liou and Wang (1992b) have shown how to drastically reduce the complexity of the calculation: starting from $l_0 = f(a_1 \ldots a_n)$ as a first estimation of the lower bound of $f(y_1 \ldots y_n)$, one looks for index set $I = \{i: c_i < f(a_1 \ldots a_n)\}$. If $i \in I$, then the authors prove that $f(a_1 \ldots a_{i-1}, b_i, a_{i+1} \ldots a_n) < f(a_1 \ldots a_n)$. Then let $i \in I$ be the index for which the substitution of a_i by b_i gives the greatest decrease of f. Substitute a_i by b_i in $f(a_1 \ldots a_n)$ and obtain the new bound $l_1 < l_0$, look for the new index set $I = \{i: c_i < l_1\}$ and so on. The process stops in a finite number of steps (until I = Ø) and the correct lower bound of $f(y_1 \ldots y_n)$ is found. A similar process works for the upper bound. Lee and Park (1997) noticed that the lower bound of $f(y_1 \ldots y_n)$ is of the form $f(a_1 \ldots a_i, b_{i+1} \ldots b_n)$ provided that the c_i's are ranked in non-decreasing order and that index i is the unique one such that $f(a_1 \ldots a_i, b_{i+1} \ldots b_n) - c_i > 0$ and $f(a_1 \ldots a_{i+1}, b_{i+2} \ldots b_n) - c_{i+1} \leq 0$. A similar result holds for the upper bound. This remark leads to an exact procedure that is of complexity O(n·logn). Guh *et al.* (1996) have further improved Liou and Wang (1992b)' method, resulting in an exact procedure of complexity O(n). Some authors (Juang *et al.* 1991) have suggested using a Monte-Carlo method for computing the fuzzy weighted average. This approach requires that membership functions be changed into cumulative distribution functions; both the approach and the results are debatable (see the discussion by Chen *et al.* 1998).

The level-cutting approach combined with the vertex method has many advantages: it is general, in particular it copes with multiple appearance of variables in the expression of the function; it comes down to standard interval computation; it applies to many functions; exact points of the resulting membership function are obtained, twice as many as chosen level-cuts. Using the support, the core and the 0.5-level-cut, it comes down to using a six parameter piecewise linear fuzzy interval as used by Steyaert *et al.* (1995). Some researchers (e.g., Schmucker, 1984) have been tempted to discretize the support of the fuzzy intervals instead of their range, and perform the extension principle explicitly, searching for a maximum on a finite set. However this method gives rise to computational anomalies (Dubois and Prade, 1979a; Dong and Wong, 1987, for instance) such as a loss of convexity of the obtained piecewise linear approximation.

The limitations of the vertex method have been discussed in detail by Yang *et al.* (1993). These authors define a set of remarkable points e_k in the already introduced hypercube $H = \times_{i=1,\ldots,n} [a_i, b_i]$. They are called "poles" and contain the possible local extrema of the function f on H. In H there may be some particular zones of dimension n –1 where a partial derivative f'_i of f vanishes. Yang *et al.* distinguish between uniform spaces, that is a set H containing no particular zone, quasi-uniform spaces containing particular zones parallel to the corresponding axis (x_i for $f'_i = 0$), and non-uniform spaces otherwise. A vertex of H that is a pole is called a normal pole and it is a particular pole otherwise. But not all vertices are poles. Any local extremum of f on H is a pole. It is then enough to apply the vertex method to poles so as to get the interval $f(H)$. In a uniform space there are only two poles that are vertices of H. In a quasi uniform space, there are no particular poles, only vertices. Linear functions f always induce uniform spaces. Rational fractions whose numerators and denominators are sums of products of independent variables, as well as weighted averages, induce quasi-uniform spaces. In the non-uniform case, poles may be interior extrema. In that case, discontinuous membership functions may result, due to the presence of such extrema as shown by Woods et al, (1992). Moreover the vertex method does not apply to the case when the hypercube H contain singularities of the function f (for instance sin 1/x, which is discussed by these authors). Generic programming libraries for computing with fuzzy intervals using the vertex method and its improvements have been described by Anile *et al.* (1995).

10.5 ALTERNATIVE FUZZY INTERVAL CALCULI

In the two previous sections, the fuzzy quantities which serve as operands in mathematical expressions to be evaluated supposedly correspond to unrelated variables. When this is no longer true, the extension principle must be modified so as to include supplementary constraints which express relationships between variables. Such relationships may be just functional links or inequalities which come from physical considerations; but they may also stem from weaker definitions of the Cartesian product appearing in the extension principle involving several fuzzy quantities as in (10.19). The assumption of noninteractivity of variables prevents error compensation in fuzzy interval analysis, but implies that the results of fuzzy interval calculations may become very imprecise; including interactivity in the form of supplementary constraints, or by choosing a triangular norm stricter than minimum, may produce more precise results, although at the expense of increased computational difficulties. The calculus of fuzzy quantities, whether interactive or not, is 'pessimistic' since it propagates all the imprecision that bears on the data. The last point of this section deals with an ' optimistic ' view of fuzzy interval analysis. This view can be regarded as a tool for coping with the lack of inverses in fuzzy arithmetics. It has indeed emerged in connection with the problem of solving equations where some coefficients are fuzzy. This time, we compute the fuzzy ranges of parameters under a constraint of maximal allowed imprecision. It is important to note that all of these alternative treatments of fuzzy quantities remain within the range of possibility theory.

10.5.1 Fuzzy Interval Calculations with Linked Variables

Assume a link between two variables x and y exists under the form of a domain $D \subseteq \mathbb{R}^2$ which acts as the actual range of (x, y); if moreover, x and y have respective fuzzy ranges M and N, then in order to calculate the range of f, the extension principle (10.19) must be adapted as follows: $\forall$ z,

$$[f(M, N \mid D)](z) = \sup\{\min(M(x), N(y)) : z = f(x, y); (x, y) \in D\} \qquad (10.37)$$

Notice that the calculation of f(M, N) from an improper representation can be carried out using (37). For instance the square of the fuzzy interval M, i.e., M^2, is obtained by letting ((x,y) = x·y and D = {(x, y): x = y } in (37). Such constrained fuzzy interval analysis compensates for some of the defects of fuzzy arithmetics such as the lack of inverse M(–)M ≠ 0, M⊗(1/M) ≠ 1, etc. Indeed these equalities are restored if the constraint x + y = 0 (resp. x.y = 1) is added to the constraints involved in the definition of fuzzy subtraction (resp. division); more discussion along that line is in (Klir, 1997). More generally, D can be a fuzzy set of $\mathbb{R}^2$ and equation (37) thus becomes, in the scope of possibility theory,

$$[f(M, N \mid D)](z) = \Pi_{M \times N}(f^{-1}(\{z\}) \cap D) = \sup\{\min(M(x), N(y), D(x, y)): z = f(x, y)\}.$$

The study of properties, and of course the calculation of fuzzy quantities of the form $f(M,N|D)$, can be very difficult, and belongs to the field of nonlinear optimisation. Here we consider only crisp forms of interaction between involved fuzzy variables. The case of fuzzy interaction is considered in the two subsequent subsections.

Linear Interaction. One interesting and easy-to-tackle example of non-fuzzy interaction is that of linear interaction between variables, i.e., when D is described by a linear equality. This case is addressed in Dubois and Prade (1981). Let $x_1 \ldots x_n$ be variables, each associated with a fuzzy range $M_1 \ldots M_n$, respectively. We want to calculate the possibility distribution associated with the variable $z = \sum_{i=1,\ldots n} a_i \cdot x_i$, where the a_i's are real numbers and the x_i's are linked through the linear normalisation constraint $\sum_{i=1,\ldots n} x_i = 1$ which defines the domain D. Let N be the fuzzy range of z. The following results are valid, see Dubois and Prade (1981, 1982a):

- $\mathrm{hgt}(N) = \sup N(z) = [M_1 \oplus \ldots \oplus M_n](1)$. When this value is less than 1, then the M_i's can be truncated consistently with Proposition 3, i.e., it is possible to change $M_i(x)$ into $\min(\mathrm{hgt}(N), M_i(x))$ without altering the result. When $\mathrm{hgt}(N) = 1$ it means that there are n values $b_1 \ldots b_n$ summing up to 1, such that $\forall$ i, $b_i \in \mathrm{core}(M_i)$.

- If the M_i's are u.s.c. fuzzy intervals with bounded support, then N is also an u.s.c. fuzzy interval with bounded support, and its level-cuts can be found from the level-cuts of the M_i's for levels at most equal to hgt(N). Namely, assume that $0 \le a_1 \le a_2 \le \ldots \le a_n$ and $M_{i\lambda} = [b_i, B_i]$ $\forall$i. Then the following results hold, see Dubois and Prade (1981, 1982a):

$$\inf N_\lambda = \inf\{\sum_{i=1}^{n} a_i \cdot x_i: x_i \in [b_i, B_i]\ \forall i; \sum_{i=1}^{n} x_i = 1\}$$
$$= \max_{k=1,n}\{\sum_{j=1}^{k-1} B_j \cdot a_j + (1 - \sum_{j=1}^{k-1} B_j - \sum_{j=k+1}^{n} b_j)a_k + \sum_{j=k+1}^{n} b_j \cdot a_j \quad (10.38)$$

$$\sup N_\lambda = \sup\{\sum_{i=1}^{n} a_i \cdot x_i: x_i \in [b_i, B_i]\ \forall i; \sum_{i=1}^{n} x_i = 1\}$$
$$= \min_{k=1,n}\{\sum_{j=1}^{k-1} b_j \cdot a_j + (1 - \sum_{j=1}^{k-1} b_j - \sum_{j=k+1}^{n} B_j)a_k + \sum_{j=k+1}^{n} B_j \cdot a_j\}. \quad (10.39)$$

These results are valid for unbounded support fuzzy intervals if all λ–cuts are bounded as indicated in Section 10.3.2. In equations (10.38) and (10.39) summations of the form $\sum_{j=1}^{k-1}$ (respectively $\sum_{j=k+1}^{n}$) are cancelled when k = 1 (respectively, k = n).

• If the M_i's are fuzzy intervals of type L-L, then N is also of type L-L when hgt(N) = 1 (otherwise it is a truncated L-L fuzzy interval, and subnormal). It is then enough to apply (10.38-10.39) to the cores and supports of the M_i's as described by the parameters of the L-L representation.

• When the x_i's are linked by a general linear equality, this equality can be reduced to $\Sigma_{i=1,\ldots n}\, y_i = 1$, changing the x_i's into y_i's of the form $y_i = k_i \cdot x_i$.

• When the a_i's are themselves variables bounded by fuzzy intervals A_i, then under the assumptions of the corollary of Proposition 1, N_λ can be obtained by means of the level-cuts$\{(A_{i\lambda}, M_{i\lambda})$: i=1,...,n$\}$. More precisely, inf$N_\lambda$ (respectively, supN_λ) is found using (10.38) (respectively, (10.39)), with a_i = inf$A_{i\lambda}$ (respectively, sup$A_{i\lambda}$), see Dubois and Prade (1982a).

The attractiveness of the calculated expression lies in its interpretation as a fuzzy discrete mathematical expectation with imprecise probability values (the M_i's), where the underlying random variable may itself take fuzzy values (the A_i's). The results encompass those of Nahmias (1979) or Kwakernaak (1978, 1979), where only the a_i's are variables with fuzzy ranges, the x_i's being constant probabilities (hence no interactivity constraint is involved). The case of imprecise probability values is also encompassed, the latter being understood as fuzzy intervals lying in [0, 1], associated with variables x_i which satisfy the normalisation condition of probabilities. Dong and Wong (1989) note that instead of using the above explicit calculation of this discrete fuzzy mathematical expectation, the vertex method can be applied to the convex polyhedron $\{(x_1 \ldots x_n)$: $x_i \in [b_i, B_i]$ for i = 1...n; $\Sigma_{i=1,\ldots n}\, x_i = 1\}$.

Interactive fuzzy interval analysis more generally concerns the possibility of adding pieces of knowledge to fuzzy arithmetics. For instance performing the sum of fuzzy variables that are known to be close to each other. Then the interactivity domain can be chosen as a crisp or fuzzy proximity relation (see Yager and Kelman (1996), Klir (1997), and more recently Dubois *et al.* (1998) for an application to data fusion of "close" values). Another use of interactive fuzzy arithmetics is made by Pan and Yuan (1997) who apply Bayes theorem when prior probabilities or likelihoods are fuzzy intervals.

10.5.2 Additions of Fuzzy Intervals in the Sense of a Triangular Norm

In the regular formulation of the extension principle involving unrelated variables, the underlying possibility distribution is the Cartesian product of the fuzzy ranges of the variables, as defined in Section 10.2.3. A more general model of a Cartesian product (and also intersection) of fuzzy intervals is obtained using a triangular norm (Dubois and Prade, 1981). It corresponds to a weak form of interaction since the core of the generalized fuzzy Cartesian product is still a standard Cartesian product. Interaction occurs only for smaller levels of membership.

Definitions. A triangular norm (t-norm in short, see, e.g., Schweizer and Sklar (1963, 1983), Klement *et al.* (1999), and Chapter 2 by Fodor and Yager in this volume) is a binary operation $*$ in the unit interval, such that $([0, 1], *)$ is a totally ordered Abelian semigroup with neutral element 1. Therefore $*: [0, 1]^2 \to [0, 1]$ is a t-norm if:

i) $*$ is commutative
ii) $*$ is associative
iii) $*$ is non-decreasing in both places (it is enough in one place due to the commutativity)
iv) $\alpha * 1 = \alpha$, for all $\alpha \in [0, 1]$. (Consequently 0 is a zero element, that is $0 * \alpha = 0$ for all $\alpha \in [0, 1]$).

The four basic triangular norms are the minimum operation, the product, the Lukasiewicz t-norm $\alpha *_L \beta = \max(0, \alpha + \beta - 1)$ (also called bounded, or linear, product) and the least t-norm denoted $*_W$ such that $\alpha *_W \beta = 0$ on the half-open square $[0, 1)^2$, i.e., except for condition (iv). For each t-norm $*$, we have $*_W \leq * \leq \min$. All continuous t-norms can be built upon the three basic t-norms min, product and $*_L$. A specific role is played by continuous Archimedean t-norms (this is when $\alpha * \alpha < \alpha$ on the open unit interval (0, 1)), which are transforms of either the product or the bounded product $*_L$. A continuous Archimedean t-norm always admits a representation of the form:

$$\alpha * \beta = \phi^{-1} (\min(\phi(0), \phi(\alpha) + \phi(\beta))), \alpha, \beta \in [0, 1]$$

where $\phi: [0, 1] \to [0,+\infty)$ is a strictly decreasing continuous function with $\phi(1) = 0$. The function ϕ is called an additive generator of $*$ and it is unique up to a positive multiplicative constant (Ling, 1965; Schweizer and Sklar, 1983). A continuous t-norm $*$ is called strict if $\phi(0) = +\infty$ and then it is a transform of the product. An Archimedean t-norm $*$ is called nilpotent if $\phi(0) < +\infty$ and then it is a transform of $*_L$.

On the other hand, both limit t-norms min and can be approximated by the continuous Archimedean t-norms, either nilpotent or strict ones (Dombi, 1986). In fact any continuous t-norm, as well as the least one $*_W$, can be approximated by continuous Archimedean t-norms (see Jenei and Fodor, 1998; Nguyen *et al.*, 1998, Mesiar, 1998).

Triangular norm expressions $\alpha * \beta$ are a multiple-valued extension of intersection operations when $\alpha, \beta \in [0, 1]$ are membership grades. A more restrictive formulation of the extension principle can thus be obtained, substituting $*$ to 'min' in equation (19). The restriction on the values of operands is all the more drastic as the distance between the t-norm $*$ and the minimum operation is bigger. In particular, if $* = *_W$ then only one of the involved variables may leave the core of its range at a time. Hence a weak interaction between the variables exists when (x, y) is restricted by $\pi(x, y) = M(x) * N(y)$. The addition of M and N is then defined by extending the expression of the fuzzy addition (28) using a t-norm (Dubois and Prade, 1981):

$$[M \oplus_* N](c) = \sup\{M(c - a) * N(a) : a \in \mathbb{R}\}.$$

This is again a Sugeno integral-like convolution (Wang and Klir, 1992; Markova-Stupnanova, 1997b) with respect to a possibility measure. Only the corresponding possibility measure involved in the convolution is built up using $*$ instead of min. The following result is a starting point for the study of such operations:

Proposition 6: *(Dubois and Prade, 1983a) If M and N are u.s.c. fuzzy intervals with bounded supports then $M \oplus_* N$ is a fuzzy interval for any $*$. Moreover the core of $M \oplus_* N$ is always $[\underline{m} + \underline{n}, \overline{m} + \overline{n}]$.*

The obtained fuzzy interval is not necessarily u.s.c., unless the t-norm is right continuous. Note that $\oplus_*$ is associative over u.s.c. fuzzy intervals whenever $*$ is lower semicontinuous on the half-open square $[0, 1)^2$, this case including all four basic t-norms. The extension principle with t-norms has been first studied by Schweizer (1975) when M and N are probability distribution functions, in the framework of stochastic geometry. It has been also proposed by Klement (1981) for fuzzy numbers in the shape of probability distribution functions. When $*$ is Lukasiewicz t-norm, the addition of fuzzy intervals is very akin to the inf-convolution in convex analysis (Rockafellar, 1970), and to the usual dynamic programming equation (Bellman, 1957).

T-norm based additions do not have the same properties as the min-based standard one. For instance, the only idempotent element with respect to the addition of fuzzy intervals based on the standard extension principle is the crisp 0. However, for a general t-norm–based addition this is no longer true. So, e.g., any fuzzy number M with core {0} is a $\oplus_{*W}$ - idempotent for the weakest t-norm. For a continuous

Archimedean t-norm $*$ with an additive generator ϕ, the corresponding $\oplus_*$ – idempotents are just the fuzzy numbers $M = (0, s, t)_{LR}$, with core $\{0\}$, for which both $\phi \circ L$ and $\phi \circ R$ are subadditive functions on $[0, +\infty)$ (Markova–Stupnanova, 1997a, Markova, 1997a). The idempotents play the role of a relative zero. Take, e.g., $* = *_L$ the Lukasiewicz t-norm, and $I = (0, 1, 1)_{LL}$, $L(x) = \max(0, 1 - x)$. The triangular fuzzy number I is a $\oplus_{*L}$ - idempotent and $M \oplus_{*L} I = M$ holds for a continuous fuzzy interval M if and only the absolute value of the derivative $|M'(x)| \leq 1$ in all points x where the corresponding derivative existsC

Closed Form Results and Shape Invariance Properties. A closed form solution of the fuzzy addition in the case of fuzzy intervals was provided by Dubois and Prade (1981) for the three basic triangular norms other than the minimum. Several other results were obtained since then by Fullér and Keresztfalvi (1992), Kawaguchi and Da-te (1993, 1994), Hong and Hwang (1994a), Hong(1995), Markova (1995, 1997b), Mesiar (1995, 1996a,b, 1997a,b). More specifically the simple following result

$$\text{if } x \leq \underline{m} + \underline{n}, \quad [M \oplus_* N](x) = \max(M(\underline{n} - x), N(\underline{m} - x)) \tag{10.40}$$

$$\text{if } x \geq \overline{m} + \overline{n} \quad [M \oplus_* N](x) = \max(M(x - \overline{n}), N(x - \overline{m})) \tag{10.41}$$

obtains in the following cases:

i) $* = *_W$

ii) $*$ is a continuous Archimedean t-norm with additive generator ϕ such that the functions $\phi \circ M^+$, $\phi \circ M^-$, $\phi \circ N^+$, and $\phi \circ N^-$ are concave (on the supports of M and N).

Recall that an additive generator of the bounded product $*_L$ is $\phi_L(x) = 1 - x$. Then the convexity of M^+, M^-, N^+, N^- (on the supports of M and N) ensures the validity of (10.40-10.41). For the product the corresponding additive generator is, e.g., $\phi_P(x) = -\log x$. Then (10.40-10.41) is fulfilled with $*$ = product if $\log M^+$, $\log M^-$, $\log N^+$, $\log N^-$ are convex (on the supports of M and N). When $M = (\underline{m}, \overline{m}, s, t)_{LR}$, $N = (\underline{n}, \overline{n}, u, v)_{LR}$, (10.40-10.41) reads

$$M \oplus_* N = (\underline{m} + \underline{n}, \overline{m} + \overline{n}, \max(s,u), \max(t, v))_{LR}. \tag{10.42}$$

in contrast with the noninteractive case (10.35).

For a general continuous Archimedean t-norm $*$ with additive generator ϕ, the following closed form result was obtained for fuzzy intervals of the same type and same spreads, by Mesiar (1996b), improving a result of Hong and Hwang (1994a).

Proposition 7: *For the u.s.c. fuzzy intervals* $M_i = (\underline{m}_i, \overline{m}_i, s, t)_{LR}$, *for* $i = 1,\ldots,n$, *let functions* $\phi \circ L$, $\phi \circ R$ *be convex. Let* $\underline{m} = \sum_{i=1}^{n} \underline{m}_i$ *and* $\overline{m} = \sum_{i=1}^{n} \overline{m}_i$. *Then*

$$\text{if } x \le \underline{m}, \quad M_1 \oplus_* \ldots \oplus_* M_n(x) = \phi^{-1}(\min(\phi(0),\ n \cdot (\phi \circ L(\frac{\underline{m} - x}{ns}))))$$

$$\text{if } x \ge \overline{m} \quad M_1 \oplus_* \ldots \oplus_* M_n(x) = \phi^{-1}(\min(\phi(0),\ n \cdot (\phi \circ R(\frac{x - \overline{m}}{nt})))).$$

For two fuzzy intervals M and N having the same spreads: $M = (\underline{m}, \overline{m}, s, t)_{LR}$, $N = (\underline{n}, \overline{n}, s, t)_{LR}$, this result reads $M \oplus_* N = (\underline{m} + \underline{n}, \overline{m} + \overline{n}, 2s, 2t)_{L2R2}$, with $L2(x) = \phi^{-1}(\min(\phi(0), 2\phi \circ L(x)))$. See Mesiar (1997b) for further generalizations.

Proposition 7 applies for the linear product $*_L$ case, e.g., when shape functions are of the form $\max(0, 1 - x^p)$, $p > 0$. Then $\phi_L(\max(0, 1 - x^p)) = \min(1, x^p)$ is convex on the support of $\max(0, 1 - x^p)$ for $p \ge 1$, and it is concave for $0 < p \le 1$ (and then (10.40-10.41) can be applied). Similarly, for the product, the condition $\phi \circ L(x) = x^p$ leads to the shape functions of the form e^{-x^p}, for which Proposition 7 can be applied in the case $p \ge 1$ and (10.40-10.41) holds when $0 < p \le 1$.

In standard fuzzy arithmetics, due to (10.35), the shape invariance (shape-preservation) for addition of fuzzy intervals when $* = \min$ holds for any type of shapes. The same property holds for the $*_W$-based addition of fuzzy intervals. Only these limit t-norms can preserve all types of shapes (Mesiar, 1996a). However, for a given $*$, some particular shapes can be preserved, and vice versa, a given shape can be preserved in the case of some special t-norms (Kolesarova, 1997, 1998).

The choice of shapes of the fuzzy intervals and of the t-norm interact via the extension principle. The problem of finding 3-tuples (M, N, $*$) such that $M \oplus_* N$ is invariant has been discussed by Gebhardt (1995). A useful tool in t-norm-based fuzzy arithmetics is the transformation principle (Mesiar 1996a). For the increasing automorphism ψ: $[0, 1] \to [0, 1]$, and u.s.c. fuzzy interval M, N:

$$M \oplus_* N = \psi((\psi^{-1} \circ M) \oplus_{*_\psi} (\psi^{-1} \circ N))$$

where ψ^{-1} is the inverse automorphism and $*_\psi$ is the ψ– transform of $*$ (Schweizer and Sklar, 1983), $x *_\psi y = \psi^{-1}(\psi(x) * \psi(y))$. The transformation principle permits:

- the modification of the shape of fuzzy intervals and hence proving, e.g., the equivalence of linear and bell-shapes;

• the modification of the t-norm in current use and therefore reducing the application of continuous Archimedean t-norms to the basic t-norms $*_L$ and product.

Linear shapes (triangular and trapezoidal fuzzy quantities) are closed under the addition based on t-norms $*$ weaker than $*_L$, i.e., $* \le *_L$, (see Dubois and Prade, 1981; Kolesarova, 1995, 1998):

$$M \oplus_* N = (\underline{m} + \underline{n},\ \overline{m} + \overline{n},\ \max(s, u),\ \max(t, v))_{LL},\ \text{if } * \le *_L$$

with $L(x) = \max(0, 1 - x)$ (trapezoidal fuzzy intervals). Recall that for $0 < p \le 1$, the Yager t-norms $\alpha * \beta = \max(0, 1 - ((1 - \alpha)^p + (1 - \beta)^p)^{1/p})$ are weaker than $*_L$ and thus the above result applies. For any $p > 0$, the corresponding additive generator of a Yager t-norm is $\phi_p(\alpha) = (1 - \alpha)^p$. The following closure result (Keresztfalvi and Kovacs, 1992) holds for Yager t-norms with $p \ge 1$, and more generally for t-norms with additive generator $\phi_p(\alpha) = g(\alpha)^p$ where g is a continuous decreasing function from [0,1] to $[0, +\infty)$ with $g(1) = 0$. Let $M = (\underline{m}, \overline{m}, s, t)_{LL}$, $N = (\underline{n}, \overline{n}, u, v)_{LL}$ with $L = g^{-1}$. Then

$$M \oplus_* N = (\underline{m} + \underline{n},\ \overline{m} + \overline{n},\ (s^q + u^q)^{1/q},\ (t^q + v^q)^{1/q})_{LL},$$

if $*$ is a t-norm with generator g^p, and where $q = p/(p-1)$. The choice of the parameter p can be applied to control the increase of uncertainty, varying the output spreads between the minimal (max of input spreads) and the maximal (sum of input spreads). The above result also holds, taking $*$ as the linear product $*_L$, for $L(x) = \max(0, 1 - x^p)$ with $p > 1$ (Dubois and Prade, 1981).

Exponential shapes e^{-kx} are preserved when the addition of fuzzy intervals is based on:

• any t-norm $*$ weaker than the product $\alpha * \beta \le \alpha \cdot \beta$;

• any t-norm $*$ with additive generator $\phi(x) = (-\log x)^p$, $p > 0$ (so-called Aczél-Alsina t-norms).

Similar observations with respect to the resulting spreads as in the case of linear shapes are still valid.

Due to the transformation principle, for any continuous shape K strictly decreasing on its support, the shape K is preserved when the addition of fuzzy intervals is based on:

• any t-norm $*$ weaker than $*_K$ generated by $K*$, such that $K*(x) = \inf\{y, K(y) = x\}$;

• any t-norm $*$ generated by an additive generator $(K*)^p$, where $p > 0$.

Conversely, $*_L$ preserves any convex shape and shapes of the form $\max(0, 1 - x^p)$, $p > 0$, while product preserves any shape L with log L convex, and shapes of the form e^{-x^p}, $p > 0$ (Badard, 1982; Dubois and Prade, 1981). Analytical expressions for the addition of fuzzy intervals for a general continuous t-norm $*$, including ordinal sum, are described in more detail by De Baets and Markova-Stupnanova (1997). They notice that if λ is $*$-idempotent for a continuous t-norm $*$ ($\lambda * \lambda = \lambda$) then $(M \oplus_* N)_\lambda = M_\lambda \oplus N_\lambda$.

Infinite Sums. It is also tempting, in the scope of comparing fuzzy interval analysis and the calculus of random numbers, to look for fuzzy counterparts of well known limit theorems of probability theory involving infinite sums of independent fuzzy variables (in the sense of a t-norm). For the minimum operation, this program is not exciting since $M \oplus M = 2M$ generally, and especially for fuzzy intervals (see Section 10.3.4) For $* \neq \min$ and fuzzy intervals, this property is no longer valid (except with crisp intervals), and sums of the form

$$s^*{}_n(M) = \frac{M \oplus_* \ldots \oplus_* M}{n}$$

that are the fuzzy counterparts of the Law of Large Numbers (LLN), are no longer invariant. It is easy to see that, for instance, when equation (42) is valid, it turns out that $\lim_{n \to +\infty} s^*{}_n(M)$ is the core of M. This form of convergence for the LLN was extensively studied by Badard (1982), Williamson (1991), Fullér (1992), Fullér and Triesch (1993), Hong and Kim(1996, 1998), Salakhutdinov and Salakhutdinov (1998). For some criticisms and remarks see also Markova-Stupnanova (1998). Fullér and Triesch give an example of convergence to non-crisp intervals. Some fuzzy counterparts of the Central Limit Theorem of probabilities can be contemplated via converging infinite sums of fuzzy intervals; first steps in this direction are made by Dishkant (1981). Proposition 7 can be extended even to the infinite series case (Triesch 1993, Hong and Hwang 1994a, Mesiar 1997b), if the requirements of Proposition 7 hold for More recently, Hong (1996) extended and generalized this result for the case of arbitrary continuous Archimedean t-norms. For Lukasiewicz t-norm, the search for limit theorems comes close to recent works on the analogy between probability theory and dynamic programming (Quadrat, 1990).

Approximate Calculations. Fullér and Keresztfalvi (1991) have generalised Nguyen's reduction of the extension principle (10.19) to a set-valued calculation to the case of weakly interactive variables in the sense of a t-norm. Namely, they proved that

$$f(Q_1, Q_2)_\lambda = \bigcup_{\alpha * \beta \geq \lambda} f((Q_1)_\alpha, (Q_1)_\beta)$$

holds for an u.s.c. t-norm-based modification of the extension principle (applied to a continuous function f); it is equivalent to the existence of a pair (a, b) depending upon the choice of the image c for which $\sup\{Q_1(x) * Q_2(y): f(x, y) = c\}$ is attained. It is however not clear if this result may lead to efficient approximate computations. Williamson (1991) has proposed a different computational scheme based on support discretization similar to probabilistic computation schemes. It would be interesting to compare fuzzy interval analysis with the t-norm product with the calculus of random intervals with a view to check whether the former is or not an approximation of the latter.

Upper and lower analytical approximations for the result of t-norm-based–fuzzy additions of fuzzy intervals are provided by Hong and Hwang (1997). A very different type of approximate calculation of additions of fuzzy intervals based on the product t-norm is proposed by Kosheleva *et al.* (1997): the idea is to exploit the well known result stating that for $a, b \in [0, +\infty)$, $\max(a,b) = \lim_{n \to +\infty} (a^n + b^n)^{1/n}$. Then the extension principle can be approximated by a sum of products, to which a fast Fourier transform can be applied.

Lastly, Otto *et al.* (1993) consider the computation of functions $f^{©}(M, N)$ of fuzzy variables under the form

$$f^{©}(M, N)(v) = \sup\{E_1(u_1) © E_2(u_2): f(u_1, u_2) = v\}$$

where © is a monotonic idempotent operation such that $0 © 1 = 1 © 0 = 0$. They give conditions under which applying the vertex method to level-cuts yields the level-cuts of $f^{©}(M, N)$ which then coincide with the result of the standard extension principle. This is for instance when f is continuous and monotonic increasing, and

$$(M © N))_\lambda \cap f^{-1}(\, f(\underline{m}_\alpha, \underline{n}_\alpha)\,) = \emptyset \;\forall \alpha > \lambda;$$
$$(M © N))_\lambda \cap f^{-1}(\, f(\overline{m}_\alpha, \overline{n}_\alpha)\,) = \emptyset \;\forall \alpha > \lambda.$$

10.5.3 Multidimensional Fuzzy Quantities

In this subsection we only briefly sketch the developments in the generalisations of fuzzy intervals to multidimensional spaces. On the one hand, this generalization captures the concept of interactive fuzzy variables, and makes it possible to compute with them. On the other hand some mathematical notions such as complex fuzzy numbers can be defined.

Modelling Fuzzy Interaction. Sarna (1989, 1998) has provided exact analytical expressions for interactive calculi of the form $\sup_{x,y}\min(M(x), M(y), \frac{\min(x, y)}{\max(x, y)})$ when M and N are triangular fuzzy numbers or trapezoidal fuzzy intervals. Buckley (1989a, 1990) has considered negatively and positively associated fuzzy variables x and y restricted with fuzzy intervals M and N. Variables x and y are said to be positively (resp. negatively) associated if the core of their joint fuzzy domain D is restricted to belong to a crisp domain containing the line $y = ax + b$ with $a > 0$ (resp $a < 0$) on supp(MxN). Buckley shows conditions under which this kind of interactivity does not affect sums, products, maximum or minimum of discrete and continuous fuzzy intervals. For instance if core(D) contains the set $\{x: M^+(x) = N^+(x)\} \cup \{x: M^-(x) = N^-(x)\}$, which is a form of positive association, then the interactivity constraint does not affect the result of these operations, since the maximum of min(M(x), N(y)) is attained on this set.

This kind of positive or negative association is also addressed by Nakamura (1990). He defines a joint possibility distribution from a basic fuzzy number M with core {0} and support (–1, 1) by rotating $R(x, y) = \min(M(x), M(y))$ by $\pi/4$ radians around its core (m, n), which gives a fuzzy relation $R^{\pi/4}$. Then Nakamura uses a correlation coefficient $\rho \in [-1, +1]$ that acts on the shape of the fuzzy relation and rotates again the result by $-\pi/4$, namely

$$R_\rho(x, y) = (R^{\pi/4}(x, \frac{y}{1-\rho}))^{-\pi/4} \quad \text{for } 1 > \rho \geq 0;$$

$$R_\rho(x, y) = (R^{\pi/4}(\frac{x}{1+\rho}, y))^{-\pi/4} \quad \text{for } 0 \geq \rho > -1.$$

When $1 > \rho \geq 0$ there is positive association and $\mathrm{supp}(R_\rho)$ is a lozenge whose main diagonal is along the line $y = x$, while when $0 > \rho \geq -1$ there is negative association and $\mathrm{supp}(R_\rho)$ is a lozenge whose big diagonal is along the line $y = -x$. When $\rho = 1$ and -1, $R_\rho(x, y)$ is taken as the limit of the above expressions and its support reduces to a segment. The non interactive case is retrieved when $\rho = 0$. Such a canonical nonseparable fuzzy relation has projections equal to M on both axes and can be used as a two dimensional shape function to define a family of fuzzy relations parameterized by the core and spread values.

This construct is systematized by Ramik and Nakamura (1993). In this work, a canonical fuzzy quantity of dimension 2 is of the form

$$D(x, y) = \min(M(a_1x + b_1y), M(a_2x + b_2y))$$

where M is a symmetric fuzzy number with core {0} and support (–1, 1) This fuzzy set is a joint canonical fuzzy quantity of dimension 2 if and only if its projection on the two axes are $M(\cdot)$ and $M(\cdot)$, respectively. Ramik and Nakamura (1993) give necessary and sufficient conditions on the parameters a_1, b_1, a_2, b_2 for D to have M(x) and M(y) as projections. For instance, if the product a_1a_2 is positive then the condition for $\sup_x D(x,y) = M(x)$ is $|a_1b_2 - b_1a_2| = |a_1 + a_2|$. See Miyakawa *et al.* (1993) for dimension n fuzzy quantities.

Some works aim at quantifiying the level of interactivity between fuzzy variables linked by the fuzzy relation D by means of some kind of interaction index, like for stochastic variables. Buckley (1989a, 1990) proposes to use the interaction index of the joint uniform probability distribution on supp(D). Some authors have envisaged interactivity between fuzzy intervals in the spirit of statistical correlation coefficients. The proposal of Murthy *et al.* (1985) comes down to a probabilistic view of the membership function as done in the center of gravity defuzzification method. More recently Yu (1993) has built up a correlation coefficient for bounded support fuzzy numbers M and N based on the integral of $M(x)\cdot N(x) + (1 - M(x)) \cdot (1- N(x))$ over a bounded referential. However problems with this approach are pointed out by Hong and Hwang (1996). Ramik and Nakamura (1993) also introduce a degree of interaction for their canonical fuzzy numbers of dimension 2. It is defined as a simple function of the parameters a_1, b_1, a_2, b_2 that characterize the fuzzy domain D.

Additions and subtractions of interactive fuzzy numbers whose joint distribution is a canonical fuzzy number of dimension 2 can be analytically calculated, just as for LR non-interactive fuzzy intervals (see Ramik and Nakamura (1993)). In particular, a property such as $M(-)M = 0$ can be retrieved by assuming negative association ($\rho = -1$)

Fuzzy Vectors. Another view of higher dimensional fuzzy intervals consists in devising a calculus of vectors with fuzzy components. Like any general structure can be fuzzified, the fuzzy complex number plane can be studied, as well as n-dimensional reals $\mathbb{R}^n$, etc. Any such fuzzification is all the more acceptable as it includes (as a natural embedding) the underlying structure and preserves as many properties as possible.

An n-dimensional fuzzy number is a normal fuzzy subset M of $\mathbb{R}^n$ such that each level-cut $M_\lambda = \{\mathbf{x} \in \mathbb{R}^n;\ M(\mathbf{x}) \geq \lambda\}$ is a non-empty convex compact subset of $\mathbb{R}^n$. For the discussion about several types of 2 – and more dimensional fuzzy numbers we refer to Kaufmann and Gupta (1985). Two common operations on n-dimensional numbers on $\mathbb{R}^n$, namely the multiplication by a real constant and the addition of vectors $\mathbf{x}$ on $\mathbb{R}^n$ are defined by means of the extension principle. Namely, let M, N be n-dimensional fuzzy numbers and c be a real constant. Then

$$c\cdot M(\mathbf{x}) = M(\mathbf{x}/c),\ \mathbf{x} \in \mathbb{R}^n,\ c \neq 0,$$
$$0\cdot M(\mathbf{x}) = 0,\ \mathbf{x} \neq \mathbf{0},\ 0\cdot M(\mathbf{0}) = 1,$$

and, unsurprizingly rewriting the extension principle for vectors as,

$$(M \oplus N)(\mathbf{z}) = \sup\{\min(\ M(\mathbf{x}), N(\mathbf{y})\ ): \mathbf{x} + \mathbf{y} = \mathbf{z}\ \}.$$

Properties similar to the ones of upper semi-continuous fuzzy intervals arithmetics also hold in the n-dimensional case, as well as the calculations with linked variables (e.g., Bouchon–Meunier *et al.*, 1997). For some applications see e.g., Subrahmanyam and Sudarsanam (1994).

Fuzzy Complex Numbers. A fuzzy complex number Z can be most simply defined as a unimodal fuzzy subset of the complex plane $\mathbb{C}$ such that each level-cut $Z_\lambda = \{z \in \mathbb{C};\ Z(z) \geq \lambda\}$ is a convex compact subset of $\mathbb{C}$. However, this definition excludes the fuzzy goniometric complex numbers introduced in Kaufmann and Gupta (1991), $Z = (R,\Phi)$, where R is a continuous positive fuzzy interval and Φ is a continuous fuzzy interval with the support contained in an interval $[a, a + 2\pi]$, and

$$Z(z) = \min(R(r), \Phi(\phi)), \text{ for } z = r \cdot e^{i\phi}.$$

The lack of the convexity of level-cuts of goniometric fuzzy complex numbers does not violate a more general definition of a fuzzy complex number which is due to Buckley (1989c). Namely, a fuzzy subset Z of the complex plane $\mathbb{C}$ is a fuzzy complex number if

- $Z(\cdot)$ is continuous;
- The strict level-cut $Z_{>\lambda} = \{z \in \mathbb{C};\ Z(z) > \lambda\}$ is open bounded, connected and simply connected for all $\lambda \in [0,1)$;
- The core $Z_1 = \{z \in \mathbb{C},\ Z(z) = 1\}$ is non-empty, compact, arcwise connected and simply connected.

If M and N are continuous fuzzy intervals, we can introduce (Buckley, 1989c) a rectangular fuzzy complex number $Z = M + i\,N$ defining

$$Z(z) = \min(M(x), N(y)), \text{ for } z = x + iy.$$

Based on the extension principle, the obvious operation on complex numbers can be extended to the fuzzy complex numbers (Buckley, 1989c). So, e.g., the sum $Z \oplus V$ of two fuzzy complex numbers is unsurprizingly defined by

$$(Z \oplus V)(z) = \sup\{\min(\ Z(z_1), V(z_2)): z_1 + z_2 = z\ \}.$$

The addition of fuzzy complex numbers is associative, commutative, etc., and $(Z \oplus V)_\lambda = Z_\lambda \oplus V_\lambda,\ \lambda \in (0, 1]$. Similarly, the multiplication $Z \otimes V$ is introduced and $(Z \otimes V)_\lambda = Z_\lambda \otimes V_\lambda,\ \lambda \in (0, 1]$.

The complex conjugate Z^* is defined by $Z^*(z) = Z(\bar{z})$, where $\bar{z} = x - i\,y$. The modulus $|Z|$ of a fuzzy complex number Z is a positive continuous fuzzy interval such that

$$|Z|(x) = \sup\{Z(z): |z| = x\} \text{ for } x \geq 0$$

Proposition 8:
i) $|Z|_\lambda = |Z_\lambda|$
ii) $|Z_1 \oplus Z_2| \subseteq |Z_1| \oplus |Z_2|$
iii) $|Z_1 \otimes Z_2| = |Z_1| \otimes |Z_2|$.

For rectangular fuzzy complex numbers,

$$(x_1 + i\,y_1) \oplus (x_2 + i\,y_2) = (x_1 \oplus x_2) + i\,(y_1 \oplus y_2),$$

while for the goniometric fuzzy complex numbers Z, denoted (R, Φ), such that $Z(z) = \min(R(r), \Phi(\phi))$ where $z = r\,e^{i\phi}$:

$$(R_1, \Phi_1) \otimes (R_2, \Phi_2) = (R_1 \otimes R_2, \Phi_1 \oplus \Phi_2) \text{ and } |(R, \Phi)| = R.$$

A product-based extension principle was applied to the sum of fuzzy complex numbers by Kim (1993). For applications, we refer to (Buckley, 1991, 1992). A somewhat modified approach to fuzzy complex numbers analysis can be found in (Wu and Qiu, 1998).

10.5.4 Fuzzy Equations and the Optimistic Calculus of Fuzzy Intervals

A natural way of interpreting an equality relationship between two mathematical expressions A and B involving fuzzy quantities is to state the identity of the membership functions of A and B corresponding to each side of the equality. This kind of equality between fuzzy quantities really implies that the imprecision pervading A and B is the same, and not only that they have the same order of magnitude. Using fuzzy interval analysis, addition and subtraction of fuzzy terms usually increase the level of imprecision. Hence transferring a term from expression A to expression B deteriorates the imprecision of B but reduces the imprecision of A; such a transfer is not valid even using standard interval calculus. For instance a fuzzy equation of the form:

$$X\,(-)\,M = N$$

where M, N, and X (the unknown fuzzy range of the unknown parameter x) are fuzzy intervals cannot be solved by letting X be $M \oplus N$, since in this equation, X is implicitly more precise than N (whereas $M \oplus N$ is not). This fact is related to the lack of inverses in the calculus of fuzzy quantities (Yager, 1980a), e.g., $\oplus$ and $\otimes$ are no longer group operations. Solving this type of equations may be useful when a fuzzy quantity expresses a requirement of maximal tolerance (say N) on a performance index (say y) which is itself the result of a computation involving unknown quantities (say x) whose value is implicitly constrained by this computation (say via x – a = y, where a is fuzzily constrained by M).

Optimistic Addition and Subtraction. A way around the problem of fuzzy equations must be searched for outside of the body of standard fuzzy operations. Preliminary discussions and attempts along this line appear in Yager (1979b) and Dubois and Prade (1980b). A canonical approach for this problem has been found in Dubois and Prade (1983a, 1984), Sanchez (1984), by relating fuzzy quantity equations to the more general problem of fuzzy relational equations, see Chapter 6 by De Baets in this volume.

For instance the simple linear equation X (–) M = N may or may not have solutions, since it translates into the following functional equation with unknown X:

$$\forall\, b \in \mathbb{R},\ N(b) = \sup_{x,\,z}\{\min(X(x), M(z)) \colon x - z = b\}.$$

Let $\alpha \wedge^{-1} \beta$ be the greatest solution of equation $\min(\alpha, \xi) = \beta$, when it exists. Namely

$$\begin{aligned} \alpha \wedge^{-1} \beta &= 1 && \text{if } \alpha = \beta \\ &= \beta && \text{if } \alpha > \beta. \end{aligned}$$

The operation $\wedge^{-1}$ is closely related to the so-called Gödel implication in many-valued logics:

$$\begin{aligned} \alpha \rightarrow \beta &= 1 && \text{if } \alpha \le \beta \\ &= \beta && \text{if } \alpha > \beta. \end{aligned}$$

For x = a, the solution X(a) of the equation must then satisfy:

$$\forall\, a,\ X(a) \le M(a - b) \wedge^{-1} N(b).$$

Hence the greatest solution of the functional equation is defined, when it exists, by

$$\forall\, a,\ X(a) = \inf\{M(a - y) \wedge^{-1} N(y) \colon y \in \mathbb{R}\}.$$

Using the above remarks, we find an alternative definition for the addition of fuzzy intervals M and N, denoted M)+(N, as follows

$$[M\ )\!+\!(\ N](z) = \inf_x\{M(z - x) \rightarrow N(x)\} \tag{10.43}$$

which contrasts with the usual addition of fuzzy intervals (10.28). Clearly, if the equation X (–) M = N has a solution then, this solution is X = N)+(M. More generally, N)+(M is always the greatest solution to the relaxed fuzzy equation of the form X (–) M ⊆ N. The property stating that M)+(– M = 0 holds, but)+(is not a commutative operation. It is easy to check (Dubois and Prade, 1983a, 1984) that if $M = (m, s, t)_{LR}$, $N = (n, u, v)_{RL}$, then, provided that $u \le s$ and $v \le t$ it holds that

$$M\ )\!+\!(\ N = (\ m + n,\ s - v,\ t - u\ )_{LR}$$

In other words, M)+(N is more precise than M, in that the imprecision of M is eroded by that of N through operation)+(. Note that M)+(N does not necessarily exist (i.e., it may be the empty set), especially if M is more precise than N; for instance, if $m \in \mathbb{R}$, m)+(N does not exists (= Ø) while N)+(m = N ⊕ m. Moreover, the fact that M)+(N exists does not imply that a solution of X (–) M = N exists.

The nestedness of the level-cuts computed from the interval equations X (–) $M_\alpha = N_\alpha$ may fail to hold. For instance consider the equation X(–)M = N. Let $\alpha < \beta$ be two membership grades. The solution to $X(-)M_\alpha = N_\alpha$ is $N_\alpha)\!+\!(M_\alpha$ if it exists. However the inclusion $N_\beta)\!+\!(M_\beta \subseteq N_\alpha)\!+\!(M_\alpha$ is not warranted by $\alpha < \beta$. The use of the representation of fuzzy intervals by pairs of functions (μ^-, μ^+) such that $\mu^-(\lambda) = \inf M_\lambda$ and $\mu^+(\lambda) = \sup M_\lambda$ due to Goetschel (1997) only apparently copes with this problem since if N is represented by (v^-, v^+) then the optimistic addition leads to consider $(v^-, v^+) + (\mu^-, \mu^+) = (\mu^+ + v^-, \mu^- + v^+)$ which is always defined but may not underlie any genuine fuzzy interval.

The imprecision-eroding property of)+(justifies the name 'optimistic fuzzy interval mathematics'. The duality between the pessimistic operation ⊕ and the optimistic one,)+(, is not self-evident when Equations (10.43) and (10.28) are compared. However when M and N are crisp intervals, then (Dubois and Prade, 1984)

$$M \oplus N = \bigcup_{y \in N} M \oplus y$$
$$M\ )\!+\!(\ N = \bigcap_{y \in N} M \oplus y$$

where $M \oplus y = \{x + y : x \in M\}$. The two above identities are known as the 'Minkowski addition' and the 'Minkowski subtraction' respectively. It is obvious that operation)+(assumes a maximal compensation of errors, since M)+(N is the smallest set S such that $\forall z \in S, \exists\, y \in N, z - y \in M$.

N.B. When applied on $\mathbb{R}^n$, the two 'Minkowski operations' M ⊕ (–N) and M)+((–N) enable the definition of dilation and erosion operations of a set M by another set N, which are at the basis of mathematical morphology (Matheron, 1975). Their fuzzy extensions (using different implications in (10.43)), have given birth to fuzzy mathematical morphologies which have been developed, especially on $\mathbb{R}^2$, by Bloch and Maitre (1994, 1995), De Baets *et al.* (1994a,b) and Sinha and Dougherty (1995) and successfully applied in image processing.

General Optimistic Interval Analysis. Clearly the above approach does not depend upon the nature of the involved operation (here addition). Optimistic counterparts for any function f have been defined (Dubois and Prade, 1984); the case of the product operation and equation of the form $X \otimes M = N$ are studied in the cited paper. Optimistic interval analysis are one way of handling and solving any fuzzy equation of the form $f(X, M) = N$ (Sanchez, 1984), or simultaneous equations of the form $\{f(X, M_i) = N_i: i = 1, p\}$ (Gottwald, 1984, 1993). General results on such fuzzy equations are given by Biacino and Lettieri (1989). Let X* be defined as

$$X^*(a) = \inf_x \{ M(x) \rightarrow N(f(a, x))\}.$$

The following results can be established:

- If f is monotone, M and N are convex then so is X*.
- If M and N are normal then so is X* if there is $b \in \mathbb{R}$ such that $M(b) \leq N(f(a,b))$ for all $a \in \mathbb{R}$.
- if f is continuous and M, N are u.s.c. fuzzy intervals then so is X*.
- If f is monotone, M and N are bounded-support u.s.c. fuzzy intervals and $\exists a \in \text{supp}(M)$, $\exists\, c < d \in \text{supp}(N)$ such that both $f(a,y) = c$ and $f(a,y) = d$ admit solutions, then X* is bounded. Moreover $f(X^*, M) = N$ if and only if $f(X^*_\alpha, M_\alpha) = N_\alpha$ for all $\alpha \in (0, 1]$.

The latter result leads Biacino and Lettieri (1989) to conclude that under the prescribed assumptions, a solution to equation $f(X, M) = N$ exists if and only if solving the level-cut equations $f(X_\alpha, M_\alpha) = N_\alpha$ give a family of closed intervals $\{X_\alpha: \alpha \in (0, 1]\}$that are the level-cuts of a fuzzy interval. Some related results can be found in Wasowski (1997). Zhao and Govind (1991a,b) establish similar results for simple fuzzy equations of one variable based on the four arithmetic operations in terms of level-cuts only. They establish their results assuming that the upper and lower bounds of level-cuts are differentiable functions of the membership grades. Buckley and Qu (1990) solve quadratic fuzzy equations of the form $MX^2 + NX = Q$. Vrba (1992) proposes algorithms for finding inverses of arithmetic operations with fuzzy intervals. Wang and Ha (1994) solve systems of fuzzy linear equations of the form $AX \oplus U = X$, where X is a vector of fuzzy variables, and U is a vector of fuzzy intervals and A is a scalar matrix.

In the preceding lines, fuzzy equations $f(X, M) = N$ are based on the sup-min extension principle (19). Fuzzy equations based on a sup-triangular norm principle can also be solved using results by Pedrycz (1985), on general fuzzy relational equations. Some material on this topic appears in Dubois and Prade (1983a, 1984) for addition; the more general equation $f(X, M) = N$ has been also considered (Dubois and Prade, 1984) using triangular norms. When the t-norm $*_L$ is used, fuzzy equations are very much related to some earlier works in path algebras by Cuninghame-Greene (1979) and more recently Cuninghame-Greene and Cechlarova (1995).

Discussion. Although the area of ' optimistic ' fuzzy interval mathematics seems to be very promising at a theoretical level, its practical appeal may be limited by the lack of flexibility of the fuzzy-set equality which defines the equations. First, membership functions are never known with such precision as to enable the equality M = N between two fuzzy quantities to be checked at the operational level. This remark makes the fact that even simple equations such as X(–)M = N may not have any solution very frustrating, if not counter-intuitive: some may argue that equations which always have solutions in deterministic case should have solutions (at least approximate ones) in a fuzzy environment. Noticeably Buckley and Qu (1991a) complain on the lack of solutions (even fuzzy complex ones) to fuzzy quadratic equations. Similarly, systems of fuzzy linear equations in the sense defined above are much more restrictive than crisp equations as already shown by Negoita (1978), and Dubois and Prade (1980b). This is discussed also in Buckley and Qu (1991b). Necessary and sufficient conditions for the existence of fuzzy solutions are found in Jiang (1986). Changing equation X(–)M = N into X(-)M $\subseteq$ N, although slightly improving this situation, is not satisfactory, since conditions of the form M(x) $\leq$ N(x) are still difficult to obtain and contradict the actual precision of membership functions.

However the problem is to understand in the right way what a fuzzy equation means. As pointed out earlier, tight fuzzy equations f(X, M) $\subseteq$ N makes perfect sense if N is viewed as a flexible constraint, M represents the uncertainty pervading an uncontrollable parameter y, and the problem is to choose the value a of a decision variable x in such a way as $f(a$, M) $\subseteq$ N. Then X visualizes the possible range of values for x, X(a) being the degree of feasibility of choosing x = a. A stronger form of this type of equation is supp(f(X, M)) $\subseteq$ core(N). It contrasts with the alternative view of fuzzy equations in the form $f(a$, M) $\cap$ N $\neq \emptyset$, where both M and N represent fuzzy constraints on decision variables y and z, respectively. Given a solution x = g(m, n) of the crisp equation f(x, m) = n obtained by deciding y = m and z = n, a solution to the loose fuzzy equation $f(a$, M) $\cap$ N $\neq \emptyset$ is X = g(M, N) in the sense of the extension principle. The difference between these two views has been discussed by Dong and Wong (1989) in the framework of linear systems of fuzzy equations with fuzzy coefficients. They propose to solve such equations on the basis of level-cuts using the vertex method. Fullér (1990) studies the influence of the change of membership functions on the fuzzy solution to systems of loose fuzzy linear equations. Buckley and Qu (1991b) also address fuzzy linear equations systems, and emphasize the difference between the solutions of equations represented by membership function equalities (tight), and solutions consisting in applying the extension principle to the crisp solutions to an equation (loose). They also notice that the solution of the form

$$X(a) = \sup\{ \min(M(b), N(c)) : f(a, b) = c\}$$

may differ from X = g(M, N) if fuzzy interval analysis is blindly applied to explicit analytical expressions g(b, c) of the solutions where one variable may appear several times, as pointed out earlier in Section 10.3.4.

Fuzzy equations are encountered in decision analysis problems, for instance when extending Saaty 's method for evaluating the absolute weight of criteria from an expert estimation of their relative importance. This is discussed in van Laarhoven and Pedrycz (1983), Boender *et al.* (1989), and Kwiesielewicz (1998). Most of the proposed approaches to this problem that come down to solving a set of fuzzy linear equations look ad hoc so far. This topic also pertains to decision-making in a fuzzy environment (see the special volume of the Handbooks of Fuzzy Sets Series on this topic (Slowinski, 1998)). In particular the meaning of this type of fuzzy equation is not always the same in all situations. This question also touches on the comparison of fuzzy quantities, as clear from above. There are several points of view in the comparison of fuzzy quantities; this is the topic of the next section.

10.6 COMPARISON OF FUZZY INTERVALS

The question of comparing fuzzy quantities has received attention from many researchers. Extensive comparative surveys are in Bortolan and Degani (1985), Li and Lee (1987), Tseng and Klein (1988), Hirota and Pedrycz (1989), Zhu and Lee (1992), Chang and Lee (1994). Indeed this problem is important in practice: when using fuzzy interval analysis, very often the results take the form of fuzzy intervals that must be compared for the purpose of making a decision (e.g., Baas and Kwakenaak, 1977; Tong and Bonissone, 1980). This problem is ill-posed and not so simple. Indeed due to the imprecision pervading fuzzy intervals it is sometimes difficult to claim in the absolute that a fuzzily-known value is greater or smaller than another, just as for random variables. The way to address this problem partially depends on the problem at hand, on the mathematical setting in which the fuzzy intervals are cast. If it is crucial to find a definite answer, one attitude is to defuzzify the fuzzy intervals and to use the corresponding ordering. If one insists on supplying a faithful picture of the relative positions of the fuzzy intervals at the expense of decisiveness, then relational techniques, some of which rely on interval orders are more suitable. Besides the comparison can be done on purely possibilistic grounds or may take advantage of the connection of fuzzy intervals to random intervals so as to exploit probabilistic comparison indices. Wang (1997) has recently carried out a systematic investigation of the problem of ranking fuzzy quantities, and especially fuzzy intervals. In this thesis, over forty methods are reviewed. A classification is adopted for better legibility. In the first class of approaches, each fuzzy quantity is changed into a real number, and is thus independently evaluated. In the second class fuzzy intervals are compared to a reference set. In the last class of approaches, fuzzy relations are constructed so as to carry out pairwise comparisons. For this class,it is necessary to further process the fuzzy relation in order to get a final ranking. See the volume edited by Slowinski (1998) in the Handbooks of Fuzzy Sets Series for decision-making applications.

This section gives an overview of methods organized partly along these lines. We restrict ourselves to fuzzy intervals with bounded support (and normal) for the sake of simplicity.

10.6.1 Positioning a Number with Respect to a Fuzzy Quantity

Before comparing two fuzzy quantities it is useful to consider how a precise number can be located with respect to a variable x with a fuzzy range Q. In the framework of possibility theory, it is easy to define the sets of numbers which are possibly greater than or equal to x, and necessarily greater than x. They are denoted by $[Q, +\infty)$ and $(Q, +\infty)$, respectively, and defined using possibility and necessity measures (Prade, 1982; Dubois and Prade, 1983b):

$$[Q, +\infty)(a) = \Pi_Q((-\infty, a]) = \sup\{Q(x): x \le a\}; \quad (10.44)$$

$$(Q, +\infty)(a) = N_Q((-\infty, a)) = \inf\{1 - Q(x): x \ge a\} \quad (10.45)$$

Note that $[Q, +\infty)$ and $(Q, +\infty)$ are convex, i.e., fuzzy intervals, and their membership functions are, up to discontinuity points, the upper and lower distribution functions of Q, respectively (as per equations (10.44) and (10.45)). See Figure 1. Note further that when Q is a real number r, then $[r, +\infty)$ and $(r, +\infty)$ are respectively recovered. Similarly, the fuzzy intervals $(-\infty, Q]$ and $(-\infty, Q)$ can be defined from (10.44) and (10.45) by complementation, e.g., $(-\infty, Q) = \overline{[Q, +\infty)}$. When Q is a fuzzy interval, then $(-\infty, Q] \cap [Q, +\infty) = Q$ (standard min intersection). In the general case, the convex hull of Q is obtained. Similarly $(-\infty, Q) \cap (Q, +\infty) = \emptyset$ for all fuzzy intervals $Q \in [0, 1]^{\mathbb{R}}$. Moreover $Q \cup (Q, +\infty) \subseteq [Q, +\infty)$, without equality generally. This is linked to the lack of an excluded middle law in fuzzy set theory with the usual max-min connectives.

Lastly, it is possible to define intervals with fuzzy boundaries, denoted [P, Q], [P, Q), (P, Q], and (P, Q) respectively, by intersecting fuzzy intervals defined by (44-45) (Dubois and Prade, 1983c). For instance, $[P, Q] = (-\infty, Q] \cap [P, +\infty)$, $[P, Q) = (-\infty, Q) \cap [P, +\infty)$ … etc…

Delgado *et al.* (1988) have generalized this approach to comparing a fuzzy interval and a scalar value. Instead of defining $[Q, +\infty)(a) = \Pi_Q((-\infty, a])$ using a possibility measure, they use a Sugeno λ-measure $m_{(\lambda)}((x, y]) = \dfrac{[Q, +\infty)(y) - [Q, +\infty)(x)}{1 + \lambda[Q, +\infty)(x)}$ for $\lambda > -1$ and define a modified semi-infinite interval $[Q, +\infty)^{\lambda}(a) = m_{(\lambda)}((-\infty, y])$; and similarly using $m^*_{(\lambda)}(A) = 1 - m_{(\lambda)}(A^c)$, $(Q, +\infty)^{\lambda}(a) = m^*_{(\lambda)}((-\infty, y])$. This change affects $[Q, +\infty)$ and $(Q, +\infty)$ on their increasing parts. For $\lambda = 0$ one recovers the usual definition.

10.6.2 Ranking Fuzzy Intervals via Defuzzification

A simple method for ranking fuzzy intervals consists in turning a fuzzy interval into a precise value and then use the ranking induced by these values. Let r denote the ranking function and $>_r$ denote the associated ordering, such that M $>_r$ N if and only if r(M) > r(N). The advantage of this approach is that it does provide a ranking eventually. The drawback is that it may not reflect the actual undecisiveness pervading a decision situation in case of imprecise data. Moreover the basic question is: what should be the ranking function? In the case of random numbers for instance the usual ranking method is by the expected value, as advocated by utility theory for instance. In the case of fuzzy intervals, this is typically a so-called defuzzification problem.

A number of empirical proposals for ranking fuzzy intervals exist, based on defuzzification procedures. Many of them exploit the characteristic values described in Section 10.2.2, although some early approaches are very rudimentary; for instance Adamo (1980) proposes r(M) = sup M_λ for a prescribed level λ.

Yager (1979c) has focused on fuzzy sets of the unit interval. He proposed to take r(M) as the center of gravity GR(M) or its possibilistic counterpart, $r_\Pi(M) = \sup_{x \in [0, 1]} \min(M(x), x)$. As a defuzzification method, the latter makes sense only in the unit interval. The center of gravity defuzzification is really the expected value of classical utility theory where the membership function is renormalized in a simple way. Later, Chang (1981) has used the numerator of the center of gravity formula as a ranking function. The center of gravity of the area limited by the fuzzy interval is more generally used for ranking purposes by Murakami *et al.*(1983). They exploit both the average value and the average membership grade of the fuzzy interval. Li and Lee (1987, 1988) exploit the probabilistic connexion suggested by the center of gravity method and propose the associated standard deviation as a complementary information. They also propose a more general precise equivalent value akin to the center of gravity: they assume a probability density function on the support of M that they introduce both in the numerator and denominator of the center of gravity expression. They study the case of uniform distributions (this is the center of gravity method) and the case of a distribution proportional to the membership function M(·) (which comes down to squaring the membership function in the center of gravity method). Yoon (1996) is another work that uses the center of gravity as a ranking function. Cheng (1998) introduces minor modification to Murakami *et al.* (1983) and to Li and Lee (1987, 1988)'s methods: he proposes to rank fuzzy quantities by the Euclidean distance of their geometrical center of gravity to the origin in the first method, and to use a relative standard deviation instead of an absolute one for improved discrimination.

Yager (1981) very early proposed to use the middle point of the expected value $r(M) = \bar{E}(M)$ defined by (10), and studied the properties of this index in more details, noticing that the linearity property $r(M \oplus N) = r(M) + r(N)$ holds, which ensures a good consistency of this ranking procedure with fuzzy arithmetics (since r(aM) =

a·r(M) holds too in that case). Indeed we have that if M $\geq_r$ N and M' $\geq_r$ N' then M⊕M' $\geq_r$ N⊕N'. It is strange that this index is known for ranking purposes and is seldom used in fuzzy control for defuzzification. The above properties indeed indicate a remarkable invariance of the induced rankings under affine transformations of the referential.

De Campos Ibañez and Gonzalez Muñoz (1989) point out that these properties remain under the following Hurvicz-like defuzzification

$$r_\alpha(M) = \alpha E^*(M) + (1 - \alpha)E_*(M) = \int_0^1 (\alpha \sup M_\lambda + (1 - \alpha)\inf M_\lambda)d\lambda$$

where the parameter α is an optimism index that can be tuned by the decision-maker. Liou and Wang (1992a) propose later on exactly the same method. Gonzalez (1990) proposes a more general defuzzification scheme using the s-mean interval instead of E(M). Campos and Gonzalez (1994) use a general fuzzy measure on [0, 1] instead of a Lebesgue measure in the above integral.

Choobineh and Li (1993) advocate a ranking function that is coherent with Yager's $\overline{E}(M)$ and that applies to any fuzzy bounded quantities be they normal or not. The approach heavily relies on the choice of a referential interval [a, b] that contains the fuzzy intervals. The ranking function takes the difference between areas between M and the upper bound and between M and the lower bound of [a, b].

Noticing that $r_\Pi(M) = \sup_{x \in [0, 1]} \min (M(x), x)$ is a Sugeno integral with respect to a possibility measure, Gonzalez (1990) considered the Sugeno integral with respect to the dual necessity measure $r_N(M) = \inf_{x \in [0, 1]} \max (M(x), 1 - x)$ as a pessimistic ranking index and again proposed to calculate a weighted average of both, so as to tune the level of optimism.

10.6.3 Goal-Driven Ranking Methods

This family of methods consists of comparing fuzzy intervals $\{M_1,\ldots,M_n\}$ to be ranked and some reference fuzzy set. There are two views of the reference fuzzy set: it can be understood as an ideal fuzzy interval; then the problem is to minimize some distance to this fuzzy set. Or it can be viewed as a fuzzy constraint and the problem is to compute the degree of satisfaction of this fuzzy constraint by the fuzzy intervals. The rankings are done according to these evaluations. The advantage of this kind of approach is that the decision-maker can interfere in the ranking process while absolute defuzzification methods look more objective, hence sometimes more arbitrary.

The oldest method of this type is due to Jain (1977). It uses a fuzzy goal G whose meaning is 'as great as possible'. It is a fuzzy set of the convex closure I of the union of the supports of the M_i's, with an increasing membership function. The fuzzy quantities are then compared with this fuzzy goal and the ranking index is

$$r(M_i) = \sup_x \min (G(x), M_i(x)) = \Pi_i(G). \qquad (10.46)$$

where $\Pi_i(G)$ is the possibility that G is attained by M_i. Jain (1977) assumed $G(x) = \left(\frac{x}{\sup I}\right)^k$, for some $k > 0$. One of Yager (1979c) proposals is of that form, when $I = [0, 1]$, and $k = 1$. The problem is that only the right side of the fuzzy intervals is exploited. Clearly one is tempted to use the necessity measure $N_i(G) = 1 - \Pi_i(G^c))$ as well, so as to account for the left side of the fuzzy intervals (with $G^c(x) = 1 - G(x)$). Chen (1985)'s method is similar; however he uses a maximising set $G(x) = \left(\frac{\sup I - x}{\sup I - \inf I}\right)^k$ and a minimising set base $G'(x) = \left(\frac{\inf I - x}{\inf I - \sup I}\right)^k$ and uses an average ranking index $r(M_i) = \frac{\Pi_i(G) + 1 - \Pi_i(G')}{2}$ which takes care of both sides of the fuzzy intervals. In fact in Chen 's method, $G(x) = 1 - G'(x)$ when $k = 1$, so that $r(M_i) = \frac{\Pi_i(G) + N_i(G)}{2}$. Kim and Park (1990) improved Chen's method with an index of optimism to reflect the decision maker attitude and used a weighted average of $\Pi_i(G^c)$ and $N_i(G)$. However, when their index of optimism is 0.5, the corresponding outcome is the same as by Chen's method with $k = 1$. It is clear that in the above methods $G(\cdot)$ plays the role of a utility function and the fuzzy numbers that of lotteries in the same spirit as for decision theory. Hence the ranking indices $\Pi_i(G)$ and $N_i(G)$ play the roles of the expected utility; they have been axiomatized as such (Dubois and Prade, 1995), (Dubois, Prade and Sabbadin, 1998).

Baas and Kwakernaak (1977) proposed a ranking index of the form $\Pi_i(G)$ where $G = (-\infty, \widetilde{\max}(M_1,\ldots,M_n)]$. Baldwin and Guild (1979a) refine the index using a fuzzy ordering relation P and compute the ranking index as $\min_{j \neq i} \Pi_i(M_j \circ P)$. Ammar (1989) also proposes the fuzzy set $G = \widetilde{\max}(M_1,\ldots,M_n)$ as a reference set and defines two optimality indices of M_i: $r_1(M_i) = \max\{\lambda, (M_i)_\lambda = G_\lambda\}$; $r_2(M_i) = \max\{\lambda, (M_i)_\lambda \neq G_\lambda \text{ and } (M_i)_\lambda \cap G_\lambda \neq \emptyset\}$; he proposes finally to rank the fuzzy intervals according to $\max(r_1(M_i), r_2(M_i))$.

Instead of using possibilistic like comparisons between a fuzzy interval and a fuzzy goal, Yager (1980b) computes the Hamming distance between a fuzzy subset of the unit interval M_i and the fuzzy set G, namely $d(M_i, G) = \int_0^1 |G(x) - M_i(x)|\, dx$ where $G(x) = x$. The same author (Yager, 1979c) also suggested a probabilistic counterpart of the Jain's method under the form $r(M) = \dfrac{\int_{-\infty}^{+\infty} G(x)M(x)dx}{am(M)}$, where am(M) is the area under M(·) (see Section 10.2.2). This is an extension of the center of gravity defuzzification using the goal function G.

10.6.4 Fuzzy Ordering Relations Induced by Fuzzy Intervals

If the aim of the comparison between fuzzy intervals is to provide an accurate picture of their relative positions, it is more natural to build a fuzzy ordering on the set of fuzzy intervals and exploit this fuzzy relation for the purpose of ranking or selecting. In the presence of imprecision on the data, the faithfulness of the obtained picture is somewhat incompatible with the decisiveness property that may be needed in some circumstances. However the fuzzy relational approach to fuzzy interval comparison offers a rich basis for discussion pertaining to which fuzzily rated decision is the best.

Possibilistic Interval Graphs. Let $M = [\underline{m}, \overline{m}]$ and $N = [\underline{n}, \overline{n}]$ be two closed intervals restricting the values of variables x and y, respectively. The question of ranking such intervals can be viewed as describing their respective locations on the real line. Basically there are four relative positions between M and N:

i) $M \succ N$ if and only if $\forall x \in M, \forall y \in N, x > y$,
which is equivalent to $\underline{m} > \overline{n}$

ii) $M \succ^{+} N$ if and only if $\forall x \in M, \exists y \in N, x \geq y$,
which is equivalent to $\underline{m} \geq \underline{n}$

iii) $M \succ^{-} N$ if and only if $\forall y \in N, \exists x \in M, x > y$,
which is equivalent to $\overline{m} > \overline{n}$

iv) $M \succcurlyeq N \equiv \neg (N \succ M)$ if and only if
$\exists (x, y) \in M \times N, x \geq y$, which is equivalent to $\overline{m} \geq \underline{n}$.

These relations form what is known as an interval ordering structure (Fishburn, 1985). $M \succ N$ is the strongest relation between intervals and in terms of variables, it means that x is necessarily greater than y, while (iv) is the weakest relation, asserting that $x \geq y$ is only possible. (ii) and (iii) are intermediary statements dealing with the least and greatest values of x and y, respectively. Indeed (i) implies both (ii) and (iii); and any of these three statements implies (iv). Denoting $M \approx N$ the conjunction of $M \succcurlyeq N$ and $N \succcurlyeq M$ (intersecting intervals, it can be shown that $M \succ^{+} N$ if and only if $M \succ A \approx N$ for some interval A, and $M \succ^{-} N$ if and only if $M \approx A \succ N$ for some interval A.

One way of using this interval ordering structure on fuzzy intervals is to select a membership level λ and compare the level-cuts M_λ and N_λ accordingly. For instance, Buckley and Chanas (1989) use the relation $M_\lambda \succ N_\lambda$. Another approach is to fuzzify the four relations according to the extension principle, as done in Dubois and Prade (1983b), by means of four dominance indices:

$$d(M \succ N) = \inf_{x,\, y} \{\max(1 - M(x), 1 - N(y)) : x \le y\} \tag{10.47}$$
$$d(M \succ^{+} N) = \inf_{x} \sup_{y} \{\max(1 - M(x), N(y)) : x \ge y\} \tag{10.48}$$
$$d(M \succ^{-} N) = \sup_{x} \inf_{y} \{\max(M(x), 1 - N(y)) : x \le y\} \tag{10.49}$$
$$d(M \succcurlyeq N) = \sup_{x,\, y} \{\min(M(x), N(y)) : x \ge y\} . \tag{10.50}$$

When M and N are standard intervals, then each of the four relations hold if and only if the corresponding index is 1. In the case of crisp intervals theses indices are 'all or nothing', i.e., range on {0, 1}; with fuzzy intervals the four relations become fuzzy as assessed by the corresponding index. In the framework of possibility theory, the four indices are interpreted as degrees of possibility or necessity of fuzzy events, namely:

$$d(M \succ N) = N_M((N, +\infty)); \qquad d(M \succ^{+} N) = N_M([N, +\infty));$$
$$d(M \succ^{-} N) = \Pi_M((N, +\infty)); \qquad d(M \succcurlyeq N) = \Pi_M([N, +\infty)).$$

In terms of level-cuts the values of the indices can be interpreted as follows:

$$d(M \succ N) \ge \lambda \text{ if and only if } \inf M_{1-\lambda} > \sup N_{1-\lambda}$$
$$d(M \succ^{+} N) \ge \lambda \text{ if and only if } \inf M_{1-\lambda} \ge \inf N_\lambda$$
$$d(M \succ^{-} N) \ge \lambda \text{ if and only if } \sup M_\lambda > \sup N_{1-\lambda}$$
$$d(M \succcurlyeq N) \ge \lambda \text{ if and only if } \sup M_\lambda \ge \inf N_\lambda.$$

Historically, the index $d(M \succcurlyeq N)$ was first suggested by Baas and Kwakernaak (1977), and $d(M \succ^{-} N)$ by Watson *et al.* (1979). If used alone, this index lacks discrimination power in some situations; to cope with this fact, Baldwin and Guild (1979a) inserted a fuzzy inequality relation in its expression instead of the crisp relation $x \ge y$. Watson *et al.* (1979) suggested the index $d(M \succ N)$, which is equal to $1 - d(N \succcurlyeq M)$. Further, Tsukamoto *et al.* (1981) proposed a set of indices which basically contain the same information as the above ones. Ramik and Nakamura (1993) extend the index $d(M \succcurlyeq N)$ to interactive fuzzy intervals.

In spite of their complicated expressions, these indices have some intuitive appeal and are easily calculated as intersecting points between membership functions of M, N, and their fuzzy complements, as pictured in Figure 5. Moreover it is easy to check that $d(M \succcurlyeq N)$ and $d(M \succ N)$ can be retrieved by positioning the fuzzy difference M(–)N with respect to 0 as follows:

$d(M \succcurlyeq N) = (-\infty, M(-)N]\,(0)$; $d(M \succ N) = (-\infty, M(-)N)\,(0)$.

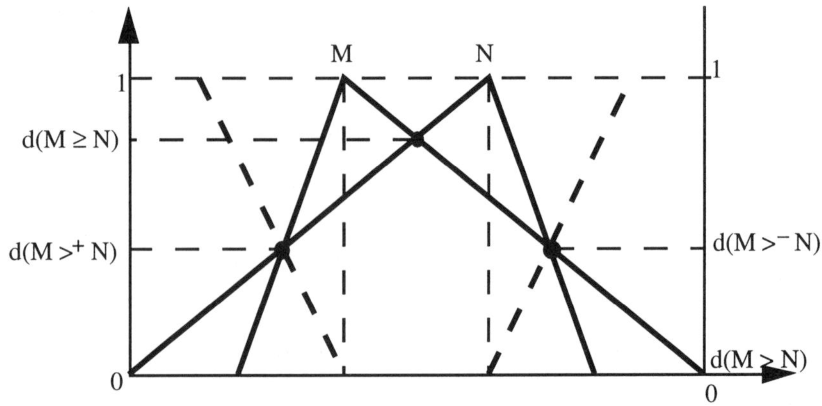

Figure 10.5

Let us point out some properties relating the introduced indices (see Dubois and Prade, 1983b):

$$d(M \succcurlyeq N) \geq \max(d(M \succ^- N), d(M \succ^+ N));$$
$$\min(d(M \succ^- N), d(M \succ^+ N)) \geq d(M \succ N);$$
$$d(M \succ N) = 1 - d(N \succcurlyeq M).$$

The strong dominance index has the following remarkable properties (Ovchinnikov and Migdal, 1987; Roubens and Vincke, 1988; Dubois and Prade, 1991b)

i) $d(M \succ M) = 0$ (irreflexive)
ii) $\min(d(M \succ N), d(N \succ M)) = 0$ (antisymmetry)
iii) $d(M \succ N) \geq \min(d(M \succ N'), d(N' \succ N))$ (transitivity)
iv) $\min(d(M \succ N), d(M' \succ N')) \leq \max(d(M \succ N'), d(M' \succ N))$ (Ferrers relation)

These properties are the fuzzy extension of the characteristic properties of interval orders. The two other indices are tournament relations in most cases:

$$d(M \succ^+ N) = 1 - d(N \succ^+ M);$$
$$d(M \succ^- N) = 1 - d(N \succ^- M).$$

These equalities may fail to hold when the extrema in the expressions defining these indices coincide with discontinuity points of both $M(\cdot)$ and $N(\cdot)$, see (Dubois and Prade, 1983b) for a complete discussion. The comparison indices $d(M \succcurlyeq N)$ and $d(M \succ N)$ have been extended by Delgado using t-norms instead of minimum in the expression of $d(M \succcurlyeq N)$, and using their modified interval $[Q, +\infty)^{\lambda}$ defined in Section 10.6.1. They proved that these λ-modified relations are insensitive to scalar multiplication of the fuzzy intervals and that they are consistent with fuzzy additions:

$$d(M \oplus M' \succ N \oplus N') \geq \min(d(M \succ N), d(M' \succ N')).$$

The two relations $\succ^{+}$ and $\succ^{-}$ can be used jointly on standard intervals and it yields another stronger natural ordering:

$M \gtrsim N$ if and only if $M \succcurlyeq^{+} N$ and $M \succcurlyeq^{-} N$

if and only if $\widetilde{\min}(M, N) = N$
if and only if $\widetilde{\max}(M, N) = M$.

This method can be applied to fuzzy intervals using the extended minimum and maximum. This approach for ranking fuzzy intervals has been proposed by Dubois and Prade (1978b) and subsequently by Freeling (1980), Ramik and Rimanek (1985), Lowen (1985). Despite its well-foundedness, this criterion is of the ' all – or – nothing ' type and yields a partial ordering since $\widetilde{\max}(M, N)$ may happen to be neither M nor N (see also Baldwin and Guild, 1979b). Notice that $M \gtrsim N$ if and and only if $\forall \lambda \in [0, 1]$, $M_{\lambda} \succcurlyeq^{+} N_{\lambda}$ and $M_{\lambda} \succcurlyeq^{-} N_{\lambda}$. Hence it is enough that one of these inequalities fail for a single membership grade for violating this criterion. It is thus very easy to find situations of undecisiveness, where intuitively one would expect that M should be declared greater than N because they are very different, and other situations where M and N are arbitrarily close pointwise, but $M > N$ is valid. However the method is general enough to be extended to more general random intervals (without nestedness property for level-cuts) and to the VSOP fuzzy numbers of Tamura and Horiuchi (1993), consisting of pairs of functions from the reals to the unit interval. Gonzalez and Vila (1991, 1992) consider a discrete version of the ordering $M \gtrsim N$ where only a finite set of level-cuts is envisaged. This view reduces the problem of comparing fuzzy intervals to one of comparing vectors. The authors study the Pareto ordering (a discrete version of $M \gtrsim N$) and a lexicographic ordering giving priority to higher membership grades.

Probabilistic Relational Comparison Methods. A random set extension of the interval order notion has been proposed by Chanas *et al.* (1993). Namely consider the three cases of random intervals mentioned in Section 10.2.1. Denoting $M_{\lambda} = [\underline{m}_{\lambda}, \overline{m}_{\lambda}]$ the level-cut of M and x^{-} and x^{+} random variables depending on two parameters α and β that generate the random interval $I(M) = [x^{-}, x^{+}]$ with the following assumptions:

i) consonant case:

$$\alpha = \beta = \lambda: I(M) = [x^{-}(\lambda), x^{+}(\lambda)] = [\underline{m}_{\lambda}, \overline{m}_{\lambda}];$$

ii) asymmetric case:

$$\alpha = \lambda = 1 - \beta\text{: } I(M) = [x^-(\lambda), x^+(\lambda)] = [\underline{m}_\lambda, \overline{m}_{1-\lambda}];$$

iii) independent case:

$$I(M) = [x^-(\alpha), x^+(\beta)] = [\underline{m}_\alpha, \overline{m}_\beta].$$

Assume the two fuzzy numbers M and N depend on the same random parameters α and β that generate the random interval I(M) and I(N). Define the probabilistic dominance indices between M and N as follows:

$$d_p(M \succ N) = \text{Prob}(I(M) \succ I(N));$$
$$d_p(M \succ^+ N) = \text{Prob}(I(M) \succ^+ I(N));$$
$$d_p(M \succ^- N) = \text{Prob}(I(M) \succ^- I(N));$$
$$d_p(M \succcurlyeq N) = \text{Prob}(I(M) \succcurlyeq I(N)).$$

Actually, there are three families of indices reflecting the three assumptions. In the consonant case, $d_p(M \succ N)$ and $d_p(M \succcurlyeq N)$ coincide with their possibilistic counterparts (not the other ones). The property $d_p(M \succ N) = 1 - d_p(N \succcurlyeq M)$ obviously holds, as well as

$$d_p(M \succ N) \le \min(d_p(M \succ^+ N), d_p(M \succ^- N), d_p(M \succcurlyeq N));$$
$$d_p(M \succcurlyeq N) \ge \max(d_p(M \succ^+ N), d_p(M \succ^- N));$$
$$d_p(M \succ N) > 0 \text{ implies } d_p(M \succcurlyeq N) = 1.$$

Chanas *et al.* (1993) prove the max-min transitivity of $d_p(M \succ N)$ in all cases.

In a similar spirit, but with a much more debatable rationale, Lee *et al.* (1994)'s relational approach is very similar to the natural probabilistic relation obtained when comparing two standard random variables x and y according to their joint distributions p. Then

$$\text{Prob}(x \ge y) = \int_{u \ge v} p(u, v)\,du\,dv.$$

Lee *et al.* (1994) apply this approach to fuzzy interval ranking, where $p(u, v) = \frac{M_i(u) * M_j(v)}{am(M_i * M_j)}$. However they do it on fuzzy intervals with discretized support so as to simplify calculations. They give an iterative ranking technique, close to Orlovsky (1978)'s method, that exploit the constructed fuzzy relation. This approach is really the comparative counterpart to the ranking approach by the center of gravity defuzzification. Note that the relation $>_p$ such that $M_i >_p M_j$ if and only if $\text{Prob}(x \ge y) > \text{Prob}(y \ge x)$ is not transitive.

Nakahara (1998) recently tried to connect the possibilistic pairwise dominance indices of Dubois and Prade (1983) and probabilistic indices of the form Prob($x \geq y$) for comparing intervals interpreted as uniformly distributed random variables. Namely, conditions such as d($M \succ N$) $\geq \lambda$ can be related to Prob($x \geq y$) for $x \in M_\lambda$ and $y \in N_\lambda$.

Relational Methods Based on Area Compensation. Instead of exploiting the coordinates of intersection points between the membership functions of fuzzy intervals or their complements for estimating their respective positions, it also makes sense to compare them by means of areas delimited by the membership functions.

When comparing two fuzzy intervals M_i and M_j, Kerre (1982, 1987) compares the Hamming distances between each of M_i and M_j, and $\widetilde{\max}(M_i, M_j)$ so as to decide which is the greatest. Referring to Figure 6, and denoting δ the Hamming distance, it is easy to check that

$$\delta(M_j, \widetilde{\max}(M_i, M_j)) = S_1 + S_2; \quad \delta(M_i, \widetilde{\max}(M_i, M_j)) = S_3 + S_4.$$

It yields a binary relation on the set of fuzzy intervals to be compared whereby $M_i > M_j$ if $S_1 + S_2 > S_3 + S_4$. Nakamura (1986) refines this approach either using a normalization factor or comparing the left and right sides separately and performing a weighted average of the corresponding indices, still using Hamming distances. Kolodziejczyk (1986) proposes more complex indices along the same lines of thought. ,

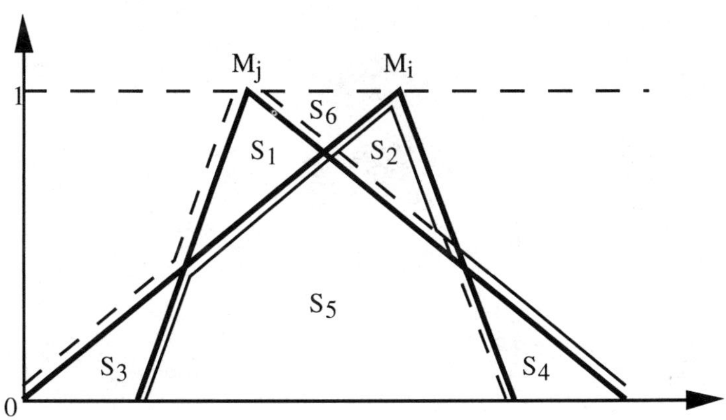

Figure 10.6

McCahon and Lee (1990) compute a dominance index of M_i over M_j by means of the proportions of area PA(M_i) and pa(M_i) common to M_i and each of $\widetilde{\max}(M_i, M_j)$ and $\widetilde{\min}(M_i, M_j)$.Then $M_i > M_j$ if M_i overlaps $\widetilde{\max}(M_i, M_j)$ more than M_j, and M_i overlaps $\widetilde{\min}(M_i, M_j)$ less than M_j. In terms of surfaces in Figure 6, it means that $M_i > M_j$ if $S_2 > S_4$ and $S_1 > S_3$ since $PA(M_i) = S_2 + S_5$ and $pa(M_i) = S_3 + S_5$. In case of conflict, a similar logic is proposed involving terms of the form $PA(M_i)/(pa(M_i) + PA(M_i))$, called proportion of the optimum, as well as $pa(M_i) + PA(M_i)$. Especially the authors wish to make a strict ranking decision when M_i and M_j are symmetric fuzzy intervals with the same core, on the basis of specificity.

The Hamming distance operates a vertical comparison of the membership functions. Tseng and Klein (1989) use an horizontal comparison to compute a pairwise dominance index based on the analysis of areas delimited by the membership functions of M_i and M_j. Namely there are three kinds of areas:

- those where M_i dominates M_j; this is when $\underline{m}_{i\lambda} > \underline{m}_{j\lambda}$ or $\overline{m}_{i\lambda} > \overline{m}_{j\lambda}$; this is S_1, S_2 and S_6.
- the one where M_i and M_j overlap; this is in the intersection area; this is S_5.
- those where M_j dominates M_i; this is S_3 and S_4.

Summing the contributions over the membership grades leads to computing (see Fig. 6)

$$A(M_i > M_j) = \int_0^1 \max(0, \sup M_{i\alpha} - \sup M_{j\alpha})\,d\alpha + \int_0^1 \max(0, \inf M_{i\alpha} - \inf M_{j\alpha})\,d\alpha$$

$$= S_1 + S_2 + 2\,S_6 \qquad \text{(see Figure 6).}$$

$$A(M_i = M_j) = am(M_i \cap M_j) = S_5;\ A(M_j > M_i) = S_3 + S_4.$$

The weak dominance index is $d_{TK}(M_i \geq M_j) = \dfrac{A(M_i > M_j) + A(M_i = M_j)}{am(M_i) + am(M_j)}$.

Moreover $d_{TK}(M_i \geq M_j) + d_{TK}(M_j \geq M_i) = 1$ and $d_{TK}(M_i \geq M_i) = 0.5$. Note the difference of the index with the one obtained by Kerre (1982)'s method. The Hamming distance does not account for the horizontal distance between M_i and M_j embodied here by the area S_6. Hence Tseng and Klein's approach better describes the relative positions of the fuzzy intervals on the metric scale.

Yuan (1991) has proposed an area-based dominance index based on comparing the fuzzy difference of the form $M_i(-)M_j$ to 0. If A^+_{ij} is the area of the fuzzy interval

$[0, M_i(-)M_j]$ and A^-_{ij} denotes the area of the fuzzy interval $[M_i(-)M_j, 0]$, then he computes $d_Y(M_i \geq M_j) = \dfrac{A^+_{ij}}{A^+_{ij} + A^-_{ij}}$. However it is not difficult to check that $A^+_{ij} = A(M_i>M_j) + A(M_i=M_j)$ and $A^-_{ij} = A(M_j>M_i) + A(M_i=M_j)$ so that the two methods are the same. Yuan(1991) proves the transitivity of the relation defined by $M_i \geq_{TK} M_j$ if and only if $d_{TK}(M_i \geq M_j) \geq 0.5$.

These basic area evaluations were also used by Roubens (1990) with dominance index based on area compensation:

$$d_R(M_i \geq M_j) = \max(0, \tfrac{1}{2}(A(M_i>M_j) - A(M_j>M_i))).$$

This dominance index is min-transitive and satisfies the Ferrers property already encountered in the previous subsection. Fortemps and Roubens (1996) suggested to drop the positivity assumption and studied

$$\begin{aligned} d_{FR}(M_i \geq M_j) &= \tfrac{1}{2}(A(M_i>M_j) - A(M_j>M_i)) \\ &= \tfrac{1}{2}(S_1 + S_2 + 2\,S_6 - S_3 - S_4) \text{ in Figure 6.} \end{aligned}$$

The same basic area evaluations also appear in Saade and Schwartzlander (1992) However like Mc Cahon and Lee (1990), they separate the comparisons between the right sides of M_i and M_j and the comparisons between the left sides, namely they use dominance indices $A^r(M_i>M_j)$ and $A^l(M_i>M_j)$ denoting the two terms appearing in the above expression of $A(M_i>M_j)$. These two terms can be expressed as Hamming distances between $(-\infty, M_i]$ and $(-\infty, \widetilde{\max}(M_i, M_j)]$, as well as $[M_i, +\infty)$ and $[\widetilde{\max}(M_i, M_j), +\infty)$ respectively. However $A(M_i>M_j)$ differs from the Hamming distance between M_i and $\widetilde{\max}(M_i, M_j)$. They also consider the ordering $\geq_{TK}$ induced by comparing $A(M_i > M_j)$ and $A(M_j > M_i)$. Interestingly Saade and Schwartzlander (1992) (also Fortemps and Roubens (1996)) proved that

$$A(M_i>M_j) - A(M_j > M_i) = 2(\bar{E}(M_i) - \bar{E}(M_j))$$

so that the relational method based on $A(M_i>M_j)$ comes down to Yager (1981)'s method based on defuzzification by the middle point of the mean interval. They also show that this approach is robust to small changes in the membership functions. The transitivity of the relation $\geq_{TK}$ then should not be surprizing.

Jimenez (1996) studies the property of yet another dominance index of the same family (where $E_*(M)$ and $E^*(M)$ are defined in Section 10.2.2):

$$d_J(M_j > M_i) = \min\left(1, \max\left(0, \frac{E^*(M_i) - E_*(M_j)}{E^*(M_i) + E^*(M_j) - E_*(M_i) - E_*(M_j)}\right)\right).$$

A different kind of dominance index related to $\bar{E}(M_i)$ is proposed by Campos and Gonzalez (1994). When ranking fuzzy intervals using the Hurvicz-like defuzzification $r_\alpha(M)$, it is clear that the ranking may depend on the choice of the optimism parameter α. The dominance region of M_i over M_j is defined as the set

$$D(M_i \geq M_j) = \{\alpha \in [0, 1]: r_\alpha(M_i) \geq r_\alpha(M_j)\}.$$

$D(M_i \geq M_j)$ is an interval of the form $[\alpha*, 1]$ or $[0, \alpha*]$ or $[0, 1]$ or $\emptyset$. $D(M_i \geq M_j) = [0, 1]$ when $E^*(M_i) \geq E^*(M_j)$ and $E_*(M_i) \geq E_*(M_j)$. Moreover $D(M_i \geq M_j) \cap D(M_i \geq M_j) = I(M_i, M_j)$ is reduced to a singleton or is empty. This set-valued dominance notion has nice properties similar to the ones expected in preference modeling:

$$D(M_i \geq M_j) \cup D(M_i \geq M_j) = [0, 1]$$
$$D(M_i \geq M_i) = [0, 1]$$
$$D(M_i \geq M_j) = (D(M_j \geq M_i))^c \cup I(M_i, M_j)$$
$$D(M_i \geq M_j) \cap D(M_j \geq M_k) \subseteq D(M_i \geq M_k).$$

The width of $D(M_i \geq M_j)$ is yet another pairwise dominance index.

From Pairwise Comparison Indices to a Ranking. The next step is to derive a ranking of n fuzzy quantities from valued pairwise comparisons. See Dubois and Prade (1983b) for the possibilistic indices where generic suggestions are made. There are basically three options. We can derive global ranking indices, which yield linear orderings (one per index) comparing each fuzzy interval with their fuzzy maximum or minimum (we are then back to the goal-driven approach; see the previous Section 10.6.3). This is the simplest choice at the computational level. We may also adopt a more combinatorial approach, computing $d(M_{\sigma(1)} > M_{\sigma(2)} > \ldots > M_{\sigma(n)})$ for some choice of ordering index and all permutations as suggested by Buckley (1989b). It presupposes an extension of the pairwise comparison indices to n-ary comparison indices. This extension is obvious in the approaches that extend interval orders such as the possibilistic relations and the random interval view. It is far less obvious for the area compensation methods. Moreover it is clearly difficult at the computational level due to the high combinatorial complexity. In contrast one may as well process the fuzzy relations obtained from the valued pairwise comparison indices using appropriate exploitation techniques (surveyed in Fodor *et al.* (1998) in the volume edited by Slowinski (1998) in the Handbooks of Fuzzy Sets Series). A generic method is described in Wang (1997), that extracts a total order relation from a fuzzy relation. This procedure is proposed for ordering fuzzy intervals on the basis of an acyclic fuzzy relation. Acyclicity is a rather weak property that is verified by most pairwise comparison indices. The key issue is whether some transitivity exists or not (see also Wang and Da Ruan, 1995; Wang *et al.*, 1995b). The recent works on fuzzy orderings by Bodenhofer (1998) seem to offer yet another research line towards a more systematic approach to the problem of ranking fuzzy numbers.

10.6.5 Fuzzy Dominance Indices and Linguistic Methods

Another type of description of the relative position of fuzzy intervals M_i and M_j consists in using a distribution instead of a scalar value for the dominance index. This view was suggested rather early and is in full accordance with the spirit of fuzzy logic, although it has been seldom exploited since then.

A first idea (Tong and Bonissone, 1980) is to compute a fuzzy dominance index of M_i over the other M_j's. What is obtained is a fuzzy quantity representing a global dominance profile $\Delta(M_i)$ for M_i. For instance the extension principle can be applied to

$$\Delta(x_i) = x_i - \frac{\sum_{j \neq i} w_j x_j}{\sum_{j \neq i} w_j} \quad \text{where } w_j = \min_k d(M_j \geqslant M_k)$$

using the Baas and Kwakernaak (1977) index of possibility of dominance. Then the decision-maker provides linguistically meaningful dominance profiles $L_1, \ldots, L_p$. They are fuzzy intervals whose meaning can be " approximately equal to average", "marginally better", "definitely better"..., etc. The next step is one of linguistic approximation, by which the most suitable label is found for $\Delta(x_i)$, by computing a distance between $\Delta(x_i)$ and the meaning of each label. The merit of such an approach is its user-friendliness and the non-commitment to arbitrary precision. One may even think of asking the user to estimate the relative worth of decisions by means of the labels of dominance profiles (Efstathiou and Bonissone, 1979; Efstathiou and Tong, 1982). The obtained linguistic preference relation is then exploited so as to get a ranking. However this kind of approach is beyond the scope of this chapter.

The use of fuzzy-valued dominance indices is also suggested by Mabuchi (1988) who bases his analysis on the visualization of the length distribution of the level-cuts of the fuzzy differences $M_i(-)M_j$. Namely, he considers the length of level-cuts of $[0, +\infty) \cap M_i(-)M_j$. and $(-\infty, 0] \cap M_i(-)M_j$., the positive and negative parts of $M_i(-)M_j$. These are denoted $d^+_\lambda = \max(0, \overline{m}_{i\lambda} - \underline{m}_{j\lambda})$ and $d^-_\lambda = \max(0, \overline{m}_{j\lambda} -$

$\underline{m}_{i\lambda}$). Let $J_{ij}(\lambda) = \frac{d^+_\lambda - d^-_\lambda}{d^+_\lambda + d^-_\lambda}$. Then plotting $J_{ij}(\lambda)$ against λ gives a faithful view of the relative positions of M_i and M_j. the more $J_{ij}(\lambda)$ overlaps [0,1] the more M_i dominates M_j, the more $J_{ij}(\lambda)$ overlaps [–1, 0] the more M_j dominates M_i; this is symmetric because $J_{ij}(\lambda) = -J_{ji}(\lambda)$. See Figure 7.

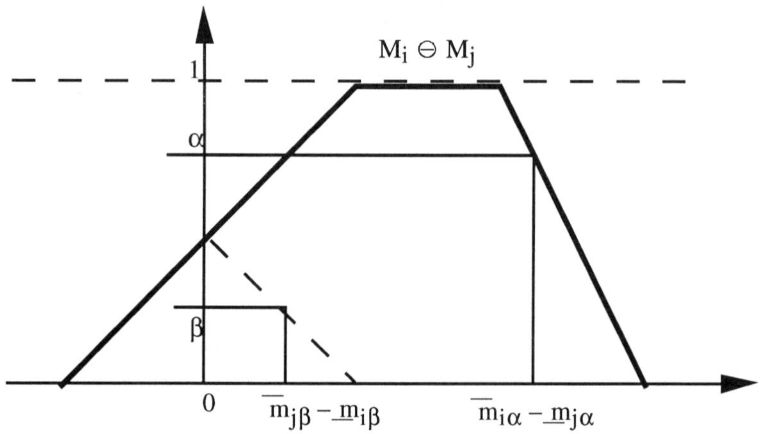

Figure 10.7

When $M_{i\lambda} \cap M_{j\lambda} = \emptyset$, $J_{ij}(\lambda) = 1$ if $\underline{m}_{i\lambda} > \overline{m}_{j\lambda}$. When $M_i = M_j$, $J_{ij}(\lambda) = 0\ \forall\lambda$. When $M_i = m \in \mathbb{R}$ and $M_j = n \in \mathbb{R}$, and $m > n$, $J_{ij}(\lambda) = 1$ by convention, and 0 if $m = n$. Unless both $J_{ij}(\lambda)$ and $J_{jk}(\lambda)$ are negative, transitivity holds in the sense of the inequality $J_{ij}(\lambda) \geq \min(J_{ij}(\lambda), J_{jk}(\lambda))$. The profile J_{ij} can be used either in a linguistic approach to comparing M_i and M_j: linguistic terms are expressed by template profiles and J_{ij} is matched against them. Another exploitation consists in computing a scalar pairwise dominance index $\overline{J}_{ij} = 2\int_0^1 \lambda J_{ij}(\lambda)d\lambda$ where more importance is given to higher membership grades. This approach is sensitive to the case when $M_i \subset M_j$ and M_i is much more precise than M_j and moves inside M_j. For instance if M_j is close to the inner right side of M_i and still in its core, then M_j is considered on the right of M_i.

10.6.6 Criteria for Ranking Fuzzy Intervals

How to compare fuzzy interval rankings methods? This question has been considered by several authors (Bortolan and Degani (1985), Li and Lee (1987), Tseng and Klein (1988), Hirota and Pedrycz (1989), Yuan (1991), Zhu and Lee

(1992), etc., basically on the basis of examples. For instance in the first experimental study by Bortolan and Degani (1985), an extensive set of ranking techniques was tested. Most ranking methods were found to be counter-intuitive for some situations, but the set of 4 indices (47 – 50) always provided a reasonable answer when normal fuzzy quantities were compared for the purpose of describing their relative positions, although a clear ranking was not always obtained in this case, due to possible conflict between the four fuzzy relations. In fact, although there are various approaches, many of them are not really distinct. Some relations existing between some methods are pointed out above, and more work can be found in Wang and Kerre (1996, 1998 part I) for defuzzification methods, Wang and Kerre (1998 part II) for relational methods, and Wang (1997) more generally. Many methods yield the same final rankings and some rankings are related to one another.

On the other hand, the intuition is widely used to judge whether an ordering procedure is reasonable or not. In order to adopt a more scienfific point of view, Yuan (1991) points out four sensible criteria for comparing ordering methods: the mode of presentation, the rationality, the discrimination power, and the robustness.

Decisive vs. Descriptive Comparison Methods. The classification in this chapter is driven by the mode of presentation, and is more precisely goal-driven since the mode of presentation of the results depend on what kind of use is made of these results. Our classification also goes by types of techniques. There are two dimensions to this classification: whether the decision-maker is supposed or not to influence the final ranking, and whether the aim is to decide or to describe. Descriptive methods tend to be as objective as possible, and accept to postpone decisions. This is the case with relational methods that extend the interval order structure. However they provide a complex description of the relative positions of the fuzzy intervals, either because it is based on several order relations that reflect several possibly antagonistic points of view on the ranking, or it is based on a valued relation that is not obvious to exploit because it merges conflicting points of view (as often the case with multiple-criteria problems). For instance many difficulties for devising ranking methods are due to the fact that a fuzzy interval has two sides and that the information conveyed by comparing the right sides may contradict the information conveyed by the left sides. A comparison method can either describe correctly the two phenomena (as Dubois and Prade, 1983b)'s family of possibilistic indices do) or let them compensate each other (like many area-based techniques, or defuzzification methods, do) with a risk of getting counterintuitive results. Finally, one may resort to ask the decision-maker whether he finds the right sides more important than the left side, or whether he accepts risk or not.

Decisive methods tend to let the decision-maker slip in the choice of the ranking. This is unsurprizing because, due to the imprecision of the fuzzy intervals, the extraction of a clear ranking may sound arbitrary. It is better to ask the decision-maker about his attitude in front of uncertainty (via the introduction of the optimism coefficient in the defuzzification methods) or about his expectation levels (via the specification of one or several fuzzy goal sets as in the methods that stem from Jain (1976)'s proposal), or both (Kim and Park's (1990) improvement of Chen (1985)'s method).

Rationality. Regarding rationality, not much work has been done so far at the formal level. Many techniques are ad hoc and the benchmarks run so far emphasize coherence with intuition but lack generality. Clearly, everybody accepts the idea that if M is on the right of N and they do not overlap, then M > N should be obtained. This is again a natural starting point. However, beyond some obvious consensual requirements, the intuition of some authors differs from the intuition of others, especially if their purposes differ. Rationality can only be envisaged inside groups of methods. The principles of rationality may differ for a ranking method based on defuzzification, and for a descriptive relational method. The state of the art is still weak from the point of view of axiomatic foundations, compared to utility theory that basically ranks random numbers. However some progress has been made recently. Among the defuzzification methods, the ones based on the mean interval of a fuzzy interval, initiated by Yager (1981) sound the most promising as the natural counterpart of expected utility, due to their close relationship to Choquet integrals. An axiomatisation of this type of method sounds possible because of existing works on the axiomatisation of such integrals in the scope of decision theory (Schmeidler, 1989). Similarly, recent works on qualitative utility theory based on possibility theory (Dubois and Prade, 1995; Dubois, Prade and Sabbadin, 1998) can pave the way to an axiomatics for the goal-driven ranking methods of Section 10.6.3, taking the function G as a utility function, and fuzzy intervals as uncertainty distributions. As for the relational methods, the study of rationality presupposes a strong anchoring of the comparison indices in the theory of crisp preference modelling (after Fishburn (1985), Roubens and Vincke (1985)) and fuzzy preference modelling (see the companion volume in the Handbooks of Fuzzy Sets Series edited by Slowinski (1998); Fodor and Roubens, 1994). The rationality of the area compensation methods again stem from their strong links with the Choquet-type mean-value of fuzzy intervals. However it would be worth studying the links between area compensation methods and random interval orders of Chanas *et al.* (1993).

Discrimination Power. The discrimination power is an important criterion. Clearly some methods lack discrimination, like the possibilistic $d(M \succcurlyeq N)$ index of Baas and Kwakernaak (1977). However this is not because it has some intrinsic flaw. This is because it is only part of a bigger picture that is a generalization of the well known interval order. Taking the whole picture into account (the 4 indices) restores discrimination. However discrimination does not imply decisiveness. Some authors seem to look for a ranking function that would result in maximal discrimination, and maximal decisiveness. It is not clear that it makes sense. For instance it is not clear that two symmetric fuzzy intervals with identical cores and different supports should be objectively discriminated from the point of view of their ranking. They have the same defuzzified value according to most objective methods. The only ways of discriminating them are:

i) to compare the fuzzy intervals on the basis of specificity (as done by Mc Cahon and Lee, 1990). However it does not mean that one fuzzy interval is greater, strictly speaking, than the other as these authors suggest. One is less risky than the other.

ii) to ask the decision-maker for his attitude in front of risk.

Robustness. Robustness is defined by Yuan (1991) as follows: A fuzzy dominance index $d(M \geq N)$ is robust if for any threshold ε, and approximation M' of M there is another threshold η such that if the distance $\delta(M, M')$ between M and M' satisfies $\delta(M, M') < \varepsilon$ then $|d(M \geq N) - d(M' \geq N)| < \eta$. The robustness is an important issue, and should be checked systematically as proposed by Yuan (1991) who claimed that some methods proposed in the literature are not robust, such as some indices of Nakamura (1986). Fortemps and Roubens (1996) have proved the robustness of their method, hence of the Choquet integral based defuzzifications. However there is also here a need for a systematic study on existing methods.

Learning. Rather than looking for a priori criteria for ranking fuzzy intervals one may envisage a data-driven approach whereby an expert ranks a set of prototype fuzzy intervals and the ranking relation is learned from this information. For instance Pedrycz (1986) proposes to learn the fuzzy relation "greater than" in Baas and Kwakernaak (1977) method by solving a set of fuzzy relational equations of the form $r(M_i) = (M_1 x \ldots x M_n) \circ R$, where $R(x_1,\ldots,x_n) = 1$ if $x_i \geq x_j$ for all $j \neq i$. Requena *et al.* (1994, 1995) apply a neural network technique to the learning of a subjective ranking method.

Equality of fuzzy values. Beside ranking, and related to it, is the checking of equality. The equality of fuzzy variables x and y restricted by quantities M and N can be assessed in the possibilistic style using the index of consistency

$$\mathrm{Pos}(x = y) = \sup_{a \in \mathbb{R}} \{\min(M(a), N(a))\}$$

introduced by Zadeh (1979). It evaluates to what extent it is possible to find a common value for x and y. It is also a degree of intersection between M and N (= $\Pi_M(N)$ or = $\Pi_N(M)$ in terms of possibility measures). This is a very weakly demanding notion but the only meaningful one in practice. Indeed the dual necessity index $\mathrm{Nes}(x = y) = 1 - \mathrm{Pos}(x \neq y)$ is very often equal to 0 even if M = N. This is because, due to imprecision, there is generally no certainty that two fuzzy variables take the same value.

In this section a fuzzy interval represents the distribution of an ill-known quantity. It is thus very important to distinguish between identical fuzzy variables (x = y) and fuzzy variables with identical possibility distributions (M = N). In the latter case nothing enforces the condition x = y. The quantity

$$N_M(N) = 1 - \Pi_M(N^c) = \inf_{a \in \mathbb{R}} \max(1 - M(a), N(a))$$

evaluates the condition of containment of the support of M in the core of N since $N_M(N) = 1$ if and only if $\mathrm{supp}(M) \subseteq \mathrm{core}(N)$. It does not test the equality of variables. Note that $N_M(N) = \min(d(M \succ^+ N), 1 - d(M \succ^- N))$. A natural measure of equality of fuzzy intervals is $d(M = N) = \min(N_M(N), N_N(M))$ which is 1 only if N = M and is a regular interval. For equal fuzzy sets d(M = N) cannot be larger than 0.5. However this measure is not specific to fuzzy intervals and makes sense for general fuzzy sets as well. Area-based measures of containment and equality of fuzzy sets could be used to compare fuzzy quantities, although with the same proviso as above; see Willmott (1981) for this point of view.

In particular the problem of comparing two uncertain quantities, dealt with here, differs from the one of comparing two membership functions and from the one of checking if an ill-known object attribute (described by a possibility distribution) satisfies a fuzzy condition (Dubois and Prade, 1982c). The first of these last two problems is encountered in linguistic approximation when a linguistically meaningful fuzzy set is used to interpret another one that comes from some computation. The second problem is one of fuzzy pattern matching (Cayrol *et al.*, 1982). For the first of these problems, the rationality principles that govern the choice of comparison indices may be totally different from the ones that underlie the ranking of fuzzy variables, which aims as a tool for decision under uncertainty.

10.7 CONCLUSION: APPLICATIONS OF FUZZY NUMBERS AND INTERVALS

The brief survey in this section makes no claim to being exhaustive; but in our opinion, indicates the major trends which have been developed in applying fuzzy intervals to mathematical or engineering areas.

Fuzzy Analysis. The integration of point-to-set maps, also called correspondences, has been studied by Aumann (1965) among others. An extension of Riemann's integration to fuzzy interval-valued mapping consistent with Aumann's theory is proposed in Dubois and Prade (1982b), who also investigate the differentiation problem when the fuzzy mapping has level-cuts whose boundaries are differentiable. A different definition is proposed by Puri and Ralescu (1981), who extend the classical notion of the derivative $\lim_{x \to x_0} \frac{f(x) - f(x_0)}{x - x_0}$ using (implicitly) the 'optimistic' fuzzy subtractions as in Section 10.5.4. These two definitions coincide when they simultaneously make sense, but they do not make sense for the same sets of fuzzy mappings. Further works in the analysis of fuzzy functions including n-dimensional case, can be found, e.g., in Kaleva (1987), Buckley (1991, 1992), Liu (1994), Burgin and Sostak (1994), Burgin (1995), Kalina (1997). See the book by Diamond and Kloeden (1994) and the next chapter by these authors. Dually, a fuzzy-valued integral of a real-based real-valued mapping over a fuzzy interval can be defined (Dubois and Prade, 1982b). Further works dealing with the integration of fuzzy valued functions appear in several papers (Klement, 1987, Zhang and Wang, 1993, etc).

Fuzzy Statistics. Fuzzy numbers had some impact in statistics. The calculus of fuzzy valued real mappings is basic for introducing random fuzzy sets. They extend random sets, where only point to set mappings are used (Dempster, 1967), to fuzzy mappings. Several proposals have been made along this line by Féron (1976), Kwakernaak (1979, 1978), Puri and Ralescu (1986), Kruse and Meyer (1987), and Miyakoshi and Shimbo (1984a, b) to quote only the earliest. Computing the expected value of a fuzzy random variable is a matter of fuzzy arithmetics. Many results in random fuzzy sets have appeared since then, such as Ban (1990, 1991), Puri and Ralescu (1991), Stojakovic (1994). Applications of fuzzy intervals to the gathering of statistical data seems promising (Kruse, 1987; Kruse and Meyer, 1987; Viertl, 1996). They are surveyed in the paper of Gebhardt, Gil and Kruse (1998) in the Handbooks of Fuzzy Sets Series, in Slowinski (1998)'s volume. Besides, fuzzy time-series have recently been studied by Song and Chissom (1993 a, b, 1994) and Song *et al.* (1995). Related to the probabilistic literature is the fuzzy-valued probability of fuzzy events proposed by Yager (1979a). Fuzzy probabilities in Markov processes are considered by Stanford (1982).

Computer Sciences. The attractiveness of fuzzy quantities for the representation of imprecise, subjectively originated information is self evident. Hence fuzzy intervals are of interest in database research. The use of fuzzy intervals in information systems has been proposed rather early by Oftedal (1981). Umano (1982), Buckles and Petry (1984), Prade(1984), and Prade and Testemale (1984) introduce fuzzy intervals as attribute values in relational data bases (see the chapter by Bosc *et al.* (1999) in the volume edited by Bezdek; Dubois and Prade (1999) in the Handbooks of Fuzzy Sets Series. The truth value of a fuzzy statement with respect to an imprecisely described fact stored in the data base can be viewed as a fuzzy quantity on the unit interval representing a fuzzy set of truth values (Zadeh, 1979). Operations on fuzzy intervals become useful for dealing with patterns of reasoning where fuzzy quantifiers appear (Zadeh, 1983, Dubois and Prade, 1988b, Sanchez, 1996). Note that the compositional rule of inference applied in several areas often reduces to fuzzy interval computations (Dubois and Prade, 1989a; Fullér and Zimmerman, 1992; Hellendoorn, 1990; Hong and Hwang, 1994b). More generally, operations on fuzzy intervals enable calculations with numerical scale-based linguistic categories to be performed by the computer (see Mandic and Mamdani, 1980 for an early example of linguistic computation package).

Systems Engineering. However fuzzy interval analysis methods have had so far little impact on the huge domain of fuzzy control where fuzzy intervals are used to represent fuzzy rules. However, fuzzy rule systems serve in representing non-linear non-fuzzy functions and are used with a precise input. A counterexample is Procyk and Mamdani (1979) whose rule tuning method consists in modifiying the fuzzy rule conclusions using fuzzy arithmetic operations. Fuzzy numbers were also found useful for environment modelling in robotics (Kim *et al.*, 1994).

Operations Research. The potential role of fuzzy interval analysis in operations research has been perceived quite early (Prade, 1980). In this field, fuzzy quantities have been used in graph problems as well as linear programming. In the case of directed graphs (such as PERT-CPM networks), where the length of arcs is only imprecisely known, the extension of shortest path algorithms and critical path methods has received early attention in Dubois and Prade (1978c), and Chanas and Kamburowski (1981). Maximum flow problems in networks with fuzzy arc capacities are considered by Chanas and Kolodziejczyk (1982). Specific applications of fuzzy shortest-path techniques are by Prade (1979) (scheduling) and Dubois (1983) (bus networks). The use of fuzzy intervals in scheduling under imprecise task durations is exemplified in Fortemps (1997), and Dubois *et al.* (1995). An overview of the application of fuzzy intervals in discrete optimization, including graph problems is proposed by Chanas and Kuchta (1998), in the two companion volumes of the Handbooks of Fuzzy Sets edited by Slowinski (1998) and Zimmermann (1999). Fuzzy quantities have been introduced in linear programming by Negoita (1978), Dubois and Prade (1980b). The main problem is to derive deterministic equivalent formulations of fuzzy programs through the use of dominance indices for the comparison of fuzzy intervals (Tanaka and Asai,1984 a,b; Dubois, 1987b; Campos and Verdegay, 1989; Fullér and Zimmermann (1993); Ramik and Rommelfanger 1993, 1996)). Applications of fuzzy intervals in optimization techniques include regression analysis with fuzzy models and fuzzy data (Kacprzyk and Fedrizzi, 1992; Tanaka and Diamond, 1998); this includes the introduction of fuzzy intervals in neural networks (Ishibuchi *et al.*, 1995). For overview monographs see Lai and Hwang (1992) and Sakawa (1993). Slowinski (1998) contains various surveys dealing with fuzzy quantities in operations research problems.

Decision Analysis. Decision analysis seems to be one of the most natural field of application of fuzzy quantities. Very often opinions provided by decision-makers about the value of choices or the occurrence of events are linguistic rather than numerical, although these linguistic terms generally refer to numerical scales. Hence, there are many instances when opinions are satisfactorily modeled by a fuzzy quantity over an appropriate scale. This idea has been used by Watson *et al.* (1979) in the classical framework of expected utility theory, where the utility of identified consequences as well as the probabilities of occurrence of these consequences are imprecise. These authors only consider binary decision trees. N-ary decision trees have been subsequently studied by the same authors (Dubois and Prade, 1982a) using results which are summarised in Section 10.5.1. Interesting discussions about this approach can be found in Freeling (1980, 1984) and French (1984). This approach is of course very much related to the question of sensitivity analysis in classical decision models (e.g., Sage, 1981). For an overwiew of fuzzy decision analysis, see Chen *et al.* (1992) and Slowinski (1998).

Multiple-objective decision models where fuzzy intervals assess individual's opinions about criteria and a fuzzy weighted average is performed, are proposed by Baas and Kwakernaak (1977), Baptistella and Ollero (1980), Buckley (1985). Sakawa (1993) describes computerized interactive optimization procedures where the user is supposed to improve upon available efficient solutions by describing his (evolving) pattern of preference, viewed as the gradient of an overall utility function. This pattern is described by means of fuzzy quantities. Further, Saaty's methodology for decision under pairwise comparison (Saaty, 1980) has been generalized to imprecise pairwise relative evaluations of efficient solutions in multicriteria optimization problems by van Laarhoven and Pedrycz (1983), see also Wagenknecht and Hartmann (1983), Lootsma (1985), Boender *et al.*(1989), Wang *et al.* (1995a).

Engineering Design. The areas of application of fuzzy interval analysis also overlap those of standard interval analysis, which used to be applied to problems of error propagation in mathematical programming and numerical analysis (Moore, 1979) and also constraint propagation with interval constraints. Indeed, the community of people in fuzzy interval analysis is more and more concerned with the classical interval analysis approach and the constraint propagation techniques (Tsang, 1993). Noticeably, the area of engineering design offers a natural field of application where fuzzy intervals are useful in the representation of design imprecision. An early example of application of fuzzy intervals and their operations in design is the choice, under-several goals, of cutting parameters on the various stages of a machining transfer line, presented in Dubois (1987a). The works of Antonsson and colleagues build on both interval mathematics and fuzzy intervals. See Sebastian and Antonsson (1996) for a representative volume, and Zimmermann (1999) that contains a survey article. These authors use non compensating idempotent operations (the minimum and the maximum) between local preference profiles, and more recently include compensating idempotent operations (that are neither t-norms not t-conorms); see Antonsson and Otto (1995), Scott and Antonsson (1995) and Otto and Antonsson (1991). This is probably one interesting alley of future research beyond the now classical maxmin fuzzy interval mathematics.

Acknowledgements

The authors are grateful to Erik Antonsson for his useful comments and suggestions on a preliminary draft of this chapter. One of the authors acknowledges the support of grant VEGA 1/4046/97.

REFERENCES

Adamo J. M. (1980). Fuzzy decision trees, *Fuzzy Sets and Systems*, **4**, 207-219.

Ammar S. (1989). Determining the 'best' decision in the presence of imprecise information, *Fuzzy Sets and Systems*, **29**, 293-302.

Anile A. M., Deodato S. and Privitera G. (1995). Implementing fuzzy arithmetic, *Fuzzy Sets and Systems*, **72**, 239-250.

Antonsson E. and Otto K. (1995). Imprecision in engineering design, *ASME J. of Mechanical Design*, **117**(B), 25-32.

Ambrosio R. and Martini G. B. (1984). Maximum and minimum between fuzzy symbols in non-interactive and weakly non-interactive situations, *Fuzzy Sets and Systems,* **12**, 27-35.

Aumann R. J. (1965). Integrals of set-valued functions, *J. Math. Anal. Appl.*, **12**, 1-22.

Baas S. M. and Kwakernaak H. (1977). Rating and ranking of multiple aspect alternatives using fuzzy sets, *Automatica,* **13**, 47-58.

Badard R. (1982). The law of large numbers for fuzzy processes and the estimation problem, *Information Sciences*, **28**, 161-178.

Badard R. (1984). Fixed point theorems for fuzzy numbers, *Fuzzy Sets and Systems,* **13**, 291-302.

Baekeland R. and Kerre E. (1988). Piecewise linear fuzzy quantities: A way to implement fuzzy information into expert systems and fuzzy databases, *Uncertainty and Intelligent Systems* (Bouchon B., Saitta L. and Yager R. R., eds.), Lecture Notes in Computer Sciences, vol. 313, Springer-Verlag, Berlin, 119-126.

Baldwin J. F. and Guild N. C. F. (1979a). Comparison of fuzzy sets on the same decision space, *Fuzzy Sets and Systems,* **2**, 213-231.

Baldwin J. F. and Guild N. C. F. (1979b). Comments on the fuzzy max operator of Dubois and Prade, *Int. J. Systems Science*, **10**, 1063-1064.

Ban J. (1990). Radon – Nikodym theorem and conditional expectation of fuzzy-valued measures and variables, *Fuzzy Sets and Systems,* **34**, 383-392.

Ban J. (1991). Ergodic theorems for random compact sets and fuzzy variables in Banach spaces, *Fuzzy Sets and Systems,* **44**, 71-82.

Baptistella L.F.B. and Ollero A. (1980). Fuzzy methodologies for interactive multicriteria optimization, *IEEE Trans. Syst. Man Cybern.*, **10**, 355-365.

Bellman R. (1957). *Dynamic Programming*, Princeton University Press, Princeton, N.J.

Bertoluzza C. and Bodini A. (1998). A new proof of Nguyen's compatibility theorem in a more general context, *Fuzzy Sets and Systems*, **95**, 99-102.

Bertoluzza C., Corral N. and Salas A (1995). On a new class of distances between fuzzy numbers, *Mathware and Soft Computing,* **2**, 71-84.

Bezdek J.C., Dubois D. and Prade H. (Eds.) (1999). *Fuzzy Sets in Approximate Reasoning and Information Systems.* The Handbooks of Fuzzy Sets Series (Dubois D. and Prade H., eds.), Kluwer Acad. Publ., Boston.

Biacino L. and Lettieri A. (1989). Equations with fuzzy numbers, *Information Sciences,* **47**, 63-76.

Bilgic T. and Türksen I.B. (1999). Measurement of membership functions: Theoretical and empirical work, *Fundamentals of Fuzzy Sets* (Dubois D. and Prade H. eds.), Kluwer Acad. Publ., 1999. This volume.

Bloch I. and Maitre H. (1994). Fuzzy mathematical morphology, *Annals of Mathematics and Artificial Intelligence*, **10**, 55-84.

Bloch I. and Maitre H. (1995). Fuzzy mathematical morphologies: A comparative study, *Pattern Recognition*, **28**, 1341-1387.

Bodenhofer U. (1998). *A similarity-based generalisation of fuzzy orderings*, PhD Thesis, J. Kepler University, Linz, Austria.

Boender C., De Graan J. G and Lootsma F. (1989). Multicriteria decision analysis with fuzzy pairwise comparisons, *Fuzzy Sets and Systems,* **29**, 133-144.

Bortolan G. and Degani R. (1985). A review of some methods for ranking fuzzy subsets, *Fuzzy Sets and Systems,* **15**, 1-19.

Bosc P., Buckles B.B., Petry F. E. and Pivert O. (1999). Fuzzy databases, *Fuzzy Sets in Approximate Reasoning and Information Systems* (Bezdek J.C., Dubois D. and Prade H., eds.), Kluwer Acad. Publ., New York, 403-468.

Bouchon-Meunier B., Kosheleva O., Kreinovich V. and Nguyen, H. T. (1997). Fuzzy numbers are the only fuzzy sets that keep invertible operations invertible, *Fuzzy Sets and Systems,* **91**, 155-163.

Braae M. and Rutherford D.A. (1978). Fuzzy relations in a control setting, *Kybernetes*, **7**, 185-188.

Buckles B. P. and Petry F. E. (1984). Extending the fuzzy data base with fuzzy numbers, *Information Sciences*, **34**, 145-155.

Buckley J. J. (1985). Ranking alternatives using fuzzy numbers, *Fuzzy Sets and Systems,* **15**, 21-31.

Buckley J. J. (1989a). On the algebra of interactive fuzzy numbers, *Fuzzy Sets and Systems,* **32**, 291-306.

Buckley J. J. (1989b). A fuzzy ranking of fuzzy numbers, *Fuzzy Sets and Systems,* **33**, 119- 122.

Buckley J. J. (1989c). Fuzzy complex numbers, *Fuzzy Sets and Systems,* **33**, 333-345.

Buckley J. J. (1990). On the algebra of interactive fuzzy numbers: The continuous case, *Fuzzy Sets and Systems,* **37**, 317-326.

Buckley J. J. (1991). Fuzzy complex analysis I: Differentiation, *Fuzzy Sets and Systems,* **41**, 269-284.

Buckley J. J. (1992). Fuzzy complex analysis II: Integration, *Fuzzy Sets and Systems,* **49**, 171-179.

Buckley J. J. and Chanas S. (1989). A fast method of ranking alternatives using fuzzy numbers, *Fuzzy Sets and Systems,* **30**, 337-338.

Buckley J. J. and Qu Y. (1990). Solving linear and quadratic fuzzy equations, *Fuzzy Sets and Systems,* **38**, 43-61.

Buckley J. J. and Qu Y. (1991a). Solving fuzzy equations: a new solution concept, *Fuzzy Sets and Systems,* **39**, 291-303.

Buckley J. J. and Qu Y. (1991b). Solving systems of linear fuzzy equations, *Fuzzy Sets and Systems,* **43**, 33-44.

Burgin M. (1995). Nonclassical analysis: Fuzzy continuity and convergence, *Fuzzy Sets and Systems,* **75**, 291-299.

Burgin M. and Sostak M. (1994). Fuzzyfication of the theory of continuous functions, *Fuzzy Sets and Systems,* **64**, 71-81.

Campos L. and Gonzalez (1994). Further contributions to the study of average value for ranking fuzzy numbers, *Int. J. Approximate Reasoning*, **10**, 135-163.

Campos L. and Verdegay J. L. (1989). Linear programming problems and ranking of fuzzy numbers, *Fuzzy Sets and Systems,* **32**, 1-12.

Cayrol M., Farreny H. and Prade H. (1982). Fuzzy pattern matching, *Kybernetes,* **11**, 103-116.

Chanas S., Delgado M., Verdegay M. and Vila M.A. (1993). Ranking fuzzy intervals in the setting of random sets, *Information Sciences,* **69**, 201-217.

Chanas S. and Kamburovski J. (1981). The use of fuzzy variables in PERT, *Fuzzy Sets and Systems,* **3**, 11-19.

Chanas S. and Kolodziejczyk W. (1982). Maximum flow in a network with fuzzy arc capacities, *Fuzzy Sets and Systems,* **8**, 165-173.

Chanas S. and Kuchta D. (1998). Discrete fuzzy optimization, *Fuzzy Sets in Decision Analysis Operations Research and Statistics* (Slowinski R., ed.), The Handbooks of Fuzzy Sets Series, Kluwer, Boston, USA.

Chanas S. and Nowakowski M. (1988). Single value simulation of fuzzy variable, *Fuzzy Sets and Systems*, **25**, 43-57.

Chang P. T. and Lee E. S. (1994). Fuzzy arithmetics and comparison of fuzzy numbers, *Fuzzy Optimization: Recent Advances* (Delgado M., Kacprzyk J., Verdegay J. L., Vila M. A., eds), Physica-Verlag, Heidelberg, Germany, 69-81.

Chang W. (1981). Ranking of fuzzy utilities with triangular membership functions. *Proc. Int. Conf. on Policy Analysis and Information Systems,* Taipei, 263-271.

Chen S. H. (1985). Ranking fuzzy numbers with maximizing sets and minimizing sets, *Fuzzy Sets and Systems,* **17**, 113-129.

Chen S. J., Hwang, C. L. and Hwang, F. P. (1992). *Fuzzy Multiple Attribute Decision Making – Methods and Applications*, Lecture Notes in Economics and Mathematical Systems, vol. 375, Springer Verlag, Berlin.

Chen H. K., Hsu W. K. and Chiang W. L. (1998). A comparison of the vertex method and with JHE method, *Fuzzy Sets and Systems,* **95**, 201-214.

Cheng C. H. (1998). A new approach for ranking fuzzy numbers by distance method, *Fuzzy Sets and Systems,* **95**, 307-317.

Choobineh F. and Li H. (1993). An index for ordering fuzzy numbers, *Fuzzy Sets and Systems,* **54**, 287-295.

Cuninghame-Green R. A. (1979). *Minimax Algebra*, Lecture Notes in Economical and Mathematical Systems, vol. 166, Springer-Verlag, Berlin.

Cuninghame-Green R. A. and Cechlarova V. (1995). Residuation in fuzzy algebra and some applications, *Fuzzy Sets and Systems,* **71**, 227-239.

Czogala E. and Hirota K. (1986). *Probabilistic Sets: Fuzzy and Stochastic Approach to Decision Control and Recognition Processes*, ISR 91, Verlag TUV Rheinland, Köln.

De Baets B. (1999). Analytical solution methods for fuzzy relational equations, *Fundamentals of Fuzzy Sets* (Dubois D. and Prade H., eds.), Kluwer Acad. Publ., 1999. This volume.

De Baets, B., Kerre E. and Gupta M. M. (1994a). The fundamentals of fuzzy mathematical morphology — Part 1: Basic concepts, *Int. J. General Systems*, **23**, 155-171.

De Baets, B., Kerre E. and Gupta M. M. (1994b). The fundamentals of fuzzy mathematical morphology — Part 2: Idempotence, convexity and decomposition, *Int. J. General Systems*, **23**, 307-322.

De Baets, B., Mares, M. and Mesiar, R. (1997). T-partitions of the real line generated by idempotents shapes, *Fuzzy Sets and Systems,* **91**, 177-184.

De Baets B, and Markova-Stupnanova A. (1997). Analytical expressions for the addition of fuzzy intervals, *Fuzzy Sets and Systems,* **91**, 203-213.

De Campos Ibañez L.M. and Gonzalez-Muñoz (1989). A subjective approach for ranking fuzzy numbers, *Fuzzy Sets and Systems*, **29**, 145-154.

Delgado M., Verdegay J. L. and Vila M. A (1988). A procedure for ranking fuzzy numbers using fuzzy relations, *Fuzzy Sets and Systems*, **26**, 49-62.

Delgado M., Verdegay J. L. and Vila M. A (1994). Fuzzy numbers, definitions and properties, *Mathware and Soft Computing,* **1**, 31-43.

Delgado M., Vila M. A. and Voxman W. (1998a). On a canonical representation for fuzzy numbers, *Fuzzy Sets and Systems*, **93**, 125-135.

Delgado M., Vila M. A. and Voxman W. (1998b). A fuzziness measure for fuzzy numbers: Applications, *Fuzzy Sets and Systems*, **94**, 205-216.

Dempster A. P. (1967). Upper and lower probabilities induced by a multivalued mapping, *Ann. Math. Stat.*, **38**, 325-339.

Denneberg D. (1994). *Nonadditive Measure and Integral*, Kluwer Academic, Dordrecht, The Netherlands.

Diamond P. and Kloeden P. (1994). *Metric spaces of fuzzy sets*, World Scientific, Singapore.

Dijkman J. G., van Haeringen H. and de Lange S. T. (1983). Fuzzy numbers, *J. Math. Anal. Appl.*, **92**, 301-341.

Dishkant H. (1981). About membership functions estimation, *Fuzzy Sets and Systems,* **5**, 141-147.

Dombi J. (1986). Properties of the fuzzy connectives in the light of general representation theory, *Acta Cybernetica*, **7**, 313-321.

Dong W. and Shah H. (1987). Vertex method for computing functions of fuzzy variables, *Fuzzy Sets and Systems*, **24**, 65-79.

Dong W., Shah H. and Wong F. (1985). Fuzzy computations in decision and risk analysis, *Civ. Eng. Syst.*, **2**, 201-208.

Dong W. and Wong F. (1987). Fuzzy weighted averages and the implementation of the extension principle, *Fuzzy Sets and Systems*, **21**, 183-201.

Dong W. and Wong F. (1989). Interactive variables and fuzzy decisions, *Fuzzy Sets and Systems*, **29**, 1-19.

Dubois D. (1982). A law of large numbers for nonconvex fuzzy sets of the real line, *Busefal* (IRIT, Université P. Sabatier, Toulouse), **9**, 31-38.

Dubois D. (1983). A fuzzy heuristic, interactive approach to the optimal network problem, *Advances in Fuzzy Sets and Possibility Theory and Applications* (Wang P. P., ed.), Plenum Press, New York, 253-276.

Dubois D. (1987a). An application of fuzzy arithmetics to the optimization of industrial machining processes, *Mathematical Modelling,* **9**, 461-475.

Dubois D. (1987b). Linear programming with fuzzy data, *Analysis of Fuzzy Information* (Bezdek J. C., ed.), Vol. 3, CRC Press, Boca Raton, Fl., 241-264.

Dubois D., Fargier H., and Prade (1995). Fuzzy constraints in job-shop scheduling, *J. Intellig. Manufacturing*, **6**, 215-234.

DuboisD., Nguyen H.T. and Prade H. (1999). Possibility theory, probability and fuzzy sets: Misunderstandings, bridges and gaps, *Fundamentals of Fuzzy Sets* (Dubois D. and Prade H., eds.), Kluwer Acad. Publ., 1999. This volume.

Dubois D., Ostasiewicz W. and Prade H. (1999). Fuzzy sets: History and basic notions, *Fundamentals of Fuzzy Sets* (Dubois D. and Prade H., eds.), Kluwer Acad. Publ., 1999. This volume.

Dubois D. and Prade H. (1978a). Operations on fuzzy numbers, *Int. J. Systems Science*, **9**, 613-626.

Dubois D. and Prade H. (1978b). Comment on 'Tolerance analysis using fuzzy sets' and 'A procedure for multiple aspect decision making', *Int. J. Systems Science.*, **9**, 357-360.

Dubois D. and Prade H. (1978c). Algorithmes de plus court chemin pour traiter des donnees floues, *RAIRO, Serie R. O.*, **12**, 213-227.

Dubois D. and Prade H. (1979a). Fuzzy real algebra: Some results, *Fuzzy Sets and Systems,* **2**, 327-348.

Dubois D. and Prade H. (1979b). Operations in a fuzzy-valued logic, *Inf. & Control*, **43**, 224-240.

Dubois D. and Prade H. (1980a). *Fuzzy Sets and Systems: Theory and Applications*, Academic Press, New York.

Dubois D. and Prade H. (1980b). Systems of linear fuzzy constraints, *Fuzzy Sets and Systems,* **3**, 37-48.

Dubois D. and Prade H. (1981). Additions of interactive fuzzy numbers, *IEEE Trans. Automatic Control,* **26**, 926-936.

Dubois D. and Prade H. (1982a). The use of fuzzy numbers in decision analysis, *Fuzzy Information and Decision Processes* (Gupta M. M. and Sanchez E., eds.), North-Holland, Amsterdam, 309-321.

Dubois D. and Prade H. (1982b). Towards fuzzy differential calculus, *Fuzzy Sets and Systems*, **8**, Part I. Integration of fuzzy mappings, 1-17; Part II. Integration on fuzzy intervals, 105-116; Part III. Differentiation, 225-233.

Dubois D. and Prade H. (1982c). What does 'convergence' mean for fuzzy numbers?, *Proc. IFAC Symp. on Theory and Applications of Digital Control* (Mahalonabis A. K., ed.), Pergamon Press, New York, 433-438.

Dubois D. and Prade H. (1983a). Inverse operations for fuzzy numbers, *Proc. IFAC Symp. on Fuzzy Information, Knowledge Representation and Decision Processes* (Sanchez E. and Gupta M. M., eds.), Pergamon Press, Oxford, 391-396.

Dubois D. and Prade H. (1983b). Ranking fuzzy numbers in the setting of possibility theory, *Information Sciences,* **30**, 183-224.

Dubois D. and Prade H. (1983c). Twofold fuzzy sets: An approach to the representation of sets with fuzzy boundaries, based on possibility and necessity measures, *Fuzzy Math.* (Huazhong, China), **3**(4), 53-76.

Dubois D. and Prade H. (1983d). On distances between fuzzy points and their use for plausible reasoning, *IEEE Int. Conf. on Systems Man and Cybernetics,* Bombay and New Delhi, IEEE, Pistacaway, NJ, 300-303.

Dubois D. and Prade H. (1984). Fuzzy set-theoretic differences and inclusions and their use in the analysis of fuzzy equations, *Control Cybern.* (Warsaw), **13**, 129-146.

Dubois D. and Prade H. (1985). Fuzzy cardinality and the modeling of imprecise quantification, *Fuzzy Sets and Systems,* **16**, 199-230.

Dubois D. and Prade H. (1986). Fuzzy sets and statistical data, *Europ. J. Op. Res.*, **25**, 345-356.

Dubois D. and Prade H. (1987a). Fuzzy numbers: An overview, *Analysis of Fuzzy Information*, Vol. I (Bezdek J., ed.), CRC Press, Boca Raton, FL, 3-39.

Dubois D. and Prade H. (1987b). The mean value of a fuzzy number, *Fuzzy Sets and Systems,* **24**, 279-300.

Dubois D. and Prade H. (eds.) (1987c). Special Issue on Fuzzy Numbers. *Fuzzy Sets and Systems,* **24**(3).

Dubois D. and Prade H. (1988a). Possibility Theory. An Approach to Computerized Processing of Uncertainty. Plenum Press, New York.

Dubois D. and Prade H. (1988b). On fuzzy syllogisms, *Computational Intelligence,* **4**, 171-179.

Dubois D. and Prade H. (1989a). Processing fuzzy temporal knowledge, *IEEE Trans. Syst. Man and Cybern.*, **19**, 729-744.

Dubois D. and Prade H. (1989b). Order-of-magnitude reasoning with fuzzy relations, *Revue d'Intelligence Artificielle* (Hermes, Paris), **3**(4), 69-94.

Dubois D. and Prade H. (1991a). Random sets and fuzzy interval analysis, *Fuzzy Sets and Systems,* **42**, 87-101.

Dubois D. and Prade H. (1991b). On the ranking of ill-known values in possibility theory, *Fuzzy Sets and Systems,* **43**, 311-317.

Dubois D. and Prade H. (1995). Possibility theory as a basis for qualitative decision theory. *Proc. 24th Int. Joint Conf. on AI*, Montreal, Canada, 1924-1930.

Dubois D., Prade H. and Sandri S. (1993). On possibility/probability transformations. In: Fuzzy Logic. State of the Art, (R. Lowen, M. Roubens, eds.), Kluwer Acad. Publ., Dordrecht, 103-112.

Dubois D., Prade H. and Sabbadin R. (1998). Qualitative decision theory with Sugeno integrals. Proc. of the 14th Conf. on Uncertainty in Artificial Intelligence, Madison, July 24-26, 1998, (G. Cooper, S. Moral, eds.), Morgan Kaufmann, San Francisco, 121-128.

Dubois D., Prade H. and Yager R. R. (1998). Computation of intelligent fusion operations based on constrained fuzzy arithmetic, *Proc. IEEE Int. Conf. on Fuzzy Systems*, Anchorage, Al., 767-772.

Efstathiou J. and Bonissone P. (1979). Ranking fuzzy sets using linguistic preference relations, *Proc. IEEE Int. Conf. on Systems Man and Cybern.* Denver, Co, USA.

Efstathiou J. and Tong R. (1982). Ranking fuzzy sets: A decision-theoretic approach, *IEEE Trans. Syst. Man. Cybern.*, **12**, 655-659.

Eklund P. and Gähler W. (1988). Basic notions for fuzzy topology, *Fuzzy Sets and Systems,* **26**, 333-356.

Féron R. (1976). Ensembles aléatoires flous, *C.R. Acad. Sci. Ser. A.*, **282**, 903-906.

Fishburn P. (1985). *Interval Orders and Interval Graphs.* Wiley, New-York.

Jenei S. and Fodor J. (1998). On continuous triangular norms, *Fuzzy Sets and Systems*, **100**, 273-282.

Fodor J., Orlovski S.A., Perny P. and Roubens M. (1998). The use of fuzzy preference models in multiple criteria choice, ranking and sorting, *Fuzzy Sets in Decision Analysis, Operations Research and Statistics* (Slowinski R., ed.), Kluwer Acad. Publ., Boston, 69-101.

Fodor J. and Roubens M. (1994). *Fuzzy Preference Modelling and Multicriteria Decision Support*, Kluwer Acad. Publ., Dordrecht.

Fodor J. and Yager R.R. (1999). Fuzzy set-theoretic operators and quantifiers, *Fundamentals of Fuzzy Sets* (Dubois D. and Prade H., eds.), Kluwer Acad. Publ., 1999. This volume.

Fortemps P. (1997). Jobshop scheduling with imprecise durations: A fuzzy approach, *IEEE Trans. Fuzzy Systems*, **5**, 557-569.

Fortemps P. and Roubens M. (1996). Ranking and defuzzification methods based on area compensation, *Fuzzy Sets and Systems,* **82**, 319-330.

Freeling A. N. S. (1980). Fuzzy sets and decision analysis, *IEEE Trans. Syst. Man. Cybern.*, **10**, 341-354.

Freeling A. N. S. (1984). Possibilities versus fuzzy probabilities – Two alternative decision-aids, *Fuzzy Sets and Decision Analysis* (Zimmermann H. J., Zadeh L. A. and Gaines B. R., eds.), TIMS Studies in the Management Sciences, vol. 20, North-Holland, Amsterdam, 67-71.

French S. (1984). Fuzzy decision analysis: some criticisms, *Fuzzy Sets and Decision Analysis* (Zimmermann H. J., Zadeh L. A. and Gaines B. R., eds.), TIMS Studies in the Management Sciences, vol. 20, North-Holland, Amsterdam, 29-44.

Fullér R. (1990). On stability in possibilistic linear equality systems with Lipschitzian fuzzy numbers, *Fuzzy Sets and Systems*, **34**, 347-354.

Fullér R. (1991). On product-sum of triangular fuzzy numbers, *Fuzzy Sets and Systems*, **41**, 83-87.

Fullér R. (1992). A law of large numbers for fuzzy numbers, *Fuzzy Sets and Systems,* **45**, 299-303.

Fullér R. and Keresztfalvi T. (1991). On generalization of Nguyen's theorem, *Fuzzy Sets and Systems,* **41**, 371-374.

Fullér R. and Keresztfalvi T. (1992). T-norm-based addition of fuzzy intervals, *Fuzzy Sets and Systems,* **51**, 155-159.

Fullér R. and Mesiar R. (eds.) (1997). Special Issue on Fuzzy Arithmetic, *Fuzzy Sets and Systems,* **91**(2).

Fullér R. and Triesch E. (1993). A note on the law of large numbers for fuzzy variables, *Fuzzy Sets and Systems,* **55**, 235-236.

Fullér R. and Zimmermann H. J. (1992). On computation of the compositional rule of inference under triangular norms, *Fuzzy Sets and Systems,* **51**, 267-275.

Fullér R. and Zimmermann H. J. (1993). Fuzzy reasoning for solving fuzzy mathematical programming problems. *Fuzzy Sets and Systems,* **60**, 121-133.

Gähler S. and Gähler W. (1994). Fuzzy real numbers, *Fuzzy Sets and Systems,* **66**, 137-158.

Gähler W. (1992). Fuzzy topology in Topology, Measures and Fractals, *Math. Research*, **67**, Academia-Verlag, Berlin, 188-197.

Gebhardt A. (1995). On types of fuzzy numbers and extension principles, *Fuzzy Sets and Systems,* **75**, 311-318.

Gebhardt J., Gil M. A. and Kruse R. (1998). Fuzzy set-theoretic methods in statistics, *Fuzzy Sets in Decision Analysis, Operations Research and Statistics*, The Handbook of Fuzzy Sets Serie, Kluwer Academic Publ., Dordrecht, The Netherlands, 311-348.

Giachetti R. E. (1996). *The Mathematics of Triangular Fuzzy Numbers to support a Model of Imprecision in Design,* Ph.D. Dissertation, Industrial Engineering, North-Carolina State University, US.

Giachetti R. E. and Young R. E. (1997a). Analysis of the error in standard approximation for multiplication of triangular and trapezoidal fuzzy numbers, and the development of a new approximation , *Fuzzy Sets and Systems,* **91**, 1-15.

Giachetti R. E. and Young R. E. (1997b). A parametric representation of fuzzy numbers and their arithmetic operators, *Fuzzy Sets and Systems,* **91**, 185-202.

Gil M. A. (1992). A note on the connection between fuzzy numbers and random intervals, *Statistics and Probability Lett.*, **13**, 311-319.

Goetschel R. H. (1997). Representations with fuzzy darts, *Fuzzy Sets and Systems,* **89**, 77-106.

Goetschel R. Jr. and Voxman W. (1983). Topological properties of fuzzy numbers, *Fuzzy Sets and Systems,* **10**, 87-99.

Gonzalez A. (1990). A study of the ranking function approach through mean values, *Fuzzy Sets and Systems,* **35**, 29-43.

Gonzalez A. and Vila M. A. (1991). A discrete method for studying indifference and order relations between fuzzy numbers, *Information Sciences,* **56**, 245-258.

Gonzalez A. and Vila M. A. (1992). Dominance relations on fuzzy numbers, *Information Sciences,* **64**, 1-16.

Gottwald S. (1984). On the existence of solutions of systems of fuzzy equations, *Fuzzy Sets and Systems,* **12**, 301-302.

Gottwald S. (1993). *Fuzzy Sets and Fuzzy Logic*, Vieweg, Braunschweig.

Guh Y.Y., Hong C.C., Wang K.M. and Lee E.S. (1996). Fuzzy weighted average: A max-min paired elimination method, *Computers Math. Applic.,* **32**, 115-123.

Harman B. (1992a). On associativity of the product of modified real fuzzy numbers, *Tatra Mount. Math. Publ.*, **1**, 45-50.

Harman B. (1992b). Sum and product of the modified real fuzzy numbers, *Kybernetika* (Prague), **Suppl. 28**(1/6), 37-40.

Hellendoorn H. (1990). Closure properties of the compositional rule of inference, *Fuzzy Sets and Systems,* **35**, 163-184.

Heilpern S. (1992). The expected value of a fuzzy number, *Fuzzy Sets and Systems,* **47**, 81-87.

Heilpern S. (1997). Representation and application of fuzzy numbers, *Fuzzy Sets and Systems,* **91**, 259-268.

Herencia J. A. (1997). Graded numbers and graded convergence of fuzzy numbers, *Fuzzy Sets and Systems,* **88**, 183-194.

Hirota K. and Pedrycz W. (1989). Interpretation of results of ranking methods with the aid of probabilistic sets, *Fuzzy Sets and Systems,* **32**, 263-274.

Höhle U. (1981). Representation theorems for L-fuzzy quantities, *Fuzzy Sets and Systems,* **5**, 83-107.

Höhle U. (1987). Fuzzy real numbers as Dedekind cuts with respect to a multiple-valued logic, *Fuzzy Sets and Systems,* **24**, 263-278.

Hong D. H. (1995). A note on t-norm-based addition of fuzzy intervals, *Fuzzy Sets and Systems,* **75**, 73-76.

Hong D. H. (1996). A note on the convergence of T-sum series of L-R fuzzy numbers, *Fuzzy Sets and Systems,* **77**, 253-254.

Hong D. H. and Hwang S. Y. (1994a). On the convergence of T-sum of L-R fuzzy numbers, *Fuzzy Sets and Systems,* **63**, 175-180.

Hong D. H. and Hwang S. Y. (1994b). On the compositional rule of inference under triangular norms, *Fuzzy Sets and Systems,* **66**, 25-38.

Hong D. H. and Hwang S. Y. (1996). A note on the correlation of fuzzy numbers, *Fuzzy Sets and Systems,* **79**, 401-402.

Hong D. H. and Hwang S. Y. (1997). A T-sum bound of LR-fuzzy numbers, *Fuzzy Sets and Systems,* **91**, 239-252.

Hong D. H. and Kim H. (1996). A law of large numbers for fuzzy numbers in a Banach space, *Fuzzy Sets and Systems*, **77**, 349-354.

Hong D. H. and Kim H. (1998). A note to the sum of fuzzy variables, *Fuzzy Sets and Systems*, **93**, 121-124.

Ishibuchi H., Kwon K. and Tanaka H. (1995). A learning algorithm of fuzzy neural networks with triangular fuzzy weights, *Fuzzy Sets and Systems,* **71**, 277-293.

Jacas J. and Recasens J. (1993). Fuzzy numbers and equality relations, *Proc. IEEE Int. Conf. on Fuzzy Systems*, San Francisco, 1298-1301.

Jain R. (1976). Tolerance analysis using fuzzy sets, *Int. J. Systems Science*, **7**, 1393-1401.

Jain R. (1977). A procedure for multiple aspect decision-making, *Int. J. Systems Science*, **8**, 1-7.

Jiang H. (1986). The approach to solving simultaneous linear equations that coefficients are fuzzy numbers, *J. Nat. Univ. Defence Technology (Chinese)*, **3**, 96-102.

Jimenez M. (1996). Ranking fuzzy intervals throgh the comparison of its expected intervals. *Int. J. of Uncertainty, Fuzziness and Knowledge-Based Systems*, **4**, 379-388.

Jimenez M. and Rivas J. A (1998). Fuzzy number approximation. *Int. J. of Uncertainty, Fuzziness and Knowledge-Based Systems*, **6**, 69-78.

Juang C.-H., Huang, X. H. and Elton D. J. (1991). Fuzzy information processing by Monte Carlo simulation technique, *J. Civil Eng. Systems*, **8**, 19-25.

Kacprzyk J. and Fedrizzi M. (1992). *Fuzzy Regression Analysis*, Physica-Verlag, Heidelberg.

Kaleva O. (1985). On the convergence of fuzzy sets, *Fuzzy Sets and Systems,* **24**, 53-65.

Kaleva O. (1987). Fuzzy differential equations, *Fuzzy Sets and Systems,* **24**, 301-317.

Kalina M. (1997). Derivatives of fuzzy functions and fuzzy derivatives, *Tatra Mount. Math. Publ.*, **12**, 27-34.

Kaufmann A. (1975). *Introduction to the Theory of Fuzzy Subsets*, Academic Press, New York.

Kaufmann A (1980). La simulation des ensembles flous, *CNRS Round Table on Fuzzy Sets*, Lyon, France (unpublished proceedings).

Kaufmann A. (1981). Hybrid Convolution, a way to combine fuzzy numbers and random numbers, *Fuzzy Math.* (Huazhong, China), **1**(2), 1-12.

Kaufmann A. and Gupta M. M. (1985). *Introduction to Fuzzy Arithmetic– Theory and Applications*, Van Nostrand Reinhold, New York.

Kaufmann A. and Gupta M. M. (1988). *Fuzzy Mathematical Models in Engineering and Management Science.* North-Holland, Amsterdam.

Kawaguchi M. F. and Da-Te T. (1993). A calculation method for solving fuzzy arithmetic equations with triangular norms, *Proc. 2d IEEE Int. Conf. on Fuzzy Systems (FUZZ-IEEE)*, San Francisco, 470-476.

Kawaguchi M. F. and Da-Te T. (1994). Some algebraic properties of weakly non-interactive fuzzy numbers, *Fuzzy Sets and Systems,* **68**, 281-291.

Keresztfalvi T. and Kovacs M. (1992). g, p-fuzzification of arithmetic operations, *Tatra Mount. Math. Publ.*, **1**, 65-71.

Kerre E. (1982). The use of fuzzy numbers in electrocardiological diagnosis, *Approximate Reasoning in Decision Analysis* (Gupta M. M. and Sanchez E., eds.), North-Holland, Amsterdam, 277-282.

Kerre E. (1987). Fuzzy approach to ECG diagnosis, *Encyclopedia of Systems and Control* (Singh, M. ed.), Pergamon Press, Oxford, UK, 1405-1407.

Kerre E. (1993). *Introduction to the Basic Principles of Fuzzy Set Theory and Some of its Applications*. Communication & Cognition, Gent, Belgium.

Kerre E. and Van Schooten A. (1988). A deeper look on fuzzy numbers from a theoretical as well as from a practical point of view, *Fuzzy Logic in Knowledge-Based Systems, Decision and Control* (M. M. Gupta, T. Yamakawa, eds.), North-Holland, Amsterdam, 173-196.

Kim J. B. (1993). On product-sum of fuzzy complex numbers of an elliptic type, *J. Fuzzy Math.*, **1**(3), 611-617.

Kim K. and Park K. S. (1990). Ranking fuzzy numbers with index of optimum, *Fuzzy Sets and Systems,* **35**, 143-150.

Kim W. J., Ko J. H. and Chung M. J. (1994). Uncertain robot environment modelling using fuzzy numbers, *Fuzzy Sets and Systems,* **61**, 53-62.

Klawonn F. and Kruse R. (1993). Equality relations as a basis for fuzzy control, *Fuzzy Sets and Systems,* **54**, 147-156.

Klement E. P. (1981). Operations on fuzzy sets and fuzzy numbers related to triangular norms, *Proc. 11th IEEE Int. Symp. on Multiple-Valued Logic*, Oklahoma City, 218-225.

Klement E. P. (1985). Integration of fuzzy valued functions, *Revue Roumaine Math. Pures Appl.*, **30**, 375-384.

Klement E. P. (1987). Strong law of large numbers for random variables with values in the fuzzy real line, *IFSA Commun. Math. Chapt.*, 7-11.

Klement E. P., Mesiar R. and Pap E. (1999). *Triangular Norms*, a book to appear.

Klir G. J. (1997). Fuzzy arithmetic with requisite constraints, *Fuzzy Sets and Systems,* **91**, 165-175.

Klir G. J. (1999). Measures of uncertainty and information, *Fundamentals of Fuzzy Sets* (Dubois D. and Prade H., eds.), Kluwer Acad. Publ., 1999. This volume.

Kolesarova A. (1995). Additions preserving the linearity of fuzzy intervals, *Tatra Mount. Math. Publ.*, **6**, 75-81.

Kolesarova A. (1997). Similarity preserving t-norm based additions of fuzzy numbers, *Fuzzy Sets and Systems,* **91**, 215-229.

Kolesarova A. (1998). Triangular norm-based addition preserving linearity of T-sums of linear fuzzy intervals, *Mathware and Soft Computing,* **5**, 91-98.

Kolesarova A. and Riecan B. (1993). T_∞-fuzzy observables, *Int. J. Theor. Physics*, **32**, 1691-1707.

Kolodziejczyk W. (1986). Orlovsky's concept of decision-making with a fuzzy relation — Further results. *Fuzzy Sets and Systems*, **20**, 11-20.

Kosheleva O., Cabrera S.D., Gibson G.A. and Koshelev M. (1997). Fast implementations of fuzzy arithmetic operation using fast Fourier transform, *Fuzzy Sets and Systems*, **91**, 269-277.

Kruse R. (1987). On a software tool for statistics with linguistic data, *Fuzzy Sets and Systems,* **24**, 377-383.

Kruse R. and Meyer K. D (1987). *Statistics with Vague Data*, Kluwer Academic, Dordrecht, Netherlands.

Kwakernaak H. (1978). Fuzzy random variables — Vol. I : Definitions and theorems, *Information Sciences,* **15**, 1-29.

Kwakernaak H. (1979). Fuzzy random variables — Part II : Algorithms and examples in the discrete case, *Information Sciences,* **17**, 253-278.

Kwiesielewicz M. (1998). A note on the fuzzy extension of Saaty's priority theory, *Fuzzy Sets and Systems,* **95**, 161-173.

Lai Y. J. and Hwang C. L. (1992). *Fuzzy Mathematical Programming – Methods and Applications*, Lecture Notes in Economics and Mathematical Systems, vol. 394, Springer-Verlag, Berlin.

Lee D. H. and Park D. (1997). An efficient algorithm for fuzzy weighted average, *Fuzzy Sets and Systems,* **87**, 39-45.

Lee K. M., Cho C. H. and Lee-Hwang H. (1994). Ranking fuzzy numbers with satisfaction function, *Fuzzy Sets and Systems,* **64**, 295-311.

Li R.J. and Lee E. S. (1987). Ranking fuzzy numbers: A comparison, *Proc. North-American Fuzzy Inf. Processing Soc. Workshop (NAFIPS'87),* Purdue University, West Lafayette, IN, 169-204.

Li E. S. and Lee R. J. (1988). Comparison of fuzzy numbers based on the probability measure of fuzzy events, *Comp. & Math. with Appl.*, **15**, 887-896.

Ling C. H. (1965). Representation of associative functions, *Publ. Math. Debrecen*, **12**, 189-212.

Liou T. S. and Wang M. J. J. (1992a). Ranking fuzzy numbers with integral value, *Fuzzy Sets and Systems,* **50**, 247-255.

Liou T. S. and Wang M. J. J. (1992b). Fuzzy weighted average: An improved algorithm. *Fuzzy Sets and Systems,* 307-317.

Liu X. (1994). On the continuity of fuzzy number valued function, *Fuzzy Sets and Systems,* **68**, 245-247.

Lootsma F. A. (1985). Performance evaluation of nonlinear optimization methods via pairwise comparison and fuzzy numbers, *Math. Prog.*, **33**, 93 et seq.

Lowen R. (1980). Convex fuzzy sets, *Fuzzy Sets and Systems,* **3**, 291-310.

Lowen R. (1983a). On ($\mathbb{R}$(L),⊕), *Fuzzy Sets and Systems,* **10**, 203-209.

Lowen R. (1983b). Hyper-spaces of fuzzy sets, *Fuzzy Sets and Systems,* **9**, 287-311.

Lowen R. (1985). The order aspect of the fuzzy real line, *Manuscripta Math.,* **39**, 293-309.

Lowen R. (1996). *Fuzzy Set Theory*, Kluwer, Dordrecht.

Mabuchi S (1988). An approach to the comparison of fuzzy subsets with an α-cut dependent index, *IEEE Trans. Syst. Man. Cybern.*, **18**, 264-272.

Mandic N. J. and Mamdani E. H. (1980). A linguistic fuzzy calculator. Res. Rep. No. 9, Queen Mary College, University of London, UK.

Mares M. (1977a). How to handle fuzzy quantities, *Kybernetika* (Prague), **13**, 23-40.

Mares M. (1977b). On fuzzy quantities with real and integer values, *Kybernetika* (Prague), **13**, 41-56.

Mares M. (1993a). Algebraic equivalences over fuzzy quantities, *Kybernetika* (Prague), **29**(2), 121-132.

Mares M. (1993b). Remarks on fuzzy quantities with finite support, *Kybernetika* (Prague), **29**(2), 133-143.

Mares M. (1994). *Computation Over Fuzzy Quantities*, CRC Press, Boca Raton, FL.

Mares M. and Mesiar R. (1996). Processing of sources of fuzzy quantities, *Proc. IPMU'96*, Granada, 359-363.

Markov S. M. (1995). On directed interval arithmetics and its applications, *J. Universal Computer Science*, **1**(7), 510-521.

Markov S. M. (1996). On the foundations of interval mathematics, *Scientific Computing and Validated Numerics (Proc. SCAN – 95)* (Alefeld G. and Frommer, A., eds.), Akademie-Verlag, Berlin, 307 et seq.

Markova A. (1995). Additions of L-R fuzzy numbers, *Busefal* (IRIT, Université P. Sabatier, Toulouse), **63**, 25-29.

Markova A. (1997). Idempotents of the T-addition of fuzzy numbers, *Tatra Mount. Math. Publ.,* **12**.

Markova A. (1998). T-sum of L-R fuzzy numbers, *Fuzzy Sets and Systems,* **85**, 379-384.

Markova-Stupnanova A. (1997a). A note to the addition of fuzzy intervals based on the continuous Archimedean t-norm, *Fuzzy Sets and Systems,* **91**, 253-258.

Markova-Stupnanova A. (1997b). Pseudo-convolutions and their idempotents. Proc. 7th Inter. Fuzzy Syst. Assoc. World Cong. (IFSA'97), Prague, June 25-29, 1997, Vol. 1, Academia, 484-487.

Markova-Stupnanova A. (1998). A note on recent results on the law of large numbers for fuzzy numbers, *BUSEFAL* (IRIT, Univ. P. Sabatier, Toulouse), **76**, 12-18.

Matheron G. (1975). *Random Sets and Integral Geometry*, John Wiley & Sons, New York.

McCahon C. and Lee E. S. (1990). Comparing fuzzy numbers: the proportion of the optimum method, *Int. J. Approximate Reasoning*, **4**, 159-181.

McCain R. A. (1983). Fuzzy confidence intervals, *Fuzzy Sets and Systems,* **10**, 281-290.

Mesiar R. (1993). Fuzzy measurable functions, *Fuzzy Sets and Systems,* **59**, 35-42.

Mesiar R. (1995). Computation over LR-fuzzy numbers, *Proc. CIFT'95*, Trento, 165-176.

Mesiar R. (1996a). LR-fuzzy numbers, *Proc. 6th Int. Conf. On Information Processing and Management of Uncertainty in Knowledge-based Systems (IPMU'96)*, Granada, Spain, 337- 342.

Mesiar R. (1996b). A note on the T-sum of L-R fuzzy numbers, *Fuzzy Sets and Systems,* **79**, 259-261.

Mesiar R. (1997a). Shape preserving additions of fuzzy intervals, *Fuzzy Sets and Systems,* **86**, 73-78.

Mesiar R. (1997b). Triangular norm-based additions of fuzzy intervals, *Fuzzy Sets and Systems,* **91**, 231-237.

Mesiar R. (1998). Approximations of continuous t-norms by strict t-norms with smooth generators, *BUSEFAL* (IRIT, Université Paul Sabatier, Toulouse), n° **75**, 72-79.

Miyakawa M., Nakamura K., Ramik J. and Rosenberg I. G. (1993). Joint canonical fuzzy numbers, *Fuzzy Sets and Systems,* **53**, 39-49.

Miyakoshi M. and Shimbo M. (1984a). A strong law of large numbers for fuzzy random variables, *Fuzzy Sets and Systems,* **12**, 133-142.

Miyakoshi M. and Shimbo M. (1984b). An individual ergodic theorem for fuzzy random variables, *Fuzzy Sets and Systems,* **13**, 285-290.

Mizumoto M. and Tanaka K. (1976a). The four operations of arithmetic on fuzzy numbers, *Syst. Comput. Controls*, **7**(5), 73-81.

Mizumoto M. and Tanaka K. (1976b). Algebraic properties of fuzzy numbers, *Proc. Int. Conf. On Cybernetics and Society*, Washington, DC, 559-563.

Mizumoto M. and Tanaka K. (1976c). Some properties of fuzzy sets of type 2, *Inf. Control*, **31**, 312-340.

Mizumoto M. and Tanaka K. (1979). Some properties of fuzzy numbers, *Advances in Fuzzy Set Theory and Applications* (Gupta M. M., Ragade R. K. and Yager R. R., eds.), North-Holland, Amsterdam, 153-165.

Moore R. (1966). *Interval Analysis*, Prentice-Hall, Englewood Cliffs, NJ.

Moore R. (1979). *Methods and Applications of Interval Analysis*, SIAM Studies on Applied Mathematics, Society for Industrial and Applied Mathematics, Philadelphia.

Murakami S., Maeda H. and Imamura S. (1989). Fuzzy decison analysis on the development of centralized regional energy control systems. *Proc. IFAC Symp. on Fuzzy Information, Knowledge Representation and Decision Processes* (Sanchez E. and Gupta M. M., eds.), Pergamon Press, Oxford, 363-368.

Murthy C. A, Pal S. K. and Dutta Majumder D. (1985). Correlation bet ween two membership functions, *Fuzzy Sets and Systems,* **17**, 23-38.

Nahmias S. (1978). Fuzzy variables, *Fuzzy Sets and Systems,* **1**, 97-110.

Nahmias S. (1979). Fuzzy variables in a random environment, *Advances in Fuzzy Set Theory* (Gupta M.M., Ragade R. and Yager R. R., eds.), North-Holland, Amsterdam, 165-180.

Nakahara Y. (1998). User oriented ranking criteria and application to fuzzy mathematical programming problems, *Fuzzy Sets and Systems,* **94**, 275-286.

Nakamura K. (1986). Preference relations on a set of fuzzy utilities as a basis for decision-making, *Fuzzy Sets and Systems*, **20**, 147-162.

Nakamura K. (1990). Canonical fuzzy number of dimension two and fuzzy utility difference for understanding preferential judgements, *Information Sciences,* **50**, 1-22.

Nanda S. (1989). On sequences of fuzzy numbers, *Fuzzy Sets and Systems,* **33**, 123-126.

Negoita C. V. (1978). *Management Applications of Systems Theory*, Birkhauser Verlag, Basel.

Nguyen H. T. (1978). A note on the extension principle for fuzzy sets, *J. Math. Anal. Appl.*, **64**, 369-380.

Nguyen H. T. and Kreinovitch V., Nesterov V. and Nakamura M. (1997). On hardware support for interval computations and for soft computing: theorems, *IEEE Trans. on Fuzzy Systems*, **5**, 108-127.

Nguyen H. T., Kreinovitch V. and Wojciechowski P. (1998). Strict Archimedean t-norms and t-co-norms as universal approximators, *Int. J. Approximate Reasoning*, **18**, 239-249.

Oftedal H. (1981). Imprecision of specification of information systems parameters: A study of decision point probabilities, *Inform. Syst.*, **6**, 101-109.

Orlovsky S.A. (1978). Decision-making with a fuzzy preference relation, *Fuzzy Sets and Systems*, **1**, 155-168.

Otto K. and Antonsson E. (1991). Trade-off strategies in engineering design, *Research in Engineering Design* (Springer Verlag), **3**, 87-104.

Otto K. and Antonsson E. (1994). Design parameter selection in the presence of noise, *Research in Engineering Design* (Springer Verlag), **6**, 234-246.

Otto K., Lewis A. D. and Antonsson E. (1993). Approximating α-cuts with the vertex method, *Fuzzy Sets and Systems,* **55**, 43-50.

Ovchinnikov S. and Migdal M. (1987). 0n ranking fuzzy sets, *Fuzzy Sets and Systems,* **24**, 113-117.

Pal N. and Bezdek J.C. (1999). Quantifying different facets of fuzzy uncertainty, *Fundamentals of Fuzzy Sets* (Dubois D. and Prade H., eds.), Kluwer Acad. Publ., 1999. This volume.

Pan Y. and Yuan B. (1997). Bayesian inference of fuzzy probabilities, *Int. J. of General Systems*, **26**(1-2), 73-90.

Papoulis A. (1965). *Probability, Random Variables and Stochastic Processes*, McGraw-Hill, New York.

Parratt L. G. (1971). *Probability and Experimental Errors in Science*, Dover, New York.

Pedrycz W. (1985). On generalized fuzzy relation equations and their applications, *J. Math. Anal. Appl.,* **107**, 520-536.

Pedrycz W. (1986). Ranking multiple aspect alternatives — Fuzzy relational approach, *Automatica*, **22**, 251-253.

Pedrycz W. (1994). Why triangular membership functions?, *Fuzzy Sets and Systems,* **64**, 21-30.

Prade H. (1979). Using fuzzy set theory in a scheduling problem, *Fuzzy Sets and Systems,* **2**, 153- 165.

Prade H. (1980). Operations research with fuzzy data, *Fuzzy Sets: Theory and Applications to Policy Analysis and Information Systems* (Wang P. P. and Chang S. K., eds.), Plenum Press, New York, 155-170.

Prade H. (1982). Modal semantics and fuzzy set theory. In: Fuzzy Set and Possibility Theory. Recent Developments. (R.R. Yager, ed.), Pergamon Press, New York, 232-246.

Prade H. (1984). Lipski's approach to incomplete information data bases restated and generalized in the setting of Zadeh's possibility theory, *Inf. Syst.*, **9**, 27-42.

Prade H. and Testemale C. (1984). Generalizing database relational algebra for the treatment of incomplete/uncertain information and vague queries, *Information Sciences,* **34**, 115-143.

Procyk T. S. and Mamdani M. M. (1979). A linguistic self organizing process controller, *Automatica*, **15**, 15-30.

Puncochar J., Drahos M. and Vrba J. (1996). Fuzzy number as a product of geometrical construction, *Fuzzy Sets and Systems,* **83**, 43-50.

Puri M. and Ralescu D. (1981). Differentielle d'une fonction floue, *C. R. Acad. Sci. Paris, Ser. I,* **293**, 237-239.

Puri M. and Ralescu D. (1983). Differentials of fuzzy functions, *J. Math. Anal. Appl.*, **91**, 552-558.

Puri M. and Ralescu D. (1986). Fuzzy random variables, *J. Math. Anal. Appl.*, **114**, 409-422.

Puri M. L. and Ralescu D. (1991). Convergence theorems for fuzzy martingales, *J. Math. Anal. Appl.*, **160**, 107-122.

Quadrat J. P. (1990). Théorèmes asymptotiques en programmation dynamique, *C. R. Acad. Sci. Paris*, Série I, **311**, 745-748.

Ramik J. (1986). Extension principle in fuzzy optimization, *Fuzzy Sets and Systems*, **19**, 29-37.

Ramik J. and Nakamura K. (1993). Canonical fuzzy numbers of dimension two, *Fuzzy Sets and Systems,* **54**(2), 167-181.

Ramik J. and Rimanek J. (1985). Inequality relation between fuzzy numbers and its use in fuzzy optimization. Fuzzy Sets and Systems, 16, 123-138.

Ramik J. and Rommelfanger H. (1993). A single- and multi-valued order on fuzzy numbers and its use in linear programming with fuzzy coefficients, *Fuzzy Sets and Systems,* **57**, 203-288.

Ramik J. and Rommelfanger H. (1996). Fuzzy mathematical progamming based on some new inequality relations, *Fuzzy Sets and Systems,* **81**, 77-87.

Rao M. B. and Rashed A. (1981). Some comments on fuzzy variables, *Fuzzy Sets and Systems,* **6**, 285-292.

Requena I., Delgado M. and Verdegay J. L. (1994). Automatic ranking of fuzzy numbers with the criterion of a decision-maker learnt by an artificial neural network, *Fuzzy Sets and Systems,* **64**, 1-21.

Requena I., Blanco A., Delgado M. and Verdegay J. L. (1995). A decision personnal index of fuzzy numbers based on neural network, *Fuzzy Sets and Systems,* **73**, 185-201.

Rockafellar R. T. (1970). *Convex Analysis*, Princeton Univ. Press, Princeton, NJ.

Rodabaugh S. E. (1982). Fuzzy addition in the L-fuzzy real line, *Fuzzy Sets and Systems,* **8**, 39-51.

Roubens M. (1990). Inequality constraints between fuzzy numbers and their use in mathematical programming, *Stochastic vs. Fuzzy Approaches to Multiobjective Mathematical Programming* (Slowinski R. and Teghem J., eds.), Kluwer Academic, Dordrecht, Netherlands, 321-330.

Roubens M. and Vincke P. (1985). *Preference Modelling*, Springer Verlag, Berlin.

Roubens M. and Vincke P. (1988). Fuzzy possibility graphs and their application to ranking fuzzy numbers, *Non-Conventional Preference Relations in Decision Making* (Kacprzyk J. and. Roubens M., eds.), 119-128.

Saade J. J. and Schwarzlander H. (1992). Ordering fuzzy sets over the real line: An approach besed on decision making under uncertainty, *Fuzzy Sets and Systems,* **50**, 237-246.

Saaty T. L. (1980). *The Analytic Hierarchy Process: Planning, Priority Setting, Resource Allocation*, McGraw-Hill, New York.

Sage A. P. (1981). Sensitivity analysis in systems for planning and decision support, *J. Franklin Inst.*, **312**, 265-291.

Sakawa M. (1993). *Fuzzy Sets and Interactive Multiobjective Optimization*, Plenum Press, New York.

Salakhutdinov R.R., Salakhutdinov R.Z. (1998). The law of large numbers and T-stability fuzzy numbers, *BUSEFAL* (IRIT, Univ. P. Sabatier), **74**, 64-67.

Sanchez E. (1984). Solution of fuzzy equations with extended operations, *Fuzzy Sets and Systems,* **12**, 237-248.

Sanchez E. (1996). Truth-qualification and fuzzy relations in natural languages, application to medical diagnosis, *Fuzzy Sets and Systems,* **84**, 155-167.

Sarna M. (1989). Fuzzy relations on fuzzy and non fuzzy numbers — Fast computation formulas, *Fuzzy Sets and Systems,* **29**, 155-163.

Sarna M. (1998). Fuzzy relations on fuzzy and non fuzzy numbers; fast computation formulas: II. *Fuzzy Sets and Systems,* **93**, 63-74.

Schweizer B. (1975). Multiplications on the space of probability distribution functions, *Aeq. Math.*, **12**, 156-183.

Schweizer B. and Sklar A. (1963). Associative functions and abstract semi-groups, *Publ. Math. Debrecen*, **10**, 69-81.

Schweizer B. and Sklar A. (1983). *Probabilistic Metric Spaces*, North Holland, New York.

Scott M. and Antonsson E. (1995). Aggregation functions for engineering design trade-offs, *Proc. of the 9th Int. Conf. on Design Theory and Methodology*, Vol. 2, 389-396.

Sebastian H.-J. and Antonsson E. (eds.) (1996). *Fuzzy Sets in Engineering Design and Configuration*, Kluwer Academic, Dordrecht, The Netherlands.

Shafer G. (1976). *A Mathematical Theory of Evidence*, Princeton University Press, Princeton.

Sinha D. (1990). A general theory of fuzzy arithmetic, *Fuzzy Sets and Systems*, **36**, 339-364.

Slowinski R (1998). Fuzzy sets in decision analysis operations research and statistics, *The Handbooks of Fuzzy Sets Series*, Kluwer, Boston, USA.

Schmeidler, D. (1989)Subjective probability and expected utility without additivity. Econometrica, 57, 571-587.

Schmucker K. J. (1984). *Fuzzy Sets, Natural Language Conputations and Risk Analysis*, Computer Science Press, Rockville, MD.

Sinha D., Dougherty, E.R. (1995). A general axiomatic theory of intrinsically fuzzy mathematical morphologies, *IEEE Trans. on Fuzzy Systems*, **3**,389-403.

Slowinski R. (Ed.) (1998). *Fuzzy Sets in Decision Analysis, Operations Research and Statistics*. The Handbooks of Fuzzy Sets Series (D. Dubois, H. Prade, eds.), Kluwer Acad. Publ., Boston.

Song Q. and Chissom B. S. (1993a). Fuzzy time series and its models, *Fuzzy Sets and Systems,* **54**, 269-277.

Song Q. and Chissom B. S. (1993b). Forecasting enrollments with fuzzy time series: Part I, *Fuzzy Sets and Systems,* **54**, 1-9.

Song Q. and Chissom B. S. (1994). Forecasting enrollments with fuzzy time series: Part II, *Fuzzy Sets and Systems,* **62**, 1-8.

Song Q., Leland R. L. and Chissom B. S. (1995). A new fuzzy time-series model of fuzzy number observations, *Fuzzy Sets and Systems,* **73**, 341-348.

Stanford R. E. (1982). The set of limiting distributions for a Markov chain with fuzzy transition probabilities, *Fuzzy Sets and Systems,* **7**, 71-78.

Stein W. E. (1983). A note on convexity and fuzzy random variables, BUSEFAL (IRIT, Université P. Sabatier, Toulouse), **14**, 43-46.

Stein W. E. and Talati K. (1981). Convex fuzzy random variables, *Fuzzy Sets and Systems,* **6**, 271-283.

Steyaert H., Van Parys F., Baekeland R. and Kerre E. (1995). Implementation of piecewise linear fuzzy quantities, *Int. J. Intelligent Systems*, **10**, 1049-1059.

Stojakovic M. (1994). Fuzzy valued measure, *Fuzzy Sets and Systems,* **65**, 95-104.

Stojakovic M. and Stojakovic Z. (1996). Additions and series of fuzzy sets, *Fuzzy Sets and Systems,* **83**, 341-346.

Subrahmanyam P. V. and Sudarsanam S. K. (1994). On some fuzzy functional equations, *Fuzzy Sets and Systems,* **64**, 333-338.

Tamura N. and Horiuchi K. (1993). VSOP fuzzy numbers and fuzzy comparison relations, *Proc. FUZZ-IEEE*, San Francisco, 1287-1292.

Tanaka H. and Asai K. (1984a). Fuzzy linear programming with fuzzy numbers, *Fuzzy Sets and Systems,* **13**, 1-10.

Tanaka H. and Asai K. (1984b). Fuzzy solution in fuzzy linear programming problems, *IEEE Trans. Syst. Man. Cybern.*, **14**, 325-328.

Tanaka H. and Diamond P. (1998). Fuzzy regression analysis, *Fuzzy Sets in Decision Analysis, Operations Research and Statistics* (Slowinski, R., ed.), The Handbook of Fuzzy Sets Series, Kluwer Academic Publ., Dordrecht, 349-390.

Tong R. M. and Bonissone P. P. (1980). A linguistic approach to decision making with fuzzy sets, *IEEE Trans. Syst. Man. Cybern.*, **10**, 716-723.

Triesch E. (1993). On the convergence of product-sum series of L-R fuzzy numbers, *Fuzzy Sets and Systems,* **53**, 189-192.

Tseng T. Y. and Klein C. (1988). A survey and comparative study of ranking procedures in fuzzy decision making. Working Paper n° 8812101, Dept. of Industrial Engineering, University of Missouri-Columbia, USA.

Tseng T. Y. and Klein C. (1989). New algorithm for the ranking procedure in fuzzy decision-making, *IEEE Trans. Syst. Man. Cybern.*, **19**, 1289-1296.

Tsukamoto Y., Nikiforuk P. N. and Gupta M. M. (1981). On the comparison of fuzzy sets using fuzzy chopping, *Proc. 8th Triennial World Congress IFAC,* Vol. 5, Pergamon Press, Oxford, 46-52.

Tsang E. (1993). *Foundations of Constraint Satisfaction,* Academic Press, New York.

Umano M. (1982). Fredom-0: A fuzzy data base system, *Fuzzy Information and Decision Processes* (Gupta M. M. and Sanchez E., eds.), North-Holland, 339-347.

Van Laatroven P. J. M. and Pedrycz W. (1983). A fuzzy extension of Saaty's priority theory, *Fuzzy Sets and Systems,* **11**, 229-241.

Van Leekwijk W. and Kerre E. (1998). Defuzzification: criteria and classification. University of Gent, Belgium. To appear in *Fuzzy Sets and Systems.*

Viertl R. (1996). *Statistical Methods for Non Precise Data.* CRC Press, Boca Raton, FL.

Vrba J. (1992). A note on inverses in arithmetics with fuzzy numbers, *Fuzzy Sets and Systems,* **50**, 267-278.

Wagenknecht M. and Hartmann K. (1983). On fuzzy rank-ordering in poly-optimization, *Fuzzy Sets and Systems,* **11**, 253-264.

Wang P. Z. (1983). From the fuzzy statistics to the falling random subsets, *Advances in Fuzzy Sets, Possibility Theory and Applications* (Wang P. P., ed.), Plenum Press, New York, 81-96.

Wang X. (1997). *A comparative study of the ranking methods for fuzzy quantities*, PhD. Thesis, University of Gent, Belgium.

Wang X. and Ha M. (1994). Solving a system of fuzzy linear equations, *Fuzzy Optimization: Recent Advances* (Delgado M., Kacprzyk J., Verdegay J. L. and Vila M. A., eds.), Physica-Verlag, Heidelberg, Germany, 102-108.

Wang X. and Kerre E. (1996). On the classification and the dependencies of the ordering methods, *Fuzzy Logic Foundations and Industrial Applications* (Rua D., ed.), Kluwer Academic Publishers, Dordrecht, The Netherlands, 73-88.

Wang X. and Kerre E. (1998). Reasonable properties for the ordering of fuzzy quantities, *Fuzzy Sets and Systems*, to appear.

Wang X., Kerre E. and Da Ruan. (1995a). Consistency and weights of judgment matrix in fuzzy AHP, *Int. J. of Uncertainty, Fuzziness and Knowledge-based Systems*, **3**, 35-46.

Wang X., Kerre E. and Da Ruan. (1995b). Transitivity of fuzzy orderings based on pairwise comparisons, *Int. J. Fuzzy Math.*, **3**, 455-463.

Wang X. and Da Ruan. (1995). On the transitivity of fuzzy preference relations in ranking fuzzy numbers, *Fuzzy Set Theory and Advanced Mathematical Applications* (Rua D., ed.), Kluwer Academic Publishers, Dordrecht, The Netherlands, 155-173.

Wang Z. and Klir G. J. (1992). *Fuzzy Measure Theory,* Plenum Press, New York.

Wasowski J. (1997). On solutions to fuzzy equations, *Control and Cybern.*, **26**, 653-658.

Watson S. R., Weiss J. J. and Donnell M. (1979). Fuzzy decision analysis, *IEEE Trans. Syst. Man. Cybern.*, **9**, 1-9.

Williamson R. C. (1991). The law of large numbers for fuzzy variables under a general triangular norm extension principle, *Fuzzy Sets and Systems,* **41**, 55-81.

Willmott R. (1981). Mean measures of containment and equality between fuzzy sets, *Proc. 11th Int. Symp. on Multiple Valued Logic*, Oklahoma City, 183-190.

Woods K., Otto K. and Antonsson E. (1992). Engineering design calculations under uncertainty, *Fuzzy Sets and Systems*, **52**, 1-20.

Wu C and Qiu J. (1998). Some remarks for fuzzy complex analysis, *Fuzzy Sets and Systems,* to appear.

Yager R. R. (1979a). A note on probabilities of fuzzy events, *Information Sciences,* **18**, 113-129.

Yager R. R. (1979b). On solving fuzzy mathematical relationships, *Inf. Control,* **41**, 29-55.

Yager R. R. (1979c). Ranking fuzzy subsets over the unit interval, *Proc. 17th IEEE Int. Conf on Decision and Control*, San Diego, CA, 1535-1437.

Yager R. R. (1980a). On the lack of inverses in fuzzy arithmetic, *Fuzzy Sets and Systems,* **4**, 73-82.

Yager R. R. (1980b). On choosing between fuzzy subsets, *Kybernetes,* **9**, 151-154.

Yager R. R. (1981). A procedure for ordering fuzzy subsets of the unit interval, *Information Sciences,* **24**, 143-161.

Yager R. R. (1982). Measuring tranquility and anxiety in decision-making: An application of fuzzy sets. *Int. J. General Systems,* **9**, 249-260.

Yager R. R. (1986). A characterization of the extension principle, *Fuzzy Sets and Systems*, **18**, 205-219.

Yager R. R. and Filev D. (1993). On the issue of defuzzification and selection based on a fuzzy set, *Fuzzy Sets and Systems*, **55**, 255-271.

Yager R. R. and Filev D. (1994). *Essentials of Fuzzy Modeling and Control,* Wiley, New York.

Yager R. R. and Kelman (1996). A. fusion of fuzzy information with consideration of compatibility, partial aggregation and reinforcement, *Int. J. Approximate Reasoning*, **15**, 93-122.

Yang H. Q, Hua Y. and Jones J. D. (1993). Calculating functions of fuzzy numbers, *Fuzzy Sets and Systems,* **55**, 273-283.

Yoon K. P. (1996). A probabilistic approach to rank complex fuzzy numbers, *Fuzzy Sets and Systems,* **80**, 167-176.

Young R. C. (1931). The algebra of many-valued quantities, *Math. Ann.*, **104**, 260-290.

Yu C. (1993). Correlation of fuzzy numbers, *Fuzzy Sets and Systems*, **55**, 303-307.

Yuan Y. (1991). Criteria for evaluating fuzzy ranking methods, *Fuzzy Sets and Systems*, **44**, 139-157.

Zadeh L. A. (1965). Fuzzy sets, *Inf. Control,* **8**, 338-353.

Zadeh L. A. (1975a). The concept of a linguistic variable and its application to approximate reasoning, *Information Sciences,* Part I: **8**, 199-249; Part II: **8**, 301-357; Part III: **9**, 43-80.

Zadeh L. A. (1975b). Calculus of fuzzy restrictions, *Fuzzy Sets and Their Applications to Cognitive and Decision Processes* (Zadeh L. A., Fu K. S., Shimura M. and Tanuka K., eds.), Academic Press, New York, 1-39.

Zadeh L. A. (1978). Fuzzy sets as a basis for a theory of possibility, *Fuzzy Sets and Systems,* **1**, 3-28.

Zadeh L. A. (1979). A theory of approximate reasoning, *Machine Intelligence*, Vol. 9 (Hayes J. E., Michie D. and Mikulich L. I., eds.), John Wiley & Sons, New York, 149-194.

Zadeh L. A. (1983). A computational approach to linguistic quantifiers in natural language, *Comp. Math. Appl.*, **9**, 149-184.

Zadeh L. A. (1996). Fuzzy logic = computing with words, *IEEE Trans. Fuzzy Systems*, **4**, 103-111.

Zhang G. Q. (1993). The convergence of a sequence of fuzzy integrals of fuzzy number-valued functions on the fuzzy set, *Fuzzy Sets and Systems,* **59**, 43-57.

Zhang D. and Wang Z. (1993). Fuzzy integrals of fuzzy-valued functions, *Fuzzy Sets and Systems,* **54**, 63-67.

Zhao R. and Govind R. (1991a). Algebraic characteristics of extended fuzzy numbers, *Information Sciences*, **54**, 103-130.

Zhao R. and Govind R. (1991b). Solutions of algebraic equations involving generalized fuzzy numbers, *Information Sciences,* **56**, 199-243.

Zhao R. and Govind R. (1991c). Defuzzification of fuzzy intervals, *Fuzzy Sets and Systems,* **43**, 45-56.

Zhu Q. and Lee E. S. (1992). Comparison and ranking of fuzzy numbers, *Fuzzy Regression Analysis* (Kacprzyk J. and Fedrizzi M., eds.), Omnitech Press, Warsaw, Poland, 21-44.

Zimmermann H. J. (Ed.) (1999). *Practical Applications of Fuzzy Technologies*, The Handbooks of Fuzzy Sets Series (D.Dubois, Prade H., eds.), Kluwer Academic Publ., New York.

11 METRIC TOPOLOGY OF FUZZY NUMBERS AND FUZZY ANALYSIS

Phil Diamond
Peter Kloeden

Abstract: This chapter gives an overview of distances between fuzzy numbers and the topology that these metrics induce. The metric structure allows the development of fuzzy analysis and various applications to interpolation, approximation and differential equations.

11.1 INTRODUCTION

Real analysis and calculus of functions of a real variable form a highly developed and very important theory for mathematics, engineering and applied science. This theory involves concepts of convergence, limits, continuity, differentiation and integration. These techniques are applied to a vast number of areas: approximation, differential equations, optimization and control theory, to name but a few.

Central to and implicit in these ideas are topological notions like open and closed sets, neighbourhoods, completeness, compactness, separability and so on. In turn, these concepts are either defined in terms of, or are consequences of metric distance between points in the real field $\Re$ and real Euclidean spaces

$\Re^n$. Further generalizations extend metric distance to infinite dimensional spaces such as the space $C([a,b])$ of continuous functions on the real interval $[a,b]$ and the space $L^p([a,b])$ of $p-$summable functions on $[a,b]$. Definitions are so framed that these spaces inherit analogous topological structure and concepts to those in $\Re^n$, or the definitions are themselves abstracted further in formulations of general topological spaces.

Consequently, when formulating a calculus of fuzzy-valued functions and a topology of fuzzy numbers, it is natural first to develop metrics on spaces of fuzzy sets and fuzzy numbers. Then, definitions of limit and convergence will allow the development of limit operations such as differentiation and integration, and formulation of a calculus and its applications.

However, fuzzy sets and fuzzy numbers are more complicated objects than real vectors or real numbers, and so metric distance and the subsequent topological development is closer to that of function spaces like $C([a,b])$ or $L^p([a,b])$ than for $\Re^n$ or $\Re$. The class of fuzzy sets has to be restricted to some extent but still includes those that are commonly used in both theory and applications. In particular, the space of upper semicontinuous, normal fuzzy convex fuzzy sets $\mathcal{E}^n$ has properties which provide rich topological structure and powerful tools for a theory of fuzzy functions.

The level sets $[X]^\beta$ of $X \in \mathcal{E}^n$ are nonempty compact convex subsets of $\Re^n$. Thus, many of the techniques of convex set-valued analysis can be used. These techniques, together with the Negoita-Ralescu "Stacking Theorem" provide one of the two main features of analysis in $\mathcal{E}^n$.

The other prominent feature is that, again via convex compact level sets and the Stacking Theorem, there is an isometric correspondence between $\mathcal{E}^n$ and a positive cone in the space of continuous functions on the $(n-1)-$sphere $S^{n-1} \subset \Re^n$. Consequently, sophisticated functional analytic techniques can be applied to problems of fuzzy analysis. The systematic application of these two basic principles to fuzzy topology and fuzzy analysis is the dominant theme of this survey. Since many of its results will be incorporated into later sections, there is quite a long section on analysis with convex compact subsets of $\Re^n$.

This chapter is structured as follows: Section 2 introduces $\mathcal{K}^n$ and $\mathcal{K}^n_C$, respectively the spaces of nonempty compact and compact convex subsets of $\Re^n$, endowed with the Hausdorff topology. The support function s_A of $A \in \mathcal{K}^n_C$ is defined. Section 3 defines the space $\mathfrak{D}^n$ of upper semicontinuous fuzzy sets on $\Re^n$ with bounded support and its subset $\mathcal{E}^n$ of fuzzy convex elements in $\mathfrak{D}^n$. Algebraic operations are defined via the Extension Principle and the support function s_U of $U \in \mathcal{E}^n$ introduced. Elements of $\mathfrak{D}^n$ and $\mathcal{E}^n$ are characterized in terms of their family of level sets, which is the Stacking Theorem. Some useful classes of fuzzy sets and fuzzy numbers are discussed. Section 4 introduces the notion of metric distances between elements of $\mathfrak{D}^n$ and of $\mathcal{E}^n$ and studies completeness, separability and convergence relationships for $\mathcal{E}^n$ in the various metric topologies. Section 5 characterizes the compact sets of $\mathcal{E}^n$ in the metric topologies defined in Section 4. All these results hold for fuzzy numbers simply by putting $n = 1$. Section 6 develops a calculus for fuzzy set valued functions of a real variable. The two last sections deal with applications, Section 7 is devoted to interpolation and Bernstein approximation, and Section 8 to fuzzy differential equations. The property of possibilistic irreversibility for fuzzy DEs, which detracts from their practical applicability, is discussed and a recent interpretation in terms of differential inclusions, which overcomes many of the drawbacks is examined.

Many results that are true for fuzzy real numbers, that is in $\mathcal{E}^1$, are formally identical to those in $\mathcal{E}^n$ and have the same statement6 and proof. When this is the case, the general result is given. For brevity, $\mathcal{E}^1$ is usually written as $\mathcal{E}$. Special subspaces of $\mathcal{E}$, such as LR-fuzzy numbers, are discussed in §3.2. Proposition 4.3.2 refers to the second result in §4.3 but sequential numbering is used for referring to equations. Bibliographical notes are given at the end of each section.

Much of the teatment has been drawn from the book by Diamond and Kloeden, 1994, but proofs have been omitted. All proofs and much, much more material, to flesh out the skeleton of this survey, can be found in that monograph and especially in the references given at the end of each section below.

11.2 CALCULUS OF COMPACT CONVEX SUBSETS IN $\Re^n$

11.2.1 Subsets and Algebraic Operations

Attention will be focussed on the following two spaces of *nonempty* subsets of $\Re^n$:

(i) $\mathcal{K}^n$ consisting of all nonempty *compact* (that is, closed and bounded) subsets of $\Re^n$,

(ii) $\mathcal{K}^n_C$ consisting of all nonempty *compact convex* subsets of $\Re^n$.

Note the strict inclusion $\mathcal{K}^n_C \subset \mathcal{K}^n$. Recall that a nonempty subset A of $\Re^n$ is convex if for all $a_1, a_2 \in A$ and all $\lambda \in [0,1]$ the point

$$a = \lambda a_1 + (1 - \lambda) a_2 \tag{1}$$

belongs to A. For any nonempty subset A of $\Re^n$ denote by $\operatorname{co} A$ its *convex hull*, that is the totality of points a of the form (1) or, equivalently, the smallest convex subset containing A. Clearly $A \subseteq \operatorname{co} A = \operatorname{co}(\operatorname{co} A)$, with $A = \operatorname{co} A$ if A is convex. Moreover $\operatorname{co} A$ is closed (compact) if A is closed (compact).

Let A and B be two nonempty subsets of $\Re^n$ and let $\lambda \in \Re$. Define (Minkowski) addition and scalar multiplication respectively by $A + B = \{a + b : a \in A, b \in B\}$, $\lambda A = \{\lambda a : a \in A\}$. Clearly then

Proposition 2.1.1 *$\mathcal{K}^n$ and $\mathcal{K}^n_C$ are closed under the operations of addition and scalar multiplication.*

In fact, these two operations induce a *linear structure* on $\mathcal{K}^n$ and $\mathcal{K}^n_C$ with zero element $\{0\}$. The structure is that of a cone rather than a vector space because, in general, $A + (-1)\,A \neq \{0\}$.

Example 2.2.2. Let $A = [0,1]$ so that $(-1)\,A = [-1,0]$, and so

$$A + (-1)\,A = [0,1] + [-1,0] = [-1,1]\,.$$

Thus, adding -1 times a set does not constitute a natural operation of subtraction. Instead, define the *Hukuhara difference* $A \sim_h B$ of nonempty sets A and B, provided it exists, as the nonempty set C satisfying $A = B + C$.

Example 2.1.3. From the preceding example,

$$[-1,1] \sim_h [-1,0] = [0,1] \text{ and } [-1,1] \sim_h [0,1] = [-1,0]\,.$$

Clearly, $A \sim_h A = \{0\}$ for all nonempty sets A. From (2.3) an obvious necessary condition for the Hukuhara difference $A \sim_h B$ to exist is that some translate of B is a subset of A, $B + \{c\} \subseteq A$ for some $c \in \Re^n$. When it exists, $A \sim_h B$ is unique. However, the Hukuhara difference need not exist, as is seen from the following example

Example 2.1.4. $\{0\} \sim_h [0,1]$ does not exist, since no translate of $[0,1]$ can ever belong to the singleton set $\{0\}$.

11.2.2 The Hausdorff Metric

Let x be a point in $\Re^n$ and A a nonempty subset of $\Re^n$. Define the *distance* $d(x,A)$ *from* x *to* A by

$$d(x,A) = \inf\{||x-a|| : a \in A\}\,.$$

Thus $d(x,A) = d(x,\overline{A}) \geq 0$ and $d(x,A) = 0$ if and only if $x \in \overline{A}$, the *closure* of A in $\Re^n$. could be infinite if A is unbounded.

The ε*-neighbourhood of* A is the subset

$$S_\varepsilon(A) = \{x \in \Re^n : d(x,A) < \varepsilon\}.$$

Its closure is the subset $\overline{S}_\varepsilon(A) = \{x \in \Re^n : d(x,A) \leq \varepsilon\}$. In particular, denote by $\overline{S_1^n}$ the *closed unit ball* in $\Re^n$, that is $\overline{S_1^n} = \overline{S}_1(\{0\})$, which is obviously a compact subset of $\Re^n$. Note that $\overline{S}_\varepsilon(A) = A + \varepsilon\overline{S_1^n}$ for any $\varepsilon > 0$ and any nonempty subset A of $\Re^n$. Sometimes, $S(A,\varepsilon)$ and $\overline{S}(A,\varepsilon)$ will be written for $S_\varepsilon(A)$ and $\overline{S}_\varepsilon(A)$ where this is typographically more convenient.

Now let A and B be nonempty subsets of $\Re^n$. Define the *Hausdorff separation* of B from A by $d_H^*(B,A) = \sup\{d(b,A) : b \in B\}$ or, equivalently, $d_H^*(B,A) = \inf\{\varepsilon > 0 : B \subseteq A + \varepsilon\overline{S_1^n}\}$. Thus $d_H^*(B,A) \geq 0$, with $d_H^*(B,A) = 0$ if and only if $B \subseteq \overline{A}$. In addition, the *triangle inequality*

$$d_H^*(B,A) \leq d_H^*(B,C) + d_H^*(C,A)$$

holds for all nonempty subsets A, B and C of $\Re^n$. In general, however $d^*_H(A, B) \neq d^*_H(B, A)$. This is schematically shown in Figure 2.1.

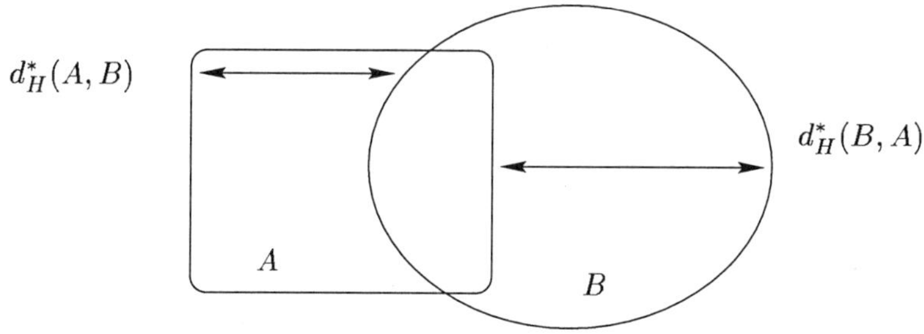

Fig. 2.1. The Hausdorff separations of A and B.

Define the *Hausdorff distance* between nonempty subsets A and B of $\Re^n$ by

$$d_H(A, B) = \max\{d^*_H(A, B), d^*_H(B, A)\} . \tag{2}$$

This is now symmetric in A and B. Consequently,

$$\begin{aligned} &\text{(a)} \quad d_H(A, B) \geq 0 \text{ with } d_H(A, B) = 0 \text{ if and only if } \overline{A} = \overline{B} \, ; \\ &\text{(b)} \quad d_H(A, B) = d_H(B, A) \, ; \\ &\text{(c)} \quad d_H(A, B) \leq d_H(A, C) + d_H(C, B) \, , \end{aligned} \tag{3}$$

for any nonempty subsets A, B and C of $\Re^n$. Restricting attention to the subspaces $\mathcal{K}^n$, $\mathcal{K}^n_C$, the Hausdorff distance (2) is a metric, the *Hausdorff metric.*

Proposition 2.2.1 *Both* $(\mathcal{K}^n, d_H)$ *and* $(\mathcal{K}^n_C, d_H)$ *are complete separable metric spaces.*

11.2.3 Compact Subsets of $\Re^n$

A nonempty subset of $\Re^n$ is compact if and only if it is closed and bounded. For such subsets the distance from a point to a set, the Hausdorff separation and the Hausdorff metric are always finite. Moreover, the infima and suprema are always attained, and so can be replaced by the corresponding minima and maxima.

The following properties of the Hausdorff metric will be useful in the sequel.

Proposition 2.3.1 *If* $A, A', B, B' \in \mathcal{K}^n$ *then*

$$\text{(i)} \quad d_H(tA, tB) = t \, d_H(A, B) \text{ for all } t \geq 0 \, , \tag{4}$$

$$\text{(ii)} \quad d_H(A + B, A' + B') \leq d_H(A, A') + d_H(B, B') \, , \tag{5}$$

$$\text{(iii)} \quad d_H(\text{co } A, \text{co } B) \leq d_H(A, B) \, . \tag{6}$$

Proposition 2.3.2 *If $A, B \in \mathcal{K}^n_C$ and $C \in \mathcal{K}^n$ then*

$$d_H(A + C, B + C) = d_H(A, B) .$$

Define the *magnitude* of a nonempty subset A of $\Re^n$ by

$$||A|| = \sup\{||a|| : a \in A\}$$

or, equivalently,

$$||A|| = d_H(\{0\}, A) . \tag{7}$$

Here, $||A||$ is finite and the supremum is attained when $A \in \mathcal{K}^n$. From (4) it obviously follows that $||tA|| = t\,||A||$ for all $t \geq 0$. In addition, from (3b, c) and (7) obtain

$$\big|\, ||A|| - ||B|| \,\big| \leq d_H(A, B)$$

for all $A, B \in \mathcal{K}^n$. The magnitude (7) can thus be thought of as a Lipschitz function $||\cdot|| : \mathcal{K}^n \rightarrow \Re^+$.

Many applications involve compact subsets of the metric spaces $(\mathcal{K}^n, d_H)$ and $(\mathcal{K}^n_C, d_H)$. Say that a subset $\mathcal{A}$ of $\mathcal{K}^n$ (or $\mathcal{K}^n_C$) is *uniformly bounded* if there exists a finite constant $c(\mathcal{A})$ such that

$$||A|| \leq c(\mathcal{A}) \text{ for all } A \in \mathcal{A} .$$

The following is a simple characterization of compactness:

Proposition 2.3.3 *A nonempty subset $\mathcal{A}$ of the metric space $(\mathcal{K}^n, d_H)$, or $(\mathcal{K}^n_C, d_H)$, is compact if and only if it is closed and uniformly bounded.*

Set inclusion induces a partial ordering on $\mathcal{K}^n$. Write $A \leq B$ if and only if $A \subseteq B$, where $A, B \in \mathcal{K}^n$. Then

$$\mathcal{L}(B) = \{A \in \mathcal{K}^n : B \leq A\} \quad , \quad \mathcal{U}(B) = \{A \in \mathcal{K}^n : A \leq B\}$$

are closed subsets of $\mathcal{K}^n$ for any $B \in \mathcal{K}^n$. In fact, from Proposition 2.4.3, $\mathcal{U}(B)$ is a compact subset of $\mathcal{K}^n$. These assertions remain true with $\mathcal{K}^n_C$ replacing $\mathcal{K}^n$ everywhere.

11.2.4 Support Functions

The metric space $(\mathcal{K}^n_C, d_H)$ of nonempty compact convex subsets of $\Re^n$ with the Hausdorff metric d_H is a separable metric space with the linear structure of a cone, but is not a vector space. A theorem of Rådström shows that this space embeds as a cone in a Banach space. While of considerable theoretical interest, this result is not especially easy to apply. More useful information can be gained directly from the underlying mathematical concept of the *support function* of a nonempty subset.

Let A be a nonempty subset of $\Re^n$. The support function of A is defined for all $p \in \Re^n$ by

$$s(p, A) = \sup\{\langle p, a\rangle : a \in A\},$$

where $\langle \cdot, \cdot \rangle$ denotes the inner product in $\Re^n$. This may take the value $+\infty$ when A is unbounded. However, when A is a compact, convex subset of $\Re^n$ the supremum is always attained and the support function $s(\cdot, A) : \Re^n \to \Re$ is well defined. Indeed, it satisfies the bound $\left|s(p, A)\right| \leq ||A|| \, ||p||$ for all $p \in \Re^n$ and is uniformly Lipschitz with

$$\left|s(p, A) - s(q, A)\right| \leq ||A|| \, ||p - q||$$

for all $p, q \in \Re^n$. In addition, for all $p \in \Re^n$, $s(p, A) \leq s(p, B)$ if $A \subseteq B$ and $s(p, \mathrm{co}\,(A \cup B)) \leq \max\{s(p, A), s(p, B)\}$.

The support function $s(p, A)$ is uniquely paired to the subset A in $\mathcal{K}_C^n$ in the sense that $s(p, A) = s(p, B)$ for all $p \in \Re^n$ if and only if $A = B$ when A and B are restricted to $\mathcal{K}_C^n$. It also preserves set addition and nonnegative scalar multiplication. That is, for all $p \in \Re^n$,

$$s(p, A + B) = s(p, A) + s(p, B)$$

and, in particular

$$s(p, A + \{x\}) = s(p, A) + \langle\, p, x \,\rangle$$

for any $x \in \Re^n$, and

$$s(p, tA) = t\, s(p, A) \ , \ t \geq 0 \, .$$

In fact, the Hausdorff metric is related to the support function for subsets $A, B \in \mathcal{K}_C^n$ by

$$d_H(A, B) = \sup\left\{|s(p, A) - s(p, B)| : p \in S^{n-1}\right\} \tag{8}$$

where $S^{n-1} = \{x \in \Re^n : ||x|| = 1\}$ is the unit sphere in $\Re^n$.

Let $C(S^{n-1})$ denote the Banach space of continuous functions $f : S^{n-1} \to \Re$ with the supremum norm

$$||f|| = \sup\{||f(p)|| : p \in S^{n-1}\} \, .$$

The support function can be used to embed the metric space $(\mathcal{K}_C^n, d_H)$ isometrically as a positive cone in $C(S^{n-1})$. For this define $j : \mathcal{K}_C^n \to C(S^{n-1})$

by $j(A)(\cdot) = s(\cdot, A)$ for each $A \in \mathcal{K}^n_C$. From the properties of the support function, j is a univalent mapping satisfying

$$j(A+B) = j(A) + j(B), \quad j(tA) = t\, j(A)\,,\ t \geq 0\,,$$

$$\|j(A) - j(B)\| = d_H(A, B)$$

for all $A, B \in \mathcal{K}^n_C$. The desired positive cone is the image $j(\mathcal{K}^n_C)$ in $C\left(S^{n-1}\right)$. Obviously j is continuous, as is its inverse $j^{-1} : j(\mathcal{K}^n_C) \to \mathcal{K}^n_C$.

Now, return to the support function $s(p, A)$ considered as being defined for all $p \in \Re^n$. For a fixed $A \in \mathcal{K}^n_C$, it is *positively homogeneous*:

$$s(tp, A) = t\, s(p, A)\,,\ t \geq 0$$

for all $p \in \Re^n$, and *subadditive*:

$$s(p_1 + p_2, A) \leq s(p_1, A) + s(p_2, A)$$

for all $p_1, p_2 \in \Re^n$. Moreover, combining these shows that $s(\cdot, A)$ is a convex function, that is,

$$s(\lambda p_1 + (1-\lambda)p_2, A) \leq \lambda s(p_1, A) + (1-\lambda) s(p_2, A)$$

for all $p_1, p_2 \in \Re^n$ and $\lambda \in [0, 1]$.

The nonempty compact convex subsets of $\Re^n$ are in fact uniquely characterized by such functions.

Proposition 2.4.1 *For every continuous, positively homogeneous and subadditive function $s : \Re^n \to \Re$ there exists a unique nonempty compact convex subset*

$$A = \{x \in \Re^n : \langle\, p, x\,\rangle \;\leq\; s(p) \text{ for all } p \in \Re^n\}$$

which has s as its support function.

11.2.5 L^p-Metrics

In some situations, particularly when investigating random sets, it is advantageous to use an L^p metric on $\mathcal{K}^n_C$ instead of the Hausdorff metric. For $1 \leq p < \infty$ define the *L^p-metric* δ_p on$\mathcal{K}^n_C$ by

$$\delta_p(A, B) = \left(\int_{S^{n-1}} |s(x, A) - s(x, B)|^p\, \mu(dx)\right)^{1/p} \tag{9}$$

for all $A, B \in \mathcal{K}^n_C$, where $s(\cdot, A)$ is the support function of A and μ is Lebesgue measure on the unit sphere S^{n-1}. This is certainly a metric and

Proposition 2.5.1 $(\mathcal{K}^n_C, \delta_p)$ *is a complete, separable metric space for each* $1 \leq p < \infty$.

Like the Hausdorff metric d_H (see (3) in §2.2), the δ_p metric also enjoys the following properties:

$$\begin{aligned}\delta_p(tA, tB) &= t\,\delta_p(A, B)\ ,\ t \geq 0\ ,\\ \delta_p(A+C, B+C) &= \delta_p(A, B)\ ,\\ \delta_p(A+B, A_1+B_1) &\leq \delta_p(A, A_1) + \delta_p(B, B_1)\end{aligned}$$

for all $A, A_1, B, B_1, C \in \mathcal{K}_C^n$. Moreover,

$$\delta_p(A, \{0\}) = \left(\int_{S^{n-1}} |s(x, A)|^p\,\mu(dx)\right)^{1/p}\ ,$$

the L^p norm on the space of Lebesgue integrable functions $s(\cdot) : S^{n-1} \to \Re$.

From (8) and (9),

$$\delta_p(A, B) \leq d_H(A, B) \tag{10}$$

for all $A, B \in \mathcal{K}_C^n$ and any $1 \leq p < \infty$. Strict equality holds here if and only if the sets A and B are parallel forms. Consequently, strict inequality is the typical case. There is, however, an inequality in the opposite direction, though not with a universal constant.

Proposition 2.5.2. *For all $A, B \in \mathcal{K}_C^n$ and $1 \leq p < \infty$*

$$c_p(A, B)\big(d_H(A, B)\big)^{\frac{p+n-1}{p}} \leq \delta_p(A, B) \tag{11}$$

where

$$c_p(A, B) = \left(\frac{\beta(p+1, n-1)}{\beta(1/2, (n-1)/2)\,(\mathrm{diam}\,(A \cup B))^{n-1}}\right)^{1/p}$$

and $\beta(\cdot, \cdot)$ is the beta function.

Here diam (C) denotes the *diameter* of a nonempty compact (not necessarily convex) subset C of $\Re^n$, that is

$$\mathrm{diam}\,(C) = \max\{||c_1 - c_2|| : c_1, c_2 \in C\}\ .$$

The multiplier $c_p(A, B)$ in Proposition 2.5.2 is unbounded and becomes infinite when A and B are the same singleton subset of $\Re^n$. Nevertheless, from the inequalities it follows that the Hausdorff metric d_H (sometimes written δ_∞) and the δ_p metrics, for any $1 \leq p < \infty$, induce the same topology on $\mathcal{K}_C^n$.

Proposition 2.5.3. *All of the metrics δ_p, for $1 \leq p \leq \infty$, induce the same topology on $\mathcal{K}_C^n$.*

Then Proposition 2.3.3 implies

Proposition 2.5.4. *For any $1 \leq p \leq \infty$ a nonempty subset $\mathcal{A}$ of the metric space $(\mathcal{K}_C^n, \delta_p)$ is compact if and only if it is closed and uniformly bounded.*

11.2.6 Continuity and Measurability

Consider mappings F from a domain T in $\Re^k$ into the metric space $(\mathcal{K}_C^n, d_H)$. The usual definition of continuity of mappings between metric spaces applies here. A set valued mapping $F : \Re^k \to \mathcal{K}_C^n$ is *continuous* at t_0 in T if for every $\epsilon > 0$ there exists a $\delta = \delta(\epsilon, t_0) > 0$ such that

$$d_H\left(F(t), F(t_0)\right) < \epsilon \tag{12}$$

for all $t \in T$ with $||t - t_0|| < \delta$. Equivalently, this can be stated in terms of convergence of sequences, that is as

$$\lim_{t_n \to t_0} d_H\left(F(t_n), F(t_0)\right) = 0$$

for all sequences $\{t_n\}$ in T with $t_n \to t_0$.

Using the Hausdorff separation d_H^* and neighbourhoods, observe that (12) is equivalent to both

$$d_H^*\left(F(t), F(t_0)\right) < \epsilon, \quad \text{and} \quad d_H^*\left(F(t_0), F(t)\right) < \epsilon,$$

and the first of these can be rewritten as

$$F(t) \subset S_\epsilon(F(t_0)) := F(t_0) + \epsilon\, S_1^n\,, \tag{13}$$

for all $t \in T$ with $||t - t_0|| < \delta$, in which case the mapping F is said to be *upper semicontinuous* at t_0. Similarly, the second is just

$$F(t_0) \subset S_\epsilon(F(t)) := F(t) + \epsilon\, S_1^n\,, \tag{14}$$

for all $t \in T$ with $||t - t_0|| < \delta$, in which case F is said to be *lower semicontinuous* at t_0. Clearly F is continuous at t_0 if and only if it is both upper semicontinuous and lower semicontinuous at t_0. A set valued mapping can be lower semicontinuous without being upper semicontinuous, and vice versa.

Example 2.6.1. The set valued mapping F from $\Re$ into $\Re$ defined by

$$F(t) = \begin{cases} \{0\}, & \text{for } t = 0 \\ [0,1], & \text{for } t \in \Re \setminus \{0\} \end{cases}$$

is lower semicontinuous, but not upper semicontinuous, at $t_0 = 0$. On the other hand, $F : \Re \to \Re$ defined by

$$F(t) = \begin{cases} [0,1], & \text{for } t = 0 \\ \{0\}, & \text{for } t \in \Re \setminus \{0\} \end{cases}$$

is upper semicontinuous, but not lower semicontinuous, at $t_0 = 0$.

Let $\mathcal{B}(\Re^k)$ and $\mathcal{B}(\mathcal{K}^n_C)$ denote the σ-algebras of Borel subsets of $\Re^k$ and $(\mathcal{K}^n_C, d_H)$ respectively. Adopting the usual definition of Borel measurability of a mapping between metric spaces, if T is a subset of $\Re$, a mapping $F : T \to \mathcal{K}^n_C$ is *measurable* if

$$\{t \in T : F(t) \in B\} \in \mathcal{B}(\Re^k) \text{ for all } B \in \mathcal{B}(\mathcal{K}^n_C) .$$

Write $F^{-1}(A) = \{t \in T : F(t) \cap A \neq \emptyset\}$ for any subset A of $\Re^n$. Then

Proposition 2.6.2 *The following assertions are equivalent :*
(i) $F : T \to \mathcal{K}^n_C$ *is measurable;*
(ii) $F^{-1}(B) \in \mathcal{B}(\Re^k)$ *for all* $B \in \mathcal{B}(\Re^k)$*;*
(iii) $F^{-1}(\mathcal{O}) \in \mathcal{B}(\Re^k)$ *for all open subsets* $\mathcal{O}$ *of* $\Re^n$*;*
(iv) $F^{-1}(C) \in \mathcal{B}(\Re^k)$ *for all closed subsets* C *of* $\Re^n$*;*
(v) $d(\cdot, F(\cdot)) : T \to \Re$ *is measurable for each* $x \in \Re^n$*;*
(vi) $||F(\cdot)|| : T \to \Re$ *is measurable;*
(vii) $s(x, F(\cdot)) : T \to \Re$ *is measurable for each* $x \in \Re^n$*.*

In (v)-(vii) measurability of the single valued mapping $T \mapsto \Re$ is with respect to the Borel σ-algebras $\mathcal{B}(\Re^k)$ and $\mathcal{B}(\Re^1)$. Such mappings are measurable if they are continuous. For set valued mappings,

Proposition 2.6.3 $F : T \to \mathcal{K}^n_C$ *is measurable if it is upper semicontinuous or lower semicontinuous, and hence if it is continuous.*

For set valued mappings, measurability is also preserved on taking limits.

Proposition 2.6.4 *Let* $F_i : T \to \mathcal{K}^n_C$ *be measurable for* $i = 1, 2, 3, \ldots$ *and suppose that* $\lim_{i\to\infty} d_H(F_i(t), F(t)) = 0$ *for almost all* $t \in T$*. Then* $F : T \to \mathcal{K}^n_C$ *is measurable.*

A *selector* of a set valued mapping F from T into $\Re^n$ is a single valued mapping $f : T \to \Re^n$ such that

$$f(t) \in F(t) \text{ for all } t \in T .$$

Proposition 2.6.5 *If* $F : T \to \mathcal{K}^n_C$ *is measurable, then it has a measurable selector* $f : T \to \Re^n$*.*

The following result, known as the *Castaing Representation Theorem*, gives an additional characterization of measurability of a set valued mapping.

Theorem 2.6.6 $F : T \to \mathcal{K}^n_C$ *is measurable if and only if there exists a sequence* $\{f_i\}$ *of measurable selectors of* F *such that*

$$F(t) = \overline{\bigcup\{f_i(t) : i = 1, 2, \ldots\}}$$

for each $t \in T$*.*

11.2.7 Differentiation

There are many different definitions of differentiation for a set valued mapping F from a domain $T \subset \Re^k$ into $\Re^n$. The definition due to de Blasi, which is considered here, has the advantage of being equivalent to Fréchet differentiability of the support function $s(\cdot, F(\cdot)) : T \to C(S^{n-1})$.

Say that a mapping $F : T \to \mathcal{K}^n_C$ is (de Blasi) *differentiable* at $t_0 \in T$ if there exists a positively homogeneous and upper semicontinuous mapping $D_{t_0} : \Re^k \to \mathcal{K}^n_C$ such that

$$\lim_{||\Delta t|| \to 0} \frac{1}{||\Delta t||} d_H \left(F(t_0 + \Delta t), F(t_0) + D_{t_0}(\Delta t)\right) = 0$$

for Δt with $t + \Delta t \in T$. Call D_{t_0} the (de Blasi) *differential* of F at t_0 and write it sometimes as $D^{(F)}_{t_0}$ to indicate its relationship with a particular mapping F.

Proposition 2.7.1 *The de Blasi differential D_{t_0} is unique if it exists.*

Proposition 2.7.2 *Suppose that $F, G : T \to \mathcal{K}^n_C$ are de Blasi differentiable at $t_0 \in T$ with differentials $D^{(F)}_{t_0}$, $D^{(G)}_{t_0}$. Then $F + G$ and λF, for any $\lambda \in \Re$, are de Blasi differentiable at t_0 with differentials $D^{(F+G)}_{t_0} = D^{(F)}_{t_0} + D^{(G)}_{t_0}$ and $D^{(\lambda F)}_{t_0} = \lambda D^{(F)}_{t_0}$, respectively.*

Proposition 2.7.3 *If $F : T \to \mathcal{K}^n_C$ is de Blasi differentiable at $t_0 \in T$ then it is continuous at t_0.*

Differentiable set valued mappings also enjoy the following basic property of single valued differentiable mappings.

Proposition 2.7.4 *Let $F : T \to \mathcal{K}^n_C$ be differentiable at each $t_0 \in T$, where T is open and convex. Then $F(t) \equiv$ constant for all $t \in T$ if and only if its de Blasi differential satisfies $D_{t_0}(\Delta t) \equiv \{0\}$ for all $t_0 \in T$ and $\Delta t \in \Re^k$.*

The de Blasi differential also vanishes at an interior $\preceq$–maximum or $\preceq$–minimum. That is, if $\preceq$ is the partial order on $\mathcal{K}^n_C$ induced by set inclusion and $F(t) \preceq F(t_0)$ for all $t \in T$, F is said to have a $\preceq$ –maximum at t_0 and similarly for a $\preceq$ –minimum.

Proposition 2.7.5 *Let $F : T \to \mathcal{K}^n_C$ be differentiable at each $t_0 \in T$ and have a $\preceq$–maximum or $\preceq$–minimum at $t_0^* \in \operatorname{int} T$. Then $D_{t_0^*}(\Delta t) \equiv \{0\}$ for all $\Delta t \in \Re^k$.*

The support function $s(\cdot, F(t)) : S^{n-1} \to \Re^n$ of $F(t) \in \mathcal{K}^n_C$ for $t \in T$ can be considered as a mapping $s(\cdot, F(\cdot)) : T \to C(S^{n-1})$. Then the following fundamental relationship holds

Theorem 2.7.6 *$F : T \to \mathcal{K}^n_C$ is de Blasi differentiable at $t_0 \in T$ if and only if $s(\cdot, F(\cdot)) : T \to C(S^{n-1})$ is Fréchet differentiable at t_0. Moreover*

$$s(\cdot, F(t_0))'(\Delta t) = s(\cdot, D_{t_0}(\Delta t))$$

for all $\Delta t \in \Re^k$ where $s(\cdot, F(t_0))'$ is the Fréchet differential of $s(\cdot, F(\cdot))$ at t_0 and D_{t_0} is the de Blasi differential of F at t_0.

Example 2.7.7. $F : (0, 2\pi) \to \mathcal{K}_C^1$ defined by $F(t) = (2 + \sin t)\, \overline{S}_1^1$ for $t \in (0, 2\pi)$ is de Blasi differentiable at all $t_0 \in (0, 2\pi)$ with de Blasi differential $D_{t_0}(\Delta t) = \cos t_0.\Delta t.\overline{S}_1^1$. The support function $s(p, F(t)) = 2 + \sin t$ for $p \in S^{n-1}$ has Fréchet differential

$$s(\cdot, F(t_0))'(\Delta t) = \cos t_0.\Delta t = s\left(\cdot, \cos t_0.\Delta t.\overline{S}_1^1\right) .$$

If $F : T \to \mathcal{K}_C^n$ is de Blasi differentiable at all $t_0 \in T$. Then $s(\cdot, F(\cdot)) : T \to C(S^{n-1})$ is Fréchet differentiable at all $t_0 \in T$, and hence is absolutely continuous on T.

For mappings $F : T \to \mathcal{K}_C^n$, where $T \subseteq \Re^1$, there is another definition of differentiation of F which involves Hukuhara difference quotients of F. Suppose for all $\Delta t > 0$ sufficiently small that the Hukuhara differences

$$F(t_0 + \Delta t) \sim_h F(t_0) \ , \ F(t_0) \sim_h F(t_0 - \Delta t) \tag{15}$$

exist with $t_0 \pm \Delta t \in T$. Then F is sid to be *Hukuhara differentiable* at $t_0 \in T$ if there exists an $F'(t_0) \in \mathcal{K}_C^n$, called the *Hukuhara derivative* of F at t_0, such that

$$\lim_{\Delta t \to 0+} d_H\left((F(t_0 + \Delta t) \sim_h F(t_0))/\Delta t, F'(t_0)\right) = 0,$$

$$\lim_{\Delta t \to 0+} d_H\left((F(t_0) \sim_h F(t_0 - \Delta t))/\Delta t, F'(t_0)\right) = 0.$$

Note that the Hukuhara difference $F(t_0 - \Delta t) \sim_h F(t_0)$ does not exist for the simple linear mapping $F(t) = t\,A$, where $A \in \mathcal{K}_C^n$, for any $t_0 > 0$. This is the reason for taking the limit as in the last equation above.

Since a Hukuhara difference $A \sim_h B$ only exists if a translate of B is contained in A, it follows from (15) that

Proposition 2.7.8 *If $F : T \to \mathcal{K}_C^n$ with $T \subseteq \Re^1$ is Hukuhara differentiable at $t_0 \in T$, then* diam $(F(t))$ *is nondecreasing at t_0.*

For the set valued mapping F in Example 2.7.7, diam $(F(t)) = 2 + \sin t$ is not nondecreasing for $-\pi/2 \leq t \leq 3\pi/2$, so F is not Hukuhara differentiable for any $t \in (\pi/2, 3\pi/2)$, although it is de Blasi differentiable there. However, when a set valued mapping is Hukuhara differentiable it is also de Blasi differentiable.

Proposition 2.7.9 *If $F : T \to \mathcal{K}_C^n$ with $T \subseteq \Re^1$ is Hukuhara differentiable at $t_0 \in T$ then it is de Blasi differentiable at t_0 and $F'(t_0)\Delta t = D_{t_0}(\Delta t)$ for*

all $\Delta t \in \Re^1$, where $F'(t_0)$ is the Hukuhara derivative of F at t_0 and D_{t_0} is the de Blasi differential of F at t_0.

A continuously differentiable mapping $F : T \to \mathcal{K}_C^n$, where T is an open convex subset of $\Re^k$, also satisfies a generalization of the Mean Value Theorem for derivatives.

Proposition 2.7.10 *Let $F : T \to \mathcal{K}_C^n$ have de Blasi differential $D_{t_0}(\Delta t)$, for each $t_0 \in T$, which is continuous in $t_0 \in T$ for all $\Delta t \in \Re^k$. Then*

$$d_H\left(F(t_1)\,,\,F(t_0)\right) \leq ||t_1 - t_0|| \max_{t \in \overline{t_0 t_1}} ||D_t(\tau)||$$

for all $t_0, t_1 \in T$ with $t_0 \neq t_1$, where $t \in \overline{t_0 t_1}$ means that $t = t_0 + s\tau$ for $0 \leq s \leq ||t_1 - t_0||$ and $\tau = (t_1 - t_0)/||t_1 - t_0||$.

An analogous result for $T \subseteq \Re^1$ and the Hukuhara derivative can be derived more easily from a Fundamental Theorem of Calculus identity which is introduced in the next subsection.

11.2.8 Integration

Let $F : [0, 1] \to \mathcal{K}_C^n$ and let $\mathbb{S}(F)$ denote the set of integrable selectors of F over $[0, 1]$. Then the *Aumann integral* of F over $[0, 1]$ is defined as

$$\int_0^1 F(t)\,dt = \left\{ \int_0^1 f(t)\,dt : f \in \mathbb{S}(F) \right\}.$$

If $\mathbb{S}(F) \neq \emptyset$, then the Aumann integral exists and F is said to be *Aumann integrable.*

Say that F is *integrably bounded* on $[0, 1]$ if there exists an integrable function $g : [0, 1] \to \Re^1$ such that

$$||F(t)|| \leq g(t) \quad \text{for almost all} \quad t \in [0, 1].$$

If such an F has measurable selectors, then they are also integrable and $\mathbb{S}(F)$ is nonempty.

Theorem 2.8.1 *If $F : [0, 1] \to \mathcal{K}_C^n$ is measurable and integrably bounded then it is Aumann integrable over each $[a, s] \subset [0, 1]$ with $\int_a^s F(t)\,dt \in \mathcal{K}_C^n$ for all $s \in [a, 1]$.*

For a set valued mapping F as in Theorem 2.8.1, the Castaing Representation Theorem 2.6.5 applies and provides a sequence $\{f_i\}$ of integrable selectors which are pointwise dense in F. Moreover,

$$\int_0^1 F(t)\,dt = \overline{\left\{ \int_0^1 f_i(t)\,dt : i = 1, 2, \ldots \right\}}$$

so only these selectors need be considered to evaluate $\int_0^1 F(t)\,dt$.

The Aumann integrability of a mapping $F : [0,1] \to \mathcal{K}_C^n$ is fundamentally related to the Bochner integrability of its support function $s(\cdot, F(\cdot)) : [0,1] \to C(S^{n-1})$.

Theorem 2.8.2 *Suppose that $F : [0,1] \to \mathcal{K}_C^n$ is measurable. Then F is Aumann integrable if and only if $s(\cdot, F(\cdot)) : [0,1] \to C(S^{n-1})$ is Bochner integrable, in which case*

$$s\left(\cdot, \int_0^1 F(t)\,dt\right) = \int_0^1 s(\cdot, F(t))\,dt \tag{16}$$

where the integral on the right is the Bochner integral.

From (16), obtain the pointwise equality

$$s\left(p, \int_0^1 F(t)\,dt\right) = \int_0^1 s(p, F(t))\,dt$$

for all $p \in S^{n-1}$, where the integral on the right is now the Lebesgue integral.

From Theorem 2.8.2 the Aumann integral satisfies

$$\int_0^1 (F(t) + G(t))\,dt = \int_0^1 F(t)\,dt + \int_0^1 G(t)\,dt, \tag{17}$$

$$\int_a^c F(t)\,dt = \int_a^b F(t)\,dt + \int_b^c F(t)\,dt\ ,\ 0 \le a \le b \le c \le 1, \tag{18}$$

$$\int_0^1 \lambda F(t)\,dt = \lambda \int_0^1 F(t)\,dt\ ,\ \lambda \in \Re, \tag{19}$$

for all Aumann integrable $F, G : [0,1] \to \mathcal{K}_C^n$, with

$$\int_0^1 F(t)\,dt \subseteq \int_0^1 G(t)\,dt \text{ if } F(t) \subseteq G(t) \text{ for all } t \in [0,1]\,.$$

In addition, the Aumann integral uniquely determines its integrand.

Proposition 2.8.3 *If $F, G : [0,1] \to \mathcal{K}_C^n$ are Aumann integrable with $\int_0^1 F(t)\,dt = \int_0^1 G(t)\,dt$, then $F(t) = G(t)$ for almost all $t \in [0,1]$.*

Similarly, the following convergence properties can be established

Theorem 2.8.4 *Let $F_i, F : [0,1] \to \mathcal{K}_C^n, i = 1, 2, \ldots,$ be measurable and uniformly integrably bounded. If $F_i(t) \to F(t)$ for all $t \in [0,1]$ as $i \to \infty$, then*

$$A_i = \int_0^1 F_i(t)\,dt \to A = \int_0^1 F(t)\,dt \text{ as } i \to \infty\,.$$

Theorem 2.8.5 *Let $F_i : [0,1] \to \mathcal{K}_C^n, i = 1, 2, \ldots$, be measurable and uniformly integrably bounded and suppose that $A_i = \int_0^1 F_i(t)\,dt \to A \in \mathcal{K}_C^n$ as $i \to \infty$. Then there exists a measurable mapping $F : [0,1] \to \mathcal{K}_C^n$ such that $A = \int_0^1 F(t)\,dt$.*

Theorem 2.8.6 *If $F, G : [0,1] \to \mathcal{K}_C^n$ are integrable, then so also is*

$$d_H\left(F(\cdot), G(\cdot)\right) : [0,1] \to \Re^1$$

and

$$d_H\left(\int_0^1 F(t)\,dt\,,\ \int_0^1 G(t)\,dt\right) \leq \int_0^1 d_H\left(F(t), G(t)\right)\,dt\,.$$

Integration and differentiation of set valued mappings $F : [0,1] \to \mathcal{K}_C^n$ are essentially inverse operations.

Proposition 2.8.7 *Let $F : [0,1] \to \mathcal{K}_C^n$ be measurable and integrably bounded. Then*

$$\lim_{\Delta t \to 0+} \frac{1}{\Delta t} \int_{t_0}^{t_0+\Delta t} F(t)\,dt = F(t_0) \tag{20}$$

for almost all $t_0 \in [0,1]$. In particular, (3.12) holds at all $t_0 \in [0,1)$ where F is continuous.

Theorem 2.8.8 *Let $F : [0,1] \to \mathcal{K}_C^n$ be measurable and integrably bounded. Then $A : [0,1] \to \mathcal{K}_C^n$ defined by*

$$A(t) = \int_0^t F(s)\,ds \tag{21}$$

for all $t \in [0,1]$ is de Blasi differentiable for almost all $t_0 \in (0,1)$ with de Blasi differential $D_{t_0}^{(A)}$ satisfying

$$D_{t_0}^{(A)}(\Delta t) = F(t_0)\,\Delta t$$

for all $\Delta t \in \Re^1$.

In fact, the mapping A in (21) is Hukuhara differentiable for almost all $t_0 \in (0,1)$, with Hukuhara derivative $A'(t_0) = F(t_0)$.

A counterpart of the first Fundamental Theorem of Calculus,

$$F(t_1) = F(t_0) + \int_{t_0}^{t_1} F'(t)\,dt\,,\ 0 \leq t_0 < t_1 \leq 1\,, \tag{22}$$

holds for a Hukuhara differentiable $F : [0,1] \to \mathcal{K}_C^n$, with continuous Hukuhara derivative F' on $[0,1]$. In general, (22) does not hold for a de Blasi differentiable mapping with $D_t(1)$ replacing $F'(t)$ in the equation, since the resulting expression implies that F is Hukuhara differentiable, whereas this need not be the case. Consider the mapping F in Example 2.7.7, with $t_0 < \pi/2 < t_1$. However, from (18) it is always true that

$$\int_a^{t_1} D_t(1)\, dt = \int_a^{t_0} D_t(1)\, dt + \int_{t_0}^{t_1} D_t(1)\, dt$$

for any $0 \leq a \leq t_0 \leq t_1 \leq 1$.

11.2.9 Bibliographical Notes

Kuratowski, 1996, is a standard reference for the topological properties of spaces of subsets of a topological space, including $\Re^n$. More specific details for $\Re^n$ can be found in Hausdorff, 1957, Michael, 1951, 1956 and Berge, 1963. Propositions 2.2.1 and 2.3.1 are proved in Lay, 1982. See also Castaing and Valadier, 1970.

The addition and scalar multiplication of subsets of $\Re^n$ were introduced by Minkowski, 1967. The Hukuhara difference is due to Hukuhara, 1967.

The assertions of Propositions 2.3.1 and 2.3.2 are of mixed origin: (4) is obvious, (5) and Proposition 2.3.2 are due to de Blasi (de Blasi, 1976; Lemma 2.2) and Rådström, 1952, while (6), which is known as Price's inequality, is proved in Rådström, 1952. A proof of Proposition 2.3.3 can be found in Diamond and Kloeden, 1989. See Castaing and Valadier, 1970, for a proof of Proposition 2.3.5.

General references for §2.4 include Aubin and Frankowska, 1990, Castaing and Valadier, 1970, Lay, 1982 and Rockafellar, 1970. Proposition 2.4.1 is adapted from Rockafellar (Rockafellar, 1970; Theorem 13.2,Corollary 13.2.1).

Shephard and Webster, 1985 and Vitale, 1985, investigate L^p metrics for $\mathcal{K}^n_C$. The propositions in §2.5 are allfrom Vitale, 1985.

Propositions 2.6.1-2.6.3 are proved in Castaing and Valadier (Castaing and Valadier, 1970, Chapter III).

The de Blasi differential is based on an idea of Bridgeland, 1970. Propositions 2.7.1–2.7.4 are proved in de Blasi 1976, whereas Theorem 2.7.5 is due to Banks and Jacobs, 1970. See Hukuhara, 1967, for more details on the Hukuhara derivative and a proof of Proposition 2.7.8. Proposition 2.7.9 is adapted from Banks and Jacobs, 1970, and is proved using the equivalence of de Blasi differentiation and the Fréchet differentiability of support functions, cf. Theorem 2.7.6.

Theorem 2.8.1 is proved in Aumann, 1965. See also Castaing and Valadier, 1970, and Aubin and Frankowska, 1990. Theorem 2.8.2 is due to Debreu, 1967, who defined the integral of a compact convex set valued mapping in terms of the Bochner integral of its support function. See Artstein, 1974, as well. The remaining propositions and theorems are adapted from Artstein, 1974, Banks and Jacobs, 1970, and Hermes, 1968.

11.3 THE SPACE $\mathcal{E}^n$

11.3.1 Definitions and Basic Properties

A fuzzy subset of $\Re^n$ is defined in terms of a membership function which assigns to each point $x \in \Re^n$ a grade of membership in the fuzzy set. Such a membership function $U : \Re^n \to I = [0, 1]$ is used synonomously to denote the corresponding fuzzy set. Denote by $\mathcal{F}^n$ the set of all fuzzy sets of $\Re^n$.

For each $\alpha \in (0, 1]$ the *α-level set* $[U]^\alpha$ of a fuzzy set u is the subset of points $x \in \Re^n$ with membership grade $U(x)$ of at least α, that is

$$[U]^\alpha = \{x \in \Re^n : U(x) \geq \alpha\} .$$

The *support* $[U]^0$ of a fuzzy set is then defined as the closure of the union of all its level sets, that is

$$[U]^0 = \overline{\bigcup_{\alpha \in (0,1]} [U]^\alpha} .$$

It immediately follows that

Property 3.1.1 *For all* $0 \leq \alpha \leq \beta \leq 1$ $[U]^\beta \subseteq [U]^\alpha \subseteq [U]^0$.

In general, some level sets of a fuzzy set may be empty. Indeed, in the trivial case of $U(x) \equiv 0$ for all $x \in \Re^n$, even the support is empty; U here is the *empty fuzzy set*. Attention will be restricted to *normal fuzzy sets* which satisfy

Assumption 1 *U maps $\Re^n$ onto I.*

Clearly $[U]^1 \neq \emptyset$, which is often used as an alternative definition of normality. In view of Property 3.1.1,

Property 3.1.2 $[U]^\alpha \neq \emptyset$ for all $\alpha \in I$.

For practical and theoretical purposes it is convenient to further restrict the membership functions under consideration so that

Assumption 2 *$[U]^0$ is a bounded subset of $\Re^n$.*

Assumption 3 *The function U is upper semicontinuous.*

Hence, each level set $[U]^\alpha$, and also $[U]^0$ by definition, is a closed subset of $\Re^n$. Moreover, they are all bounded since they are subsets of $[U]^0$, which is bounded, and so

Property 3.1.3 *$[U]^\alpha$ is a compact subset of $\Re^n$ for all $\alpha \in I$.*

Another immediate consequence of Assumption 3 is

Property 3.1.4 *For any nondecreasing sequence $\alpha_i \to \alpha$ in I*

$$[U]^\alpha = \bigcap_{i \geq 1} [U]^{\alpha_i} .$$

The totality of fuzzy sets satisfying the three Assumptions 1–3 will be denoted by $\mathfrak{D}^n$. In most cases, however, attention will be restricted to a subset of $\mathfrak{D}$ with the following additional property: a fuzzy set U is said to be *fuzzy convex* if

$$U(\lambda x + (1-\lambda)y) \geq \min\{U(x), U(y)\} \tag{23}$$

for all $x, y \in [U]^0$ and $\lambda \in [0,1]$. As a final restriction on the membership functions, the following allows powerful methods of convex analysis to be used.

Assumption 4 *U is fuzzy convex.*

Let U be fuzzy convex and $x, y \in [U]^\alpha$ for some $\alpha \in (0,1]$, so $U(x) \geq \alpha$ and $U(y) \geq \alpha$. Then

$$U(\lambda x + (1-\lambda)y) \geq \min\{U(x), U(y)\} \geq \alpha$$

for any $\lambda \in [0,1]$, so $\lambda X + (1-\lambda)y \in [U]^\alpha$. Hence $[U]^\alpha$ is a convex subset of $\Re^n$ for any $\alpha \in (0,1]$. The support $[U]^0$ is also convex, which follows from the fact that

$$d_H\left([U]^\alpha, [U]^0\right) \to 0 \text{ as } \alpha \to 0+ \tag{24}$$

and the completeness of the metric space $(\mathcal{K}_C^n, d_H)$. Thus

Property 3.1.5 *If U is fuzzy convex, then $[U]^\alpha$ is convex for each $\alpha \in I$.*

Denote by $\mathcal{E}^n$ the space of all fuzzy subsets U of $\Re^n$ which satisfy Assumptions 1-4; that is, normal, fuzzy convex, upper semicontinuous fuzzy sets with bounded supports. When $n = 1$, elements of $\mathcal{E}$ are called *fuzzy numbers* and the compact convex level sets are compact real intervals.

Combining Properties 3.1.1-3.1.5 gives

Proposition 3.1.6 *Let $U \in \mathcal{E}^n$ and write $C_\alpha = [U]^\alpha$ for $\alpha \in I$. Then*
(i) C_α is a nonempty compact convex subset of $\Re^n$ for each $\alpha \in I$;
(ii) $C_\beta \subseteq C_\alpha$ for $0 \leq \alpha \leq \beta \leq 1$; and
(iii) $C_\alpha = \bigcap_{i=1}^\infty C_{\alpha_i}$ for any nondecreasing sequence $\alpha_i \to \alpha$ in I.
Or, equivalently, $d_H(C_{\alpha_i}, C_\alpha) \to 0$ as $\alpha_i \uparrow \alpha$.

The converse of Proposition 3.1.6 also holds and is the *Negoita-Ralescu Stacking Theorem.*

Proposition 3.1.7 *Let $\mathcal{C} = \{C_\alpha : \alpha \in I\}$ be a family of subsets of $\Re^n$ satisfying* (i),(ii) *and* (iii) *of Proposition 3.1.6, and define $U : \Re^n \to I$ by*

$$U(x) = \begin{cases} 0, & \text{if } x \notin C_0, \\ \sup\{\alpha \in I : x \in C_\alpha\}, & \text{if } x \in C_0. \end{cases} \tag{25}$$

Then $U \in \mathcal{E}^n$ with $[U]^\alpha = C_\alpha$ for $\alpha \in (0,1]$ and

$$[U]^0 = \overline{\bigcup_{\alpha \in (0,1]} C_\alpha} \subseteq C_0 \,. \tag{26}$$

Remark 3.1.8 If the condition of convexity on the C_α is dropped, Propositions 3.1.6 and 3.1.7 are still true, but with $\mathfrak{D}^n$ replacing $\mathcal{E}^n$. The results are also true if more general base spaces than $\Re^n$ are used, such as a Banach space.

An advantage of only requiring the $U \in \mathcal{E}^n$ to be upper semicontinuous and not continuous is that the nonempty compact convex subsets of $\Re^n$ can then also be included in $\mathcal{E}^n$ by identification with their characteristic functions. Recall that the characteristic function χ_A of a subset A of $\Re^n$ is defined by

$$\chi_A(x) = \begin{cases} 1, & \text{if } x \in A \\ 0, & \text{if } x \notin A \,. \end{cases}$$

A straightforward application of Proposition 3.1.7 gives

Proposition 3.1.9 *If $A \in \mathcal{K}^n_C$, then $\chi_A \in \mathcal{E}^n$.*

In the context of fuzzy sets, a subset A of $\Re^n$, or more precisely its characteristic function χ_A, is called a *crisp subset* of $\Re^n$.

Addition and scalar multiplication of fuzzy sets in $\mathcal{E}^n$ will be defined level setwise, that is, for $U, V \in \mathcal{E}^n$ and $c \in \Re \setminus \{0\}$

$$[U+V]^\alpha = [U]^\alpha + [V]^\alpha \tag{27}$$
$$[c\,U]^\alpha = c\,[U]^\alpha \tag{28}$$

for each $\alpha \in I$.

Proposition 3.1.10 *$\mathcal{E}^n$ is closed under addition (27) and scalar multiplication (28).*

These definitions are compatible with the Zadeh Extension Principle for addition and scalar multiplication of fuzzy sets. That is,

$$(U+V)(z) = \sup_{z=x+y} \min\{U(x)\,,\, V(y)\}$$
$$(c\,U)(x) = U(x/c).$$

In $\mathcal{E}^n$, these are equivalent to the level set definitions (27) and (28) respectively.

The concept of support function of a nonempty compact convex subset of $\Re^n$, introduced in §2.4, can be importantly generalized to fuzzy sets in $\mathcal{E}^n$. Let $U \in \mathcal{E}^n$ and define $s_U : I \times S^{n-1} \to \Re$ by

$$s_U(\alpha, p) = s(p, [U]^\alpha) = \sup\{\langle p, a\rangle \,:\, a \in [U]^\alpha\} \tag{29}$$

for $(\alpha, p) \in I \times S^{n-1}$, where $s(\cdot, [U]^\alpha)$ is the support function of $[U]^\alpha$. Call s_U the *support function* of the fuzzy set U. Note that the supremum in (29) is actually attained, since the level set $[U]^\alpha$ is compact, and so can be replaced by the maximum. Moreover, for $U, V \in \mathcal{E}^n$, $U = V$ if and only if $s_U = s_V$, since the support function on $\mathcal{K}_C^n$ uniquely characterizes the elements of $\mathcal{K}_C^n$; see Proposition 2.4.1.

The following properties follow directly from those of support functions of the level sets.

Proposition 3.1.11 *Let $U \in \mathcal{E}^n$. Then the support function s_U is*

(i) *uniformly bounded on $I \times S^{n-1}$;*

(ii) *Lipschitz in $p \in S^{n-1}$ uniformly on I;*

(iii) *for each $\alpha \in I$*

$$d_H([U]^\alpha, [V]^\alpha) = \sup\{|s_U(\alpha, p) - s_V(\alpha, p)| : p \in S^{n-1}\}. \qquad (30)$$

Furthermore, there is the following dependence on membership grade.

Proposition 3.1.12 *Let $U \in \mathcal{E}^n$. Then $s_U(\cdot, p)$ is nonincreasingand left continuous in $\alpha \in I$ for each $p \in S^{n-1}$.*

Example 3.1.13 Let $U \in \mathcal{E}$ have level sets $[U]^\alpha = [-1+\alpha, 1-\alpha]$ for $\alpha \in I$. Then, as $S^0 = \{-1, +1\}$, the support function s_U is given by $s_U(\alpha, \pm 1) = 1 - \alpha$ for all $\alpha \in I$.

11.3.2 Useful Subsets of $\mathfrak{D}^n$ and $\mathcal{E}^n$

In many papers in the literature, additional restrictions have been imposed on the membership functions of fuzzy sets in $\mathcal{E}^n$, in effect defining subsets of $\mathcal{E}^n$ which are useful in different practical and theoretical contexts. This is particularly true for the fuzzy numbers $\mathcal{E}$.

A fuzzy set $U \in \mathfrak{D}^n$ or $U \in \mathcal{E}^n$ is called a *Lipschitzian fuzzy set* if it is a Lipschitz function of its membership grade in the sense that

$$d_H([U]^\alpha, [U]^\beta) \leq K\,|\alpha - \beta| \qquad (31)$$

for all $\alpha, \beta \in I$ and some fixed, finite constant K. In view of Proposition 3.1.11, this is equivalent to the support function $s_U(\cdot, p)$ being Lipschitz uniformly in $p \in S^{n-1}$. Denote the respective subsets of Lipschitzian fuzzy sets by $\mathfrak{D}^n_{\text{Lip}}$ and $\mathcal{E}^n_{\text{Lip}}$.

The *triangular fuzzy numbers* are those fuzzy sets in $\mathcal{E}$ which are characterized by an ordered triple $(x_\ell, x_c, x_r) \in \Re^3$ with $x_\ell \leq x_c \leq x_r$ such that $[U]^0 = [x_\ell, x_r]$ and $[U]^1 = \{x_c\}$, for then

$$[U]^\alpha = [x_c - (1-\alpha)(x_c - x_\ell),\, x_c + (1-\alpha)(x_r - x_c)] \qquad (32)$$

for any $\alpha \in I$. Hence their graph is a triangle. Let $\mathcal{E}_{TG}$ denote the subset of all triangular fuzzy numbers. Note that

$$d_H([U]^\alpha, [U]^\beta) = |\alpha - \beta| \max\{x_c - x_\ell, x_r - x_c\}$$

so all the triangular fuzzy numbers are Lipschitzian and $\mathcal{E}_{TG} \subset \mathcal{E}_{\text{Lip}}$.

Slightly more general are the *trapezoidal fuzzy numbers*, that is those $U \in \mathcal{E}$ whose graph is a trapezium. These are uniquely characterized by an ordered quadruple $(x_\ell, x_{c\ell}, x_{cr}, x_r)$, with $x_\ell \leq x_{c\ell} \leq x_{cr} \leq x_r$ such that $[U]^0 = [x_\ell, x_r]$ and $[U]^1 = [x_{c\ell}, x_{cr}]$. Denote the subset of trapezoidal fuzzy numbers by $\mathcal{E}_{TP}$. Then $\mathcal{E}_{TG} \subset \mathcal{E}_{TP} \subset \mathcal{E}_{\text{Lip}}$.

The LR-*fuzzy numbers* generalize the trapezoidal fuzzy numbers by allowing the lateral edges of the graphs to be those of strictly monotonic, continuous functions. Thus, if $[U]^0 = [x_\ell, x_r]$ and $[U]^1 = [x_{c\ell}, x_{cr}]$, where $x_\ell \leq x_{c\ell} \leq x_{cr} \leq x_r$, then U is continuous on $\Re$, strictly increasing on $[x_\ell, x_{c\ell}]$ and strictly decreasing on $[x_{cr}, x_r]$. Denote the subset of LR-fuzzy numbers by $\mathcal{E}_{LR}$ and note that if $U \in \mathcal{E}_{LR}$ then U need not be Lipschitzian.

Any fuzzy number $U \in \mathcal{E}$ has $[U]^1 = [x_{c\ell}, x_{cr}] \subseteq [U]^0 = [x_\ell, x_r]$ with U upper semicontinuous, nondecreasing on $[x_\ell, x_{c\ell}]$ and nonincreasing on $[x_{c\ell}, x_r]$. The *θ-crisp fuzzy number*, where $0 < \theta < \pi/2$, are those $U \in \mathcal{E}$ with

$$|U(s_1) - U(s_2)| \leq \tan\,\theta\,|s_2 - s_1|$$

for any s_1, s_2 with $x_\ell \leq s_1 \leq s_2 \leq x_{c\ell}$ or $x_{cr} \leq s_1 \leq s_2 \leq x_r$. In fact (2.3) must then be satisfied for any $s_1, s_2 \in \Re$, so that a θ-crisp fuzzy number is Lipschitzian with $\tan\,\theta$ as the Lipschitz constant. Thus $\mathcal{E}_\theta \subset \mathcal{E}_{\text{Lip}}$, where $\mathcal{E}_\theta$ denotes the subset of all θ-crisp fuzzy numbers. Since $\tan\,\theta$ is strictly increasing in $\theta \in (0, \pi/2)$, it follows that $\mathcal{E}_{\theta'} \subset \mathcal{E}_\theta$ if $0 < \theta' < \theta < \pi/2$.

Finally, for any nonempty subset K of $\Re^n$ define the subset $\mathcal{E}^n(K)$ by

$$\mathcal{E}^n(K) = \{U \in \mathcal{E}^n : [U]^0 \subset K\}\,, \tag{33}$$

with analogous definitions for $\mathcal{E}^n_{\text{Lip}}(K)$, $\mathcal{E}^n_\theta(K)$ and the like.

11.3.3. Bibliographical Notes

The original idea of fuzzy convexity goes back to Zadeh, 1965; Nguyen, 1978, exploited the idea in $\mathcal{E}$ without using normality, which was introduced by Puri and Ralescu, 1983. Level sets were first comprehensively used by Mizumoto and Tanaka, 1976. The Stacking Theorem 3.1.7 was first proved by Negoita and Ralescu, 1975. Propositions 3.1.10 and equations (27), (28) were proved in Nguyen, 1978,, in the general context of extending a continuous function $f : \Re \times \Re \to \Re$ to $f : \mathcal{E} \times \mathcal{E} \to \mathcal{E}$ using the Zadeh Extension Principle. Note that this does not apply in $n > 1$ dimensions unless f maps compact convex subsets into compact convex subsets – that is, f is a convex and continuous mapping.

Support functions have long been used in the theory of convex analysis, and the first application to fuzzy set theory seems to be Puri and Ralescu, 1986. Proposition 3.1.12 is due to Diamond and Kloeden, 1989.

Lipschitz fuzzy sets of Section 3.2 were introduced by Puri and Ralescu, 1986. Triangular fuzzy numbers are part of the general LR-fuzzy methodology introduced by Dubois and Prade, 1982. Goetschel and Voxman, 1981, used the notion of θ-crisp fuzzy numbers.

11.4 METRICS ON $\mathcal{E}^n$

11.4.1 Definitions and Basic Properties

The most commonly used metrics on $\mathcal{E}^n$ involve the Hausdorff distance between the level sets of the fuzzy sets. They are, in fact, function space metrics applied to functions $\phi : I \to \Re^+$ defined by

$$\phi(\alpha) = d_H\left([U]^\alpha, [V]^\alpha\right) \tag{34}$$

for $\alpha \in I$, where $U, V \in \mathcal{E}^n$. In view of (30) and Proposition 3.1.12, these functions are left continuous and hence measurable on I.

The *supremum metric* d_∞ on $\mathcal{E}^n$ is defined by

$$d_\infty(U, V) = \sup\left\{d_H\left([U]^\alpha, [V]^\alpha\right) : \alpha \in I\right\} \tag{35}$$

for all $U, V \in \mathcal{E}^n$ and is obviously a metric on $\mathcal{E}^n$. The supremum in (35) need not be attained, so cannot be replaced by the maximum.

Example 4.1.1 Let $U, V \in \mathcal{E}^1$ be defined on level sets by

$$[U]^\alpha = [V]^\alpha = [0, 1] \text{ for } 0 \leq \alpha \leq 1/2,$$

$$[U]^\alpha = \{0\},\ [V]^\alpha = [0, 2(1-\alpha)] \text{ for } 1/2 < \alpha \leq 1,$$

and so

$$\phi(\alpha) = d_H\left([U]^\alpha, [V]^\alpha\right) = \begin{cases} 0 & \text{for } 0 \leq \alpha \leq 1/2 \\ 2(1-\alpha) & \text{for } 1/2 < \alpha \leq 1. \end{cases}$$

Then $\sup\{\phi(\alpha) : \alpha \in I\} = 1$, but this is not attained.

The $L^{p,\infty}$*-metrics* d_p on $\mathcal{E}^n$ are a class of metrics for $1 \leq p < \infty$ defined by

$$d_p(u, v) = \left(\int_0^1 d_H\left([u]^\alpha, [v]^\alpha\right)^p \, d\alpha\right)^{1/p} \tag{36}$$

for all $u, v \in \mathcal{E}^n$.

Proposition 4.1.2 *$(\mathcal{E}^n, d_p)$ is a metric space for $1 \le p < \infty$.*

It is not immediately obvious that these d_p are metrics, but symmetry and the triangle inequality follow from corresponding properties of the d_H metric and the Minkowski inequality. To prove $d_p(U,V) = 0$ only if $U = V$ requires extending $d_H([U]^\alpha, [V]^\alpha) = 0$ for almost all α to all $\alpha \in I$ and this follows from the upper semicontinuity of U, V.

From the properties of the Hausdorff metric listed in Propositions 2.3.1 and 2.3.2,

$$d_p(cU\,, cV) = |c|\, d_p(U,V) \tag{37}$$

$$d_p(U+W\,, V+W) = d_p(U,V) \tag{38}$$

$$d_p(U+W\,, V+W') \le d_p(U,V) + d_p(W,W') \tag{39}$$

for all $c \in \Re \setminus \{0\}$, all $U, V, W, W' \in \mathcal{E}^n$ and $1 \le p \le \infty$. Properties (37)–(39) are also satisfied by d_∞.

In view of the identity (30) relating the Hausdorff metric distance between level sets and the distance between their support functions, an alternative expression for d_∞ is given by

$$d_\infty(U,V) = \sup\left\{|s_U(\alpha,q) - s_V(\alpha,q)| : (\alpha,q) \in I \times S^{n-1}\right\}$$

for all $U, V \in \mathcal{E}^n$. Analogous expressions also hold for the d_p metrics , that is

$$d_p(U,V) = \left(\int_0^1 \left(\sup\left\{|s_U(\alpha,q) - s_V(\alpha,q)| : q \in S^{n-1}\right\}\right)^p \, d\alpha\right)^{1/p}$$

for all $U, V \in \mathcal{E}^n$ and each $1 \le p < \infty$.

The L^p–metrics δ_p on $\mathcal{K}_C^n$ which are directly defined in terms of support functions, give rise to another class of metrics on $\mathcal{E}^n$. These are the $L^{p,p}$–*metrics* ρ_p, $1 \le p < \infty$, defined by

$$\rho_p(U,V) = \left(\int_0^1 \delta_p\left([U]^\alpha, [V]^\alpha\right)^p \, d\alpha\right)^{1/p} \tag{40}$$

for all $U, V \in \mathcal{E}^n$. From the definition (9) of the metric δ_p on $\mathcal{K}_C^n$, (40) can be expressed more transparently as

$$\rho_p(U,V) = \left(\int_0^1 \int_{S^{n-1}} |s_U(\alpha,q) - s_V(\alpha,q)|^p \, \mu(dq)\, d\alpha\right)^{1/p}, \tag{41}$$

where μ is unit Lebesgue measure on the unit sphere S^{n-1} in $\Re^n$. That ρ_p is a metric on $\mathcal{E}^n$ follows similarly to Proposition 4.1.2. It is not hard to show that properties (37)–(39) above are also satisfied by the ρ_p metrics.

11.4.2 Completeness

In this section it is shown that $\mathcal{E}^n$ is complete with respect to the metric d_∞, whereas only $\mathcal{E}^n(K)$, where K is a nonempty compact subset of $\Re^n$ is complete with respect to the d_p and ρ_p metrics. That $\mathcal{E}^n$ itself is not complete with respect to these last metrics follows from the fact that it is not closed under convergence in these metrics.

Example 4.2.1 Fix p, $1 \le p < \infty$ and define $U : \Re \to I$ and $U_k \in \mathcal{E}$ by

$$U(x) = \begin{cases} 0 & \text{if } x < 1 \\ x^{-2p} & \text{if } x \ge 1 \,. \end{cases}$$

$$U_k(x) = \begin{cases} 0 & \text{if } x \notin [1,k] \\ x^{-2p} & \text{if } x \in [1,k] \end{cases}$$

for $k = 1, 2, 3, \ldots$. These functions have level sets

$$[U_k]^\alpha = \begin{cases} \left[1, \alpha^{-1/2p}\right] & \text{if } k^{-2p} \le \alpha \le 1 \\ [1,k] & \text{if } 0 \le \alpha \le k^{-2p} \end{cases},$$

$$[U]^\alpha = \begin{cases} \left[1, \alpha^{-1/2p}\right] & \text{if } 0 < \alpha \le 1 \\ [0,\infty) & \text{if } \alpha = 0 \,. \end{cases}$$

Since $[U]^0$ is unbounded, $U \notin \mathcal{E}$. For the $L^{p,\infty}$–metric d_p, $\{U_k\}$ is Cauchy and

$$d_p(U_k, U) = \left(2\, k^{-p}\right)^{1/p} = 2^{1/p}\, k^{-1} \to 0$$

as $k \to \infty$. Hence $\mathcal{E}$ is not closed under the d_p metric. Finally, for the $L^{p,p}$–metric ρ_p

$$\begin{aligned} s_{U_k}(\alpha, -1) = s_U(\alpha, -1) &= -1 \qquad \text{for} \quad \alpha \in (0,1] \\ s_U(\alpha, +1) &= \alpha^{-1/2p} \quad \text{for} \quad \alpha \in (0,1] \\ s_{U_k}(\alpha, +1) &= \begin{cases} \alpha^{-1/2p} & \text{for } k^{-2p} \le \alpha \le 1 \\ k^{-1/2p} & \text{for } 0 \le \alpha \le k^{-2p} \end{cases}, \end{aligned}$$

so that

$$\rho_p(U_k, U) \le 2^{1/p}\, k^{-1} \to 0 \text{ as } k \to \infty$$

since $S^0 = \{\pm 1\}$ and integration over S^0 is just summation. Hence $\mathcal{E}$ is not closed under the ρ_p metric.

In each case where completeness holds, the proof involves first finding a limit u and then using the Stacking Theorem 3.1.7 for the level sets to show that $u \in \mathcal{E}^n$, or $\mathcal{E}^n(K)$, as appropriate.

Proposition 4.2.2 *$(\mathcal{E}^n, d_\infty)$ is a complete metric space.*

From Example 4.2.1, $\mathcal{E}^n$ is not complete with respect to the d_p or ρ_p metrics. However, for any nonempty compact subset K of $\Re^n$ the subset

$$\mathcal{E}^n(K) = \{u \in \mathcal{E}^n : [u]^0 \subseteq K\}$$

is closed with respect to the metrics d_p and ρ_p.

Proposition 4.2.3 *$(\mathcal{E}^n(K), d_p)$ is a complete metric space for each nonempty compact subset K of $\Re^n$ and each $1 \leq p < \infty$.*

It will be discussed in §4.4 that $(\mathcal{E}^n(K), d_p)$ and $(\mathcal{E}^n(K), \rho_p)$ have the same topologies. In view of this fact,

Proposition 4.2.4 *$(\mathcal{E}^n(K), \rho_p)$ is a complete metric space for each nonempty compact subset K of $\Re^n$ and each $1 \leq p < \infty$.*

11.4.3 Separability

Recall that a subset A of a metric space X is said to be *dense* in X if its closure $\bar{A} = X$. If X has a countable dense subset, it is said to be *separable*. The concept is important for approximation theory and is very often crucial in a wider context where an approximation argument is used.

It was seen in the previous section that the d_∞ metric is both useful and simple to use. However, its major failing is that

Proposition 4.3.1 *$(\mathcal{E}^n, d_\infty)$ is not a separable space.*

However, the d_p metrics do induce separable topologies on $\mathcal{E}^n(K)$ for compact subsets K of $\Re^n$.

Proposition 4.3.2 *For $1 \leq p < \infty$, $(\mathcal{E}^n, d_p)$ is separable.*

11.4.4 Convergence Relationships

There are relationships between the convergences with respect to the different metrics on $\mathcal{E}^n$ defined in §4.1. These will be established in this section. For instance, the d_p and ρ_p metrics induce the same topology on $\mathcal{E}^n$ for each $1 \leq p < \infty$, whereas d_∞–convergence implies d_p–convergence.

Example 4.4.1 Let $U = \chi_{[0,1]} \in \mathcal{E}^1$ and define $U_k \in \mathcal{E}^1$ on level sets by

$$[U_k]^\alpha = \begin{cases} [0,1] & \text{if } 1/k < \alpha \leq 1 \\ [0,2] & \text{if } 0 \leq \alpha \leq 1/k \end{cases}$$

for $k = 1, 2, 3, \ldots$. Then

$$d_p(U_k, U) = (1/k)^{1/p}, \quad d_\infty(U_k, U) = 1.$$

From the boundedness of I and $I \times S^{n-1}$, it follows that

$$d_p(U,V) \leq d_q(U,V), \quad \rho_p(U,V) \leq \rho_q(U,V) \tag{42}$$

for all $U, V \in \mathcal{E}^n$ and $1 \leq p \leq q$. Moreover, from inequalities (10) and (11), relating the Hausdorff metric d_H and the L^p–metric δ_p on $\mathcal{K}_C^n$,

$$\rho_p(U, V) \leq d_p(U, V) \leq d_\infty(U, V), \tag{43}$$

$$c_p\left([U]^0, [V]^0\right) d_{p+n-1}(U, V)^{(p+n-1)/p} \leq \rho_p(U, V),$$

for all $U, V \in \mathcal{E}^n$. Hence,

Proposition 4.4.2 *d_p and ρ_p induce the same topology on $\mathcal{E}^n$.*

From inequality (43) is also obtained

Proposition 4.4.3 *d_∞–convergence implies d_p–convergence, and thus ρ_p–convergence, for all $1 \leq p < \infty$.*

Example 4.4.1 shows that the converse of Proposition 4.4.3 is false. However, from elementary functional analysis it follows that $d_\infty(U, V) = \lim_{p \to \infty} d_p(U, V)$ for all $U, V \in \mathcal{E}^n$ and, in view of Proposition 4.4.2, hence also that

$$d_\infty(U, V) = \lim_{p \to \infty} \rho_p(U, V)$$

for all $U, V \in \mathcal{E}^n$.

11.4.5 Bibliographical Notes

§4.1 Other metrics have been discussed in Goetschel and Voxman, 1981, 1983, and Kloeden, 1982. The first recorded use of the d_∞ metric appears to be Heilpern, 1981. Puri and Ralescu, 1983, 1985, 1986, exploited its metric properties in a metric space context, and introduced the isometrical embedding. Klement, Puri and Ralescu, 1986, used the d_1 metric for a strong law of large numbers, and make reference to more general d_p metrics. The ρ_p metrics were first introduced by Diamond and Kloeden, 1990, who showed their equivalence to d_p metrics and investigated their topological characteristics.

§4.2 Proposition 4.2.2 is due to Puri and Ralescu, 1985. All other results in this section come from Diamond and Kloeden, 1994, as does Example 4.2.1.

§4.3 Klement, Puri and Ralescu, 1986, showed that $(\mathcal{E}^n, d_\infty)$ is not separable, but that $(\mathcal{E}^n, d_1)$ is. Diamond and Kloeden, 1990, extended this result to the d_p $(1 \leq p < \infty)$ metrics, and Diamond, 1990, investigated the separability for spaces of starshaped fuzzy sets.

§4.4 Propositions 4.4.2, 4.4.3 appear in Diamond and Kloeden, 1990.

11.5 COMPACTNESS CRITERIA

11.5.1 Introduction

In this section compact sets of $\mathcal{E}^n$ are characterized with respect to various metric topologies. By *compact* is meant *sequentially compact*; that is, $\mathcal{U} \subset$

$(\mathcal{E}^n, \nu)$ is said to be compact if any sequence $\{U_k\}$ in $\mathcal{U}$ has a subsequence converging to a point in $\mathcal{U}$ in the metric ν.

The condition of uniform boundedness is common to all the compactness criteria of subsequent sections. The subset $\mathcal{U} \subset \mathcal{E}^n$ is said to be *uniformly support–bounded* if there is a constant $L > 0$ such that the support sets of all $U \in \mathcal{U}$ lie within a ball of radius L in $\Re^n$, that is if $||[U]^0|| \leq L$ for all $U \in \mathcal{U}$.

The most convenient way of approaching the question of compactness is to first embed $\mathcal{E}^n$ as a cone in an appropriate Banach space. Recall from §3.1 that each $U \in \mathcal{E}^n$ was characterized by its support function s_U. It will be convenient to use the notation $U^*(\alpha, x) := s_U(\alpha, x),\ \alpha \in I,\ x \in S^{n-1}$. Let $\mathfrak{v}(y)$ be the total variation on I of a function $y : I \to \Re$. Since for any $U \in \mathcal{E}^n$ and $x \in S^{n-1}$, $U^*(., x)$ is nonincreasing in $\alpha \in I$ it has total variation on I

$$\mathfrak{v}(U^*(., x)) = U^*(0, x) - U^*(1, x). \tag{44}$$

Hence

$$\sup_{x \in S^{n-1}} \mathfrak{v}(U^*(., x)) \leq 2 \sup_{(\alpha, x) \in I \times S^{n-1}} |U^*(\alpha, x)| \leq 2 \sup_{a \in [u]^0} ||a||,$$

which implies that $U^* \in \mathbf{B}_\infty(S^{n-1}, B\mathfrak{v}(I))$ where $B\mathfrak{v}(I)$ is the Banach space of functions of bounded variation on I with norm $\mathfrak{v}(y)+|y(0)|$. Here $\mathbf{B}_\infty(X, Y)$ denotes the space of bounded mappings f from a compact space X into a normed space Y, endowed with the norm $||f||_{\mathbf{B}_\infty} = \sup_{x \in X} ||f(x)||_Y$. From Proposition 3.1.12 (ii) it follows that $U^* \in \mathbf{B}_\infty(I, C(S^{n-1}))$. Hence for any $U \in \mathcal{E}^n$, U^* belongs to the space

$$\mathfrak{X} = \mathbf{B}_\infty(S^{n-1}, B\mathfrak{v}(I)) \cap \mathbf{B}_\infty(I, C(S^{n-1}))$$

which is a Banach space with the norm

$$||y|| = \sup_{x \in S^{n-1}} \mathfrak{v}(\chi(., x)) + \sup_{\alpha \in I} \sup_{x \in S^{n-1}} |\chi(\alpha, x)|. \tag{45}$$

Thus there is an embedding j of the whole space $\mathcal{E}^n$ to $\mathfrak{X}$, defined by $U^* = j(U)$, $j : \mathcal{E}^n \to \mathfrak{X}$. Besides preserving the linear structure of $\mathcal{E}^n$, j also satisfies

$$d_\infty(u, v) \leq ||j(u) - j(v)|| \leq 3d_\infty(u, v) \tag{46}$$

which follows from equation (30), (44) and the definition of the norm (45). Consequently j is not an isometry, but by (46) it is a homeomorphism, which is sufficient for our requirements here.

The nature of the metric, whether it be d_∞ or d_p $(1 \leq p < \infty)$, will determine the character of the extra conditions required for compactness.

A family of support functions $\mathcal{U}^* = \{U^* : U \in \mathcal{U}\}$ is said to be *equi-leftcontinuous* in $\alpha \in I$ uniformly in $x \in S^{n-1}$ if for each $\epsilon > 0$ there exists $\delta = \delta(\epsilon) > 0$ such that $U^*(\beta, x) \leq U^*(\alpha, x) < U^*(\beta, x) + \epsilon$ for all $\beta - \delta < \alpha \leq \beta$, $x \in S^{n-1}$ and $U^* \in \mathcal{U}^*$. In addition, a set $\mathcal{U} \subset \mathcal{E}^n$ is said to have the *Blaschke property* if and only if it is uniformly support–bounded and $\mathcal{U}^*$ is equileftcontinuous.

The insufficiency of closedness and support–boundedness for compactness in the d_∞ metric is illustrated by the following example, which suggests that some further condition like equileftcontinuity is required.

Example 5.1.1 Define $\mathcal{U} = \{U_r \in \mathcal{E}^1 : r \in I\}$, where U_r is defined by its α–level sets as

$$[U_r]^\alpha = \begin{cases} [0,1] & \text{if } 0 \leq \alpha \leq r\,, \\ \{1\} & \text{if } r < \alpha \leq 1\,, \end{cases}$$

for every $r \in I$. It can be shown that $\mathcal{U}$ is not totally bounded and hence cannot be compact. Here, $U_r^* \in \mathcal{U}^*$ satisfies

$$U_r^*(\alpha, x) = \begin{cases} +1 & \text{if } x = +1\,,\, 0 \leq \alpha \leq 1\,, \\ 0 & \text{if } x = -1\,,\, 0 \leq \alpha \leq r\,, \\ -1 & \text{if } x = -1\,,\, r < \alpha \leq 1\,, \end{cases}$$

for each $r \in I$, since $S^0 = \{\pm 1\}$. Observe that the set $\mathcal{U}$ is uniformly support–bounded since $[U_r]^0 = I$ for each $r \in I$. However, the set $\mathcal{U}^*$ is not equileft-continuous in $\alpha \in I$ uniformly in $x \in S^0$. To see this, take any $0 < \epsilon < 1$ and $x = -1$. Then $U_r^*(1, -1) = -1$ and for any $\delta > 0$ there exist α with $1 - \delta < \alpha \leq 1$ for which

$$-1 = U_r^*(1,-1) \leq U_r^*(\alpha, -1) < U_r^*(1,-1) - \epsilon = -1 + \epsilon$$

is not satisfied for *all* $U_r^* \in \mathcal{U}^*$. In fact, for those U_r^* with $1 - r < \delta$, there exist α , $1 - \delta < \alpha < r$ for which $U_r^*(\alpha, -1) = 0 > -1 + \alpha$.

The Blaschke property is necessary and sufficient for compactness in $(\mathcal{E}^n, d_\infty)$:

Proposition 5.1.2 *A closed set of $\mathcal{U}$ of $(\mathcal{E}^n\,,\, d_\infty)$ is compact if and only if $\mathcal{U}$ is uniformly support–bounded and $\mathcal{U}^*$ is equileftcontinuous in $\alpha \in I$ uniformly in $x \in S^{n-1}$.*

11.5.2 Compact Subsets in $(\mathcal{E}^n\,,\, d_p)$

The Blaschke property is sufficient for compactness of a closed set in the d_p metric topology, because each d_p topology, $1 \leq p < \infty$, embeds continuously into $(\mathcal{E}^n, d_\infty)$, since $d_p \leq d_\infty$. Consequently, the compact sets in the latter space, which are precisely the closed Blaschke sets (Proposition 5.1.2), are compact in $(\mathcal{E}^n, d_p)$.

However, the Blaschke property is too strong to be also necessary. This is because $d_p \leq d_\infty$ and a more appropriate condition is desirable for L_p type spaces, related to equileftcontinuity in the stronger topology. Let $U \in \mathcal{E}^n$. If for each $\epsilon > 0$ there exists $\delta = \delta(\epsilon, U) > 0$ such that for all $0 \le h < \delta$

$$\int_h^1 \delta_\infty \left([U]^\alpha, [U]^{\alpha-h}\right)^p \, d\alpha < \epsilon^p$$

say that U is *p-mean leftcontinuous,* $1 \le p \le \infty$. If for nonempty $\mathcal{U} \subset \mathcal{E}^n$ this holds uniformly in $U \in \mathcal{U}$, say $\mathcal{U}$ is *p-mean equileftcontinuous.* If, in addition, $\mathcal{U}$ is uniformly support–bounded, then $\mathcal{U}$ is said to have the *p-Blaschke property.* Observe that for the corresponding family of support functions, this property translates as

$$\int_h^1 (U^*(\alpha - h, x) - U^*(\alpha, x))^p \, d\alpha < \epsilon^p$$

for all $0 \le h < \delta$, $x \in S^{n-1}$ and $U^* \in \mathcal{U}^*$, and that in the limit $p = \infty$ this concept is just the Blaschke property of the previous section. However, p-Blaschke does not imply Blaschke (although the converse is true), as the following example shows.

Example 5.2.1 Define $\mathcal{U} \subset \mathcal{E}^n$ by the sequence $\mathcal{U} = \{\widetilde{U}, U_1, U_2, \cdots\}$ where

$$U_k^*(x) = \begin{cases} x^k & \text{if } 0 \le x \le 1 \\ 0 & \text{if } x \notin [0,1] \end{cases}$$

and $\widetilde{U}(x) = 1$, if $x = 1$, and 0 otherwise. Clearly $\mathcal{U}$ is uniformly support–bounded. Now $\mathcal{U}$ is 1-mean equileftcontinuous, and thus 1-Blaschke, but not equileftcontinuous. To see this, for $0 < h < 1$

$$\begin{aligned} \int_h^1 d_\infty \left([U_k^*]^{\alpha-h}, [U_k^*]^\alpha\right) d\alpha &= \int_h^1 \left(\alpha^k - (\alpha - h)^k\right) d\alpha \\ &= (k+1)^{-1} \left(1 - (1-h)^{k+1} - h^{k+1}\right) \\ &\le \left(1 - (1-h)^2 - h^2\right)/2 \\ &= h - h^2 \le h \end{aligned}$$

for all $0 \le h \le 1$ and all $k \ge 1$. Thus $\int_h^1 d_\infty \left([U_k^*]^{\alpha-h}, [U_k^*]^\alpha\right) d\alpha \le \epsilon$ if $0 < h < \delta(\epsilon) = \epsilon$ and this is the required condition. On the other hand,

$$\sup_{h \le \alpha \le 1} d_\infty \left([U_k]^{\alpha-h}, [U_k]^\alpha\right) = \sup_{h \le \alpha \le 1} \left|\alpha^k - (\alpha - h)^k\right| = 1 - (1-h)^k$$

and for $h \ge 0$ this has supremum 1 as $k \to \infty$, while for $h = 0$ the supremum is 0. Consequently, $\mathcal{U}$ is not equileftcontinuous, and thus not Blaschke.

Proposition 5.2.2 *A closed set $\mathcal{U}$ of $(\mathcal{E}^n, d_p)$, $1 \le p \le \infty$, is compact if and only if $\mathcal{U}$ has the p-Blaschke property.*

Recall that a space is locally compact at a point x if there is some neighbourhood N of x whose closure $\overline{N}$ is compact. The space itself is locally compact if it is locally compact at each of its points. From the result above it follows that

Corollary 5.2.3 *The space $(\mathcal{E}^n, d_p)$, $1 \leq p < \infty$, is locally compact. Moreover a subset $\mathcal{U}$ is locally compact if and only if every uniformly support–bounded and closed subset of $\mathcal{U}$ is p-Blaschke.*

Remark 5.2.4. In contrast, the space $(\mathcal{E}^n, d_\infty)$ is not locally compact.

11.5.3 Bibliographical Notes

Example 5.1.1 is due to Kaleva, 1985. The idea of the Blaschke property for a class of fuzzy sets $\mathcal{U}$ appears in Diamond and Kloeden, 1989, where Proposition 5.1.1 was proved. The proof in parts uses ideas of compacteness in bounded variation associated with Helly's Selection Theorem, which may be found in Graves, 1946. Blaschke's name has been used in the definition because he showed the completeness of the space $(\mathcal{K}_C^n, d_H)$.

The p–Blaschke concept, proofs and examples of §5.2 are taken from Diamond and Kloeden, 1990.

11.6 FUZZY SET VALUED MAPPINGS OF REAL VARIABLES

11.6.1 Continuity and Measurability

In this section mappings F from a domain T in $\Re^k$ into the space $\mathcal{E}^n$ of fuzzy sets on $\Re^n$ are considered. Unless otherwise indicated, the metric d_∞ on $\mathcal{E}^n$ is used. Thus $F : T \to \mathcal{E}^n$, that is $F(t) \in \mathcal{E}^n$ for all $t \in T$. Such a mapping is called a *fuzzy set valued mapping of k real variables.*

The usual definition of continuity of mappings between metric spaces will be used. A fuzzy set valued mapping $F : T \to \mathcal{E}^n$ is *continuous* at $t_0 \in T$ if for every $\epsilon > 0$ there exists a $\delta = \delta(t_0, \epsilon) > 0$ such that

$$d_\infty(F(t),\ F(t_0)) < \epsilon \tag{47}$$

for all $t \in T$ with $||t - t_0|| < \delta$. Alternatively, (47) can be written in terms of the convergence of sequences, that is as

$$\lim_{t_k \to t_0} d_\infty(F(t_k),\ F(t_0)) = 0$$

for all sequences $\{t_k\} \subset T$ with $t_k \to t_0$.

From the definition of the metric d_∞, (47) is equivalent to the continuity of the level set mappings $[F(\cdot)]^\alpha : T \to \mathcal{K}_C^n$ at $t_0 \in T$ with respect to the Hausdorff metric d_H, uniformly in $\alpha \in I$. The uniformity in α is essential for the converse implication.

Example 6.1.1 Define $F : (-1, 1) \to \mathcal{E}$ level setwise as $[F(t)]^0 = [0, 1]$, $[F(t)]^1 = \{1\}$ for all $t \in (-1, 1)$ and

$$[F(t)]^\alpha = \begin{cases} \{1\} & \text{for } t \in (-1, 0] \\ [1 - (1 - \alpha)^{1/t}, 1] & \text{for } t \in (0, 1) \end{cases}$$

when $0 < \alpha < 1$. Then $d_H([F(t)]^0, [F(0)]^0) = d_H([F(t)]^1, [F(0)]^1) = 0$ for all $t \in (-1, 1)$ and

$$d_H([F(t)]^\alpha, [F(0)]^\alpha) = \begin{cases} 0 & \text{for } t \in (-1, 0] \\ (1 - \alpha)^{1/t} & \text{for } t \in (0, 1) \end{cases}$$

when $0 < \alpha < 1$. So each level set mapping $[F(\cdot)]^\alpha$ is continuous at $t_0 = 0$, but

$$d_\infty(F(t), F(0)) = \sup_{\alpha \in I} (1 - \alpha)^{1/t} = 1$$

for all $t \in (0, 1)$ and F is not continuous with respect to the d_∞ metric on $\mathcal{E}$. Note, however, that F is continuous at $t_0 = 0$ with respect to the d_p metric since

$$d_p(F(t), F(0)) \leq \left| \frac{t}{t + p} \right|^{\frac{1}{p}}, \quad \text{for all} \quad t \in (-1, 1).$$

Upper and lower semicontinuity of a fuzzy set valued mapping $F : T \to \mathcal{E}^n$ is defined level setwise uniformly in $\alpha \in I$. Say that F is *upper semicontinuous* at $t_0 \in T$ if for every $\epsilon > 0$ there exists a $\delta(t_0, \epsilon) > 0$ such that

$$d_H^*([F(t)]^\alpha, [F(t_0)]^\alpha) < \epsilon \tag{48}$$

for all $\alpha \in I$ and $t \in T$ with $\|t - t_0\| < \delta$, and *lower semicontinuous* at t_0 if for every $\epsilon > 0$ there exists a $\delta = \delta(t_0, \epsilon) > 0$ such that

$$d_H^*([F(t_0)]^\alpha, [F(t)]^\alpha) < \epsilon \tag{49}$$

for all $\alpha \in I$ and $t \in T$ with $\|t - t_0\| < \delta$. Note that (48) and (49) can be written respectively as

$$[F(t)]^\alpha \subseteq [F(t_0)]^\alpha + \epsilon S_1^n, \quad \text{and} \quad [F(t_0)]^\alpha \subseteq [F(t)]^\alpha + \epsilon S_1^n$$

for all $\alpha \in I$ and $t \in T$ with $\|t - t_0\| < \delta$, where $S_1^n = \{x \in \Re^n : \|x\| < 1\}$.

Measurability of a mapping $F : T \to \mathcal{E}^n$ can also be defined with respect to the d_∞ metric on $\mathcal{E}^n$. The mapping $F : T \to \mathcal{E}^n$ is *measurable* if $\{t \in T;\ F(t) \in \mathcal{B}\}$ is a Borel subset of $\Re^k$ for all Borel subsets $\mathcal{B}$ of the metric space $(\mathcal{E}^n, d_\infty)$, or equivalently if $\{t \in T;\ d_\infty(F(t), U) \leq \epsilon\}$ is a Borel subset of $\Re^k$ for all $\epsilon > 0$ and $U \in \mathcal{E}^n$.

Defining the integral of a fuzzy set valued mapping requires a stronger form of measurability. A mapping $F : T \to \mathcal{E}^n$ is *strongly measurable* if the level set mappings $[F(\cdot)]^\alpha : T \to \mathcal{K}_c^n$ are measurable (cf. Proposition 2.6.1) for all $\alpha \in I$.

Proposition 6.1.2 *If $F : T \to \mathcal{E}^n$ is strongly measurable, then it is measurable.*

A measurable fuzzy set valued mapping need not, however, be strongly measurable. Note that a continuous mapping $F : T \to \mathcal{E}^n$ is strongly measurable. In fact

Proposition 6.1.3 *If $F : T \to \mathcal{E}^n$ is upper or lower semicontinuous at each $t_0 \in T$, then F is strongly measurable on T.*

Strong measurability is also preserved on taking limits, which follows from Proposition 2.6.4 applied to level set mappings

Proposition 6.1.4 *Let $F_i : T \to \mathcal{E}^n$ be strongly measurable for $i = 1, 2, , 3, \ldots$ and suppose that $\lim_{i\to\infty} d_\infty(F_i(t),\ F(t)) = 0$ for almost all $t \in T$. Then $F : T \to \mathcal{E}^n$ is strongly measurable.*

In conclusion, note that a continuous mapping $F : T \to \mathcal{E}^n$ need not attain a maximum or minimum (with respect to $\leq$) on a nonempty compactsubset of T as they need not exist. The mapping $F : [0, 1] \to \mathcal{E}$ defined by $F(t) = \chi_{[t,1+t]}$ for $0 \leq t \leq 1$ is a simple counterexample.

11.6.2 Differentiation

The de Blasi and Hukuhara definitions of differentiation of set valued mappings generalize in a natural way to fuzzy set valued mappings. In this section mappings $F : T \to \mathcal{E}^n$ with domain $T \subseteq \Re^k$ are considered and T is usually assumed to be open and convex and, in particular, an open interval when $k = 1$. Unless otherwise indicated, the d_∞ metric on $\mathcal{E}^n$ is always used.

A mapping $F : \Re^k \to \mathcal{E}^n$ is *homogeneous* if

$$F(\lambda t) = \lambda F(t) \text{ for all } \lambda > 0,\ t \in \Re^k, \tag{50}$$

recalling that scalar multiplication of fuzzy sets is defined level setwise. That is, (50) is equivalent to

$$[F(\lambda t)]^\alpha = \lambda[F(t)]^\alpha \text{ for all } \lambda > 0,\ t \in \Re^k, \alpha \in I.$$

The upper semicontinuity of such mappings was defined in the previous section.

A mapping $F : T \to \mathcal{E}^n$ is (de Blasi) *differentiable* at $t_0 \in T$ if there exists an upper semicontinuous, homogeneous mapping $D_{t_0} : \Re^k \to \mathcal{E}^n$ such that

$$\lim_{||\Delta t|| \to 0} \frac{1}{||\Delta t||} d_\infty(F(t_0 + \Delta t),\, F(t_0) + D_{t_0}(\Delta t)) = 0. \tag{51}$$

The mapping D_{t_0} here is called the (de Blasi) *differential* of F at t_0. It will sometimes be written as $D_{t_0}^{(F)}$ to indicate its relationship with a particular mapping F.

In view of the definition of the d_∞ metric and the level set addition of fuzzy sets, (51) can be rewritten as

$$\lim_{||\Delta t|| \to 0} \frac{1}{||\Delta t||} d_H([F(t_0 + \Delta t)]^\alpha,\, [F(t_0)]^\alpha + [D_{t_0}(\Delta t)]^\alpha) = 0 \tag{52}$$

uniformly in $\alpha \in I$. Thus the level set mapping $[F(\cdot)]^\alpha : T \to \mathcal{K}_C^n$ is de Blasi differentiable at $t_0 \in T$ with de Blasi differential

$$D_{t_0}^{([F]^\alpha)}(\Delta t) = [D_{t_0}^{(F)}(\Delta t)]^\alpha, \text{ for all } \Delta t \in \Re^k, \tag{53}$$

for each $\alpha \in I$. From Propositions 2.7.1 and 2.7.2 on de Blasi differentiable set valued mappings it immediately follows that

Proposition 6.2.1 *The de Blasi differential $D_{t_0}^{(F)}$ of a mapping $F : T \to \mathcal{E}^n$ is unique if it exists.*

Proposition 6.2.2 *Suppose that $F, G : T \to \mathcal{E}^n$ are de Blasi differentiable at $t_0 \in T$ with differentials $D_{t_0}^{(F)}$, $D_{t_0}^{(G)}$. Then $F + G$ and λF, for any $\lambda \in \Re$, are de Blasi differentiable at $t_0 \in T$ with differentials*

$$D_{t_0}^{(F+G)} = D_{t_0}^{(F)} + D_{t_0}^{(G)}, \quad D_{t_0}^{(\lambda F)} = \lambda D_{t_0}^{(F)},$$

respectively.

A fuzzy set valued mapping need not, however, be de Blasi differentiable at a point $t_0 \in T$ when each of its level set mappings is de Blasi differentiable at t_0 uniformly in the membership grade $\alpha \in I$. The problem is that the de Blasi differentials of the level set mappings may not be the level sets of a fuzzy set.

$D_{t_0}^\alpha(\Delta t)$ for $D_{t_0}^{([F]^\alpha)}(\Delta t)$ in following theorem and proof.

Theorem 6.2.3 *Let $F : T \to \mathcal{E}^n$ and suppose that its level set mappings $F^\alpha := [F(\cdot)]^\alpha : T \to \mathcal{K}_C^n$ are de Blasi differentiable at $t_0 \in T$ uniformly in $\alpha \in I$ with differentials $D_{t_0}^\alpha := D_{t_0}^{([F]^\alpha)}$ satisfying*

$$D_{t_0}^\beta(\Delta t) \subseteq D_{t_0}^\alpha(\Delta t) \tag{54}$$

for all $\Delta t \in \Re^k$ *and* $0 \leq \alpha \leq \beta \leq 1$. *Then* F *is de Blasi differentiable at* $t_0 \in T$ *and its differential* $D_{t_0}^{(F)}$ *has level sets*

$$[D_{t_0}^{(F)}(\Delta t)]^\alpha = D_{t_0}^\alpha(\Delta t)$$

for all $\Delta t \in \Re^k$ *and* $\alpha \in I$.

Differentiable fuzzy set valued mappings inherit many of the properties of differentiable crisp set valued mappings since these hold for their level set mappings. Propositions 2.7.3, 2.7.4, 2.7.5 and 2.7.10 carry over immediately if

$$\|D_{t_0}^{(F)}(\Delta t)\| = \sup_{\alpha \in I} \|[D_{t_0}^{(F)}(\Delta t)]^\alpha\|$$

is written for the differential $D_{t_0}^{(F)}$ of a fuzzy set valued mapping $F : T \to \mathcal{E}^n$. Hence

$$\|D_{t_0}^{(F)}(\Delta t)\| = \|[D_{t_0}^{(F)}(\Delta t)]^0\| \longrightarrow 0$$

as $\|\Delta t\| \to 0$. Note that $\chi_{\{0\}} \in \mathcal{E}^n$ is the fuzzy analogue of the crisp zero set $\{0\} \in \mathcal{K}_C^n$. In addition, by a constant in $\mathcal{E}^n$ is meant a fixed fuzzy set in $\mathcal{E}^n$.

Proposition 6.2.4 *If* $F : T \to \mathcal{E}^n$ *is de Blasi differentiable at* $t_0 \in T$, *then it is continuous at* t_0.

Proposition 6.2.5 *Let* $F : T \to \mathcal{E}^n$ *be a de Blasi differentiable at each* $t_0 \in T$ *where* $T \subseteq \Re^k$ *is open and convex. Then* $F(t) \equiv$ const. *for all* $t \in T$ *if and only if its differentials satisfy* $D_{t_0}^{(F)}(\Delta t) \equiv \chi_{\{0\}}$ *for all* $\Delta t \in \Re^k$ *and all* $t_0 \in T$.

Proposition 6.2.6 *Let* $F : T \to \mathcal{E}^n$ *have a* $\leq$*-maximum or* $\leq$*-minimum at* $t_0 \in T$ *and be de Blasi differentiable at* t_0. *Then* $D_{t_0}^{(F)}(\Delta t) = \chi_{\{0\}}$ *for all* $\Delta t \in \Re^k$.

A mapping $F : T \to \mathcal{E}^n$ is *continuously differentiable* on T if $D_{t_0}^{(F)}$ exists for all $t_0 \in T$ and $D_{t_0}^{(F)}(\Delta t)$ is continuous on T as a function of t_0 for each fixed $\Delta t \in \Re^k$. The following analogue of the Mean Value Theorem for derivatives holds.

Proposition 6.2.7 *Let* $F : T \to \mathcal{E}^n$ *be continuously differentiable on* T, *where* $T \subseteq \Re^k$ *is open and convex. Then*

$$d_\infty(F(t_1), F(t_0)) \leq \|t_1 - t_0\| \max_{t \in \overline{t_0 t_1}} \|D_t^{(F)}(e)\| \tag{55}$$

for all $t_1, t_0 \in T$ *with* $t_1 \neq t_0$, *where* $e = (t_1 - t_0)/\|t_1 - t_0\|$ *and* $t \in \overline{t_0 t_1}$ *means* $t = t_0 + se$ *for* $0 \leq s \leq \|t_1 - t_0\|$, *that is,* t *is on the line segment joining* t_0 *and* t_1.

Recall the notation of the support function of a crisp and fuzzy set, in particular

$$U^*(\alpha, x) = s(x, [U]^\alpha) = \max_{a \in [U]^\alpha} \langle a, x \rangle$$

for $\alpha \in I$, $x \in S^{n-1}$. Let $F : T \to \mathcal{E}^k$ be de Blasi differentiable at $t_0 \in T$. From the properties of support functions in §2.4

$$\begin{aligned}
&\sup_{(\alpha,x) \in I \times S^{n-1}} \frac{1}{\|\Delta t\|} \left| F(t_0 + \Delta t)^*(\alpha, x) - F(t_0)^*(\alpha, x) - D_{t_0}^{(F)}(\Delta t)^*(\alpha, x) \right| \\
&\sup_{(\alpha,x) \in I \times S^{n-1}} \frac{1}{\|\Delta t\|} \left| s(x, [F(t_0 + \Delta t)]^\alpha) - s(x, [F(t_0) + D_{t_0}^{(F)}(\Delta t)]^\alpha) \right| \\
&\frac{1}{\|\Delta t\|} d_\infty(F(t_0 + \Delta t),\ F(t_0) + D_{t_0}^{(F)}(\Delta t)) \\
\to\quad & 0 \text{ as } \|\Delta t\| \to 0,
\end{aligned}$$

so $F(\cdot)^*(\alpha, x) : T \to \Re$ is differentiable at $t_0 \in T$ (uniformly in $(\alpha, x) \in I \times S^{n-1}$) with differential $D^*_{(\alpha,x)}(\Delta t) = D_{t_0}^{(F)}(\Delta t)^*(\alpha, x)$ for all $\Delta t \in \Re^k$ and each $(\alpha, x) \in I \times S^{n-1}$.

For $T \subseteq \Re$, differentiation can be defined in terms of support functions. A mapping $F : T \to \mathcal{E}^n$ is *Bobylev differentiable* at $t_0 \in T \subseteq \Re$ if the support function $F(\cdot)^*(\alpha, x)$ is differentiable at $t_0 \in T$ uniformly in $(\alpha, x) \in I \times S^{n-1}$ and there exists a fuzzy set $U \in \mathcal{E}^n$ such that

$$U^*(\alpha, x) = \frac{d}{dt} F(t)^*(\alpha, x) \Big|_{t=t_0} \tag{56}$$

for all $(\alpha, x) \in I \times S^{n-1}$. Call U the *Bobylev derivative* of F at t_0 and write it as $F'(t_0)$. Since

$$\begin{aligned}
& \frac{1}{|\Delta t|} d_\infty(F(t_0 + \Delta t),\ F(t_0) + F'(t_0)\Delta t) \\
= \quad & \sup_{(\alpha,x) \in I \times S^{n-1}} \frac{1}{|\Delta t|} |F(t_0 + \Delta t)^*(\alpha, x) - F(t_0)^*(\alpha, x) \\
& \qquad - F'(t_0)^*(\alpha, x)\Delta t| \\
\to\ & 0 \text{ as } \Delta t \to 0
\end{aligned}$$

and since the de Blasi differential is unique if it exists, a Bobylev differentiable fuzzy set valued mapping is de Blasi differentiable and

$$D_{t_0}^{(F)}(\Delta t) = F'(t_0)\Delta t \tag{57}$$

for all $\Delta t \in \Re$. In particular, $F'(t_0)$ is unique if it exists.

Example 6.2.8 Let $F : (0, 2\pi) \to \mathcal{E}$ be defined on level sets by

$$[F(t)]^\alpha = (1-\alpha)(2+\sin t)[-1,1],$$

so that

$$F(t)^*(\alpha, x) = (1-\alpha)(2+\sin t)$$

for all $\alpha \in I$, $x \in \{\pm 1\}$ and $t \in (0, 2\pi)$. Then F is Bobylev differentiable at all $t_0 \in (0, 2\pi)$ with derivative $F'(t_0) \in \mathcal{E}$ given by

$$[F'(t_0)]^\alpha = (1-\alpha)\cos t_0\,[-1,1]$$

for $\alpha \in I$, so

$$F'(t_0)^*(\alpha, x) = (1-\alpha)\cos t_0$$

for all $\alpha \in I$ and $x \in \{\pm 1\}$.

Note that the differentiability of the support functions alone in (56) is insufficient to guarantee the existence of a fuzzy set $u \in \mathcal{E}^n$ with U^* coinciding with the derivatives. The situation is similar to that in Theorem 6.2.3 and in particular condition (54).

Another useful definition of differentiation of a fuzzy set valued mapping of a single real variable is based on Hukuhara difference quotients. A mapping $F : T \to \mathcal{E}^n$ is *Hukuhara differentiable* at $t_0 \in T \subseteq \Re$ if for some $h_0 > 0$ the Hukuhara differences

$$F(t_0 + \Delta t) \sim_h F(t_0), \quad F(t_0) \sim_h F(t_0 - \Delta t) \tag{58}$$

exist in $\mathcal{E}^n$ for all $0 < \Delta t < h_0$ and if there exists an $F'(t_0) \in \mathcal{E}^n$ such that

$$\lim_{\Delta t \to 0+} d_\infty((F(t_0 + \Delta t) \sim_h F(t_0))/\Delta t,\; F'(t_0)) = 0$$

and

$$\lim_{\Delta t \to 0+} d_\infty((F(t_0) \sim_h F(t_0 - \Delta t))/\Delta t,\; F'(t_0)) = 0$$

The fuzzy set $F'(t_0)$ is called the *Hukuhara derivative* of F at t_0.

Recall that $U \sim_h V = W \in \mathcal{E}^n$ is defined on level sets, that is as $[U \sim_h V]^\alpha = [U]^\alpha \sim_h [V]^\alpha = [w]^\alpha$ for all $\alpha \in I$. In view of the definition of the metric d_∞, all the level set mappings $[F(\cdot)]^\alpha$ are Hukuhara differentiable at t_0 with Hukuhara derivatives $[F'(t_0)]^\alpha$ for each $\alpha \in I$ when $F : T \to \mathcal{E}^n$ is Hukuhara differentiable at t_0 with Hukuhara derivative $F'(t_0)$.

When the Hukuhara differences (58) exist the support functions satisfy

$$s(p, [F(t_0 + \Delta t)]^\alpha) - s(p, [F(t_0)]^\alpha) = s(p, [F(t_0 + \Delta t)]^\alpha \sim_h [F(t_0)]^\alpha)$$

for all $p \in S^{n-1}$, $\alpha \in I$. Hence

$$\begin{aligned} &d_H(([F(t_0 + \Delta t)]^\alpha \sim_h [F(t_0)]^\alpha)/\Delta t\,,\, [F'(t_0)]^\alpha) \\ &\quad = \sup_{p \in S^{n-1}} \left|(s(p, [F(t_0 + \Delta t)]^\alpha) - s(p, [F(t_0)]^\alpha))\Delta t - s(p, [F'(t_0)]^\alpha)\right| \end{aligned}$$

for all $\alpha \in I$, so F is Bobylev differentiable at t_0 whenever it is Hukuhara differentiable there and the Bobylev and Hukuhara derivatives coincide. Since a Bobylev differentiable function is de Blasi differentiable, F is also de Blasi differentiable at t_0 and its de Blasi differential and Hukuhara derivative are related by equation (58).

A fuzzy set valued mapping can, however, be Bobylev or de Blasi differentiable at t_0 without being Hukuhara differentiable at t_0, essentially because the Hukuhara differences in (58) may not exist. From Proposition 2.7.8 follows

Proposition 6.2.9 *If $F : T \to \mathcal{E}^n$ is Hukuhara differentiable at $t_0 \in T \subseteq \Re$, then diam $([F(t)]^\alpha)$ is nondecreasing in t at t_0 for each $\alpha \in I$.*

The fuzzy set valued mapping in Example 6.2.8 is Bobylev differentiable at $t_0 = \pi/2$ but is not Hukuhara differentiable there because the Hukuhara differences (58) do not exist for $t_0 = \pi/2$ and $\Delta t > 0$. In fact, diam $([F(t)]^\alpha) <$ diam $([F(\pi/2)]^\alpha)$ here for $t > \pi/2$ and all $0 \leq \alpha < 1$.

Hukuhara and Bobylev differentiable mappings inherit the following properties from their de Blasi differentiability.

Proposition 6.2.10 *If $F : T \to \mathcal{E}^n$ is Hukuhara (or Bobylev) differentiable at $t_0 \in T \subseteq \Re$, then*
(i) its derivative $F'(t_0)$ is unique;
(ii) it is continuous at t_0.

Proposition 6.2.11 *If $F, G : T \to \mathcal{E}^n$ are Hukuhara (or Bobylev) differentiable at $t_0 \in T \subseteq \Re$, then $F + G$ and cF, for all $c \in \Re$, are Hukuhara (or Bobylev) differentiable at t_0 and*

$$(F + G)'(t_0) = F'(t_0) + G'(t_0), \quad (cF)'(t_0) = cF'(t_0).$$

Proposition 6.2.12 *Let $F : T \to \mathcal{E}^n$ be Hukuhara (or Bobylev) differentiable at point $t_0 \in T \subseteq \Re$. Then $F(t) \equiv$ const. for all $t \in T$ if and only if $F'(t_0) = \chi_{\{0\}}$ for all $t_0 \in T$.*

Proposition 6.2.13 *Let $F : T \to \mathcal{E}^n$ have a $\leq$-maximum or $\leq$-minimum at $t_0 \in T \subseteq \Re$ and be Hukuhara (or Bobylev) differentiable at t_0. Then $F'(t_0) = \chi_{\{0\}}$.*

Proposition 6.2.14 *Let $F : T \to \mathcal{E}^n$ be Hukuhara (or Bobylev) differentiable at each point $t_0 \in T \subseteq \Re$ with derivative $F' : T \to \mathcal{E}^n$ continuous at each*

$t_0 \in T$. *Then for all* $t_0, t_1 \in T$ *with* $t_0 < t_1$

$$d_\infty(F(t_1), F(t_0) \leq |t_1 - t_0| \max_{t_0 \leq t \leq t_1} \|F'(t)\|.$$

Finally, note that if $r : T \to \Re^n$ is differentiable at t_0 with derivative $r'(t_0)$, then $F : T \to \mathcal{E}^n$ defined by $F(t) = \chi_{\{r(t)\}}$ is Hukuhara differentiable at t_0 with derivative $F'(t_0) = \chi_{\{r'(t_0)\}}$. Moreover, if $F : T \to \mathcal{E}$ is Hukuhara differentiable at t_0 and $[F(t)]^\alpha = [f_\alpha(t), g_\alpha(t)]$ for $\alpha \in I$, then f_α and g_α are differentiable at t_0 for each $\alpha \in I$ and the Hukuhara derivative $F'(t_0)$ has level sets

$$[F'(t_0)]^\alpha = [f'_\alpha(t_0), g'_\alpha(t_0)] \tag{59}$$

for each $\alpha \in I$. The converse need not apply, that is the differentiability of the end point mappings f_α, g_α need not imply the Hukuhara differentiability of F since the intervals in (59) need not be the level sets of a fuzzy set in $\mathcal{E}$.

11.6.3 Integration

Integration of a fuzzy set valued mapping of a single real variable is defined level setwise in terms of the Aumann integrals of its level set mappings. Without loss of generality, restrict attention to mappings defined on the unit interval $[0, 1]$, that is $F : [0, 1] \to \mathcal{E}^n$.

A mapping $F : [0, 1] \to \mathcal{E}^n$ is said to be *integrably bounded* if there exists an integrable function $h : [0, 1] \to \Re$ such that

$$\|F(t)\| \leq h(t) \quad \text{for all} \quad t \in [0, 1].$$

Since $\|F(t)\| = \|[F(t)]^0\| \geq \|[F(t)]^\alpha\|$ for all $\alpha \in I$, each level set valued mapping $[F(\cdot)]^\alpha : [0, 1] \to \mathcal{K}_C^n$ is integrably bounded as defined in §2.8.

Suppose that $F : [0, 1] \to \mathcal{E}^n$ is strongly measurable (see §6.1) and integrably bounded. By Theorem 2.8.1 each level set mapping $[F(\cdot)]^\alpha : [0, 1] \to \mathcal{K}_C^n$ is Aumann integrable on $[0, 1]$. If there exists a fuzzy set $U \in \mathcal{E}^n$ such that $[U]^\alpha = \int_0^1 [F(t)]^\alpha dt$ for all $\alpha \in I$, say that F is *integrable* over $[0, 1]$ and call U, which is denoted by $\int_0^1 F(t)dt$, its *integral* over $[0, 1]$. Thus, for each $\alpha \in I$

$$\left[\int_0^1 F(t)dt\right]^\alpha = \int_0^1 [F(t)]^\alpha dt.$$

By the following theorem such a $U \in \mathcal{E}^n$ always exists.

Theorem 6.3.1 *If* $F : [0, 1] \to \mathcal{E}^n$ *is strongly measurable and integrably bounded, then it is integrable over* $[a, b]$ *for any* $[a, b] \subseteq [0, 1]$ *with* $a < b$.

A continuous mapping $F : [0,1] \to \mathcal{E}^n$ is strongly measurable by Proposition 6.1.4. Since $\|A\| = d_H(\{0\}, A)$,

$$\begin{aligned} \big| \|[F(s)]^0\| - \|[F(t)]^0\| \big| &\leq d_H([F(s)]^0, [F(t)]^0) \\ &\leq d_\infty(F(s), F(t)) \end{aligned}$$

for all $s, t \in [0,1]$, so $\|F(\cdot)\| = \|[F(\cdot)]^0\| : [0,1] \to \Re$ is itself continuous and hence integrable on $[0,1]$. This gives

Corollary 6.3.2 *A continuous mapping $F : [0,1] \to \mathcal{E}^n$ is integrable.*

The integral of mapping $F : [0,1] \to \mathcal{E}^n$ inherits many basic properties from the Aumann integrals of its level set mappings. From (17)–(19)

Proposition 6.3.3 *If $F, G : [0,1] \to \mathcal{E}^n$ are integrable, then $F + G$ and λF for any $\lambda \in \Re$ are integrable with*

$$\begin{aligned} \int_0^1 \{F(t) + G(t)\} dt &= \int_0^1 F(t) dt + \int_0^1 G(t) dt, \\ \int_0^1 \lambda F(t) dt &= \lambda \int_0^1 F(t) dt \end{aligned}$$

Proposition 6.3.4 *If $F, G : [0,1] \to \mathcal{E}^n$ are integrable with $F(t) \leq G(t)$ for each $t \in [0,1]$, then*

$$\int_0^1 F(t) dt \leq \int_0^1 G(t) dt$$

The convergence theorem, Theorem 2.8.4, carries over level setwise to fuzzy set valued mappings.

Theorem 6.3.5 *Let $F_i, F : [0,1] \to \mathcal{E}^n$, $i = 1, 2, ...,$ be strongly measurable and uniformly integrably bounded. If $F_i(t) \to F(t)$ in the d_∞ metric as $i \to \infty$ for all $t \in [0,1]$, then*

$$U_i = \int_0^1 F_i(t) dt \to u = \int_0^1 F(t) dt \quad \text{as} \quad i \to \infty$$

The integral inequality in Theorem 2.8.6 also has a counterpart for fuzzy set valued mappings

Proposition 6.3.6 *If the functions $F, G : [0,1] \to \mathcal{E}^n$ are integrable, then $d_\infty(F(\cdot), G(\cdot)) : [0,1] \to \Re$ is integrable and*

$$d_\infty\left(\int_0^1 F(t) dt, \int_0^1 G(t) dt\right) \leq \int_0^1 d_\infty(F(t), G(t)) dt. \tag{60}$$

Example 6.3.7 If $F : [0,1] \to \mathcal{E}$ is integrable with level sets $[F(t)]^\alpha = [f_\alpha(t), g_\alpha(t)]$, then $f_\alpha, g_\alpha : [0,1] \to \Re$ are integrable and

$$\left[\int_0^1 F(t)dt\right]^\alpha = \int_0^1 [F(t)]^\alpha dt = \left[\int_0^1 f_\alpha(t)dt, \int_0^1 g_\alpha(t)dt\right] \qquad (61)$$

for each $\alpha \in I$. Conversely, if the f_α, g_α are integrable for each $\alpha \in I$, then F is integrable and (61) holds. In particular, if $F(t) = \chi_{\{r(t)\}}$ where $r : [0,1] \to \Re$ is integrable, then F is integrable and

$$\int_0^1 F(t)dt = \int_0^1 \chi_{\{r(t)\}}dt = \chi_{\{\int_0^1 r(t)dt\}}.$$

From Theorem 2.8.2 the support functions $F(\cdot)^*(\alpha, x) : [0,1] \to \Re$ of an integrable mapping $F : [0,1] \to \mathcal{E}^n$ are integrable for each $(\alpha, x) \in I \times S^{n-1}$ with

$$\left(\int_0^1 F(t)dt\right)^*(\alpha, x) = \int_0^1 F(t)^*(\alpha, x)dt. \qquad (62)$$

Bobylev, in fact, defined integration of a mapping F through the relationship (62), requiring the existence of a fuzzy set $u = \int_0^1 F(t)dt \in \mathcal{E}^n$.

From Theorem 6.3.1 an integrable mapping $F : [0,1] \to \mathcal{E}^n$ is also integrable on any subinterval $[a,b] \subseteq [0,1]$. The additivity of integrals over adjacent subintervals $[a,b]$, $[b,c] \subseteq [0,1]$, that is

$$\int_a^c F(t)dt = \int_a^b F(t)dt + \int_b^c F(t)dt, \qquad (63)$$

follows level setwise from the corresponding property (18) for set valued mappings. When the upper end point of the integration subinterval $[a,t] \subseteq [0,1]$ is considered as a variable the integral

$$G(t) = \int_a^t F(s)\, ds \qquad (64)$$

defines a mapping $G : [a,1] \to \mathcal{E}^n$.

Example 6.3.9. If $F(t) \equiv U_0 \in \mathcal{E}^n$ for all $t \in [0,1]$, then $G(t) = \int_0^t F(s)\, ds = tU_0$. This is easily verified using (62) and the properties of support functions.

From the additivity property (63), obtain

$$\int_a^{t'} F(s)\, ds = \int_a^t F(s)\, ds + \int_t^{t'} F(s)\, ds \qquad (65)$$

for all $0 \leq a \leq t \leq t' \leq 1$, so the Hukuhara difference

$$\int_a^{t'} F(s)\, ds \sim_h \int_a^t F(s)\, ds = \int_t^{t'} F(s)\, ds \tag{66}$$

exists. Consequently, from Proposition 2.7.8,

Proposition 6.3.10 *If $F : [0,1] \to \mathcal{E}^n$ is integrable,* diam $([\int_0^t F(s)\, ds]^\alpha)$ *is nondecreasing in t for each $\alpha \in I$.*

A Mean Value Theorem for Integrals also holds

Theorem 6.3.11 *If $F : [0,1] \to \mathcal{E}^n$ is continuous, then*

$$d_\infty \left(\int_a^{t'} F(s)\, ds, \int_a^t F(s)\, ds \right) \leq |t' - t| \max_{t \leq s \leq t'} \|[F(s)]^0\|$$

for all $0 \leq a \leq t \leq t' \leq 1$.

The above theorem is in fact a special case of the Mean Value Theorem for derivatives, Theorem 6.2.13, applied to the integral mapping G defined in (64) which turns out to be Hukuhara differentiable with Hukuhara derivative $G' = F$. This is a consequence of the additivity property of integrals (65) and the implied existence of the Hukuhara differences (66). This result provides a Fundamental Theorem of Calculus for fuzzy set valued mappings of a single real variable.

Theorem 6.3.12 *If $F : [0,1] \to \mathcal{E}^n$ is continuous, then for all $t \in (a,1) \subseteq [0,1]$ the integral $G(t) = \int_a^t F(s)\, ds$ is Hukuhara differentiable with Hukuhara derivative $G'(t) = F(t)$.*

Corollary 6.3.13 *If $F : [0,1] \to \mathcal{E}^n$ is Hukuhara differentiable and its Hukuhara derivative F' is integrable over $[0,1]$, then*

$$F(t) = F(t_0) + \int_{t_0}^t F'(s)\, ds \tag{67}$$

for all $0 \leq t_0 \leq t \leq 1$.

The proof follows from its counterpart for the level set mappings $[F(\cdot)]^\alpha$ for each $\alpha \in I$. The result need not, however, hold for the Bobylev derivative or for the de Blasi differential $D_t^{(F)}(1)$. In this sense the Hukuhara derivative appears to be a natural inverse operation of integration for fuzzy set valued mappings.

11.6.4 Bibliographical Notes

There are many papers which refer to or use fuzzy set valued mappings of a single real variable, in particular those taking values in $\mathcal{E}$. Measurability, differentiation, integration and other basic properties are considered in the papers, amongst others, of Dubois and Prade, 1982a, 1982b, 1982c, Goetschel and Voxman, 1981, 1983, Kaleva, 1987, 1990a, 1990b, 1991, Kloeden, 1980, 1982, 1987, 1991, Puri and Ralescu, 1983, 1985, 1986 and Seikkala, 1987.

§6.1 is based on Kaleva, 1990a, 1990b. An equivalent definition to strong measurability was proposed by Puri and Ralescu, 1986. The results onde Blasi differentiation in §6.2 are from Diamond and Kloeden, 1994, though many are motivated by analogous results for Hukuhara differentiation in Kaleva, 1987, 1990a, Puri and Ralescu, 1983 and in the work of Bobylev, 1985, 1990a. Proposition 6.2.4 is from Diamond and Kloeden, 1994. Puri and Ralescu, 1986, defined the expectation of a fuzzy random variable using the Aumann integral level setwise. The generalization to integrals of fuzzy set valued mappings of a single real valuable presented here was carried out by Kaleva, 1987, 1990a. See also Bobylev, 1985. Theorem 6.3.5 is from Diamond and Kloeden, 1994.

11.7 INTERPOLATION AND APPROXIMATION

11.7.1 Interpolation and Splines

The problem of interpolation for fuzzy sets is as follows: suppose that at various time instants t information $F(t)$ is presented as a fuzzy set. The aim is to approximate the function $F(t)$, for all t in the domain of F, by simple linear approximants of the form

$$\lambda_0(t)F_0 + \lambda_1(t)F_1 + \cdots + \lambda_n(t)F_n$$

where the F_j are fixed fuzzy sets and the $\lambda_j(\cdot)$ are scalar valued mappings.

Let $t_0 < t_1 < \cdots < t_n$ be $n+1$ distinct points in $\Re$ and let $U_0, U_1, \ldots, U_n$ be $n+1$ fuzzy sets in $\mathcal{E}$. A *fuzzy polynomial interpolation* of the data is a fuzzy–valued function $F : \Re \to \mathcal{E}$ satisfying:

(i) For all $i = 0, 1, \ldots, n$, $F(t_i) = U_i$;

(ii) F is continuous;

(iii) If $U_i = \chi_{\{x_i\}}$, $i = 0, 1, \ldots, n$, are crisp real numbers, and if f is the unique polynomial of degree $\leq n$ such that $x_i = f(t_i)$, $i = 0, 1, \ldots, n$, then $F(t) = \chi_{f(t)}$. That is, if the data is crisp, then the interpolation F is a crisp polynomial.

A function F fulfilling these conditions may be constructed as follows. For each $\alpha \in [0, 1]$ and $i = 0, 1, \ldots, n$, let $C_\alpha^i = [U_i]^\alpha$. Since each $U_i \in \mathcal{E}$, C_α^i , $i = 0, 1, \ldots, n$, is a nonempty compact interval and satisfies the conditions of Proposition 3.1.6.

For each $\mathbf{x} = (x_0, x_1, \ldots, x_n) \in \Re^{n+1}$, denote by $P_\mathbf{x}$ the unique polynomial of degree $\leq n$ such that

$$P_\mathbf{x}(t_i) = x_i \ , \ i = 0, 1, \ldots, n \, .$$

That is, by the crisp Lagrange interpolation formula

$$P_{\mathbf{x}}(t) = \sum_{i=0}^{n} x_i \left(\prod_{i \neq j} \frac{t - t_j}{t_i - t_j} \right) .$$

Finally, for each $t \in \Re$ and all $\xi \in \Re$ define $F(t) \in \mathcal{E}$ by

$$(F(t))(\xi) = \sup \left\{ \alpha \in [0,1] : \exists\, \mathbf{x} \in C_\alpha^0 \times \cdots \times C_\alpha^n \text{ such that } P_{\mathbf{x}}(t) = \xi \right\} .$$

Theorem 7.1.1 *$F(t) \in \mathcal{E}$ for each $t \in \Re$. Moreover,*

1. *$F(t_i) = U_i$, $i = 0, 1, \ldots, n$;*

2. *F is continuous;*

3. *If for all $i = 0, 1, \ldots, n$, $U_i = \chi_{[a_i, b_i]}$, $a_i \leq b_i$, and for each $t \in \Re$*

   $$J(t) = \{\xi \in \Re : \exists\, \mathbf{x} \in [a_0, b_0] \times \cdots \times [a_n, b_n] \text{ such that } P_{\mathbf{x}}(t) = \xi\} ,$$

then F reduces to the function $F(t) = \chi_{J(t)}$.

Corollary 7.1.2 *The interpolation polynomial can be written level setwise as*

$$[F(t)]^\alpha = \{y \in \Re : y = P_{\mathbf{x}}(t),\ x_i \in [U_i]^\alpha,\ i = 1, 2, \ldots, n\} ,$$

for $0 \leq \alpha \leq 1$.

When the data U_i presents as triangular fuzzy numbers, values of the interpolation polynomial are also triangular fuzzy numbers. Then $F(t)$ has a particularly simple form that is well suited to computation. Denote by $L_i(t)$, $i = 1, 2, \ldots, n$, the polynomials

$$L_i(t) = \prod_{j \neq i}^{n} \frac{t - t_j}{t_i - t_j} .$$

Theorem 7.1.3 *Let (t_i, U_i), $i = 1, 2, \ldots, n$ be the observed data and suppose that each of the $U_i = (U_i^\ell, U_i^c, U_i^r)$ is an element of $\mathcal{E}_{TG}$. Then for each $t \in [t_0, t_n]$, $F(t) = (F_\ell(t), F_c(t), F_r(t))$ is in $\mathcal{E}_{TG}$ and*

$$F_\ell(t) = \textstyle\sum_{L_i(t) \geq 0} L_i(t) U_i^\ell - \sum_{L_i(t) < 0} L_i(t) U_i^r ,$$

$$F_m(t) = \textstyle\sum_{i=0}^{n} L_i(t) U_i^c ,$$

$$F_r(t) = \textstyle\sum_{L_i(t) \geq 0} L_i(t) U_i^r - \sum_{L_i(t) < 0} L_i(t) U_i^\ell .$$

It is well known that Lagrange interpolation can suffer poor interpolatory approximation, for example the Runge effect for equally spaced $t_0, t_1, \ldots, t_n$. To overcome these potential disadvantages, interpolation can be done using splines.

Let S_k denote the *family of splines* of order k with *knots* t_i, $i = 0, 1, \ldots, n$. That is $s \in \mathrm{S}_k$ if

(i) s is C^{k-2} on the interval $[t_0, t_n]$,

(ii) s is a polynomial of order k on the interval $[t_i, t_{i+1}]$ for $i = 0, 1, \ldots, n-1$.

As in the case of fuzzy interpolation polynomials, which by Corollary 7.1.2 are expressed by level sets in terms of crisp interpolation polynomials, define fuzzy spline interpolation in terms of level sets and crisp splines. Let $\mathbf{x} = (x_0, x_1, \ldots, x_n)$ and let $S_{\mathbf{x}}(t)$ be a spline in S_k which interpolates the data (t_i, x_i), $i = 0, 1, \ldots, n$. Then the *interpolating fuzzy spline of order* k which interpolates the data (t_i, U_i), $U_i \in \mathcal{E}$, $i = 0, 1, \ldots, n$ is defined level setwise by $[F(t)]^\alpha = \{y \in \Re : y = P_{\mathbf{x}},\ x_i \in [U_i]^\alpha\}$, for all $0 \leq \alpha \leq 1$.

If $s_i \in \mathrm{S}_\mathbf{k}$ interpolates the data (t_j, f_j), $j = 0, 1, \ldots, n$, where $f_j = 1$ if $j = i$ and zero otherwise, then $S_{\mathbf{x}}(t) = \sum_{i=0}^{n} x_i s_i(t)$ and similarly to Corollary 7.1.2, $F(t) = \sum_{i=0}^{n} s_i(t) U_i$. Consequently, F is continuous on $[t_0, t_n]$ and if all the $U_i \in \mathcal{E}_{TG}$, then so also is $F(t) \in \mathcal{E}_{TG}$ for all $t \in [t_0, t_n]$.

Example 7.1.4 Consider the case when $k = 2$, which gives *piecewise linear* interpolants. Since $s_i(t) > 0$ if $t \in (t_{i-1}, t_{i+1})$ and $s_i(t) = 0$ for $t \notin [t_{i-1}, t_{i+1}]$, then

$$F(t) = \frac{(t_{i+1} - t)}{(t_{i+1} - t_i)} U_i + \frac{(t - t_i)}{(t_{i+1} - t_i)} U_{i+1}$$

for all $t \in [t_i, t_{i+1}]$, $i = 0, 1, \ldots, n-1$. Hence, in terms of level sets, writing $[U_i]^\alpha = [a_i^\alpha, b_i^\alpha]$ and $[F(t)]^\alpha = [F_\ell^\alpha(t), F_r^\alpha(t)]$,

$$F_\ell^\alpha(t) = \frac{(t_{i+1} - t)}{(t_{i+1} - t_i)} a_i^\alpha + \frac{(t - t_i)}{(t_{i+1} - t_i)} a_{i+1}^\alpha \quad \text{for all} \quad t \in [t_i, t_{i+1}]$$

$$F_r^\alpha(t) = \frac{(t_{i+1} - t)}{(t_{i+1} - t_i)} b_i^\alpha + \frac{(t - t_i)}{(t_{i+1} - t_i)} b_{i+1}^\alpha \quad \text{for all} \quad t \in [t_i, t_{i+1}] \,.$$

For $k = 4$ a cubic spline is obtained. A similar result to Theorem 7.1.3 obtains.

Theorem 7.1.5 *Let (t_i, U_i), $i = 1, 2, \ldots, n$ be the observed data and suppose that each of the $U_i = (U_i^\ell, U_i^c, U_i^r)$ is an element of $\mathcal{E}_{TG}$. Then for each $t \in$*

$[t_0, t_n]$, $F(t) = (F_\ell(t), F_c(t), F_r(t))$ *is in* $\mathcal{E}_{TG}$ *and*

$$F_\ell(t) = \sum_{s_i(t)\geq 0} s_i(t)U_i^\ell - \sum_{s_i(t)<0} s_i(t)U_i^r ,$$

$$F_m(t) = \sum_{i=0}^{n} s_i(t)U_i^c ,$$

$$F_r(t) = \sum_{s_i(t)\geq 0} s_i(t)U_i^r - \sum_{s_i(t)<0} s_i(t)U_i^\ell .$$

Remark 7.1.6 The sign of L_i is easy to compute and thus the representation of Theorem 7.1.3 is readily computable. The same is not exactly true for the s_i. Also, the interpolation conditions are not sufficient to determine the splines s_i uniquely. Two extra conditions are required. Many such conditions are available, but only the "not–a–knot" condition will be considered. It turns out that the crisp spline equations can be manipulated so that $s_0 = s_1$ and $s_{n-1} = s_n$ and this supplies two additional conditions which are suitable for our purposes.

Proposition 7.1.7 *Assume that* $s_0, s_1, \ldots, s_n$ *satisfy the not–a–knot condition. Then for any* i

1. s_i *is not identically zero on any subinterval* $[t_j, t_{j+1}]$, $j = 0, \ldots, n-1$;
2. s_i *is of constant sign on any given* $[t_j, t_{j+1}]$, $j = 0, \ldots, n-1$;
3. s_i *changes sign at* t_j *for all* $j \neq i$.

Hence the formulas of Theorem 7.1.5 are more readily computable if the additional conditions are supplied by the not–a–knot property.

Finally, consider the case when the fuzzy data is $U_i = y_i + A$, $i = 0, 1, \ldots, n$, $y_i \in \Re$, where A is a constant fuzzy set in $\mathcal{E}$. Then if $\mathbf{y} = (y_0, y_1, \ldots, y_n)$,

$$G(t) = S_{\mathbf{y}}(t) + A$$

is a fuzzy spline interpolation with level curves which are C^{k-2} on $[t_0, t_n]$. That is, the fuzzy spline interpolate is just a translation of the crisp spline, by the constant fuzzy set A which perturbs each of the data points. Here, C^{k-2} denotes that the curves are $k-2$ times continuously differentiable.

11.7.2 Bernstein Approximation

For $F : [0, 1] \to \mathcal{E}^p$, define the n-th *Bernstein polynomial approximant* $B_n F$ to be

$$B_n F(t) = \sum_{j=0}^{n} \binom{n}{j} t^j (1-t)^{n-j} F(j/n) .$$

It is easy to see that $B_n F$ is a continuous fuzzy valued mapping on $[0, 1]$ with values in $\mathcal{E}^p$.

Theorem 7.2.1 *Let* $F : [0,1] \to \mathcal{E}^p$ *be continuous. Then* $\lim_{n\to\infty} B_n F = F$ *uniformly in* $[0,1]$ *in the sense that* $d_\infty(B_n F(t), F) \to 0$ *uniformly in* $0 \leq t \leq 1$ *as* $n \to \infty$.

If F is not fuzzy convex valued, then a technical difficulty arises and certain inequalities involving sums of nonconvex sets are no longer valid. However, the Bernstein approximants can still be constructed and a partial approximation result is true. Recall that $\mathfrak{D}^p$ was the totality of all *normal, upper semicontinuous* fuzzy sets on $\Re^p$ and that $\mathcal{E}^p \subset \mathfrak{D}^p$ because members of $\mathcal{E}^p$ have the extra condition of fuzzy convexity.

If K is a compact subset of $\Re^p$, denote its *convex hull* by co K. For $u \in \mathfrak{D}^p$, define its convexification co $u \in \mathcal{E}^p$ in terms of level sets by

$$[\mathrm{co}\, u]^\alpha = \mathrm{co}\left([u]^\alpha\right).$$

Clearly, if $u \in \mathcal{E}^p$, then $\mathrm{co}\, u = u$. For a mapping $F : [0,1] \to \mathfrak{D}^p$, define the mapping $\mathrm{co}\, F : [0,1] \to \mathcal{E}^p$ by $(\mathrm{co}\, F)(t) = \mathrm{co}\,(F(t))$, $0 \leq t \leq 1$.

The effect of the Bernstein construction is to convexify F. To make this observation more precise, a fuzzy version of the Shapley-Folkman Lemma, estimating the distance between sums and their convexification, is needed.

Lemma 7.2.2 *Let* $U_j \in \mathfrak{D}^p$, $j = 0, 1, \ldots, n$ *be such that* $||U_j|| = d_\infty(U_j, \chi_{\{0\}}) \leq M$. *Then*

$$d_\infty\left(\sum_{j=0}^{n} U_j, \sum_{j=0}^{n} \mathrm{co}\, U_j\right) \leq M p^{1/2}.$$

This then gives

Theorem 7.2.3 *Let* $F : [0,1] \to \mathfrak{D}^p$ *be continuous. For any* ϵ, $0 < \epsilon < 1/2$, $B_n F$ *converges to* co F *uniformly in* $\epsilon \leq t \leq 1 - \epsilon$.

11.7.3 Bibliographical Notes

§7.1 Lagrange interpolation for fuzzy data was explored by Lowen, 1990. Kaleva avoided the well known computational problems associated with crisp Lagrange interpolation and used spline approximation (Kaleva, 1991). Theorem 7.1.1 is due to Lowen, 1990, while the related Corollary 7.1.2 was noted by Kaleva, 1991. Example 7.1.4 and all other results appear in Kaleva, 1991.

§7.2 Bernstein approximation for compact multifunctions was first treated by Vitale, 1979. The results of this section are due to Diamond and Ramer, 1993. The crisp Shapley-Folkman convexifying Lemma is in Arrow and Hahn, 1971. The approximation process can be extended to Korovkin systems, which are much more general than and include the Bernstein polynomials (Diamond and Ramer, 1993).

Classical results on splines, interpolation and Bernstein approximation can be found in Davis, 1963 and de Boor, 1998, to name but two from an enormous literature. Other forms of approximation of fuzzy sets themselves have been discussed in Collings and Kloeden, 1991 and Diamond, Kloeden and-Vladimirov, 1998.

11.8 FUZZY DIFFERENTIAL EQUATIONS

11.8.1 Introduction

Evolutionary mechanisms of many important dynamical processes are often modelled by differential equations, which need to be solved for specific initial conditions. For the most basic situation involving first order autonomous ordinary differential equation in $\Re^n$ this *initial value problem* takes the form

$$\frac{dx}{dt}(t) = f(x(t)), \quad x(0) = x_0, \tag{68}$$

for which it is required to find a *solution*, that is a differentiable function $x = x(t)$ with $x(0) = x_(0)$ that satisfies the DE in some interval (t_1, t_2) for some $t_1 < 0 < t_2$, that may be infinite. When the mapping f on the right hand side of DE is continuous, such an initial value problem is equivalent to solving the *integral equation*

$$x(t) = x_0 + \int_{t_0}^{t} f(x(s))ds$$

for a continuous function $x = x(t)$.

Crisp differential equations often represent an idealization of real situations where imprecision may in fact play a significant role. Random differential equations have long been used to incorporate the effects of statistical fluctuations. Imprecision due to uncertainty or vagueness suggests the introduction of *fuzzy differential equations*. In this case *interpret* the right hand side of the DE in (68) as a mapping $f : \mathcal{E}^n \to \mathcal{E}^n$ with the solutions taking values $X(t) \in \mathcal{E}^n$ for all appropriate t. Traditionally the d_∞ metric on $\mathcal{E}^n$ and the Hukuhara derivative have been used for fuzzy DE.

Example 8.1.1 Let $A \in \mathcal{E}$ have level sets $[A]^\alpha = [A_1^\alpha, A_2^\alpha]$ for $\alpha \in I$ and suppose that a solution $X : [0, T] \to \mathcal{E}$ of the fuzzy DE

$$X'(t) = AX \tag{69}$$

on $\mathcal{E}$ has level sets $[X(t)]^\alpha = [X_{1,\alpha}(t), X_{2,\alpha}(t)]$ for $\alpha \in I$ and $t \in [0,T]$. This the fuzzy DE is equivalent to the coupled system of ordinary DE

$$\begin{aligned} X'_{1,\alpha}(t) &= \min\{A_1^\alpha X_{1,\alpha}(t), A_2^\alpha X_{1,\alpha}(t), A_1^\alpha X_{2,\alpha}(t), A_2^\alpha X_{2,\alpha}(t)\} \\ X'_{2,\alpha}(t) &= \max\{A_1^\alpha X_{1,\alpha}(t), A_2^\alpha X_{1,\alpha}(t), A_1^\alpha X_{2,\alpha}(t), A_2^\alpha X_{2,\alpha}(t)\} \end{aligned} \tag{70}$$

for $\alpha \in I$, since by Theorem 6.3.4 the Hukuhara derivative $X'(t)$ has level sets $[X'(t)]^\alpha = \left[X'_{1,\alpha}(t), X'_{2,\alpha}(t)\right]$ for $a \in I$ and $t \in [0,T]$ and by the Extension Principle the fuzzy set $f(X(t)) = AX(t)$ has level sets

$$\begin{aligned} [AX(t)]^\alpha = & \ [\min\{A_1^\alpha X_{1,\alpha}(t), A_2^\alpha X_{1,\alpha}(t), A_1^\alpha X_{2,\alpha}(t), A_2^\alpha X_{2,\alpha}(t)\}, \\ & \ \max\{A_1^\alpha X_{1,\alpha}(t), A_2^\alpha X_{1,\alpha}(t), A_1^\alpha X_{2,\alpha}(t), A_2^\alpha X_{2,\alpha}(t)\}] \end{aligned}$$

for all $\alpha \in I$ and $t \in [0,T]$.

Example 8.1.2 For $A = \chi_{\{-1\}} \in \mathcal{E}$ the fuzzy differential equation (69) becomes

$$X'(t) = -X(t)$$

and the system of ordinary DE (70) reduces to

$$X'_{1,\alpha}(t) = -X_{2,\alpha}(t), \quad X'_{2,\alpha}(t) = -X_{1,\alpha}(t)$$

for $\alpha \in I$, with solution corresponding to an initial value $X_0 \in \mathcal{E}$ with $[X_0]^\alpha = [X_{01,\alpha}, X_{02,\alpha}]$ for $\alpha \in I$ given by

$$\begin{aligned} X_{1,\alpha}(t) &= \frac{1}{2}(X_{01,\alpha} - X_{02,\alpha})\,e^t + \frac{1}{2}(X_{01,\alpha} + X_{02,\alpha})\,e^{-t} \\ X_{2,\alpha}(t) &= \frac{1}{2}(X_{02,\alpha} - X_{01,\alpha})\,e^t + \frac{1}{2}(X_{01,\alpha} + X_{02,\alpha})\,e^{-t} \end{aligned}$$

for $\alpha \in I$ and all $t \geq 0$. Thus for $X_0 = \chi_{\{c_0\}}$ the solution $X(t) = \chi_{\{c_0 e^{-t}\}} \to \chi_{\{0\}}$ as $t \to \infty$. In contrast, when $[X_0]^\alpha = [\alpha - 1, 1 - \alpha]$ for $\alpha \in I$ the solution has level sets

$$[X(t)]^\alpha = \left[(\alpha - 1)e^t, (1-\alpha)e^t\right] = (1-\alpha)e^t[-1,1]$$

for all $\alpha \in I$ and $t \geq 0$. In particular, $\text{diam}[X(t)]^\alpha = 2(1-\alpha)e^t$, so the solution becomes fuzzier with increasing time.

11.8.2 Existence and Uniqueness of Solutions

The existence and uniqueness of such a solution of an initial value problem (68), which is often written as $x = x(t; x_0)$, is important for both mathematical and modelling considerations. Classical existence and uniqueness theorems assume that the mapping f on the right hand side of the DE satisfies a Lipschitz condition and the proof requires the completeness of the space on which the DE is defined. A fuzzy DE

$$X'(t) = f(X), \quad X \in \mathcal{E}^n, \tag{71}$$

on the complete metric space $(\mathcal{E}^n, d_\infty)$ satisfies a *global Lipschitz condition* if there exists a finite constant $K > 0$ such that

$$d_\infty(f(X), f(Y)) \leq K\, d_\infty(X, Y) \quad \text{for all} \quad X, Y \in \mathcal{E}^n. \tag{72}$$

Theorem 8.2.1 *For any finite $T > 0$ and $x_0 \in \mathcal{E}^n$, a fuzzy differential equation (71) with Hukuhara derivative on $(\mathcal{E}^n, d_\infty)$, which satisfies a global Lipschitz condition (72) has a unique solution $X : [0, T] \to \mathcal{E}^n$ corresponding to the initial value $X(0) = X_0$.*

Many variations of this theorem are possible. For example, the global Lipschitz condition (72) could be weakened to a local Lipschitz condition, that is with a different constants for each bounded subset of $\mathcal{E}^n$. Then existence is guaranteed only on a subinterval depending on the given initial value and possibly very small, unless a *growth bound* such as

$$\| f(X) \| \leq K\,(1 + \| X \|), \qquad X \in \mathcal{E}^n,$$

is also satisfied. Other metrics on $\mathcal{E}^n$ could also be used provided the resulting metric space is complete, or the differential equation is restricted to a subspace which is complete. An example of the latter is a subspace of fuzzy sets in $\mathcal{E}^n$ with supports contained in a given compact subset of $\Re^n$ where a d_p metric is used.

11.8.3 Reinterpreting Fuzzy DEs

Although fuzzy DE have been proposed as a way of modelling uncertain and incompletely specified systems, the approach outlined in previous sections incurs significant difficulties due to the nature of the fuzzy derivatives that have been used. It was mentioned in §6.2 that Hukuhara differences need not exist in the definition of that derivative, while the Bobylev and de Blasi derivatives, even when they exist at each $\alpha-$level, need not lie in the space of fuzzy sets. This places serious restrictions on the class of models that can be used and is a constraint on modelling uncertain systems by systems of the type (71).

However, the most critical problem is that posed by Theorem 6.2.9. In the context of fuzzy DE, this means that any solution $X(t)$ to a *fuzzy initial value problem* (FIVP)

$$X'(t) = G(t, X), \quad X(t_0) = X_0, \tag{73}$$

has diam $([X(t)]^0)$ nondecreasing in $t \geq t_0$, which has been called a *possibilistic irreversibility property.* This was seen in Example 8.1.2, where an ordinary DE in which the origin was asymptotically stable had an unstable solution for the fuzzy counterpart, whenever the initial condition was a nondegenerate (that is, not crisp) fuzzy number $X_0 \in \mathcal{E}$. In that example, $\|X(t)\| \to \infty$ as $t \to \infty$, in stark contrast to the behaviour of the ordinary DE. Consequently, this formulation of fuzzy DEcannot sensibly participate in any of the rich behaviour of ordinary DE such as periodicity, stability, bifurcation and the like. This means that the whole approach is ill-suited for modelling in that it cannot reflect real situations even if they are vague and ill-defined.

Recently, Hüllermeier, 1997, suggested a different formulation of the FIVP based on a family of differential inclusions at each $\alpha-$level, $0 \leq \alpha \leq 1$,

$$x'(t) \in [G(t, x(t))]^\alpha, \quad x(0) \in [X_0]^\alpha, \tag{74}$$

where now $[G(\cdot, \cdot)]^\alpha : \mathbb{R} \times \mathbb{R}^n \to \mathcal{K}_C^n$, the space of nonempty convex compact subsets of $\mathbb{R}^n$. The idea is that the set of all such solutions $\Sigma_\alpha(X_0, T)$ would be the $\alpha-$level of a fuzzy set $\Sigma(X_0, T)$, in the sense that all attainable sets $\mathcal{A}_\alpha(X_0, t)$, $0 < t \leq T$, are levels of a fuzzy set $\boldsymbol{\mathcal{A}}(X_0, t)$ on $\mathbb{R}^n$. Considering $\Sigma(X_0, T)$ to be the solution of the fuzzy DE $X' = G(t, X)$, $X(0) = X_0$ thus captures both uncertainty and the rich properties of differential inclusions in one and the same technique. Note that this approach is not a family of level set *differential equations* but deals with *differential inclusions*. It is this distinction which leads to fundamentally different behaviour of the system from that of the traditional fuzzy DE. Both the solution set and attainability set are fuzzy sets, and under weaker conditions on G than the theorems of §8.2.

Example 8.3.1 Returning to Example 8.1.2, interpreted as a family of differential inclusions for $t \geq 0$, $0 \leq \alpha \leq 1$,

$$X'_\alpha(t) = -\lambda X_\alpha(t), \quad X_\alpha(0) \in [X_0]^\alpha = (1 - \alpha)[-1, 1], \tag{75}$$

a much more intuitive and attractive outcome emerges. Since $-X_\alpha = \{-x_\alpha\}$ is a singleton set in $\mathcal{K}_C^1$, (75) becomes

$$x'_\alpha(t) = -x_\alpha(t), \quad x_\alpha(0) \in (1 - \alpha)[-1, 1],$$

which has solution set $\Sigma_\alpha(X_\alpha, t)$ on $[0, t]$ comprising the functions

$$x_\alpha(t) = x_\alpha(0)e^{-\lambda t}, x_\alpha(0) \in [X_0]^\alpha.$$

Consequently, $\Sigma_\alpha(X_\alpha, t) = (1 - \alpha)e^{-\lambda t}[-1, 1]$. So $X(t)$ is interpreted to be the symmetric triangular fuzzy number-valued function whose $\alpha-$level is

$[X(t)]^\alpha = \Sigma_\alpha(X_\alpha, t) = (1-\alpha)e^{-\lambda t}[-1,1]$. This result is much more intuitive than the classical fuzzy solution given in Example 8.1.2 since now diam (supp $X(t)) \to 0$ as $t \to \infty$.

Let $\Omega \subset \mathbb{R}\times\mathbb{R}^n$ be an open subset containing $(0, x_0)$ and let $H : \Omega \to \mathcal{K}^n_C$. The *differential inclusion*

$$x'(t) \in H(t, x(t)), \quad x(0) = x_0, \tag{76}$$

is said to have a *solution* $y(t)$ on $[0,T]$ if $y(\cdot)$ is absolutely continuous, $y(0) = x_0$ and $y(\cdot)$ satisfies the inclusion a.e. in $[0,T]$.

The *boundedness assumption* is said to hold if there exist $b,\ T,\ M > 0$ such that

- the set $Q = [0,T] \times (x_0 + (b+MT)\overline{S}_1^{;n}) \subset \Omega$, where $\overline{S}_1^{;n}$ is the closed unit ball of $\mathbb{R}^n$;
- H maps Q into the ball of radius M.

Denote the set of all solutions or *solution set* of (76) on $[0,\tau]$ by $\Sigma(x_0,\tau)$, the *attainable set* by $\mathcal{A}(x_0,\tau) = \{x(\tau) : x(\cdot) \in \Sigma(x_0,\tau)\}$ and write $\mathcal{Z}_T(\mathbb{R}^n) = \{x(\cdot) \in C([0,T];\mathbb{R}^n) \ :\ x'(\cdot) \in L^\infty([0,T];\mathbb{R}^n)\}$. It is known that for every $x_1 \in x_0 + b \operatorname{int} \overline{S}_1^{;n}$, $\Sigma(x_1, T)$ exists and is a compact subset of $\mathcal{Z}_T(\mathbb{R}^n)$, and each attainable section $\mathcal{A}(x_1,\tau)$, $0 < \tau \le T$, is a compact subset of $\mathbb{R}^n$. In fact, although these sets are not in general convex, they are acyclic which is stronger than simply connected.

Let $\Omega \subset \mathbb{R} \times \mathbb{R}^n$ be open and let J be a real compact interval. A mapping $F : \Omega \times J \to \mathcal{K}^n$ is said to be *regularly quasi-concave on* J if

- For all $(t,x) \in \Omega$ and $\alpha,\ \beta \in J$,

$$F(t,x;\beta) \supseteq F(t,x;\alpha) \quad \text{whenever} \quad \beta \le \alpha. \tag{77}$$

- If $\{\alpha_k\}$ is a nondecreasing sequence in J converging to α, $\alpha_k \uparrow \alpha$, then for all $(t,x) \in \Omega$

$$\bigcap_k F(t,x;\alpha_k) = F(t,x;\alpha).$$

The definition is adapted from the usual definition for real-valued functions but drops the convexity of level sets. Note the similarity with conditions of the Stacking Theorem 3.1.7.

Now, consider the family of differential inclusions

$$x'_\alpha(t) \in F(t,x;\alpha), \quad x(0) = x_0,\ \alpha \in J,$$

with attainability set $\mathcal{A}_\alpha(x_0, T)$ and solution set $\Sigma_\alpha(x_0, T)$ on $[0, T]$.

Theorem 8.3.2 *Let $F : \Omega \times J \to \mathcal{K}_C^n$ be upper semicontinuous on Ω, regularly quasi-concave on J and suppose that the boundedness assumption holds. Then the mapping $\alpha \mapsto \mathcal{A}_\alpha(x_0, T)$ is a regularly quasi-concave map from J to $\mathcal{K}^n$ and $\alpha \mapsto \Sigma_\alpha(x_0, T)$ is a regularly quasi-concave mapping from J to $\mathcal{Z}_T(\mathbb{R}^n)$.*

Theorem 8.3.3 *Let $X_0 \in \mathcal{E}^n$ and let Ω be an open set in $\mathbb{R} \times \mathbb{R}^n$ containing $\{0\} \times \text{supp}\,(X_0)$. Suppose that $G : \Omega \to \mathcal{E}^n$ is upper semicontinuous and write $F(t, x; \alpha) = [G(t, x)]^\alpha \in \mathcal{K}_C^n$ for all $(t, x, \alpha) \in \mathbb{R}^{n+1} \times [0, 1]$. Let the boundedness assumption, with constants b, M, T, hold for all $x_0 \in \text{supp}\,(X_0)$ and for the inclusion*

$$x'(t) \in F(t, x; 0), \quad x(0) \in \text{supp}\,(X_0). \tag{78}$$

Then the attainable sets $\mathcal{A}_\alpha(X_\alpha, T)$, $\alpha \in [0, 1]$, of the family of inclusions

$$X'_\alpha(t) \in F(t, X_\alpha; \alpha), \quad X_\alpha(0) \in [X_0]^\alpha, \ \alpha \in [0, 1], \tag{79}$$

are the level sets of a fuzzy set $\boldsymbol{\mathcal{A}}(X_0, T) \in \mathfrak{D}^n$. The solution sets $\Sigma_\alpha([X_0]^\alpha, T)$ of (79) are the level sets of a fuzzy set $\boldsymbol{\Sigma}(X_0, T)$ defined on $\mathcal{Z}_T(\mathbb{R}^n)$.

Remark 8.3.4 If the condition that G be bounded on $[0, \infty) \times \Gamma$, where $\Gamma \subseteq \mathbb{R}^n$ is open, be added to the conditions of the theorem, the interval of existence and consequences extend to $[0, \infty)$.

Suppose that G is bounded on $[0, \infty) \times \Gamma$. It is a natural question to ask as to what is the asymptotic behaviour of $\boldsymbol{\Sigma}(X_0, T)$ as $T \to \infty$. The set Γ is said to be *globally viable* under G if for any $X_0 \in \mathcal{E}^n(\Gamma)$ $\boldsymbol{\mathcal{A}}(X_0, t) \subseteq \mathfrak{D}^n(\Gamma)$ for all $0 \le t < \infty$. In this case, say that $W \in \mathfrak{D}^n(\Gamma)$ is an *absorbing set* for the system if for every bounded subset $\mathcal{V} \subset \mathcal{E}^n(\Gamma)$ there is a $t_\mathcal{V} > 0$ such that $\boldsymbol{\Sigma}(\mathcal{V}, t) \subseteq W$ for all $t > t_\mathcal{V}$. Here $\boldsymbol{\Sigma}(\mathcal{V}, t)$ denotes the fuzzy set with $\alpha-$levels given by

$$[\boldsymbol{\Sigma}(\mathcal{V}, t)]^\alpha = \bigcup_{X_0 \in \mathcal{V}} [\boldsymbol{\Sigma}(X_0, t)]^\alpha.$$

and by $\boldsymbol{\Sigma}(\mathcal{V}, t) \subseteq W$ is meant that

$$\boldsymbol{\mathcal{A}}(X_0, t) \subseteq W, \quad X_0 \in \mathcal{V}, \ t > t_\mathcal{V},$$

in the sense of fuzzy sets, where $X \subseteq Y$ if and only if $X(\xi) \le Y(\xi)$ for all $\xi \in \Re^n$. Note that while $\boldsymbol{\Sigma}(\mathcal{V}, t)$ and $\boldsymbol{\mathcal{A}}(\mathcal{V}, t)$ are certainly fuzzy sets, their level sets need not be compact, since an infinite union of compact sets need not be compact. However, this does not affect the $\subseteq$ relation in the last equation, nor does it affect the validity of the previous and following discussion.

Example 8.3.5 In Example 8.3.1, any triangular fuzzy number $U \in \mathcal{E}_{TG}$ with 0 in the interior of supp(U) is absorbing. In this case, $\chi_{\{0\}}$ is a global attractor.

Example 8.3.6 Consider the differential inclusion on $\mathbb{R}^+$,

$$y'(t) \in -y(t) + E\cos t, \quad E = [-\varepsilon, \varepsilon],\ \varepsilon > 0. \tag{80}$$

The solution set is given for $t \geq 0$ by

$$\begin{aligned} y(t) &\in \frac{1}{2}(\sin t + \cos t)E + \left(y(0) - \frac{1}{2}E\right)e^{-t} && (81)\\ &= \Sigma(t, y(0)), && (82)\end{aligned}$$

because every integrable selection ϕ from $E\cos t$ satisfies $e_1(t) \leq \phi(t) \leq e_2(t)$, where for $i = 1, 2$,

$$e_i(t) = \left((-1)^i \mathrm{sgn}\,(\cos t)\right)\cos t, \quad t \geq 0.$$

Write the solution set with initial condition $y(0) = y_0$ as $\Sigma(t, y_0)$. Note that $x \in \Sigma(0, x) = -\frac{1}{2}E + (x + \frac{1}{2}E)$ and that

$$\Sigma(t + \tau, x) = \Sigma(t, \Sigma(\tau, x)).$$

Put $c(t) = \frac{\sqrt{2}}{2}\sin(t + \pi/4)$ and note that E is an absorbing set for the system, since for any bounded interval $D = [a, b]$,

$$\Sigma(t, D) = c(t)E + [a, b]e^{-t} - c(0)e^{-t}$$

and this is certainly a subset of E for $t > t_D$, where

$$Ke^{-t_D} := \left(\max\{|a|, |b|\} + \sqrt{2}\varepsilon\right)e^{-t_D} \leq \left(1 - \sqrt{2}/2\right)\varepsilon$$

Moreover, $\rho(\Sigma(t, D), c(t)E) \leq Ke^{-t} \to 0$ as $t \to \infty$, so in this sense the trajectory in $\mathcal{K}_C^1$ is asymptotically stable for the system.
Now let $U \in \mathcal{E}$ be the fuzzy number with level sets $[U]^\alpha = (1 - \alpha)[-1, 1]$ and transfer this analysis to the fuzzy DE

$$Y'(t) = -Y(t) + U\cos t,$$

via the family of differential inclusions (80) with $\varepsilon = 1 - \alpha$. From (81), clearly

$$\Sigma(X, t) = c(t)U + (X - c(t)U)\,e^{-t},$$

U is an absorbing set and $c(t)U$ is an asymptotically stable trajectory of the system in $\mathcal{E}$.

11.8.4 Bibliographical Notes

Fuzzy differential equations have been considered by Bobylev, 1990a, Kaleva, 1990b, 1991 and Seikkala, 1987, amongst others. Example 8.1.1 and 8.1.2 are from Diamond and Kloeden, 1994. The existence and uniqueness theorem, Theorem 8.2.1 is taken from Kaleva, 1990b. See Kaleva, 1991 and Kloeden, 1991, for an analogue of the Peano theorem for fuzzy differential equations. See Kloeden, 1982, for an abstract approach to fuzzy dynamical systems. Hüllermeier, 1997, introduced the new interpretation discussed in §8.3. That attainability and solution sets are fuzzy sets is proved rigorously in Diamond and Watson, 1999, where Theorems 8.3.2 and 8.3.3 can be found. The notion of attractive sets and Examples 8.3.5 and 8.3.6 are from Diamond, 1999a. Ideas of Lyapunov stability are explored in Diamond, 1999b. The basic principles and definitions regarding differential inclusions and their solution and attainability sets can be found in Aubin and Cellina, 1984, which also discusses viability theory.

11.9 CONCLUSION

This chapter has surveyed the more commonly used types of fuzzy numbers and metric distances between elements of spaces of fuzzy numbers. This gives a metric space topology and some topological properties were studied, such as characterization of compact sets in various metric spaces. By far the most important space for practical purposes is $(\mathcal{E}, d_H)$ and its subspaces, for example LR-fuzzy numbers. In this space, a calculus of fuzzy functions is possible and an overview was given of integration, differentiation and other methods typical of mathematical analysis.

Two broad areas of applications were considered: approximation and interpolation, and fuzzy differential equations. In both, the idea of a distance between fuzzy numbers is central to development of theory and techniques.

Throughout the whole discussion, two general mathematical procedures have been emphasised. First, to study fuzzy sets and fuzzy functions by considering their level sets, obtaining results by methods of set-valued analysis and afterwards "gluing" the level sets back into fuzzy objects via the Stacking Theorem 3.1.7 is an extremely powerful technique. Second, fuzzy metric spaces can be embedded, via support functions, as cones in Banach spaces and sophisticated functional analytic techniques can be applied. Both these procedures treat fuzzy objects as classical mathematical elements with their own particular structures and are not mere fuzzifications of existing crisp results.

References

K.J. Arrow and F.H. Hahn (1971) *General Competitive Analysis*, Holden–Day, San Francisco.

Z. Artstein (1974) On the calculus of closed set–valued functions, *Indiana Univ. J. Math.*, **24**, 433-441.

J.–P. Aubin and A. Cellina (1984) *Differential Inclusions,* Springer Verlag, New York.

J.–P. Aubin and H. Frankowska (1990) *Set–Valued Analysis*, Birkhäuser, Basel.

R.J. Aumann (1965) Integrals of set–valued functions, *J. Math. Anal. Applns.*, **12**, 1–12.

H.T. Banks and M.Q. Jacobs (1970) A differential calculus of multifunctions, *J. Math. Anal. Applns.*, **29**, 246–272.

C. Berge (1963) *Topological Spaces: including a treatment of multi–valued functions, vector spaces and convexity*, Oliver & Boyd, Edinburgh.

F.S. de Blasi (1976) On the differentiability of multifunctions, *Pacific J. Math.*, **66**, 67–81.

V.N. Bobylev (1985) Support functions on a fuzzy set and its characteristic properties, *Math. Notes USSR*, **37**, 281–285.

V.N. Bobylev (1990a) A possibilistic argument for irreversibility, *Fuzzy Sets and Systems*, **34**, 73–80.

V.N. Bobylev (1990b) Cauchy problem under fuzzy control, *BUSEFAL*, **21**, 117–126.

P. Bridgeland (1970) Trajectory integrals of set valued functions, *Pacific J. Math.*, **53**, 43–68.

Ch. Castaing and M. Valadier (1970) *Convex Analysis and Measurable Multifunctions*, Springer Lecture Notes in Mathematics Vol. 580, Springer–Verlag, Heidelberg.

I.L. Collings and P. E. Kloeden (1995) Continuous approximation of fuzzy sets, *J. Fuzzy Math.*, **3**, 449–453.

P.J. Davis (1963) *Interpolation and Approximation*, Blaisdell Publishing, Waltham, Mass.

C. de Boor (1978) *A Practical Guide to Splines*, Applied Mathematical Sciences **27**, Springer–Verlag, Berlin.

G. Debreu (1967) Integration of correspondences, *Proc. 5th. Berkeley Symp. Math. Stat. Probab.*, Vol. 2, 351–376.

P. Diamond (1990) A note on fuzzy starshaped fuzzy sets, *Fuzzy Sets and Systems*, **37**, 193–199.

P. Diamond (1999a) Time-dependent differential inclusions, cocycle attractors and fuzzy differential equations, submitted for publication.

P. Diamond (1999b) Stability and Lyapunov functions for fuzzy differential equations, submitted for publication.

P. Diamond and P.E. Kloeden (1988) A note on compact sets in spaces of subsets, *Bull. Austral. Math. Soc.*, **38**, 393–395.

P. Diamond and P.E. Kloeden (1989) Characterization of compact sets of fuzzy sets, *Fuzzy Sets and Systems*, **29**, 341–348.

P. Diamond and P.E. Kloeden (1990) Metric spaces of fuzzy sets, *Fuzzy Sets and Systems*, **35**, 241–249; Corrigendum, **45** (1992), 123.

P. Diamond and P. Kloeden (1994) *Metric Spaces of Fuzzy Sets: Theory and Applications*, World Scientific, Singapore.

P. Diamond, P. Kloeden and A. Vladimirov (1998) Spikes, broken planes and the approximation of convex fuzzy sets, *Fuzzy Sets and Systems*, **99**, 225–232.

P. Diamond and A. Ramer (1993) Approximation of knowledge and fuzzy sets, *Cybernet. Syst.*, **24**, 407–417.

P. Diamond and P. Watson (1999) Regularity of solution sets for differential inclusions quasi-concave in a parameter, *Appl. Math. Letters*, to appear.

D. Dubois and H. Prade (1982a) Towards fuzzy differential calculus. Part 1: Integration of fuzzy mappings, *Fuzzy Sets and Systems*, **8**, 1–17.

D. Dubois and H. Prade (1982b) Towards fuzzy differential calculus. Part 2: Integration on fuzzy intervals, *Fuzzy sets and Systems*, **8**, 105–116.

D. Dubois and H. Prade (1982c) Towards fuzzy differential calculus. Part 3: Differentiation, *Fuzzy Sets and Systems*, **8**, 225–233.

R. Goetschel and W. Voxman (1981) A pseudometric for fuzzy sets and certain related results, *J. Math. Anal. Applns.*, **81**, 507–523.

R. Goetschel and W. Voxman (1983) Topological properties of fuzzy numbers, *Fuzzy Sets and Systems*, **10**, 87–99.

S. Gottwald (1986) Characterization of the solvability of fuzzy operators, *Elektronische Informat. Kybernetik*, **22**, 67–91.

L. M. Graves (1946) *The Theory of Functions of Real Variables*, McGraw–Hill, New York.

F. Hausdorff (1957) *Set Theory*, Chelsea Press, New York.

S. Heilpern (1981) Fuzzy mappings and fixed point theorem, *J. Math. Anal. Applns.*, **83**, 566–569.

H. Hermes (1968) Calculus of set–valued functions and control, *J. Math. Mech.*, **18**, 47–59.

M. Hukuhara (1967) Integration des applications measurables dont la valeur est un compact convexe, *Funkc. Ekvacioj.*, **10**, 205–223.

E. Hüllermeier (1997) An approach to modelling and simulation of uncertain dynamical systems, *Int. J. Uncertainty, Fuzziness & Knowledge-Based Systems*, **5**, 117–137.

O. Kaleva (1985) On the convergence of fuzzy sets, *Fuzzy Sets and Systems*, **17**, 53–65.

O. Kaleva (1987) Fuzzy differential equations, *Fuzzy Sets and Systems*, **24**, 301–317.

O. Kaleva (1990a) The Cauchy problem for fuzzy differential equations, *Fuzzy Sets and Systems*, **35**, 389–396.

O. Kaleva (1990b) On the calculus of fuzzy valued mappings, *Appl. Math. Lett.*, **3**, 55–59.

O. Kaleva (1994) Interpolation of fuzzy data, *Fuzzy Sets and Systems*, **61**, 63–70.

E.P. Klement, M.L. Puri and D.A. Ralescu (1986) Limit theorems for fuzzy random variables, *Proc. Roy. Soc. London, A*, **407**, 171–182.

P.E. Kloeden (1980) Compact supported sendographs and fuzzy sets, *Fuzzy Sets and Systems*, **4**, 193–201.

P.E. Kloeden (1982) Fuzzy dynamical systems, *Fuzzy Sets and Systems*, **7**, 275–296.

P. E. Kloeden (1987) Chaotic mappings of fuzzy sets, Proc. 2nd. IFSA, Vol 1, 368–371.

P.E. Kloeden, *Chaotic iterations of fuzzy sets*, Fuzzy Sets and Systems, **42** (1991), 37–42.

P.E. Kloeden (1991) Remarks on Peano theorem for fuzzy differential equations, *Fuzzy Sets and Systems*, **44**, 161–163.

K. Kuratowski (1966) *Topology, Vols. I & II*, Academic Press, New York.

S.R. Lay (1982) *Convex Sets and Their Applications*, John Wiley and Sons, New York.

R. Lowen (1990) A fuzzy Langrange interpolation theorem, *Fuzzy Sets and Systems*, **34**, 33–38.

E. Michael (1951) Topologies on spaces of subsets, *Trans. Amer. Math. Soc.*, **71**, 152–182.

E. Michael (1956) Continuous selections: I, Ann. Math. **63**, 361–381; II, Ann. Math. **64**, 562–580; III, Ann. Math. **65**, 375–390.

H. Minkowski (1967) Theorie der konvexen Körper, insbesondere Begründung ihres Oberflächen Begriffs, (Collected Papers), Chelsea Publishing Company, New York, vol. 2, 131–229.

M. Mizumoto and K. Tanaka (1976), The four operations of arithmetic on fuzzy numbers, *Systems, Computers Controls*, **7**, 73–81.

C.V. Negoita and D.A. Ralescu (1975) *Applications of Fuzzy Sets to Systems Analysis*, John Wiley and Sons, New York.

H.T. Nguyen (1978) A note on the extension principle for fuzzy sets, *J. Math. Anal. Applns.*, **64**, 369–380.

M.L. Puri and D.A. Ralescu (1983) Differentials of fuzzy functions, *J. Math. Anal. Applns.*, **91**, 552–558.

M.L. Puri and D.A. Ralescu (1985) The concept of normality for fuzzy random variables, *Ann. Probab.*, **13**, 1373–1379.

M. L. Puri and D.A. Ralescu (1986) Fuzzy random variables, *J. Math. Anal. Applns.*, **114**, 409–422.

H. Rådström (1952) An embedding theorem for spaces of convex sets, *Proc. Amer. Math. Soc.*, **3**, 165–169.

R.T. Rockafellar (1970) *Convex Analysis*, Princeton University Press, Princeton, N.J.

S. Seikkala (1987) On the fuzzy initial value problem, *Fuzzy Sets and Systems*, **24**, 319–330.

G.C. Shephard and R.J. Webster (1985) Metrics for sets of convex bodies, *Mathematika*, **12**, 73–88.

R.A. Vitale (1985) L_p metrics for compact, convex sets, *J. Approx. Theory*, **45**, 280–287.

R.A. Vitale (1979) Approximation of convex set–valued functions, *J. Approx. Theory*, **26**, 301–316.

L.A. Zadeh (1965) Fuzzy sets, *Informat. Control*, **8**, 338–353.

Index